Spatial Biases in Perception and Cognition

Our experience of the world is influenced by numerous spatial biases, most of which influence us without our being aware of them. These biases are related to illusions and asymmetries in our perception of space, relationships between space and other qualities, dynamics of moving objects, dynamics of scene configuration, and dynamics related to perception and action. Consideration of these biases provides insight into how we perceive, remember, and navigate space, as well as how we interact with objects and people in space. This volume introduces and reviews numerous spatial biases and provides descriptions and examples of each bias. The contributors discuss historical and current theories for many biases and provide new explanatory theories. Providing a "one-stop shop" for information on such a key aspect of our experience in the world, this volume will interest anyone curious about our understanding of space.

Timothy L. Hubbard is an adjunct faculty member at Arizona State University and Adjunct Doctoral Dissertation Chair at Grand Canyon University, and was previously a full professor at Texas Christian University. He has published 90 papers in peer-reviewed scientific journals, 12 chapters in various academic books, and more than 100 scientific conference presentations. He is a consulting editor for *Attention, Perception, & Psychophysics* and *Journal of Experimental Psychology: Human Perception and Performance*, and an associate editor for *Auditory Perception & Cognition* and *Frontiers in Psychology*. He has published scholarly reviews and empirical findings on several spatial biases and is a fellow of the Association for Psychological Science and of the Psychonomic Society.

Spatial Biases in Perception and Cognition

Edited by

Timothy L. Hubbard
Arizona State University and Grand Canyon University

CAMBRIDGE
UNIVERSITY PRESS

CAMBRIDGE
UNIVERSITY PRESS

University Printing House, Cambridge CB2 8BS, United Kingdom

One Liberty Plaza, 20th Floor, New York, NY 10006, USA

477 Williamstown Road, Port Melbourne, VIC 3207, Australia

314–321, 3rd Floor, Plot 3, Splendor Forum, Jasola District Centre, New Delhi – 110025, India

79 Anson Road, #06–04/06, Singapore 079906

Cambridge University Press is part of the University of Cambridge.

It furthers the University's mission by disseminating knowledge in the pursuit of education, learning, and research at the highest international levels of excellence.

www.cambridge.org
Information on this title: www.cambridge.org/9781107154988
DOI: 10.1017/9781316651247

First published 2018

Printed and bound in Great Britain by Clays Ltd, Elcograf S.p.A.

A catalogue record for this publication is available from the British Library.

Library of Congress Cataloging-in-Publication Data
Names: Hubbard, Timothy L., editor.
Title: Spatial biases in perception and cognition / edited by Timothy L. Hubbard, previously of Texas Christian University.
Description: Cambridge, United Kingdom ; New York, NY : Cambridge University Press, 2018. | Includes bibliographical references.
Identifiers: LCCN 2018008153 | ISBN 9781107154988 (hardback) | ISBN 1107154987 (hardback) | ISBN 9781316607077 (pbk.) | ISBN 1316607070 (pbk.)
Subjects: LCSH: Space perception. | Spatial behavior. | Cognition.
Classification: LCC QP491 .S73 2018 | DDC 152.14/2–dc23
LC record available at https://lccn.loc.gov/2018008153

ISBN 978-1-107-15498-8 Hardback

Contents

Part IV Perception and Action

The plate section can be found between pp 260 and 261

Figures

Tables

Contributors

Jose Antonio Aznar-Casanova
University of Barcelona, Spain

Jiuyang Bai
Illinois State University, USA

Paolo Bartolomeo
Inserm U1127 and Sorbonne Université, Paris, France

Leonardo Gomes Bernardino
Universidade Federal de Uberlandia, Brazil

Ivana Bianchi
University of Macerata, Italy

Mario Bonato
University of Padua, Italy

Pom Charras
University Paul Valery Monpellier, France

Vincent Cialdella
Illinois State University, USA

Krzysztof Cipora
University of Tübingen, Germany

Ophelia Deroy
University of London, UK

Irune Fernandez-Prieto
Hospital Sant Joan de Deu, Barcelona, Spain

Kristin M. Gagnier
Johns Hopkins University, USA

Souta Hidaka
Rikkyo University, Japan

Timothy L. Hubbard
Arizona State University and Grand Canyon University, USA

Helene Intraub
University of Delaware, USA

J. Scott Jordan
Illinois State University, USA

Dirk Kerzel
University of Geneva, Switzerland

Raymond M. Klein
Dalhousie University, Canada

Alexander Kranjec
Duquesne University, USA

Juan Lupianez
University of Granada, Spain

Michael K. McBeath
Arizona State University, USA

Tobias Meilinger
Max Planck Institute for Biological Cybernetics, Tübingen, Germany

Michael Morgan
City University of London, UK

Jochen Müsseler
RWTH Aachen University, Germany

Jordi Navarra
Hospital Sant Joan de Deu, Barcelona, Spain

John G. Neuhoff
College of Wooster, USA

Nora S. Newcombe
Temple University, USA

Hans-Christoph Nuerk
University of Tübingen, Germany

Katarzyna Patro
Ulm University, Germany

Ralph S. Redden
Dalhousie University, Canada

Elena Rusconi
University of Trento, Italy

Ugo Savardi
University of Verona, Italy

Thomas W. Schubert
University of Oslo, Norway

Daniel S. Schloesser
Illinois State University, USA

Charles Spence
University of Oxford, UK

Yoichi Sugita
Waseda University, Japan

Caterina Suitner
University of Padua, Italy

Wataru Teramoto
Kumamoto University, Japan

Barbara Tversky
Columbia University, USA

Carlo Umiltà
University of Padua, Italy

Jan M. Wiener
Bournemouth University, UK

Jessica K. Witt
Colorado State University, USA

Myrka Zago
IRCCS Santa Lucia Foundation, Rome, Italy

Introduction

Spatial Biases: What Are They and Why Are They Important?

Timothy L. Hubbard

What is a spatial bias? This can be most easily understood if we consider each of the words in the term. First, "spatial" refers to configurations of or distances between objects in the environment. Its use in the term "spatial bias" fits our everyday conception of what we mean by "space," but also goes considerably beyond our everyday conception to include stimuli that might not initially be considered spatial. For example, we might think of different auditory pitches as being located in an auditory space (e.g., height on a musical staff) or different numerical quantities as providing a framework for mapping space (e.g., reflecting different extents or different coordinates in a graph). Second, "bias" refers to a consistent pattern of perceptual judgment or motor action. In some cases, there might be a consistent direction of error in judgment or action away from the veridical response (e.g., as when observers consistently overshoot the actual final location of a moving target in their judgment of the final location); in other cases, there might be a consistent pattern of judgment or action in the absence of a clear veridical response (e.g., associating higher auditory pitches with higher locations in the picture plane). Thus, a "spatial bias" is a consistent pattern of perceptual or cognitive judgments or motor actions involving configuration or distance in response to a stimulus when those judgments or actions do not correspond to the veridical stimulus or when there is no clear veridical stimulus to compare with those judgments or actions.

Although various spatial biases have long been of interest in psychology and related disciplines, information on such biases has been scattered across many different sources. One aim of this book is to provide a catalog of spatial biases and descriptions of the different domains and situations within which such biases have been found. A second aim is to highlight the potential relevance of spatial biases in a wide range of domains and situations. Although historical examination of spatial biases often focused on various types of static illusions (e.g., Müller-Lyer illusion, horizontal–vertical illusion) or spatial biases in cognitive mapping (e.g., effects of hierarchical organization of knowledge), recent decades have seen the discovery of new spatial biases associated with scenes (e.g., boundary extension), invariant physical principles (e.g., momentum, gravity), moving objects (e.g., flash-lag effect, looming), numeric cognition (e.g., SNARC), musical cognition (e.g., SMARC), and social cognition (e.g., group membership, status). The extent to which spatial processing can influence other types of cognition or behavior (e.g., the Simon effect, numerical processing in SNARC) is also important. The book is divided into four main sections. Part I deals with various anisotropies (i.e., asymmetries) and illusions. Part II deals with biases that arise from the dynamics of objects. Part III expands the view from dynamics of objects to dynamics of scenes, and Part IV examines dynamics arising from actions within scenes.

Part I: Anisotropies and Illusions

Part I begins with a consideration of perceptual biases in geometry, which is evidenced in many classical visual illusions. Morgan (Chapter 1) suggests the retina provides a coarse topographic map, and that measurements on the map are carried out by "coincidence detectors" and "collector units" that have wide spatial tuning and are unable to provide precise positional information for single positions independent of the surrounding context. Aznar-Casanova and Bernardino (Chapter 2) widen the focus to consider differences between physical space and visual space, and they consider perceptual anisotropies in the frontoparallel (coronal), medial (sagittal), and horizontal (transverse) planes including the horizontal–vertical illusion, oblique effect, foreshortening, and effects of the observer's eye level and ground surface. Aznar-Casanova and Bernardino note that anisotropies are influenced by properties of the stimuli (e.g., spatial frequencies, size, orientation), the type of task (e.g., search, recognition, magnitude estimation), and within-observer variables (e.g., hemispheric specialization, laterality, age).

Anisotropies are related to numerical magnitude by Cipora, Patro, and Nuerk (Chapter 3), who consider findings on spatial–numerical associations. They distinguish between spatial–numerical associations involving extension (i.e., larger numbers associated with large extents) or directionality (i.e., larger numbers associated with a particular direction or region in space, e.g., the SNARC effect). They highlight that the relationship between numbers and space is bidirectional, review several variables that influence such spatial–numerical associations, and conclude that such associations are highly flexible, situated, triggered by different types of informational content, and operate at different levels of information processing and in different temporal frames. The relationship of space to an individual's perceptual judgments or motor actions is further considered by Umiltà, Bonato, and Rusconi (Chapter 4), who overview the Simon effect (responding is facilitated when the response is located in the same part of space as the stimulus), SMARC effect (higher pitches are associated with higher locations in space), and STEARC effect (different parts of space are associated with the past or the future). Umiltà and colleagues consider these effects as examples of a broader S–R (i.e., stimulus–response) compatibility involving spatial coding of nonspatial information.

The relationship between auditory pitch and spatial location is further considered by Deroy, Fernandez-Prieto, Navarra, and Spence (Chapter 5), who review findings that higher-frequency sounds facilitate responding to stimuli in the upper part of visual space and lower-frequency sounds facilitate responding to stimuli in the lower part of visual space. Deroy and colleagues suggest that spatial attributes of auditory pitch result from a crossmodal correspondence between vision and audition that is linked to the connection between auditory perception of sounds and visual perception of sound sources. In the last chapter of Part I, Kranjec (Chapter 6) examines connections between space and language, that is, how verbal and nonverbal representations of spatial relationships interact. A case study of a patient with simultagnosia reveals that exposure to nonlinguistic spatial information can improve perception of spatial relations between stimuli that are otherwise not perceived. Kranjec also suggests that high-level categorical information about space can influence perceptual representation in ways that language does not.

Part II: Dynamics of Objects

Part II begins with a consideration of mislocalization of the onset position of a moving stimulus. Müsseler and Kerzel (Chapter 7) examine two such biases: the Fröhlich effect and the onset repulsion effect (judged onset location is displaced slightly forward or backward, respectively, from the actual onset location). They review several possible mechanisms (e.g., sensation time, metacontrast, lateral inhibition, attentional mechanisms) and suggest how the conflict between the Fröhlich effect and onset repulsion effect might be resolved. Mislocalization of the offset position of a moving target is reviewed by Hubbard (Chapter 8), who focuses on representational momentum (judged final location is displaced slightly forward of the actual final location). Hubbard reviews effects of several variables that influence representational momentum and suggests several properties that characterize representational momentum. The flash-lag effect (a perceived misalignment of a moving object with a briefly presented stationary target) is discussed by Hubbard (Chapter 9), who reviews effects of several variables that influence the magnitude of the flash-lag effect, its properties, and its major theories.

The influence of an actual or implied gravitational attraction on perceptual and motor responses is considered by Zago (Chapter 10), who reviews several behavioral tests for perception of the direction of gravitational attraction and acceleration. Zago suggests that the representation of gravity involves a multiplicity of representations, and that although estimates of the gravity vector are generally accurate under typical ecological conditions, they can be biased under unusual conditions. Influences of auditory information on visual motion perception are discussed by Teramoto, Hidaka, and Sugita (Chapter 11), who present findings challenging the traditional view that vision is dominant over other modalities in spatial processing and supporting a view that auditory information can have a strong effect on visual motion perception even when those sounds do not contain explicit spatial information. The dynamics of looming objects are considered by Neuhoff (Chapter 12), who notes that visual and auditory looming objects have priority in processing and bring about responses not evoked by motion in any other direction. Neuhoff points out that looming objects are often perceived as closer than those objects really are, and that this bias might have been shaped by evolution to enhance the probability of survival and reproduction.

Part III: Dynamics in Scenes

Part III begins with a consideration of boundary extension (a previously viewed scene is remembered as including information that was not actually viewed, but that would likely have been present just beyond the edges of the actually viewed scene) by Intraub and Gagnier (Chapter 13), who consider a variety of cognitive factors (e.g., viewpoint, attention, medium of presentation, neuropsychological effects, developmental effects) supporting a multisource model of boundary extension. Spatial contraries (the organizing of space along opposing dimensions, e.g., up–down, left–right, etc.) and the experience of mirror reflections are examined by Bianchi and Savardi (Chapter 14), who suggest that spatial contraries are an essential aspect of the framework within which people gesture and interact and are integral to the relationship between an object/body and its perceived reflection. Aesthetics and preferences in scene composition are discussed by Hubbard (Chapter 15), who reviews findings on aesthetic judgments and preferences regarding relative location, facing direction (orientation), contour shape, and ratio on objects in a scene. Additionally, theories for some of these preferences, as well as general properties of spatial bias in scenes, are considered.

A review of spatial biases in the classic cognitive mapping literature is provided by Tversky (Chapter 16), who presents a new Spatial Reference theory to account for a wide range of such biases. This theory involves spatial biases related to grouping and hierarchical organization of spatial information, use of canonical axes, use of reference points (landmarks), perspectives, paths, schemas, and route planning. The importance of hierarchical and categorical information in spatial bias is underscored by Newcombe (Chapter 17), who overviews data on the effects of spatial categories (e.g., different quadrants defined by cardinal axes) and presents the Category Adjustment Model (CAM) in which fine-grained spatial information is nested within larger spatial categories. CAM is a Bayesian model based on unbiased information that is weighted by category, and it generates predictions consistent with several previously observed spatial biases.

Part IV: Perception and Action

Part IV begins with a consideration of the consequences of brain damage, and Charras, Lupianez, and Bartolomeo (Chapter 18) review findings involving brain damage and visual neglect. The authors present evidence for a general left–right imbalance in visuospatial and representational processes, review evidence that pre-attentive processing is relatively spared in neglect patients, and argue for an attentional account of neglect. An evolutionary account for several spatial biases is provided by McBeath (Chapter 19), who argues that many biases reflect predictive expectations. McBeath suggests that observers and actors exhibit an array of spatial biases that vary in robustness from universal invariant properties (e.g., increases in visual angle of a looming object) to socially learned tendencies (e.g., American pedestrians who are approaching each other usually veer to their respective right sides). One spatial bias that is related to visual search is inhibition of return (IOR), and Klein and Redden (Chapter 20) argue for the existence of two different types of IOR: one type involves a reduction in signal strength in an early salience map, and the other type involves a response bias criterion shift in a later map.

The effect of action on perception is discussed by Witt (Chapter 21), who reviews data suggesting that a perceiver's ability to perform an intended action affects his or her perception of the salient stimuli. Witt provides a historical overview of the action-specific approach to perception, addresses whether such effects reflect perception or judgment, and discusses unresolved issues regarding action-specific perception. Biases involved in navigating from an initial origin point to a desired destination are discussed by Wiener and Meilinger (Chapter 22), who review studies on route choice in which navigators have different levels of knowledge regarding that environment. Social aspects of space related to group membership, stereotyping, and status are addressed by Suitner and Schubert (Chapter 23), who review the links between social cognition and spatial cognition from both an applied and a theoretical perspective. Finally, on a more philosophical note, Jordan, Cialdella, Schloesser, and Bai (Chapter 24) discuss two forms of bias within cognitive science (efficient-cause bias, realism bias) and how those biases influence theories of perception, cognition, and action in general and spatial representation in particular.

Why Spatial Biases?

As the preceding brief introduction to each of the chapters shows, spatial biases are relevant in a wide range of domains and situations. Although for many years such biases had been of limited interest, appealing primarily to researchers interested in perception and psychophysics, the chapters in this book show that spatial biases impact many different areas of our everyday experience. In many cases, spatial biases are not errors per se, but reflect adaptations based on expectations (e.g., Chapters 8, 9, 10, 13, 19). In some cases, spatial biases might have been selected for by evolution (e.g., Chapters 8, 10, 12, 19, 23) or reflect recent experience or context (e.g., Chapters 7, 14, 15, 16, 21, 24). In other cases, spatial biases reflect the hierarchical or linguistic organization of spatial knowledge (e.g., Chapters 6, 15, 16, 17, 22, 23) and contributions of language (e.g., Chapter 6) and attention (e.g., Chapters 7, 9, 18, 21). In still other cases, spatial biases reflect crossmodal or multisensory (e.g., Chapters 5, 11) or motor (Chapter 4) influences and an interplay of multiple mechanisms (e.g., Chapters 2, 3, 4, 20). Spatial biases reflect consequences of our neural structure and development (e.g., Chapters 1, 2, 18, 20, 22) and our responses (e.g., Chapters 2, 4, 15, 24), as well as cultural practices (e.g., Chapters 19, 23, 24).

Several different themes emerge across the chapters in this book. One theme is that different spatial biases arise from different causes or combinations of causes, and these causes can range from low-level physiological mechanisms to high-level organization of semantic knowledge and culturally specific learning. A second theme is that some spatial biases appear to provide at least some adaptations for daily functioning; even so, whether other biases provide as yet undiscovered adaptations or were too harmless to be selected against is not clear. A third theme is that consideration of spatial biases can provide useful tools that can yield important insight into the functioning of perception and cognition. A fourth theme is that spatial biases do not just reflect illusions based on a few laboratory artefacts, but permeate much of our everyday perception and cognition. Indeed, such biases are ubiquitous, even if unnoticed, in our everyday experience. Overall, an understanding of the roles and mechanisms of spatial biases is important in a variety of domains and situations in our everyday experience, and the chapters in this book represent important steps toward such an understanding.

Part

I

Anisotropies and Illusions

Chapter 1

Perceptual Biases in Elementary Geometry

Michael Morgan

To quote Weintraub and Schneck (1986, p. 147), 'The term *illusion*, because of its connotations of magic and deception,[1] neither of which is germane to the topic at hand, shall be restricted to the title.' We shall use it sparingly. The implication of a departure between perception and reality has little to no justification in post-Kantian scientific discourse (for an extended justification of this position see Morgan, 1996). To see that the term is inept, it suffices to point out that both cinematography and colour metamerism have been in the past called illusions, but now that their mechanism is understood we know that no trickery or fooling of the senses are involved. The limitations on sensory acuity necessarily imposed by cone-action spectra or low-pass temporal frequency filtering are no more illusions than our failure to see UV radiation.

The perceptual biases seen in the Müller-Lyer, Poggendorff, Ebbinghaus, Fraser 'twisted cord' and many other figures (Robinson, 1972) are biases in basic perceptual geometry. That is, they involve perceptually based judgements of angle and distance. It is immediately apparent why centuries of research have failed to produce a comprehensive theory of these biases: It is because we have no theory of elementary visual geometry in the first place. Before embarking on an analysis of biases in this chapter, we shall review what little is known about basic visual geometry. There will be no attempt to produce a comprehensive list of geometrical biases: there are several excellent reviews already (Robinson, 1972; Coren & Girgus, 1978; Gillam, 1988).

The Concept of Visual Direction

The starting point for visual geometry is conventionally the idea of *visual direction*. If we examine a single-point light source in the dark, it has a definite location in space relative to our body. We can therefore make a saccadic eye movement to it or make a pointing movement. These movements are not necessarily without error. Saccades often fall short of the target (saccadic undershoot) but generally have the same mean polar angle relative to the fovea as does the target. By definition, a point light source imaged on the retina has a *visual direction* defined in a polar co-ordinate framework with the fovea as its origin. The visual direction is a two-dimensional vector with dimensions of polar angle and distance from the fovea.

Two otherwise identical objects can co-exist in visual space if they have different visual directions. There are reasons to think that the sense of visual direction is innate; Schlodtmann pressed on the eyeball of congenitally blind patients and had them point to

[1] The German word for a perceptual illusion, '*Täuschung*', also means a deception or deceit.

the resulting phosphene. Pressure on the temporal retina of the left eye resulted in pointing to the right-side of the body, just as it does in normally sighted subjects, because of the mirror reversal of the retinal image. The blind patients would have been unlikely to have acquired this knowledge through experience.

If two objects in space have two different visual directions, it follows that there is a vector describing the angle between them and their separation, or equivalently, two vectors, each joining one of the objects to the fovea. We shall say that this vector describes their *relative spatial position.* This must be the basis for elementary visual geometry. But how is this vector made explicit, and what are its properties?

Basic Visual Geometry

Relative Spatial Position

We begin with a necessarily brief review of acuity for relative spatial position, which reveals the underlying resources for more complex visual geometry tasks. Relative spatial acuity has sometimes been called *hyperacuity* because in some cases, notably vernier acuity, it can be finer than the acuity for resolving two adjacent bars or dots (Westheimer, 1981), a fact readily explained by low-pass filtering in the optics of the eye affecting the latter more than the former (Barlow, 1980; for review, see Morgan, 1990). However, in the case of widely separated features, thresholds for relative spatial acuity can be greater than the 30 arcmin of foveal spatial resolution. We shall therefore use the more general term *relative spatial acuity.* The brief review that follows will be confined to the 2D case, although many of the remarks apply equally to stereoscopic acuity. The most basic kinds of acuity are found with just two or three points in the visual field. Between any two points there is a vector that gives the distance (a scalar magnitude) and angle between them, in a coordinate system with its origin at the fovea. Distance and angle are thus coded separately. Since the biases in the Müller-Lyer, Poggendorff and other figures are essentially biases of angle and distance, these basic mechanisms are clearly relevant.

Acuity for Distance

Observers can compare the distance between two points to a remembered standard, or the mean of a set (Westheimer & McKee, 1977; Morgan, Watamanuik, & McKee, 2000), with an accuracy approximately proportional to 5% of the standard distance. Larger standard distances involve increasing retinal eccentricities and thus larger receptive fields, although whether this is the main cause for the Weber relationship is disputed (Levi, Klein, & Yap, 1988; Morgan & Watt, 1989). Alternatively, if there are three points, the observer can compare them pairwise. One version of this task is bisection, in which the distance AB is compared to BC, where ABC are collinear. Usually in distance tasks, the angle between the points is held constant over trials. If it is varied, acuity falls slightly but not drastically (Morgan, unpublished observations), possibly because of angle-dependent bias (the horizontal–vertical bias). An important fact about both distance and angle acuity is that they are not reduced by placing irrelevant features in the vicinity of the two targets (Morgan & Ward, 1985; Morgan, Hole, & Ward, 1990), provided all the elements in the scene are clearly resolved. This means that the magnitude of the vector between two points can be encoded, ignoring other points in the scene – a process termed *selective attention.*

Acuity for Size and Area

It is an important question whether there are special mechanisms for encoding the overall size of an object. Biases of perceived size are seen in the Ames Room, the Harvest Moon phenomenon and the Ebbinghaus circles. The evidence is that, when possible, the visual system uses 1D measures of size, because these are more accurate. For example, when comparing the size of two circles, observers compare their diameters (Morgan, 2005; Nachmias, 2008; Solomon, Morgan, & Chubb, 2011). When they compare the shape of ellipses, they compare the ratios of minor to major axes, and are thus less accurate than comparing size of circles, because two noisy measurements are needed rather than one. When comparing the area of irregular polygons, observers are less accurate than with circles (Raphael & Morgan, 2016). However, irregular shapes can still be compared, so there must still be some mechanism. One possibility is overall energy, which has also been proposed for the similar problem of computing the number of dots in an extended dot cluster (Dakin, Tibber, Greenwood, Kingdom, & Morgan, 2011; Morgan, Raphael, Tibber, & Dakin, 2014). It may be noted in passing that dot number biases the estimation of cluster size, and vice versa, but these are not normally considered geometrical biases, so we shall not recapitulate the discussion by Dakin et al. (2011) and Raphael and Morgan (2016).

The Role of Constancy in Causing Biases

In the Ames room, the further-away person looks dramatically smaller than the nearer one. This has claims to be called a true 'illusion' because it results from the deliberately misleading manipulation of perspective to give the impression that the room is square rather than trapezoidal. Misplaced constancy has frequently been evoked to account for 2D as well as 3D biases (for history, see Ross & Plug, 2002). It may or may not be responsible for the Harvest Moon phenomenon. It seems particularly strange to call the latter an illusion when we never see the Moon at its 'true' size, whatever that would mean.

It is not always appreciated that constancies themselves are an example of perceptual bias, resulting in the shift of the P50 point of the psychometric function rather than its shape (Burbeck, 1987; McKee & Welch, 1992). An artificial constancy based on training was demonstrated by requiring observers to compare the distance between two parallel lines to that of a remembered standard (Morgan, 1992). The standard was different for different orientations of the stimulus: to wit, the standard size increased with CW rotation, a fact that the observer had to discover with the help of auditory feedback. After training, observers were just as accurate in this task as they were with a single standard. They could even use a number presented on the screen instead of angle to perform the scaling. If we were going to describe this task in the language of 'illusion', we would be tempted to say that observers had the 'illusion' that the size of the figure *decreased* with CW rotation, but of course the alternative description is that the observer has learned a new decision rule based on context. The extent to which perceptual and decision biases can be disentangled is a difficult one, with a long history in relation to adaptation level theory (Helson, 1964; Restle & Levinson, 1971; Morgan, Dillenburger, Raphael, & Solomon, 2012), and has to be considered whenever some new 'illusion' is claimed.

Acuity for Angle

This is often measured by presenting two points and asking the observer whether the imaginary line joining them is tilted left or right of vertical (or of horizontal). Alternatively, a three-dot array can be used as in bisection, but the observer is asked whether the center dot is shifted left or right of the axis defined by the two outer dots. In ideal conditions the thresholds for angle can be even less than 1 degree of visual angle (DVA). It is a striking fact that acuity is similar for separated dots, lines and even gratings (review: Morgan, 1990). However in the case of dots, acuity falls off rapidly if the reference angle is not one of the cardinal axes, even after extensive learning (Morgan, 1990). In the case of gratings, the angular threshold increases as the spatial frequency decreases, in other words, as they get more 'fuzzy' (Burr & Wijesundra, 1991). This parallels the finding that acuity thresholds for both distance and angle increase with blur (Stigmar, 1970; for review, see Morgan, 1990). Also as in the case of distance, acuity does not increase by placing irrelevant features between the two dots (Morgan & Ward, 1985; Morgan et al., 1990).

Comparison between Distance and Angle Acuity

This may seem like comparing the proverbial apples and oranges, because distance is measured in degrees of visual angle (DVA) while angle acuity is measured in radians. However, both DVA and Radians are SI-derived units based on ratios of lengths, so all is not lost. If we use the three-dot bisection array, a kind of comparison is possible using the measure of distance of the middle dot from the null position, either in the direction of the outer dots (distance deviation) or in the orthogonal direction (angular deviation). If this is done with a vertical or horizontal array, acuity for angle is greater than that for distance (Morgan et al., 1990; for review, see Morgan, 1990). It is interesting to note that saccadic eye movements show a similar effect: the variable area for angle is less than that for distance (van Opstal & van Gisberen, 1987). However, the reason may be unconnected to saccades. Bisection acuity involves comparing two distances, and thus twice the sensory noise of the angle noise, which can be done with just one pair of dots, provided there is a relatively less noisy internal reference. Moreover, the angle advantage disappears if the dots are not on a cardinal axis, or if the axis is varied over trials (for review, see Morgan, 1990).

Acuity for Curvature

In the Orbison figure (Robinson, 1972), the bias is one of local curvature in a circle, so curvature may be relevant. The curvature of an arc is the derivative abs(dS/dθ) where S is the length of the arc from a given point and θ is the tangential angle. Thus curvature involves both a distance and an angle measure. Watt (1984) showed that a length limit of 30 min arc applies to the highly accurate discrimination of curvature difference. Within this constraint and the previously reported orientational restriction (Watt & Andrews, 1982), curvature may determine the level of performance.

Acuity for Average Position

Of particular relevance to geometrical biases may be the ability of observers to locate the centroid of collections of neighbouring features with high relative efficiency. By efficiency here we mean that relative to an ideal observer who takes every feature into account, the real observer does not take a small sub-sample but includes a large fraction, perhaps as high as 100% (Morgan & Glennerster, 1991). Eye movements towards a cluster of dots tend to go to its centroid (Findlay, 1982). Observers comparing the distance between two dot clusters to a standard distance tend to use the separation between the centroids, even if they attempt to do otherwise (Morgan & Glennerster, 1991; Harris & Morgan, 1993).

Acuity for Collinearity

To determine that two lines are collinear we need to establish (a) that they have the same orientation and (b) that this orientation is the same as that of the virtual line joining them. Biases in this computation produce the Poggendorff effect. Morgan and Dillenburger (2016) have recently conjectured that the largest cause of the bias is in the construction of the virtual line.

How Are Biases Related to Acuity?

Fechner (see Solomon, 2011) called the inverse of acuity the 'variable error' and any bias the 'fixed error'. A simple analogy is the 'grouping' of darts thrown at the bull's eye of a dartboard. The standard deviation of the landing positions is the variable error; any deviation of the mean from 'bull's eye' is the fixed error, or bias. In practice, the two kinds of error are usually measured from a psychometric function, based on a large series of trials in which the observer is shown two dots with a varying tilt from the vertical and has to classify the stimulus as clockwise or anti-clockwise. (This is for two-dot angle acuity; the procedure for other acuities is easily imagined.) A point on the psychometric function has as its ordinate p, the probability of a CW response, and as its abscissa it has the actual angle. Psychometric functions are usually well fitted by a Cumulative Gaussian with two parameters σ and μ. The value of σ gives us Fechner's 'variable error' and is otherwise called the 'threshold' or 'JND', while μ is the difference of its fitted 50% point (ordinate) from 0 (abscissa) and is otherwise called the 'bias'. Morgan, Hole, and Glennerster (1990) found that the bias (σ) in the outgoing-arrowhead version of the Müller-Lyer figure had no effect on the JND (σ), meaning that the arrowheads did not increase sensory noise. Similar findings have been reported for the Poggendorff (Morgan & Dillenburger, 2016) and Judd (Morgan et al., 1990) biases, and for the rod-and-frame bias (Morgan, Grant, Melmoth, & Solomon, 2015, Fig. 3). There are, however, counterexamples, such as the decrease in angular acuity for a grating when it is subjected to a bias by a tilted surround. Solomon and Morgan (2006) dubbed this effect 'stochastic recalibration', and suggested that it arose from a change in bias over trials masquerading as a variable error (Wickelgren, 1968).

When bias and acuity are independent, there is a strong argument for expressing the bias as the ratio σ/μ and calling this the standardized bias. When this is done for the Poggendorff effect (Morgan & Dillenburger, 2016), the standardized bias is close to unity. In the rod-and-frame case, the bias in 'frame-sensitive' observers is about two times that of the JND. There are not enough measured cases to state a general rule, but where the standardized bias is unity (for example, in the Poggendorff figure), it is salutary to reflect that the striking perceptual bias is comparable in size to the sensory noise.

Theories of Positional Coding

Qualitative Local Sign (*Localzeichen*)

Lotze (see Morgan, 1977) proposed that each nerve fiber coming from the retina carried a *local sign* to the central nervous system (CNS) that coded for its retinal position. Thus, by examining which 'local sign' was being transmitted along the optic nerve, the CNS could determine which point on the retina was stimulated and thus the visual direction of the source object. Lotze was not specific about the nature of local sign but seems to have thought it was a quality. He did not think of it in terms of position on a map. Indeed, his starting point is that CNS maps are useless because they would need another eye inside the skull to read them (in this he was following Descartes, who thought that the image would have to collapse to a point before it could be interpreted; thus, the pineal hypothesis). The problem with Lotze's theory is that it gives no clue to how local sign is used for geometry. How do we know how far apart two local signs are if the difference between them is qualitative rather than quantitative? Granted, one could have an enormous set of rules for pairwise comparison, such that the distance and angle are defined by every pairwise combination of local sign. This amounts to a huge lookup table, and if its purpose is to recover a quantitative signal, one might wonder why one does not start with a quantitative signal in the first place. The only clue to mechanism that Lotze gives us in the *Medizinische Psychologie* is the idea that every local sign is reflexively linked to an eye movement that brings the object into the fovea. It is the covert or overt eye movement that gives us our sensation of space, Lotze proposes, presumably by what we now call efference copy. This could explain how we know that two points have a different visual direction, but it gives no obvious basis for calculating the distance and angle between two points without a lookup table. For this reason, Lotze's version of local sign has not survived, although the term has, without necessarily being understood. Instead, attention has shifted to maps in the CNS.

Topographic Maps

It is known that the retina has an orderly projection to several 2D maps in the central nervous system, including the primary visual cortex (V1) and the superficial layers of the superior colliculus (King, 2004). These are maps in the sense that neighbourhood relations on the retina are preserved in the projection. Locally the 'neural image' looks like the retinal image, although over greater distances neither angle nor distance are accurately preserved (for further discussion of *topographic homeomorphism*, see Goodhill, Finch, & Sejnowski, 1995). In such maps every point on the retina excites a different group of cells on the map. All very well, but how is this translated into a geometry? If the task is merely to determine the visual direction of a single point and to unleash a motion response in that direction, an analogue mechanism such as the mapping of the superficial layer of the SC on to a deeper pre-motor map and thus to the eye muscles can, and does, suffice (King, 2004). Thus, electrical stimulation of the superficial SC can evoke a saccade in the visual direction corresponding to the position on the map. This is the kind of link between local sign and visual direction conjectured by Lotze, and it suffers from the same weakness when it comes to *relative visual direction.* It cannot explain directly how we can compare the magnitude or direction of two vectors. Topographic maps are probably a necessary condition for visual geometry, but they are not sufficient.

Local Filters

Following the discovery of orientation-tuned and spatial frequency–tuned single neurones in V1, it has been suggested several times (e.g. Wilson, 1986) that these neurons could form the basis for relative position acuity. Compelling evidence for this idea comes from masking studies of abutting-line vernier acuity, where gratings oriented at small angles relative to the main axis of the vernier figure cause more masking that do parallel gratings (Findlay, 1973). This has suggested an opponent mechanism in which the output of two neurons tuned to orientations on either side of the target are compared. In addition, Parker and Hawken (1985) found that the statistical reliability of single V1 cells in discrimination changes with position of a grating was similar to that of human observers. However, abutting line vernier acuity may not be the ideal condition to consider because it is not obviously a relative position task. Nor does phase sensitivity tell us anything directly about *relative* spatial position acuity. The discontinuity of the line at the vernier 'break' may well provide sufficient information for a population of cells to decode the direction of the 'break'. Orientation acuity for widely separated dots would be much harder to account for in this way, particularly when there are irrelevant intervening features in the way (Morgan et al., 1990). For this and other reasons, such as the insensitivity of acuity to random changes in contrast of the dots (Morgan & Regan, 1987), a consensus has emerged (e.g. Morgan, 1990; Akutsu, McGraw, & Levi, 1992) that a local filter model may be adequate for local tasks, such as vernier acuity, but not for 'long range' positional acuity tasks.

Coincidence Detectors

In an attempt to extend the local filter model to account for long-range acuity, Morgan and Regan (1987) proposed filters (coincidence detectors, or CDs) that summed the outputs from two widely separated local filters (subfields). A further opponent stage determines which of two optimally placed CDs, tuned to different separations on either side of the standard, is more active, and thus whether the separation between the two test stimuli is greater than the standard separation. A similar scheme of CDs and subfields tuned to angular separation is easily envisaged.

A common objection to CDs is the computational explosion that would result if every pair of neurones on a topographical map had to have a different CD (Bock & Goode, 1994). However, this would depend on how many neurons there were on the map. Clearly, a separate CD for every pair of neurones in V1 would be neither feasible nor desirable. However, a subfield summing the outputs of all cells in a hypercolumn would be more feasible, and could explain why we can randomize the colour, contrast and probably many other attributes of the two dots in a relative positional acuity task without (much) loss of acuity. Recall that the map is not responsible for spatial resolution. The beauty of an opponent process is that it is actually advantageous to have widely tuned components because this enhances the signal/noise ratio (Hinton, McClelland, & Rummelhart, 1986). A topographic map of just a few hundred units (a matrix of 10×10), each with Gaussian profile–receptive field covering several degrees of visual angle, could probably do reasonably well, although modelling would be needed to see just how well, in comparison to the human observer. A map with 100 subfields needs only 4,950 CDs to span the space – hardly a formidable fraction of the $>20 \times 10^9$ neurones in human cortex. Where this map might be located in the brain is unknown at present, although at least one fMRI study points to the lateral occipital cortex (LOC) as a possible site (Tibber, Anderson, Melmoth, Rees, & Morgan, 2009).

A deeper problem, shared by all filter theories, is that the observer would have to know which CD to use for the task. Suppose the task were 2AFC, in which the observer has to decide whether the pair of dots in the first interval is tilted more or less CW from vertical than in the second. Suppose, moreover, that the angle of the first pair is randomly changed over trials. To perform the task efficiently, the observer would have to select the most active CD in the first interval and put it opposite the most active CD in the second interval. Some top-down selective process would be needed to do this. Classical evidence that the first stimulus in a 2AFC task can select the filter to be used in the second is the relatively high slope for detection versus discrimination (see Solomon, 2009). Consider the thought experiment of instructing the observer to decide whether the pair of red dots is tilted CW or ACW, ignoring dots of all other colours. This is not far from the experiment of Morgan et al. (1990) with irrelevant flankers or dots, and we can thus anticipate little interference with acuity from the irrelevant features, provided the dots are not too close together. Note also that this exactly what we ask the observer in the case of the Müller-Lyer figure: 'ignore the arrows'.

Other 'Collector Units' and FRFs

The coincidence detector is just one of a menagerie of imaginary creatures called 'collector units' or FRF (Filter-Rectify-Filter) units. These are envisaged as filters that combine the outputs of lower-level filters after a stage of rectification (e.g. Bock & Goode, 1994; Kingdom, Prins, & Hayes 2003). Suppose the task is to locate the center of a roughly circular cloud of dots, half of which are black and half of which are white, on a mean-luminance background. The collector unit that collected the most energy could encode the position of the cluster. The FRF model envisages a set of low-level detectors that respond with negative output to black dots and positive output to white dots. The sign of the output is made positive before their outputs are collected (hence Filter-RECTIFY-Filter). This is merely a mathematical formalism, since real neurons do not have negative firing rates. The real process is one in which some neurons respond positively to white dots and some respond positively to black dots (Half Wave Rectification), and their outputs are then collected without regard for their original polarity.

Collector units (CUs) can be envisaged for a worryingly large number of tasks. 'Worryingly' because we once again risk a computational explosion, even if the CUs are widely tuned. Consider the task of determining the orientation of a line of spaced dots, half of which are white and half black. A CU that collects along radial lines on our coarse topographic map would do the trick. But consider another task: Which of two horizontal lines of dots with vertical position varying sinusoidally along the x-axis has the higher frequency of sinusoidal modulation (Tyler, 1973)? There is unlikely to be a specific CU for this task, or if there is, the number of CUs is limited only by our imagination. Rather, the observer would have to decide what basic collector unit is best adapted to the task: for example, a CU that collects energy along the horizontal axis at a particular vertical position in the row would get any increasing amount of energy as the spatial frequency of the sinusoid increases. In other words, an apparently complicated task can often be solved by a simple method. An existence proof for this is that if observers are asked to compare the area of two circles, they actually compare their diameters (Morgan, 2005; Nachmias 2008; Solomon, Morgan, & Chubb, 2011). We do not need CU 'templates' for every possible shape. They do not exist, and therefore, observers are relatively poor at comparing the areas of two ellipses when they have different aspect ratios (Morgan, 2005) or the areas of differently shaped polygons (Raphael & Morgan, 2016). If observers choose a simple heuristic to solve a difficult task, they are clearly vulnerable to biases, as we shall see later.

Classification and Mechanisms of Geometrical Bias

Using the toolkit of concepts introduced earlier in the chapter, and a few others, the main causes of geometric biases will now be introduced with selected examples. The list of examples does not pretend to do any more than round up the usual suspects, and I apologise to the reader whose favorite biases may not have been included.

The 'Perfect Storm' Principle

We should not expect to find a single explanation of a bias. The longer a figure has been in the literature, the more it mutates to accrete new biases to make it newsworthy. Illusions have evolved in the literature not to be informative, but to be conspicuous. Striking variants leave progeny in the textbooks, while their less successful but possibly more comprehensible relatives die out (Morgan & Casco, 1990). An example is the Poggendorff figure, which combines many separate biases (Hotopf & Ollerearnshw, 1972; Hotopf, Ollerearnshw, & Brown, 1974; Day & Dickenson, 1976; Hotopf & Hibberd, 1989). For example, if the normal vertical version of the figure is rotated so that the previously oblique points are vertically aligned, only a small bias remains. The figure is normally shown in a vertical configuration, because the bias with the horizontal figure is considerably less impressive. Morgan and Dillenburger (2016) have described four different biases in the Poggendorff, including a small effect of near-orientation inhibition (Blakemore, Carpenter, & Georgeson, 1970) and a larger bias involved in the construction of virtual lines.

Müller-Lyer's 1889 paper was entitled "*Optische Urteilstauschungen*" (Basic Optical Illusions) and described many variants of his eponymous figure (Figure 1.1), but the arrowhead version has proved the most popular because of its conspicuousness and impenetrability. The first effect he describes is the 'bent line' effect (Müller-Lyer, 1889;

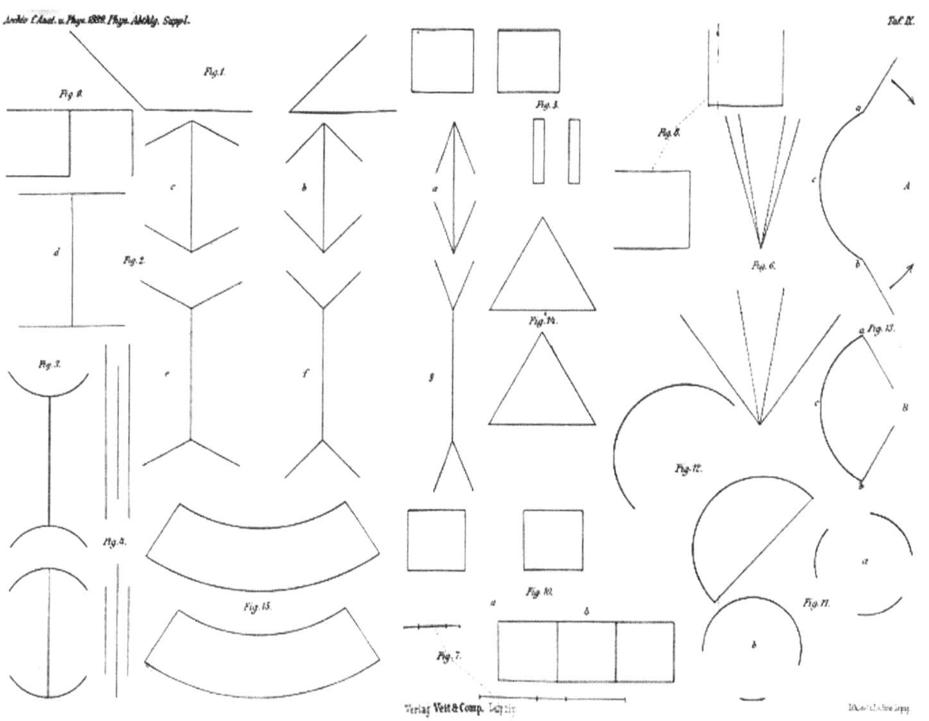

Figure 1.1 Various configurations investigated by Müller-Lyer, from his 1889 paper 'Optische Urteilstauschigen'.

Figure 1), in which the lines composing an obtuse (135 degree) angle look longer than those composing a 45 degree angle. They also seem longer than a single line on its own (Müller-Lyer, 1896a, 1896b). Müller-Lyer ascribes this effect to assimilation: the line segments are assimilated in part to the larger figure of which they form a part. The principle of assimilation is further illustrated by the 'parallel lines' effect in Müller-Lyer's Figure 4, which Pressey (2014) in his 'assimilation theory' was later to consider the prototype of all assimilation biases (see also Jordan & Schiano, 1986, on this point).

Müller-Lyer considered his Figures 2 and 4 to show a basic effect. However, he remarks that the effect in his Figure 1, while being clear and repeatable, is not very strong (*"zwar deutlich und constant, doch nicht sehr stark"*). Therefore, he invents Figure 2 in which the angles are turned into arrowheads, and this is the origin of the textbook version. One cannot help feeling that a lot of later speculation about perspective cues and depth could have been avoided if Müller-Lyer had stuck with his original version, or if his Figure 3 had outcompeted the arrowheads.

Biases in Cue Combination

In the Fraser (1908) 'twisted cord' effect, the lines are locally tilted but globally vertical (Figure 1.2). The local and global tilts are independent cues to orientation, which would normally be in agreement. In the Fraser figure, they are not. If the observer is asked whether the lines are parallel, the answer is 'no'. Presumably then, the perceived orientation combines the two different cues (local and global) in some manner. It is questionable whether we should call a cue combination a bias at all. The implicit assumption seems to be that the observer ought to obey the instruction to look at the global orientation and ignore the local. But this may be impossible. Nothing could be more natural, or more mistaken, than to assume that the visual system can obey verbal instructions in the manner intended by the experimenter (Merleau-Ponty, 1945). To take another example, in the Müller-Lyer figure we ask the observer to measure the shaft length and to ignore the arrowheads. But, as Merleau-Ponty (1945) insisted, no mechanism for isolating the shaft exists: There is no simple 'sensation' corresponding to the length of the shaft. Similarly, in the Fraser figure,

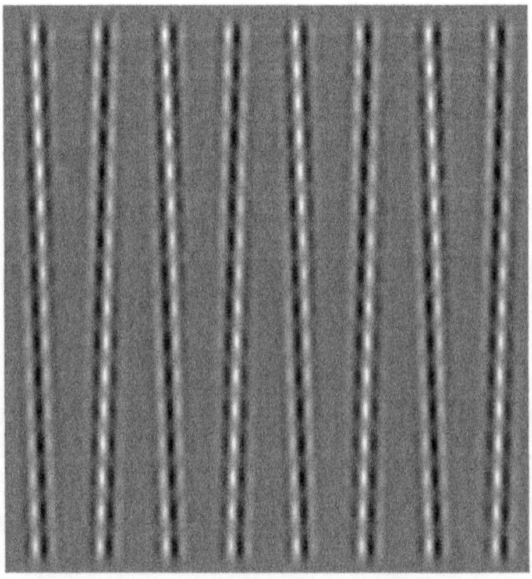

Figure 1.2 The Fraser twisted cord. Each vertical line is composed of Gabor wavelets, phase-shifted as they go up the page. The vertical lines do not look parallel, although the separation between them looks constant. Evidently, acuity for parallelism does not depend upon distance acuity.

to determine global orientation we would need some form of collector unit that is insensitive to local orientation. Again, such a mechanism may not exist. Morgan and Baldassi (1997) proposed a model in which CUs weakly tuned to a particular line of global orientation collected outputs of filters tuned locally to that orientation, which would normally be a good idea. The important point is that CUs tuned to one global orientation and different local orientation do not exist. The so-called Fraser bias is thus a constraint on visual processing.

The logical principles underlying the Fraser effect apply with very few further assumptions to the Munsterberg and Café-Wall figures (Morgan & Moulden, 1986), and arguably also throw light on the appearance of 'spider webs' in grids and lattices (Morgan & Hotopf, 1989).

The general principle that you will get a surprising result if you ask the visual system to use a mechanism that it does not possess may be of wide application. It is like asking the visual system to report the content of a single movie frame, ignoring all others, or to say what three wavelengths are in a mixture.

Bayesian Theory

Where cue combination is involved, a Bayesian analysis may be appropriate. The essence of the Bayesian approach to perception (Knill & Richards, 1996) is that sensory evidence is useful but not completely reliable because of sensory noise and missing information. Sensory information is therefore supplemented by a *Bayesian prior*. The prior is a *likelihood function*, which gives the likelihood of different states of the outside world, in the absence of sensory evidence. The sensory evidence also gives rise to a likelihood function, expressing the likelihood of different states of the world given the sensory evidence. These two likelihood functions multiplied together give the *posterior distribution*, which expresses the likelihood of different states of the world taking into account both the Bayesian prior and the sensory evidence. The observer must then decode the posterior distribution to make a decision – for example, one based on the position of the maximum.

An application of Bayesian theory to the motion analogue of the Fraser twisted cord (Lisi & Cavanagh, 2015; Morgan, 2015) has been described by Kwon, Tadin, and Knill (2015). The effect in the motion domain is that a Gabor patch with its envelope moving in one direction and its carrier in another is seen to move in a direction dominated by the carrier. (Interpret Figure 3 as having both vertical position and time on the y-axis). Kwon and colleagues utilise the physics of a moving object with texture on its surface to suggest a Bayesian account of the phenomenon, using Kalman filtering.

The usual way of testing a Bayesian account of cue combination is to reduce the reliability of the sensory evidence, with the prediction that this will make the prior dominate. For example, Tomassini, Morgan, and Solomon (2010) showed that the tendency to assimilate oriented textures to a cardinal axis is increased when their orientation variance is increased. There have been few systematic attempts to enhance geometrical biases by adding noise. However, it is known that the Poggendorff bias is increased by optical blur (Coren, Ward, Porac, & Fraser, 1978, Glass 1970), and Morgan (1999) showed that this was true for the acute-angle version of Poggendorff but not for the obtuse-angle version, suggesting a different mechanism for the two cases. Since localization accuracy decreases with Gaussian blur (Watt & Morgan, 1984), the finding for acute angles supports a Bayesian view. However, Morgan (1999) offered an alternative account based on second-stage filtering. Coren et al. (1978) investigated the effects of blur on a variety of geometrical figures and found increases for the Poggendorff, Müller-Lyer and Parallel-Lines biases but, puzzlingly, none for the Zöllner figure. There is clearly room for further investigations, including the use of added pixel noise as an alternative to blur.

Second-Stage Filtering

The reasoning in this section is directly inspired by the discussion of hierarchical positional coding by Watt (1988, 1991). Earlier in the chapter, a model for positional coding was set out that had only a coarse sampling of positions, with large and overlapping positional units (PUs) at each point. This was not to be confused with the detailed topographic map in V1. Precision in such a map is obtained by taking the differential outputs of PUs from different positions (CUs). The map codes for relative positions, but we may predict that it would be poor at absolute accuracy for a single point. All the evidence is in favour of the last prediction, since acuity for the relative position of two features is exceptionally poor if the features are presented with a time interval between them (Westheimer, 1981). The advantages of a differential map are computational simplicity, since only a small number of positions are involved, and an improvement in the signal/noise ratio (Hinton et al., 1986; Watt, 1988). The disadvantage is the inability to encode the micro-position of small features independent of features nearby. We may instruct the observer to tell us the distance between the ends of the target line in the Müller-Lyer figure and to ignore the arrowheads, but there is no machinery to do this or any simple sensation corresponding to the position of the intersection (Merleau-Ponty, 1945). We find this difficult to realize because when we inspect one end of the figure, we clearly see a Y intersection composed of three lines. But this is a matter of pattern recognition, not of positional accuracy. The idea that we see the *position* of the intersection clearly is a cognitive delusion.

A particularly clear case of this principle is seen in Müller-Lyer's Figures 5 and 6 from his 1896 paper and in the follow-up by Morgan and Glennerster (1991). Observers were asked measure the distance between two markers on a line. If the line continued for a short distance beyond the markers, they were biased towards seeing it as longer relative to a neutral line.

Similar reasoning has been used to account for the Poggendorff bias (Morgan, 1999) and the Judd bias (Judd, 1899; Morgan & Casco, 1990). In the Poggendorff case, the proposal is that the position of the intersections between pointers and parallels is shifted by coarse second-stage filtering. The Judd effect is that in an '*H*' shape with the verticals tilted, the horizontal line is tilted in same direction, and also appears shorter than the horizontal in an upright 'H'. A model based on coarse second-stage filtering accounts for this effect. The Judd effect can also be considered the essential element of the Zöllner effect, where many '*H*' components are put into an *HHHHHHHHH* configuration. This is not to say that coarse filtering is the only cause of the Zöllner bias; other factors, such as near-orientation repulsion, may also be involved. The Zöllner bias is probably another 'perfect storm'.

The CD model predicts that a particular dot in a dense dot cluster would not have a unique positional signal. This was verified by Morgan and Glennerster (1991) and by Harris and Morgan (1993). The observer's task was to compare the distance between two white dots to a standard distance. The white dots were embedded in a cluster of black dots, which the observer was instructed to ignore. Both the distance between the white target dots and the centroid of the surrounding clusters were varied independently over trials. Results showed that the target dots were judged closer together when the centroids of the surround were closer together than when the centroids were further apart. The similarity to the Müller-Lyer effect is rather obvious.

Near-Orientation Repulsion

Using a configuration like that in the top row of Figure 1.3, Blakemore et al. (1970) showed that the top line of the acute angle was perceptually repulsed from the lower line. This was shown by adjusting the orientation of the top line to match that of the middle line. Morgan and Dillenburger (2016) confirmed this effect with a 2AFC procedure in which the task was to decide in which of two configurations (left and right in Figure 1.3) the upper two lines were more parallel. Both experiments (Blakemore et al., 1970 and Morgan, 2016) agreed that the bias was in the region of 1–2 degrees DVA. A small gap at the angle apex reduced the effect (Experiment 3.2 in Morgan, 2016), as did making the angle a T junction (Experiment 3.3), but the effect was equally strong when the task was to align the pointer with a distant dot (Experiment 3.4).

Carpenter and Blakemore (1973) suggest that angle expansion caused by near-orientation inhibition could be involved in the Hering (bowed lines), Zöllner and Poggendorff biases, which they illustrate in their Figure 2. A word of caution is needed here. 'Angle expansion' is an incorrect description of the effect of near-orientation repulsion, since the two lines in a 45 degree acute angle have the same orientations as the lines in an obtuse 135 degree angle, but repulsion would lead to an angle expansion in the first case

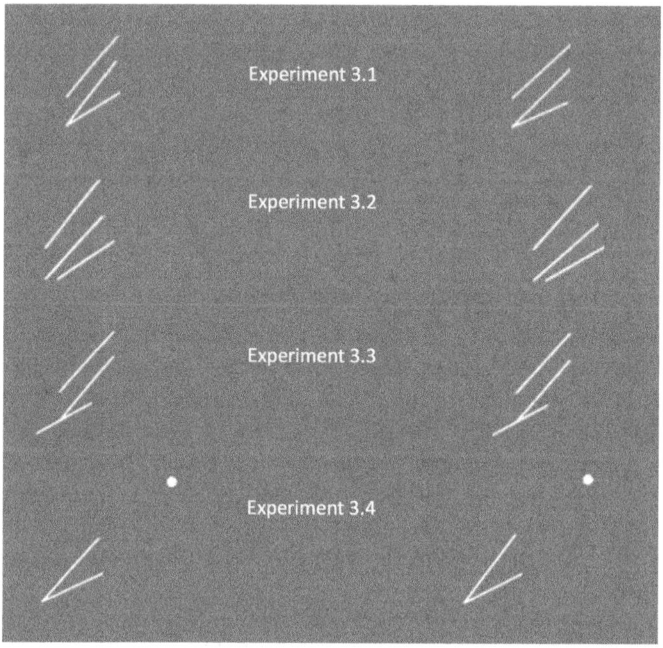

Figure 1.3 Configurations used by Morgan and Dillenburger (2016) to test the orientation-repulsion effect reported by Blakemore et al. (1970). The task (2AFC) was to decide whether the upper two lines in the configuration on the left were more or less parallel than those on the right. In Experiment 3.4 the task was to decide in which of the two figures the line and the dot were more collinear. For further explanation, see the text.

and an angle contraction in the second. In the Poggendorff configuration, these two effects would go in the same direction. Thus it is incorrect to describe the Poggendorff bias (hereafter, the 'P-bias') as an angle expansion, but this is not an objection to orientation repulsion as the explanation. On the other hand, the absence of significant repulsion in Morgan's (2016) Experiment 3.3 (Figure 1.3) does not support the equivalence of the acute and the obtuse angle in causing repulsion. Another objection to the orientation repulsion explanation is that the biases found by Blakemore et al. (1970) and by Morgan and Dillenburger (2016) were no greater than 2 degrees, whereas the P-bias, even for a single pointer and a dot target, is at least 5 degrees (e.g. Morgan, 1999; Hamburger, Hanson, & Gegenfurtner, 2007; Morgan & Dillenberger, 2016). Moreover, Hotopf and Ollerearnshaw (1972) measured the apparent orientation of the pointer (which they call the 'traversal') by matching to a neutral line, and found that even with a 30 degree traversal, the bias was less than half of that of the P-bias measured by adjusting the angle *between* the traversals. Morgan (1999) measured the apparent orientation of a 30 arcmin pointer using a 2-dot comparison stimulus and found no significant bias, although there was a P-bias with a very short (6 arcmin) pointer, which was proposed to be due to neural blurring. Finally, Wenderoth, White, and Beh (1978) used a pointer-dot alignment task but varied the position of the dot on either side of the pointer. Their Figure 10 shows that there was no significant misalignment when the dot was on the unattached side of the pointer, but the normal P-bias when it was on the attached side. All these findings suggest that although the Poggendorff bias may indeed be due to 'mispointing' (Hotopf & Hibberd, 1989; Ninio, 2014; Ninio & O'Regan, 1999), it is not primarily due to orientation repulsion.

To supplement the meagre repulsion effect, Morgan and Dillenburger (2016) provided evidence for a much larger bias in the construction of virtual lines across the virtual surface provided by the two parallels. These constructed lines tend towards taking the shortest path across the gap. This account is consistent with the presence of a P-bias in a figure where the parallels are 'cognitive contours' (Tibber, Melmoth, & Morgan, 2008), and it has something in common with Zanutti's (1976) 'amodal completion' mechanism for the P-bias, although it does not share her prediction that the distance between the parallels will be perceived as compressed.

Contrast and Assimilation (*Kontrast und Konfluxion*)

These are words, not mechanisms, but they are useful in drawing attention to the opposite directions of different classes of bias. In the Parallel Lines figure, the target line is *assimilated* into the length of the flanking lines. Thus a line flanked by longer lines looks longer than one flanked by shorter lines. None of the theories we have considered so far accounts for this, except possibly an extended version of the Bayesian Theory, which we consider below, in which the average of the two lines becomes a prior. By a little 'stretch' of imagination, the Müller-Lyer arrowhead effect also becomes an assimilation (to the overall size of the figure), as Müller-Lyer himself conjectured. The Fraser effect is also an assimilation, of the global orientation to the local. On the other hand, the Ebbinghaus circles and angle-repulsion are clear examples of contrast. The opposite direction of the Ebbinghaus and Parallel lines effect is particularly striking. In the Ebbinghaus figure the central circle is seen as smaller when surrounded by *larger* circles. In the parallel-lines figure the central line is seen as smaller when surrounded by *smaller* lines. Matters are further complicated by the finding (Jordan & Uhlarik, 1985) that assimilation in the lines effect changes to contrast when the context lines are presented before the targets instead of simultaneously. It would be interesting to know whether the same manipulation changes contrast in the Ebbinghaus figure to assimilation.

Müller-Lyer himself grappled with this problem of '*Kontrast und Konfluxion*' in his first 1896 paper. He notes that there is assimilation of line length to parallel flankers, but that a segment of a long line appears shorter than the same-sized segment of a shorter line (contrast). However, the configurations in his Figures 5 and 6 (see Figure 1.1 in this chapter) bear a strong similarity to those used by Morgan et al. (1990) in a study of the Arrowhead bias. The perceived length of the distance between the vertical markers in the comparison stimulus was measured relative to the standard stimulus, and was found to be longer when the distance from the arrowheads *D* was small, but not when it was large. Morgan et al. (1990) proposed that large second-stage filters were unable to resolve a small distance *D*, and thus overestimated the distance between the markers. The effect reported by Müller-Lyer may thus have nothing to do with contrast in his Figure 6, but only assimilation in Figure 5. However, the existence of contrast cannot be denied in the Ebbinghaus circles figure.

The *Kontrast und Konfluxion* problem also crops up in the orientation domain. In the Fraser twisted cord, the global line orientation is shifted in the direction of its local orientation, but if the local orientation is >30 degrees, the effect changes to contrast-repulsion (Morgan & Baldassi, 1997; Tyler & Nakayama, 1984). In parafoveal vision, a Gabor patch appears tilted in the same direction as the surrounding patches, but in central vision it appears tilted in the opposite direction (Mareschal, Morgan, & Solomon, 2010).

Assimilation is related to the idea that the visual system has a known ability to average over small numbers of stimuli. This has been shown for position (Morgan & Glennerster, 1991), orientation (Dakin, 2001) and size (Gorea, Belkura, & Solomon, 2014; Solomon et al., 2011). Parkes, Lund, Andellucci, Solomon, and Morgan (2001) suggested that in parafoveal vision, averaging is obligatory. A Bayesian account might suggest that the average size of a set of elements can serve as a prior for individual elements, which are therefore attracted to the mean. This is not a superficially plausible account of the parallel-lines assimilation effect, but it is the best we have right now.

Conclusions

The naïve theory of vision conceives of the perceived visual image as like a physical image, on which measurements can be made directly with a ruler. Every point in this image is thought of as an object, the distance and angle of which from other points can be accurately measured. When this expectation is disappointed, believers in the naïve theory complain that they have been 'deceived' or 'fooled'. In reality, there is no such image (Merleau-Ponty, 1945). Measurements have to be carried out by neurons, which have limited resolution in time and space. As a first step towards a theory of Geometrical Measurement, this review has postulated a coarse topographic map of the retina, with only a limited number of 'positional units'. Measurements on this map are carried out by differential units termed 'coincidence detectors' or, more generally, 'collector units'. A map having only a few positional units, each with wide spatial tuning, is computationally efficient, and enhances the signal/noise ratio, but it is unable to provide positional information for single points independently of their surrounding context. This can have occasionally surprising results.

Chapter 2

Perceptual Anisotropies in Visual Space

Jose Antonio Aznar-Casanova and
Leonardo Gomes Bernardino

"Space and time are modes in which we think,
not conditions in which we exist."
Albert Einstein

This morning you turned off the alarm clock, got out of bed, and walked around your home without bumping into the furniture. During the first hours of your day you performed several similar activities with little or no difficulty, which wouldn't have been possible without a representation of the environment surrounding you. How does this come about?

Understanding the nature of sensory and perceptual experience is one of the fundamental problems in human thought. Our visual awareness of the external world is the end result of processes initiated in our photoreceptors, which perform sensory transduction and send electric signals to the brain along the optic nerve. Although we are not conscious of this, these signals carry data about basic perceptual properties (direction of movement, position, size, color, shape), which are hierarchically processed in different levels and structures of the central nervous system.

In the 1980s, studies that performed ablations on the brains of monkeys revealed two main and specific routes of visual processing: the *ventral pathway* and the *dorsal pathway* (Ungerleider & Mishkin, 1982; Mishkin, Ungerleider, & Macko, 1983). The ventral pathway leads from the primary visual area, and data travels along it through the temporal lobe. Its function is related to the recognition of objects, and therefore it is also known as the "what" pathway. In the dorsal pathway, the processing of spatial information occurs in the parietal lobe, and due to its function, it is also known as the "where" pathway. Goodale and Milner (1992) subsequently argued that it would be more appropriate to call the dorsal stream the "how" pathway, given its role not only in locating but also in planning and executing tasks (for a review, see Norman, 2002). More recently, several authors have found evidence to suggest that spatial information is also processed along the ventral pathway, thus challenging the notion of a strict functional division between these visual pathways (Carlson, Hogendoorn, Fonteijn, & Verstraten, 2011; Macevoy & Epstein, 2007; McKyton & Zohary, 2007).

One conclusion to be drawn from all this is that our perception is action-orientated (Cutsuridis, Hussain, & Taylor, 2011), and most of our behavior is primarily but not exclusively driven by visual information. Based on perception–action coupling, Cutting and Vishton (1995) proposed a division of space into three zones or layers: (1) *peripersonal space*, also called near space and referring to distances within arm's reach (up to 2 m); (2) *action space*, distances from 2 m up to 30 m; and (3) *space in perspective or vision space*, involving distances beyond 30 m. Clearly, therefore, understanding how we represent space

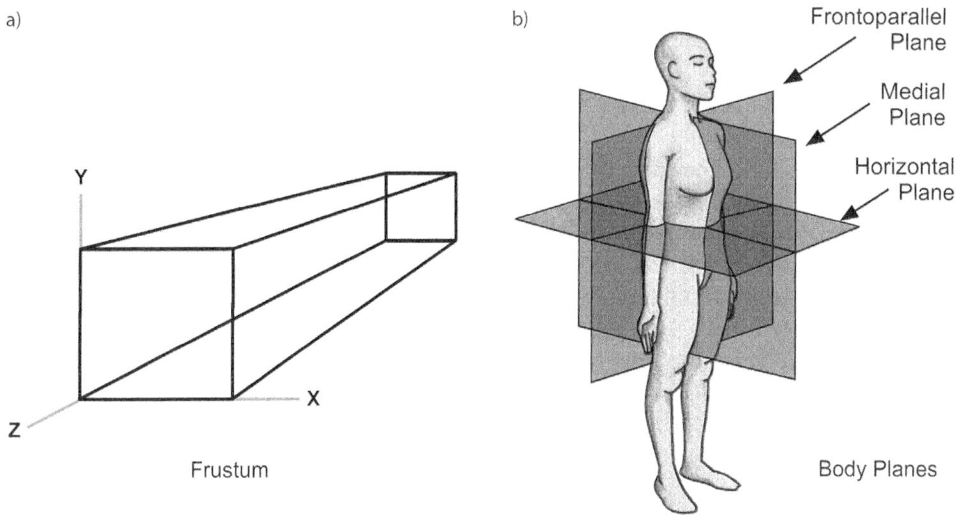

Figure 2.1 (a) Three-dimensional spatial configuration defined by orthogonal axes (X, Y, and Z), with VS represented as an implicit frustum. (b) Spatial planes starting from the egocentric frame of reference. Note that all of them are determined by some spatial embodiment.

is far from a trivial problem, and in addressing it we need to clarify what we mean by *space*. One way of doing so is to distinguish between *physical space* (PS) and *visual space* (VS).

PS is constituted by objects and by surfaces and backgrounds that we consider to exist independently of us. By contrast, VS is an internal representation of our phenomenological experience with PS, which is organized into a three-dimensional configuration defined by orthogonal axes (X, Y, and Z), as indicated in Figure 2.1a. This representation therefore has a spatial structure: in other words, there are certain underlying distance relationships between the objects themselves, as well as between the objects and the beholder (Indow, 2004).

Starting from the spatial configuration of objects, surfaces, and backgrounds, the visual system builds an *exocentric* or *allocentric* frame of reference, that is, one external to the observer. According to O'Keefe (1991, 1993), this reference frame is mapped by pyramidal cells in the hippocampus, cells whose representational ability is limited to three dimensions. This does not, however, mean that PS is necessarily three-dimensional. Furthermore, the convergence of visual, proprioceptive, vestibular, and somesthetic data produces a corporal representation from which VS itself stretches. What is established, therefore, is another reference for mediating our interaction with the environment. Because this frame of reference is body-centered, it is called *egocentric*.

In the context of research on VS perception there are two key questions that remain unanswered: Which properties of PS are preserved when we map it? And how reliable is our representation of PS? The biochemical, physiological, and psychophysiological processes that originate our perception in general, and specifically our VS, are intrinsically ambiguous. There is not necessarily an exact equivalence between the physical stimulus and its perceptual representation; in other words, the end result of spatial representation (i.e., VS) is not necessarily faithful to the exterior world (i.e., PS). First of all, we do not build a point-to-point representation of our visual field. Therefore, saying that vision is our window to the world or comparing it to a photographic camera is not accurate, as such metaphors are only partially applicable. Second, the context and our knowledge, experience, and expectations also affect perceptual processes, which means that perception is a constructive process, it is active and creative. Visual illusions are good examples of the constructive nature of visual processing.

In recent decades, countless studies involving a wide variety of experimental tasks and paradigms have shown that our estimates regarding perceived dimensions in space are less accurate and precise than would be the case were PS and VS to be coincident (Cuijpers, Kappers & Koenderink, 2000a, 2000b; Hershenson, 1999; Koenderink & van Doorn, 2008; Morgan, 2003; Sedgwick, 1986, 2001; Wagner, 1985, 2006; Wagner & Gambino, 2015). We can conclude, therefore, that VS is neither homogeneous nor *isotropic*. This latter term is borrowed from physics and is used to refer to a substance or object whose properties remain constant regardless of the direction considered. The contrasting term is *anisotropic*. Thus, the observations according to which human performance varies considerably depending on the orientation of stimuli (cardinal or oblique, radial or tangential) and the position of those stimuli in space (to the left or right, up or down, near or distant) are referred to as *VS anisotropies*.

In light of the above, our goal in this chapter is to provide an overview of perceptual anisotropies in VS and to discuss theoretical and methodological perspectives for future research. Perceptual anisotropies are generally investigated from three spatial planes (Figure 2.1b): the *frontoparallel*, the *medial* (or sagittal), and the *horizontal* (or transverse) (Sedgwick, 2001). These planes constitute the VS and are determined by an egocentric frame of reference that in our species is body-centered or head-centered. Based on the results of psychophysical and neurophysiological research, we will present and discuss the processes that originate and modulate the multiple phenomena related to perceptual anisotropies on these planes.

Perceptual Anisotropies in the Frontoparallel Plane

The frontal (or coronal) plane is any longitudinal plane that divides the body into anterior and posterior halves. Our biped posture and the frontal position of our eyes define our visual direction and impose a frontal plane, separating the space in front of and behind us. Here we are interested in the frontoparallel plane, which can be defined as a plane in front of and at any distance from the observer and that contains objects without difference of depth between them. Observing monocularly a static point of fixation, we have access to a two-dimensional oculocentric system with two orthogonal directions, horizontal and vertical, axes X and Y, respectively (Figure 2.1a).

There is a long tradition of research into perceptual anisotropies in the frontoparallel plane (see Blumenfeld, 1913; Helmholtz, 1886/1962; Hillebrand, 1902), and this has shown how our representation of this plane is not a flat surface, despite what its name suggests. For example, in a paradigm known as distance alley (D-alley) experiments, participants are asked to adjust points' positions in such a way that they appear to have a constant lateral distance between them, i.e., equally separated in a frontoparallel plane. The result is a curved surface and the direction of the curve is concave in the peripersonal space and convex in the extrapersonal space. Notably, there is also evidence that the structure of the frontoparallel plane is better described by Euclidean geometry in the peripersonal space (Indow & Watanabe, 1988; Indow, 2004). Let us now consider some of the phenomena related to perceptual anisotropies in the frontoparallel plane.

Oblique Effect

Several psychophysical studies have observed better performance on measures of accuracy and precision when task stimuli are vertically or horizontally orientated, as compared with other orientations (Campbell & Kulikowski, 1966; Howard, 1982; Howard & Rogers, 2002; Mansfield & Ronner, 1978). This preference for horizontal and vertical spatial orientations is known as the *cardinal effect*. Accordingly, worse performance in the judgment of stimuli aligned to non-cardinal or oblique orientations can be observed in many species. This perceptual anisotropy of VS is referred to as the *oblique effect* (Appelle, 1972). It should be noted that this effect is not limited to two-dimensional representations, since there is evidence of its occurrence in three-dimensional VS in a Vernier acuity task (Aznar-Casanova, Torrents, & Torro-Alves, 2008).

In terms of explaining the oblique effect, optical properties, unequal distribution of cells in the retina, and oculomotor factors have all been proposed as being responsible for this anisotropy. However, there is considerable evidence to suggest that the answer to its origin and modulation lies in the neurophysiology of the visual system. The first findings regarding initial levels of processing came from the classic single-cell studies of Hubel and Wiesel (e.g., 1959, 1962). Their results showed that in the cat's visual cortex the number of brain cells that selectively respond to cardinal orientations is greater than the number that are sensitive to oblique orientations. Similar results were also found in macaque monkeys (De Valois, Yund, & Hepler, 1982), animals whose visual system is very similar to that of humans. Li, Peterson and Freeman (2003) extended previous findings showing that the brain cells with a preference for oblique meridians are less selective – in other words, they exhibit broader orientation tuning widths. However, this prevalence was observed only in simple cells in the fovea and in the periphery of the visual field, which respond to greater spatial frequencies. These authors suggest that the oblique effect originates in and is modulated by intracortical connections.

Following the results obtained in animals, a study using fMRI showed that the human striate cortex responds differently to cardinal and obliquely orientated stimuli (45° and 135°), although no such difference was observed among areas of the extrastriate cortex (Furmanski & Engel, 2000). Taken together, these findings confirm the hypothesis that the oblique effect is the result of a greater number of brain cells with high selectivity and which respond preferentially to horizontal and vertical orientations. Thus, orientation is a characteristic represented in the striate cortex within hypercolumns in pinwheel-like patterns and centered in brain cells with a preference for stimuli aligned to the orientations of 0° and 90° (Das & Gilbert, 1997; Peterhans, von der Heydt, & Baumgartner, 1986). However, the representation of orientation is not made only on this fine spatial scale but also in topographic maps formed by neighboring brain cells that share the preference for a specific orientation (Freeman, Brouwer, Heeger, & Merriam, 2011).

More recent research has shown that the explanation for this phenomenon is not circumscribed to the initial levels of visual processing. For example, a study conducted with different animals (owls, cats, and marmoset monkeys) and using optical imaging techniques revealed that areas of the extrastriate cortex also have a smaller proportion of brain cells that are selective to stimuli in orientations of 30°, 60°, 120°, and 150° (Liu & Pettigrew, 2003). These findings are supported by evidence for the involvement of superior visual areas in the oblique effect in humans when performing an orientation perception task (Buchanan-Smith & Heeley, 1993), as well as in a relative depth judgment task with single-image random-dot stereograms (Lupón-Bas et al., 2014). These three-dimensional percepts are generated by the activity of the extrastriate visual pathways, mainly in the dorsal stream (Stidwill & Fletcher, 2011).

In a similar vein, a study that combined a psychophysical task with a brain-imaging technique (fMRI) showed a preference for cardinal orientations (be they indoor and outdoor scenes or non-scene stimuli) in an area of the medial temporal cortex (ventral stream) known as the parahippocampal place area (PPA) (Nasr & Tootell, 2012). This difference in PPA activity shows that the oblique effect is not simply a result of activity in the striate cortex, but that it also involves superior areas such as the dorsal and ventral pathways of visual processing. Interestingly, a study with ferrets indicated that orientation selectivity might be present early in the life of these animals (Chapman & Bonhoeffer, 1998). This finding suggests that initial visual experiences have little influence on the development of this prevalence, even if they are important to its maintenance. However, although this research provides an interesting clue about genetic and environmental interaction to produce the oblique effect, we still have to determine the exact role of ontogeny in the development and maintenance of the human visual mechanisms involved in this anisotropy.

In addition to neurophysiological explanations, the influence of gravitational force has also been proposed to be a factor in the origin of the oblique effect. Gravity establishes an aligned vertical directional axis that is coincident with the body axis. It therefore contributes to better performance in perceptual judgments made in the vertical and horizontal meridians, which are parallel and perpendicular, respectively, to its force. In fact, there is evidence that visual, vestibular, proprioceptive, and gravitational information are combined to define the vertical direction (Vingerhoets, De Vrijer, van Gisbergen, & Medendorp, 2009). However, the egocentric frame of reference plays a major role in the oblique effect, which has been observed even in microgravity environments (Lipshits, Bengoetxea, Cheron, & McIntyre, 2005; Lipshits & McIntyre, 1999).

Another proposed explanation is that visual scenes in the environment are more likely to involve cardinal than oblique data. This hypothesis finds support in studies that have analyzed the cortical activation to horizontal, vertical, and oblique-oriented gratings (Coppola, White, Fitzpatrick & Purves, 1998) and the statistical properties of visual scenes, whose results show a greater proportion of horizontal contents, followed by vertical, and lastly by oblique contents (Keil & Cristobal, 2000). It was also observed that asymmetries in the vertical and horizontal representations of visual scenes are more commonly verified in low and high spatial frequencies, respectively.

Interestingly, some studies have found biases that are diametrically opposed to those observed in the oblique effect, that is, better performance with obliquely oriented images. This phenomenon is referred to as the *horizontal effect* (see Essock, DeFord, Hansen, & Sinai, 2003; Hansen & Essock, 2004, 2006). As in the oblique effect, the name came from the stimuli direction associated with worse performance on perceptual tasks. To explain these results the authors suggested that the representation of orientations correlates negatively with their availability in visual scenes. Hence, the widespread availability of horizontally aligned information would "compensate" for a poorer representation of this horizontal content by means of a neural normalization process. By the same token, oriented rare contents would be overrepresented and, therefore, better perceived, as in the case of oblique orientations.

At first glance, these results suggest that the oblique effect and the horizontal effect have different underlying neural mechanisms. However, both are field-independent orientation anisotropies, that is, their occurrence depends on the processing of stimulus orientation information, regardless of their position in the visual field. Therefore, in order to elucidate these controversial findings, we still need to determine the exact relationship between behavioral measures and neural activity (Mannion, McDonald, & Clifford, 2010).

It should also be noted that many human studies in recent decades have found behavioral evidence for the existence of anisotropies in meridian-relative orientation. This means that the representation of orientation depends on its angular position in the visual field. For instance, some studies observed better performance when the orientation and the angular meridian coincided, especially in the case of major eccentricities (Westheimer, 2003, 2005). The neural substrates for this radial bias have already been reported in cats and monkeys (Leventhal, 1983, and Schall, Perry, & Leventhal, 1986, respectively). Based on the findings of fMRI studies, this bias is now being associated with activity in the striate and extrastriate cortices, and the results also demonstrate the existence of a direct relationship between orientation tuning and retinotopic maps (Freeman et al., 2011; Mannion et al., 2010; Sasaki et al., 2006).

The oblique effect is clearly a robust behavioral phenomenon, even if it is still not completely understood. One question that remains to be answered is whether anisotropy for different tasks has a common origin or whether, for each particular task, it is the result of changes in the underlying properties of neural mechanisms at a single level of processing (Heeley, Buchanan-Smith, & Cromwell, 1997). Of course, the main difficulty lies in the fact that we are yet to understand exactly how orientation is encoded by the visual system.

In addition to perceptual anisotropies between cardinal and non-cardinal orientations, many psychophysical and neurophysiological studies also indicate the existence of asymmetries between the horizontal and vertical orientations.

Horizontal-Vertical Illusion (HVI)

One manifestation of this anisotropy is known as the *horizontal-vertical illusion* (HVI). This illusion occurs when, in an arrangement involving two lines of the same length, one horizontally oriented and the other vertically oriented, the length of the latter is overestimated in relation to the former. A well-known example of this is the Shepard illusion, in which the surfaces of the boards of two tables are perceived as being different. Although they are in fact identical, the board of the table on the left seems longer and narrower in comparison with that on the right (Figure 2.2a). It is important to note that this is not the common interpretation of Sheppard's illusion (Tyler, 2011).

The HVI has been extensively investigated over the last decades and it is a robust phenomenon (for recent discussion, see Hamburger & Hansen, 2010). The magnitude of the illusion varies between 2% and 6% for L-shape arrangements (Figure 2.2b) and reaches 9% for inverted-T arrangements (Figure 2.2c) in pen-and-paper and computerized tasks

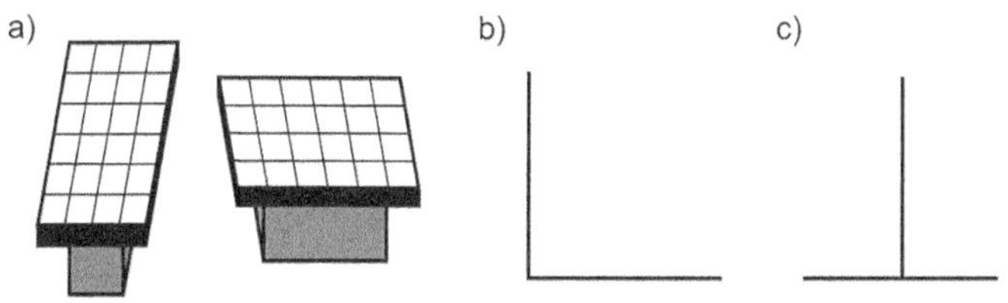

Figure 2.2 The Shepard illusion (a), in which the boards of the tables are perceived as being different despite being identical, and the two more common arrangements of the horizontal-vertical illusion (HVI), in L-shape (b) and inverted T-shape (c).

with line drawings and pictures. The larger magnitude of the second configuration is related to the bisection of the horizontal line by the vertical, and hence a model to explain the HVI must take into consideration two components: orientation anisotropy and length bisection (Mamassian & de Montalembert, 2010; Mikellidou & Thompson, 2013). It should also be noted that the magnitude of the HVI can reach very high values, approximately 20% in virtual reality environments using a head-mounted display, while in real-world, outdoor settings it may exceed 50% in-depth extents on the ground (Chapanis & Mankin, 1967; Higashiyama, 1996; Higashiyama & Ueyama, 1988; Yang, Dixon, & Proffitt, 1999).

Many explanations have been proposed for this perceptual anisotropy and they can be divided into four groups (for a review, see Higashiyama, 1996): (1) gravity, (2) physiology, (3) frame of reference, and (4) perspective. Gravitational force is an important reference system insofar as it strongly influences our sense of movement, making us experience the horizontal and vertical dimensions as unequal, i.e., an equal distance in the vertical direction is perceived as longer than in the horizontal direction.

For instance, any healthy adult is able to walk a distance of 300 m in a flat terrain without much effort; however, the same distance is not easy to accomplish in a slanted terrain (e.g., a hill). Thus, VS is affected by the energy employed in locomotion, which is related to the effort to cope with gravity (Howard & Templeton, 1966). From this point of view, the HVI would result from a misperception of the vertical axis due to the influence of gravity. Although this commonsense explanation has not been extensively tested, there are studies that support it (e.g., Bhalla & Proffitt, 1999; Yang et al., 1999).

The physiological explanations are based on the notion of an unequal distribution of cells in the retina, for instance, a greater number of cells dedicated to horizontal direction processing. This prevalence is also found in the cortical organization of humans and other animals such as cats, ferrets, and monkeys (Chapman & Bonhoeffer, 1998; Coppola et al., 1998; Furmanski & Engel, 2000; Li et al., 2003, Liu & Pettigrew, 2003). Thus, these asymmetries in the representation of meridians would result in a magnification of the vertical representation. Oculomotor factors are another source of asymmetries, since the fact that the execution of ocular movements along the vertical axis is more complex would also generate these vertical biases. Furthermore, there is evidence that the HVI is affected by small changes in the position of the eyes (Richter, Wennberg, & Raudsepp, 2007).

It is important to note that while observing a static point, the visual field has an elliptical shape and comprises a visual angle of roughly 160° in the horizontal meridian and 135° in the vertical meridian (Ruch & Fulton, 1960). The fact that our visual field is horizontally elongated is the basis for the frame-of-reference explanation (Künnapas, 1955, 1957). According to this explanation, vertically oriented stimuli would be overestimated because they are closer to the boundaries imposed by the shape of our visual field.

This framing effect yields certain predictions about the HVI. Its magnitude, for example, should be reduced or eliminated under monocular vision conditions and in complete darkness, because the visual field would, respectively, become less asymmetric and have no distinctive limits. In a series of experiments, Prinzmetal and Gettleman (1993) confirmed these predictions. In the same vein, a more recent study investigated the horizontal/vertical aspect ratio of rectangles and found evidence that the HVI is a two-dimensional phenomenon, which also supports the framing effect (van Ee & Erkelens, 2000).

This evidence is contrary to the perspective explanation of the HVI. According to the latter, a vertically oriented line would be perceived as slanted away from the observer, which

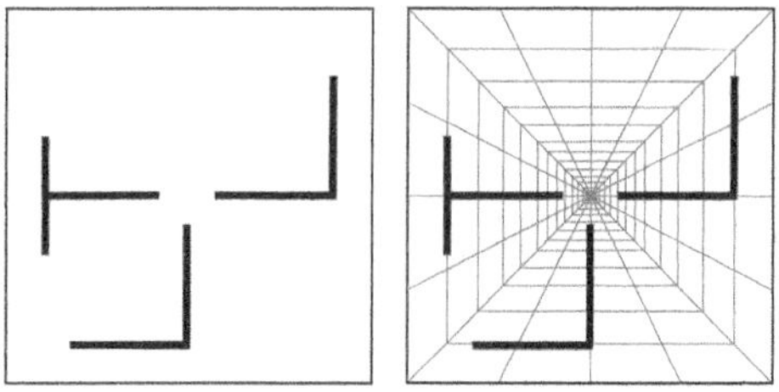

Figure 2.3 Depth interpretation and HVI. According to the perspective explanation, observers tend to perceive the L-shape arrangements as receding in depth, which results in the overestimation of the vertical lines because of size scaling. Thus, our representation of space would be more akin to the figure on the right, in which a bigger perceptual distortion in comparison to the figure on the left (no depth cues) can be observed.

would result in its overestimation (Figure 2.3). This occurs because our visual system size-scales to compensate for the perspective effect. Thus, it was proposed that HVI is the result of inappropriate size-constancy scaling (Gregory, 1963; Woodworth, 1938). It has also been suggested that the interpretation of line arrangements in the frontoparallel plane as flat projections of 3D scenes reflects a preference for the way in which we represent visual scenes (Mamassian & Landy, 2001). However, this claim is controversial insofar as the participants in a study did not report making use of this three-dimensional interpretation (Ward, Porac, Coren, & Girgus, 1977).

The perspective explanation presupposes that the HVI is modulated by the configuration of the stimuli because this provides depth cues that might be used for size scaling. For example, the greater magnitude of the HVI in inverted-T as compared with L-shape arrangements probably occurs because of the presence of the T-junction occlusion cue. Although this prediction has not been confirmed (Wolfe, Maloney, & Tam, 2005), other studies have found that the addition of a surrounding context affects this illusion. Specifically, the size distortions observed in the HVI were differentially modulated depending on the texture gradients into which the stimuli were positioned (Williams & Enns, 1996). These results nonetheless indicate that this depth interpretation interacts with and adds to the frame of reference to produce the perceptual distortions observed in the HVI.

The results discussed in this section show that the HVI is observed across a wide range of stimuli, tasks, experimental paradigms, and settings, highlighting its complexity and multi-determined nature. It is clear, therefore, that simple models based on just one of the aforementioned explanations will be incapable of describing, explaining, and predicting the phenomena related to the HVI.

Line-Bisection Asymmetry

A well-known spatial task in which perceptual anisotropies can also be observed is the line-bisection task. In this task, participants indicate the midpoint of a horizontal line, that is, the spatial location that divides it into two equal parts. At first glance, this may appear to be a task that involves only the dorsal pathway and, therefore, that it is related to an egocentric reference system (Fink, Marshall, Weiss, & Zilles, 2001). However, there is behavioral and neurophysiological evidence that participants make comparisons between the segments that go beyond the division of the line, comparisons that are characteristic of object recognition and whose processing takes place in the ventral pathway and which depend on an allocentric reference system (Riestra, Womack, Crucian, & Heilman, 2002; Foxe, McCourt, & Javitt, 2003).

In general, participants indicate a midpoint to the left of the real physical midpoint. In persons with no neuropsychological deficits this leftward bias is called *pseudoneglect*

(Bowers & Heilman, 1980). This term alludes to those patients with cerebral lesions who manifest attention deficits in the visual field contralateral to the lesion hemisphere, a neuropsychological condition known as *hemispatial neglect*. This clinical condition tends to be observed in patients with lesions in the right cerebral hemisphere, which is involved in the mobilization of global attention in spatial tasks. In support of this hypothesis, a meta-analysis of more than 70 studies concluded that modifications of the cerebral organization associated with aging affects attention modulation, which would explain the observation of leftward bias in younger participants and rightward bias in older people (Jewell & McCourt, 2000). Besides age, many studies indicate that the magnitude and even the direction of the deviations depend on other within-subject variables (e.g., gender and hand dominance), as well as on stimuli features (Brodie, 2010; van Vugt, Fransen, Creten, & Paquier, 2000).

The line bisection task with vertically oriented stimuli also reveals an asymmetry, in this case an upward shift that is referred to as *altitudinal pseudoneglect* (Nielsen, Intriligator, & Barton, 1999; Suavansri, Falchook, Williamson, & Heilman, 2012). A similar pattern is observed in bisection tasks involving radially oriented lines, where results show biases above the real midpoint (Shelton, Bower, & Heilman, 1990; Toth & Kirk, 1996). Some authors argue that this bias is due to an interpretation of VS as receding in depth, indicating a departure from the egocentric reference. This interpretation received support from studies showing that obliquely oriented stimuli produce vertical disparities that are involved in the perception of distance and depth (Brenner, Smeets, & Landy, 2001; Chowdhury, Christiansen, Morgan, & DeAngelis, 2008; Rogers & Bradshaw, 1993).

This upward bias has been consistently reported, indicating that this anisotropy is affected by attention- and context-related factors related to the stimuli (Chieffi, 1996; Fink et al., 2001), as well as by hand dominance, since this kind of bias was observed only in right-handed people (Scarisbrick, Tweedy, & Kuslansky, 1987). There is also evidence that this asymmetry is not correlated with the task of horizontally oriented bisection or with the HVI (Post, O'Malley, Yeh, & Bethel, 2006, van Vugt et al., 2000), suggesting that distinct and dissociated processes underlie the biases observed in these phenomena.

Perceptual Asymmetries of Visual Fields

The three perceptual anisotropies (oblique effect, HVI, and line bisection asymmetry) discussed so far are mainly observed in arrangements of stimuli presented in the central visual field. However, the presentation of stimuli in different positions in the visual field also results in asymmetric representations. Based on the horizontal meridian (X-axis), the frontoparallel plane can be divided into the right and left visual fields. In many behavioral tasks, research has found differences in the performance of participants depending on the hemifield in which stimuli are presented. Verbal stimuli, for example, are processed more efficiently when presented in the right visual field (Scott & Hellige, 1998), whereas in tasks requiring judgment of stimuli brightness, numerosity, and size, greater accuracy and shorter reaction times were observed when stimuli were presented in the left visual field (Charles, Sahraie, & McGeorge, 2007; Nicholls, Bradshaw, & Mattingley, 1999).

Many studies on the horizontal meridian have used the divided visual field paradigm in individuals with or without neurological problems, and their results have been interpreted as evidence of hemispheric specialization (Bourne, 2006; Lindell & Nicholls, 2003; Sergent, 1982, 1995). In other words, differences in performance indicate an advantage in the processing of information by the contralateral hemisphere. According to Sergent (1982), a general rule would be that the left and right hemispheres are specialized in the processing of high and low spatial frequencies, respectively. This hypothesis has not been confirmed by other studies (Boeschoten, Kemner, Kenemans, & Engeland, 2005). However, these discrepancies can be explained by the coarse-to-fine hypothesis of visual information processing (see Peyrin et al., 2010).

Alongside explanations based on structural mechanisms, the main hypothesis proposed to account for these phenomena concerns the involvement of the right cerebral hemisphere, especially the parieto-occipital cortex. For instance, studies have observed a preferential distribution of both transient attention and sustained attention toward the left visual field (Brodie, 2010; Corbetta & Shulman, 2002; Du & Abrams, 2010; Fink et al., 2001; Flöel et al., 2005; Halligan & Marshall, 1994; Mao, Zhou, Zhou, & Han, 2007). In addition to structural mechanisms of attention, we cannot rule out the influence of the scanning direction of visual information, which is strongly affected by our reading habits. In Western languages, for example, reading is performed from left to right and top to bottom. This pattern of organization would result in a greater allocation of attention to the left visual field, even in nonverbal tasks.

We can also divide the frontoparallel plane by taking the Y-axis as our reference. Many studies also indicate the existence of perceptual anisotropies between the *upper visual field* (UVF) and the *lower visual field* (LVF) (for a review, see Christman & Niebauer, 1997). Some studies report a processing advantage when stimuli are presented in the UVF, mainly in tasks with verbal stimuli, although this advantage is also described in spatial tasks (Fukusima & Faubert, 2001) and for masked face processing (Quek & Finkbeiner, 2016). In a great variety of tasks, however, participants were observed to perform better when stimuli were presented in the LVF, including moving targets (Hubbard, 2001). The *vertical meridian asymmetry* is a particular case of this lower visual hemifield advantage (Carrasco, Talgar & Cameron, 2001; Talgar & Carrasco, 2002), and recently it has beenshown that this asymmetry is not observed at intercardinal locations beyond 30° of polar angle from the vertical meridian (Abrams, Nizam & Carrasco, 2012).

This downward asymmetry has also been observed in eye movement studies. For instance, researchers have observed a greater number of fixations (Parkhurst & Niebur, 2003) and of saccadic movements to the LVF (Foulsham & Kingstone, 2010). Interestingly, these authors demonstrated that this predominance remained even for inverted images (180° rotations), in which case most saccades were directed upward. By contrast, there is evidence of a dissociation between sensory, proprioceptive, and oculomotor information in the representation of the UVF and LVF. For example, Richter et al. (2007) found more accurate judgments of L-shaped arrangements in the LVF, independently of whether or not the stimulus was presented through prismatic lenses, which invert the image.

Two hypotheses have been proposed to explain these perceptual anisotropies. Based on the evidence that attention shares neural substrates with eye movements (see Awh, Armstrong & Moore, 2006), the first hypothesis involves the differential allocation of attention to the LVF (He, Cavanagh, & Intriligator, 1996). This idea is supported by the finding that the V1 region representing the LVF sends more projections to the dorsal pathway, the system involved in attention, than to the corresponding part of the UVF (Gazzaniga & Ladavas, 1987). However, clinical studies with hemispatial neglect patients suggest that the LVF and the UVF possess distinct attentional systems that depend on the posterior parietal cortex/dorsal pathway and on the inferior temporal cortex/ventral pathway, respectively (Rapcsak, Cimino, & Heilman, 1988; Shelton et al., 1990).

The second hypothesis concerns differences in visual processing between the UVF and the LVF. In a series of psychophysical studies, Carrasco and colleagues found that asymmetry is modulated by sensory factors such as the spatial frequency of the stimuli (e.g., Carrasco et al., 2001). Other evidence for a LVF advantage in object processing comes from neurophysiological studies. For instance, researchers have observed a higher density of photoreceptors in the superior retina (which represents the inferior part of VS), larger sensory magnitudes for lower field stimuli, and greater activation of the visual cortex for stimuli in the LVF (Curcio & Allen, 1990; Hagler Jr., 2014; Portin, Vanni, Virsu, & Hari, 1999).

A further point of note is that the representation of the visual field is described as a retinotopical map because it preserves the spatial arrangement of information from the retina. Based on this principle, information from the LVF and UVF is processed and represented in different regions in both the striate cortex (above and below the calcarine fissure, respectively) and the extrastriate cortex (dorsally and ventrally, respectively) (Sereno et al., 1995). Further support for differences between the LVF and UVF in the initial stages of visual processing comes from electrophysiological (Qu, Song, & Ding, 2006) and brain imaging (Liu, Heeger, & Carrasco, 2006) studies.

According to Previc (1990), this asymmetry is a result of our evolutionary process. In this context it is worth noting the influence of our bipedal posture and gravitational force on locomotion and manipulation of objects. It is more likely, therefore, that the information most relevant to our survival will be found in the LVF. Hence, the LVF is related to peripersonal space and visuomotor coordination, and its processing occurs in the magnocellular and dorsal pathways. On the other hand, the UVF is related to action space and to visual search and object recognition tasks, and its processing occurs in the parvocellular and ventral pathways. This association between the LVF and near space, as well as that between the UVF and far space, may seem odd, because we presuppose no depth difference between objects in the frontoparallel plane. However, for stimuli presented in the vertical meridian, their height in the visual field is a monocular depth cue. In other words, even if there are no other depth cues available, objects presented in the UVF are perceived as farther away than objects in the LVF (Farran, Whitaker, & Patel, 2009; Tozawa, 2010).

In short, the studies discussed earlier in the chapter suggest that horizontal biases are related to hemispheric specialization, and that vertical biases occur due to differences between the ventral and dorsal pathways. Furthermore, the leftward bias is explained by attention mechanisms, which are affected by our experience of scanning for visual information, especially the reading direction in our native language. Thus, we can conclude that the perceptual anisotropies in the horizontal and vertical meridians have different underlying mechanisms.

In a study that supports this dissociation, Nicholls, Mattingley, Berberovic, Smith, & Bradshaw (2004) found no strong correlation between the horizontal and vertical anisotropies for oriented visual stimuli. In addition, they observed greater upward biases in the left spatial hemifield on a vertical bisection task, a result that was later reported by Suavansri et al. (2012). Based on these findings, they claim that oblique biases arise from the interaction between the cognitive and neural mechanisms involved in the vertical and horizontal biases. It should also be noted that an earlier study revealed greater leftward biases in the UVF than in the LVF on a horizontal bisection task (McCourt & Jewell, 1999). It was suggested that this interaction takes place in the lateral intra-parietal area (Nicholls et al., 2004; Fink et al., 2001).

One way of understanding better the interaction between vertical and horizontal biases is to analyze participants' performance based on visual field quadrants, namely upper right, upper left, lower left, and lower right. For example, judgments made in the lower right quadrant result in more accurate size estimates (Raudsepp, 2001b), even though greater perceptual distortions have been described in this quadrant (Fukusima & Faubert, 2001).

Using this approach, Hagenbeek and Van Strien (2002) found processing advantages for the upper right and lower left quadrants in a face-matching task, results similar to those reported by Christman and Niebauer (1997) and Thomas and Elias (2011). The close relationship between these quadrants was also described in an interesting study whose aim was to determine the spatial distribution of hallucinations associated with sleep paralysis (Girard, Martius, & Cheyne, 2007). In the experimental paradigm used by these authors, the inside VS representation is directly accessed without being affected by the type of stimulus or the task. Their results showed a greater concentration of projected hallucinations in the upper right and lower left visual fields.

More research is needed to understand orientation processing in VS and, consequently, the mechanisms underlying these perceptual anisotropies. Furthermore, and as we have seen, a common hypothesis proposed to explain these anisotropies is the depth interpretation of VS, even in the frontoparallel plane. Thus, further investigation is also required of the more distal extra-personal space, as well as of the three-dimensional representation of VS, which is directly related to the horizontal and medial planes and which includes the Z-axis.

Perceptual Anisotropies in the Medial and Horizontal Planes

The medial plane is composed of the Y and Z axes (Figure 2.1a), dividing space into left and right halves. The organization of the central nervous system itself follows a bilateral symmetry in the Y-axis. The brain, for example, is divided into two hemispheres, each controlling the motor and sensory contralateral systems. There is psychophysical and neurophysiological evidence that brain cells in the caudal intraparietal area respond selectively to stimuli that stretch longitudinally in depth, downward and vice versa – that is, in the medial plane. These cells are called axis-orientation-selective neurons and they are directly implicated in 3D shape representation (Sakata, Tsutsui, & Taira, 2005).

The horizontal plane is composed of the X and Z axes (Figure 2.1a), dividing space into upper (above) and lower (below) halves. One horizontal plane is that determined by the eye-level height of the observer. Another very important horizontal plane is the ground surface, where our locomotion and interaction with the environment take place. The medial and horizontal planes are closely related because depth (Z-axis) is intrinsically present in both of them. Thus, it can be stated that these planes are of a different nature to the frontoparallel plane, in which depth is not embedded.

Since the Renaissance (e.g., the work of Della Francesca, Da Vinci, or Dürer), many painting techniques have made use of monocular depth cues to create a three-dimensional impression on what is actually a two-dimensional flat surface (see Kubovy, 1986). To achieve this effect, the lost depth information is substituted by the violation of certain properties of Euclidian geometry. In this context, Rock (1983) suggested that in the case of an observer experiencing a convincing illusion of depth when examining such a natural scene in a photograph or painting, the effect is probably due to the fact that the array of retinal photoreceptors captures two-dimensional images (due to projective geometry) with very similar characteristics to those in three dimensions.

More recently, Saunders and colleagues (Saunders & Knill, 2001; Saunders & Backus, 2006, 2007) investigated the importance of linear perspective for perceiving depth from monocular images. They used 2D perspective images of a slanted rectangular object, which were sufficient to create a strong 3D percept, and demonstrated that participants could infer inclination from the "skew" of the contour (angle subtended between the diagonals and the horizontal side of the rectangle). Subsequently, they concluded that both parallelism and orthogonality constraints are used to perceive depth from a linear perspective.

We also recover 3D information through binocular depth cues, for example, convergence and binocular disparity. The latter is especially important. It should be remembered that there are relatively large areas of the visual field that are perceived by both eyes because of their frontal position. However, the lateral distance between the eyes results in slightly different images, especially in the horizontal coordinates. These are called horizontal disparities, and they provide powerful depth information for our representation of space. There is evidence that the mechanisms for detecting horizontal disparities are related to a stereo anisotropy (for more details, see Serrano-Pedraza, Brash, & Read, 2013; Serrano-Pedraza, & Read, 2010).

It is now evident that the visual system uses single monocular and binocular depth cues and their integration to build a general representation of three-dimensional space. The mechanisms underlying this process have been widely investigated in the context of size and distance judgments made under full and reduced vision conditions (Epstein, Park, & Casey, 1961; Tozawa & Oyama, 2006). The results indicate the existence of anisotropies in distance perception both between the observer and an object (*egocentric distance*) and also between two objects (*exocentric distance*).

Anisotropies of Perceived Egocentric Distances

It has been known for several decades that the perceived egocentric distance does not correspond exactly to the physical distance. Moreover, this asymmetry is modulated by factors such as context and experimental task, revealing the complex nature of the relationship between perceived distance and physical distance. Even though some studies have found an accelerated linear function relating physical to perceived distance in internal environment (Teghtsoonian & Teghtsoonian, 1969) and other studies a decelerated nonlinear function (Gilinsky, 1951), the most common result has been a decelerated function between them (Da Silva, 1985). That is, the perceived distance is linearly compressed in external environments. This egocentric distance underestimation is larger under reduced-cue conditions, but it also occurs under full-cue conditions (Foley, 1972; Foley, Ribeiro-Filho, & Da Silva, 2004; Higashiyama, 1996; Loomis & Philbeck, 2008).

Most of these results were obtained through verbal reports and visual matching tasks, the so-called direct methods. These have been criticized for introducing cognitive biases into perceptual judgments, thus compromising their veridicality. According to Gogel and Da Silva (1987), size and distance perception involves two levels of processing: (1) perceptual, information immediately available to the observer, based on size-distance invariance hypothesis; and (2) cognitive, which depends on the former but which is mediated by memory or past experiences (e.g. familiar size). In the view of Gogel and Da Silva, we are aware of the compression of egocentric distance, and try to correct it through cognitive mechanisms.

Since the 1990s, egocentric distance has also been investigated using indirect methods, by asking observers to perform an action, for example, throwing an object or walking directly or indirectly (triangulation task) toward a target. These actions may be carried out under continuous visual control (visually guided task) or initiated with visual feedback and finished without it (visually directed task). In this context, many authors have provided evidence that egocentric distance judgments under full-cue conditions are accurate and reliable (little variability) for physical distances of up to approximately 20 meters (Fukusima, Loomis & Da Silva, 1997; Loomis, Da Silva, Fujita, & Fukusima, 1992, 1996; Loomis & Philbeck, 1999). Although several studies have reported different estimates when using direct and indirect methods, Philbeck and Loomis (1997) found very similar egocentric distance estimates when using verbal report and visually directed walking. In addition to the task and the availability of depth cues, these authors also manipulated the position of targets, which were presented at eye level or ground level. The results revealed the fundamental role of angular elevation in the representation of egocentric distances (see also Foley, 1977; Sedgwick, 1986).

The study by Ooi, Wu, and He (2001) found that *eye level* is used as a reference to compute angular elevation for stimuli presented below the horizon. These stimuli are then perceived as being closer and more elevated in the visual field as physical distance increases. The importance of eye level to the three-dimensional representation of VS has been acknowledged, especially in terms of its close links to the horizon reference line (Bertamini, Yang, & Proffitt, 1998; Ozkan & Braunstein 2010). Eye level can be considered to coincide with the horizon, which is defined as "the infinitely distant limit of a ground plane" (Rogers, 1996, p. 142), it being a characteristic of the optical arrangement and not of the environment. Recently, it was observed that the judged eye-level strongly affects the perceived egocentric distance in outdoor and virtual reality settings. This result is consistent with a model based on angular elevation bias (see Durgin & Li, 2011; Li, Phillips, & Durgin, 2011).

Another factor that is fundamental to these asymmetries is the *ground surface.* In classical ground theory, Gibson (1950) described this surface as an important reference for extracting the general spatial configuration and establishing relationships of distance between objects and the observer. This theory was corroborated by Sinai, Ooi, and He (1998). These authors showed that when the surface texture information is available and it is homogeneous, then space representation is fairly accurate, and this affects the perception of absolute distance. Thus, any interruption (e.g., spatial gap) or incongruence (e.g., textural change) in the surface diminishes considerably the accuracy of distance perception. Another study found anisotropies when the terrain is disrupted by a dip or a lump; in this case, objects placed on the ground are perceived as being farther away (dip) or closer (lump) in relation to the actual physical distance (Meng & Sedgwick, 2001). These contextual effects of the structure of the surrounding environment have been reported in other studies and probably explain, at least partially, the great diversity in results regarding egocentric distance (Lappin, Shelton, & Rieser, 2006; Wu, Ooi, & He, 2004).

There is also evidence that observers predominantly use ground surface information for spatial representation (Bian, Braunstein, & Andersen, 2005). In a series of experiments, they presented computer-generated 3D scenes depicting a textured room. There were two texture scales (small and large elements), one displayed at ground and ceiling surfaces and another at sidewalls, which was counterbalanced across trials. Thus, any difference cannot be attributed to specific texture or surface. When the observers were exposed to contradictory information between surfaces, they mainly rely on the ground information to represent the scene. The authors referred to this as the *ground dominance effect*, which occurs due to the intrinsic features of the surface and not its position in the visual field (Bian, Braunstein, & Andersen, 2006). Thus, this effect cannot be explained by better performance of participants when the stimuli are presented in the lower visual field (see section Perceptual Asymmetries of Visual Fields).

Similar results have also been found in the binocular rivalry condition (Ozkan & Braunstein, 2009), in size estimation of three-dimensional objects (Champion & Warren, 2008), and in change detection of inclined planes (Ozkan & Braunstein, 2010). Bian and Andersen (2010) reported better performance in change detection on the ground surface with an exposure time of 500 msec. These results indicate that the ground surface advantage occurs in the early stages of visual processing (bottom-up). On the other hand, a decrease in reaction time was observed for conditions in which the ground and ceiling surfaces were presented in the LVF and UVF, respectively. This result also suggests the involvement of top-down mechanisms in the perceptual organization of visually perceived space.

The processing advantage of ground surface information probably has its origin in the ecology of the human species, as pointed out by Gibson. It is on this surface that we move and where most of the objects with which we interact daily are placed. In support of this hypothesis, Yang and Purves (2003) recognized the importance of ground surface information in the occurrence of perceptual anisotropies. However, they argue that these asymmetries are the result of a probabilistic strategy of the visual system that has been selected during the evolution of our species. Thus, VS would be statistically built, generating percepts from ambiguous stimulation based on the probability distribution of PS geometry. This explanation is useful when it comes to understanding in-depth perceptual anisotropies, whether egocentric distance judgments or exocentric distance judgments.

Anisotropies of Perceived Exocentric Distances

Since the work of Hillebrand (1902), it has been known that perceived exocentric distance does not correspond exactly to physical distance, i.e., there is no exact correspondence between VS and PS, as discussed throughout this chapter. Hillebrand asked observers to arrange pairs of luminous points extending toward them so that they appeared as two straight parallel lines, an experimental paradigm known as parallel alley (P-alley). However, the observers produced neither straight nor parallel lines: the data showed curved lines that tended to converge on the midline of their eyes. These results were complemented by the work of Blumenfeld (1913), indicating that P-alleys do not coincide with D-alleys, with the latter lying outside the former. Taken together, these findings indicate an underestimation of in-depth distances and, therefore, VS compression. It is important to note that these exocentric distance judgments were made in complete darkness, that is, in the absence of any depth cues. However, similar results were later reported under less restricted viewing conditions, showing a distorted representation of VS even in natural settings (Haber, 1985; Levin & Haber, 1993; Wagner, 1985).

More recently, exocentric pointing tasks have been extensively used to investigate the perception of exocentric distance. In these tasks, the observer controls a pointer via remote control and has to direct it toward a target (see Koenderink & van Doorn, 1998). In a laboratory setting, with distances smaller than 6 meters and reduced visual conditions, Cuijpers et al. (2000a) found systematic and significant biases in the adjustment of a line segment toward a specific point. The results suggest that the biases depend linearly on the orientation of and angle between objects. Similar perceptual anisotropies have been described in collinearity tasks in the laboratory (Cuijpers et al., 2002), as well as in an open field (Kelly, Loomis & Beall, 2004), in parallelism tasks (Cuijpers, Kappers, & Koenderink, 2000b), and in the horizontal and medial planes in a 3D exocentric pointing task (Doumen, Kappers, & Koenderink, 2006).

These studies also revealed that exocentric relations are scaled by perceived egocentric distance. Thus, in order to estimate exocentric distance we first make a rough estimate of the egocentric distance (Aznar-Casanova et al., 2008; Foley et al., 2004). This implies that the exocentric distance estimate is less accurate than that of the egocentric distance. By contrast, studies using other tasks (e.g., parallelism tasks) have shown that egocentric and exocentric distance perceptions are independent and determined by different mechanisms (Loomis, Da Silva, Fujita, & Fukusima,1992; Li et al., 2011).

These contradictory results may be due to differences in the tasks used (verbal reports or exocentric pointing tasks), in the environmental context (reduced or full vision conditions), in the experimental setting (natural open field or indoor laboratory), and in the distances involved (peripersonal or action space). What is clear is that the representation of exocentric (as well as egocentric) distance is affected by multiple variables, highlighting the dynamic and adaptive nature of VS. Thus, it is difficult experimentally to determine (1) whether VS has a geometry that best describes it, and (2) in the event that it does, what that geometry would be.

Although several studies have sought to answer these two questions, the debate remains unsettled. For instance, there is evidence that VS is constructed based on Euclidean geometry (Haber, 1985; Wagner, 1985); on a Riemannian geometry of constant and negative curvature, that is, hyperbolic (Indow, 2004; Indow & Watanabe, 1988; Luneburg, 1947, 1950); on a Riemannian geometry of variable geometry (Koenderink, van Doorn & Lappin, 2000; Koenderink, van Doorn, Kappers & Lappin, 2002; Koenderink, van Doorn, Kappers, Doumen, & Todd, 2008); and on a non-Riemannian geometry (Cuijpers et al., 2000a, 2000b, 2002). There are also those who argue that it is not possible to determine a geometry inherent to VS (Foley et al., 2004), or who approach this issue from the probabilistic and statistical point of view (Yang & Purves, 2003). In the face of this complexity, we can conclude that VS does not possess a well-structured geometry because it is multi-determined and depends on a series of factors (e.g., task, visual conditions, experimental settings, etc.), which also modulate perceptual egocentric and exocentric distance anisotropies.

Conclusions

In this chapter we have discussed the extent to which our representation of PS is precise and accurate. This is an important question, because most of our behavior is visually dependent, emphasizing the close relationship between perception and action. We have reviewed classical and recent research into a wide variety of perceptual phenomena, research that reveals the anisotropic nature of VS. The representation of VS is modulated by a multiplicity of variables, since it is contextual (what is in it), momentary (where the observer is), and task-dependent. Understanding VS is therefore challenging, although not impossible, despite the inherent difficulties and complexity. Indeed, considerable effort has been made and significant advances have been achieved by psychophysical and neurophysiological approaches in recent decades. Nevertheless, we still do not know, for example, what role phylogeny and ontogeny play in the representation of VS, and what the relevance of their role would be for clinical conditions that involve perceptual alterations (e.g., ADHD and schizophrenia). Research on perceptual anisotropies is important, therefore, because it can help us understand the mechanisms underlying VS representation and, consequently, the interaction and integration of perception, attention, and memory, and the planning and execution of movement.

Acknowledgments

This work was funded through a grant awarded by the Spanish Ministry of Economy and Competitiveness (MINECO) to J. A. Aznar-Casanova (reference PSI2012–35194). We thank Manuel Moreno Sanchez for his help in improving the figures in this chapter. Thanks also go out to Nelson Torro Alves and Bruno Marinho de Sousa for their comments and suggestions.

Chapter 3

Situated Influences on Spatial–Numerical Associations

Krzysztof Cipora, Katarzyna Patro, and Hans-Christoph Nuerk

Spatial Influences Are Ubiquitous and Flexible

In the Western culture, numerical information can be organized in multiple ways. For instance, rulers and modern computer keyboards adopt a default left-to-right direction of number ordering, which we also use when writing down consecutively increasing numbers. However, humans are much more flexible in their usage of numerical information. We invented negative numbers, whose direction on a conventional axis reverses the direction for natural numbers. We can also memorize or process numbers randomly distributed in space or make systematic use of circular, vertical, or diagonal directionality while using mobile phones, clocks, or thermometers. How does our mind flexibly adopt not only one but multiple different formats for representing numbers in space?

The ability to form *spatial–numerical associations* (SNA) constitutes one of the basic traits of the human mind (e.g., Cipora & Nuerk, 2013; Cipora et al., 2016; de Hevia, Vallar, & Girelli, 2008a; Fischer M. & Shaki, 2014; Hubbard, Piazza, Pinel, & Dehaene, 2005; for reviews, see Cipora, Patro, & Nuerk, 2015 and Patro, Nuerk, Cress, & Haman, 2014 for a taxonomy of different SNAs). While the origin of this ability is a highly contested issue (for biological innate views, see de Hevia, Izard, Coubart, Spelke, & Streri, 2014; Rugani, Vallortigara, Priftis, & Regolin, 2015; for more culturally oriented views, see Nuerk et al., 2015; Núñez & Fias, 2017), the literature is fairly consistent in that SNAs are directly or indirectly modulated by the direction of our reading habits (for review, see Fischer M. & Shaki, 2015; also, Shaki, Fischer M., & Petrusic, 2009; Zebian, 2005). The reading habit account offers an explanation for default organization of numerical sequences being ordered from left to right in Western cultures but not in Arab or Hebrew cultures. In cognitive experiments, participants from Western cultures are usually more likely to associate small numbers with the left-hand side and larger numbers with the right-hand side, while participants from Middle Eastern cultures show the reverse or null effect (see Fischer M. & Shaki, 2014; Shaki, Fischer M., & Göbel, 2012).

While there is a general default for cultural spatial habits, it is also obvious that we live in a world that provides us with a multiplicity of directional spatial instructions and arrangements in different situations. For instance, bilingual speakers of English and Hebrew or English and Arab can switch between different reading directions quickly and with great ease. Is such spatial-directional flexibility also relevant for SNAs? When considering this question, it is essential to acknowledge that spatial influences and their cultural modulations are not restricted to reading (Nuerk et al., 2015). Many other situations or spatial activities strongly influence the way that we mentally associate numbers with space and learn SNAs (Newcombe, Levine, & Mix, 2015). Even though many of these spatial activities are not directly related to numbers (e.g., shifting an object from left to right), they can determine in which direction spatial–numerical information will be processed in a particular situation (Patro, Fischer U., Nuerk, & Cress, 2016a for a controlled training study). The processes underlying such number line formation have been described by Nuerk et al. (2015) and Patro, Nuerk, and Cress (2016b) in detail.

Most current research, however, is concerned with fixed SNAs rather than with their flexibility. Here, we focus on flexibility, which is perhaps the more interesting and the more telling attribute of how human number cognition works. We will argue that *the situated influences on spatial–numerical associations* may be a key factor in understanding human SNAs and thus deserves more attention.

A General Distinction of Spatial–Numerical Processing: Extension and Directionality

Situated influences operate in a particular framework of spatial–numerical processing, which will be introduced here first as a theoretical background. Generally, at least two types of SNA can be distinguished: one based on extension and another on directionality (Cipora et al., 2015; Patro et al., 2014). The first one refers to the phenomenon wherein larger numbers are more readily associated with larger extensions in space (length, surface, spatial intervals; see also Walsh, 2003 for ATOM theory). Such an association can be observed already in infants and schoolchildren, and it can be used for embodied training of basic arithmetic skills in kindergarten and school children (e.g., de Hevia & Spelke, 2010; Fischer M. & Brugger, 2011; Fischer U., Moeller, Bientzle, Cress, & Nuerk, 2011; Link, Moeller, Huber, Fischer U., & Nuerk, 2013; Link, Nuerk, & Moeller, 2014; Moeller, Neuburger, Kaufmannn, Landerl, & Nuerk, 2009; Ramani & Siegler, 2008; Siegler & Opfer, 2003). Extensive reviews about this relationship can be found elsewhere (Dackermann, Huber, Bahnmueller, Nuerk, & Moeller, 2015; Newcombe et al., 2015).

However, our multidirectional spatial experience, to which we just referred above, appears to be more relevant for the second type of SNA – a directional relation between numbers and space. One of the most thoroughly investigated SNAs from this category is the *Spatial Numerical Association of Response Codes* (SNARC effect; Dehaene, Bossini, & Giraux, 1993; Wood, Willmes, Nuerk, & Fischer M., 2008 for meta-analysis). In the most typical experimental procedure (parity judgment task), participants classify numbers presented on the screen as odd or even using lateralized keys. After half of the experiment, the assignment of response keys is changed, so that for each number there is a response with the left-hand key and one with the right-hand key. In Western cultures (cf. Shaki et al., 2009), participants typically respond more quickly to small numbers using the left key and to large numbers with the right key. Behaviorally similar results are also observed in a magnitude comparison task. There are several variants of the SNARC effect. The most common variant (described earlier) is referred to as horizontal SNARC because numbers are associated with the horizontal (left–right) dimension. However, numerical magnitudes may also be mapped onto other spatial dimensions (e.g., the vertical SNARC; faster responses to small/large-magnitude numbers in the lower/upper space). The latter is considered to be more universal, consistent across cultures, and less shaped or modulated by cultural experience (Shaki & Fischer M., 2012). The SNARC effect has also been described in the sagittal dimension (i.e., small-magnitude numbers are associated with space close to the body and large-magnitude numbers with space farther from the body; see Winter, Matlock, Shaki, & Fischer M., 2015 for review).

The horizontal SNARC effect is quite ubiquitous. It has been observed in unimanual setups (i.e., binary responses with two fingers of the same hand), in tasks requiring responses with one finger, feet, directional saccades, and for stimuli other than symbolic numbers, such as letters, days of the week, nonsymbolic number magnitudes and object sizes (Ren, Nicholls, Ma, & Chen, 2011 for object sizes; Fischer M. & Shaki, 2014 for extensive review), and even in Go–NoGo tasks in which no spatial response location is assigned for different responses (Fischer M. & Shaki, 2016). So far, SNARC research has largely been descriptive or demonstrative, i.e., it has shown that the SNARC effect can be observed for another stimulus choice, another response choice, another experimental setting. However, in our review, we will show that the SNARC effect seems not to be modulated equally for all conditions. In particular, we will show that there are situated constraints driving or interfering with directional SNAs.

Bidirectional Relations between Numbers and Space

A standard question for SNAs is whether spatial processing influences number processing or vice versa, i.e., where links between numbers and space originate from. An overview of the possible patterns of directional relationships between numerical and spatial processing is depicted in parts I and II of Figure 3.1 (see Fischer M. & Shaki, 2014 for current overview).

First, processing numerical magnitudes causes biases in spatial processing. Regarding directionality, it was shown that perceiving small-magnitude numbers causes covert shifts of attention to the left side, while perceiving large-magnitude numbers causes attention shifts to the right side (Fischer M., Castel, Dodd, & Pratt, 2003; but see Fattorini, Pinto, Rotondaro, & Doricchi, 2015). Shaki and M. Fischer (2014; Exp. 1) showed that random numbers generated while walking in an empty room influenced the direction of subsequent turns (left/right turns were more likely after small/large-magnitude numbers). In sum, representing number magnitude causes spatial shifts of attention or motion in the direction predicted by SNA directionality. Such numerical influences have also been documented for processing nondirectional spatial extension, such as, for instance, estimation of a line's midpoint flanked by smaller or larger numbers (Fischer, 2001; de Hevia, Girelli, Bricolo, & Vallar, 2008b; de Hevia Girelli & Vallar, 2006; de Hevia & Spelke, 2009).

As we have mentioned earlier, spatial processing (e.g., performing motion in space) might also influence numerical processing. For instance, random number generation can be influenced by the direction of head movements. While performing leftward head movements, participants generate smaller numbers than when performing rightward movements (Loetscher, Schwarz, Schubiger, & Brugger, 2008). Additionally, changes in body posture

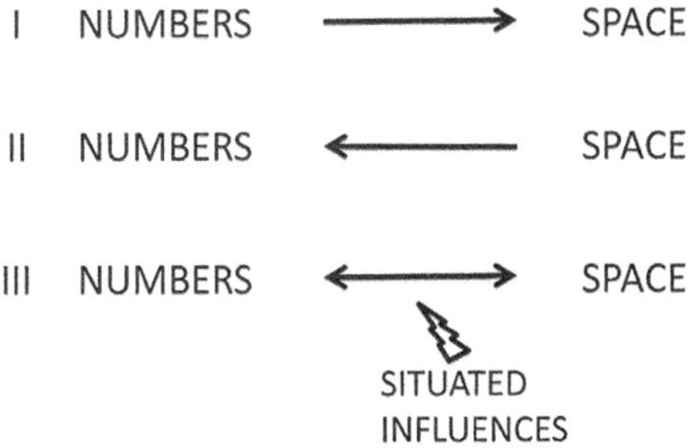

Figure 3.1 Overview of relationships between numbers and space.

and leftward/rightward turns while walking influence estimation of numerical values accordingly (Eerland, Guadalupe, & Zwaan, 2011; Shaki & Fischer M., 2014, Exp. 2).

To summarize, there is now ample evidence that SNAs represent a bidirectional link between numbers and space.

A Framework of Embodiment Influences on SNAs: Groundedness, Embodiedness, and Situatedness

The evidence we presented suggests that SNAs, including their exact directionalities, are hard-wired. This claim is congruent with typical accounts of SNA. On the other hand, much evidence suggests that SNAs can also be influenced by situational factors. These influences (see part III of Figure 3.1) are the main point of interest for the current work.

M. Fischer (2012; see also M. Fischer & Brugger, 2011) proposed a general theoretical framework on how our experience may shape numerical processing. This hierarchical framework includes groundedness, embodiedness, and situatedness, which are hypothesized to play a role in establishing SNAs. *Groundedness* refers to universal constraints of cognition, such as gravity, that express invariance in the external world. In regards to numerical cognition, smaller numbers would be associated with lower space and larger numbers with upper space (vertical SNARC), since larger piles that are comprised of more elements are typically higher than smaller piles. *Embodiedness* refers to performance signatures reflecting bodily (sensory and motor) constraints of our bodies. These sensory and motor experiences are incorporated into the structure of knowledge and can be evoked by activation of several concepts (even abstract ones). In terms of numerical cognition, such influences include universal cultural experiences such as reading in one direction (Wasner, Moeller, Fischer U., & Nuerk, 2014), which may affect horizontal SNARC (which also falls into this category). *Situatedness*, the last level of this hierarchy, refers to the situational constraints on the availability of knowledge that are exerted by the state of one's body and its environment. They may also refer to perceptual and bodily (i.e., somatosensory and motor) influences of the ongoing (e.g., experimental) situation. Such a modulation of the SNARC effect is not due to long-lasting internalized bodily experiences such as reading direction, but rather to the sensory and bodily constraints of the experimental situation in that moment, which can be referred to as *situated influences*. As M. Fischer (2012) proposes, situated influences may easily interfere with embodiedness. Altogether, this distinction, followed by empirical evidence, emphasizes that associations of numbers and space do not obligatorily originate from stable representations, but rather may be considered as effects of interaction between long-term cognitive structures and the current situation or task demands. Nevertheless, data collected up to date shows such interactions come from varied experimental procedures, which possibly influence different stages of processing and still need to be systematized.

A Taxonomy of Situated Influences on Directional SNAs

Typically, the SNARC effect was considered to be a signature of stable mapping of number representations on spatial locations (points on a directional mental number line; Dehaene et al., 1993). However, to date, the literature has documented several instances of situated influences on the SNARC effect. Surprisingly, studies have found that even very short manipulations may lead to prominent changes in the SNARC effect, showing that

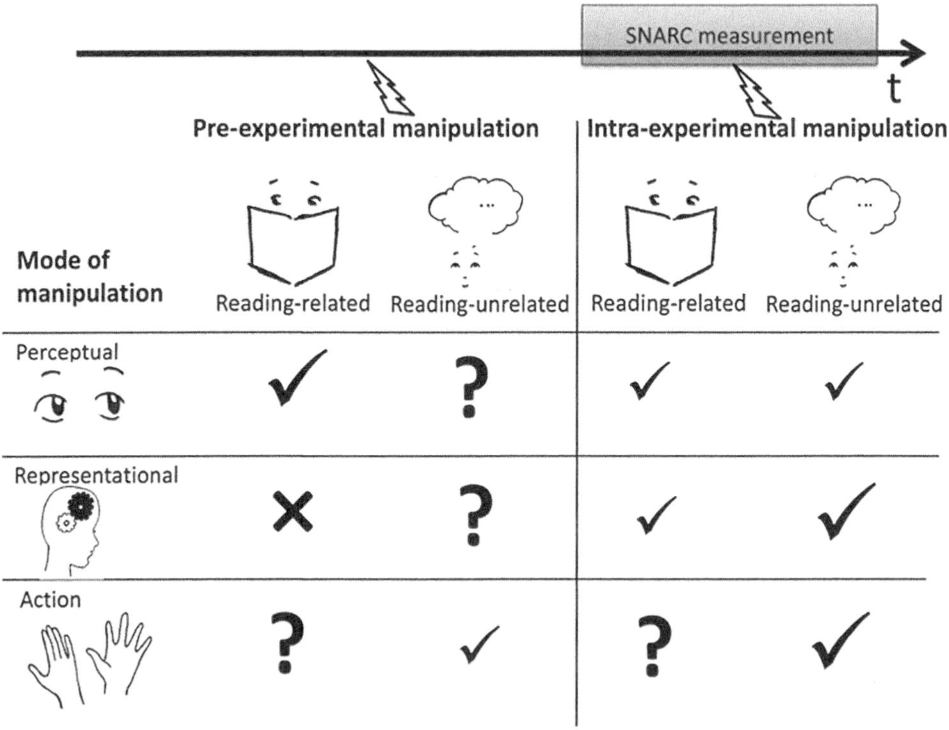

Figure 3.2 Overview of proposed types of situated influences on the SNARC effect. Large tics (✓) represent strong evidence for effects of given manipulation type. Small tics (✓) represent poor evidence for effect of given manipulation type. Question mark (?) represents manipulation types that are not yet tested, whereas x sign (×) denotes manipulation types that were not successful.

embodied and situated factors can dominate even long-standing sensorimotor habits. Nevertheless, manipulations that lead to temporal changes in typically observed SNAs varied largely at both technical and, even more importantly, conceptual levels. Therefore, we wish to go beyond a review of all available situated influences and present a new systematization and taxonomy of situated influences (see Figure 3.2 and Table 3.1). We hope that this taxonomy will clarify our understanding of the SNARC effect and guide further studies to improve our understanding of how situational factors may change typical associations between abstract and concrete concepts. Note that we restrict our taxonomy to situated influences on horizontal SNARC. We decided not to include the vertical SNARC effect for the following theoretical and practical reasons in this review. The theoretical reason is that vertical SNARC is, as we already mentioned, classified as a signature of groundedness in cognition. It is theoretically postulated (Fischer M., 2012) that this phenomenon is more stable than the horizontal effect, and therefore, situated influences are likely more difficult to detect; however, there is still not much evidence supporting that claim. The practical reason is related and may have its source in the above theoretical considerations: Published studies of situated influences on vertical SNA mapping are very scarce. Nevertheless, we must make a note here that there is some evidence that vertical spatial mapping of numbers may also be modified by situated linguistic factors. Göbel (2015) has recently demonstrated that counting direction in a vertical array may be

Table 3.1 Overview of situational influences on the SNARC effect

Mode of manipulation	Time of manipulation	
	Pre-experimental	Intra-experimental
Perceptual	How to cook SNARC – large/small magnitude numbers placed at the beginning / end of text lines **(R-R)** [Fischer M. et al., 2010] Bilingual participants reading a paragraph of a text written from left to right / right to left before a parity judgment task **(R-R)** [Shaki & Fischer M., 2008; Exp. 1]	Bilingual participants reading number words written from left to right / right to left **(R-R)** [Fischer M. et al., 2009; Exp. 1] Simon compatibility of stimulus-response location influences the SNARC effect **(R-U)** [Keus & Schwarz, 2005; Gevers et al., 2005; Tlauka, 2002#; Mapelli et al., 2003#] SPARC-SNARC interaction: auditory numbers are superadditively slow when both SNARC and SPARC are incompatible (e.g., large number is in low pitch and on the left side) **(R-U)** [Weis et al., 2016]
Representational	Bilingual participants listening to a passage of text in a given language **(R-R)** [Shaki & Fischer M., 2008; Exp. 2] #	Bilingual participants being auditorily presented with number words in a given language **(R-R)** [Fischer M. et al., 2009; Exp. 2] Imaging numbers as placed on the clock face versus on the ruler **(R-U)** [Bächtold et al., 1998] Neglect patients asked to imagine numbers as positions on a clock face process large magnitude numbers – which are placed on the left-hand side – less efficiently **(R-U)** [Vuilleumier et al., 2004; Exp. 4]. The SNARC effect depends on number range present in the task **(R-U)** [Dehaene et al., 1993; Exp 3; Fias et al., 1996; Exp. 1] The SNARC effect depends on the magnitude of criterion number the number is to be compared with in magnitude classification **(R-U)** [Nathan, Shaki, Salti, & Algom, 2009] Distortion of the SNARC effect caused by the presentation of descending number sequences **(R-U)** [Lindemann et al., 2008] Distortion of the SNARC effect caused by working memory load **(R-U)** [Herrera et al., 2008; van Dijck et al. 2009]

Table 3.1 *(cont.)*

Mode of manipulation	Time of manipulation	
	Pre-experimental	**Intra-experimental**
		SNARC for order in sequences stored in working memory **(R-U)** [van Dijck & Fias, 2011; Ginsburg et al., 2014; Ginsburg & Gevers, 2015; Huber et al., 2016] Verbal labels instead of spatial locations determine the direction of the SNARC effect **(R-U)** [Gevers et al., 2010; Imbo et al., 2012; Georges et al., 2015]
Action	Rapid construction of SNA in preschoolers **(R-U)** [Patro et al., 2016a]	Varied response key distance does not influence the SNARC effect **(R-U)** [Schiller et al., 2016] # Direction of unimanual SNARC depends on hand orientation: palm up versus palm down **(R-U)** [Riello & Rusconi, 2011] SNARC effect with crossed hands depends on reference frames **(R-U)** [Dehaene et al., 1993; Exp. 6, Fischer M. & Hill, 2004; Wood et al., 2006a #; Müller & Schwarz, 2007; Viarouge et al., 2014] Direction of the SNARC effect changed by spatial compatibility of stimuli successful only when the same effectors are used for both tasks – spatial and numerical **(R-U)** [Notebaert et al., 2006] SNARC disappears [Tan & Dixon, 2011] or is significantly reduced [Pfister et al., 2013] when calculated from RTs to number stimuli that were preceded by identical numbers. **(R-U)**. SNARC is weaker when it is calculated for RTs to number stimuli that were preceded by SNARC-incompatible response **(R-U)** [Pfister et al., 2013]

R-R – reading related manipulation
R-U – reading-unrelated manipulation
sign denotes non-successful manipulations

influenced by reading direction. In her study, Cantonese speakers (who read from left to right as well as from top to bottom) counted objects arranged in a square in the direction depending on the directionality of a text presented to them just before. Note, however, that this study did not test the vertical SNARC effect per se, only counting direction. This one study might suggest that situated influences may also play an important role for vertical SNAs. Furthermore, M. Fischer (2013) reports unpublished studies showing modulations of the vertical SNARC effect. However, given the lack of literature, it is too early to make a definitive judgment.

To theoretically and practically structure situated influences on this specific instance of SNA, we organize our review around three categorical distinctions that are illustrated in Figure 3.2. More detailed description of each type of situated influences is presented in Table 3.1. All these distinctions refer directly to the SNARC effect. However, we believe they can be, at least partly, extended to other directional SNA types.

(i) ***Mode of manipulation:*** This is a primary distinction in our model because it directly refers to stages of processing that seem to be involved in SNA formation. We distinguish three modes of manipulation: perceptual, representational, and action-related.

a. *Perceptual manipulations* refer to alterations of stimuli that are biased so that perception itself includes spatial–directional component.
b. *Representational manipulations* are those in which a spatial–directional component is to be produced or processed mentally (i.e., perceptual input does not involve directionality). For instance, when a reading material is presented auditorily, the directional component may only be activated at a representational but not perceptual level.
c. *Action-related manipulations* refer to changes in the type of motor/effector responses that are performed prior to the task or within a given task. Action and perceptual components are often congruently interconnected. Nevertheless, under some conditions, they may be contrasted against each other. To illustrate this distinction, one may imagine typing text on a typical computer keyboard. The motor activity during typing on the keyboard is by no means directional. However, the perceptual properties observed by the person who is typing are directional (i.e., letters appear in a sequence from left to right).[1]

(ii) ***Time of manipulation:*** Our secondary distinction specifies that within each mode, applying manipulations may vary in time. We will see that this is an important factor for the reliability of situated influences.

a. *Pre-experimental manipulation:* The manipulation is introduced before the task measuring the SNARC effect is administered. For instance, a text in a given direction (e.g., an instruction) has to be read before the SNARC task with Arabic numerals is administered.
b. *Intra-experimental manipulation:* The manipulation is applied during the task that is used to measure the SNARC effect. For instance, bilingual participants are tested with number words of languages with different reading directions within the experiment itself.

[1] However, evidence exists that emotional valence of words can be biased by the position of letters of which a given word consists on the QWERTY keyboard (so-called QWERTY effect; Jasmin & Casasanto, 2012). We would like to thank Martin Fischer for pointing this to us.

(iii) ***Reading relatedness:*** After categorizing the reviewed manipulations regarding mode and time, we propose to further divide them into two additional types, depending on whether they are reading related or not.

a. *Reading related:* The manipulation refers to reading, for instance, the reading direction of presented number words.
b. *Reading unrelated:* The manipulation does not refer to reading, such as the presentation of an analog clock face.

Our taxonomy is not based on similarity of groups (e.g., bilinguals) or task procedures. We might assume that converging results on SNARC changes due to a given manipulation type could shed some light on the nature of underlying processes. Although most of the published studies report successful situated manipulations of SNARC, in our overview, we also include null results (i.e., failures to manipulate the SNARC effect within a given manipulation type), which we mark with the # sign. Such null results may be due to technical details of utilized procedures as well as the fact that a given processing stage/representation form is less prone to changes caused by situational factors. The latter would be more likely if null results were found using multiple tasks (i.e., task-overlapping construct validity).

The manipulations we discuss can be generally categorized in two groups. First, there might be competing representations that may be activated concurrently. For instance, when the SNARC effect is examined with crossed hands, the right responding hand is on the left side while the right response code (i.e., the response key) is on the left side. One question is whether directional SNAs follow the response codes or the response hands, or both (cf. Wood, Nuerk, & Willmes, 2006b). Second, the existing space-to-number mapping may be distorted. For instance, the standard mapping of larger numbers to the right space may be disturbed if numbers in a cooking recipe appear on the left or right side of the text (Fischer M., Millis, & Shaki, 2010).

Characteristics of Situated Influences on the SNARC Effect

Perceptual Manipulations

Pre-Experimental

Manipulations present at the perceptual level, applied directly before the SNARC-relevant task, may strongly influence the magnitude of the effect or even cause it to disappear. Several such manipulations can be found in the literature, and the majority of them are reading related.

M. Fischer et al. (2010) exposed participants to numbers placed within the lines of cooking recipes in the way that was either congruent or incongruent with the left-to-right reading direction (e.g., small numbers on the left side vs. on the right side). This kind of manipulation was pre-experimental, since the reading task was irrelevant to subsequent parity judgment. The placement of numbers in the text lines was also irrelevant to both reading and parity judgment tasks. This manipulation was perceptual, because a text had to be perceived, while motor actions were not modulated. It led to significant changes in SNARC: the incongruent condition reduced the effect in monolingual English speakers and induced its reversal in monolingual Hebrew speakers.

In another experiment, Shaki and M. Fischer (2008; Exp. 1) tested bilingual Russian–Hebrew speakers. The participants were pre-experimentally primed with a short paragraph written in Hebrew or in Russian. After reading a paragraph in Hebrew (i.e., right-to-left script) the SNARC effect was significantly reduced.

In sum, pre-experimental situated perceptual manipulations can influence the SNARC effect, at least when they are reading-related. By the term *reading-related* we mean that the studies refer directly or indirectly to the direction of the reading process in a given culture or a given subject in a certain experimental condition. We do not claim that reading-related effects stem from the reading process itself.[2] Rather they could also stem from attentional function directly or indirectly influenced by the reading process. This is evident in cultural habits and activities that are performed mostly in a reading-consistent direction (see Nuerk et al., 2015 for such examples).

Intra-experimental

M. Fischer, Shaki, and Cruise (2009; Exp. 1) have demonstrated that the SNARC effect can be effectively reduced by intra-experimental perceptual manipulation. They tested bilingual participants (Russian and Hebrew speakers). Each trial consisted of two sequentially presented stimuli. First, a number word written in either Russian (script from left to right) or Hebrew (script from right to left) was presented, and the participants were asked to assess its parity status. Subsequently, an Arabic number was presented, the parity status of which was also to be assessed. Having applied such manipulation, the authors observed that the SNARC effect was significantly reduced in trials containing a Hebrew number word and in the following digit trials. This result suggests that to alter the SNARC effect in bilinguals (i.e., in case there are concurrent, relatively stable number to space mappings), pre-experimental warm-up in a given scanning/reading direction is not needed.

The SNARC effect is thus susceptible to manipulations applied intra-experimentally on a perceptual level, and can be changed by reading direction even on a trial-by-trial basis. Nevertheless, to the best of our knowledge, this is the only reading-related study that falls into this category.

However, there are reading-unrelated studies for both the visual and auditory modality. As regards visual modality, Keus and Schwarz (2005) observed that the SNARC effect is present in Simon incongruent conditions (i.e., number presented on the left and response given on the right) but not in Simon congruent conditions (see also Gevers, Caessens, & Fias, 2005 see Umiltà, et al., chapter 4 in this volume). On the other hand, some studies did not demonstrate the interaction between SNARC and Simon (Mapelli, Rusconi, & Umiltà, 2003; Tlauka, 2002). In the same vein, Weis, Estner, van Leeuwen, and Lachmann (2016) observed a superadditive interaction between the SPARC effect (Spatial Pitch Association of Response Codes, i.e., higher pitches are associated with the right response code and low pitches with the left response code; Lijdl, Kolinsky, Lochy, & Morais, 2007) and the SNARC effect. The SNARC effect was more pronounced for incongruent SPARC mappings. So, when experimental situations are complex and require resolution of perceptual incongruences, the SNARC effect is more pronounced.

Summary – Perceptual Manipulations

Perceptual manipulations can strongly influence the SNARC effect irrespective of whether they are applied prior to or during the task measuring the SNARC effect. We wish to note

[2] We would like to thank Martin Fischer for raising this important issue.

that perceptual manipulations may or may not be related to mental changes at the conceptual or representational level. Nevertheless, the results of the studies discussed in the preceding sections suggest that directional perceptual activity (usually but not necessarily related to reading; Padakannaya, Devi, Zaveria, Chengappa, & Vaid, 2002; see Nuerk et al., 2015 for diverging views) plays an important (and most likely causal) role in the formation of the SNARC effect.

Representational Manipulations

A representational manipulation was intended to influence directionality at a representational level without directional stimuli (e.g., script in left-to-right reading) being actually perceived. A good example is a study from Shaki and M. Fischer (2008), in which verbal material was presented auditorily. The verbal material was from two languages with different reading directions, but the reading direction was only mentally represented, because no written material was presented.

Pre-experimental

Experiment 2 from the study by Shaki and M. Fischer (2008) is, to the best of our knowledge, the only study in which a representational manipulation was administered before the task aimed at measuring the horizontal SNARC effect. Bilingual participants (Russian and Hebrew speakers) listened to a short text passage in Hebrew or in Russian. Afterwards, the participants performed a typical parity judgment task. There was no difference in the magnitude of the SNARC effect between two language conditions. Thus, an attempt to prime the concurrent reading-related mapping associated with given language, but without explicit reference to scanning direction, failed to influence the SNARC effect.

Thus, pre-experimental manipulation of reading direction representations was *not sufficient* to modulate subsequent SNARC experiments. This does not rule out the possibility that reading-unrelated but number-related pre-experimental representational manipulations (e.g., direction of the number line) may modulate the SNARC effect. This should be examined in future studies.

Intra-experimental

Most of situated influences on the SNARC effect we identified belong to this category. Furthermore, the vast majority of these manipulations are not related to reading. The only study we identified that referred to reading direction, but in an implicit way, was Experiment 2 by M. Fischer et al. (2009). Number words were presented binaurally in Hebrew or in Russian, in a randomized order, and participants were to assess their parity status. The SNARC effect differed significantly between trials presented in the two languages. As predicted, the effect was significantly reduced for Hebrew number words. This result clearly indicates that the bias caused by represented reading direction in a given language may significantly influence spatial mapping of number representations. Contrary to pre-experimental manipulation, where explicit reference to reading direction is necessary, in intra-experimental manipulations, implicit references to the representation of reading direction are sufficient to induce the SNARC effect. What is more, these data are a nice demonstration that the distinction between pre-experimentally cued direction and intra-experimentally activated direction is important for whether SNARC manipulations can be observed.

Several representational manipulations not related to reading also caused changes in the SNARC effect, providing several insights into the nature of the SNARC effect. Bächtold, Baumüller, and Brugger (1998), in the training phase, asked one group of participants to *imagine* numbers as placed on a clock face (where large numbers are placed to the left of small numbers). Another group was asked to imagine these numbers as placed on the ruler (where large numbers are placed to the right of small numbers). Afterwards, all participants performed magnitude classification and had to imagine the clock face or the ruler throughout the whole experiment.[3] The instruction stated that they were to assess whether each presented number referred to earlier or later than 6 o'clock / longer or shorter than 6 centimeters. Contrary to the ruler condition where the regular SNARC effect was found, in the clock face condition, the SNARC effect was reversed. Therefore, situated factors that were intended to evoke alternative number-space mapping successfully influenced the observed SNARC effect. Note that in this experiment, the manipulation was again aimed at evoking concurrent and stable spatial representations rather than online building of such. This result was further confirmed by Vuilleumier, Ortigue, and Brugger (2004; Exp. 4). The authors demonstrated that neglect patients, who normally neglect the left side of representational (including numerical) space (Zorzi, Priftis, & Umlità, 2002), processed large numbers less efficiently when required to imagine them on a clock face. Note that like in the famous Piazza del Duomo study in neglect patients (Bisiach & Luzatti, 1978), clock faces were not perceived in this experiment, but only imagined. Therefore, such experiments show that – as in neglect – not only perception but also mental representation of directional stimuli modulates human cognition.

A seminal study by Dehaene et al. (1993, Exp. 3; see also Fias, Bryspaert, Geypens, & d'Ydewalle, 1996, Exp. 1) demonstrated that the SNARC effect depends on the context in which a given number appears. Therefore, the spatial mapping may be created during an ongoing task instead of being permanent. The researchers used two numerical intervals (0–5 and 4–9) presented in separate blocks. They were most interested in results for numbers 4 and 5 because these were present in both intervals. The response pattern for these numbers was characteristic of large-magnitude numbers (i.e., right-hand advantage) when the numbers appeared in the interval 0–5. The opposite was true when the same numbers appeared in the interval 4–9 (i.e., left-hand advantage). Therefore, the task set largely modifies the pattern of the SNARC effect. Similar to this approach, Nathan, Shaki, Salti, and Algom (2009) examined trial-to-trial changes in the SNARC effect. They used a magnitude comparison task with a varied criterion value changing from trial to trial. In each trial, the participant was first instructed about the criterion value (e.g., 4/7) to which the target number (e.g., 6) was to be compared. Depending on the criterion, each target number could be classified as larger or smaller. The pattern of results (i.e., right-hand or left-hand advantage) for a given number depended on the relative magnitude of a number (i.e., whether it was smaller or larger than the criterion; see also Vuilleumier et al., 2004) rather than its absolute value. In sum, this shows that the SNARC effect is not absolute, but rather relative to the mental set within an experiment.

[3] Note that the manipulation started pre-experimentally here because the participant had to imagine the clock face or the ruler before the experiment, but it persisted intra-experimentally because the participant had to imagine the clock face or the ruler throughout the whole experiment. Therefore, it was not classified as a pre-experimental-only manipulation. In fact, the manipulation was both pre- and intra-experimental.

Lindemann, Abolafia, Pratt, and Bekkering (2008) presented participants with ascending, descending, or disordered (i.e., with no monotone order) number sequences and asked them to memorize these sequences. Subsequently, the participants were to assess the parity status of a presented number and, at the end, to recall the memorized sequences. Interestingly, SNARC was only present after ascending and disordered sequences, but disappeared after descending sequences. Moreover, an effect of SNARC alternation was observed only when all sequences in a block were of the same type. This observation is complementary to bilingual studies reported earlier, when activation of a directional mapping caused changes in the SNARC effect. The difference between these two tasks (apart from the presence of a reading component) is that here, the directionality was created (or rather disturbed) online and had to be kept in active working memory, whereas in bilingual experiments, the prime word evoked a reading-related directionality that was already established in long-term memory in the participants.

This early experiment of Lindemann et al. (2008) converges with another line of research, which demonstrated that loading working memory while performing a SNARC task modulated SNARC effects. Herrera, Macizo, and Semenza (2008) demonstrated that the SNARC effect measured with a magnitude classification task is reduced when participants have to maintain visuospatial information (i.e., Corsi block-tapping task sequences) in their memory. This reduction in the SNARC effect was not present in the parity judgment task. Van Dijck, Gevers, and Fias (2009) replicated this finding and provided evidence for further dissociation between parity judgment and magnitude classification by showing that the SNARC effect measured with the former is reduced when verbal working memory is loaded. These results provide evidence for the claim that working memory is important for the situated mapping of numbers to space. This would explain why the pattern of the SNARC effect depends largely on the task set.

A spectacular demonstration of the role of working memory resources for the appearance of the SNARC effect was shown by van Dijck and Fias (2011). They asked participants to memorize a sequence of numbers. Subsequently, the participants performed parity judgment on this material (and were to refrain from assessing parity of numbers that were not present in the sequence). At the end of such a block, they were to reproduce the sequence. The observed SNARC effect could theoretically be dependent on either the number's magnitude or its position in the sequence. For instance, in the sequence *9 2 4 6 1*, the largest magnitude number is placed in a leftmost (first) location and the smallest magnitude number is placed in the rightmost location. The results show that the SNARC effect depended on number position, not on number magnitude. In a subsequent experiment, van Dijck and Fias (2011) used sequences of vegetable and fruit names. Participants were to memorize these sequences and to perform binary fruit–vegetable classification. The authors again observed a significant SNARC-like effect depending on the element's position in a sequence. This implies that order can drive SNARC-like effects not only for numbers, which have ordinal characteristics, but even for naturally unordered categories like fruits and vegetables.

The working memory account may also explain the appearance of SNARC in typical tasks, when there are no situational manipulations. The basic assumption behind that explanation is that the numbers of an experimental set are also stored in working memory. However, ordered discrete items are stored in the most salient serial order. For numbers, this means that the smallest number is stored at the first position and the largest at the last position. For letters and other ordered sequences, the first element is mentally stored at the first serial order position. As a general principle, the first element is – in our culture – always associated with the leftmost space and the last element with the rightmost space. Therefore, in line with this account, the SNARC effect could be considered as a signature of the working memory function (Fias & van Dijck, 2016) and is a special case of broader phenomenon of spatial coding of ordinal position in a working memory sequence (Abrahamse, van Dijck, & Fias, 2016).

A subsequent study by Ginsburg, van Dijck, Previtali, Fias, and Gevers (2014) demonstrated that the effect of ordinal sequence (instead of SNARC effect depending of number magnitude) is present only when participants are required to retrieve the sequence from memory. Therefore, WM retrieval, and not information maintenance, seems to be related to the SNARC effect. The authors also conclude that there might be multiple mechanisms underlying effects for numerical magnitude and ordinal position. They also speculate that the SNARC effect may originate from temporal activation of long-term memory representations. Similar conclusions can also be drawn from later studies by Ginsburg and Gevers (2015) and Huber, Klein, Moeller, and Willmes (2016). Therefore, although the WM account can explain SNARC and its situated changes to some extent, it is not the only possible mechanism responsible for these effects.

Another linguistic representational manipulation modulating the SNARC effect was described by Gevers et al. (2010). According to these authors, the SNARC effect can be explained by verbal coding of response sides. Namely, small-magnitude numbers are associated not with the left side of space, but rather with the verbally coded concept "left," whereas large-magnitude numbers are associated with the "right" concept. Effects depending on spatial location and abstract verbal concepts can be contrasted and tested against each other. Gevers et al. (2010; Exp. 3) instructed participants to perform parity judgment by pressing the keys labeled with "right" and "left." The keys could be placed on the right- or left-hand side of the display. Interestingly, the observed SNARC effect depended on the verbal codes ascribed to the response keys, but not to the spatial locations of keys. This was also replicated in 9–11-year-old children (Imbo, De Brauwer, Fias, & Gevers, 2012). Nevertheless, a recent study shows that when spatial aspects are emphasized in the task instructions (i.e., "press the button on the left/right side of the display"), the number–space associations depend both on verbal-spatial and visuospatial mechanisms (Georges, Schiltz, & Hoffmann, 2015). However, Georges et al. (2015) used a magnitude classification task that is thought to rely more on visuospatial than verbal processing in general (van Dijck et al., 2009). In sum, linguistic influences might be dominant in the parity judgment task, while spatial aspects may be more salient in the magnitude comparison task. However, this depends on the focus of the task instructions.

Summary – Representational Manipulations

As the SNARC effect is thought to originate from a representational level, manipulations on this level seem to be a natural way of influencing this effect. Note that only manipulations that were also applied intra-experimentally (c.f. Bächtold et al., 1998) were found to be successful. However, there was a large variety of these manipulations, including imagery, manipulating the task set, working memory load, etc. One of these manipulations was reading related, but it referred to text directionality in an associative way (meaning of the word as well as its phonological properties associated with a given language and its reading direction). The other did not involve a reading component.

Unsurprisingly, it seems that the SNARC effect originates mostly from a representational level, and multiple components seem to modulate or even trigger it in a rather flexible, possibly even trial-by-trial basis. In fact, the effect seems to largely originate from dynamic changes and representations that are actively constructed during task performance. Evidence for such online construction of number representation (i.e., on a trial-by-trial basis) was found in neglect patients as well (Hoeckner et al., 2008). Typically, when a regular task set (whole range of one-digit numbers) and instructions are presented, the online construction seems to lead to what we observe and ascribe as a stable SNARC effect. Otherwise, the construction leads to observed distortions in the typical pattern of results.

Action Manipulations

Action manipulations refer to experimental settings where motor actions or body postures per se are manipulated. Obviously, there is a close link between perception and action, which we do not wish to ignore. Thus, when we change perceptual stimuli (e.g., written number words of different reading directions), we might also trigger particular actions of participants. Vice versa, if participants are instructed to perform particular motor actions, they might also perceive them visually or somatosensorily. Here, we focus on manipulations that target motoric or bodily movements or postures per se. Determining the extent to which such manipulations work in part or as a whole because they are also perceived via multiple modalities remains an important question for the future.

Pre-Experimental

Patro et al. (2016a) trained preschool children to play a nonnumerical spatial game in which they had to move an object leftward or rightward. Just a few minutes of this directional training evoked SNAs: children trained with the rightward movement formed left-to-right SNA, and those trained with leftward movement formed right-to-left SNA. However, as outlined above, although motor action was the agent of manipulation here, children also perceived accompanying congruent spatial changes (i.e., the spatial trajectory of movement observed on the screen; cf. Patro et al., 2016b). Future experiments must clarify whether the motor action manipulation is still successful when visual perception is prohibited or even interfered with (e.g., in appropriately designed virtual reality experiments).

One must also keep in mind that the only evidence for this type of manipulation comes from preschoolers. It remains unknown whether action manipulations would cause the same effects in adults. We believe this also should be addressed in future studies.

Intra-Experimental

In several situations, the SNARC effect is not susceptible to situated influences related to changes in effectors. As we have already mentioned, this can be observed in unimanual setups. Typically, participants asked to assess parity status of presented numbers by using two fingers of the same hand respond faster with the finger located on the right/left to large/small magnitude numbers (Priftis, Zorzi, Meneghello, Marenzi, & Umlità, 2006). Riello and Rusconi (2011) show that when hands are in a prone position, a significant SNARC effect can be observed only in case of right-hand responses. In contrast, when hands are in a supine position, a significant SNARC can be observed only in left-hand responses. The explanation for this effect is that SNARC may be observed only when the direction of the Mental Number Line is congruent with individual's finger-counting routine. The authors propose that finger series provide a relevant frame of reference in the external space that may be necessary for the SNARC effect to appear. Furthermore, as it has been recently demonstrated, in bimanual setups the distance between response keys does not influence the SNARC effect (Schiller, Eloka, & Franz, 2016).

On the other hand, several studies have shown that modifying the way the participants respond in the SNARC task leads to changes in the observed pattern of results. In general, these manipulations are based on body frame adjustments by contrasting relative (depending on individual's body/limb) and absolute positions or directions. In the first study demonstrating the SNARC effect, Dehaene et al. (1993; Exp. 6) found that when participants respond with their hands crossed, the effect depends on the position of a key, not on the responding hand. A similar procedure was utilized by M. Fischer and Hill (2004). Parity judgment of auditorily presented numbers was performed both with crossed and uncrossed (parallel) hands. Furthermore, in one condition, the participants were allowed to look at their hands and in the other condition they were not. In the first condition, the SNARC effect depended on the position of responses in external space (i.e., small/large numbers associated with given left/right spatial location) irrespective of whether hands were crossed or uncrossed. However, when the participants were blindfolded, the effect was present only in the crossed hands condition. The authors' interpretation was that in the crossed-hands condition, the proprioceptive input was stronger, which led to the building of a stronger frame of reference that produced the SNARC effect. Müller and Schwarz (2007) also support the claim that the SNARC effect depends on spatial location of the response button and not on the responding hand. They found that SNARC depended on spatial location even when the instruction emphasized parity-to-hand mapping (which was constant across session), whereas hand-to-button mapping was altered (crossed and parallel hands blocks).

Wood et al. (2006a) attempted to replicate this finding by testing a larger sample and using different materials (Arabic numbers, written and auditory number words, and dice patterns). The participants were instructed to respond with the right or left key. The authors failed to replicate the effect in any condition. According to the interpretation by Wood et al. (2006b), both body-centered and hand-based associations may be activated when a subject performs the task. These associations are contrasted against each other in the crossed-hands condition. In the study by Wood et al. (2006a), they referred more to the responding hand (left/right hand) than to the response side (left/right side), because blocks with crossed hands were preformed after blocks with parallel hands. In the study by M. Fischer and Hill (2004), the stimuli were presented auditorily, and this (especially in the blindfolded condition) did not induce external frames of reference. The authors provide evidence that individuals may rely on different frames of reference in terms of their SNAs. These frames are stable intra-individually but differ across individuals. Therefore, the effects of motor manipulations aimed at changing the SNARC effect may also depend on frames of reference, which a given participant preferentially activates.

Viarouge, Hubbard, and Dehaene (2014) provide further support for the claim that multiple reference frames provide situated influences on the SNARC effect when addressed intra-experimentally. They differentiate between global, body-centered, and object-related frames. Influence of each frame may be manipulated by the experimental context (constituted by response locations, response set or visual feedback provided by the response set). The authors argue that in a typical situation, a global left-right reference frame is activated. They also postulate that several situated influences on the SNARC effect may be caused by activations of concurrent reference frames (i.e., focusing the instructions on participants' hands instead of focusing on directions in space).

Action manipulations may also take other forms, which do not depend on altering body-centered frame of references to absolute ones. Notebaert, Gevers, Verguts, and Fias (2006; Exp. 1) found that the SNARC effect might be influenced by a concurrent task in which participants are to respond in a SNARC-incompatible way. The SNARC effect in the orientation judgment trials (upright vs. tilted) was largely influenced by the type of inducer task (e.g., it was reversed in participants who were to respond to small/large magnitude numbers with right/left hand) within a concurrent task. In the second experiment, Notebaert et al. (2006) demonstrated similar results with a spatial inducer task. However, this kind of action manipulation is effective only if the same effectors are used in both inducer trials and trials that are used to measure the SNARC effect (Notebaert et al., 2006; Exp. 4). Therefore, an action defined as performing a concurrent task that requires SNARC-incompatible bimanual decisions may lead to distortion in the observed SNARC effect. Action manipulation does not need to involve concurrent inducer tasks. The effect varies largely from trial to trial. Namely, the SNARC effect calculated based on trials that were preceded by the same numbers disappears (Tan & Dixon, 2011) or is significantly reduced (Pfister, Schroeder, & Kunde, 2013). Furthermore, a significant reduction in SNARC is also observed when it is calculated on trials preceded by trials requiring spatially incompatible (left/right to large/small magnitude number) responses (Pfister et al., 2013). Apart from overall reduction in the SNARC effect, the effect also seems to be more disorganized. Namely, there was a larger error term in the regression model indicative of SNARC (i.e., the data points were more dispersed). It seems that action (i.e., type of preceding response given by the participant) was sufficient to impact the SNARC effect.

Summary – Action Manipulations

Regarding action manipulations, the pattern of results must be carefully differentiated. The distance or extension of left and right response codes does not alter the SNARC effect. Whether responses are bimanual or unimanual, and whether there are small or large distances between response hands, does not seem to play a major role in the effect. Like for the mental set of numbers in an experiment in the representational manipulations, the spatial reference frames for different physical distances again seem to be relative rather than related to absolute distance.

However, the picture changes dramatically if action reference frames are reversed or induced. When hands are crossed, the SNARC effect can be eliminated, possibly because the body-centered action reference frame and the default representational reference frame for numerical sequences point into opposite directions. In such scenarios, instruction seems to be crucial, i.e., whether it focuses on external response keys or body-centered effectors such as response hands. Reversal of action direction can even pre-experimentally reverse the SNARC effect, at least in preliterate children.

To summarize, absolute distance of action parameters does not modulate the SNARC, but the reversal of directionality of action modulates the SNARC at least with appropriate instructions.

Conclusions

Our review suggests that SNAs and, in particular, the SNARC effect consist of a highly flexible, situated spatial mapping of numbers. It operates at different levels of human information processing (perceptual, representational, action), in different temporal frames (pre-experimentally activated mappings or online activated mappings during the experiment) and is triggered by different informational content (reading direction, reading-unrelated action direction, mental numerical set). We will briefly elaborate on the most important points of this conclusion.

- ***Flexibility:*** All the phenomena described earlier in the chapter consistently show that our cognitive system can easily adjust to different ways of mapping numbers to space. Depending on the situation, the system can immediately switch between concurrent mappings, suppress dominant representations, or create a new representation just from the beginning. Flexibility can be triggered before the experiment (e.g., by reading texts before or performing actions in particular direction) or during the experiment (e.g., by presenting number words in different scripts). Importantly, it cannot only change for one experiment, but also on a trial-by-trial basis.
- ***Situatedness:*** The SNARC effect is situated in the context of the experiment or even of a given trial. We hope to have convinced the reader that the default SNARC effect is just a base for SNARC research because it can be modulated by multiple situational contexts within or even before the experiments. Perceptions, representations, and actions as well as reading-related and reading-unrelated content can all modify or reverse the SNARC effect.
- ***Different levels of information processing:*** The variety of situated influences on the SNARC seems to suggest that this effect cannot be considered only as stimulus–response compatibility effect, present at the response selection stage of processing (e.g., Gevers, Ratinckx, De Baene, & Fias, 2006). The SNARC effect can be modulated by manipulations at a perceptual, representational, or action level. Moreover, most of these manipulations do not only modulate the SNARC when the response is given but also modify the reference frames in which space and numbers are mapped even before the experiment. Since the SNARC can be induced even without directional response selection (Fischer M. & Shaki, 2016), a pure response selection account cannot explain all results. However, this does not rule out the possibility that SNARC operates at multiple levels, response selection being one of them.
- ***Different temporal frames:*** The SNARC effect can be modulated both prior to and during SNARC measurement procedure. Moreover, effects of manipulations vary depending on the time they are applied. This is theoretically important because it seems to suggest that different temporal reference frames with different durations can be activated. When we outlined the flexibility above, we highlighted that spatial reference frames can be flexibly altered over a very short duration, even less than a second on a trial-to-trial basis. However, spatial reference frames do not need to be altered or to dissolve during such a short period of time. Pre-experimental, representational and action-related manipulations trigger directionality of spatial reference frames, which at least persist during the subsequent experiment without any additional updating of those reference frames. Whether different temporal reference frames can operate in parallel or whether just one reference frame can be activated over different durations remains an open question for future research.
- ***Different informational content:*** When embodied and situated modulations of the SNARC effect are discussed, it is mostly discussed with reference to reading direction. However, we hope to have shown that script and written words are just one of many types of information that can alter the SNARC effect. Mental imagery,

reversal of motoric codes, and training of directional action can all modulate the SNARC effect in a similar way as particular reading-related content. We conclude that spatial–numerical mappings can be modulated and induced (cf. Nuerk et al., 2015) without reading-related content.

To conclude: Gevers and Lammertyn (2005) titled their review paper on the SNARC effect "The hunt for SNARC," emphasizing the difficulties in examining this phenomenon. Our review shows that successful SNARC hunting requires operating at several information-processing levels. Apart from isolating and investigating separate mechanisms that influence the SNARC effect, future effort should focus on testing the interplay between these mechanisms. It seems that what we observe as a typical SNARC effect should be considered not as a unitary phenomenon, but rather as an effect of a situated interplay between very specific configurations of relevant factors. Without such a multi-componential approach, the SNARC effect averts our understanding just like the Snark averts being found in Lewis Caroll's "The Hunting of the Snark" poem.

Acknowledgments

We would like to thank Martin Fischer for his helpful comments. We are also grateful to Shadi Akbari for her help with preparing figures and to Barbara Peysakhovich for English proofreading. KC and HCN are supported by funding from the German Research Foundation (DFG NU 265/3–1) on "Linguistic Influences on Numerical Cognition: A Cross-Cultural Investigation Using Natural Specificities of Polish and German languages."

Chapter 4

S-R Compatibility with Physical and Representational Locations

The Simon, SMARC, and STEARC Effects

Carlo Umiltà, Mario Bonato, and Elena Rusconi

The concept of stimulus-response (S-R) compatibility has a rich legacy. In this chapter we offer a primer on a popular instance of spatial S-R compatibility, the Simon effect. We highlight empirical findings and theoretical frameworks from the extensive literature on this effect. We then show how these protocols and models have more recently been applied to the study of mental representations, whose spatial dimension cannot be directly observed but may be evoked by sensory and symbolic stimuli. In particular, we focus on sound pitch (see SMARC effect section) and time, encompassing a broad range of operational definitions – from stimulus duration to memory for ordered events (see STEARC effect section). In either case, the discovery of spatial compatibility effects between nonspatial stimuli and spatial responses boosted speculations on the existence of a stereotypical mental spatial representation and prompted empirical studies on potential interactions between sound processing and space and between time processing and space. Although it is plausible that these spatial compatibility effects may share some basic mechanism(s), different hypotheses have been put forward for the origin of their spatial stimulus code. This chapter offers an overview and pinpoints essential literature on phenomena that are united by spatial stimulus coding whereby the code originates in perceptual or conceptual characteristics, is hard-wired, spontaneously acquired, or culturally driven.

The Simon Effect

The notion of stimulus-response (S-R) compatibility is in principle very simple: The process of selecting the correct response based on the features of the command stimulus can be of different degrees of difficulties, depending on the ease with which relevant stimulus features and relevant response features match. Kornblum, Hasbroucq, and Osman (1990) called this (physical or conceptual) similarity *dimensional overlap* between stimulus and response sets – that is, *dimensional overlap* would be "the degree to which a stimulus set and a response set, or two or more aspects of a stimulus set or a response set, are perceptually, structurally, or conceptually similar" (Kornblum & Lee, 1995, p. 875). That happens, for example, when stimuli shown on the left or right side are assigned to response devices located on the left or the right side. When the relation between stimuli and responses is direct and natural, it is described as being compatible (for example: left stimulus / left response and right stimulus / right response). When the relation is indirect and unnatural, it is described as incompatible (for example, left stimulus / right response and right stimulus / left response). If the relevant stimulus feature and the relevant response feature are both spatial in nature, as in the example just provided, we can speak of spatial S-R compatibility. The speed and accuracy with which one can first select and then execute a selected response to an event (a stimulus) vary as a function of many factors, with compatibility being only one, and spatial compatibility being just one of the many forms of compatibility.

Simon Effect vs. S-R Spatial Compatibility

Compatibility effects were first noted in the 1860s by the Dutch physiologist/ophthalmologist Franciscus Donders (1868/1969). His seminal work on the use of reaction times (RTs) to study the duration of information-processing stages included experiments that employed compatibility manipulations. The concept of S-R compatibility was, however, formalized only later by the American psychologist Paul Fitts (Fitts & Seeger, 1953; Fitts & Deininger, 1954).

Simon conducted a series of studies in the late 1960s investigating compatibility effects that occur when stimulus location is irrelevant to performing the task but responses are defined along spatial dimensions (see Simon, 1990 for a review). Hedge and Marsh (1975) called the effect of irrelevant location correspondence the "Simon effect," and this is the name by which it has continued to be referred to. Umiltà and Nicoletti (1990) stressed the distinction between compatibility effects that occur when stimulus location is task-relevant and compatibility effects that occur when stimulus location is task-irrelevant (but see Buetti, Lleras, & Moore, 2014 for a critique of the notion of "task-irrelevancy"), referring to the former as "spatial S-R compatibility proper" and to the latter as the "Simon effect."

Kornblum et al. (1990) proposed a comprehensive taxonomy, which was based on the notion of dimensional overlap. They distinguished different types of S-R compatibility, depending on the type of task from which they emerge. There are four categories of tasks according to the dimensions on which the overlap occurs. Ensembles of type 1 include tasks for which there is no dimensional overlap between stimulus dimension (relevant or irrelevant) and response dimension – for example, when letters of the alphabet are shown at fixation and the subject is instructed to press the left key for vowels and the right key for consonants. In ensembles of type 2, the tasks are characterized by a dimensional overlap between the relevant stimulus dimension and the response dimension. The spatial S-R compatibility task, which produces the classical spatial S-R compatibility effect proper, belongs to this category. As already mentioned, there is compatibility when the subject is instructed to respond to a stimulus that appears on one side by pressing the key located on the same side (i.e., right–right and left–left), whereas there is incompatibility when the instructions are to press the key on the opposite side (i.e., right–left and left–right). In ensembles of type 3, the irrelevant stimulus dimension overlaps with the response dimension. The Simon task, which produces the classical Simon effect, belongs to this category. Typically, the stimulus-relevant dimension is color, and the subject is instructed to select the left or right response on the basis of stimulus color. Because the stimuli appear to the left or right of fixation, here too there is compatibility (right stimulus / right response, left stimulus / left response) or incompatibility (right stimulus / left response, left stimulus / right response). Ensembles of type 4 have dimensional overlap between relevant and irrelevant stimulus dimensions but no overlap of either stimulus dimension with the response dimension. For example, there is compatibility if a stimulus (e.g., an H) that, based on task instructions, requires a response with the right key appears on the right. In contrast, there is incompatibility if that same stimulus appears on the left. Tasks in this category include the Eriksen flanker task (e.g., Eriksen & Eriksen, 1974). Kornblum and Lee (1995) added four more categories to the taxonomy (ensembles of types 5 through 8) to include tasks with dimensional overlap on two or more dimensions.

The present chapter deals exclusively with the Simon effect. A point that needs to be clarified is whether it is admissible to make a sharp distinction between the Simon effect and spatial compatibility proper. From now on, we will dispense with the notation "S-R", which is redundant. The critical point is that, in the case of a task that gives rise to spatial compatibility proper, task instructions explicitly refer to the spatial location (often right or left) of both the stimuli and the responses. For example, if the stimulus appears on the right side, the correct response is to be executed by pressing the button on the right side (compatible condition) or the button on the left side (incompatible condition). In contrast, in the case of the Simon task, which gives rise to the Simon effect, only responses are defined with reference to their locations. For example, if a red dot appears, the correct response is the one on the left. If the red dot appears on the left, the condition is compatible (or corresponding). If the red dot appears on the right, the condition is incompatible (or non-corresponding). Note that it is immaterial whether locations are defined in absolute terms – that is, with reference to the observer's body midline – or in relative terms – that is with reference to the other stimulus or the other response (e.g., Nicoletti, Anzola, Luppino, Rizzolatti, & Umiltà, 1982; Umiltà & Nicoletti, 1985).

The obvious difference between the Simon task and the spatial compatibility task (see, e.g., Proctor & Vu, 2006; Umiltà & Nicoletti, 1990) is that, in the former, the stimulus spatial codes are relevant for performing the task and must be formed intentionally. In the Simon task, instead, the stimulus spatial codes are not relevant for performing the task and are formed automatically. However, the most compelling reason for keeping apart the Simon effect and spatial compatibility proper is that these effects are attributable to different mechanisms. A widely accepted model of the Simon effect is a dual process one, which distinguishes between different response selection routes (e.g., Hommel, 2011a; Proctor & Vu, 2006). One route is automatic, in the sense of producing response activation that is independent of the intended response. The other route is intentional and involves a translation process from the stimulus to the correct response. Various terms are used to refer to these processing routes. The automatic route is often called direct or unconditional, and is thought to rely on innate (or overlearned) long-term associations between stimuli and responses. The intentional route is called indirect or conditional, and is thought to rely on short-term associations for the specific task. The Simon effect is attributed to the fact that the spatially corresponding response is automatically activated, regardless of task instructions. If the required response is not the spatially corresponding one, it must be inhibited, with a cost in terms of time and accuracy.

Some models invoke an identical mechanism for explaining spatial compatibility proper (e.g., Proctor & Vu, 2006): The spatially corresponding response is automatically activated and must be inhibited if it is not the one required. In contrast, other models hypothesize only a single, intentional response selection route. They attribute the spatial compatibility effect to the number of transformational operations, or processing stages, which have to be performed, from input to output (e.g., Deininger & Fitts, 1955). Performance is more efficient when fewer operations are required (e.g., Rosenbloom & Newell, 1987).

The strongest evidence that the Simon effect and spatial compatibility proper must be kept apart comes from the observation that their time course differs. This difference emerges when one examines RT distributions (e.g., De Jong, Liang, & Lauber, 1994; Rubichi, Nicoletti, Iani, & Umiltà, 1997), in which RTs for each condition are ranked from fastest to slowest and partitioned into bins. Distributional analyses have shown that the Simon effect is largest at the shortest bin and decreases at longer bins. Although alternative interpretations of the distribution functions can be provided (Zhang & Kornblum, 1997), the most straightforward interpretation is that the Simon effect decreases as RT increases. Because the stimulus spatial codes are presumed to be formed automatically and to be irrelevant to the task, it is logical to assume that their effects are greatest shortly after stimulus onset and then decay with the passage of time and/or with slower RTs (De Jong et al., 1994; Hommel, 2011a; Van der Lubbe & Verleger, 2002). Simon, Acosta, Mewaldt, and Speidel (1976) showed that there is an initial advantage for the corresponding RTs, which decreases over time. Similarly, the lateralized readiness potential (LRP), which is a measure of differential activation of the lateral motor areas of the cortex, is initially asymmetric, favoring the motor area contralateral to the responding hand (e.g., Eimer, 1998). This asymmetry then disappears.

In contrast, in the case of the spatial compatibility proper (but also of crossed hands and vertical variants of the Simon task), the effect remains unchanged or even increases as RTs become slower (Proctor, Vu, & Nicoletti, 2003; Wiegand & Wascher, 2005, 2007). Wascher et al. (2001; see also Buhlmann, Umiltà, & Wascher, 2007; Pellicano, Lugli, Baroni, & Nicoletti, 2009; Van der Lubbe & Abrahamse, 2011; Van der Lubbe, Abrahamse, & De Kleine, 2012) proposed that the decreasing function is a unique consequence of automatic activation through privileged visual pathways (through the dorsal pathway, perhaps). These observations have originated the distinction between a *visuospatial* Simon effect, which decreases, and a *cognitive* Simon effect, which increases or remains unchanged (e.g., Pellicano et al., 2009; Wiegand & Wascher, 2007).

Coding of the Irrelevant Stimulus Location

The critical operation from which the Simon effect originates is the coding of stimulus location. If the task instructions require pressing the right-side button when the command stimulus is red, whether it appears on the left or on the right is irrelevant. Clearly, explaining why and how that happens is necessary for explaining the Simon effect, but it says very little about its mechanisms (e.g., Hommel, 2011a, 2011b). This section of the chapter is concerned with how the stimulus spatial code is formed. Possible mechanisms of the Simon effect will be discussed in the next section.

Two widely accepted accounts for why, in a Simon task, the stimulus spatial code is formed are referred to as “attention shifting” and “referential coding” (Hommel, 2011a, 2011b). The former is based on the so-called premotor theory of spatial attention (PMTA), put forward by Rizzolatti, Riggio, Dascola, and Umiltà (1987; more recent versions of PMTA can be found in Smith & Schenk, 2012; Van der Lubbe & Abrahamse, 2011; Van der Lubbe et al., 2012). It maintains attention shifts in space when a congruent motor program for moving the eyes is formed. The critical point is that, once the oculomotor program is formed, two parameters become available: the direction and the amplitude of the intended (but not necessarily executed) movement. The direction parameter (i.e., right, left, above, or below) becomes the spatial code of the stimulus toward which the programmed movement is directed.

Two implications of the attention shift hypothesis are that the spatial code will be relative to where attention is positioned at the onset of the stimulus – that is, at the moment when the motor program is formed – and that, if an attention shift does not occur, no Simon effect is observed. Consistent with the hypothesis, no Simon effect was found when no attention shift was needed to detect the target stimulus, and the Simon effect, when present, was observed with reference to the direction of the attention shift needed to detect the target stimulus (e.g., Nicoletti & Umiltà, 1989, 1994; Rubichi et al., 1997).

Thus, the attention shift explanation attributes spatial coding to the frame of reference provided by the location to which attention is shifted. In contrast, the referential coding hypothesis assumes that the spatial code is generated because the stimulus location is represented relative to the position of a reference object (e.g., Hommel, 2011a, 2011b). An implication of the referential coding hypothesis is that the Simon effect should occur on the condition that a reference object is present in the display. The problem with the referential coding hypothesis is that it does not envisage a mechanism by which the stimulus spatial code is formed. It confines itself to stating that the code is formed with reference to another object, which explains the term "referential coding." However, it is not specific about how the coding operation takes place. Van der Lubbe and Abrahmse (2011) linked the referential coding hypothesis to the theory of event coding (TEC) proposed by Hommel, Müsseler, Aschersleben, and Prinz (2001), but Hommel himself appears to be uncertain about this link (Hommel, 2011b). In contrast, the attention shift hypothesis is very specific about how the stimulus spatial code is formed: The code is the directional parameter in the oculomotor program.

Hommel et al. (2001) proposed a framework for perception and action planning, TEC, that is intended to explain many aspects of perception–action relations, including the Simon effect. TEC can be invoked for explaining why the corresponding response is automatically activated in the Simon task. It is based on the notion of common coding, according to which perception and action are represented in the same coding system. Particularly relevant here is the concept of an event code. It consists of the integrated feature codes that represent an event. These codes receive input from the sensory systems and modulate the activities of motor systems. Importantly, when a stimulus possessing particular stimulus features is perceived, it primes the actions that produce those same features.

Evidence that the relation between a response and the effect it produces is important to action has been demonstrated in studies of compatibility effects between a response and its effect on the environment (e.g., Kunde, 2001). Tucker and Ellis (1998; also, Phillips & Wards, 2002) applied the concept of action affordances to Simon effects obtained for object properties. They showed that the specific actions an object affords are automatically facilitated.

Another hypothesis that deals with the emergence of the stimulus spatial code is the salient features coding hypothesis (Weeks & Proctor, 1990; also Proctor & Cho, 2006; Proctor & Vu, 2006). This hypothesis is not concerned with the mechanisms by which the spatial codes are formed. It is primarily concerned with intentional S-R translation processes and, therefore, is primarily applicable to spatial compatibility proper. It maintains that stimuli and responses are coded asymmetrically, with one as salient, or positive polarity, and the other as nonsalient, or negative polarity. Weeks and Proctor (1990) attributed the up-right/down-left mapping advantage obtained in spatial compatibility tasks to the fact that "up" and "right" are coded as salient and "down" and "left" as nonsalient, and that mapping maintains correspondence between the salience, or polarity, of the codes.

A Computational Model

The effects of correspondence of an irrelevant spatial stimulus code with the response are usually attributed to a direct, or automatic, activation route rather than to intentional translation. Zorzi and Umiltà (1995) developed a computational model of the Simon effect that is based on this notion. Their model is implemented within a connectionist network of nodes connected by weighted excitatory and inhibitory links, which determine the activation values associated with the nodes. Stimuli are represented by position nodes, which code left or right position, and feature nodes, which code the relevant stimulus dimension. Responses are instantiated as nodes representing left or right response. Activation of a position node primes the corresponding response and inhibits the non-corresponding response through an inhibitory link between the two responses. The relevant feature nodes are "learned" through presenting each of the four possible input patterns along with the required response to simulate what occurs when a subject receives instructions about how to perform the task. At stimulus onset, the input node for the irrelevant stimulus position is assigned a value of 1, and this activation then dissipates over time. Because discrimination of the relevant stimulus feature is assumed to take longer than discrimination of location, the input node for the relevant feature does not receive its activation value of 1 until several cycles later. After stimulus onset, the network continues to cycle until the activation of a response exceeds a threshold value.

Zorzi and Umiltà (1995) performed simulations showing that the model generates the Simon effect. The simulations show facilitation when stimulus location corresponds with the response location and interference when it does not. Their model also generates the decreasing of the Simon effect as RTs become longer. Tagliabue, Zorzi, Umiltà, and Bassignani (2000) modified the original model to examine how the Simon effect might be eliminated or reversed after practice with an incompatible spatial mapping (see discussion further on). In the modified model, a distinction is made between short-term memory associations (STM links) and long-term memory associations (LTM links). STM links are established by instructions and are in effect for the specific task that is to be performed. LTM links are preexisting associations between stimulus and response locations that are relatively permanent. The modified model simulated the basic Simon effect, with both facilitatory and inhibitory components. It also produced the customary pattern of the Simon effect decreasing as a function of time after stimulus onset. Tagliabue et al. (2000) also investigated whether the reduction of the Simon effect after practice with an incompatible spatial mapping could be attributed to changes in LTM links or STM links between the stimulus position nodes and the response nodes. Their conclusion was that the practice effects were due to changes in STM links. The modified model is thus able to account for basic aspects of the Simon effect and the influence on it of prior practice with certain spatial mapping.

Reversals of the Simon Effect

Corresponding responses are not always faster than non-corresponding ones: The Simon effect can be negative. There are four types of reversals (see Proctor & Vu, 2006, for a comprehensive discussion): (1) the Hedge and Marsh reversal, which occurs when the task is characterized by an incompatible relevant S-R mapping; (2) a relative frequency reversal, which occurs when non-corresponding trials are more frequent than corresponding trials; (3) a mixed-task reversal, which occurs when location-irrelevant trials are intermixed with incompatibly mapped location-relevant trials; and (4) a practice-induced reversal, which occurs after practicing with an incompatible spatial mapping.

Hedge and Marsh (1975) found a positive Simon effect when the S-R color mapping was compatible (subjects responded by pressing the button of the same color as the stimulus), and a reverse Simon effect when it was incompatible (subjects responded by pressing the button of the opposite color as the stimulus). They attributed both the regular and the reverse Simon effects to a strategy of misapplying the response rule appropriate for the relevant color–response dimension to the irrelevant spatial dimension. This is the logical recoding account.

De Jong et al. (1994) proposed a model based on two processes with different time courses. A stimulus would automatically activate its spatially corresponding response shortly after onset, and then this activation would decay. Simultaneously, a task-defined S-R translation (identity or reversal) would operate on the relevant stimulus feature, and would generalize to the irrelevant stimulus location. For a compatible S-R mapping of the relevant dimension, both components produce activation of the corresponding response, which leads to a Simon effect. For an incompatible mapping, the two components counter each other, and the conditional, reversed component comes to dominate as the unconditional, automatic activation of the corresponding response decays.

By varying the relative numbers of corresponding and non-corresponding trials, corresponding responses can be made more likely than non-corresponding responses, and vice versa. When non-corresponding responses are more frequent, they tend also to be faster. Stürmer, Leuthold, Soetens, Schröter, and Sommer (2002) attributed the change in magnitude of the Simon effect as the percentage of corresponding trials varied to the formation of STM links in the conditional (intentional) response–selection route. That is, when a certain trial type predominates, subjects receive more practice with that trial type, which can cause a response bias through STM links.

Stürmer et al. (2002; also Hommel, Proctor, & Vu, 2004; Notebaert, Soetens, & Melis, 2001) showed that strong sequential effects occur in Simon tasks, that is, that performance on the current trial is a function of the preceding trial. The Simon effect manifests itself on those trials that immediately follow trials where the location relation was corresponding, whereas it vanishes or reverses following trials where it was non-corresponding. The interpretation of these results is that of modulation (or gating), on a trial-by-trial basis, of the direct, or automatic, response–selection route. It would be suppressed after a non-corresponding trial. In contrast, the suppression would be released following a corresponding trial. The suppression hypothesis, however, predicts a reduction of the Simon effect when non-corresponding trials predominate but not a reversal of the Simon effect. Hommel et al. (2004) pointed out that there is a perfect confound between the compatibility sequence and the feature combination sequence, which opens the possibility that what looks like an effect of cognitive control processes is, in fact, an effect of repetitions and alternations of S-R features.

When trials with an incompatible location-relevant mapping are intermixed with trials for which stimulus location is irrelevant, the mapping has an influence on direction and magnitude of the Simon effect. Proctor and Vu (2002; also, Proctor, Vu, & Marble, 2003) showed that when the location-relevant mapping was compatible, a positive Simon effect was obtained, but when the location-relevant mapping was incompatible, the Simon effect reversed.

The Simon effect is influenced, not only by an incompatible spatial mapping that is currently in effect, as when location-relevant and location-irrelevant trials are mixed, but also by an incompatible spatial mapping that was previously in effect. These findings indicate that the incompatible mapping from the previous spatially incompatible task continued to exert an influence on response selection for the current Simon task. Tagliabue et al. (2000; also Tagliabue, Zorzi, & Umiltà, 2002) showed that practice with an incompatible mapping of only 72 trials is sufficient to eliminate the Simon effect when a Simon task was performed immediately after the practice session. Most accounts of the Simon effect emphasize the role of automatic LTM links, which are innate or acquired through years of experience. However, the reversal and elimination of the Simon effect through relatively little practice with an incompatible spatial mapping before performing the Simon task, or when Simon trials are intermixed with incompatibly mapped location-relevant trials, indicates that task-defined STM links can produce automatic activation of their associated responses as well.

Conclusions

It is apparent that the Simon effect should be investigated by applying the ancient principle of "divide and conquer." That is, in the Simon effect there are several issues that promise to be better elucidated if they are tackled separately. The first is that the Simon effect and spatial compatibility proper belong to different conceptual families (e.g., Hommel, 2011a, 2011b; Umiltà & Nicoletti, 1990). The former is a close relative of the Stroop effect and the Eriksen flanker effect. Its explanation is to be found in the competition between two processing routes. An explanation for the latter is to be found in the different number of processing stages that occur in a single processing route. In the case of the Simon task, there are then at least three specific issues, which are orthogonal and need to be investigated independently. The first is the mechanism by which the task-irrelevant stimulus spatial code is formed. The two main hypotheses are the attention shifting hypothesis and the referential coding hypothesis (e.g., Hommel, 2011a). We favor the attention shifting explanation because it provides a specific mechanism based on which the spatial code is formed, that is, the code would be provided by the direction parameter in the program for a saccade. This explanation is often invoked in the context of PMTA. Note, however, that the link between the attention shifting explanation and PMTA does not necessarily refer to its classical version (Rizzolatti et al., 1987). Other, more sophisticated versions (e.g., van der Lubbe et al., 2011; Smith & Schenk, 2012) can be invoked as well. A second issue concerns why the corresponding response is automatically elicited, regardless of whether it is the one required by task instructions. Here one can invoke TEC (Hommel et al., 2001), which maintains that perceived events (i.e., stimuli) and produced events (i.e., responses) are represented through common codes. In other words, actions are represented in terms of sensory-motor bindings. Therefore, the stimulus code would automatically elicit the identical response code. Finally, it is necessary to explain the so-called sequential effects, that is why the Simon effect diminishes, or even disappears, when, in a sequence of trials, the foregoing trial is non-corresponding. One possibility is to invoke the intervention of control mechanisms. Alternatively, one can make recourse to the more or less complete repetition of features in a sequence of trials (e.g., Hommel, 2011a). Similarly, control mechanisms or spontaneous decay can be invoked to explain why the Simon effect decays as RT becomes slower or the S-R interval lengthens.

The SMARC Effect

This section describes how the spatial compatibility[1] rationale can be used to investigate the mental representation of pitch height.[2] We start by looking at the earliest studies with classical speeded-choice tasks and finding initial evidence for a vertical spatial representation of pitch height in Simon's work; we then present in some detail our studies on the SMARC effect and identify five hypotheses about the origin of spatial pitch codes. Finally, we highlight relevant research threads and open issues, including the place of the SMARC effect among other spatial compatibility effects.

Early Investigations

In one of the first demonstrations of the Simon effect, using monoaural pure tones intermixed with binaural tones, Simon and Small (1969) conveyed imperative information with pitch height (400 Hz and 1,000 Hz). Half of their participants were instructed to press the left key in response to the low-pitch tone and the right key to the high-pitch tone; the other half received opposite instructions. Simon and Small's (1969) report focused only on the compatibility effect between task-irrelevant stimulus location and response side; however, in a later study, Simon, Mewaldt, Acosta, and Hu (1976) stated that no compatibility effects were found between pitch (high/low) and response side (left/right) in those data. Simon and Craft (1971) instructed participants to respond with lateralized keypresses to the location of a monoaural tone of intermediate pitch, to the pitch (high/low) of a binaural tone, or to the pitch and location (left/right) of a monoaural tone. In the latter condition, half of their participants were instructed to respond with a right keypress to a high-pitch tone in their right ear and with a left keypress to a low-pitch tone in their left ear. The other half was given the opposite assignment; in either case, pitch provided redundant information. Simon and Craft (1971) reported performance superiority with the redundant cue, followed by the fixed-pitch monoaural tone location, followed by the variable-pitch binaural tone. They suggested that while compatible directional cues (location) are more effective than symbolic cues (pitch) to convey simple right/left commands, the presence of redundant information facilitates information processing. Also in this case, no compatibility effect was detected between pitch and response side (Simon et al., 1976).

Earlier, Mudd (1963, 1965; Mudd & McCormick, 1960; find earlier authors in Rusconi et al., 2006a) had reported spatial stereotypes in association with pitch, especially along the vertical, but also in the horizontal, dimension, and demonstrated the effectiveness of pitch for cueing visual search in visual displays. Simon et al. (1976) picked this up and tested compatibility effects between pitch (200 Hz or 500 Hz) versus location of an imperative tone from a high/low source on the vertical meridian, and up/down movements of a response-switch. They found "an interaction between two potent population stereotypes, a frequency-space stereotype and a tendency to respond toward the source of stimulation" (p. 354), with participants responding to the pitch of the tone. Performance in the

[1] The term "compatibility" is used in the present section and in the following one to indicate both cases in which the spatial stimulus attribute is relevant to the task and cases in which it is irrelevant; see Figure 4.1.

[2] Pitch height is a perceptual attribute of auditory stimuli that is primarily determined by soundwave frequency (Moore, 2003); all of the frequencies reported here were played at comfortable listening levels.

compatible condition (i.e., requiring up responses to the high-pitch tone and down responses to the low-pitch tone) was significantly faster than performance in the incompatible condition (i.e., with opposite S-R mappings). Additionally, a significant Simon effect was found, but only in the pitch–response compatible trials. Simon and colleagues suggested that processing the symbolic content of a tone under instructions that conflict with a stereotype (i.e., pitch–response associations) may overload the information-processing apparatus and prevent the other stereotype (i.e. location–response associations) from influencing performance. However, it could also be that the irrelevant location code had decayed (see previous section) by the time a response was selected in the incompatible condition.

More recently, Walker and Ehrenstein (2000) used a selective-listening paradigm to examine interactions between auditory dimensions (the starting pitch of a dynamic stimulus or the direction of pitch change) and manipulated S-R mappings in pitch classification tasks with unimanual keypresses. They used six high- and six low-pitched tones between 501 Hz and 1,585 Hz to start an ascending or a descending glide, with an end-pitch ranging between 316 Hz and 2,512 Hz. With vertical responses, they found the expected compatibility effect for both starting pitch and glide directions. In the incompatible condition, the task-irrelevant attribute (i.e., starting pitch if participants were judging glide direction, or glide direction if they were judging initial pitch) did not influence performance. No preferential mapping of pitch or glide direction on spatial responses was found in the horizontal dimension. Walker and Ehrenstein's (2000) concluded that only the hypothesis that pitch is represented in vertical space was supported.

The Spatial-Musical Association of Response Codes (SMARC) Effect

In early studies, either space or pitch height or both space and pitch height were task-relevant. Because similar words are used to describe pitch height and vertical spatial locations, such compatibility effects may arise from linguistic coincidence rather than (or in addition to) spatial compatibility. Thus, Rusconi, Kwan, Giordano, Umiltà, and Butterworth (2005, 2006a) tested whether pitch height can influence spatial (vertical/horizontal) responses when tones are not to be localized in space and pitch height is task-irrelevant. They included a crossed-hand manipulation to test whether the putative effects followed responding hand or response location. They also introduced musicianship as a relevant factor, by assuming that musical expertise would increase pitch processing automaticity and boost compatibility effects, if any, when pitch height is task-irrelevant.

In a first experiment, Rusconi et al. (2006a) reported an advantage for upper responses to high pitch and lower responses to low pitch, and an advantage for right responses to high pitch and left responses to low pitch, when non-musicians were asked to judge the pitch of pure tones, ranging from 164.81 Hz to 415.30 Hz (E3, F3#, G3#, A3#, D4, E4, F4#, G4#), relative to a fixed reference tone (261.63 Hz, C4). This compatibility effect followed response location rather than responding hand, and was not influenced by distance from the reference pitch. Rusconi and colleagues then introduced a task requiring participants to decide whether a sound stimulus was played by wind or percussion instruments (i.e., a timbre judgment task). Pitch (range: 174.61–466.16 Hz; F3#, G3#, A3#, C4, E4, F4#, G4#, A4#, centered on 293.66 Hz, D4) was varied orthogonally to timbre and no reference tone was presented. Only the vertical compatibility effect survived in non-musicians and this time it interacted with distance, being significant only for the highest- and lowest-pitched sounds, which were separated by more than an octave. In musicians, however, both vertical (and also with pitches separated by less than an octave) and horizontal compatibility effects were found in the same task. Because the effect was more pronounced in musicians, Rusconi and colleagues named it the Spatial-Musical Association of Response Codes (a.k.a. SMARC) effect. At the other end of the musicality spectrum, Douglas and Bilkey (2007) reported that individuals with congenital amusia – a pitch-processing deficit – show significantly smaller SMARC effects than matched non-musicians, who show smaller SMARC effects than matched musicians. Because the horizontal SMARC effect was only present in musicians when the pitch was task-irrelevant and it was overall weaker than the vertical effect, Rusconi and colleagues proposed that the vertical spatial representation was more natural and suggested that the horizontal SMARC effect could be the product of orthogonal remapping (Cho & Proctor, 2003). They dismissed the hypothesis that in their study it could derive from a separate representation linked to familiarity with musical keyboards, because the main instrument of their participants did not always map pitch left to right (e.g., violin vs. guitar). A study by Lidji, Kolinsky, Lochy, and Morais (2007) produced evidence for the SMARC effect, confirming its relation with musicianship. They tested the keyboard hypothesis by correlating individual number of years of piano experience with the horizontal SMARC effect, and they concluded that keyboard experience is not related in any simple way to the SMARC effect.

Rusconi, Giordano, Casey, Umiltà, and Butterworth (2006b) tested the keyboard hypothesis by collecting behavioral measures of keyboard familiarity (i.e., piano playing performance) along with measuring the SMARC effect in a timbre judgment task. They arranged response keys either horizontally or along the real vertical rather than transversally as in their previous study (Figure 4.1). This tested their assumption of cognitive

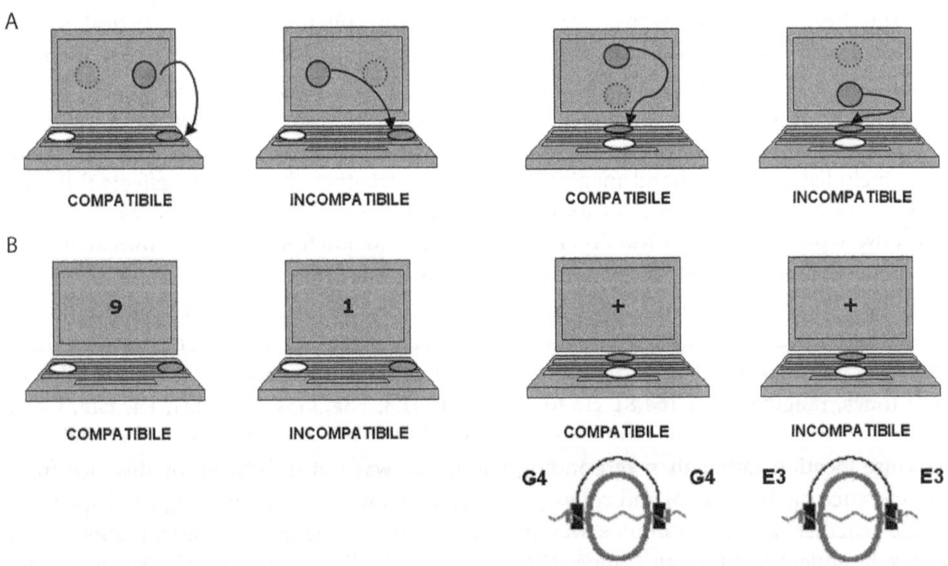

Figure 4.1 Examples of compatibility effects when stimuli and responses share a spatial dimension. Stimuli and correct responses are shown in white. (A) Physical compatibility along the horizontal (left half of the figure) and the vertical dimension (right half). (B) Representational compatibility: the Spatial–Numerical Association of Response Codes (SNARC) is shown on the left and the Spatial–Musical Association of Response Codes (SMARC) is shown on the right. Adapted from Rusconi et al. (2006a) with permission.

equivalence between the transversal and the vertical arrangement of response keys (Vu, Proctor, & Pick, 2000, for visual stimuli) for the SMARC effect. Further, Rusconi et al. (2006b) recruited Western instead of Chinese musicians as in their previous study, to assess whether the SMARC effect could be generalized to nontonal language speakers. Their results highlighted both a vertical and a horizontal SMARC effect. However, the horizontal SMARC effect did not correlate with individual keyboard skills, leading the authors to retain the orthogonal remapping hypothesis as the most parsimonious explanation for the horizontal SMARC effect. Finally, Rusconi et al. (2006b) manipulated the spatial location of the sound source either along the vertical or along the horizontal meridian during the timbre classification task. Along the vertical meridian, they found a SMARC effect but no Simon effect; along the horizontal meridian, they found a Simon effect but no SMARC effect, similar to early studies. They speculated that the vertical component of the SMARC effect might have overridden the Simon effect due to the weakness of sound source localization abilities along the vertical meridian (see, e.g., Yost, 2007). This prompts the question of where the spatial code of the stimulus generating a vertical SMARC effect comes from.

The Origin of the Spatial Code

A first hypothesis, the motor coding hypothesis, relates this association to the spatial nature of the actions of a performer. It, however, is logically jeopardized by the extreme variability across musical instruments of the relationship between pitch and the spatial nature of a performer's actions. In a cello, the hand has to move downward to produce high tones. In a guitar, right-handed players move the left hand to the right and the right hand downward, left-handed players moving the right hand to the left and the left downward. In other instruments, the mapping is essentially horizontal. For instance, with a transverse flute, pitch becomes higher by opening keys in a leftward direction, on a piano pitch becomes higher by playing keys in a rightward direction. Given this variability in the relationship between musical pitch and the effects of spatial action, explanations of spatial coding based on learned motor patterns are unwarranted. Although it is reasonable to expect influences from differential musical training and music-reading habits, the robustness of the vertical coding in non-musicians suggests that the motor coding hypothesis, which could be compatible with Hommel et al.'s (2001) TEC, is not a plausible primary source.

The proprioceptive coding hypothesis (Stumpf, 1883) posits that the association of pitch with vertical space emerges from the systematic variation of proprioceptive sensations associated with vocal performance. When making low sounds, the chest resonates, and when making high sounds, the head resonates. The "up" and "down" of musical pitch may thus correlate with the spatial "up" and "down" of our bodies (Zbikowski, 1998). It could also be related with rise and descent of the human larynx while speaking (Parkinson, Kohler, Sievers, & Wheatley, 2012). In line with the proprioceptive coding hypothesis, this association would be universal and expected to emerge in all cultures that practice singing or where verbal communication embeds pitch modulations. Evidence of cultural invariance is absent for linguistic associations between pitch and space, though. In ancient Greece, music theorists described pitch as "sharp" or "heavy"; in Bali and Java cultures, they describe it as "small" and "large"; and the Suyá in Amazonia as "young" and "old." However, evidence of an implicit association between pitch height and vertical space has been reported even in the absence of spatial words to describe pitch (Parkinson et al., 2012).

A hypothesis proposed by Pratt (1930) evokes a perceptual bias in the processing of spatial sound. He observed that, "as soon as one has his attention directed to the spatial property of a tone, the phenomenon of pitch-locality becomes very real and unmistakable" (p. 284). Interestingly, human acoustic localization ability differs across the main spatial dimensions. Along the horizontal meridian, for example, cues such as interaural time, phase, and level difference are available to locate a sound source. Along the vertical meridian, instead, interaural difference cues are unavailable, since the signal will be the same at the two ears. Less effective cues, such as spectral shape cues, will guide localization (Yost, 2007). Because the auditory modality is not as well equipped as the visual modality, pitch height might be naturally hijacked as an acoustic extra-cue in case of uncertain sound localization.

A further hypothesis points to environmental regularities in everyday acoustical environment: larger objects, which are more likely to be located near the ground than smaller objects, tend to produce lower pitches when struck than smaller objects (Stumpf, 1883). Evidence of a statistical association between the location of natural sounds sources and sound frequency has been recently provided by Parise, Knorre, and Ernst (2014). Also, the environmental regularities hypothesis would predict cultural invariance in the use of pitch as a localizing heuristic (i.e. a nonspatial attribute of the sound stimulus, which may implicitly/explicitly help localize stimuli along the vertical meridian because of its statistical association with the position of natural sound sources). This hypothesis could complement the perceptual bias hypothesis by providing a reason why pitch may assist localization and intersensory associations between visual elevation and pitch height. It could alternatively provide the basis for an attentional orienting hypothesis analogous to that advanced for the Simon effect (see previous section). Indeed, both exogenous and endogenous orienting would occur more frequently toward higher positions in space for higher pitch sounds and lower positions in space for lower pitch sounds, thus creating a long-term association between the direction of attention orienting and sound pitch.

Conclusions (and Current Directions)

Thirteen years on, studies on the SMARC effect (Rusconi et al., 2005; 2006a) or those exploring spatial biases in connection with pitch height have expanded. The most noteworthy research threads include investigations linking pitch processing with crossmodal attention and those exploring relations between the SMARC effect and other known compatibility effects. These investigations promise to clarify to what extent structurally similar effects such as the Simon (see previous section), Spatial-Numerical Association of Response Codes (SNARC; see Cipora, Patro, & Nürk, Chapter 3 in this volume), Spatial-Temporal Association of Response Codes (STEARC; see the following section), and SMARC effects can all be subsumed under a general theoretical framework or attributed to common neural mechanisms.

Crossmodal attention studies have especially focused on audiovisual interactions or interactions with tactile/haptic modalities (see Deroy et al., Chapter 5 in this volume). Overall they indicate that sound pitch cues select spatial attention across modalities, and vice versa. For example, pitch influences visual and haptic (in musicians, at least) bisection performance with horizontal lines or rods when it is either task-relevant or task-irrelevant, leading to the suggestion that it could facilitate rehabilitation interventions on visuospatial neglect (e.g., Ishihara et al., 2013); preliminary evidence on the use of sound pitch in the rehabilitation of acalculia, via facilitation of connections between space and numerical magnitude, is also available (Domhas, Zamarian, & Delazer, 2008). In contrast, it was reported that visuospatial display could be used to cue pitch processing in congenital amusia (Albouy et al., 2015).

Recent studies have contrasted the automaticity of the SMARC effect with sung number words with that of the SNARC effect. Fischer, Riello, Giordano, and Rusconi (2013) probed the SMARC and SNARC effects against intentionality and load-sensitivity criteria of automaticity (Jonides, 1981) in a dual-task paradigm. Stimuli and response alternatives were identical, but the relevant stimulus attribute (pitch or number) differed between tasks; concomitant tasks required short-term retention of either color or location information. Their results showed that, while spatial associations of both magnitude and pitch are load-sensitive (i.e., neither spatial mapping is strictly automatic), the SMARC effect appears to be more resilient than the SNARC effect in non-musicians. Beecham, Reeve, and Wilson (2009) argued that the SNARC and SMARC effects emerge from different spatial representations. This would seem consistent with the fact that they differ in their prevalent spatial dimension (vertical/horizontal), and, likely, in the mechanisms from which their secondary spatial dimension (horizontal/vertical) originates (e.g., Holmes & Lourenco, 2012). Other studies have investigated whether the SMARC effect interacts with spatial compatibility effects generated by visual stimuli. Vu, Minakata, and Ngo (2014), for example, reported that task-irrelevant variations of sound pitch interact with stimulus coding in a spatial compatibility task by weakening the horizontal vs. the vertical dimension. However, they did not influence response selection on a trial-by-trial basis. Nishimura and Yokosawa (2009), instead, reported additive Simon and SMARC effects with task-irrelevant auditory stimuli on left/right responses, while participants judged the color of a central visual stimulus.

In conclusion, much current research on the SMARC effect is attempting to clarify its relation with the classical Simon effect and other compatibility effects originated from hypothetical mental spatial representations.

The STEARC Effect

Here we will describe how spatial compatibility tasks can provide a number of insights also on the way time is represented and processed. First, we will outline how time–space interactions *can emerge as spatial compatibility effects* (STEARC = Spatio-Temporal Association of Response Codes) *across both conceptual and experienced time formats.* Then, space-time *anisotropies* will be presented. The characteristics of time processing revealed by spatial compatibility tasks converge in suggesting that its representation *is spatial in nature.*

Spatial biases in the time domain complement Chapters 6 and 16, as well as the spatial characteristics of the numerical domain described in Chapter 4 in this volume. By the end of the section, the reason why heterogeneous time-induced spatial biases fit in a chapter that focuses on spatial compatibility should become clear.

Time–Space Interactions across Experienced and Conceptual Time

Of particular interest for implementing a parallel with the spatial compatibility domain are the tasks that use lateralized responses. These tasks show interactions between time aspects and response side, both when the time aspect is an experienced duration (short vs. long; Vallesi, Binns, & Shallice, 2008) or a conceptual order (e.g., before vs. after a specific month; Gevers, Reynvoet, & Fias, 2003; see discussion that follows). Most commonly, each participant undergoes two separate stimulus response mappings (e.g., left response associated with before or with after). The mapping leading to faster/less error-prone responses is then considered as suggestive of a specific time-space association. In analogy with the Simon effect, time–space associations often hold also when the time aspect is task-irrelevant (e.g., Gevers, Reynvoet, & Fias, 2004; Previtali, de Hevia, & Girelli, 2010; van Dijck & Fias, 2011). The association between time and space seems to be so strong as to render an analogy with the automatic coding of space, as revealed for instance by the Simon effect, discussed earlier in the chapter, not inappropriate.

Many experiments investigated space–time associations by presenting brief temporal stimuli, typically in the 100–2,000 ms interval, which had to be categorized as being longer or shorter than a reference duration. The standard finding is an association between responses occurring in the left space and relatively shorter durations and responses occurring in the right space and relatively longer durations (Vallesi et al., 2008, 2011). Time-space interactions have also been found when responding to a newly learned arbitrary sequence of items (Previtali et al., 2009; van Dijck & Fias, 2011). Even in this case of experienced order "left" associates with earlier-presented items and "right" with later-presented ones. By using videos, the same method has been applied to studying longer time durations by Santiago, Román, Ouellet, Rodríguez, and Pérez-Azor (2010). When implementing material that has an intrinsic order (e.g., digits) within an arbitrary sequence, the resulting spatial association mirrors the arbitrary order of presentation and not the intrinsic order (van Dijck & Fias, 2011).

Time can also induce spatial associations when it is prompted in abstract formats rather than experienced. It is the case, for instance, of material with an intrinsic sequential order. Tversky, Kugelmass, and Winter (1991) asked English and Arabic children and adults to arrange in space ordered items belonging to several categories, including time and quantities. They found a strong preference, for time-ordered sequences only, to arrange items according to writing direction. In Fuhrman and Boroditsky (2010), left effector-before and right effector-after associations were similar, regardless of whether the time interval encompassed seconds or years. This strongly suggests that it is the ordinal position rather than the temporal length of the sequence that subtends the effect (for months and days, see Gevers et al., 2003, 2004).

It appears that the time–space associations are flexible. For example, before is coded as left no matter whether the interval length is 10 years or 1 minute. Also words that induce the concept of past or future trigger an association with left or right side, respectively (Santiago, Lupiáñez, Pérez, & Funes, 2007). Conceptually similar effects can be found when the ordinal concepts of time are induced by asking participants to determine whether a given actor/singer became famous before or after they were born (Weger & Pratt, 2008).

In summary, similar interactions with response side seem to occur for a number of heterogeneous time categories (e.g., experienced vs. conceptual time; Bonato & Umiltà, 2014). This suggests that the coding used to process time tends to be rather general and not to be linked to a particular time interval/format.

Time–space interactions allow a different interpretation, based on the "polarity correspondence" principle put forward by Proctor and Cho (2006; also see Proctor & Vu, 2006), previously described for the Simon effect (see discussion earlier in the chapter). Recently, however, Santiago and Lakens (2015) obtained results that cast doubts on the polarity correspondence as an alternative explanation for time–space interactions (see also Bonato Saj & Vuilleumier, 2016).

Spatial Anisotropies for Time

Evidence suggests that the spatial representation of time has a direction. In Western cultures, past tends to be associated with left and future tends to be associated with right. Major determinants of this anisotropy are reading habits. Right-to-left readers present a "reversed" representation under a number of different time aspects. The effect of reading direction on the time-sequence has been described when sequential events had to be spatially arranged (see the seminal study by Tversky et al., 1991) and when using lateralized response keys (Fuhrman & Boroditsky, 2010). It also occurs when experienced durations have to be categorized (Vallesi, Weisblatt, Semenza, & Shaki, 2014) and when conceptual time is prompted by the use of different tenses (Ouellet, Santiago, Israeli, & Gabay, 2010). Flexibility in time representation because of mixed reading habits has been described in Mandarine participants by Miles, Tan, Noble, Lumsden, and Macrae (2011). Anisotropies also emerge across different axes, as is the case for the back-to-forth direction (de la Fuente, Santiago, Román, Dumitrache, & Casasanto, 2014; Torralbo, Santiago, & Lupianez, 2006).

Casasanto and Bottini (2014) provided insights on the time-to-space mapping flexibility. They showed a reversal of the left-past right-future associations by presenting right-to-left–oriented sentences. Crucially, however, this reversal took an entire experimental block to occur, as if the new mapping first had to counteract the (more spontaneous) left-to-right mapping. In contrast, this did not happen when sentences were presented according to a less common direction (top-to-bottom).

Neurological patients suffering from right-hemisphere damage and left hemispatial neglect, a disorder of spatial attention affecting contralesional hemispace (see Charras, Lupianez, & Bartolomeo, Chapter 18 in this volume), present a disrupted representation for short durations (Marin et al., 2016; Calabria et al., 2011; but see Masson, Pesenti, & Dormal, 2017). Moreover, these patients present a disrupted representation for ordered items. This disruption is evident when items occurring before a time reference are erroneously reported as occurring after it (Saj, Fuhrman, Vuilleumier, & Boroditsky, 2014), resembling transpositions in the spatial domain (Bonato & Cutini, 2015). Recently Bonato, Saj and Vuilleumier (2016) asked patients with left neglect to categorize events of a story as occurring before or after a central event. Left vs right sided responses were contrasted. Patients showed a selective slowing when responding to events occurring before the reference. Again, in left neglect patients, prismatic adaptation inducing leftward attentional deviation reduced the time underestimation they presented at baseline (Oliveri, Magnani, Filipelli, Avanzi, & Frassinetti, 2013).

From Metaphoric Structuring to Anisotropies (the Mental Time Line)

The Metaphoric Structuring (MS) is the dominant theoretical proposal accounting for cognitive interactions between time and space (Boroditsky, 2000; Casasanto & Boroditsky, 2008). Its core assumption is that language reveals and shapes cognition. This view is well supported by a wealth of spatial metaphors for time. When a more in-depth analysis is performed, however, it becomes evident that in several Western cultures time is preferentially represented according to a left-to-right orientation even in the absence of any linguistic metaphors indexing this specific spatial direction (Bonato et al., 2012; Casasanto & Bottini, 2014). The MS view, therefore, needs to take writing direction into account (Boroditsky, 2011).

A more specific possibility for framing time–space interactions consists in assuming the presence of a "mental time line" (MTL), oriented according to situated experiences, like writing direction, and implying a key modulation of spatial attention, allowing a very high flexibility (Arzy, Adi-Japha, & Blanke, 2009; Frassinetti et al., 2009; Ishihara, Keller, Rossetti, & Prinz, 2008; Vallesi et al., 2008; Ulrich & Maieborn, 2010; see Bonato et al., 2012 for review). The MTL has a narrower focus than the MS and mainly gathers evidence from spatially characterized responses and/or tasks, predicting a spatially anisometric representation of time in the spatial domain.

In analogy with the number–space interactions, the key to understanding the true nature of representation for time seems to be understanding whether its flexibility across tasks and formats (Torralbo et al., 2006) confirms or rather disconfirms its spatial nature. An alternative possibility to explain the ubiquity of spatial effects in time processing is to assume that it mirrors some supramodal mechanisms, as, for instance, those involved in retaining information in working memory (Abrahmse, van Dijck, Majerus, & Fias, 2014). It seems likely that the determinants of association with space will be better pinpointed in the future, similarly to what is happening in the numerical domain (see Figure 3.2).

Conclusions

Strong analogies seem to occur between spatial compatibility effects for lateralized stimuli and for time. Time–space interactions occur for experienced as well as conceptual time, suggesting a common spatial representation. These spatial terms have an orientation (anisotropy), which spontaneously follows the direction of writing. However, it flexibly adapts to manipulations of spatial attention and to the specific spatial layout provided.

Despite its ephemeral nature, or possibly because of it, time representation seems to be implemented by spatial means. The activation of largely overlapping networks for remembering the past and envisioning the future (Schacter & Addis, 2007) might suggest that the existence of a common (spatial) network for time representation, resembling basic spatial effects similar to those revealed by the Simon task, might not be impossible.

Chapter 5

Unraveling the Paradox of Spatial Pitch

Ophelia Deroy, Irune Fernandez-Prieto, Jordi Navarra, and Charles Spence

Introduction: The Paradox of Spatial Pitch

Try, for a moment, to imitate an orchestra conductor while listening to a familiar classical tune. Did you find your hands moving up and down in time with the high- and low-pitched notes, respectively? Or just try singing in a high-pitched and then a low-pitched voice instead: now it is your eyebrows that will likely have gone up and down (see Huron & Shanahan, 2013). This spontaneous behavior might not come as a surprise: A series of tones increasing in frequency are almost universally described as "rising"; it is also heard as a continuous ascending movement (Eitan & Granot, 2006). The connection between the notes and these spatial expressions, however, is highly paradoxical. Musical notes are not played from higher or lower locations in space, and yet are spontaneously interpreted through what seems, at best, a metaphor, and at worse a total inaccuracy. As noted by Zuckerkandl (1956, p. 270), "On the one hand music appears as the art that . . . is perceived solely in and through time, to the complete exclusion of space; on the other hand, it is full of phenomena that seem to presuppose a spatial order and that in any case are wholly incomprehensible if space is completely excluded."

The "paradox of musical motion," as it is sometimes known, has attracted the attention of many thinkers and philosophers (Larson, 2012; Nussbaum, 2007; Peacocke, 2009; Scruton, 1983). It has also been empirically investigated since Carl Stumpf's early work on the psychology of the tone (Stumpf, 1883, 1911). However, far from solving the paradox of musical motion, the scientific investigation of tonal perception has actually moved it toward different, and no less challenging problems: Cognitive scientists have shown that the spatial interpretation initially observed for music also holds for individual tones. While the musical case could be explained by a general tendency to map the temporal order of sequences onto space, no such interpretation is available to explain why pure tones should also be spontaneously mapped onto space. The mapping here seems only to come from a contrast in frequency or pitch: two sounds of higher or lower frequency played from the same static location translate into a difference between "high" and "low" (Mudd, 1963). And this turns out to be true from a very early age (Jeschonek, Pauen, & Babocsai, 2013).

In this chapter, we focus on this paradoxical spatialization of pitch, which is several steps removed from the paradox of musical motion. The paradox of musical motion questions how a sequence of notes can be perceived as *moving*. The paradox of spatial pitch asks how pitch can acquire certain spatial attributes that are at odds with the spatial properties of the resonating body producing it, and how pitch biases both spatial perception and action (Lakoff & Johnson, 1980; Parise, Knorre, & Ernst, 2014; Rusconi, Kwan, Giordano, Umiltà, & Butterworth, 2006). The paradox of musical motion is that of hearing implied or perceived motion in the absence of any physical spatial displacement – the paradox of spatial pitch, one of spatial perception divorced from the actual location of the sound-emitting body.

Many researchers assume that the spatial character of pitch contains the key to the paradox of music motion, but this optimism is something we would like to challenge here. Music is much more than pitch, and the perceived dynamism of musical motion is much more than merely "going up" or "going down" (Eitan & Granot, 2007; Walker, 1987). After all, a melody retains the same motion when played in a higher or lower key (Foster & Zatorre, 2010). As we shall see, our own approach increases the gap between the spatial interpretation given to pitch and musical listening. But why should pitch be interpreted or mapped spatially?

One of the problems with this question, nicely outlined by Ernst Mach in his book *Analysis of Sensations*, is to determine what, exactly, is meant by "spatially mapped." The space onto which pitch is mapped has little to do with actual space: "A tonal series occurs in something which is an analogue of space, but is a space of one dimension limited in both directions . . . It more resembles a vertical line" (Mach, 1959, p. 278).[1] This observation leads many to take the spatial character of pitch in a metaphorical sense of fitting within a peculiar mental line, where the differences between tones are mapped. The metaphorical interpretation, however dominant, is problematic. For one thing, it makes the paradox of spatial pitch rather trivial, as pitch would just be one of the many aspects of experience that we are inclined to represent on a mental line. More importantly, thinking about high and low pitch as a metaphor pushes the problem further away instead of resolving it: Having proposed that pitch is mapped onto a metaphorical space, we then need to understand how this metaphorical space can introduce biases in how we behave in physical space.

The metaphorical interpretations start with pitch only, and its assumed lack of spatial informativeness. Our account starts with space and the pitch of sounds that are produced in that space: How do sounds of different pitch modulate our responses when we detect, perceive, or interact with other external objects or stimuli to which these sounds relate?

Spatial pitch: Influences on Perception, Behavior, and Attention

Traditionally musical pitch has been defined as the most salient *perceptual* dimension corresponding with the *physical* dimension of sound frequency – more specifically, the fundamental frequency of a harmonic series. For instance, the fundamental frequency of the note G4 is 392 Hz, meaning that the air vibrates 392 times a second. The connection with the frequency of the resonating body emitting the sound means that pitch might vary with the displacement of the emitting source in/through space. When a sound source rapidly approaches and then passes a static observer, the frequency of the sound reaching the ear remains essentially constant at first, then falls at an increasing rate as the source gets ever closer, before dropping rapidly as the source passes. It then drops at a decreasing rate as the source recedes (Neuhoff & McBeath, 1996), the rate and magnitude of the drop being related to the distance and the speed of the source, respectively. However, as illustrated by the Doppler shift effect (Doppler, 1842), many people actually believe incorrectly that the

[1] Pitch itself is better accounted for in a multidimensional manner (e.g., Krumhansl, 2001). One might think here of Shepard's (1982) helix, Krumhansl's cone (Krumhansl, 1979), or Bharucha's MUSACT (e.g., Tillman & Bharucha, 1999).

frequency of the source *increases* as it approaches, in the same way that intensity increases. When a train passes at a railroad crossing, the pitch of the sound is perceived as rising as the train approaches and falling as it recedes. Both pitch and intensity appear to follow the same time course as the source moves. The Doppler effect reveals that our interpretation of the spatial information contained in frequency changes is already biased. More importantly for present purposes, it reminds us that changes in pitch can be linked to the actual and perceived spatial fate of the sound sources that are responsible for them.

While rising and falling pitch will be interpreted as the approaching and receding movement of its source along the horizontal plane, pitch and sound sources are also linked in *auditory localization* tasks in the vertical plane. Binaural cues play a lesser role, and spectro-temporal cues a more important one, when perceiving vertical auditory motion. Differences in frequency can play a role in the perception of moving auditory targets in the vertical plane, thanks to the filtering properties of the pinnae (Shaw, 1982). More surprisingly, people typically locate *static* sources of higher pitched sounds as originating from higher locations in space (Pedley & Harper, 1959; Pratt, 1930; Roffler & Butler, 1968). For instance, Pedley and Harper presented their participants with three different tone series. Two of the seven tones presented in each of three-tone series were the same (900 Hz and 1,400 Hz). These tones appeared as the lowest, intermediate, or highest tones in each series. The participants had to assign an elevation to each tone in the series, based on seven markers presented on the wall in front of them. While all of the sounds actually came from the same vertically arrayed pair of loudspeakers, the participants were actually misinformed that each of the sounds came from one of seven different loudspeakers. The height assigned to the source of the 900 Hz and 1,400 Hz tones was significantly influenced by their *relative* position within their respective tone series, being associated with a lower elevation when the tones were lower in frequency relative to the rest of the series of tones. Overall, though, the increasing frequency sounds in each series were generally located at higher positions.

But whereas in the Doppler case, there is a genuine correlation between the change in frequency and the approach of the sound source (which happens to be misperceived), there would appear to be no physical grounding for a given sound source emitting higher-frequency sounds if it is located higher in space.[2] So, instead of counting as a misperception of the space–pitch relation in the horizontal motion case, the vertical spatial bias generated by the pitch of sounds whose sources are static would seem to be imposing a spatial relation that simply isn't there. The paradox of spatial pitch, in other words, is taken to come from those spatial effects of pitch that cannot be accounted for as a misperception of the displacement of sound sources. As reviewed below, the key evidence for such effects comes from those behavioral paradigms in which, keeping the source of the sound constant and time irrelevant to the task, a difference in pitch is sufficient to influence performance on unrelated visual tasks. (As an aside, it is perhaps worth noting that many, but by no means all, of these tasks have been inherently spatial; see Spence, 2013 on this theme.)

[2] Though it is worth noting that a high-frequency sound will likely spread more easily when coming from a source located high in space just because the sound will be absorbed by fewer obstacles (than is the case on the ground). This is not the case for low-frequency sounds, which we are generally worse at detecting the spatial location of and which are also less likely to be absorbed by obstacles.

Speeded Classification

Demonstrations of the spatial characteristics of auditory pitch often involve variations of the speeded classification task (Ben-Artzi & Marks, 1995; Bernstein & Edelstein, 1971; Evans & Treisman, 2010; Melara & O'Brien, 1987; Patchling & Quinlan, 2002). Participants have to make speeded elevation discrimination responses to a random sequence of visual targets presented above or below fixation. On each trial, a task-irrelevant tone is presented, with the frequency of the tone varying between two predefined and easily discriminable pitches (typically referred to as high or low pitched).[3]

In their early study, Bernstein and Edelstein (1971) demonstrated that people respond more slowly when the elevation of a visual target and the relative pitch of a task-irrelevant sound are incongruent in terms of spatial pitch. The visual targets were presented diagonally to either side of fixation: upper left and lower right for some participants, lower left and upper right for the rest. The participants had to discriminate the location of the visual targets as rapidly as possible while a task-irrelevant tone was presented either simultaneous with the visual stimulus or slightly (up to 45 msec) later. Crossmodal congruency effects were observed in those blocks of trials in which the pitch of the sound varied randomly on a trial-by-trial basis, but not when the pitch of the sound was blocked. Patchling and Quinlan (2002), Evans and Treisman (2010), and a number of others have reported very similar results, though typically with the visual stimuli presented only directly above or below central visual fixation.

The Redundant-Targets Effect

Using another paradigm, Miller (1991) asked participants to make speeded responses in a go/no-go task. Visual targets were presented either above, below, or directly at central fixation, while high-pitched or low-pitched target tones were also delivered over headphones. On each trial, unimodal visual or auditory targets could be presented. There were also a number of bimodal target trials. However, regardless of the target type, the participants had to make the same simple speeded response. They were, however, instructed to refrain from responding whenever a visual stimulus was presented at fixation or when a sound with an intermediate pitch was presented. Participants responded significantly faster on when the elevation of the visual target and the pitch of the auditory target were congruent rather than incongruent. As participants made virtually no errors, this ruled out a speed–accuracy account of the congruency effect. What is more, the reaction time data violated the race model (for redundant signals reaction times, see Miller, 1982), thus arguing in favor of an explanation of this bimodal facilitation effect in terms of multisensory integration (see Harrar et al., 2014).

On the Perception of Ambiguous Visual Motion Displays

Several research groups have investigated the impact of ascending versus descending frequency sweeps on people's perception of ambiguous visual motion displays (Carnivale & Harris, 2016; Maeda, Kanai, & Shimojo, 2004; Sadaghiani, Maier, & Noppeney, 2009). In what is perhaps the first study of this type, Maeda and colleagues demonstrated that

[3] Note that in a few studies, the roles of vision and audition as target and distractor modalities, respectively, have been reversed (Melara & O'Brien, 1987).

people's perception of an ambiguous visual motion display that could be perceived as drifting upward versus drifting downward was biased by the simultaneous presentation of an ascending versus descending auditory frequency sweep (presented, once again, over headphones).[4] More recently, Carnivale and Harris (2016) replicated Maeda et al.'s (2004) findings while at the same time addressing the question of which frame of reference (i.e., gravitational or head-centered) counts as "up." The participants in the latter study had to lie on their side in order to dissociate the two frames of reference. The results revealed a contribution from both head-centered and gravitational upright. However, by far the biggest contribution seemed to be based on the head-centered upright.

Exogenous Spatial Attention Shifts

Chiou and Rich (2012) conducted a modification of Posner's now-classic exogenous attention-cuing paradigm (see Posner, 1978, 1980) in a series of studies in which the participants had to make simple speeded detection responses to a series of visual targets presented randomly from either above or below a central visual fixation point. Shortly before the presentation of each target, an auditory cue was presented from the center of the display. The idea was that if relatively higher-pitched sounds were associated with higher spatial locations, then an exogenous crossmodal spatial cuing effect might be observed: that is, participants would detect upper targets more rapidly than lower targets following the presentation of a higher-pitched than following a lower-pitched auditory cue. The results were as expected (see Figure 5.1A).

Similar results have since been reported in a study by Fernández-Prieto and Navarra (2017) in which static frequency tones were replaced by rising or falling auditory cues, presented for 200 msec prior to the onset of a visual target located in one of four positions arranged in a square around fixation (see Figure 5.2). The participants had to make a speeded detection response. While a slight (and non-significant) spatial cuing effect was observed at the shorter SOA (400 msec), a significant crossmodal spatial cuing effect was found at the longer SOA (550 msec). Intriguingly, the crossmodal cuing effect was only significant following the presentation of the rising frequency sweeps.

Concerns about a criterion-shifting account were effectively eliminated in another series of experiments conducted by Mossbridge, Grabowecky, and Suzuki (2011). These researchers used a go/no-go color matching task in which the participants had to respond to a visual target only when its color coincided with that of a reference visual stimulus presented, for 500 msec, at the start of the trial. The reference circle was always presented along with a frequency sweep (increasing vs. decreasing, with different intervals). The visual targets were either presented from one of four cardinal positions (up, down, left, or right) or from one of four diagonal locations situated around central fixation. The participants were instructed to ignore the task-irrelevant auditory cue. The direction of the frequency sweep and pitch intervals were varied orthogonally.

The results revealed that it was the direction of the frequency change (and not the height of the interval, i.e., the distance in frequency space between the initial and final pitch of the frequency sweep) that guided the direction of participants' spatial attention. The frequency

[4] Interestingly, though, Hidaka, Teramoto, Keetels, and Vroomen (2013) reported that playing low-pitched and high-pitched sounds did not influence perception of apparent motion of pairs of visual stimuli presented from higher and lower positions.

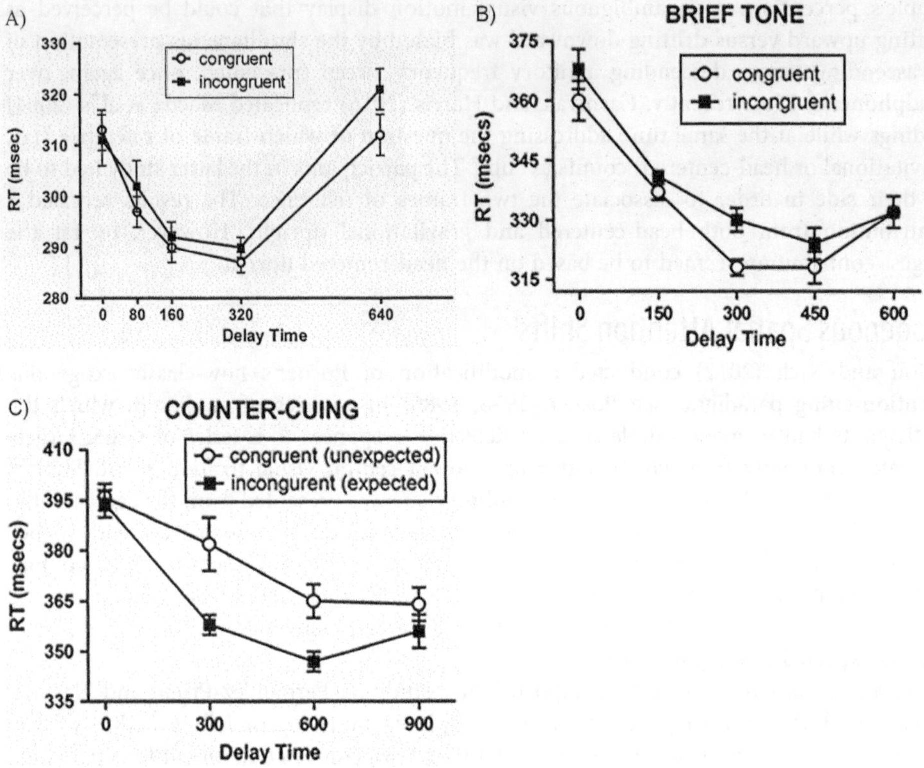

Figure 5.1 (A) Results of Chiou and Rich's (2012; Exp. 1) crossmodal cuing study. Mean reaction times (RTs) are plotted as a function of the SOA and crossmodal correspondence between the pitch of the centrally presented auditory cue and the elevation of the visual target. (Congruent = open symbols; Incongruent = filled symbols.) (B–C) Results of Chiou and Rich's (2012; Exp. 4) study of the effects of endogenous attentional orienting and crossmodal correspondence on the magnitude of spatial cuing effects observed in the vertical dimension. Mean RTs are plotted as a function of the SOA and crossmodal correspondence between the elevation of the visual target and the pitch of the centrally presented auditory cue. (Crossmodally congruent = open symbols; Crossmodally incongruent = filled symbols.) (B) The results when the pitch of the centrally presented auditory cue was nonpredictive with regard to the location (upper vs. lower) of the visual target (essentially replicating the results shown in panel A, Chiou & Rich, Exp. 1); (C) The results when the pitch of the auditory cue was made to predict of the *opposite* location (opposite in the sense of contradicting the natural crossmodal mapping). Error bars represent 1 SEM. (Reprinted with permission from Chiou & Rich, 2012, Figures 2 & 5.)

sweeps facilitated the response to targets presented immediately above or below fixation, but not to those presented in the "diagonal" positions.[5] Such results may indicate that pitch has directed attention toward a specific location rather than toward the upper or lower region of space. Given that improved RTs and/or enhanced sensitivity were reported in Mossbridge et al.'s studies, these results clearly demonstrate an *attentional* facilitation rather than a *criterion* shift that has tainted the interpretation of certain other previous studies (e.g., see Klein et al., 1987a).

Like many other researchers, Mossbridge et al. (2011) argued that their results reflected an *automatic* effect of pitch on performance/attention, since the frequency-modulated

[5] In Fernández-Prieto and Navarra's study (2017), however, global rather than highly localized shifts of spatial attention were observed.

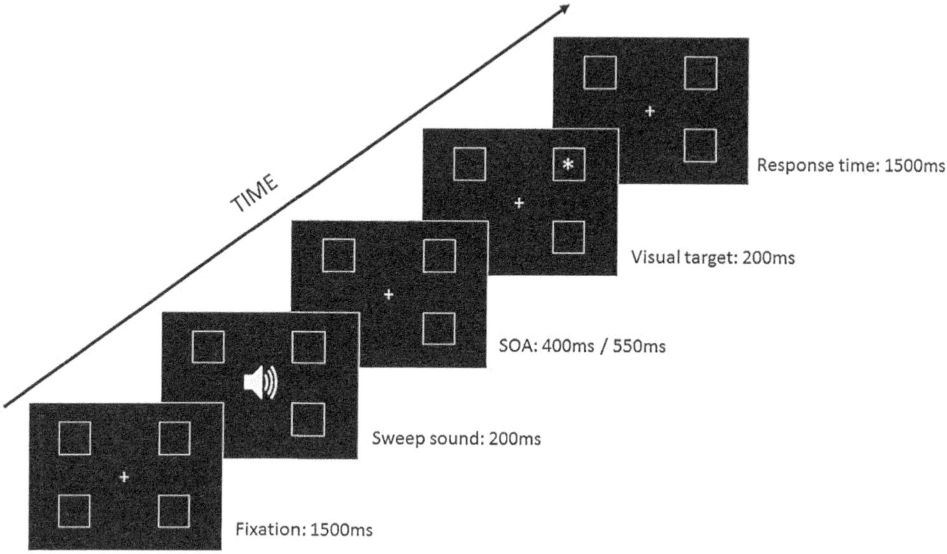

Figure 5.2 Fernández-Prieto and Navarra's (2017) experimental procedure showing the sequence and timing of events in a trial. One of two different frequency sweeps (one varying from 200Hz to 700Hz and the other from 700Hz to 200Hz) was used randomly as a "spatial" cue in an adapted version of the classical Posner paradigm. In addition, one of two different SOA's was randomly presented (400 msec and 550 msec). (Reprinted with permission from Fernández-Prieto & Navarra, 2017, Figure 1.)

auditory cues were entirely task-irrelevant and uninformative with regard to the likely location of the target, and the participants had been explicitly instructed to ignore them by the experimenter. This suggestion deserves some nuancing, though. In another of the experiments, by Chiou and Rich (2012), the pitch of the auditory cue was made informative with regard to the likely elevation of the visual target. In particular, a low-pitched (250 Hz) auditory cue predicted that the visual target would be presented from the upper target location on the majority (80%) of trials, whereas the presentation of the high-pitched auditory cue (2,500 Hz) predicted that the visual target would likely appear in the lower position instead. The participants were told about the meaning of the cue and the probabilities concerned. The results (see Figure 5.1B) revealed that the participants' attention was directed to the likely target location rather than the location associated with the pitch of the sound. It is worth noting, however, that this reversal of the spatial cuing effect took some time to emerge. Such results, therefore, imply that the processing of spatial pitch is, at least to some degree, goal dependent (Spence & Deroy, 2013).

The latter result, however, does not exclude the possibility that the exogenous, crossmodal mapping exerts an influence on people's perception/performance: To do so, one would need to demonstrate that there was no difference in the time course and magnitude of spatial cuing effects under those conditions in which the participants' exogenous and endogenous attention pulls in the same direction versus the opposite ones (see Klein et al., 1987a, 1987b). Should such a difference between these conditions be obtained, it would likely indicate a short-lasting exogenous cuing effect triggered by the pitch of the cue that is then overridden by the strategic allocation of a participant's spatial attention. In the only study to have compared performance in these two conditions directly, participants found it easier to direct their attention when the informative value of a rising or falling pitch tone matched the likely location of the subsequently presented visual target (Klein & Juckes, 1987).

SMARC and SPARC Effects

A separate line of experimental research shows that responses to higher-pitched sounds tend to be faster when given using a higher rather than a lower response button – an effect that has been labeled as the SMARC (spatio-musical association of response codes) or SPARC (spatial-pitch association of response codes) effect (Rusconi, Giordano, Casey, Umiltà, & Butterworth, 2006; see also Umiltà, Bonato, & Rusconi, Chapter 4 in this volume). These studies are inspired by the SNARC (spatio-numerical association of responses codes) effect, in which responses to small vs. large numbers are faster on the left and right side, respectively (Dehaene, Bossini, & Giraux, 1993). In these stimulus–response compatibility effects (SRC; see Proctor & Reeve, 1989), differences in response times are taken as evidence that the mental spatial representation of magnitude corresponds, or interferes, with the spatial distribution of response options: For instance, the faster responses observed to high pitch when they are associated with upper response buttons, rather than lower, suggest a vertical mapping of pitch differences, with higher pitch corresponding to the upper position.

Vertical mappings of pitch are generated consistently and automatically across musicians and non-musicians alike (Rusconi, Kwan, et al., 2006). By contrast, horizontal representations of pitch seem to be less automatically mapped and appear to be much more task-dependent (Lidji, Kolinsky, Lochy, & Morais, 2007). Moreover, subsequent research has shown that the SPARC effect on the horizontal axis is modulated by increasing the pitch range of the vocal stimuli that are used (Estner, 2014).

Temporal Ventriloquism

Finally, Orchard-Mills, Van der Burg, and Alais (2016) published a study demonstrating that the pitch of task-irrelevant auditory stimuli can modulate the temporal ventriloquism effect for pairs of visual stimuli, presented above or below fixation (cf. Parise & Spence, 2008). Specifically, in this task, the visual stimuli were presented at a range of stimulus-onset asynchronies, while one auditory stimulus was presented shortly before the first visual stimulus and the other was presented shortly after the second visual stimulus. The two sounds originated from exactly the same spatial location and provided no task-relevant information. Nevertheless, participants' visual temporal discrimination performance was significantly better when the upper visual stimulus was preceded or followed by the higher-pitch tone, and vice versa for the lower visual stimulus and lower-pitched tone.

Interim Summary

The evidence reviewed in this section clearly demonstrates that people's performance across a range of behavioral tasks can be influenced by the pitch of auditory stimuli, no matter their relevance to the task at hand. While there is undoubtedly a general tendency not to publish null results (Rosenthal, 1979), it is striking that the only studies that we are aware of where auditory pitch was found to have no effect on performance was in Carnivale and Harris's (2016) study in which the participants had to judge the subjective upright of an ambiguous letter (p or d) presented visually at fixation while tones of different pitches were presented to either ear. Perhaps, though, the fact that a null result was obtained in this study should not be seen as all that surprising as the task really involved determining the standard orientation of quite particular visual stimuli (whose identity changes when turned upside down) rather than involving a spatial task contrasting up and down.

Do these results help us determine the origin of the pitch–space relation?

Revisiting the Linguistic Hypothesis

Behavioral data certainly adds to the linguistic and phenomenological evidence that had been dominant in previous explanations. In line with the general hypothesis that language influences perception, several accounts saw the spatial contrast for pitch as coming directly from the use of movement verbs and spatial adjectives to talk about sounds. Spatial elevation is after all also linked metaphorically with many other dimensions of experience or life, such as the difference between positive and negative affect, or dominance and social power (Dolscheid & Cassasanto, 2015; Meier & Robinson, 2004). In these domains, just as for pitch, visual representations of affect or social relations will be responded to more quickly if they conform with the up-down contrast with which they are described.

Pitch was noticed as an example of quasi-universal linguistic agreement by Stumpf (1883) (though Spanish, Catalan, and French more commonly use the terms "obtuse" and "acute," and Farsi, Turkish, and Zapotec more often use "thick" and "thin"; see Shayan, Oztur, & Sicoli, 2011). The fact that English remains the language in which the majority of the psychological research in the field is conducted (Henrich et al., 2010) certainly makes the uses of the terms "high" and "low" to describe both pitch and height more salient for investigation, over and above other possible mappings.

A few years ago, Parkinson, Kohler, Sievers, and Wheatley (2012) tested the Kreung, a hill tribe from northeastern Cambodia, and one of the very few who do not use spatial language in order to describe auditory pitch. The participants in this study viewed shapes that could either rise or fall while listening to sounds that would either increase or decrease in frequency. They were required to report on the auditory change in a speeded classification task (Marks, 2004). The critical response measure was the accuracy of participants' responses in the congruent vs. incongruent trials. Intriguingly, the associations between pitch and vertical position in this group were similar to those demonstrated in the majority of other populations where pitch is described in terms of spatial height. Such results therefore support the idea that the mapping of frequency onto vertical directions does not depend on language.

The early manifestation of this mapping also suggests that language is not necessary for the association between space and pitch. Evidence in 6-month-old infants presented by Braaten (1993) has been pushed back down to 3-month-olds (Walker et al., 2010), albeit not without some controversy (Lewkowicz & Minar, 2014; Walker et al., 2014). Jeschonek et al. (2013) and Dolscheid, Hunnius, Casasanto, and Majid (2014) have also demonstrated that prelinguistic infants (7–12 months old and 4 months old, respectively) are sensitive to associations between space and pitch, or change of pitch.

Such early manifestations should not, however, necessarily lead one to conclude that language has no effect whatsoever. A recent study by Nava, Grassi, and Turati (2016) reported that the association between pitch and elevation was still quite weak in preschool children no matter whether tested with audio-visual or audio-tactile combinations of stimuli. Nava and colleagues speculated that this could perhaps be attributable to immature linguistic and auditory cues that are still developing at age 5 years.

Meanwhile, Fernandez-Prieto, Navarra, and Pons (2016) evaluated pitch-elevation correspondences in 9-to-12-year-old children using a modification of the method used previously by Rusconi, Kwan, et al. (2006). All age groups showed significant pitch–elevation association effects. However, only the older group (11–12 years old) exhibited a positive correlation between the performance in visuospatial and visuomotor tasks of the Wechsler Intelligence Scale for Children IV (WISC-IV) and a pitch discrimination auditory test. Importantly, no such correlation was observed between auditory test performance and the vocabulary subtest of WISC-IV. Such results are, then, consistent with Nava et al.'s (2016) suggestion that the association between pitch and elevation may be modulated by developmental factors.

A neuroimaging study by Sadaghiani et al. (2009) has provided additional information here. The motivation for the study was to investigate whether the biases on visual motion perception that can equally be induced by a rising or descending pitch, spatial words such as "left" or "up," or sounds actually moving from left to right, actually shared the same neural underpinnings. The study revealed not only that the three kinds of effects (pitch, words, and motion) operate at different levels of the cortical hierarchy (see Figure 5.3) but also that the influence of linguistic signals emerged primarily in the right intraparietal sulcus while the effects of pitch could be seen both in these higher-level convergence regions and in the audiovisual motion areas (hMT+/V5+), where the effects of actual auditory motion were shown (though it should be noted that some of the effects reported were actually sub-threshold).

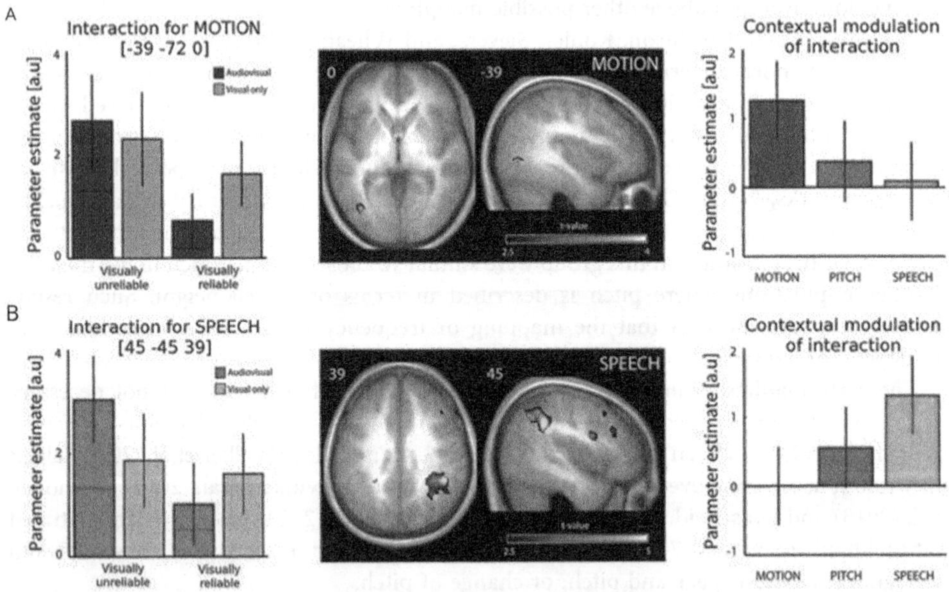

Figure 5.3 Results of Sadaghiani et al.'s (2009) neuroimaging study looking at the neural substrates of crossmodal effects of auditory stimulation on visual motion perception. Up or down could be conveyed by speech (the words "up" or "down"), or by physical motion, or by ascending or descending frequency sweep. (Reprinted with permission from Sadaghiani et al. [2009].) A black and white version of this figure will appear in some formats. For the colour version, please refer to the plate section.

Amodal Remapping and Other Conceptual Hypotheses

Conceptualizing Ordered Relations

The explanation of the spatial effects of pitch might be an instance of a broader, but not directly linguistic, tendency to map sensory magnitudes onto space. Many sensory magnitudes, such as size, weight, brightness, visual density, etc., and symbolic numbers all tend to be spontaneously mapped spatially with smaller quantities being associated to one end and larger to the other. In most left-to-right reading cultures, and in young infants, smaller quantities will be associated to the left side and larger ones to the right one. As this spatial mapping seems to be common to many different sensory attributes, and to symbolic numbers, it may be playing a key role in enabling an abstract or amodal representation of quantity, independent of the kinds of objects or domains at stake (Hubbard, Piazza, Pinel, & Dehaene, 2005; see Nieder & Dehaene, 2009 for a review).

Some theories see the use of a space as *essential* to the general representation of sensory magnitudes (e.g., Dehaene, 1997), while others consider it as a convenient, but by no means indispensable, way of representing different kinds of information on a similar format (Di Luca & Pesenti, 2011; see also Butterworth, 2005). Another debate concerns the degree of connection that theories see between the space involved in magnitudes (and, by association, for pitch) and real space.

At one end of the spectrum, Proctor and Cho (2006) consider that the spatial representation of magnitudes and pitch serves the conceptual representation of order relations. They insist, for instance, that the kind of space shown to exist for sensory magnitudes often only involves left-to-right or up-down opposites on a two-dimensional line – suggesting that space mostly acts as a practical way to represent polarity oppositions and intervals on a scale. In other words, there would be no more to the mapping of pitch contrasts into higher and lower parts of a line than an easy way to represent their difference as a "distance" between values. By contrast, the general theory of magnitude proposed by Walsh (2003) suggests that the fact that we map quantities onto space is primarily meant to help keeping track of quantities or order in the abstract, but is grounded in the demands of action: The mapping of various magnitudes onto a spatial format matters to operate and coordinate transformations for action or predictions about the immediate sensorimotor consequences of action (see also Binetti et al., 2015; Buetti & Walsh, 2009) and is supported by evidence of a neural overlap of magnitude processing in the inferior parietal cortex. Crucially, many of these mappings will reflect ecologically valid relations between time, space, magnitudes, and action: for instance, it takes more time to reach for a further object, speed of movement being fixed; an increase in brightness or loudness tends to correlate with objects coming closer in space; etc.

Why wonder whether the space at stake in the mapping (be it necessary or not to the ordered representation of pitch differences) is a conceptual, abstract space, or the motor space of movement and actions? While the latter will connect the spatial mapping of pitch to constraints about the world or action, abstract theories of magnitude converge with other accounts of the space–pitch relation, which see it as an instance of a general tendency to connect domains of experience through very general abstract concepts, such as "intense–weak" (Karwoski, Odbert, & Osgood, 1942 for an early statement; Walker, 2012; Walker & Smith, 1986). According to these theories, all domains of experience are structured by a small number of basic oppositions, which show in our ways of relating or comparing specific experiences and explain why we match pitch to elevation, but also to brightness, shape, and other dimensions.

Is Higher Pitch 'More Than' Lower Pitch?

It is fair to note that theories of magnitude that connect space and quantities have the ambition of explaining much more than the oft-documented association between pitch and space. But how good are they in explaining this mapping in particular? One primary difficulty rests with the fact that pitch is not, strictly speaking, a sensory magnitude in the sense of being linked to the amplitude of waves.[6] Auditory magnitude – strictly speaking, loudness – is shown to be mapped onto space, in a way that fits with the expected left–right mapping of quantities (Chang & Cho, 2015; Fairhurst & Deroy, 2017; Mudd, 1963). Pitch, on the other hand, corresponds to an increase in the temporal frequency but not in amplitude. It is only indirectly related to actual magnitudes, either through size of typical sources or because higher pitch tends to be perceived as louder.

Overall, lower pitched sounds tend to be emitted by larger objects, and higher pitch sounds by smaller ones (although the tension of the object's material is the critical factor here). However, this statistical regularity would predict that lower-pitched sounds would count as "more," along with larger sizes, and be on the upper part of the line used to map magnitudes. The only evidence pointing in this direction comes from Eitan, Schupak, Gotler, and Marks (2014), who showed that the association of lower-pitched sounds with larger sizes was somewhat reversed once tested dynamically, as sounds that become lower in pitch are actually congruent with shrinking size (i.e., "less").

The overall interference between pitch and loudness might also explain why pitch ends up being assimilated onto a "more or less" scale: As higher-pitched sounds tend to be perceived as louder than lower-pitched sounds played at the same decibel level (Grau & Nelson, 1988; Melara & Marks, 1990; Stevens, 1935; Verschuure & van Meeteren, 1975), they may come to correspond with "more" intensity. However, the effect is minimal: Stevens demonstrated that two tones that differ in intensity can differ in frequency by, at most, 3% (half a semi-tone) and still be perceived as equal in pitch, but subsequent studies have found a much smaller effect (Gulick, 1971).

There is still a difference between the kind of *immediate* spatial mapping of sensory magnitudes directly related to an increase in the physical amplitude of waves (i.e., "more" energy) or the actual size ("more" extension) and the indirect account needed to explain how pitch, which is first and foremost associated with temporal frequency, comes to be associated with amplitude or size. On another point, neither the general theory of magnitude nor the polarity matching account explain why the effects present themselves along the vertical dimension, at least in a privileged manner (see Rusconi, Kwan, et al., 2006), when most others (e.g., Fairhurst & Deroy, 2017; Shaki et al., 2012) are more robustly mapped onto the horizontal dimension instead. The notion of a privileged mapping remains delicate, given previous evidence that the directionality of mapping is heavily influenced by cultural factors and experience and can be easily changed. Rusconi and colleagues' results, suggesting the presence of horizontal mapping of pitch in musicians but not in

[6] It is worth noting, however, that higher pitch generates larger electrophysiological responses than does lower pitch (see Näätänen, 1990; Naatanen, Gaillard, & Mantysalo, 1978; Ruusuvirta & Astikainen, 2012).

non-musicians, show the possibility of other mappings depending on the perceiver's experience (see also Lidji et al., 2007). Still, here, the vertical mapping of pitch differences appears as early as 12 months of age, and even earlier for the dynamic version, and would therefore not depend on cultural factors.

Overall, if the interpretation of space–pitch as a conceptual mapping of "more or less" seems to face difficulties, should we accept that sensory magnitudes are matched onto space because a single format recoding of various magnitudes for action (Walsh, 2003)? To do so, the high–low contrast involved in pitch should relate to the space for action. However, as seen earlier, changes in pitch rather seem linked to approaching or receding stimuli. Here, we want to suggest that the solution to this question rests in looking at the collaboration between audition and vision in the localization of sound sources in external space.

The Crossmodal Interpretation

One alternative interpretation for the effects that have been reviewed here might be to stress not the *a*modal, but rather the *cross*modal character of the space that is involved in pitch. In other words, while *audition only* is unlikely to explain why higher sound sources are associated with higher frequency sounds, this link becomes more apparent once sound sources are considered to be tracked by vision as well as by audition.

The Role of Vision

Although the majority of studies have been conducted in the audiovisual domain, the basic phenomenon of pitch–space is rarely described as visuospatial: While Carnevale and Harris (2016, p. 113) assert that "Low- and high-pitched sounds are perceptually associated with low and high visuospatial elevations, respectively," most studies do not stress the visual character of the space involved in the testing. Parise et al. (2014) talk, for instance, of "perceived spatial elevation" in general, and Hidaka et al. (2013) of higher-pitched sounds "being 'up' in space," without specifying the modal nature of the space involved. A possible explanation for this absence of modal specification might be the largely shared assumption that space is an "amodal" representation – which is not without controversy (see Spence et al., 2013). While the spatial relations represented by each modality are certainly related in the brain and in the mind, this is insufficient to conclude that they all only pertain to a single modal space: The evidence is compatible with a multiplicity of spatial representations, with different degrees of connection depending on the hierarchy and motor goals.

In this context, it is interesting to ask what kind of space is involved in the space–pitch paradox. Does the tendency to associate higher or increasing frequency with higher location or ascending movement exist in the audio-tactile domain as well? Occelli et al. (2009) demonstrated that crossmodal congruency effects between high or low pitch and locations still exist even when these locations were given on the forearm stimulated via the participants' sense of touch. Meanwhile, Deroy, Fasiello, Hayward, and Auvray (2016) investigated whether the correspondence between frequency sweeps and visual direction of movement has analogues in the audio-tactile domain. An implicit association between frequency sweeps and direction of tactile movements on the finger was observed, though only in sighted participants: No such effect occurred for blind participants, independently of the onset of their deficit.

Although the pitch–elevation correspondence also exists between auditory pitch and tactile location, it depends on a participant's prior visual experience.[7] Note that these results are difficult to reconcile with the idea that the conceptual sense of space or metaphorical interpretation is altered in blind people, who show no difficulty in using the words "high" and "low" to describe pitch (Antović, Bennett, & Turner, 2013; Eitan, Ornoy, & Granot, 2012). It highlights the role of vision in the spatial mapping of magnitudes, in line with other studies that have shown that the spatial mappings of symbolic magnitudes are altered in congenitally blind individuals (Crollen, Dormal, Seron, Lepore, & Collignon, 2013; Pasqualotto, Taya, & Proulx, 2014).

These results reveal the close ecological link between auditory and visual space for the perceiver immersed in an environment. Can this explain why the auditory dimension of pitch finds itself mapped onto vertical visual locations?

Crossmodal Correspondences

Instead of the general idea of mapping, we want to introduce the idea that pitch–space is an instance of correspondences, with the term referring to the sometimes surprising associations that people share between different features, attributes, or dimensions of their sensory experience (Deroy & Spence, 2016; see Spence, 2011 for reviews). For example, people tend to associate higher-pitched sounds not only with locations that are higher up but also with lighter, brighter objects (Parise, 2016; Walker, 2012). While crossmodal correspondences have often been assimilated to the much more specific and idiosyncratic phenomenon of synesthesia (Deroy & Spence, 2013), they appear to be much more implicit and widespread, having now been documented between a wide variety of different stimulus attributes and dimensions across ages and cultures. The pitch–elevation correspondence is widely studied along with other audiovisual correspondences, but fits within a much wider network of associations and similarities linking pitch to various nonauditory dimensions (including taste, smell, textures, and bodily size; see Crisinel & Spence, 2009, 2012; Yau, Olenczak, Dammann, & Bensmaia, 2009; Tajadura-Jimenezet al., 2017, respectively).

Congruent with its classification as a crossmodal correspondence, the relation between pitch and elevation, like the majority of other crossmodal correspondences, would appear to be a *relative* rather than an *absolute* phenomenon (Mossbridge et al., 2011; Pedley & Harper, 1959). It is interesting to note here that seemingly random pair of tones, or frequency differences, have been used in the experiments, apparently with no pretesting being required in order to establish the appropriate pitches for the elevation of the stimuli utilized in a given study.[8] This observation provides further indirect evidence for the relative rather than absolute mapping being important. For, if it were an absolute mapping, one would have imagined that much more stimulus calibration would have been required in order to home in on that mapping prior to the conducting of a study.

[7] It would certainly be relevant to test the audiovisual correspondence after the removal of congenital cataracts (Held et al., 2011).

[8] This contrasts with the case of synesthesia – e.g., Noble et al. (2010). It would be interesting, though, in future research, given how bad most of us are at identifying pitch, to investigate whether those individuals with absolute pitch would show the same relative effects.

Internalized Statistics of the Environment?

Correspondences, such as between pitch and visual elevation, may be explained as internalization of the statistics of the environment (Spence, 2011). Instead of specific associations linked to objects, the suggestion is that, on average, high-pitched sounds likely come from sources observed higher in space – which is in line with the findings of Parise et al. (2014). The latter researchers demonstrated a statistical regularity with higher-pitched sounds originating, on average, from higher in space.[9] As sources of sounds tend overall to correspond to a position in space, and tend to be located through vision, this could explain the link between pitch and visual space.[10] This is also supported by the evidence of the Frohlich effect both for visual motion and changes in auditory pitch (see Hubbard & Ruppel, 2013). As suggested by Hubbard (2013a), auditory imagery might be involved in or activated during auditory perception and be associated to visual information regarding the object that produced the sound.

This explanation in term of the collaboration between vision and audition to locate sources also illuminates why the correspondence between pitch and space should be grounded in a bodily or head-centered frame of reference, which is the one used to coordinate our eye/head movements.

Back to Musical Motion

That the mapping between pitch and space ultimately comes from audition and vision collaborating in the localization of external objects might come as a disappointment to those who saw it as a paradox. It seems even more disappointing to those who thought that the paradoxical spatialization of pitch would be key to understanding the paradox of musical motion (see also Larson, 2012). Our account, however, still remains relevant to musical cognition.

Perceiving pitch as high or low can come to be recruited in a "pure listening" episode, where localizing sound sources is not meant to be relevant. Interestingly, amusic individuals seem unable to recruit the correspondence "offline" – or at least off-purpose (Douglas & Bilkey, 2007; Williamson, Cocchini, & Stewart, 2011). These studies open important questions regarding the extent to which musical processing involves visuospatial or audiovisual representations. Testing the spatial biases noted above with amusic participants should be crucial here in determining whether amusia comes with a general indifference to the spatial correspondence between pitch and elevation.

The link between spatial pitch and music is more complicated and obliges us to go beyond the strategic perceptual tasks through which the correspondence is evidenced. To date, the majority of the spatial biases induced by pitch, and pitch change, have involved perception, or decisions between two spatial options. Looking forward, it is also going to be important to further study, especially relevant to music, how the relative pitch of a

[9] The naturalistic recordings in this study were made via directional microphones mounted on a baseball cap designed to capture the sounds of the environment and the elevation from which they came.

[10] Note here that one other alternative explanation for the pitch–elevation mapping that does not necessarily rely on vision is based on the fact that while high-frequency sounds tend to be absorbed (or bounce) by objects on (or near) the floor, low-pitched sounds pass through them. This can explain why high-pitched sounds travel more freely above us, while low-pitched sounds can be more easily associated with near-floor positions.

sound can also bias certain bodily movements and action, or be connected to vocalization. One recent study along these lines had participants spontaneously draw a line in response to low-pitched or high-pitched notes and chords (Salgado-Montejo et al., 2016). Statistical analysis of the responses revealed that, on average, the participants drew higher lines in response to higher-pitched sounds.

Very different factors might explain the privileged association between pitch and elevation here, including the link between body size and vocal pitch (within a species, and over-and-above differences in the tension of the vocal chords), and the further link between pitch and affective expressions or social communication. While high-pitched sounds are used to attract interest in social, pacific contexts (e.g., "motherese") and are associated with smaller, nonthreatening objects, lower-pitched sounds will be associated with danger, or more negative expressions and contexts. The connection between high/low pitch and high/low locations could, then, be explained partly through another association between negative and down and positive and up. On the other hand, pitch will become painful only at the highest frequencies, making the hedonic associations of pitch less straightforward. One would still expect that the spatial biases induced by pitch would be modulated by the valence of the context or of the target of one's actions.

In our view, the contribution of pitch changes to musical motion needs to account for these links between pitch and affect, and its complex links to the space and context of actions, not to mention social interactions. Spatial biases across audition and vision have here only perhaps a role to play in speculating about the roots of instrument design and musical notation. In musical notation (something that is evolutionarily very recent and will have probably drawn on existing perceptual biases), higher pitch is indeed represented higher on the staff. It seems that musical notation has incorporated the association between pitch and elevation even since its earliest forms in ancient times (e.g., the Seikilos epitaph, around 200 BC to around AD 100). Looking at the side of musical production, not all instruments seem to be built to make sure that higher-pitch sounds are produced by manipulating the upper part of the instrument, but many of them do in the Western tradition. Explaining the vast variety of factors that have shaped the design of musical instruments across cultures is something that falls well beyond the scope of the present chapter, but it seems worth mentioning that, in recent decades, as musicians, researchers, and technology developers had tried to adapt music production to visual digital media, the exploitation of pitch–space biases is becoming more salient.

Conclusions

The paradox of spatial pitch, as has been argued here, only emerges because of the tendency to look at audition in isolation from the other senses (see also Hidaka, Teramoto, & Sugita, 2015). This unisensory perspective is also partly disembodied, forgetting the link between sensory epithelia dictated by their spatial distribution on our bodies and the way in which they collaborate to guide bodily actions and exploration. Envisaged in connection with the location of the resonating bodies in the external space, in which vision is also interested, the spatial biases generated by pitch stop looking paradoxical and start to appear more like crossmodal biases, grounded in the wider ecological statistical regularities. More than being environmental regularities, truly independent of us and our actions, the link between pitch and visual elevation integrates us and is centered on our bodies (see Mossbridge et al., 2011; Parise et al., 2014).

Shall the connection with vision, and ultimately between ears, eyes, and external objects, which seem to be grounding the space–pitch relation, suggest that it has little to say about musical motion, and musical cognition in general? The location of external sources is, after all, largely indifferent to the perception of musical motion, although it matters to the quality of the experience. Listening to a piece of music has little to do with locating objects in the vertical space. However, this leaves room for a much more important social and cultural evolutionary story, where the link between pitch, space, and our bodies might contribute to explaining the development of music in human societies, including the privileged uses of certain visual representations in music and the design of musical instruments.

Chapter 6

Representational Biases in Space and Language

Alexander Kranjec

Space is inherently relational. Space is abstract and cannot be directly seen or touched. However, organisms, objects, technologies, and the environment have certain formal regularities. It is these predictable tendencies in material structure and behavior that provide a means for grounding our understanding of spatial relations. That is to say, for a spatial representation to be *meaningful and communicable*, it requires a frame of reference (Levinson, 2003; Taylor & Tversky, 1996). Accordingly, space can be understood as a system of relations with respect to any number of embodied, perceptual, environmental, and cultural regularities – some more constrained than others (Bowerman & Choi, 2001; Clark, 1973; Levelt, 1984). For example, humans walk upright. We have bodies that are a kind of asymmetrical container with an inherent front and back, as are many of the artifacts we create, such as cars and refrigerators. We have a dominant hand (usually the right) for performing fine manual skills. Our eyes are located at the front of our bodies, and the giant primate brains behind them are generally biased toward processing visual information above and beyond input from other sensory modalities. Because spatial information is processed primarily in the right hemisphere of our brains, we are biased to attend to things on our left. In our visible environment, living things grow up, while the force of gravity tends to pull things down. The night sky rotates around a celestial pole that corresponds to North or South. People in the United States drive on the right side of the road and read from left to right.

The meanings of particular spatial words in conjunction with our knowledge of the world make understanding the examples in the previous paragraph seemingly simple. Language allows us to label spatial relations, and in the most prosaic sense, spatial language enables us to communicate the locations of things to coordinate our thoughts with others (e.g., "The door on the left"). But spatial language also supports our understanding of nonspatial domains such as time (Núñez & Cooperrider, 2013) (e.g., "The past is behind us"). In fact, beyond this kind of explicit language of verbal labels for spatial relations using prepositions, some hypothesize that language organization itself, at a higher level of event representation, possesses something like visuospatial structure (Chatterjee, 2001; Slobin, 1996; Talmy, 2000). Syntax is itself a kind of semantic reference frame providing speakers and listeners formal, structural cues for communicating and understanding knowledge about agents, actions, and objects in space.

Interactions between space and language are now well documented, having received several extensive, formal treatments over the past 20 years (Bloom, Peterson, Nadel, & Garrett, 1996; Evans & Chilton, 2007; Levinson, 2003). However, many of the details of how spatial and verbal representations interact, in the mind and brain, remain unknown. If there is no such "thing" as space (at least not in any obvious way), why might it be so important for providing a basis for meaning and communication?

Space provides an interesting case study in semantics, because interactions between space and language concern broader issues regarding the continuity of mental representation (Chatterjee, 2001, 2008; Kranjec & Chatterjee, 2010). Spatial semantics provide a sparse, paired-down system for investigating how a basic spatial concept (such as "above") may be represented by multiple systems (such as perception and language). The general idea that the same basic meaning can be represented by different systems originates with Pavio's (1986) dual coding hypothesis, which suggests that information can be coded in analog as well as symbolic formats. Theories regarding the representation of basic, abstract meanings involving both verbal and nonverbal representations have been refined in the context of more recent debates concerning embodied approaches to cognition (Louwerse & Jeuniaux, 2010; Riordan & Jones, 2011). However, many questions remain regarding the specific roles that perceptual and linguistic representations play in our understanding of space. For example, to what extent do representations best characterized as perceptual vs. verbal in nature bias spatial cognition? Another way to raise a similar question might include a discussion of the relative values of "embodied" vs. "Whorfian" explanations for such biases. For example, embodiment research suggests ways that the body, action, and perception may effect verbal representation, whereas Whorfian approaches suggest ways that language may influence nonverbal thought and perception. On the one hand, these theoretical approaches seem opposed in terms of "bottom–up" vs. "top–down" emphases. Yet on the other hand, embodied and Whorfian approaches are often portrayed as complimentary, suggesting a common representation somewhere in the middle (like an image schema) driving both kinds of effects (Holmes & Wolff, 2013; Mandler, 1992; Richardson, Spivey, Barsalou, & McRae, 2003).

Cognitive neuroscience approaches provide a means for describing how the same basic spatial meanings (depicted in perceptual or verbal forms) may be encoded by shared and/or distinct neural architectures. These approaches raise more questions regarding the nature of mental representation. For example, to what extent does the brain distinguish between the same basic spatial relation expressed in discrete formats? If the brain does distinguish between certain kinds of representations, are particular areas or networks biased toward certain forms of information? Can neuroscience data help describe potentially intermediate forms of representation whose characteristics lie somewhere between language and perception by providing evidence that some representations are not best understood as either "symbolic" or "perceptual"?

The present chapter focuses on several recent studies that examine relations between spatial and verbal representations of the same basic concepts. Although the interactions between space and language are complex, each study discussed in the current chapter focuses on the semantics of simple spatial concepts grounded in categorical figure–ground relations – the starting point for any spatial frame of reference. In English, such concepts are lexicalized using prepositions. Prepositions are small words with basic meanings but important in that they can operate on semantic and syntactic levels. Prepositions are special because they are semantically and grammatically flexible in ways that are unique in human language (Kranjec, 2013). They can serve a prescribed functional role as syntactic relational units (e.g., "on" alert vs. "in" trouble) seemingly (but not necessarily) lacking semantic content (Kranjec & Chatterjee, 2010). Semantically, spatial language, in addition to allowing us to communicate the locations of things, serves as a foundation on which more complex concepts are constructed. We can think and talk about all sorts of abstract conceptual categories such as time and morality using the language and structure of space. In this regard, prepositions may be important for grounding relational thinking more generally.

The use of basic relational language (using prepositions), grounded in fundamentally relational cognitive domains (such as space), may even make species-specific human abilities like analogy and metaphor possible (Gentner, 2003). By investigating how these particularly basic yet flexible and productive units of language are instantiated in the brain along a representational continuum, the studies presented in this chapter seek to better understand how the human brain utilizes neural structures evolved for perception and action to make meaning via language.

Each study in this chapter greatly reduces the domain of language to single words or concepts denoting simple spatial relations such as "in," "on," and "above" in order to investigate how verbal and nonverbal representations of the same meaningful spatial relations interact. The studies are aimed at addressing several theoretically related questions:

1. **How are representational formats** like words and pictures **distinguished at the neural level?** Does the brain recognize intermediate representational or iconic formats, such as *schemas*, that have both word-like and image-like qualities? Can we describe general neural-organizational biases in how certain kinds of information are represented?
2. **Are some representational formats privileged?** If the brain does distinguish between formats, are some better for representing spatial information? If so, can biasing information processing toward a particular format make spatial relations more salient?
3. **How do different representational formats interact?** To what extent do these formats interact, and in which direction? That is, do verbal labels coding for spatial relations such as "left" and "above" penetrate perception to influence thought in a neo-Whorfian sense? Or can *meaningful but nonverbal* spatial categories bias visuospatial perception without the direct influence of language?

How Are Representational Formats Distinguished at the Neural Level?

In cognitive neuroscience, research investigating the representation of basic spatial concepts has focused on prepositions. Friederici (1981) reported that Wernicke aphasics demonstrate impairments in processing locative prepositions. Landau and Jackendoff (1993) later proposed that the parietal cortex, as the terminus of the dorsal "where" pathway, also plays a role in processing prepositions. This hypothesis found support by work from Damasio and colleagues demonstrating a role for the left supramarginal gyrus and inferior frontal gyrus in the comprehension of locative prepositions (Damasio et al., 2001; Emmorey et al., 2002). Noordzij, Neggers, Ramsey, and Postma (2008) also found that understanding the meaning of categorical spatial relations expressed by locative prepositions was associated with activation in the left supramarginal gyrus. Wu, Waller, and Chatterjee (2007) found locative relations to be mediated by left inferior frontal-parietal cortices. Despite some diversity in the specific results, the literature on prepositions and that on nameable categorical spatial relations strongly implicates the left hemisphere over and above the right.

However, notwithstanding this left-hemisphere bias for processing prepositions and nameable spatial relations, spatial representation and spatial attention are more generally explained as right-lateralized functions of the inferior parietal cortex, and in particular, the right supramarginal gyrus (McFie, Piercy, & Zangwill, 1950; Vallar & Perani, 1986). In part to explain hemispheric differences in processing different kinds of spatial relations, Kosslyn

(1989) originally proposed that spatial information could be divided into two broad types: categorical relations that refer to discrete spatial relations frequently lexicalized by locative prepositions; and coordinate relations, which are finer-grained metric relations that specify the exact locations of objects in space. (See the third study discussed later in this chapter.) To summarize, after several decades of research on spatial relational cognition, the overall picture that emerges from the cognitive neuroscience literature is one in which word-like spatial relations are represented in the left hemisphere (in particular the left supramarginal gyrus), and analog spatial relations are represented in the right hemisphere (in particular the right supramarginal gyrus). However, stimuli from these studies were limited to words and pictures.

In order to refine our understanding of neural biases for distinct kinds of spatial information, Amorapanth et al. (2012) attempted to distinguish between brain areas responsible for representing spatial relations in (1) words, (2) pictures with rich perceptual detail, and at an intermediate level of iconic, or (3) schematic abstraction. *By focusing on intermediate representational formats with properties of both words (or language) and images (or perception) it is possible to test the extent to which particular brain areas reveal biases for information presented at various levels of abstraction.* For present purposes, we can call these intermediate formats *schemas*. Schemas are understood to be more concrete compared to the arbitrary letters and sounds that represent a word like "IN" and more abstract than images depicting real-world scenes in space. While the word "schema" has made several appearances in the history of ideas, here the term is used to describe any iconic representation where perceptual detail has been abstracted away from reality in order to provide a more flexible representation for cognition. Schemas occupy an intermediate position on a continuum between very concrete perceptual complexity and very abstract symbolic representation. Unlike full-fledged symbolic representations of spatial relations (i.e., prepositions), a schema (or icon) preserves some of the analog spatial qualities of the thing it stands for (Deacon, 1997; Peirce, 1955). In this way, schemas can be understood to function more like percepts. However, like words, schemas may be more likely to be processed holistically and function in a quasi-symbolic way. Thus, schemas are more abstract and regular than the thing they represent, functioning more like language. Despite cognitive neuroscience's growing interest in exploring relations between space and language, very little is known about the neural organization of such intermediate representations. More recent developmental and cognitive linguistic approaches have treated spatial schemas mainly as a theoretical construct. However, empirical support for schematic representation has a history in experimental psychology (see Peterson, Gillam, & Sedgwick, 2006, for example). Distinct yet potentially allied concepts like that of "boundary extension" (Intraub & Bodamer, 1993) have been explored using neuroscience methods (Park, Intraub, Yi, Widders, & Chun, 2007).

How might the brain represent such intermediate representations? As suggested by prior research on space and language discussed earlier here, damage to the left hemisphere in regions critical for the representation of lexicalized categorical spatial relations could, in parallel, compromise their schematic representation. Alternatively, right hemisphere areas associated with representing nonverbal spatial information may be critical for representing the same basic spatial meanings without language. The mediating role that schemas are hypothesized to play between language and perception in representing the meaning of categorical spatial relations suggests that either of the above principles of neural organization could be the case. The group study described below (Amorapanth et al., 2012) tests the validity of these two alternative hypotheses using voxel-lesion symptom mapping (VLSM) in between-format matching tasks with 34 stroke patients.

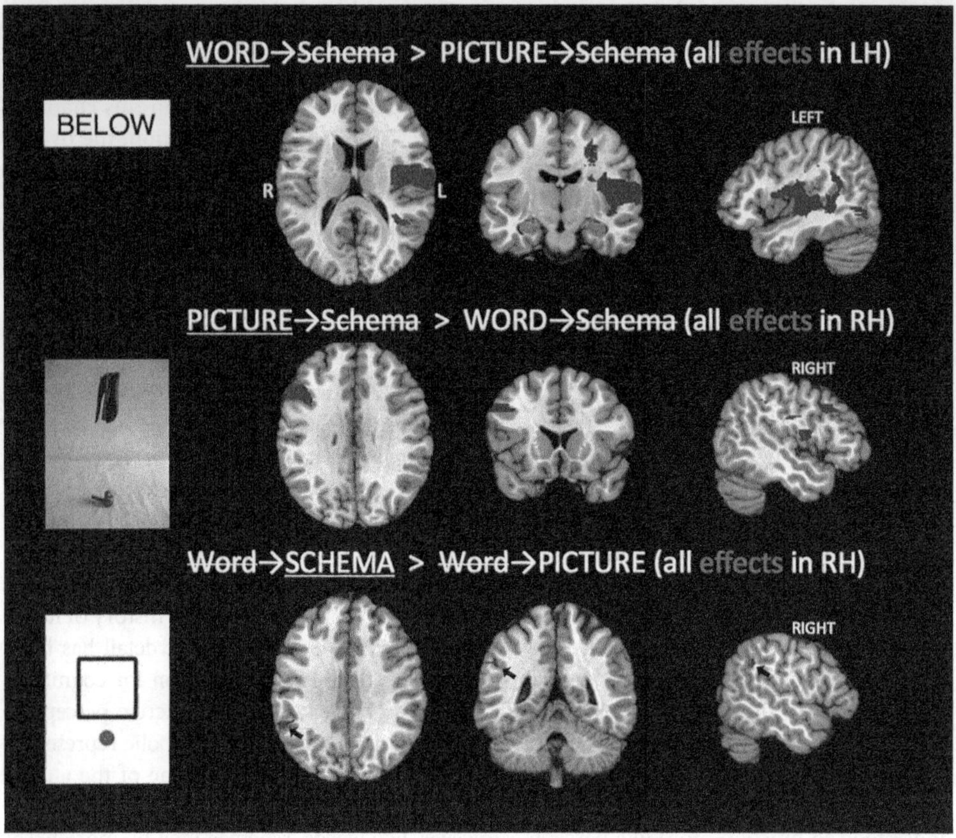

Figure 6.1 To test hypotheses regarding processing biases across hemispheres and information types, the tasks were residualized against each other in order to establish orthogonal measures for particular representational formats (Amorapanth et al., 2010). By regressing performance for one matching task onto another and plotting the residual scores, behavioral variance associated with processing within a single representational format or stimulus type (i.e., word, picture, or schema) can be isolated. In this figure, the color blue represents the isolated format in a particular analysis and the associated brain areas where damage predicts significant behavioral impairment for processing that format.

To see which brain areas were involved in processing the same spatial meanings in words vs. pictures vs. schemas, we used VLSM (Bates et al., 2003), a method that analyzes the relationship between tissue damage (lesioned brain areas) and behavior (incorrect responses on matching tasks). Brain–behavior relations are mapped on a voxel-by-voxel basis as in functional neuroimaging. Unlike fMRI, which correlates neural activity and behavior, VLSM analyses indicate brain areas where damage predicts significant behavioral impairment.

Four prepositions (IN, ON, ABOVE, and BELOW) served as the **words** in each matching task. Realistic color image stimuli served as the **pictures** designed to unambiguously depict the same spatial relations as denoted by the prepositions. The **schemas** consisted of simple lines and geometric forms. The set of four schemas varied along parameters proposed by Talmy (2000), such as containment, support, and degree of separation. Incorporating the three basic types of stimuli (words, pictures, and schemas; Figure 6.1, left), we used four matching tasks to investigate cognitive processing across representational formats. All tasks required participants to match a relation depicted in a probe item to one of four target items.

VLSM indicated those brain areas most critical for the representation of one stimulus type over and above another type between matching tasks (Figure 6.1, right). The results showed that verbal components of the matching tasks are processed in the left hemisphere (WORD > PICTURE) and pictorial components in the right hemisphere (PICTURE > WORD), and furthermore that the right hemisphere differentiates between distinct spatial formats (SCHEMA > PICTURE). Overall the results suggest that categorical spatial representation deficits in the left hemisphere are difficult to distinguish from deficits associated with labeling these relations verbally. Effects in the right hemisphere suggest that deficits processing categorical spatial relations do occur with right brain damage, and that these deficits cannot be accounted for by naming deficits. Furthermore, the right hemisphere distinguishes between different kinds of analog spatial representations (schemas vs. pictures), suggesting that the right hemisphere plays a special role in extracting schematic representations.

Are Some Representational Formats Privileged? Can Biasing Information Processing toward a Particular Format Make Spatial Relations More Salient?

VLSM results from the group lesion study suggest that the brain can distinguish between distinct formats in representing spatial relations. However, might some formats be "better" for representing spatial information compared to others? Although linguistic representations are uniquely abstract and flexible, some kinds of information are often more "efficiently encoded or retrieved using an analogue, spatial or other form of representation" (Hubbard, 2007, p. 41). Maps, pictograms, and other graphical tools have been used throughout history to represent culturally important information in such an efficient manner. Like language, they work by reducing a more complex spatial reality into a model of the world that is easier to use. But unlike language, such technologies abstract away from a more complex physical or psychological reality in order to provide a reduced or prototypical representation while still preserving something about the fundamental, analog structure of the thing it represents. These kinds of schematic representations are obviously still in use today. For example, an iconic smiley face may be better than a photograph of a smiling person for representing happiness because it can be mapped more flexibly across individuals and contexts (McCloud, 1993). And it would seem, as emoji keyboards are now standard issue on many smartphones, such iconic or schematic representations find frequent use in communication even as technology has moved from cave paintings to handheld computers.

If some representational formats are privileged, perhaps biasing information processing toward a particular format could make spatial relations more salient. In this regard, a neuropsychological case study approach is useful. In a study by Kranjec, Ianni, and Chatterjee (2013), a patient with simultanagnosia (EE555) presented an interesting case for the investigation of differences in spatial representational format. Patients with lesions in parietal cortices frequently show impaired spatial abilities. Bilateral parietal lesions can result in simultanagnosia. Patients with simultanagnosia are characteristically unable to perceive more than a single object at a time (Luria, 1959). If schemas help one abstract spatial relations from complex scenes and aid relational thinking, they might be particularly helpful for an individual with simultanagnosia. Because schemas are designed to make basic spatial relations more salient and are visually less complex than perceptually rich scenes, it was hypothesized that their inherent visual simplicity may put less attentional demand on the compromised system of someone with simultanagnosia.

Patient EE555 experienced three parietal lobe infarcts between May and June of 2004 resulting in bilateral lesions extending from the occipital lobes to middle parts of the inferior parietal sulcus. Behavioral testing indicated simultagnosia. EE555 was unable to comprehend more than one object simultaneously 30 months after her most recent stroke and showed a complete local bias with Navon Letters (Berryhill, Fendrich, & Olson, 2009).

Control subjects and EE555 performed six matching tasks similar to those in the VLSM group study. All participants were instructed to match a relation depicted in a probe item to one of four target items. (Probe-Target; for example, matching the word "IN" to one of four photographs depicting the spatial relation "IN.") Probe and target spatial relations could be depicted in Word, Picture, or Schema formats (see Figure 6.2.)

Overall, controls outperformed EE555 on all tasks. Most interestingly, on the only two tasks that did not include matching to, or from, schemas (word-to-picture [W-P] and picture-to-picture [P-P]), EE555 performed at chance levels. For all tasks with schemas (schema-to-picture [S-P], word-to-schema [W-S], schema-to-schema [S-S], picture-to-schema [P-S]), performance was significantly better than chance. Tasks were administered in the sequence as depicted so possible learning effects can be observed in Figure 6.2. EE555 performed at chance for the first task (W-P) and then appeared to improve gradually on the following two tasks (S-P and W-S). That is, initially it appears as though a learning effect could possibly explain the observed differences in accuracy from tasks 1–3. Arguing against this possibility, however, is the subsequent "recovery" of impairment in the only other matching task that did not include schemas (task 4: P-P). That EE555 is at chance performance only on the two baseline tasks without schemas (W-P and P-P) strongly suggests that the inclusion of schemas made spatial relational matching possible. Better performance for EE555 on the S-P versus W-P task suggests that schemas make spatial relations visible for EE555 in a manner that words cannot. Furthermore, EE555's spatial

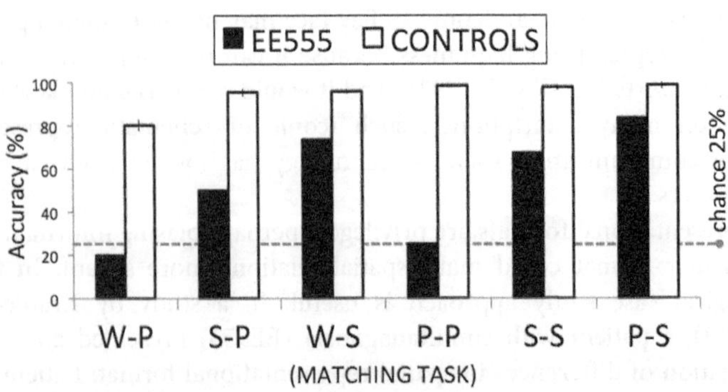

Figure 6.2 Results from six matching (Probe-Target) tasks in Kranjec, Ianni, and Chatterjee (2013). W = word; P = picture; S = schema. For instance, "W-P" refers to matching a Word (e.g., "ABOVE") to a Picture representing a figure object above a ground object.

matching performance increases with each second exposure to schemas generating the hypothesis that schemas may have a cumulative facilitatory effect.

Importantly, it is not just that EE555 could see spatial relations in schematic form where she could not see spatial relations in photographs. Rather, exposure to spatial relations in schematic format actually made seeing spatial relations between objects in more perceptually rich picture stimuli possible. (This is demonstrated by above chance performance in S-P trials.) This means that the recovery of global processing observed was not merely the result of manipulating stimulus features such as shape and proximity to make some object relations more salient (Huberle & Karnath, 2010). Additionally, verbal descriptions of spatial relations alone did not have this effect. (As demonstrated by below chance performance on W-P trials.) Access to spatial semantics was not sufficient to prime spatial attention as in a previous study (Soto & Humphreys, 2009). Rather, it seems that because schemas are "abstract but analog," they are particularly good at representing spatial relations in both a meaningful and a perceptually salient manner. Not only was EE555 able to see spatial relations when they were presented in a schematic format, schemas actually helped EE555 see spatial relations when they were presented in a more perceptually rich image format (i.e., pictures). It is possible that a relatively low-level visual process may provide the underlying mechanism for this effect. Spatial deficits associated with simultanagnosia could result from impaired processing of low-spatial frequencies in damaged areas of the dorsal visual pathway. Another study (Thomas, Kveraga, Huberle, Karnath, & Bar, 2012) found that priming simultanagnosic patients with low-spatial frequency information facilitated global processing. The authors suggest that by "biasing" low-spatial frequency processing in the intact ventral pathway, visuospatial attention and perception can be recovered.

How Do Different Representational Formats Interact? Directional Biases between Space and Language

Results from the VLSM group study showed left–right hemispheric biases in the processing of verbal-analog spatial information, respectively. Results also indicated that the right hemisphere further discriminated between distinct analog formats (schemas and pictures). The simultagnosia case study data suggests that some representational formats (schemas) may be privileged in bridging perceptual and verbalizable information about spatial relations. In order to investigate questions related to the direction of interaction between perceptual and verbal representations in neurologically intact participants, Kranjec, Lupyan, and Chatterjee (2014) conducted several behavioral experiments with healthy adults. One general question we explored concerned whether meaningful spatial categories can influence perception, and if so, whether such an effect was specifically modulated by verbal representations or by meaningful but nonverbal representations that are best described as analog in format.

Kranjec et al. (2014) made use of a well-known neuropsychological distinction. Recall that Kosslyn (1989) first suggested that spatial relations can be divided into two general kinds: (1) *categorical relations* refer to discrete spatial relations frequently lexicalized by locative prepositions like "left," "right," "above," and "below"; and (2) *coordinate relations* refer to fine-grained metric relations coding analog distances. Categorical representations specify abstract, equivalent classes of spatial relations. Coordinate representations specify the exact locations of objects in space, critical for reaching and navigation.

In our study we tested whether the discrimination of categorical and coordinate information differs as a function of meaningful spatial relations. It was predicted that categorical perceptual discrimination should be facilitated between categories that are discretely lexicalized by common locative prepositions (i.e., "easier-to-name" categories) than for categories less easily (or more ambiguously) named. This means that discriminating between two perceptually different spatial relations that are discretely lexicalized (e.g., discriminating *above* vs. *left*) should be easier as compared to discriminating between two perceptually different spatial relations that share a common spatial category between them (e.g., discriminating *above-left* vs. *below-left*). On the other hand, as much as lexicalized distinctions are associated with more robust categories (Lupyan, 2008), it was predicted that it would be more difficult to make coordinate judgments when discriminating between two perceptually different spatial relations within easier-to-name spatial categories as compared to more ambiguously lexicalized categories. In short, it was thought that "nameability" would have an opposite kind of effect on categorical and coordinate discriminations: (1) Better nameability should make locations *between* spatial categories more distinctive, while (2) making locations *within* spatial categories more alike. This relates to the phenomenon of categorical perception (see Goldstone & Hendrickson, 2010 for a review.) Previous research (Winawer et al., 2007) investigating the effects of language on color perception found that Russian-speaking participants whose language makes an obligatory distinction between shades of blue were faster to discriminate two colors when they were members of distinct verbal categories (in Russian), as compared to when they were from the same verbal category. No effects were found for English speakers using the same colors. The category advantage in Russian speakers was eliminated by verbal interference, further suggesting that verbal categories in language can affect color perception, at least as reflected by differences in reaction time, and that such effects are mediated by language online.

In the current study, neurotypical English-speaking participants had to determine if the dot-cross configurations presented to the left and right of central fixation were same or different. Only "different" trials were analyzed. They could be of four types representing each condition of interest (see Figure 6.3).

As predicted, participants were more accurate when discriminating categorical information between easier-to-name categories and more accurate at discriminating coordinate spatial information within harder-to-name categories (Experiment 1, Figure 6.4).

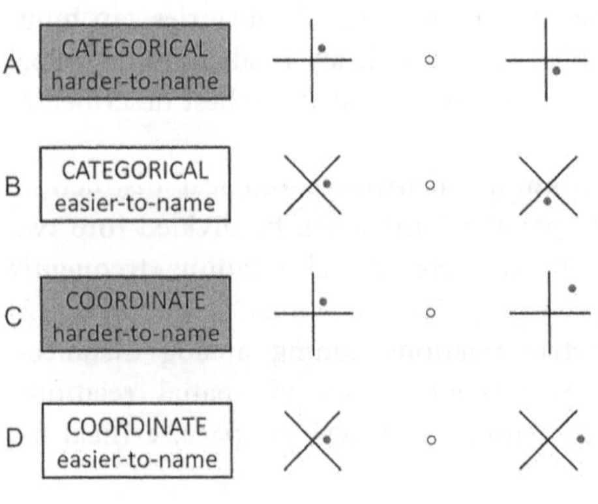

Figure 6.3 Examples of four kinds of "different" trials in Kranjec et al. (2014), Exps. 1–3.

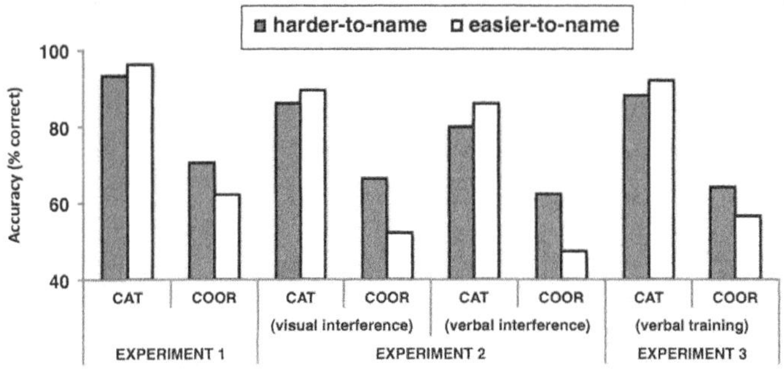

Figure 6.4 Results from Kranjec et al. (2014). Participants were more accurate when discriminating categorical information between easier-to-name categories and more accurate at discriminating coordinate spatial information within harder-to-name categories across all three experiments. Neither verbal interference nor verbal training had any effect on discrimination performance.

These results suggest that there is a relationship between discretely lexicalized spatial categories and perceptual processing of categorical and coordinate spatial relations. Subsequent experiments attempted to either down-regulate or up-regulate the involvement of language in performance of the same task.

In Experiment 2, performance on the original task was contrasted under verbal and visual interference conditions. Because verbal interference is hypothesized to down-regulate the involvement of language in task performance (Gilbert, Regier, Kay, & Ivry, 2006; Winawer et al., 2007), the advantage on the categorical easier-to-name trials and coordinate harder-to-name trials (Exp. 1) should be attenuated by verbal interference if language plays a critical role in the observed perceptual biases from Experiment 1. Alternatively, requiring participants to explicitly associate locations with distinct labels strengthens existing associations and tends to exaggerate categorical effects on perception (Lupyan, Thompson-Schill, & Swingley, 2010).

In Experiment 3, participants were trained to associate particular meaningful verbal labels (above, below, left, and right) with the spatial categories used in Experiment 1. Each of the four positions corresponding to the "×" partition was valid for a single label (above, below, left, or right), whereas each of the four positions corresponding to the "+" partition was valid for two labels (above AND right, above AND left, below AND right, or below AND left). The "up-regulation" hypothesis reasoned that if the previously observed interaction stems from the automatic involvement of category labels then strengthening the association between labels and particular spatial locations should increase the differences between easier-to-name (×) and harder-to- name (+) trials observed in Experiment 1.

Despite our best efforts however, the results from both Experiment 2 (verbal interference) and Experiment 3 (verbal training) suggest that the original spatial relation type-by-nameability interaction is resistant to online language manipulations. Neither verbal interference nor verbal training had any effect on discrimination performance. The results across all three experiments in the behavioral study suggest that although robust biases in the visual perception of spatial relations do correlate with patterns of lexicalization, they do not appear to be modulated by language online. Furthermore, on coordinate trials, we find evidence that lexicalized or "meaningful" categories predicted performance on a perceptual task that did not require participants to make a categorical discrimination.

Conclusions

Our group VLSM study provided preliminary evidence for how the brain makes gross distinctions when representing named spatial relations in different formats. More recently, fMRI methods have extended this result to other kinds of basic concepts indicating that the brains of neurotypical participants distinguish between the relative symbolic vs. analog nature of action representations, like those named by verbs vs. schematic depictions of actions (Watson, Cardillo, Bromberger, & Chatterjee, 2014). Our case study involving a patient with simultagnosia demonstrated that exposure to spatial relations in schematic format actually made seeing spatial relations between otherwise "invisible" stimuli possible, in a way that the mere naming of spatial relations could not for a patient with severely impaired visuospatial attention. A recent group patient study (Göksun, Lehet, Malykhina, & Chatterjee, 2013) showed that other kinds of externalized schematic representations, namely *gestures*, might help or compensate for difficulty with lexical access in patients with naming deficits. So et al. (2015) found that gesture as compared to language was a more effective means of encoding spatial information on a route-learning task using neurotypical participants. Both studies may suggest that, like schemas, gestures are a privileged kind of representation, or perhaps should themselves be considered a form or instantiation of schematic representations. Finally, our behavioral findings with neurotypical participants suggest that high-level categorical information about space penetrates perceptual representations in a manner that language cannot. Robust effects were found for task accuracy (not merely reaction time) over three experiments, and were impervious to both interference and training, suggesting that the mechanism for representing basic, meaningful spatial relations, such as those coded by prepositions, runs deep. These data, taken together with the group patient study, further suggest that the right hemisphere may exhibit an analog bias in representing the meaning of basic spatial categories. It is possible that within the right hemisphere, analog representations of basic spatial categories exert a modulating effect on the perception of spatial relations (even coordinate spatial relations) independent of verbal representations processed in the left hemisphere. The relatively nonverbal right hemisphere may be "semantically independent" with regard to primitive spatial concepts.

Data from all three studies presented above are consistent with the idea that mental structures such as schemas may play a role in representing meaningful patterns of basic spatial relations. Schemas capture an intermediate level of representational structure that is neither perceptual nor verbal in nature, lacking both the rich detail of percepts, and the symbolic properties of words, while encoding the necessary content for representing primitive relational semantics. Spatial schemas are distinct from complex percepts in that they are "bleached" of detail, discretely encoding the basic semantics contained within asymmetrical figure-ground relations. They are also distinguishable from language. Whereas words are symbolic in that the relation between a word and what it represents is mostly arbitrary, a schematic representation maintains the general topographic structure of the percept. In the above behavioral experiments, a schematic representation (or image schema) may provide a mechanism for representing "named" meaning at an intermediate level, preserving an aspect of the topographic perceptual representation. While not being easily modulated by verbal interference or training, schemas may play some role in linking linguistic labels and percepts. The behavioral results are consistent with the findings of both the group and single-case patient studies described above in that they suggest that meaningful, but nonverbal, spatial representations can exert influence on low-level spatial relational processing in a way that symbolic representations alone (like words) cannot. As compared to how the mind and brain represent basic concepts, space may be special in this respect. While space may play an important role in grounding semantics (in an "embodied" sense) it may also be relatively resistant to Whorfian effects because it serves such a fundamental role structuring our larger, more abstract conceptual system using analog representations. With a diverse set of cognitive neuroscience methods, space can be a model system for investigating how basic meaning is instantiated along a representational continuum.

Part II

Dynamics of Objects

Chapter 7

Mislocalizations at the Onset Position of Moving Stimuli

Jochen Müsseler and Dirk Kerzel

Discovering Mislocalizations with Moving Stimuli

A note in a book by Friedrich W. Fröhlich (1929) stated that in 1894 the Norwegian astronomer O. Pihl noticed a perceptual illusion when localizing the onset position of a moving target: Typically an observer did not notice the target at its physical onset position, but rather at some later position on its motion trajectory. In other words, a localization error in the direction of motion occurred (Figure 7.1a).

It took some time before the scientific community dealt with this phenomenon. Almost 30 years later, the same author (Fröhlich, 1923) was the first to publish systematic experiments on the mislocalization, and in the subsequent scientific debate the illusion was termed the *Fröhlich effect*. Nowadays, the Fröhlich effect is typically observed on a computer display, on which the moving target appears suddenly out of nowhere. At the time of Fröhlich (1923), such an experimental setup was difficult to implement. Instead, he used a mechanical device with a moving bar entering a window. In this case the target is not perceived at the position adjacent to the edge of the window, but at a later position within that window.

In fact, Fröhlich (1923) reported not only the mislocalization but also other phenomena. For instance, he reported that the perceived width of the moving bar appeared larger than that of the physical stimulus and that the bar looked brighter at its leading than at its trailing edge (for details, see Fröhlich, 1923; Kerzel, 2010).The mislocalization and its first explanations were amply discussed in the 1930s (e.g., by Fröhlich, 1925, 1929, 1930, 1932; Metzger, 1932; Müller, 1931; Rubin, 1930), but were neglected after World War II.

Interest in the Fröhlich effect was revived in the 1990s together with two further mislocalizations in motion direction. First, Nijhawan (1994) presented a moving target in alignment with a stationary flash and observed that the target appears to lead the flash (flash-lag effect, Figure 7.1c; for an overview, see Hubbard, Chapter 9 in this volume). In fact, Nijhawan rediscovered an observation made in follow-up studies of the Fröhlich effect (cf. Metzger, 1932; Rubin, 1930; for details, see Kerzel, 2010). Second, when observers localize the offset point of a moving target, they also tend to judge it to be ahead of the target's motion trajectory. Unlike in the previous illusions, the target never reached the judged position. This observation was termed *representational momentum* (Figure 7.1b), as it was seen as evidence for a mental analogue to the momentum of moving physical objects (Freyd & Finke, 1984; for an overview, see Hubbard, Chapter 8 in this volume).[1]

[1] Originally, representational momentum was observed with induced motions, i.e., with sequences of static stimuli. Another observation was that the offset position was judged as if the target was subject to gravity. In other words, the judged position was below the physical offset position (see also De Sá Teixeira, 2016; Zago, Chapter 10 in this volume).

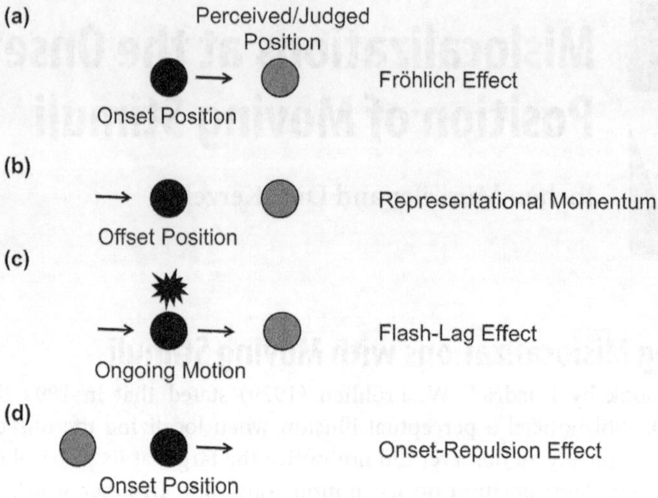

Figure 7.1 Four localization errors with moving stimuli. When the onset position (a) or the offset position (b) of a moving target is localized, observers typically make localization errors in the direction of movement. Similarly, when they judge a moving target that is presented in alignment with a flash, the target appears to lead the flash (c). These errors are known as the Fröhlich effect, flash-lag effect, and representational momentum. In the onset repulsion effect (d), the onset position is judged opposite to the direction of motion.

Beside these errors in motion direction, another error was reported at the beginning of the present century that was opposite to motion direction: In several studies, the target's onset was found to be consistently mislocalized away from the physical onset position opposite to the direction of motion (onset-repulsion effect, Figure 7.1d; Thornton, 2002; see also Actis-Grosso & Stucchi, 2003; Hubbard & Motes, 2002; Hubbard & Ruppel, 2011; Kerzel, 2002; Kerzel & Gegenfurtner, 2004). As in representational momentum, the target was never presented at the judged position, and this mislocalization has also been explained with reference to mental analogues of physical laws (Thornton, 2002).

The present chapter deals mainly with the observation of displacements of the perceived onset position. We will first give an overview about interpretations and findings of the Fröhlich effect. Then, we take into account the conditions and findings that resulted in the error opposite to motion direction (the onset-repulsion effect), and show how the conflict between the illusions might be resolved. We end the chapter with an outlook on the possible contributions of these phenomena to our understanding of perceptual processes in general.

The Mislocalization in Motion Direction: The Fröhlich Effect

Explanations of the Fröhlich effect can be roughly divided into four accounts: the sensation-time account, the metacontrast or lateral-inhibition account, the attentional account, and the mental extrapolation account.

Sensation Time

One main finding of Fröhlich was that the size of the mislocalization f increased with movement speed v, and that the ratio f/v proved to be fairly constant (Fröhlich, 1923). The fixed ratio f/v turns the spatial error into a temporal error, and Fröhlich interpreted the time constancy as an expression of the sensation time ("Empfindungszeit"). The sensation time was understood as the time between the retinal impact of light and the corresponding visual sensation (see also Kreegipuu & Allik, 2003). Fröhlich saw the finding that the sensation time varied with the brightness of stimuli as plausible evidence for his explanation. It was, as expected, on the order of 100 msec with faint stimuli and about 50 msec with bright stimuli.

At first glance, the idea of sensation time appeared attractive to explain the localization error. However, the explanation was criticized early by the contemporaries of Fröhlich, both empirically and theoretically. Rubin (1930), for instance, noted that reducing the size of the window (i.e., reducing the trajectory length) shortened the Fröhlich effect (see also Müsseler & Neumann, 1992) and thereby should also shorten the sensation time. Also, Metzger (1932) proposed that the sensation time might be longer at motion onset than at positions further along the trajectory (similar the latency differences discussed to explain other spatiotemporal phenomena; e.g., see Bachmann, 1999; Hubbard, 2014; Whitney, 2002). However, these variations are not consistent with the basic idea of sensation time and are therefore theoretically problematic. Metzger (1932) pointed out correctly that the concept of sensation time has to be applied not only to the onset of motion but to the entire motion trajectory. In this case, an observer should perceive the moving stimulus with a corresponding temporal delay, but at all positions of the trajectory.

Thus, the concept of sensation time could not provide a satisfactory answer to the question of why only the first positions were excluded from perception. Furthermore, at the time of Fröhlich it was not clear at all whether the first positions were perceptually missed or whether the first positions were simply displaced in the direction of motion.[2] Only later studies revealed that observers have indeed no access to the first positions: In experiments of Müsseler and coworkers (Müsseler & Aschersleben, 1998, Exp. 5; Müsseler & Tiggelbeck, 2013, Exp. 1), observers failed to detect brief pattern changes at motion onset but were better in detecting the change at positions further along the trajectory (see also Ansorge, Carbone, Becker, & Turatto, 2010 for evidence from reaction time studies). Thus, a mechanism is called for that prevents the very first positions from being perceived.

Metacontrast Masking and Lateral Inhibition

At first sight, accounts of the Fröhlich effect based on metacontrast masking and lateral inhibition seemed to explain the low visibility of the initial part of the trajectory (cf. Carbone & Ansorge, 2008; Geer & Schmidt, 2006; Kirschfeld & Kammer, 1999; Piéron, 1935). Metacontrast masking was first described by Stigler (1910) and refers to the observation that the visibility of a stationary, flashed target is reduced when it is followed by a mask in its spatial vicinity (visual backward masking). Backward masking is optimal when target and mask share common stimulus boundaries, for instance, when the target is a disk and the mask is a surrounding ring. Optimal stimulus onset asynchronies (SOAs) between target and mask are between 40 msec and 100 msec and are thus within the range of the temporal error in the Fröhlich effect. Piéron (1935) was the first to hypothesize the Fröhlich effect to be caused by masking mechanisms.

[2] A study of Rubin (1930) already addressed this question, but his findings remained inconclusive.

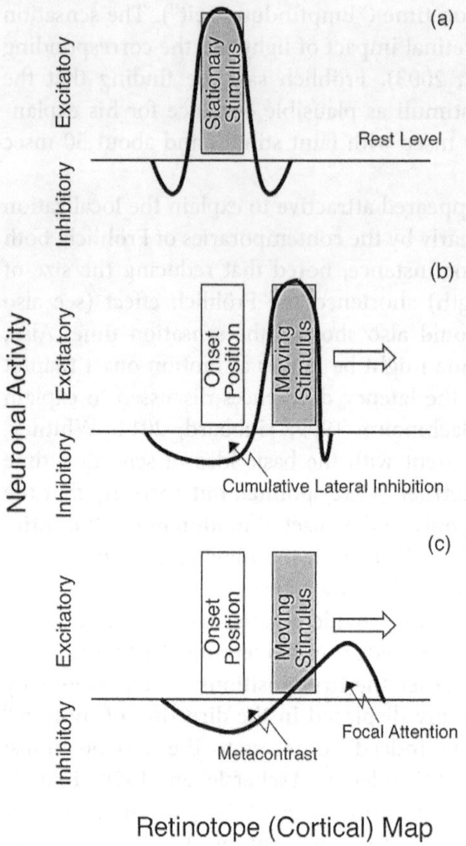

Figure 7.2 Simplified assumptions (a) of lateral inhibition with stationary stimuli, (b) cumulative lateral inhibition with moving stimuli with regard to Geer and Schmidt (2006), and (c) metacontrast and visual focal attention with regard to Kirschfeld and Kammer (1999, fig. 6). The latter figure illustrates only the additional excitatory and inhibitory neuronal activity, which is elicited by the motion of the stimulus.

Lateral inhibition has been used to interpret metacontrast masking (e.g. Bridgeman, 2006). The basic assumption is that the presentation of a stimulus elicits excitatory and inhibitory neuronal activity in a retinotopic (cortical) map, for instance in the form of a simplified Mexican-hat function (Figure 7.2a). The inhibitory parts could be crucial for the Fröhlich effect, and Geer and Schmidt (2006) proposed that multiple inhibitory connections from neighboring space have a cumulative masking effect on early parts of the target trajectory (Figure 7.2b). The authors interpreted the effect of trajectory length accordingly by assuming that inhibition from adjacent stimulus positions accumulates across the trajectory. Therefore, the Fröhlich effect increases with trajectory length.

However, if the area of lateral inhibition or any other masking mechanism just moves across a retinotopic map (cf. Figure 7.2b), there is no way to explain why only the first positions are excluded from the perception. This explanatory gap was already encountered by the sensation time account. As a matter of fact, the masking account predicts that most of the trajectory (except for the last position) should become invisible, because each stimulus presentation is masked by the subsequent stimulus presentations. Clearly, that is not the case. Therefore, an additional component is needed, which explains why the target becomes visible at all. In the subsequent accounts, this function was assigned to visual attention.

Visual Attention and Its Neuronal Implementation

At the end of the 1990s, two independently developed accounts refer to attentional mechanisms to explain the Fröhlich effect. Kirschfeld and Kammer's (1999) masking-plus-focal-attention account assumed that positions on the trajectory behind the target are masked (cf. Figure 7.2c), similar to (1935) ideas and nowadays postulated by Geer and Schmidt (2006). The new assumption was that positions on the trajectory before the target are pre-activated by the target itself (cf. Figure 7.2c). The authors associated this pre-activation with mechanisms similar to cue-induced visual focal attention. Thus, the approach combines mechanisms of metacontrast masking and visual focal attention.

With regard to Kirschfeld and Kammer (1999), focal attention ensures that masking does not occur along the entire trajectory. However, focal attention must first be shifted to the moving stimulus, and that is why the first positions of the trajectory are excluded from perception (the Fröhlich effect). Once attention has reached the moving stimulus, it becomes visible because masking is counteracted and overcome.

To support their view, Kirschfeld and Kammer (1999) presented a rotating rod that was continuously illuminated and was additionally flashed with far higher energy when it first appeared. The resulting percept was actually of two bars: a flashed bar at the correct initial position, and a moving bar that was displaced in the direction of motion (the Fröhlich effect). The interpretation was that the transient, flashed illumination of the initial orientation was strong enough to overcome masking, while the initial portion of the moving bar's trajectory was suppressed until the pre-activation (focal attention) was established. Further, Kirschfeld and Kammer concluded that perception of the moving bar had a shorter latency than perception of the stationary flashed bar, because the moving bar appeared ahead of the flashed bar and both bars appeared simultaneously (cf. the flash-lag effect; Metzger, 1932; Nijhawan, 1994; Rubin, 1930). The conclusion that moving stimuli have shorter latencies than flashed stimuli has also been confirmed in reaction-time experiments (Aschersleben & Müsseler, 1999).

The other attentional account was originally developed without reference to masking mechanisms.[3] Müsseler and coworkers (Müsseler & Aschersleben, 1998; Müsseler & Neumann, 1992, Exp. 6) simply started from three well-accepted attentional mechanisms used to explain effects with stationary stimuli (e.g., spatial cuing effects), but which should be equally applicable to situations in which stimuli are in motion (see also Ansorge et al., 2010; Carbone & Pomplun, 2007; Kerzel & Gegenfurtner, 2004; Müsseler, Stork, & Kerzel, 2002): (1) the presentation of a stimulus in the visual field elicits an attentional shift toward that stimulus; (2) an attention shift takes time; (3) a phenomenal representation of a stimulus is not available before the end of the attention shift. Applied to the Fröhlich-effect situation, this means that with the presentation of a moving stimulus, a visual focus shift is initiated, and while this shift is under way, the stimulus continues to move. The first

[3] In fact, the attentional model we started with emerged from assumptions to *address* metacontrast masking (Neumann & Müsseler, 1990; Neumann & Scharlau, 2006 [English translation of Neumann, 1987]). The basic idea was that the target in the metacontrast situation elicits an attention shift, but before this is completed, the mask appears. What the observer then perceives is the mask with shorter latencies (for similar ideas, see Bachmann, 1984, 2010; Di Lollo, Enns, & Rensink, 2000).

phenomenal representation of the stimulus is available at the end of the focus shift, and this is what is observed in the Fröhlich effect.

Both attentional accounts were able to explain the main findings observed with the Fröhlich effect, for instance, that the Fröhlich effect increases with increasing target velocity. The effect of stimulus brightness (or stimulus contrast; Fröhlich, 1923) can be plausibly addressed by assuming that establishing focal attention or eliciting an attentional shift is more effective with bright than with faint stimuli. Further, both accounts predicted that cuing the onset position with a stationary stimulus reduced the Fröhlich effect. The observed reduction effect was small but reliable (Adamian & Cavanagh, 2017; Kerzel & Müsseler, 2002; Müsseler & Aschersleben, 1998; Whitney & Cavanagh, 2000a). Finally, they explain why mislocalizations are more pronounced in the Fröhlich effect than in the flash-lag effect: At the beginning of the movement, attention is far from the moving object and a large mislocalization results. As the motion progresses, attention catches up with the moving object and the mislocalization is reduced (Müsseler et al., 2002).

However, the attentional accounts have also serious problems with some findings: for instance, that an increase of trajectory length led to an increase of the Fröhlich effect (Rubin, 1930; Müsseler & Neumann, 1992, Exp. 6). The authors of the masking-plus-focal-attention account (Kirschfeld & Kammer, 1999) might integrate Geer and Schmidt's (2006) assumption that inhibition from adjacent stimulus positions accumulates across the trajectory, which leads to larger Fröhlich effects with longer trajectories. The attention-shifting account of Müsseler and Aschersleben (1998) has more difficulties to accommodate this finding. The solution could be a modification of their third assumption, which claims that a phenomenal representation of a stimulus is not available before the end of the attention shift. This claim may be too strong. Note that shifting attention toward a stimulus implies at least coarse knowledge of the stimulus location prior to the start of the shift. What happens when an attention shift cannot be successfully completed, as would be the case when the moving target has a short trajectory and has already disappeared from the screen? Taking the third assumption literally, no stimulus should be perceived at all. It is more likely that the coarse representation of the stimulus at the beginning of the attention shift together with the incoming position information during the shift establishes what is seen. In any case, the perceived position should be closer to the starting position, i.e., the Fröhlich effect would be decreased.

Another point is that both attentional approaches use different levels of description. While Kirschfeld and Kammer (1999) selected a neuronal level of description to address masking and attentional processes, the attention shifting account of Müsseler and coworkers (Müsseler & Aschersleben, 1998; Müsseler & Neumann, 1992) was framed in a functional formal way, which leaves the neuronal processes unspecified. Therefore, Müsseler et al. (2002; see also Müsseler & Tiggelbeck, 2013) attempted to identify the attention-shifting component in the neuronal models of Kirschfeld and Kammer (1999) and particularly in the dynamic-field account of Jancke and Erlhagen (Jancke et al., 1999; Jancke & Erlhagen, 2010). The dynamic-field account, originally developed to explain the activity of neuronal populations in the primary visual cortex of the cat (Jancke et al., 1999) and then successfully applied to perceptual mislocalizations with moving stimuli (for an overview, see Jancke & Erlhagen, 2010), assumes that the presentation of a stimulus forms an activation pattern that is not restricted to the area covered by the stimulus (see also Hubbard, 1994). Rather, it spreads its activation to and integrates contextual information from the adjacent parts of the visual field. Therefore, in response to an afferent input, the activation is assumed to interact with new incoming information and, thus, modifies suprathreshold activity.

When a stimulus moves through the visual field, it can be assumed that the incoming information contributes to and modifies the activation pattern in such a way that a stimulus-driven bow wave of activity occurs, which moves continuously across the visual scene (Müsseler et al., 2002). Depending on velocity, it peaks at or even ahead of the leading edge of the stimulus. Since the Fröhlich effect emerges in the buildup phase of the bow wave, activation is accumulated starting from the resting level. The movement resulted in a skew wave that exceeds the perceptual threshold (distance between resting level and supraliminal activation), and the Fröhlich effect is observed. It has been suggested that spreading subthreshold activation constitutes a neuronal correlate of a cue-induced attentional mechanism that alters the processing of spatial information (Bocianski, Müsseler, & Erlhagen, 2008, 2010; Kirschfeld & Kammer, 1999; Müsseler & Tiggelbeck, 2013; Steinman, Steinman, & Lehmkuhle, 1995). We will return to this point later in the chapter.

Mental Extrapolation (Visual Prediction)

Mental extrapolations often occur in everyday life. When, for instance, a tennis ball flies through the visual scene, a spatial lag between the ball's position in the real world and its perceived position should emerge from neuronal transmission latencies. In order to hit the ball with a racket successfully, there must be some form of compensation, and this compensation might be in the motor system (e.g., Kerzel & Gegenfurtner, 2004). It overcomes the lag by predicting the position of the moving target forward.

However, in Nijhawan's view (1994, 2008; see also Hubbard, Chapter 9 in this volume), the lag is compensated not only by motor predictions but also by visual predictions. In this view, the flash-lag effect can be understood as the visualized percept of this prediction. The flash lags because the visual system extrapolates the position of the moving target, and this is what is seen. Thus, in the strong version of the extrapolation assumption, stimuli in motion are perceived at their real-time positions and do not lag behind. Alternative accounts assume different perceptual latencies for the flash and the moving target (e.g., Baldo & Klein, 1995; Whitney & Murakami, 1998).

The extrapolation assumption is controversially debated (see, e.g., the discussion in Baldo & Klein, 2008; Krekelberg, 2008; Nijhawan, 2008; Whitney, 2008). Although also seen as an explanation for the Fröhlich effect (e.g., Maus, Weigelt, Nijhawan, & Muckli, 2010), it is especially difficult to see how an extrapolation mechanism works at the onset position of moving stimuli. Predicting future positions of a target requires some knowledge about the target's motion direction and velocity. As there is no preceding motion trajectory at motion onset, it is unclear how extrapolation could account for the Fröhlich effect. Here one must probably recruit mechanisms that have been introduced in the previous sections.

As a last remark, it should be noted here that a predictive component also exists in the masking-plus-focal-attention account (Kirschfeld & Kammer, 1999) and in the bow-wave account or dynamic field model, respectively (Jancke & Erlhagen, 2010; Müsseler, et al., 2002). The difference is that visual prediction determines the percept while a pre-activation only prepares an area for the target to be seen.

Taking into Account the Mislocalization Opposite to the Direction of Motion: The Onset-Repulsion Effect

As already noted in the Introduction, some studies also found mislocalization opposite to the direction of motion. Note that this error is contrary to the Fröhlich effect: In the onset-repulsion effect (Thornton, 2002), the judged onset position of the target was found to be consistently mislocalized opposite to the direction of motion (Figure 7.1d; see also Actis-Grosso & Stucchi, 2003; Hubbard & Motes, 2002; Hubbard & Ruppel, 2011; Kerzel, 2002; Kerzel & Gegenfurtner, 2004).

Studies concerned with the onset-repulsion effect sometimes revealed contradictory findings. For instance, some studies found that an increment in velocity shifted the judged onset further opposite to motion direction (Kerzel, 2002; Thornton, 2002), whereas other studies did not find any effect of velocity (Actis-Grosso & Stucchi, 2003; Hubbard & Motes, 2002; Kerzel, 2002; Müsseler & Kerzel, 2004). Some authors found the onset-repulsion effect only with a relative judgment task (Kerzel, 2002), whereas others did so also with an absolute positioning task (Müsseler & Kerzel, 2004; Thornton, 2002). Further, the onset-repulsion effect seems to depend on motion type and motion direction. It is largest with smooth, continuous motions and decreases with implied motions (Kerzel, 2004; Thornton, 2002). Finally, upward motion or right-to-left motion resulted in stronger onset-repulsion effects than downward or left-to-right motion (Thornton, 2002).

It is obvious that accounts in terms of sensation time, metacontrast, or attention simply do not apply to these findings, most obviously because perceptual processes could never have been triggered at positions where the target was never presented (opposite to target motion). Instead, explanations of the onset-repulsion effect often refer to non-perceptual mechanisms. For instance, it is possible that the onset position is accurately perceived but is distorted during the delay before the judgment is made. In this case the onset-repulsion effect would originate from a memory failure, similar to the proposed mechanisms underlying representational momentum (Freyd & Finke, 1984; for an overview, see Hubbard, Chapter 8 in this volume). It is also possible that in this case observers have an imprecise percept of the onset position and tend to estimate the origin of the motion post hoc, which is subject to biases (see discussion below).

However, mislocalizations of the type of representational momentum are in the direction of motion, and a mechanism is needed to explain why the onset-repulsion effect is in the opposite direction. Therefore, it has been discussed that estimations of the onset position run backward along the observed trajectory, as this reflects more natural, physical tendencies (Thornton, 2002). In the same vein, Runeson (1974) reported that observers perceive an illusory deceleration at the onset of a motion even when the physical velocity is constant. This may result in an opposite error if the post hoc estimation of the onset position is calculated on the basis of constant motion along the entire trajectory.

The most critical point, however, is how to explain the contrary findings of the Fröhlich and onset-repulsion effects. It turned out that the experimental procedures used to measure the mislocalizations were quite different. In reports of the Fröhlich effect, observers were able to predict where the moving target would appear. For instance, the moving target always appeared at the fixed edge of a window (Fröhlich, 1923) or at two fixed eccentricities to the left or right of fixation (Müsseler & Aschersleben, 1998). In reports of the onset-repulsion effect, however, the moving target appeared randomly in a relatively large area (Hubbard & Motes, 2002; Thornton, 2002), and observers were unable to predict the onset positions.

To examine the hypothesis that the error in motion direction (the Fröhlich effect) and the error opposite to motion direction (the onset-repulsion effect) originated from the predictability of onset positions, Müsseler and Kerzel (2004; see also Müsseler et al., 2008) conducted experiments in which two different trial contexts were used. In the random context, the target appeared mostly at a random onset position in a large area of the computer screen (similar to Thornton, 2002), but in one-sixth of the trials the target appeared about 6.6° to the left or right of fixation (Figure 7.3). Only these trials of the random-context condition were compared with the trials of the constant-context condition, in which the target always appeared at the onset positions to the left or right of fixation (similar to Müsseler & Aschersleben, 1998). The judgments showed a huge difference between context conditions: The onset was localized –0.5° opposite to the direction of motion in the random-context condition (onset-repulsion effect) and 1.5° in the direction of motion in the constant-context condition (the Fröhlich effect). Thus, low predictability

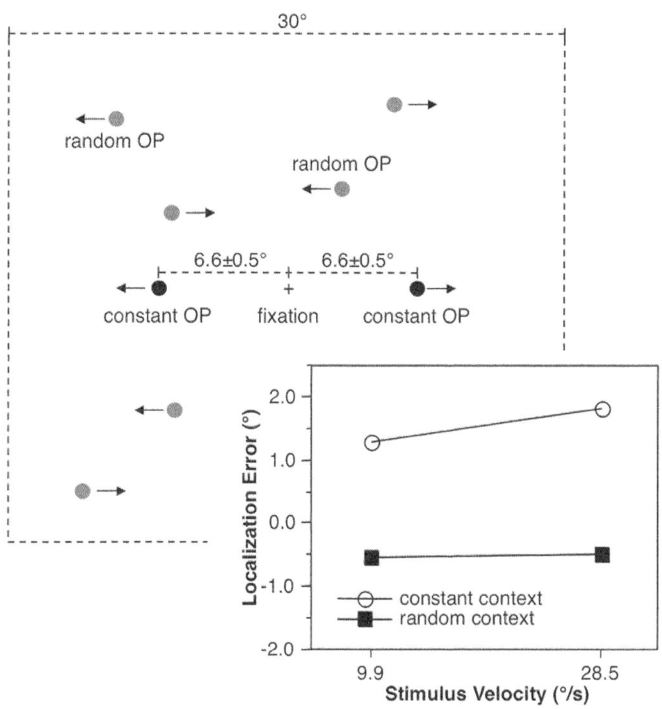

Figure 7.3 Trial contexts and findings of Müsseler and Kerzel (2004, Exp. 1). In the constant-trial context, the target always appeared at constant onset positions (OP, black dots) to the left or right from fixation. In the random-trial context, the target appeared mostly at random OPs (grey dots) in a 30 x 30° field of the computer screen, but in one-sixth of the trials also at the constant OPs. In the data analysis, only these trials were compared with the trials of the constant context. The results showed that the onset was localized opposite to the direction of motion with the random context (negative localization error of –0.5°; the onset-repulsion effect) and in the direction of motion with the constant context (positive localization error of 1.5°; the Fröhlich effect).

of onset positions led to the error opposite to the direction of motion, while high predictability of onset positions led to the error in the direction of motion.[4]

With other authors (Actis-Grosso & Stucchi, 2003; Kerzel, 2002; Kerzel & Gegenfurtner, 2004; Thornton, 2002), we assumed that the difference between context conditions originates from an error in the judgment phase. When positional predictability is low, as is the case in the random-context condition, observers may notice a target relatively late, and with every new trial they might become aware of a possible localization error. To avoid this error, they may overcompensate and point to positions opposite to motion. Consistent with strategic adjustments, differences between the random-context and constant-context conditions were visible after about 15–35 trials (Müsseler & Kerzel, 2004, Exp. 4).

However, further experiments by Müsseler and Tiggelbeck (2013) cast doubts on the overcompensation explanation. With regard to the overcompensation explanation, the error opposite to motion direction is assumed not to be a perceptual one, but to result from the tendency in the judgment phase to correct for the possible spatial error. Consequently, an overcompensation mechanism should mainly affect a localization task, but not a discrimination task. In an experiment of Müsseler and Tiggelbeck (2013, Exp. 1), moving targets either started out as squares and changed to circles at different positions on the trajectory, or appeared as circles and did not change (cf. also Ansorge et al., 2010). Observers' task was to discriminate whether or not they perceived a square during the motion of the target. The overcompensation account expected equal or worse discrimination performance in the random-context condition than in the constant-context condition. This result would point out a response bias, which compensates for a possible localization error in the judgment phase. However, the contrary finding was observed: When the squares appeared in the very first positions of the motion, better discrimination performance was found in the random-context condition than in the constant-context condition. Thereafter, the difference between context conditions vanished.

This finding was somewhat surprising, as it indicated worse perceptual performance in the constant-context condition. And this disadvantage of the constant-context condition was somewhat counterintuitive: When stimuli always appeared at predictable left/right positions, as is the case in the constant-context condition, observers could direct their attention to both positions in advance (parallel allocation of visual attention to two positions; cf. Awh & Pashler, 2000; Cave, Bush, & Taylor, 2010; Franconeri, Alvarez, & Enns, 2007; but see also Jans, Peters, & De Weerd, 2010). Directing attention to a position usually improves spatial localization in this area, as was found in several studies with stationary targets (e.g., Bocianski et al., 2008, 2010; Tsal & Bareket, 1999, 2005; Tsal, Meiran, & Lamy, 1995; Yeshurun & Carrasco, 1999). Therefore, observers in the constant-context condition of Müsseler and Tiggelbeck (2013) had not directed their attention to the left/right onset positions, or alternatively, directing attention to these positions produced worse localization performance.

---

[4] Note, however, that a mislocalization opposite to target motion has not always been observed in the random-context condition. In the study of Müsseler and Kerzel (2004), the error was in the direction opposite to target motion in three of four experiments. Further studies (Müsseler et al., 2008; Müsseler & Tiggelbeck, 2013) also found a clear difference between random-context and constant-context conditions, but the error was always in the direction of motion. The reason might be that the exploitable trials of the random-context condition were still presented above chance level at the left/right positions.

To examine the last possibility, Müsseler and Tiggelbeck (2013, Exp. 3) used an exogenous cue to direct attention in the random-context condition. The cue was presented at the onset positions, 280 msec before the moving target appeared. If, with moving stimuli, directing attention results in worse localization performance, presenting the cue should result in comparable mislocalizations in both context conditions. And this was what the results actually showed. When the cue preceded the motion onset, the localization error of the random-context condition increased in size relative to the localization error of the constant context.[5] Thus, Müsseler and Tiggelbeck's experiments delivered consistent results. When observers allocated their attention to the onset position, worse discrimination performance (Exp. 1) went hand in hand with worse localization precision (Exp. 3).

An issue that remains unexplained is why attention improves discrimination and localization performance with stationary stimuli while it seems to impair discrimination and localization precision at the onset position of moving stimuli. We speculated that, contrary to stationary stimuli, moving stimuli require fast spatial disengagement (Petersen & Posner, 2012) from the previously attended position in order to follow the stimulus, especially at the onset position. It seems plausible that this disengagement could impair processing.

How to implement this idea in the neuronal dynamic field model discussed in the previous section? Bocianski et al. (2008, 2010) already applied the model to an illusion with stationary stimuli and extended it by integrating a top–down attentional mechanism. In the empirical part of their paper, observers were confronted with blockwise presentations, similar to the random and constant context used by Müsseler and Kerzel (2004). The authors assumed that the blockwise presentation of a target at constant positions modulates the attentional baseline by arousing a peak at attended locations and by suppressing all other locations (for neuronal evidence see, e.g., Bestmann, Ruff, Blakemore, Driver, & Thilo, 2007; Smith, Singh, & Greenlee, 2000). Empirical and modeling data showed that localization precision was improved when the static target was presented in the attended area (Bocianski et al., 2010).

When, instead of stationary stimuli, a moving target is presented at the attended area, the only assumption to add is that the target might have left the region of the attentional peak already before a suprathreshold activity is reached. Moreover, the new incoming information of the target may interact within the previous activation pattern, which may additionally impair localization performance. In a sense, the postulated mechanism is similar to that accounting for effects of spatial disengagement from previously attended positions.

Note that this extension of the neuronal dynamic field model can only account for the observed differences between the random and constant context – that is, for the clear mislocalizations in motion direction in the constant-context condition and the more or less precise localizations in the random-context condition. It cannot account for the onset-repulsion effect per se – that is, for the error opposite to motion direction.

Conclusions

The present chapter focused initially on the Fröhlich effect – the localization error at the onset position in motion direction. In the nearly century-old scientific debate on this

[5] Interestingy, the opposite finding is observed with representational momentum, where a cue decreases (but does not eliminate) forward displacement (Hubbard, Kumar, & Carp, 2009).

illusion, different accounts were considered and discarded, among them the sensation-time account and the masking account. But it is worth emphasizing that these accounts did not simply disappear, but have been modified by additional findings. For instance, the sensation-time assumption is still discussed in the context of the flash-lag effect, with different latencies for the flash and the moving target (e.g., Baldo & Klein, 2008; Krekelberg, 2008; Whitney, 2008), or in the context of the Fröhlich effect, with longer latencies at the onset position than at later positions on the trajectory (e.g., Aschersleben & Müsseler, 1999; Kirschfeld & Kammer, 1999).

The underlying processing mechanisms in localizing the onset position of moving stimuli were further clarified by the discovery of the onset-repulsion effect, that is, the error opposite to motion direction. As it turned out in several studies, the localization judgments varied strongly with trial context. Perceived starting positions were in the direction of motion in constant-context conditions and opposite to motion direction (or at least essentially reduced) in random-context conditions (Müsseler & Kerzel, 2004; Müsseler & Tiggelbeck, 2013; Müsseler et al., 2008). It is likely that when stimuli always appear at predictable positions, as is the case in the constant-context condition, observers direct their attention to the positions in advance. However, one would expect that localization precision is improved, whereas the opposite was found. Localization precision and discrimination performance was worse with constant than with random context.

Trial context was also found to affect localization judgments with stationary stimuli (Bocianski et al., 2008, 2010), but here the findings were as expected (see also Tsal & Bareket, 1999, 2005; Tsal et al., 1995; Yeshurun & Carrasco, 1999). Localization precision was better with constant-context conditions than with random-context conditions. To account for these differences in the findings between stationary and moving stimuli, we speculated that moving stimuli require a spatial disengagement from the previously attended onset position in order to follow the target. It seems plausible that this attentional disengagement could impair processing. Certainly, this idea needs further confirmation, but if this conclusion proves to be true, attentional disengagement should be at the heart of explanations of the Fröhlich effect.

Acknowledgments

This research was supported by a grant from the Deutsche Forschungsgemeinschaft to Jochen Müsseler (DFG MU 1298/6). Correspondence concerning this manuscript should be addressed to Jochen Müsseler, Work and Cognitive Psychology, RWTH Aachen University, Jägerstr. 17–19, 52056 Aachen, Germany. Email: muesseler@psych.rwth-aachen.de.

Chapter 8

Influences on Representational Momentum

Timothy L. Hubbard

Memory for the final location of a moving target is often displaced in the direction of anticipated motion, and this is referred to as *representational momentum* (Freyd & Finke, 1984). Overviews of variables that influence representational momentum were presented in Hubbard (1995c, 2005b, 2014a), and comparisons of representational momentum with other types of momentum-like effect were presented in Hubbard (2014a, 2015b, 2017). This chapter consolidates earlier lists of variables that influence representational momentum into a single comprehensive catalog. Such a consolidation should be useful for future studies of representational momentum and other momentum-like effects (e.g., variables that influence representational momentum might be hypothesized to have analogous effects on other momentum-like effects). In addition, suggestions regarding a few properties of representational momentum, and the relationship of representational momentum with other spatial biases, are provided. As theories of representational momentum were addressed in Hubbard (2010), and discussion linking representational momentum with other momentum-like effects was provided in Hubbard (2014a, 2015b, 2017), those issues are not discussed here.

A common stimulus presentation method (implied motion) and response measure (probe judgment) used in studies of representational momentum are illustrated in Figure 8.1. In the top panel, three sequential presentations of a target (inducing) stimulus implying motion from left to right are shown. A fourth (probe) stimulus is shown, and the probe is located behind the final location of the target (i.e., shifted in the direction opposite to target motion), at the same location as the final location of the target, or beyond the final location of the target (i.e., shifted in the direction of target motion). Participants judge whether the probe is at the same location where the target vanished or at a different location. In the bottom panel, probability of a *same* response is plotted as a function of probe location, and representational momentum is indicated by the greater likelihood of a *same* response to a probe slightly beyond the final location than to a probe slightly behind the final location. Other methods of stimulus presentation include smooth continuous motion of a target or presentation of a single static target drawn from a longer motion sequence (e.g., a photograph of a dancer in mid-leap). Other methods of response measurement include using a computer mouse to position the cursor at the judged location or touching the display at the judged location.

Variables That Influence Representational Momentum

Hubbard (2005b) organized variables that influence representational momentum into categories related to the target, display, context, or observer. A similar scheme is followed here. Some variables have been manipulated in numerous studies (e.g., target velocity),

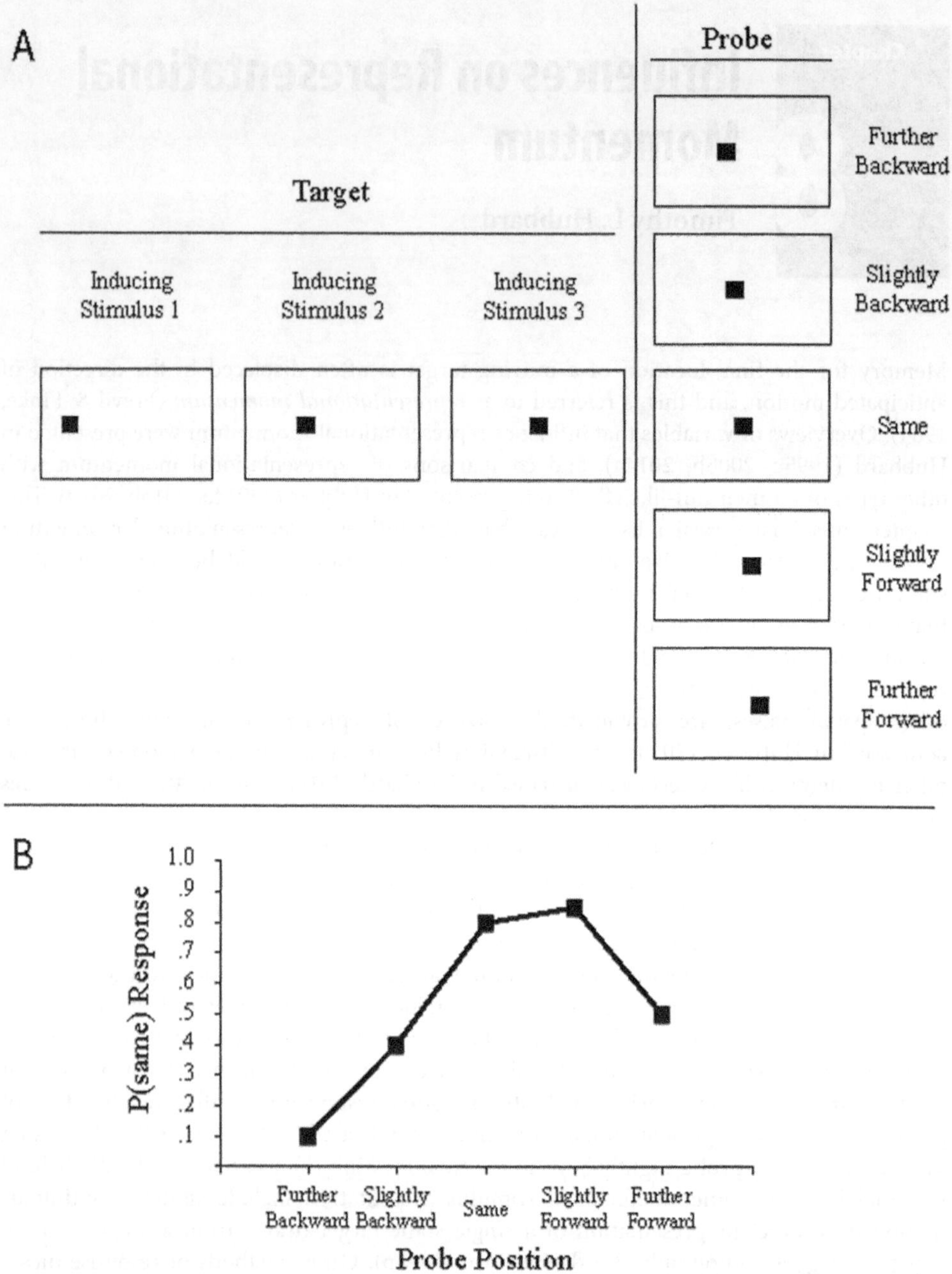

Figure 8.1 An illustration of a typical methodology and results for an experiment assessing representational momentum. In Panel A, the large rectangles indicate the outlines of the display, and the small black squares indicate the target (left) or probe (right). There are three consecutive appearances of inducing stimuli that comprise the target. In this example, the target exhibits implied rightward motion (typically, each inducing stimulus is presented for 250 msec, and there is a 250-msec interstimulus interval between successive inducing stimuli and between the final inducing stimulus and probe). A probe is presented, and position of the probe relative to the actual final position of the target varies across trials (five potential probe positions are shown in the column on the right). In Panel B, a hypothetical but typical distribution of *same* responses as a function of probe position is illustrated. The presence of representational momentum is indicated by the higher probability of *same* responses to probes forward of the final actual target location than to probes backward of the actual final target location. Adapted from "Forms of momentum across space: representational, operational, and attentional," by T. L. Hubbard, 2014a, *Psychonomic Bulletin & Review, 21*, p. 1373. Copyright 2014 by Psychonomic Society, Inc.

but discussion for each variable is limited to studies that initially demonstrated the effect or an important constraint of that variable.

Characteristics of the Target

Variables influencing representational momentum that are related to the target include (a) velocity, (b) distance traveled, (c) direction, (d) shape, (e) identity and semantic category, (f) size, (g) eccentricity, (h) visual field, (i) involvement of a human form, and (j) nonvisual stimuli.

Velocity. Faster target velocity generally leads to larger forward displacement (e.g., Freyd & Finke, 1985; Hubbard & Bharucha, 1988; de sá Teixeira, Hecht, & Oliveira, 2013). If final instantaneous velocity is constant, then a previously accelerating or decelerating target exhibits larger or smaller, respectively, forward displacement (Actis-Grosso, Bastianelli, & Stucchi, 2008; Finke, Freyd, & Shyi, 1986). Visual targets or auditory targets with an irregular velocity exhibit less representational momentum, presumably due to decreased predictability of the target (Getzmann & Lewald, 2009). Effects of velocity are diminished (a) with increases in implied friction (Hubbard, 1995b), (b) if a target is initially stationary and subsequent motion is attributed to contact from another object (Hubbard & Ruppel, 2002), (c) at very high velocities (Munger & Minchew, 2002), and (d) for single targets exhibiting continuous motion if observers cannot visually track that target (Kerzel, Jordan, & Musseler, 2001). A velocity effect has been found for changes in auditory pitch (Freyd, Kelly, & DeKay, 1990; see also Hubbard, 1995a).

Distance. In most studies of representational momentum, distance traveled by the target is not a factor. Targets travel a fixed distance on every trial (e.g., 34 degrees of rotation, Freyd & Finke, 1984) or different distances of travel are counterbalanced or randomized across trials (e.g., Hubbard & Bharucha, 1988). However, representational momentum decreases with increasing distance traveled by a launched target (Choi & Scholl, 2006; Hubbard & Ruppel, 2002).[1] Forward displacement decreases with increases in distance traveled by the target for patients with neglect or patients with right hemisphere damage, but not for control participants (McGeorge, Beschin, Colnaghi, Rusconi, & Della Sala, 2006). Relatedly, forward displacement decreases as a target approaches a boundary (Hubbard & Motes, 2005) or edge of the display (de sá Teixeira & Oliveira, 2011); however, this latter finding probably reflects expected change in target motion rather than distance traveled.

Direction. Horizontal motion in the picture plane leads to larger forward displacement than does vertical motion, and downward motion leads to larger forward displacement than does upward motion (Hubbard, 1990; Hubbard & Bharucha, 1988). Differences between leftward motion and rightward motion are usually not found (e.g., Cooper & Munger, 1993; Hubbard, 1990), but when a difference is found, rightward motion is larger (e.g., Halpern & Kelly, 1993). Forward displacement occurs for target motion in depth (e.g., Hayes, Sacher, Thornton, Sereno, & Freyd, 1996) and is larger when motion is away from the observer (Hubbard, 1996a; Nagai, Kazai, & Yagi, 2002). Forward displacement along the line of sight

[1] In the launching effect, an initially moving object contacts an initially stationary object, the initially stationary object begins to move, and the movement of the initially stationary object is attributed to contact from the initially moving object (Michotte, 1963; for review, see Hubbard, 2013d).
A launched target is thus an initially stationary stimulus that begins moving upon contact from a moving object, and whose motion is attributed to contact from the initially moving object.

occurs with motion of an observer's viewpoint through a scene (Munger, Owens, & Conway, 2005; Thornton & Hayes, 2004). Forward displacement occurs for objects rotating in depth, and displacement is larger when rotation is around an axis that corresponds to observers' and objects' coordinate systems (Munger, Solberg, Horrocks, & Preston, 1999). Differences in displacement for rotation within the picture plane are usually not reported, although one study found displacement for a target moving along a rectangular clock frame was larger for clockwise motion than for counterclockwise motion (Joordens, Spalek, Razmy, & van Duijn, 2004), and rotation is larger for targets rotating downward than rotating upward (Munger & Minchew, 2002; Munger & Owens, 2004). Displacement for a target following a circular path is forward along the tangent and inward toward the center (Hubbard, 1996b).

Shape. The majority of studies of representational momentum presented stimuli consisting of geometric shapes. If each inducing stimulus within a trial is a radically different shape, forward displacement does not occur; however, if inducing stimuli imply a consistent change of shape of a single object (e.g., growing consistently thinner, taller, larger, etc.), then forward displacement in the direction of implied change occurs (Kelly & Freyd, 1987; White, Minor, Merrell, & Smith, 1993). Implied drag does not influence forward displacement of square or rectangular translating targets or rotating pyramid shapes (Cooper & Munger, 1993), although in a more extreme case, a line moving parallel to its major axis (i.e., its smaller edge facing the direction of motion) exhibited larger forward displacement than did a line moving parallel to the direction of its minor axis (Hubbard, 2005a). Displacement is larger if an object moves in the direction it points (Freyd & Pantzer, 1995) or faces (Nagai & Yagi, 2001; Vinson & Reed, 2002).

Identity. If a target is given a verbal label suggesting a specific direction of motion, displacement is larger in that direction. For example, an ambiguous shape exhibited larger forward displacement for upward motion if that shape was labeled "rocket" than if labeled "steeple" (Reed & Vinson, 1996). This suggests displacement is influenced by object-specific information, and such information is more likely to influence displacement if the target is prototypical of its category and has a typical direction of motion (Nagai & Yagi, 2001; Vinson & Reed, 2002). Forward displacement is increased if an object moves in the direction of its typical (forward) motion (e.g., an airplane moving forward) than if that object moves in an atypical (backward) direction (Nagai & Yagi, 2001).

Size. Implied mass of pyramid-shaped objects that rotate in depth does not influence forward displacement (Cooper & Munger, 1993). Larger horizontally moving targets exhibit greater downward displacement than do smaller targets, larger downward-moving targets exhibit larger downward displacement than do smaller targets, and larger upward-moving targets exhibit smaller forward displacement than do smaller targets (Hubbard, 1997, 1998); this suggests effects of size (mass) are represented by effects of weight, as weight is experienced in the direction of gravitational attraction (i.e., the vertical axis) and not in the direction of motion. More generally, represented location of a target is influenced by implied gravitational attraction (Hubbard, 1990, 1997), and this has been referred to as *representational gravity* (see de sá Teixeira, 2014; de sá Teixeira & Hecht, 2014; Hubbard, 1995c, 2005b; Motes, Hubbard, Courtney, & Rypma, 2008; Zago, Chapter 10 in this volume). If participants estimate how long it would take to stop a moving target, more dense (massive) targets are estimated to take more effort, but not more time, to stop; this has been interpreted as suggesting mass and velocity might interact in representational momentum (de sá Teixeira, Oliveira, & Amorim, 2010).

Eccentricity. Some studies reported representational momentum for a continuously moving target did not occur if participants maintained central fixation and could not track the target (e.g., de sá Teixeira, Hecht, & Oliveira, 2013; Kerzel, 2000), and so there was little initial incentive to look for effects of eccentricity on representational momentum. However, one study reported representational momentum for visual targets and auditory targets increased as final target location moved from central to paracentral regions of the visual field and decreased as the final target location moved from paralateral to lateral regions of the visual field (Schmiedchen, Freigang, Rubsamen, & Richter, 2013). For visual targets, the initial increase probably reflects decreases in resolution acuity in paracentral and paralateral regions (cf., increase in forward displacement of blurred targets, Fu, Shen, & Dan, 2001). The reason for the pattern with auditory targets is less clear.

Visual Field. If a stimulus changing in size is presented on the midline of the visual field, and probes are presented to the left or right visual field, representational momentum for size is larger for probes in the left visual field with a retention interval of 500 msec, but this difference vanishes with longer retention intervals (White et al., 1993). Similarly, representational momentum for location in the picture plane is larger if targets are in the left than in the right visual field (Gottwald, Lawrence, Hayes, & Khan, 2015; Halpern & Kelly, 1993). When targets are lower in the picture plane, representational momentum is larger for vertically moving objects (Hubbard, 2001) and horizontally moving objects in the left visual field (Gottwald et al., 2015).

Human Form. Presentation of animated human figures (Verfaillie & Daems, 2002) or point-light walkers (Verfaillie, De Troy, & Van Rensbergen, 1994) leads to forward displacement for postures of those figures; however, one study found displacement for point-light characters limited to characters presented on a textured surface or from a static viewpoint (Jarraya, Amorim, & Bardy, 2005). Representational momentum occurs for faces changing from a neutral to a more extreme expression (Uono, Sato, & Toichi, 2010, 2014; Yoshikawa & Sato, 2006, 2008; but see Thornton, 2014). Representational momentum occurs for changes in the direction in which a face is oriented, but is reduced when gaze direction of that face differs from the direction in which the face is oriented (Hudson, Liu, & Jellema, 2009) or emotional expression was inconsistent with approach toward the observer (Hudson & Jellema, 2011). Representational momentum occurs for hand gestures in sign language; displacement is larger for gestures in the typical direction of motion than in the reversed direction, but this might reflect differences in awkwardness of motion rather than semantic meaningfulness (Wilson, Lancaster, & Emmorey, 2010). Such an awkwardness effect does not occur with a static stimulus (Munger, 2015). Representational momentum for animation of a human hand is influenced by expectations regarding whether that hand would reach toward or withdraw from an object (Hudson, Nicholson, Ellis, & Bach, 2016; Hudson, Nicholson, Simpson, Ellis, & Bach, 2016).[2]

Nonvisual Stimuli. Remembered final pitch of a stimulus changing in auditory frequency exhibits forward displacement in pitch (Freyd et al., 1990; Hubbard, 1995a; Johnston & Jones, 2006; Kelly & Freyd, 1987). Remembered location in physical space of a sound source translating in the picture plane is displaced in the direction of motion (Getzmann & Lewald, 2007; Getzmann, Lewald, & Guski, 2004). Forward displacement for location of a moving sound source occurs at the beginning and end, but not middle, of the trajectory (Getzmann & Lewald, 2007). Oculomotor behavior does not influence auditory

[2] Depictions of animate targets (e.g., a picture of a rhinoceros) do not lead to more or less forward displacement than do depictions of man-made objects (e.g., a picture of a truck; Halpern & Kelly, 1993), although in one study a larger displacement was reported when a drawing of a bug-like creature moved in the direction its "head" was pointing than when it moved in the opposite direction (Freyd & Miller, 1992).

representational momentum for a moving sound source (Getzmann, 2005). When pitch rises and falls in a predictable manner, displacement in final pitch is backward when listeners expect direction of pitch motion to reverse (Johnston & Jones, 2006). Representational momentum for a moving sound source increases as vanishing point moves from central to paracentral regions and decreases as vanishing point moves from paralateral to lateral regions (Schmiedchen et al., 2013). Changes in grasp aperture consistent with representational momentum occur when participants grasp opening or closing visual pliers (Brouwer, Thornton, & Franz, 2005) or two visual spheres moving toward or away from each other (Brouwer, Franz, & Thornton 2004).

Characteristics of the Display

Variables influencing representational momentum that are related to the display include (a) surface form, (b) whether the participant controls target motion, (c) retention interval, (d) prior probability a *same* response to a probe would be correct, (e) response measure, and (f) contrast between target and background.

Surface Form. The term "surface form" refers to the format in which the target is presented, and three formats have been used. The first is frozen-action photographs, and these are single static images drawn from a larger motion sequence (e.g., a person walking or jumping; e.g., Freyd, 1983; Futterweit & Beilin, 1994).[3] The second is implied motion (as in Figure 8.1), and this involves a series of separate static stimuli that imply motion in a specific direction (e.g., a series of rectangles at different orientations; Freyd & Finke, 1984). The third is continuous motion, and this involves a target that appears to exhibit smooth and uninterrupted motion (e.g., Hubbard & Bharucha, 1988). No difference between implied motion and continuous motion for displacement in orientation of a rotating rod (Munger & Owens, 2004) or a change in auditory frequency (Hubbard, 1995a) have been found. Kerzel (2003c) reported forward displacement for a target moving along a circular trajectory is less with continuous than with implied motion, and Faust (1990) found displacement for a horizontally moving target is less with implied than with continuous motion. Comparison of displacement across different surface forms is difficult, as direction and other characteristics of motion depicted with each type of surface form are typically different.

Control. If observers control velocity and direction (turning points) of a target, then forward displacement if that target vanishes without warning is less than if observers do not have such control (Jordan & Knoblich, 2004). If observers have control over when the target vanishes, then forward displacement decreases with increases in the latency between when observers indicate the target should vanish and when the target actually vanishes (Jordan, Stork, Knuf, Kerzel, & Musseler 2002; but see Hubbard, 2005b). Control over when the

[3] Although not explicitly measuring representational momentum, two recent studies reported data consistent with representational momentum in frozen-action photographs. Participants habituated to a set of frozen-action photographs depicting motion in a single direction. In one study, participants were more likely to interpret subsequent ambiguous display of moving dots as exhibiting motion in the opposite direction (Winawer, Huk, & Boroditsky, 2008), and in the other study, the position of a subsequent dot was shifted in the direction opposite to the direction previously depicted in a set of frozen-action photographs (Pavan, Cuturi, Maniglia, Casco, & Campana, 2011). Findings of Winawer et al. and Pavan et al. are consistent with a motion aftereffect, and this suggests the frozen-action photographs activate motion processing (cf. Kourtzi & Kanwisher, 2000; Senior et al., 2000).

target vanishes might interact with oculomotor behavior (Stork & Müsseler, 2004). If participants have previous experience controlling a target or observing another person controlling the target, then forward displacement is larger than if participants do not have such experience (Jordan & Hunsinger, 2008).

Retention Interval. Representational momentum increases during the first few hundred milliseconds after the target vanishes (e.g., Freyd & Johnston, 1987; Halpern & Kelly, 1993). This increase occurs with implied motion and continuous motion targets (but has not been examined for frozen-action photographs). Some studies report representational momentum asymptotes (e.g., Finke & Freyd, 1985; Kerzel, 2000), and other studies report declines in representational momentum with subsequent increases in retention interval (e.g., Freyd & Johnson, 1987; de sá Teixeira, Hecht, & Oliveira, 2013). However, many studies claiming representational momentum asymptotes did not examine representational momentum for extended retention intervals (e.g., Kerzel, 2000, did not examine retention intervals longer than 500 msec). Studies of retention interval typically use probe response measures (see below), as such measures do not confound time to locate and move a cursor with retention interval per se (although see de sá Teixeira, Hecht, & Oliveira, 2013).

Prior Probabilities. If error feedback suggests prior probability that a *same* response is correct is relatively small, then overall likelihood of a *same* response, but not representational momentum, is decreased (Ruppel, Fleming, & Hubbard 2009). If actual or believed prior probability that a *same* response is correct decreases, the likelihood of a *same* response, but not the magnitude of representational momentum, decreases (Hubbard & Lange, 2010). In other words, decreases in prior probability that a *same* response is correct on any given trial decreases height of the distribution of *same* responses but does not influence skew (shift) of that distribution. Decreases in actual or believed prior probability decrease false alarm rates, and decreases in believed prior probability decrease hit rates and beta but not d' (Hubbard & Lange, 2010). Prior probability is most relevant to studies using probe judgment, but it is not clear how prior probability is relevant to studies using cursor positioning or reaching response measures.

Response Measure. Three response measures have been used. The first is probe judgment, in which a stimulus similar or identical to the target is presented after the target vanishes and observers judge whether the probe is the same as or different from the target (e.g., Finke et al., 1986), or, less commonly, whether the probe is to the left or right of final target position (e.g., Kerzel, 2000). A disadvantage of probe methods is that high numbers of probes and replications of each probe type are required, and this leads to higher numbers of trials (e.g., typically 5–9 probe positions per target position). The second type is cursor positioning (referred to as *mouse pointing* by Kerzel, 2003c), in which observers use a computer mouse to position the cursor at the location in a display corresponding to the final location of the target. By clicking a button on the mouse, display coordinates of the cursor are recorded and can be compared with display coordinates of the target (e.g., Hubbard, 1990; Hubbard & Bharucha, 1988). Cursor positioning typically requires fewer trials than does probe judgment, but some stimulus dimensions are not easily adapted to such a measure (e.g., visual brightness, auditory pitch). The third method is reaching, in which participants touch (with a finger) the location in the display where the target vanished (e.g., Ashida, 2004; Kerzel & Gegenfurtner, 2003). A disadvantage of reaching is that spatial resolution is typically not as precise as probe judgment or cursor positioning. There haven't been many comparisons of response measures, but one study found probe judgments resulted in smaller estimates of representational momentum than did reaching (Kerzel, 2003c).

Contrast. Displacement in remembered luminance of a target changing in luminance is backward (in the direction opposite to representational momentum; Brehaut & Tipper, 1996) and interacts with a bias toward the level of background luminance (Favretto, 2002). If a moving target on a dark background gradually decreases in luminance (decreases in contrast), forward displacement is larger than if a target travels a shorter distance before abruptly vanishing (Maus & Nijhawan, 2006). Weaker motion signals resulting from less contrast between target and background have been suggested to result in smaller displacement (Maus & Nijhawan, 2009), although this does not appear consistent with suggestions that weaker motion signals resulting from less continuous motion result in larger displacement (Kerzel, 2003c). If a light or dark target is presented on a white or black background, representational momentum for target location is larger if contrast of target and background is high or increasing and lower if contrast of target and background is low or decreasing (Hubbard & Ruppel, 2014).

Characteristics of the Context

The context includes physical and cognitive elements. Variables influencing representational momentum that are related to the context include (a) physical surroundings, (b) landmarks, (c) shadows and shading, (d) interactions with nontarget stimuli, (e) expectations regarding future target motion, and (f) attributions regarding the source of target motion.

Physical Surroundings. If a rotating target is embedded within a larger surrounding frame, representational momentum for the target is increased if the frame rotates in the same direction or is oriented slightly beyond the final target orientation, and representational momentum for the target is decreased if the frame rotates in the opposite direction or is oriented slightly behind the final target orientation (Hubbard, 1993). Similarly, if a moving target is near a larger translating grating, representational momentum of the target is increased or decreased if the grating moves in the same or opposite direction, respectively, as the target (e.g., Whitney & Cavanagh, 2002). If a target moves between stationary distractors (Gray & Thornton, 2001) or toward a stationary barrier (Hubbard & Motes, 2005), forward displacement is decreased. If the target is expected to bounce off a barrier, forward displacement decreases as the target approaches the barrier and reverses at the moment of contact (Hubbard, 1994; Hubbard & Bharucha, 1988).

If observers view video clips from real-world scenes (e.g., train station, shopping mall), representational momentum for those scenes occurs even if different stimuli in a scene move at different velocities and in different directions (Thornton & Hayes, 2004). Forward displacement of an observer's viewpoint for motion into animated scenes (pyramid landscapes) occurs if probes tested specifically for representational momentum (Munger et al., 2005). However, if probes were more general, then responses were more consistent with boundary extension (see also DeLucia & Maldia, 2006). Forward displacement of the viewpoint occurs for observers in driving simulations (Thornton & Hayes, 2004). If viewpoint of a scene rotates to the left or right, then representational momentum for objects occurs and is larger for objects entering the scene than exiting the scene (Munger et al., 2006) even though objects are stationary and it is the viewpoint that moves. Representational momentum for objects in a scene decreases if viewpoint motion involves rotation and translation rather than just rotation (Brown & Munger, 2010).

Landmarks. Examples of context considered above were global in the sense of surrounding a target or occupying a large portion of the surrounding environment. However, context can be local in the sense of an object or landmark providing a single point of reference. If a target moves toward or away from a landmark, forward displacement of that target increases or decreases, respectively (Hubbard & Ruppel, 1999). This probably reflects a combination of representational momentum with a landmark attraction effect (cf. Bryant & Subbiah, 1994); when representational momentum and landmark attraction operate in the same direction (motion toward the landmark), they sum, and forward displacement is large, whereas when representational momentum and landmark attraction operate in opposite directions (motion away from the landmark), they partially cancel, and forward displacement is small. A target is displaced toward a distractor that is flashed at the moment the target vanishes or during the retention interval, and although consistent with a landmark effect, it has been argued a briefly presented nontarget stimulus should not be considered a landmark (Kerzel, 2002a).

Shadows and Shading. Information regarding shape or direction of motion provided by shading and shadows influences representational momentum. If the first and third (of three) inducing stimuli are convex, forward displacement is larger if the second inducing stimulus is not concave (cf. coherent motion; Kelly & Freyd, 1987), but if inducing stimuli are replaced by luminance-polarized circles (white top-half and black bottom-half for convex; black top-half and white bottom-half for concave), there is no difference in displacement (Hidaka, Kawachi, & Gyoba 2009); thus, effects of shape and shading were due to differences in perceived depth rather than luminance. If the apparent cast shadow of a horizontally moving target diverges from the path of the target, representational momentum is larger than if the cast shadow parallels or converges with the path of the target, and this suggests the diverging shadow is interpreted as suggesting motion toward the observer (Taya & Miura, 2010). If paths of the target and shadow converge, downward displacement of the target is larger, and this was interpreted as suggesting downward motion. Information from shape and shading influences perceived direction of motion, which in turn influences representational momentum.

Interactions with Nontarget Stimuli. A tone that increases or decreases in auditory frequency does not influence forward displacement of a vertically moving visual target, but downward displacement of a horizontally moving visual target is larger if the tone descends than if the tone ascends (Hubbard & Courtney, 2010). If onset of a horizontally moving visual target is accompanied by onset of a tone, representational momentum for the visual target is decreased or increased if the tone terminates slightly before or after, respectively, the visual target vanishes (Teramoto, Hidaka, Gyoba, & Suzuki 2010). A tone presented just before a visual target vanishes leads to backward displacement of that target (Chien, Ono, & Watanabe 2013; but see Teramoto et al., 2010). If a tone is presented only to the left or right ear, then backward displacement is larger if the tone is on the same side of space where visual motion originated. If observers are presented with separate visual and auditory targets moving across space, then if the nonjudged modality target vanishes before or after the judged modality target vanishes, displacement of the judged modality target is biased toward the vanishing position of the nonjudged modality target (Schmiedchen et al., 2012). However, if the difference in vanishing times was greater than 2,000 msec, then effects of the nonjudged modality target on the judged modality target only occurred if participants judged auditory targets.

Examples in the preceding paragraph involved target stimuli and nontarget stimuli in different modalities. However, effects of cast shadows, barriers, and landmarks accompanying visual targets suggest nontarget stimuli in the same modality as a target influence representational momentum for that target. In these latter cases, there is no contact between targets and nontarget stimuli (e.g., Hubbard, 1993), but contact between targets and nontarget stimuli can influence representational momentum for the target. If a target slides along a stationary surface, forward displacement of that target is decreased relative to forward displacement of an otherwise identical target presented in isolation or clearly separated from the larger stationary surface (Hubbard, 1995b, 1998). This decrease is referred to as *representational friction* (Hubbard, 1995b, 1995c), and decreases in forward displacement increase as implied friction on the moving target increases (Hubbard, 1995b, 1998). Illusory motion of a visual nontarget stimulus might (e.g., Hubbard et al., 2005) or might not (e.g., Nagai & Yagi, 2005) influence displacement of a visual target. Additional examples of nontarget influences are discussed in the section below on Attributions of the Source of Target Motion (Causality).

Expectations of Future Target Motion. If a target is expected to bounce off a barrier the target is approaching, then forward displacement decreases as the target approaches the barrier, and at the moment of contact, displacement is in the expected (new) direction rather than in the previous direction of motion (Hubbard & Bharucha, 1988). A physical reason for a direction change does not seem to be necessary, as similar patterns occur when visual targets changing in location (Verfaillie & d'Ydewalle, 1991), and auditory targets changing in frequency (Johnston & Jones, 2006), exhibit oscillatory motion in an otherwise blank display. Similarly, forward displacement is larger if a target bounces off rather than crashes through a barrier, as if elastic bouncing off preserved more momentum and rigid crashing-through depleted more momentum (Hubbard, 1994); interestingly, this difference is stronger if observers received a valid cue (e.g., a cue of "bounce" and the target then "bounced") than if observers received an invalid cue (e.g., a cue of "bounce" and the target then "crashed"). A similar combination of effects of expected target direction and actual target direction on displacement was reported by Hudson, Nicholson, Ellis et al. (2016) and Hudson, Nicholson, Simpson et al. (2016). A spoken mimetic word describing possible target behavior presented during target motion influences displacement (Gobara, Yamada, & Miura 2016).

Attribution of the Source of Target Motion (Causality). If observers attribute motion of an initially stationary target to contact from a moving launcher, forward displacement of the target is less than forward displacement of an otherwise identical target that exhibits autonomous motion (Hubbard & Favretto, 2003; Hubbard & Ruppel, 2002; Hubbard et al., 2001). The launcher does not need to exhibit physical motion, as illusory gamma motion when a launcher appears adjacent to a target results in a launching effect (Hubbard et al., 2005). Choi and Scholl (2006) replicated the decreased forward displacement for launched targets found by Hubbard et al. (2001), but they hypothesized this reflected the presence of two objects and a single motion rather than perception of causality. Hubbard (2013c) tested this by comparing displacement of a target in an entraining effect (a launcher continues in motion after contacting the target and appears to carry the target along; Hubbard, 2013d) with displacement in a launching effect; displacement of entrained targets was larger than displacement of launched targets, and this is inconsistent with Choi and Scholl's hypothesis. Increases in launcher size and velocity lead to larger representational momentum for the target (de sá Teixeira et al., 2008), and forward displacement of a launched target decreases with distance traveled (Hubbard & Ruppel, 2002).

Characteristics of the Observer

Variables influencing representational momentum that are related to the observer include (a) age, (b) allocation of attention, (c) eye movements and fixation, (d) action plans,

(e) expertise, (f) error feedback and knowledge of representational momentum, (g) affective value of the target, (h) presence of psychopathology, and (i) neurophysiological structures involved or implicated in representational momentum.

Age. Children 24–36 months of age viewed a toy car roll down an incline and vanish behind an occluder, and they could open one of four doors in the occluder to retrieve the car (Perry, Smith, & Hockema 2008). Children were more likely to open a door beyond where the car would have stopped, and this is consistent with representational momentum. Children 5 to 9 years of age viewed displays similar to those in Finke and Freyd (1985), and older children exhibited larger representational momentum than did younger children (Taylor & Jakobson, 2010). Third-grade children, fifth-grade children, and adults shown frozen-action photographs exhibited representational momentum, but there was no effect of age (Futterweit & Beilin, 1994). First-grade children, fourth-grade children, and adults shown horizontally or vertically moving geometric stimuli exhibited representational momentum, and first-grade children exhibited larger representational momentum than did fourth-grade children or adults (Hubbard et al., 1999). Older adults (more than 65 years of age) exhibited less representational momentum for an implied motion target than did younger adults (Piotrowski & Jakobson, 2011); indeed, the oldest participants did not exhibit representational momentum (which might reflect decreased ability to process implied motion). A consistent pattern of developmental changes in representational momentum has not been found, and methodological differences between studies might account for differences in findings.

Allocation of Attention. If an unrelated task (e.g., counting along with a metronome) is concurrent with target presentation, representational momentum for the target is increased (Hayes & Freyd, 2002). Similarly, larger forward (smaller backward) displacement is exhibited when participants simultaneously listen to a string of digits and respond each time three odd numbers in a row occur (Joordens et al., 2004). If attention is manipulated by having participants attend to one or two locations on a single trajectory, no effect of allocation of attention occurs (Hubbard & Motes, 2002; Kerzel, 2004). If participants are cued during the target presentation or retention interval regarding final target location, representational momentum decreases but is not eliminated (Hubbard et al., 2009). The inability of a cue to eliminate representational momentum suggests at least some portion of the displacement process is cognitively impenetrable. However, it is possible that cues presented during target presentation or during the retention interval might be displaced in the direction of motion (cf. Hubbard, 2008) and thus ineffective in cueing actual final location.

An unrelated object flashed (briefly presented) at the moment the target vanished increases representational momentum (Munger & Owens, 2004). When a flashed object is relevant to the task (i.e., participants judged target position at the time of the flash), backward displacement of the target occurs, but if a flashed object is irrelevant (i.e., observers instructed to ignore the flash), forward displacement occurs (Müsseler et al., 2002). An unrelated flash during the retention interval results in backward displacement, and this is larger if the distractor is in front of rather than behind the target (Kerzel, 2003a). These findings suggest tasks or stimuli that draw attention away from the target might influence displacement. The complementary pattern might also occur, that is, forward displacement might influence attention. For example, detection of a gap within a circle is faster and more accurate if the circle is presented slightly in front of the final location of a moving target than if the circle is presented slightly behind the final location of a moving target (Kerzel et al., 2001). Indeed, attention itself might exhibit a form of momentum (e.g., Hubbard, 2014a, 2015b; Pratt, Spalek, & Bradshaw 1999).

Eye Movements and Eye Fixation. If an observer views a continuously moving target but does not track that target, representational momentum is decreased or eliminated (de sá Teixeira, Hecht, & Oliveira, 2013; Kerzel, 2000). This initially led to suggestions that forward displacement resulted from pursuit eye movements overshooting final target position and visual persistence (e.g. Kerzel, 2000; Stork & Müsseler, 2004). However, oculomotor behavior does not influence displacement for a target undergoing implied motion (Kerzel, 2003a); similarly, pursuit movements do not occur with frozen-action photographs, which also result in representational momentum. Nonetheless, the role of eye movements has been debated (e.g., Hubbard, 2005b, 2006b; Kerzel, 2002c, 2006). Several recent studies challenge the idea that preventing visual tracking of smoothly moving targets eliminates representational momentum (e.g., Getzmann & Lewald, 2009; Schmiedchen et al., 2013; Teramoto et al., 2010). De sá Teixeira, Hecht, and Oliveira (2013) suggested constraining eye movements suppresses or conceals effects of motion representation (cf. Hubbard 2005b, 2006b, 2010, that eye movements might contribute to, but are not causal of, representational momentum). Oculomotor behavior is not related to representational momentum for auditory targets (Getzmann, 2005) or representational gravity for horizontally moving visual targets (de sá Teixeira, Hecht, & Oliveira, 2013). The presence of representational momentum in patients with schizophrenia (de sá Teixeira, Pimenta, & Raposo, 2013; Jarrett et al., 2002), who often exhibit dysfunction of pursuit eye movements, further challenges claims that pursuit eye movements are necessary for generation of representational momentum.

Action Plans. Representational momentum decreases if observers have control over target direction and velocity (Jordan & Knoblich, 2004). Similarly, displacement is influenced by whether participants trigger disappearance of the target or the target vanishes without warning (Jordan et al., 2002; Stork & Müsseler, 2004). This suggests motor plans are coupled to perceptual space, which further suggests sharing or overlap of mechanisms for perception and for action planning. Relatedly, participants with previous experience controlling a target exhibit larger forward displacement than do participants without previous experience controlling that target, and this might reflect activation of action plans (Jordan & Hunsinger, 2008). Such findings are consistent with effects of expertise on representational momentum (see discussion below), but seem inconsistent with decreases in displacement when participants control the target (Jordan & Knoblich, 2004; Jordan et al., 2002; Stork & Müsseler, 2004); however, it is possible that increased displacement occurs only during passive observation and not during active control.

Expertise. If expert drivers and novice drivers view displays involving roadway scenes filmed on board a moving automobile, all participants exhibited representational momentum, and representational momentum was larger for experts (Blättler, Ferrari, Didierjean, van Elslande, & Marmèche, 2010). If experienced pilots and non-pilots view displays based on aircraft landings, experienced pilots but not non-pilots exhibited representational momentum (Blättler, Ferrari, Didierjean, &Marmèche, 2011). Lack of displacement in non-pilots was attributed to lack of familiarity with the setting and stimuli. Experienced baseball batters exhibit larger representational momentum for a fast-moving object than do inexperienced baseball batters (Nakamoto, Mori, Ikudome, Unenaka, & Imanaka, 2015). These findings are consistent with the notion that representational momentum is related to predictability of the stimulus (cf. Kerzel, 2002d), as experts are more familiar with the stimuli, and thus more able to predict the near future of those stimuli, than are non-experts. Interestingly, this suggests that attempts to develop training to eliminate representational momentum might be futile, and this is consistent with effects of error feedback on representational momentum.

Error Feedback and Knowledge. Representational momentum was initially claimed to be impervious to error feedback (Freyd, 1987), and this was based on findings that feedback during practice trials did not influence representational momentum on subsequent experimental trials (e.g., Finke & Freyd, 1985). More recent studies challenged this claim. If participants are instructed regarding representational momentum and asked to guard against it in their responses, forward displacement on subsequent trials is reduced but not eliminated (Courtney & Hubbard, 2008). If participants are given feedback after each trial regarding whether their response was correct or incorrect, such feedback does not reduce representational momentum, although it does reduce overall probability of a *same* response to any given probe (Ruppel et al., 2009). Furthermore, reduced probability of a *same* response continues even if feedback is later withdrawn. As most experiments using probe judgment present 5–9 probe positions, the probability that any given probe will reflect the accurate target location is less than 50%, and the relatively low prior probability a *same* response would be correct might account for decreases in *same* responses (cf. Hubbard & Lange, 2010).

Affective Value. Affective value of an object (e.g., safe, painful) that a hand moved toward or away from influenced representational momentum for location of the hand; more specifically, when affective value of the stimulus matched direction of hand movement (positive affect and motion toward the object, or negative affect and motion away from an object), forward displacement of the hand was larger than when affective value of the stimulus did not match direction of hand motion (Hudson, Nicholson, Ellis et al., 2016; Hudson, Nicholson, Simpson et al., 2016). Analogously, affective value of a target moving toward a stationary object influences displacement of that target. Participants were presented with a neutral stationary object and a moving target, and they read vignettes in which the moving target was described as threatening to the stationary object, neutral, or happy; forward displacement of the moving target was larger when the situation was described as a threatening than when the situation was described as neutral or happy (Greenstein, Franklin, Martins, Sewack, & Meier 2016). Studies on representational momentum involving changes in facial expression involved different emotions depicted by a facial target, and those studies are discussed in the section on Human Form.

Psychopathology. Patients diagnosed with schizophrenia exhibit larger representational momentum than do control participants (Jarrett, Phillips, Parker, & Senior 2002), and target size, but not target velocity, influences representational momentum in such patients (de sá Teixeira, Pimenta, & Raposo, 2013). Patients diagnosed with mental retardation exhibit smaller representational momentum than do control participants (Conners, Wyatt, & Dulaney, 1998). Representational momentum is larger in patients with left spatial neglect and patients with right-hemisphere damage but no neglect (McGeorge et al., 2006), and representational momentum decreases with increases in distance traveled by the target for patients but not for control participants. Patients with left hemineglect as a consequence of right-hemisphere damage exhibited larger forward displacement than did patients with right-hemisphere damage without neglect and control participants regardless of direction of target motion (Lenggenhager et al., 2012); furthermore, patients with left hemineglect exhibited larger forward displacement for targets moving toward the left (and a patient with right hemineglect exhibited larger forward displacement for targets moving toward the right). Patients with autism spectrum disorder exhibit decreases representational momentum for changes in facial expression (Uono et al., 2014).

Neurophysiology. Several studies on neurophysiology of representational momentum involve patients diagnosed with psychopathology, and those studies were discussed in the section on Psychopathology. Retinal ganglion cells of rabbits and salamanders appear to anticipate arrival of a continuously moving stimulus (Berry, Brivanlou, Jordan, & Meister 1999), but no such patterns have been reported for implied motion or frozen-action stimuli. Representational momentum for changes in size (White et al., 1993) and location (Gottwald et al., 2015; Halpern & Kelly, 1993) are larger for stimuli in the left visual field (right hemisphere). Consistent with this, increased activity in right parietal cortex occurs when observers exhibit representational momentum (Amorim et al., 2000). Frozen-action photographs elicit activation in motion-processing areas of V5/MST (Kourtzi & Kanwisher, 2000; Senior et al., 2000). If a stimulus shown to previously elicit representational momentum is viewed after TMS over area V5/MST, representational momentum does not occur (Senior, Ward, & David 2002). Representational momentum has been linked to activity in prefrontal cortex (Rao et al., 2004).

Properties of Representational Momentum

Freyd (1987) suggested several properties of dynamic mental representation that were based on early studies of representational momentum. Subsequent research suggested additional properties (see Hubbard, 2005b, 2010, 2014a, 2015a, 2017). Four of these additional properties are considered here and involve whether representational momentum (a) reflects objective physical principles or naive physics, (b) reflects low-level processing or high-level processing, (c) is related to other spatial biases, and (d) bridges the gap between perception and action.

Physical Principles or Naive Physics?

Some researchers suggest representational momentum reflects internalization or incorporation of momentum (e.g., Finke et al., 1986), and other researchers completely reject the notion of such internalization or incorporation (e.g., Kerzel, 2000, 2003a). Displacement does not always reflect objective physical principles (e.g., representational momentum does not reflect implied mass and is influenced by expectations regarding future target behavior and by landmark attraction). What is more likely is that displacement reflects subjective consequences of such principles rather than objective principles per se (e.g., displacement is influenced by experienced weight rather than by objective mass), and that such consequences are modulated by expectations regarding the target. Consistent with this, some authors suggest representational momentum might involve naïve physics (e.g., impetus; Hubbard, 2004, 2013c; Hubbard & Ruppel, 2002; Kozhevnikov & Hegarty, 2001). An interpretation emphasizing subjective experience is consistent with contemporary emphases on embodied or grounded cognition (e.g., Barsalou, 2008; Gibbs, 2005).

Low-Level Processing or High-Level Processing?

Several strands of evidence suggest representational momentum involves at least some high-level (top-down) processes. Representational momentum is influenced by information regarding source (Hubbard, 2004, 2013c) and anticipated direction (Hubbard & Bharucha, 1988; Johnston & Jones, 2006; Verfaillie & d'Ydewalle, 1991) of target motion. Representational momentum is influenced by target prototypicality and identity (Nagai & Yagi, 2001;

Reed & Vinson, 1996; Vinson & Reed, 2002). Representational momentum is related to activity in several cortical areas (Amorim et al., 2000; Kourtzi & Kanwisher, 2000; Rao et al., 2004; Senior et al., 2000). Representational momentum occurs with different surface forms (cf. Freyd & Finke, 1984; Hubbard & Bharucha, 1988), in different modalities (cf. Johnston & Jones, 2006; Verfaillie & d'Ydewalle, 1991), and is influenced by cross-modal information (e.g., Teramoto et al., 2010). Rather than a different mechanism for each surface form or modality, it is more parsimonious to suggest a more high-level general-purpose extrapolation mechanism (cf. Hubbard, 2006b, 2014a, 2015a, 2015b, 2017). Importantly, allowing for high-level processes and influences does not mean that low-level processes cannot contribute to or modulate displacement (e.g., disrupting information normally provided by eye movements disrupts displacement; de sá Teixeira, Hecht, & Oliveira, 2103).

Relationship to Other Spatial Biases?

Representational momentum is related to other momentum-like effects (e.g., Hubbard, 2014a, 2015a, 2015b, 2017) and general perceptual or cognitive processes (e.g., mental rotation; Hubbard, 2006a; Munger, Solberg, & Horrocks, 1999; perceptual grouping; Hubbard 2011b; aesthetics; Hubbard, Chapter 15 in this volume). These relationships are beyond the scope of this chapter, but a narrower consideration of relationships of representational momentum to other biases in remembered location (Fröhlich effect, flash-lag effect, boundary extension) is presented.

Fröhlich Effect. The perceived onset (initial) location of a moving target is often displaced in the direction of target motion, and this is referred to as the *Fröhlich effect* (Kerzel, 2010; Müsseler & Kerzel, Chapter 7 in this volume). Forward displacement of the target in the Fröhlich effect appears similar to forward displacement of the target in representational momentum (cf. Hubbard, 1990; Müsseler & Aschersleben, 1988). The Fröhlich effect (e.g., Müsseler & Aschersleben, 1998) and representational momentum (e.g., Hubbard, 1990) increase with increases in target velocity. The Fröhlich effect (Whitney & Cavanagh, 2000a) and representational momentum (Hubbard et al., 2009) are decreased by presentation of a valid cue prior to stimulus presentation. Given such similarities, displacement in the Fröhlich effect might reflect representational momentum and an accurate memory for trajectory length, or displacement in representational momentum might reflect the Fröhlich effect and an accurate memory for trajectory length. However, when the Fröhlich effect and representational momentum are measured for the same trajectory, representational momentum for the vanishing point is found, but memory for initial location is displaced in the direction opposite to target motion (i.e., an onset repulsion effect; Actis-Grosso & Stucchi, 2003; Hubbard & Motes, 2002; Kerzel, 2004).

Flash-Lag Effect. If a briefly presented (flashed) stationary object is aligned with a moving target, that object appears to lag behind the moving target. This is referred to as the *flash-lag effect* (Hubbard, 2014b, Chapter 9 in this volume). Forward displacement in the flash-lag effect appears similar to forward displacement in representational momentum (cf. Munger & Owens, 2004; Shi & de'Sperati, 2008). The flash-lag effect (Wojtach et al., 2008) and representational momentum (Hubbard & Bharucha, 1988) increase with increases in target velocity. The flash-lag effect (Namba & Baldo, 2004; Vreven & Verghese, 2005) and representational momentum (Hubbard et al., 2009) are decreased by presentation of a valid cue prior to stimulus presentation. The flash-lag effect (Nagai et al., 2010; Noguchi & Kakigi, 2008) and representational momentum (Reed & Vinson, 1996) are influenced by conceptual knowledge. The flash-lag effect (Ichikawa & Masakura, 2006, 2010) and representational momentum (Hubbard, 2013c; Hubbard et al., 2001) are influenced by attributions regarding the source of target motion. The flash-lag effect (Scocchia, Actis-Grosso, de'Sperati, Stucchi, & Baud-Bovy, 2009) and representational momentum (Jordan & Hunsinger, 2008) are influenced by whether observers control the stimulus. The flash-lag effect (Sarich, Chappell, & Burgess, 2007) and representational momentum (Hayes & Freyd, 2002) increase with divided attention. The flash-lag effect (Brenner, van Beers, Rotman, & Smeets, 2006; Shi & Nijhawan, 2008) and representational momentum (Hubbard & Ruppel, 1999) increase with motion toward a landmark or fixation. The main difference is that in the flash-lag effect, displacement of the target is measured relative to the location of a nearby stationary object, whereas in representational momentum, displacement of the target is measured relative to the actual location of the target Hubbard (2013b, 2014b).

Boundary Extension. Memory for a previously viewed scene includes not just information within the initial view but also information that was not viewed but would likely have been present outside the boundaries of the initial view. This has been referred to as *boundary extension* (Hubbard et al., 2010; Intraub, 2012; Intraub & Gagnier, Chapter 13 in this volume). Boundary extension (Intraub, Daniels, Horowitz, & Wolfe, 2008) and representational momentum (Hayes & Freyd, 2002) are decreased when more attention is allocated to the target. Boundary extension (Intraub & Dickinson, 2008) and representational momentum (Freyd & Johnson, 1987) occur within tens of milliseconds after a target vanishes. Boundary extension (Intraub, Hoffman, Wetherhold, & Stoehs, 2006) and representational momentum (Jordan et al., 2002) are influenced by action plans regarding eye movements. Boundary extension (Intraub & Bodamer, 1993) and representational momentum (Courtney & Hubbard, 2008) are decreased but not eliminated if participants receive information regarding the effect and are instructed to counteract the effect. Boundary extension (Hubbard, 2011a) and representational momentum (Hubbard, 2011b) exhibit properties similar to Gestalt principles of perceptual grouping. Munger et al. (2005) suggested boundary extension for a scene occurred before representational momentum for objects in that scene.[4]

Bridging the Gap?

It takes 100 msec or so for information presented to the retina to be processed in the cortex (e.g., De Valois & De Valois, 1991); thus, there is a lag between when sensation of a stimulus begins and when information regarding that stimulus enters perceptual awareness. The flash-lag effect has been proposed to compensate for such delays (e.g., Nijhawan, 2008), but representational momentum might provide better compensation, as the flash-lag effect is defined in terms of relative position (and so a referent stationary object is necessary), and representational momentum does not require comparison to another stimulus. Indeed, Hubbard (2005b) suggested representational momentum bridged the gap between perception and action. When light from a moving stimulus stimulates the retina, a cascade of sensory, perceptual, cognitive, and perhaps motor responses is initiated. However, while this processing is occurring, the stimulus typically does not pause and wait for the observer

[4] Location and motion may be processed by the same neural mechanisms (Whitney & Cavanagh, 2002), and Intraub's (2002) suggestion that boundary extension results from continuity of spatial layout is consistent with Freyd's (1987, 1993) suggestion that representational momentum is more likely when the underlying stimulus dimension is continuous.

to finish processing, but instead continues in motion. By the time conscious awareness of the stimulus is achieved, the stimulus has already moved beyond where it was when it first stimulated the retina. If an immediate response from an observer is to be optimal, that response should be calibrated to where the stimulus is in real time and not where the stimulus was when it was initially sensed. What is needed is a bridge between perception (i.e., location at initial sensation) and action (i.e., location when a response would reach the stimulus), and representational momentum might provide such a bridge.

Summary and Conclusions

Representational momentum involves displacement of the judged location of a target in the direction of anticipated motion. This displacement can be evoked by continuous motion, implied motion, and static images that suggest a specific direction of motion, and can be measured with cursor positioning, probe judgment, and reaching responses. Representational momentum is increased if (a) velocity is faster, (b) motion is horizontal rather than vertical or downward rather than upward, (c) motion is in a prototypical direction or direction in which the target points or faces, (d) a constant object identity is maintained, (e) motion is from central to paracentral regions, (f) motion is biologically less awkward, (g) observers do not control the target, (h) retention interval is only a few hundred milliseconds long, (i) contrast of the target with the background is high or increasing, (j) a surrounding or nearby stimulus moves in the same direction, (k) motion is toward a landmark, (m) the target is entering the scene, (n) the target experiences less implied friction, (o) a nontarget stimulus that appeared when the target appeared vanishes slightly after the target vanishes, (p) motion is perceived as autonomous rather than caused by another stimulus, (q) observers divide attention between the target and another stimulus or task, (r) observers track a continuously moving target, (s) the observer has previous experience or expertise controlling the target, (t) the target is associated with threat, and (u) observers are naive regarding representational momentum. Other types of information that influence displacement more generally include (a) distance traveled by the target, (b) size of the target, (c) gaze direction of a target, (d) shape and shading, and (e) action plans of the observer.

Representational momentum exhibits several properties in addition to those initially suggested by Freyd (1987). First, representational momentum incorporates information regarding the subjective consequences of physical principles for the observer rather than information regarding objective physical principles. This is consistent with idea of embodied cognition. Second, representational momentum involves at least some high-level processing. Displacement reflects a combination of many influences, and can be found in multiple dimensions and influenced by cross-modal information. This suggests a single or small number of extrapolation mechanisms exist, although it is possible but less parsimonious that multiple modality-specific and dimension-specific extrapolation mechanisms exist. Relatedly, inability of feedback or instruction to eliminate representational momentum suggests there is at least one cognitively impenetrable (modular) component to displacement, but the influence of expectations regarding target identity and typical behavior suggest there is at least one cognitively penetrable (nonmodular) component. Third, representational momentum is related to other biases in localization such as the Fröhlich effect, flash-lag effect, and boundary extension, as well as other momentum-like processes. Fourth, rather than being an error or flaw in spatial representation, representational momentum is an adaptive solution to the problem of neural processing delay. Representational momentum helps bridge the gap between initial sensed location of a target and actual location of the target at the time information regarding initial sensation enters conscious awareness.

In representational momentum, judged final position of a moving target is displaced from actual final position, and this displacement is in the direction of anticipated or likely future movement of the stimulus. The idea of representational momentum (and displacement in general) highlights the dynamic and anticipatory nature of perception and representation and underscores important issues in visuospatial and spatiotemporal processing. Foremost among these issues is how our representational system has been tuned to anticipate what is most likely to be encountered in interactions with environmental stimuli. Such tuning potentially involved shaping of the functional architecture of the representation (see Hubbard, 2005b, 2006a). Relatedly, representational momentum might help compensate for delays in perception due to neural processing latencies. Although it might be tempting to view representational momentum as an error, such a displacement is actually a very useful adaptation. The existence of representational momentum has potential implications for other biases in localization (e.g., representational momentum of a moving target might be the basis of the flash-lag effect), and might be related to other processes in spatial representation. Representational momentum is an important and fundamental aspect of spatial representation, and may contribute to more processes than previously expected.

Chapter 9

The Flash-Lag Effect*

Timothy L. Hubbard

If an observer views a briefly presented (flashed) stationary object that is aligned with a moving target, the position of the briefly presented object seems to lag behind the position of the moving target. This is referred to as the *flash-lag effect* (Nijhawan, 1994; for review, see Hubbard, 2014b) and has implications for how we perceive and interact with stimuli in our environment (e.g., there is a delay between when the retina is stimulated and conscious perception of the stimulus [De Valois & De Valois, 1991; Nijhawan, 2008], and the flash-lag effect might reveal one way perception compensates for this delay). Several methods have been used to study the flash-lag effect, and the most common are illustrated in Figure 9.1. In Figure 9.1A, the moving target consists of a rotating bar, and the flashed object consists of two sets of dotted lines. In Figure 9.1B, the moving target consists of a revolving annulus, and the flashed object consists of a solid disk. In Figures 9.1C and 9.1D, the moving target consists of a translating bar, and the flashed object consists of a second bar presented adjacent to the translating bar or within a gap in the translating bar. The magnitude of the flash-lag effect is influenced by numerous variables, and these are briefly reviewed. Properties of the flash-lag effect, and the relationship of the flash-lag effect to other spatial biases, are discussed. Finally, some of the more common explanations for the flash-lag effect are described.

Influences on the Flash-Lag Effect

Many different variables have been manipulated in studies of the flash-lag effect. In some cases (e.g., target velocity), different values of a given variable result in different magnitudes of the flash-lag effect. In other cases (e.g., color of the moving target or flashed object), different values of a given variable reveal other aspects of the flash-lag effect.

Characteristics of the Stimulus

Numerous characteristics of the stimulus influence or reveal properties of the flash-lag effect. These include (a) timing of presentation of the flashed object relative to presentation of the moving target, (b) continuity of the target, (c) distance traveled by the target prior to presentation of the flashed object, (d) distance between the moving target and the flashed object, (e) eccentricity and visual field, (f) velocity of target motion, (g) direction of target motion, (h) binocular disparity, (i) changes in target color and lack of color mixing, (j) luminance of the moving target or of the flashed object, (k) contrast of the moving target and flashed object with the background or each other, (l) spatial frequency of the

* This chapter is an abridged and slightly updated version of Hubbard (2014b).

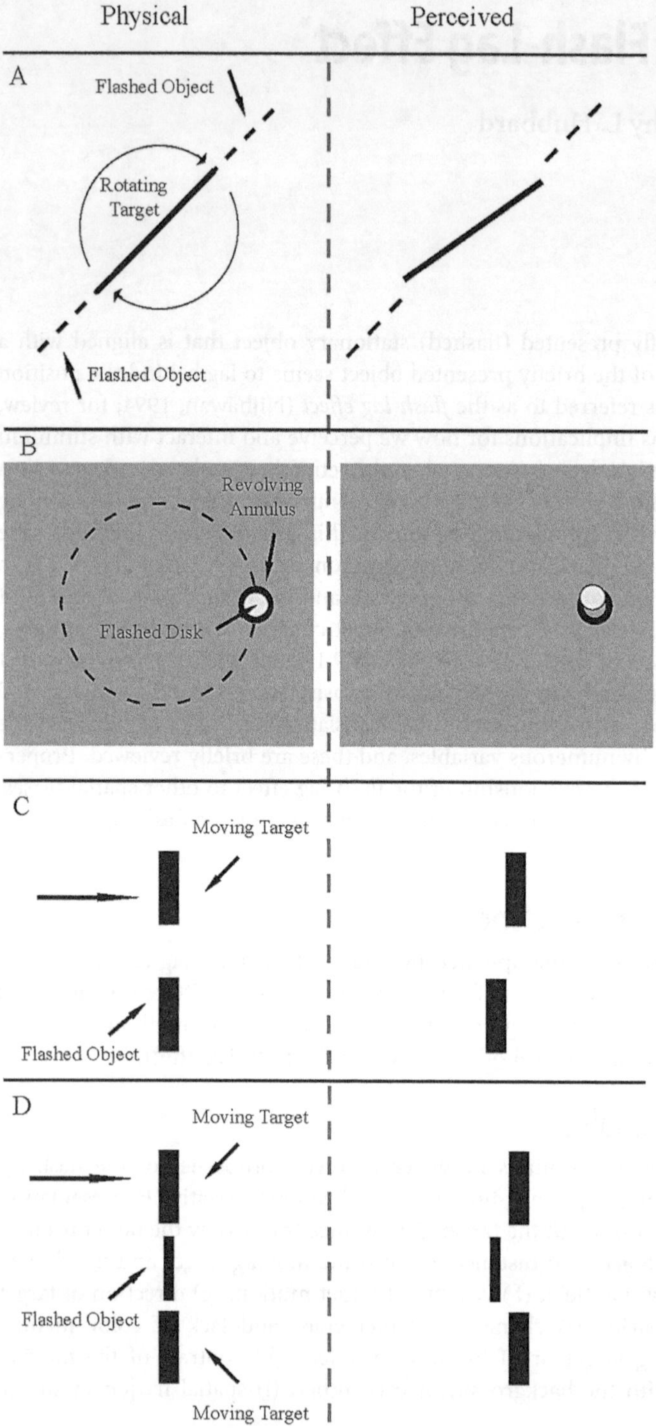

Figure 9.1 Illustrations of common stimulus displays in studies of the flash-lag effect. The actual physical stimulus is depicted on the left, and the typical perceived stimulus is depicted on the right; in each case, the position of the flashed object appears to lag behind the position of the moving target. (A) The target consists of a bar rotating clockwise, and the flashed object consists of two dashed line segments that are briefly flashed when they are in alignment with the bar. (B) The target consists of a black annulus moving clockwise on a circular path (indicated by the dashed line), and the flashed object consists of a white disk flashed within the annulus. (C) The target consists of a vertical bar moving from left to right in the upper part of the display, and the flashed object is a vertical bar in the lower part of the display that is briefly flashed when it is aligned with the moving target. (D) The target consists of two vertically aligned vertical bars moving from left to right, and the flashed object consists of a vertical bar flashed in the gap within the target. Adapted from "The Flash-Lag Effect and Related Mislocalizations: Findings, Properties, and Theories," by T. L. Hubbard, 2014b, *Psychological Bulletin*, 140, p. 309. Copyright 2014 by American Psychological Association, Inc.

moving target, (m) identity of the moving target, (n) duration of the flashed object, (o) motion of the flashed object, (p) predictability of the flashed object, and (q) presence of unrelated stimuli.

Relative Timing. The flashed object is typically presented at the onset, midpoint, or offset of target motion, and these displays are referred to as *flash-initiated*, *flash-midpoint*, and *flash-terminated*, respectively. A flash-lag effect typically occurs with flash-initiated and flash-midpoint displays, but not with flash-terminated displays (Eagleman & Sejnowski, 2000b; Khurana & Nijhawan, 1995; Nijhawan, Watanabe, Khurana, & Shimojo, 2004). Many studies that involve flash-initiated, flash-midpoint, and flash-terminated stimuli present the flashed object and moving target as aligned, but a small group of studies varied relative spatial position, and observers adjusted the stimuli so that the flashed object and moving target appear aligned (e.g., Lappe & Krekelberg, 1998).

Continuity of the Target. A flash-lag effect occurs if the moving target is presented continuously (i.e., target motion appears smooth) or if discrete and clearly separated presentations of the target (i.e., implied motion) are viewed (Rizk, Chappell, & Hine, 2009), although the effect is weaker with discrete presentations. Such a result suggests the flash-lag effect is related more to underlying continuity of the stimulus dimension along which change occurs and from which the target is drawn or sampled (e.g., location, orientation, luminance, hue, etc.) than to characteristics of the target per se. An unexpected change in color or size (Au & Watanabe, 2013; Moore & Enns, 2004) of the moving target when the flashed object is presented disrupts the flash-lag effect.

Distance Traveled by the Target. Finding a flash-lag effect with flash-initiated and flash-midpoint displays but not flash-terminated displays suggests the flash-lag effect decreases as distance traveled by the target increases. Consistent with this, increases in distance traveled by the target prior to presentation of the flashed object decreases the flash-lag effect (Vreven & Verghese, 2005). Maus and Nijhawan (2006) reported the visible threshold for moving targets was lower if the target traveled farther, and they suggested this involved larger forward displacement of the target; however, larger displacement would presumably result in an increased flash-lag effect with increases in distance traveled by the target (cf., Kanai, Sheth, & Shimojo, 2004).

Distance between Moving Target and Flashed Object. Many studies presented the flashed object embedded within the moving target (e.g., a disk within an annulus; Becker, Ansorge, & Turatto, 2009; Nijhawan, 2001; flashed objects interleaved with different parts of a moving target; Khurana & Nijhawan, 1995), but other studies varied distance between the moving target and flashed object. In the former studies, a robust flash-lag effect occurred. In the latter studies, increases in distance between the moving target and the flashed object increased the flash-lag effect (Baldo & Klein, 1995; Baldo et al., 2004). If the flashed object is embedded or interleaved within the moving target, or is spatially separate from the moving target, then there is no partial overlap between the flashed object and the moving target. However, if partial overlap occurs, the portion of the moving target inside the boundaries of the flashed object is veridically perceived, whereas the portion of the moving target outside the boundaries of the flashed object exhibits a flash-lag effect (Kanai & Verstraten, 2006). Thus, whether the moving target encloses, is spatially separated from, or partially overlaps the flashed object influences the flash-lag effect.

Eccentricity and Visual Field. When the flashed object is closer to fixation than is the moving target, the flash-lag effect is increased (Baldo, Kihara, et al. 2004; Baldo & Klein, 1995), whereas when the flashed object is farther from fixation than is the moving target, the flash-lag effect is decreased (Linares, López-Moliner, & Johnston, 2007) with increases in eccentricity of the flashed object from fixation. The flash-lag effect is more likely to occur with flash-terminated displays as eccentricity increases (Kanai et al., 2004). The flash-lag effect is increased for horizontal target motion in the left visual field relative to the right visual field (Kanai et al., 2004) and is more influenced by target direction if motion is in the right visual field (Shi & Nijhawan, 2008). However, visual field does not influence the flash-lag effect for vertically moving targets (Ishikawa & Masakura, 2006, 2010).

Velocity of Target Motion. The flash-lag effect increases with increased target velocity in the picture plane (e.g., Brenner & Smeets, 2000; López-Moliner & Linares, 2006; Wojach et al., 2008) and in depth (e.g., Lee, Khuu, Li, & Hayes, 2008). The effect of target velocity is decreased in multi-target displays relative to single-target displays (Shioiri, Yamamoto, Oshida, Matsubara, & Yaguchi, 2010). Acceleration or deceleration of the moving target results in an increased or decreased flash-lag effect, respectively (Whitney, Murakami, & Cavanagh, 2000). The effect of velocity is larger if target motion is toward fixation (Blohm et al., 2003). Random changes in velocity decrease the flash-lag effect (Vreven & Verghese, 2005). The flash-lag effect is influenced by target velocity after the flashed object is presented (Brenner & Smeets, 2000), and this is consistent with an account of the flash-lag effect based on postdiction.

Direction of Target Motion. The flash-lag effect is not influenced by whether targets ascend or descend in the picture plane (Ichikawa & Masakura, 2006, 2010). If a target reverses direction at the time of the flash, the flash-lag effect is decreased (Whitney & Murakami, 1998). If a vertically moving target turns to the left or right, a flash-lag effect occurs even if the direction of change is unpredictable (Chappell & Hinchy, 2014; Whitney, Cavanagh, & Murakami, 2000). The flash-lag effect is larger if a target moves toward fixation than away from fixation (Brenner, van Beers, Rotman, & Smeets, 2006; Kanai et al., 2004; Mateeff & Hohnsbein, 1988). A flash-lag effect also occurs for targets moving in depth (Harris, Duke, & Kopinska, 2006; Ishii, Seekkuarachchi, Tamura, & Tang, 2004). The flash-lag effect decreases if the target unpredictably changes direction near the time of the flash (Arrighi, Alais, & Burr, 2005; Murakami, 2001b; Vreven & Verghese, 2005; Whitney et al., 2000).

Binocular Disparity. A flash-lag effect occurs when the moving target or flashed object is defined solely by binocular disparity. This has been found with random-dot stimuli (Harris et al., 2006; Nieman, Nijhawan, Khurana, & Shimojo, 2006) and stereomotion (Lee et al., 2008). As noted by Nieman et al. (2006), finding a flash-lag effect with random-dot stimuli challenges hypotheses that the flash-lag effect results from retinal or low-level phenomena (e.g., as in Berry, Brivanlou, Jordan, & Meister, 1999).

Color. If a target gradually changes from red to green and a flashed object is presented during the change, a flash-lag effect for color occurs (Sheth, Nijhawan, & Shimojo, 2000). If observers view a moving green bar and a briefly flashed red bar is superimposed on the green bar, perception of a yellow bar (based on additive color mixing) might be predicted. However, a flash lag effect in which the red bar lags the green bar occurs (Nijhawan, 1997). If observers execute smooth pursuit eye movements past a stationary green bar (that appeared to move across the retina in the direction opposite to the direction of eye movement), and a red bar was briefly superimposed on the green bar, the red bar appeared to be shifted in the opposite direction; this is consistent with a flash-lag effect (Nijhawan, Khurana, Kamitani, Watanabe, & Shimojo, 1998). A flash-lag effect occurs if the moving object changes color (Kreegipuu & Allik, 2004), but the perceived color change lags perceived position (Cai & Schlag, 2001; Gauch & Kerzel, 2008b). If target color is maintained or exhibits multiple changes, a flash-lag effect occurs, but a single unexpected change at the moment of the presentation of the flashed object eliminates the flash-lag effect (Au & Watanabe, 2013).

Luminance. The flash-lag effect decreases if luminance of the flashed object increases (Öğmen, Patel, Bedell, & Camuz 2004; Purushothaman, Patel, Bedell, & Öğmen, 1998); if luminance of the flashed object is very high and luminance of the moving target is very low, the flash-lag effect reverses (i.e., a flash-lead effect occurs). A flash-lag effect is found when the moving target and flashed object are luminance modulated, equiluminant, or equiluminant-in-luminance-noise, except for the specific combination of a luminance modulated flashed object and an equiluminant or equiluminant-in-luminance-noise moving target (Chappell & Mullen, 2010). Nieman et al. (2006) found a flash-lag effect with random-dot stimuli equated in luminance with the background and distinguishable only by binocular disparity. Although for a given velocity targets with low luminance are perceived as moving faster than are targets with high luminance, and that the flash-lag effect is generally larger with faster velocities, differences in velocity related to overall luminance do not influence the flash-lag effect (Vaziri-Pashkam & Cavanagh, 2011). A flash-lag effect for luminance occurs for a moving target that is changing in luminance and a flashed object presented at the same luminance as the target (Sheth et al., 2000).

Contrast. With linearly moving targets, decreases in contrast between the flashed object and moving target increases the flash-lag effect (Kanai et al., 2004), and consistent with this, forward displacement of a moving target is increased if contrast between the target and background gradually fades (Maus & Nijhawan, 2009). However, with rotating targets, decreases in contrast between the flashed object and moving target result in a smaller flash-lag effect or a flash-lead effect (Arnold, Ong, & Roseboom, 2009). This latter finding is not completely consistent with claims the flash-lag effect results from differential latencies in processing moving targets and flashed objects.

Spatial Frequency. If a square wave grating of a given spatial frequency is briefly presented adjacent to a grating that is increasing or decreasing in spatial frequency, a flash-lag effect for spatial frequency occurs (Sheth et al., 2000). The flash-lag effect is increased as spatial frequency decreases from 1 to 0.25 cycles per degree (Cantor & Schor, 2007). Forward displacement of a blurry target (high spatial frequencies removed) is larger than forward displacement of a focused target (high spatial frequencies intact; Fu, Shen, & Dan, 2001). An increase in the flash-lag effect with loss of higher spatial frequency information does not reflect an increase in noise, as noise would be randomly distributed around the actual location and would dilute any consistent shift in a specific direction.

Target Identity. Most studies of the flash-lag effect presented geometric stimuli such as annuli, bars, and disks. If targets are chimeric stimuli comprised of the top half of one (human) face and the bottom half of another (human) face, and the top half is flashed, then observers who judge the identity of the person in the top half perform better when the bottom half is moving than when the bottom half is stationary; this was attributed to a flash-leg effect disrupting normal perceptual alignment effects in facial recognition (Khurana, Carter, Watanabe, & Nijhawan, 2006). If a single change in moving target size corresponds with presentation of the flashed object, then the flash-lag effect decreases; this has been attributed to failure to update the original object file of the moving target, and so the suddenly differently sized target is perceived as a different object from the original target (Moore & Enns, 2004). As discussed below, semantic meaningfulness regarding identity or object category attributed to the stimulus influences the flash-lag effect.

Duration of the Flashed Object. Surprisingly, many studies of the flash-lag effect do not report duration of the flashed object. Some studies report use of a single duration ranging from 5 (e.g., Shi & Nijhawan, 2008) to 70 (e.g. Moore & Enns, 2004) msec. Other studies report multiple durations ranging from 1–193 msec (Rotman, Brenner, & Smeets, 2005) or 24–200 msec (Cantor & Shor, 2007). Eagleman and Sejnowski (2000c) suggested a flash-lag effect would not occur for flash durations longer than 80 msec, but Krekelberg and Lappe (1999) reported a flash-lag effect with flash durations of 500 msec.

Motion of the Flashed Object. In most studies of the flash-lag effect, the flashed object is stationary. Indeed, Eagleman and Sejnowski (2007) claimed a flash-lag effect would turn into a flash-drag effect if motion were attributed to the flashed object. However, when a flashed object briefly moves along with the moving target, a flash-lag effect can still occur (e.g., Cravo & Baldo, 2008), but decreases with increasing motion of the flashed object (Bachmann & Kalev, 1997; Krekelberg & Lappe, 1999). It is not yet possible to disentangle effects of increased motion of the flashed object from effects of increased duration of the flashed object. A decrease in the flash-lag effect with increased motion of the flashed object might reflect an increase in perceptual processing speed for the flashed object over time (Bachmann, Luiga, Põder, & Kalev, 2003) or an increasing ability to extrapolate position of the flashed object over time. A moving flashed object results in a larger flash-lag effect than does a stationary flashed object for flash-initiated and flash-terminated, but not for flash-midpoint, displays (Gauch & Kerzel, 2008a).

Predictability of the Flashed Object. The validity of a cue presented 100 or 500 msec before the flashed object and indicating whether the flashed object would be on the left or right side of a display does not influence the flash-lag effect (Khurana et al., 2000). However, a valid cue presented 3,000–5,000 msec before the flashed object appeared results in a smaller flash-lag effect than does an invalid cue (Namba & Baldo, 2004), and a cue visible from the beginning of a trial until the flashed object appears decreases the flash-lag effect (Brenner & Smeets, 2000). The flash-lag effect decreases the second time a stimulus is shown (the first time is analogous to a cue; Rotman et al., 2005). Cueing which target in a multi-target display would be nearest the flashed object decreases the flash-lag effect (Shioiri et al., 2010). In general, the flash-lag effect is largest when unpredictability of the flashed object is highest (Baldo, Kihara, et al., 2002; Baldo & Namba, 2002; Vreven & Verghese, 2005).

Presence of Unrelated Stimuli. If an unrelated stimulus moves toward the location where a flashed object will appear, the flash-lag effect is larger than if an unrelated stimulus moves away from the location where the flashed object will appear (Maiche, Budelli, & Gómez-Sena, 2007). Similarly, if an unrelated stimulus moves toward the moving target and is occluded by the target at the time the flashed object is presented, the flash-lag effect increases regardless of whether the unrelated stimulus vanishes after being occluded or emerges from occlusion and continues its original direction of motion (Bachmann, Murd, & Põder, 2012). Bachmann et al. (2012) suggest this finding is inconsistent with many theories of the flash-lag effect; however, Bachmann et al.'s rejection of attention shift theory in particular might be overstated, as presence of an unrelated object (i.e., divided attention) is linked to increases in other spatial biases such as representational momentum (Hayes & Freyd, 2002) and boundary extension (Intraub, Daniels, Horowitz, & Wolfe, 2008).

Characteristics of the Observer

Characteristics of the observer that influence the flash-lag effect include (a) allocation of attention, (b) eye movements and eye fixations, (c) body movements of the observer, (d) perceived control of the moving target or flashed object by the observer, and (e) semantic or other meaningfulness the observer attributes to the stimulus.

Allocation of Attention. The flash-lag effect increases as distance between the moving target and flashed object increases, and this has been attributed to the time required to shift attention from the moving target to the flashed object (Baldo & Klein, 1995). In a multi-target display in which one target is cued, the flash-lag effect increases as distance of the flashed object from the cued target increases (Shioiri et al., 2010). However, embedding the flashed object within the moving target (thus removing the need for a shift of attention across space) also results in a flash-lag effect (Khurana & Nijhawan, 1995). The flash-lag effect increases if observers are engaged in a concurrent task (Sarich, Chappell, & Burgess, 2007; Scocchia, Actis-Grosso, de'Sperati, Stucchi, & Baud-Bovy, 2009). The larger flash-lag effect with random-dot stimuli in Nieman et al. (2006) supports a role of attention: tracking the target required greater attention, thus leaving less attention for detection of the flashed object, resulting in the flashed object requiring even longer to process and enter into awareness.

Eye Movements and Eye Fixation. If observers fixate a central point around which a moving target is revolving, a flash-lag effect occurs, whereas if observers track the revolving target, a flash-lag effect does not occur (Nijhawan, 1997, 2001). A flash-lag effect occurs if the flashed stimulus is presented during smooth anticipatory eye movement (Blohm et al., 2003; cf. Nijhawan, 2001). The location of a flashed object is shifted in the direction of eye movement as a moving target is tracked (Rotman, Brenner, & Smeets, 2002; Rotman et al., 2005) or in the direction of gaze after the flash (Rotman, Brenner, & Smeets, 2004). A flash-lag effect occurs during pursuit motion if a flashed object is above or below the horizontal trajectory of a moving target (van Beers, Wolpert, & Haggard, 2001). If the flash involves an object in a specific location (rather than a background flash), subsequent saccades to the location of the flashed object are not displaced, but subsequent saccades to the location of the moving target at the time of the flashed object are displaced in the direction of motion (Becker et al., 2009).

Body Movement. If observers in a darkened environment focus on an illuminated stationary line and move their heads back and forth, and another illuminated line aligned with the first line is flashed in mid-movement, then a flash-lag effect occurs (Schlag, Cai, Dorfman, Mohempour, & Schlag-Rey, 2000). If observers in a darkened environment sit in a rotating chair and a continuously illuminated line rotates with the chair, a flash-lag effect occurs if a stationary line is briefly presented near the continuously illuminated line (Cai, Jacobson, Baloh, Schlag-Rey, & Schlag, 2000). A flash-lag effect occurs if observers in a darkened environment move their (nonvisible) hands and a stationary visual line is flashed near their hand (Nijhawan & Kirschfeld, 2003), or if a vibration is applied to a finger of a hand moving past a stationary finger on the observer's other hand (a buzz-lag effect; Cellini, Scocchia, & Drewing, 2016). Thus, a flash-lag effect can occur with passive or active motion of one's own body. A flash-lag effect can also occur for judgments of another person's movements relevant to a flashed object (Kessler, Gordon, Cessford, & Lages, 2010).

Control. If observers believe they control target motion, the flash-lag effect decreases (Ichikawa & Masakura, 2006, 2013), but only if mapping between observers' movements (i.e., body movements controlling a computer mouse, which control target motion) and target movements were familiar (Ichikawa & Masakura, 2010). However, the flash-lag effect increases if observers actually control target motion, and this might reflect greater attention required for control of the target, leaving less capacity available for processing information related to the flashed object (Scocchia et al., 2009). If observers control presentation of the flashed object, then the flash-lag effect decreases (López-Moliner & Linares, 2006). This is consistent with the possibility that the flash-lag effect results at least in part from high-level attributions regarding the source of target motion.

Meaningfulness. The flash-lag effect is influenced by Gestalt perceptual grouping imposed on a stimulus, and is larger if the flashed object is nearer the leading edge of a moving target than the trailing edge of that moving target (Watanabe, 2004; Watanabe, Nijhawan, Khurana, & Shimojo, 2001). The flash-lag effect decreases if moving targets or flashed objects are meaningful (e.g., kanji segments for Japanese speakers; Noguchi & Kakigi, 2008). Relatedly, if the moving target is an object with a typical facing direction or direction of motion (e.g., a photograph of an automobile) and the flashed object is a dot presented above the target, then the flash-lag effect is larger with backward motion than with forward motion of the target (Nagai, Suganuma, Nijhawan, Freyd, Miller, & Watanabe, 2010). The complementary pattern can also occur, as the flash-lag effect can influence processing of meaningful stimuli (e.g., faces; Khurana et al., 2006).

Properties of the Flash-Lag Effect

Properties of the flash-lag effect involve whether the flash-lag effect is (a) primarily spatial or temporal, (b) a bias in perception of the moving target or the flashed object, (c) multisensory or crossmodal, (d) primarily low-level or high-level, (e) related to continuity of target behavior, (f) related to other spatial biases associated with a flashed object, and (g) related to spatial biases not associated with a flashed object.

Spatial or Temporal?

Theories based on motion extrapolation emphasize spatial aspects of the flash-lag effect (e.g., Nijhawan, 1994, 2008), whereas theories based on differential latencies emphasize temporal aspects of the flash-lag effect (e.g., Whitney & Cavanagh, 2000). Theories based on postdiction emphasize spatial (e.g., Eagleman & Sejnowski, 2002) or temporal (e.g., Whitney, 2002) aspects of the flash-lag effect. Temporal judgments (i.e., did a moving target change color prior to presentation of a flashed object?) and spatial judgments (i.e., did a moving target change color after passing a flashed object?) can be dissociated (Kreegipuu & Allik, 2004; also Gauch & Kerzel, 2008b). Such dissociations probably involve differences in processing latencies for different types of information (e.g., information about color is perceptually available earlier than information about motion; Arnold, Clifford, & Wenderoth, 2001). Other researchers (e.g., Murakami, 2001a) emphasize that the flash-lag effect involves spatiotemporal correlation. As space cannot be traversed without passing through time (and for a moving object, elapsed time relates to traversed space), it seems reasonable to consider the flash-lag effect as spatial and temporal. Indeed, interaction of spatial information and temporal information is found in other illusions (e.g., tau and kappa effects; Collyer, 1977; Jones & Huang, 1982).

Moving Target or Flashed Object?

The flash-lag effect involves bias regarding relative positions of the moving target and flashed object, but it is not initially clear if this involves bias regarding judgment of absolute location of the moving target and/or bias regarding judgment of absolute location of the flashed object. The name "flash-lag" implies veridical perception of the moving target and distortion (i.e., lagging) in perceived location of the flashed object (cf. Nijhawan, 2001), but some theories of the flash-lag effect posit veridical perception of the flashed object and distortion (displacement) in perceived location of the moving target. Displacement of absolute locations of the moving target and flashed object have been examined in two studies. In one study, a moving target followed an arc, and the flashed object was a dot in front of or behind the target; judgments of absolute locations of the dot and the target at the time the dot was presented revealed forward displacement for the dot and target, with a larger displacement for the target resulting a flash-lag effect (Shi & de'Sperati, 2008). In another study, a briefly presented stationary object was presented above or below the final location of a horizontally moving target, and the stationary object and moving target were displaced forward (Hubbard, 2008); lack of a difference in displacement is consistent with lack of a flash-lag effect in flash-terminated displays. Thus, the flash-lag effect might be a derived or second-order illusion resulting from more basic illusions involving perceived absolute positions of the moving target and flashed object (Hubbard, 2014b).

Multi-Sensory or Crossmodal?

Most research on the flash-lag effect involved visual stimuli, but evidence of auditory, haptic, and crossmodal flash-lag effects have been reported. If the moving target is an auditory frequency sweep or a single auditory frequency moving across space, and the flashed object is a briefly presented tone, an auditory flash-lag effect occurs (Alais & Burr, 2003). If a visual target is moving across space and a localized auditory tone is briefly presented, or if an auditory frequency is moving across space and a localized visual stimulus is briefly presented, then a crossmodal flash-lag effect occurs (Alais & Burr, 2003; but see Hine, White, & Chappell, 2003). If a vibration is applied to a finger of a hand moving past a stationary finger on the observer's other hand, a haptic flash-lag effect occurs (Cellini et al., 2016). If participants in a darkened room move their hand, a brief visual flash aligned with the hand is perceived as lagging behind the hand's position (Nijhawan & Kirschfeld, 2003). Crossmodal information can also decrease the flash-lag effect; if observers view a visual moving target, an auditory tone briefly presented before or after a visual flash decreases or increases, respectively, the flash-lag effect (Stekelenburg & Vroomen, 2005; Vroomen & de Gelder, 2004).

Low Level or High Level?

Several studies suggest high-level processes are involved in the flash-lag effect. The flash-lag effect does not occur prior to perceptual grouping (e.g., Watanabe, 2004; Watanabe et al., 2001), pattern recognition (e.g., Linares & López-Moliner, 2007), or tilt illusion (Arnold, Durant, & Johnston, 2003). The flash-lag effect is influenced by conceptual knowledge (e.g., Nagai et al., 2010; Noguchi & Kakigi, 2008), beliefs regarding control of the target (Ichikawa & Masakura, 2006, 2010), and predictability of the flashed object (Baldo & Namba, 2002; Vreven & Verghese, 2005). A visual flash-lag effect is influenced by presentation of an auditory tone (Vroomen & de Gelder, 2004), and a flash-lag effect occurs if the moving target and flashed object are in different modalities (Alais & Burr, 2003; Nijhawan & Kirschfeld, 2003). A flash-lag effect occurs if the moving target and flashed object are presented to different eyes (Whitney & Cavanagh, 2000b) or in random-dot stereograms (e.g., Harris et al., 2006; Nieman et al., 2006). However, Anstis (2007, 2010) found a flash-lag effect with stimuli based on a chopsticks illusion and a reversed phi illusion, and he suggested the flash-lag effect reflected physical rather than perceived motion. Also, activation patterns in retinal cells appear to extrapolate a moving target's trajectory (Gollisch &

Meister, 2010). It is likely the flash-lag effect involves a combination of low-level and high-level processes or involves top-down influences on low-level processes (e.g., Erlhagen, 2003; Gauch & Kerzel, 2009; Jancke & Erlhagen, 2010).

Continuity of Target Behavior?

A flash-lag effect occurs if the moving target exhibits continuous motion or implied motion involving discrete and distinct presentations of the target (Rizk et al., 2009), although the flash-lag effect is weaker with implied motion. The flash-lag effect decreases if the target unpredictably changes direction near the time of the flash (Arrighi et al., 2005; Murakami, 2001b; Vreven & Verghese, 2005; Whitney et al., 2000). Random changes in velocity decrease the flash-lag effect (Vreven & Verghese, 2005). If target color is constant or exhibits multiple changes, a flash-lag effect occurs, but a single color change corresponding with presentation of the flashed object eliminates the flash-lag effect (Au & Watanabe, 2013). A single change in target size corresponding with presentation of the flashed object decreases the flash-lag effect (Moore & Enns, 2004). In general, if disruption in continuity of the target can be bridged by a simple extrapolation that connects the pre-change and post-change target (e.g., implied motion using discrete and separate stimuli), then perception of a single target undergoing motion (or other change) is preserved, and a flash-lag effect occurs. However, if a disruption in continuity cannot be bridged by such an extrapolation (e.g., a single change in size or color at the moment the flashed object is presented), then the moving target is no longer perceived as a single target undergoing motion, and the flash-lag effect does not occur.

Relationship to Other Flash-Related Spatial Biases?

The flash-lag effect is one of several phenomena related to relative positions of a flashed object (or other transient change) and a moving target. Other examples include the flash-lead, flash-drag, flash-grab, and flash-jump effects.

Flash-Lead. In the flash-lead effect, judged location of a briefly flashed object leads rather than lags judged location of a moving target. Increasing luminance of a flashed object (Öğmen et al., 2004; Purushothaman et al., 1998) or ratio of luminance of the flashed object to luminance of the moving target (Baldo & Caticha, 2005) increases likelihood of a flash-lead effect. Whether absolute intensities or ratio of intensities of the moving target and the flashed object is important for flash-lag and flash-lead effects for stimulus dimensions other than luminance is not known. Additionally, precuing location of the flashed object (Hommuk, Bachmann, & Oja, 2008), and decreasing contrast between the moving target and flashed object (Arnold et al., 2009), can result in a flash-lead effect. If a moving target in a flash-terminated display vanishes at the end of target motion, a flash-lead effect can occur (Roulston, Self, & Zeki, 2006; but see Kanai et al. 2004). A flash-lead effect rather than a flash-lag effect can be observed for the first item in a perceptual stream (Bachmann & Põder, 2001) and when target motion is away from the fovea (Shi & Nijhawan, 2008). Hine et al. (2003) reported a flash-lead effect with an auditory flashed object (a click) and a visual moving target.

Flash-Drag. In the flash-drag effect, memory for the position of a briefly flashed object is shifted in the direction of nearby motion (cf. Shi & de'Sperati, 2008). For example, position of a flashed grating is shifted in the direction of motion of a larger inducer grating (Fukiage, Whitney, & Murakami, 2011). Durant and Johnston (2004) reported displacement of stationary flashed stimuli aligned with a rotating bar in the direction of rotation.

Hubbard (2008) reported displacement of a stationary object aligned with the final location of a horizontally moving target in the direction of motion. Eagleman and Sejnowski (2007) suggested the flash-lag and flash-drag effects worked in opposite directions, and increases in the flash-drag effect diluted the flash-lag effect. Curiously, a flash-drag effect (Kosovicheva et al., 2012), but not a flash-lag effect (Fukiage & Murakami, 2010), influences a tilt aftereffect; however, a flash-drag effect does not influence a position aftereffect (Fukiage & Murakami, 2013). The existence of a flash-drag effect might reflect spreading activation from the target resulting in a shift of the center of activation of the flashed object (in a network representing spatial location) toward the target (cf. Hubbard, 2008). The flash-drag effect often involves a stationary stimulus (e.g., Whitney & Cavanagh, 2000b), but a flash-drag effect could be hypothesized to occur with movement of the flashed object.

Flash-Grab. If a target moves back-and-forth, trajectory of that oscillation appears shorter than it actually is (Sinico, Parovel, Casco, & Anstis 2009). In the flash-grab effect, a flashed object at the end of an oscillating trajectory (and which occludes the target) is perceived to be at the endpoint of the perceived trajectory rather than the endpoint of the physical trajectory (Cavanagh & Anstis, 2013). Although the flash-grab effect seems similar to displacement of a flashed object in the direction of target motion in the flash-drag effect, Cavanagh and Anstis (2013) argue the flash-grab effect has a different temporal profile, requires more attention, and is larger than the flash-drag effect. The only reported instances of the flash-grab effect involve oscillating targets, and so whether the flash-grab effect is limited to such stimuli remains an open question. The flash-grab effect contributes to a tilt aftereffect (Ge, Wang, & He, 2015), might involve feedback signals from higher visual areas to early retinotopic cortex (Zhou, Ge, Wang, Zhang, & He, 2015), and occurs with visual stimuli and auditory stimuli (Krüger et al., 2015).

Flash-Jump. In the flash-jump effect, a moving object undergoes a transient change in some feature such as color or size, and that change is judged to have occurred farther along the trajectory (Cai & Schlag, 2001; Gauch & Kerzel, 2008b). The flash-jump effect has also been referred to as the *feature flash-drag effect* by Eagleman and Sejnowski (2007), as localization of a feature change is displaced (i.e., dragged) in the direction of motion. Cai and Schlag (2001) suggest the flash-jump effect results from asynchronous feature binding, and this is related to the idea of object updating in Moore and Enns (2004); specifically, different features are processed at different rates, and information relevant to different features emerges in perceptual awareness at different times. Alternatively, Eagleman and Sejnowski (2007) suggested the flash-jump effect resulted from general motion-biasing effects.

Relationship to Non-Flash Spatial Biases?

The flash-lag effect is consistent with several non-flash phenomena. These include spatial biases studied primarily in the laboratory (Fröhlich effect, representational momentum, backward referral, anisotropic distortions) as well as real-world applications (flag errors in association football).

Fröhlich Effect. The perceived onset (initial) location of a moving target is often displaced in the direction of target motion, and this is referred to as a *Fröhlich effect* (for review, Kerzel, 2010; Müsseler & Kerzel, Chapter 7 in this volume). Forward displacement in the Fröhlich effect appears similar to forward displacement of the moving target in the flash-lag effect (cf. Kreegipuu & Allik, 2003). The Fröhlich effect (e.g., Müsseler & Aschersleben, 1998) and flash-lag effect (e.g., Wojtach, Sung, Truong, & Purves, 2008) increase with increases in target velocity. The Fröhlich effect (Whitney & Cavanagh, 2000b) and flash-lag effect (Namba & Baldo, 2004; Vreven & Verghese, 2005) are decreased by presentation of a valid cue. The Fröhlich effect (Müsseler & Aschersleben, 1998) and flash-lag effect (Baldo & Klein, 1995) have been suggested to reflect the time required to shift attention. Given such similarities, displacement of a moving target in a flash-initiated display might appear as nothing more than a Fröhlich effect; indeed, the same mechanisms have been suggested to underlie the Fröhlich effect and flash-lag effect (e.g., Eagleman & Sejnowksi, 2007; Kirschfeld & Kammer, 1999; Müsseler, Stork, & Kerzel, 2002). However, the flash-lag effect (Baldo, Kihara, et al., 2002), but not the Fröhlich effect (Müsseler & Aschersleben, 1998), is influenced by eccentricity. When measured on the same set of stimuli, the flash-lag effect is larger than the Fröhlich effect (Chappell, Hine, Acworth, & Hardwick, 2006).

Representational Momentum. The perceived offset (final) location of a moving target is often displaced in the direction of target motion, and this is referred to as *representational momentum* (for review, Hubbard, 2005b, 2014a, Chapter 8 in this volume). Forward displacement in representational momentum appears similar to forward displacement of the moving target in the flash-lag effect (Munger & Owens, 2004; Shi & de'Sperati, 2008). Representational momentum (Hubbard & Bharucha, 1988) and the flash-lag effect (Wojtach et al., 2008) increase with increases in target velocity. Representational momentum (Hubbard, Kumar, & Carp, 2009) and the flash-lag effect (Namba & Baldo, 2004; Vreven & Verghese, 2005) are decreased by presentation of a valid cue. Representational momentum (Reed & Vinson, 1996) and the flash-lag effect (Nagai et al., 2010; Noguchi & Kakigi, 2008) are influenced by conceptual knowledge of the target. Representational momentum (Hubbard, 2013c; Hubbard, Blessum, & Ruppel, 2001) and the flash-lag effect (Ichikawa & Masakura, 2006, 2010) are influenced by attributions regarding the source of target motion. Representational momentum (Jordan & Hunsinger, 2008) and the flash-lag effect (Scocchia et al., 2009) are influenced by whether the observer controls the stimulus. Representational momentum (Hayes & Freyd, 2002) and the flash-lag effect (Sarich et al., 2007) are increased under divided attention. Representational momentum (Hubbard & Ruppel, 1999) and the flash-lag effect (Brenner et al., 2006; Shi & Nijhawan, 2008) increase with motion toward a landmark or fixation. Hubbard (2013b, 2014b) suggested forward displacement of the moving target in representational momentum and in the flash-lag effect are similar, but in the former, displacement of the target is measured relative to the actual position of the target, whereas in the latter, displacement of the target is measured relative to a nearby stationary object.

Backward Referral. Libet and his colleagues had experimental participants make a voluntary finger movement, and those participants noted the time on a clock face at which they decided to initiate movement (Libet, 1985; Libet, Wright, & Gleason, 1982; Libet, Gleason, Wright, & Pearl, 1983). The average reported time of the conscious decision was 200 msec prior to movement, but a readiness potential appeared in participants' EEG 500 msec prior to movement. Libet and colleagues interpreted this as suggesting that a decision to move was implemented before a participant became consciously aware of such a decision, and so the timing of the subjective conscious intent was "adjusted" backward in time to reflect the timing of the unconscious decision (cf. the notion of "postdiction" discussed below). This adjustment of the timing of conscious decision was referred to as *backward referral*. Alternatively, apparent incongruity in conscious and unconscious decision times might result from a flash-lag effect (Klein, 2002; van de Grind, 2002) or representational momentum (Joordens, Spalek, Razmy, & van Duijn, 2004); in these latter cases, represented location of the moving hand on the clock face is displaced forward from its actual position, and so timing of the conscious decision to move appeared more recent than it actually was.

Anisotropic Distortions. In studies of the flash-lag effect with visual stimuli, the flash-lag effect is larger if targets move toward fixation than away from fixation (Shi & Nijhawan, 2008) and if measured at onset of the flashed object than at offset of the flashed object (Bachmann & Kalev, 1997). The flash-lag effect is larger for targets in the left visual field relative to the right visual field and larger for targets in the upper visual field relative to the lower visual field (Kanai et al., 2004). Effects of perceptual grouping on the flash-lag effect (Watanabe, 2004; Watanabe et al., 2001) suggest anisotropic distortions associated with the flash-lag effect; indeed, Watanabe and Yokoi (2006, 2007) suggest the flash-lag effect is a special case of two-dimensional anisotropic distortion. However, it is not clear if anisotropic distortions associated with the flash-lag effect occur with nonvisual or cross-modal stimuli or are limited to visual stimuli. More critically, it is not clear whether anisotropic distortions cause the flash-lag effect or if the flash-lag effect causes anisotropic distortions (cf. Rowland & Durant, 2014). If the former, it might be possible to link other spatial biases to the flash-lag effect through mechanisms or properties of anisotropy; if the later, spatial representation is more flexible and idiosyncratic than often assumed (cf. Tversky, Chapter 16 in this volume).

Judgments of "Offsides". In association football (known as *soccer* in North America), judgment of "offsides" occurs when a player is in the opponent's half of the field, the ball is played to him by a teammate, and he is closer to the goal line at the moment the pass is initiated than both the ball and the second-to-last defender (the last one being the goalkeeper). Referees are more likely to call an offsides penalty when an offsides is not present (referred to as a *flag error*) than to not call an offsides penalty when an offsides is present (i.e., more likely to false alarm than to miss; although see Wühr, Fasold, & Memmert, 2015). Baldo, Ranvaud, and Morya (2002) suggested flag errors resulted from a flash-lag effect: The player is analogous to the moving target, passing a ball is analogous to a transient flashed object, and position of the player is displaced forward relative to position of the ball (see Catteeuw, Gilis, Wagemans, & Helsen, 2010; Giles et al., 2008). Studies of expert referees, amateur referees, and controls suggest that expert referees are more able to compensate (i.e., avoid flag errors) and that compensation involves a shift in decision-making rather than perception (Catteeuw, Helsen, Gilis, van Roie, & Wagemans, 2009; Put et al., 2012). Simulations and computer animations are useful in training referees to avoid flag errors (Catteeuw, Gilis, Jaspers, Wagemans, & Helsen, 2010).

Theories of the Flash-Lag Effect

Several theories of the flash-lag effect have been proposed, but none account for all of the data or are universally accepted. The most well-known theories include (a) motion extrapolation, (b) attention shift, (c) differential latencies, (d) postdiction, and (e) perceptual acceleration.

Motion Extrapolation

The most well-known theory of the flash-lag effect involves motion extrapolation. Nijhawan (1994, 2008) suggested the flash-lag effect occurs because the visual system extrapolates the trajectory of a moving target to compensate for delays in perception due to neural processing times. Without such compensation, perceived position of a moving target would lag behind the actual position of that target. Such an account is similar to representational momentum (see Hubbard, 2005b, 2013b, 2014a, Chapter 8 in this volume), but curiously, representational momentum is not mentioned in discussions of motion extrapolation theory. Motion extrapolation has intuitive appeal; however, several findings initially appear inconsistent with motion extrapolation. A flash-lag effect occurs with random motion (Murakami, 2001b) and unpredictable changes in target direction (Chappell & Hinchy, 2014; Eagleman & Sejnowski, 2000b; Whitney, Cavanagh, & Murakami, 2000) and target velocity (Brenner & Smeets, 2000); random or unpredictable motion would not (by definition) be predictable and able to be extrapolated. Occurrence of a flash-lag effect in flash-initiated displays (when prediction would not be possible) and lack of a flash-lag effect in flash-terminated displays (when prediction should be maximized) are problematic (but see Nijhawan, 2008).

Attention Shift

Attention shift theory suggests attention is initially focused on the moving target, and when the flashed object appears, attention shifts from the moving target to the flashed object. This shift takes time, and during this time, the moving target continues to move. By the time information regarding the flashed object reaches conscious awareness, the moving object has already moved some distance beyond where it was located when the flashed object initially appeared, and so the location of the flashed object is perceived as lagging behind the location of the moving target. An attention shift theory is consistent with effects of target velocity (Nijhawan, 1994), cueing (Namba & Baldo, 2004), and predictability of the flashed object (Baldo & Namba, 2002; Vreven & Verghese, 2005). The attention shift notion was initially proposed by Baldo and Klein (1995), but Baldo and colleagues have subsequently taken a more conservative view that distribution and allocation of attention modulate but probably do not cause the flash-lag effect (Baldo & Klein, 2010; Baldo, Kihara et al., 2002). Findings inconsistent with attention shift theory include occurrence of a flash-lag effect when the flashed object is embedded in (interleaved with) the moving target (e.g., Khurana & Nijhawan 1995) and occurrence of a flash-lag effect in flash-initiated stimuli.

Differential Latencies

Differential latencies theory suggests a moving target is processed (and reaches conscious awareness) more quickly than does a stationary flashed object. If a moving target and flashed object were aligned, information regarding the moving target would reach perceptual awareness more quickly; therefore, when information regarding the flashed object reached perceptual awareness, the moving target would already be perceived as further along its trajectory (e.g., Kafaligönül, Patel, Öğmen, Bedell, & Purushothaman, 2010; Öğmen et al., 2004; Purusothaman et al., 1998; Whitney & Murakami, 1998; Whitney, Murakami, & Cavanagh, 2000). Findings inconsistent with differential latencies theory include temporal order judgments that indicate flashed objects might be processed at the same speed (e.g., Chappell et al., 2006) or more quickly (e.g., Nijhawan et al., 2004; Raiguel, Lagae, Gulyàs, & Orban, 1989) than moving targets, effects of perceptual grouping (Watanabe, 2004; Watanabe et al., 2001), effects of semantic meaningfulness (Nagai et al., 2010; Noguchi & Kakiki, 2008), existence of visual-auditory crossmodal flash-lag (latencies for auditory stimuli are generally faster than latencies for visual stimuli; Alais & Burr, 2003; Arrighi et al., 2005), and occurrence of a flash-lag effect if the flashed object is in motion (Bachmann & Kalev, 1997).

Postdiction

Postdiction theory suggests the flash-lag effect results from information presented after the flashed object vanished (e.g., Eagleman & Sejnowski, 2000a, 2000b; Rao, Eagleman, & Sejnowski, 2001). Information regarding location is integrated over brief intervals of time for the flashed object and for the moving target,[1] and the length of this integration period ranges from 80 msec (Eagleman & Sejnowksi, 2000b) to 500 msec (Krekelberg & Lappe, 1999). For the flashed object, the average location during the integration period is the same as the individual location signals (the object does not move); however, for the moving target, the average location during the integration period beginning with the presentation of the flashed object is shifted forward as the moving target continues in motion throughout the integration period. This averaged information "adjusts" perception of the target forward at the time of the flash (cf. backward referral), and this was initially referred to as *postdiction* (Eagleman & Sejnowksi, 2000b) and subsequently referred to as *motion-biasing* (Eagleman & Sejnowski, 2007). Effects of changes in target velocity or direction of motion after presentation of the flashed object (e.g., Brenner & Smeets, 2000; Eagleman & Sejnowksi, 2000b; Rotman et al., 2005) and lack of a flash-lag effect in flash-terminated displays (e.g., Eagleman & Sejnowski, 2000b; Moore & Enns, 2004; Watanabe, 2004) are consistent with postdiction. However, postdiction theory suggests information presented prior to the flashed object should not influence the flash-lag effect, and so effects of pre-flash target behavior (Chappell & Hine, 2004) and pre-flash cues (Baldo, Kihara et al., 2002; Namba & Baldo, 2004; Vreven & Verghese, 2005) are inconsistent with postdiction theory.

Perceptual Acceleration

Perceptual acceleration theory suggests the moving target and flashed objects are processed in different perceptual streams. When a perceptual stream first appears, the first few items are processed relatively slowly, but processing latency decreases for additional stimuli in that stream; this is referred to as *perceptual acceleration* (Bachmann et al., 2003). If a flashed object is in a second stream that begins later than the stream involving the moving target, then the flashed object is processed more slowly than the moving target, and this would result in a flash-lag effect. This is similar to differential latencies theory, but differential latencies theory specifies that processing speed is a function of the type or intensity of a stimulus and not a function of the number of preceding stimuli. The perceptual acceleration theory is consistent with findings of a flash-lag effect when a moving target consists of a sequence of discrete stimuli (e.g., letters; Bachmann & Põder, 2001). However, although a moving target and a flashed object that were spatially separated might involve separate processing streams, it is less clear whether a flashed object that overlapped or interleaved the moving target (e.g., disk within an annulus) would be considered to involve separate streams. Existence of a flash-lag effect with flash-initiated stimuli, and lack of a flash-lag effect in flash-terminated displays, appears inconsistent with perceptual acceleration (but see Bachmann, 2007, 2010).

---

[1] For an early model of the flash-lag effect that involved temporal integration but did not emphasize postdiction, see Krekelberg (2001, 2003) and Krekelberg and Lappe (1999, 2000a, 2000b, 2001, 2002; but also see Shen, Zhou, Gao, Liang, & Shui, 2007).

Summary and Conclusions

The flash-lag effect is robust and occurs with many types of stimuli. A flash-lag effect occurs if target motion is linear, curvilinear, or rotary, and with continuous, sampled, or random motion. A flash-lag effect occurs with visual motion in the picture or depth plane, auditory motion in frequency or physical space, and crossmodally with visual, auditory, and haptic stimuli. A flash-lag effect occurs if the flashed object overlaps or is spatially separate from the moving target, and the flashed object can be stationary or in motion. A flash-lag effect usually occurs with flash-initiated and flash-midpoint displays but not with flash-terminated displays. The flash-lag effect is increased if (a) distance between the flashed object and moving target increases, (b) eccentricity of the flashed object or moving target increases, (c) the target moves toward fixation, (d) the target moves faster, (e) presentation of the flashed object is less predictable, (f) target position is less certain, (g) observers divide attention over multiple targets or tasks, (h) a nearby stimulus facilitates processing along the path of motion, (i) observers do not control presentation of the flashed object, (j) the flashed object is aligned with the leading rather than the trailing edge of the moving target, (k) the effect is measured at onset of flashed object motion, and (l) features of the target such as size remain constant.

The flash-lag effect exhibits several properties. First, the flash-lag effect involves spatial components and temporal components. Second, the flash-lag effect involves biases regarding perceived location of the moving target and perceived location of the flashed object. The perceived locations of the moving target and flashed object are shifted in the direction of motion, and the flash-lag effect per se presumably results from a larger shift for the moving target than for the flashed object. Third, the flash-lag effect is multisensory and crossmodal, and has been found with visual, auditory, and haptic moving targets and flashed objects. Fourth, the flash-lag effect involves high-level processing. It is possible low-level perceptual factors contribute to the flash-lag effect, but numerous studies found a role of high-level processing. Fifth, the flash-lag effect is dependent on continuity of target behavior, and discontinuities (e.g., sudden or unpredictable change in target size) disrupt the flash-lag effect. Sixth, the flash-lag effect is likely related to other flash-related spatial biases (flash-lead, flash-drag, flash-grab, flash-jump) and spatial biases not related to a flashed object (Fröhlich effect, representational momentum, backwards referral, anisotropies).

Several theories of the flash-lag effect have been proposed. Motion extrapolation theory proposes the flash-lag effect results from forward spatial displacement of the target. Attention shift theory suggests that movement of the target during the time that attention is being shifted from the moving target to the flashed object results in the target being perceived beyond the flashed object. Differential latencies theory suggests moving targets are processed more quickly than are flashed objects. Postdiction theory suggests temporal integration of location signals after the flashed object is presented shifts the represented location of the target at the time the flashed object was presented forward. Perceptual acceleration theory suggests that a flashed object is the initial stimulus in a new perceptual stream and thus is processed more slowly. However, it is unlikely a single mechanism can account for all examples of the flash-lag effect, and examples of the flash-lag effect with different types of stimuli might reflect different or multiple mechanisms (e.g., attention shifts and differential latencies appear to modulate the flash-lag effect). Some examples of the flash-lag effect might reflect representational momentum or a Fröhlich effect in which judged target position is measured relative to a nearby object rather than the target's actual position. There are data that challenge each theory, and these challenges have not been resolved.

In the flash-lag effect, position of a briefly presented object spatially aligned with a moving target is perceived to lag behind the position of that target, and research on the flash-lag effect has underscored important issues in visuospatial and spatiotemporal processing. Foremost among these issues is how our representational system compensates for delays in perception due to neural processing latencies. The answer to this question has consequences for the flash-lag effect and several other spatial biases such as the Fröhlich effect, representational momentum, and other flash-related biases. Whether the flash-lag effect reflects a fundamental bias in perception or is a derived bias based on judgment of target location relative to another object (rather than relative to actual target location) is not clear. Given that moving targets are more likely to have an immediate impact on an observer than are stationary objects, a representational system tuned to preferentially anticipate and (more quickly) process information regarding moving targets than information regarding stationary objects would be more adaptive. This would demonstrate an important way our representational system has been tuned to anticipate what is most likely to be encountered in interactions with environmental stimuli. Despite the simplicity of the flash-lag effect, such an effect might reveal important properties of our representational system and provide insight into how our representational system adapted to allow perception of and interaction with stimuli in our environment.

Chapter 10

Perceptual and Motor Biases in Reference to Gravity

Myrka Zago

Introduction

Perception, action, and imagery often involve the spatial localization of both us and objects relative to the external environment. Spatial localization, in turn, requires making reference to a frame. Because all sensory systems – along with memory – can contribute to localize a target, and because movements can involve the coordination of several body segments (as in orienting the eyes, head, and arm toward the target), a multiplicity of diverse reference frames contribute to sensorimotor transformations and internal representations (Howard, 1982; Lacquaniti, 1997; Soechting & Flanders, 1992).

Not only do the reference frames associated with each sensory channel differ between each other but they are also intrinsically unstable because they are egocentric. Thus, any time we move our eyes, the eye-centered frame shifts and rotates in parallel with the eyes. The same happens with head- and body-centered reference frames. Nevertheless, introspection as well as abundant psychophysical and behavioral evidence shows that our representations of space are generally stable.

It has long been recognized that Earth gravity helps keep multiple egocentric reference frames in spatial register (Berthoz, 2000; Paillard, 1991). One possibility is that the brain transforms initial egocentric signals into allocentric, gravity-centered representations of the world. Gravity-centered representations tend to be stable, being anchored to a physical invariant, independent of the observer's movements. In particular, the gravity reference contributes maintaining postural equilibrium, orienting ourselves, navigating through space, and recognizing objects. Indeed, the lack of a gravity reference during space flight results in cognitive, sensory, and motor derangements that take several days of adaptation to subside (Clément & Reschke, 2008).

In this chapter, I address first the issue of how we study the representation of the gravity reference behaviorally. There is a multiplicity of different tests, and each of them can reveal a different type of representation. I will show that the neural estimates of the direction of gravity and the kinematics of a target descending along the vertical are generally accurate under the most prevalent ecological conditions, but they can be biased significantly under unusual conditions. As a result, perceived gravity does not always afford a veridical representation of physical gravity. Conversely, neural estimates of gravity direction and acceleration may bias our experience of the outside world and affect our actions accordingly. The study of these two different types of bias is important insofar as it can reveal the mechanisms underlying the neural encoding of gravity effects.

Perceived Vertical

Subjective Visual Vertical (SVV). The perceived vertical direction can be assessed using a variety of tests, which do not lead necessarily to the same results. Probably the oldest and most common test is represented by the SVV, introduced by Aubert in 1861. SVV is assessed in a dark environment by asking participants to align a luminous rod to the vertical, in most cases in the fronto-parallel plane (roll tilt). Both constant (average) errors and variable errors (trial-to-trial variability measured as 1 SD) of SVV are smaller than 2° in healthy participants in the upright posture (Mittelstaedt, 1983; Tarnutzer, Bockisch, Straumann, & Olasagasti, 2009; Udo de Haes, 1970; Vingerhoets, De Vrijer, van Gisbergen, & Medendorp, 2009). One study reported small constant errors of SVV in pitch (sagittal plane), only slightly higher than those in roll (Frisén, 2010). On the other hand, both constant and variable errors become appreciable when participants lie in a tilted posture (Figure 10.1; Aubert, 1861; Clemens et al., 2011; Howard, 1982; Kaptein & Van Gisbergen, 2005; Mittelstaedt, 1983; Udo de Haes, 1970; Vingerhoets et al., 2009). When the participant is rolled in the frontal plane by a small angle (<30°), the SVV in roll tends to remain aligned with the true vertical or slightly overshoots it in the direction opposite to the body tilt (so called E-effect). Instead, for greater roll tilts, the SVV undershoots the vertical in the same direction as body tilt (so called A-effect). A-effects gradually increase for absolute roll tilts beyond 60°, and reach peak amplitudes up to 50° at about 130° roll. The SD of SVV increases to about 5–7°. A-effects are also observed in the perceptual estimates of the direction of visual motion of a patch of dots (De Vrijer, Medendorp, & Van Gisbergen, 2008). A-effects cannot be simply accounted for by an underestimation of head and body tilt, because a separate assessment of body tilt perception relative to a reference position (the subjective body tilt or SBT) shows negligible constant errors (Clemens et al., 2011; Kaptein & Van Gisbergen, 2004; Mast & Jarchow, 1996; Mittelstaedt, 1983). Notice, however, that the variable errors are generally greater for the SBT than SVV (Clemens et al., 2011; Mast & Jarchow, 1996).

Both the SVV and SBT have been tested during full immersion in water. The forces acting on the skin are greatly reduced under water because the body is almost buoyant, but the vestibular system still senses the gravitational force. In a carefully controlled study, the pattern of responses for SVV and SBT showed the same dissociation under water as on

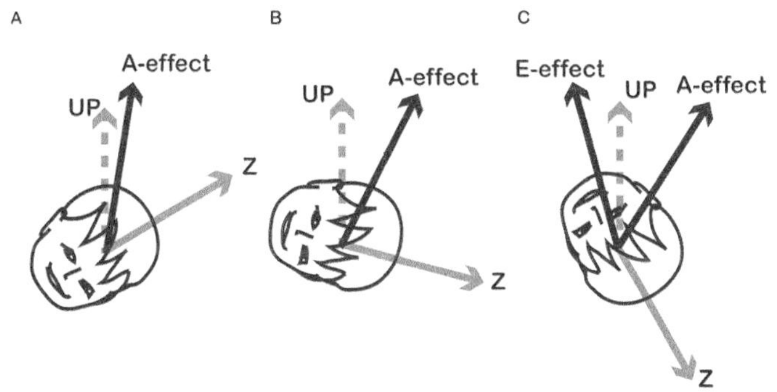

Figure 10.1 Systematic errors in the SVV when participants lie in a roll-tilted posture (60°, 105°, and 150°, from A to C, respectively). *Z* represents the orientation of the body axis. The SVV is aligned with gravity or undershoots the true vertical in the direction opposite to the body tilt (E-effects) for tilt angles smaller than 30°. The SVV overshoots the true vertical in the same direction as body tilt (A-effects) when the tilt angle is greater than 30°; this bias increase with tilt angle. An abrupt transition occurs from an A to an E effect at large tilts (≥135°).

land, but only SBT significantly differed between the two conditions, with a head-upward bias of the perceived body roll (Jarchow & Mast, 1999).

On Earth, the otoliths of the vestibular apparatus are constantly loaded by the pull of gravity. In the freefall conditions of space flight, the otoliths become unloaded. Accordingly, the SVV is typically aligned with the body axis in astronauts (Clément & Reschke, 2008; Glasauer & Mittelstaedt, 1998). Tests performed during parabolic flight showed that the SVV starts to be aligned with gravity levels above about 0.3 g (de Winkel, Clément, Groen, & Werkhoven, 2012).

In general, the biases of SVV under different environmental conditions are well accounted for by the hypothesis that the brain combines visual, vestibular, and somatosensory inputs with a prior assumption about body orientation in space, in particular the assumption that the head and the body are oriented near the vertical (see Multisensory Estimates of Upright Direction). As for the anatomo-functional substrates of SVV, a pathologic tilt of the SVV is consistently observed after lesions of the posterior insular and retroinsular cortex (Baier et al., 2013; Barra et al., 2010; Brandt, Dieterich, & Danek, 1994; Rousseaux, Honoré, Vuilleumier, & Saj, 2013). These regions receive convergent signals from the vestibular, visual, and neck proprioceptive systems and represent a putative human homologue of the parieto-insular vestibular cortex of the monkey (Brandt et al., 1994; Lopez & Blanke, 2011).

Rod-and-Frame Test. The rod-and-frame test, originally developed by Witkin (1949), measures participants' ability to align to vertical a luminescent rod inscribed within a luminescent frame in the dark. Tilting the outer frame may affect the alignment of the rod, determining deviations from the true vertical. Interestingly, there is wide inter-individual variability in the effects of the frame tilt, with larger effects in observers who rely more heavily on visual context for verticality estimates.

Subjective Haptic Vertical (SHV). Head and body tilts generally evoke vestibulo-ocular reflexes. Thus, the eyes counterroll by about 10% of a roll tilt (Miller & Graybiel, 1962). In line of principle, these reflexes might affect orientation judgments based on visual cues such as the SVV, but they would not affect the same judgments based on a manual alignment of the rod (SHV) to the perceived vertical in blindfolded participants. Bortolami, Pierobon, DiZio, & Lackner (2006) used the SHV to test the localization of the vertical for participants tilted in roll, pitch, and yaw, and found much smaller biases than in SVV. Similar results have been reported for roll by Schuler, Bockisch, Straumann, & Tarnutzer (2010). Therefore, a major component of the bias in SVV at tilted positions depends on central processing of visual information.

Perceived Upright of Objects

Oriented Character Recognition Test (OCHART). We recognize and interact more easily with objects when they are perceived as upright. Although the perceived upright of objects can be influenced by body orientation with respect to gravity, it is potentially distinct from the perceived vertical (as assessed by SVV). In the OCHART devised by Dyde, Jenkin, & Harris (2006), participants are asked to report whether they see the letter "d" or the letter "p" when the corresponding symbol is presented at various orientations. Because the letter "p" becomes the letter "d" when it is rotated by 180°, the averaged transitions "p"-to-"d" and "d"-to-"p" define a perceptual upright. Importantly, observers are not asked to make judgments relative to any reference frame, but only to identify the character.

OCHART and SVV were compared in three body postures, upright, right side down or supine, and in the presence or absence of a visual background (Dyde et al., 2006). SVV behaved as in the studies reviewed above. However, OCHART setting was close to the true vertical when the body, gravity, and visual background were aligned, but it markedly deviated when these orientations diverged among each other, the setting now tending to become aligned with the body axis. Moreover, OCHART setting depended on the orientation of the visual background much more than did the SVV.

Recognition of Faces and Scenes. The recognition of scenes, people, and actions tends to be faster and more accurate when they are aligned with the observer, whether both the scene and the observer are upright or they are both tilted relative to the vertical (Chang, Harris, & Troje, 2010; Köhler, 1940; Kushiro, Taga, & Watanabe, 2007; Troje, 2003; Yin, 1969). Thus, the ability to detect gross distortions in a modified photograph of a face is impaired when the photograph is presented upside-down relative to the observer (Lobmaier & Mast, 2007; Thompson, 1980). Similar viewer-centered effects have been described for the discrimination of inverted whole-body postures (Reed, Stone, Bozova, & Tanaka, 2003) or point-light biological motion (Chang et al., 2010; Pavlova & Sokolov, 2000; Sumi, 1984; Troje & Westoff, 2006).

However, the orientation of the scene relative to gravity also contributes. Thus, observers are sensitive to the artificial inversion of the visual effects of gravity on the motion of inanimate objects (Bingham, Schmidt, & Rosenblum, 1995; Indovina et al., 2005) or people (Zago, La Scaleia, Miller, & Lacquaniti, 2011). Moreover, the orientation relative to gravity contributes to processing global and local motion cues of point-light walker stimuli, presumably in connection with expectations about the dynamics of the body due to gravity (Chang & Troje, 2009; Shipley, 2003; Troje & Westhoff, 2006).

Conversely, a photograph or video-clip with marked "up" and "down" polarization, when rotated, can bias the perceived vertical direction (Dyde et al., 2006; Jenkin, Dyde, Jenkin, Zacher, & Harris, 2011; Jenkin, Jenkin, Dyde, & Harris, 2004). Preuss, Harris, & Mast (2013) used the York Tumbling Room to modulate the visual reference frame and the perceived orientation of the body relative to gravity. The York Tumbling Room is a full-size furnished cubic room with strong directional cues, which can be rotated around an Earth-horizontal axis independently of rotations of the visual stimuli and observers, thus disassociating various sources of information about verticality. Preuss and colleagues found that visual scenes depicting a human body with a clear visual polarity (as shown by the relative positions of the head and feet) were more accurately recognized when they were aligned with the tumbling room.

Overall, it appears that the brain does not use a unique representation of upright for the recognition of all categories of objects. Instead, a specific upright can be defined for different object categories.

Perceptual Judgments of Object Stability

When we have to predict whether an object is stable on its base of support or is going to tip over, we must implicitly determine if the vertical projection of the center of mass falls within the support surface.

To assess perceptual judgments of stability, Barnett-Cowan, Fleming, Singh, & Bülthoff (2011) presented images of objects close to a table edge to participants upright or lying on their side, and estimated the critical angle at which each object appeared equally likely to fall or right itself. Perceived gravity direction was independently tested as SVV. It was found that the perceived critical angle was significantly biased in the same direction as the SVV, consistent with a similar multisensory fusion of cues. Somewhat similar results were found by Lopez, Bachofner, Mercier, & Blanke (2009), who investigated perceptual judgments of the tendency to fall of pictures of a human figurine with implied motion that was tilted in the roll plane.

Battaglia, Hamrick, & Tenenbaum (2013) also assessed perceptual judgments of stability in upright observers, but used much more complex objects such as towers of 10 blocks arranged in randomly stacked configurations presented on a monitor. They found that the average accuracy was 66%, but interestingly the performance was well predicted by a model that uses approximate, probabilistic simulations of mechanics to make fast inferences in complex natural scenes where crucial information is missing (see discussion later in the chapter).

Multisensory Estimates of Upright Direction

As reviewed earlier, there is a potential multiplicity of different perceptual estimates of verticality depending on the specific test that is used. This multiplicity depends on the fact that the direction of gravity is assessed by the brain using not one but a multiplicity of sensory cues (visual, vestibular, somatosensory, visceral) combined with prior assumptions (see Lacquaniti et al., 2014, 2015). The major graviceptors (i.e., receptors sensitive to gravity effects) belong to the vestibular system. The otolith receptors in the maculae of the sacculus and utricle respond to a tilt of the head relative to gravity, but under dynamic conditions they do not distinguish between gravitational and inertial components (Fernandez & Goldberg, 1976). Thus, the otoliths by themselves are unable to distinguish whether the head is accelerated backward or tilted forward. However, if the inertial accelerations are short lasting, the net gravito-inertial accelerations can be disambiguated by the brain using one of two different strategies. First, the incoming otolith signals can be filtered, with the result that the low-frequency (longer-lasting) signals are interpreted as a change in the tilt angle of the head relative to gravity, whereas the high-frequency (shorter-lasting) signals are interpreted as due to an accelerated translation (Mayne, 1974). A long-lasting acceleration is a sporadic event, and when it occurs, it can give rise to a misperception of body orientation, being interpreted as a tilt relative to gravity (see the section on Spatial Disorientation).

The second strategy to disambiguate gravito-inertial accelerations consists of combining the otolith signals with those of the semicircular canals (Angelaki, McHenry, Dickman, Newlands, & Hess, 1999; Merfeld, Zupan, & Peterka, 1999). The angular head velocity signaled by the latter allows keeping track of dynamic changes in orientation of the gravity vector relative to the head. Neural correlates of the disambiguation of gravito-inertial accelerations have recently been uncovered. Thus, Purkinje cells in the caudal cerebellar vermis can estimate head tilt (Laurens, Meng, & Angelaki, 2013), and conversely neurons in the vestibular nuclei, as well as in the rostral portion of cerebellar fastigium and nodulus, are translation-selective (Angelaki, Shaikh, Green, & Dickman, 2004).

Graviceptors also are embedded in thoraco-abdominal viscera (such as the kidneys and vena cava) and provide inputs to the vestibular nuclei (where they are integrated with those from otoliths and semicircular canals; Jian, Shintani, Emanuel, & Yates, 2002). Visceral graviceptors contribute significantly to the perception of the direction of gravity, along with somatosensory afferents such as those from neck and ankle joints (Mittelstaedt, 1996).

Also vision, though not directly affected by gravity, contributes several kinds of cues to spatial orientation in naturalistic environments (Harris, Jenkin, Dyde, & Jenkin, 2011). Most visual images of our gravity-bound natural or man-made environments are anisotropic, with more image structure at orientations parallel or orthogonal to the direction of gravity in a fronto-parallel plane (Hansen & Essock, 2004). These image anisotropies are matched by corresponding anisotropies in perceptual responses, so that the direction of lines and motions is discriminated better when they are oriented vertically or horizontally than when they are oriented obliquely (Appelle, 1972; Ball & Sekuler, 1987).

The typical orientation of familiar objects (such as walls, trees, people, chandeliers, chairs), the assumption that light comes from above, the spatial relationship between static objects and their support, the direction of falling objects, all consistently point to the vertical direction of gravity effects. The visual reference to gravity is so robust that there exist attraction places ("Mystery spots") where some peculiarity of the environment (such as strong slopes of the terrain or slanted trees) provides the illusion that the gravity law is violated. Similar illusions are obtained in houses intentionally built with tilted walls.

Sensory systems provide information about the environment in egocentric reference frames, such as those centered on the eyes, head, or body. Egocentric representations are inherently unstable because they tend to shift and rotate any time the corresponding body part moves. Abundant psychophysical and behavioral evidence shows that the brain transforms initial egocentric signals into allocentric, gravity-centered representations of the world. These by definition are stable, being independent of the observer's movements. However, there is still little experimental data where gravity-centered representations are localized in the brain and how they operate. A recent study showed that neurons in the caudal intraparietal area of the monkey cerebral cortex encode visual object tilt relative to gravity direction (Rosenberg & Angelaki, 2014), but similar representations may also exist at earlier visual stages (Sauvan & Peterhans, 1999).

How are different sensory cues combined together to yield an estimate of orientation? Mittelstaedt (1983) proposed that the perceived vertical results from a weighted vector sum of the different orientation cues, the weights depending on the relative reliability of each cue and the specific task. Applying this model to their experimental results, Dyde et al. (2006) found that the influence of a visual background in the SVV test is limited, contributing about 8% of the information, compared to 77% from gravity, and 15% from the long axis of the observer's body. By contrast, vision contributes about 25% of the information for the perceptual upright in the OCHART, compared to 20% from gravity and 55% from the body. Notice that another way to assess the relative role of visual context in the estimate of verticality is provided by the rod-and-frame test mentioned in a previous section. As previously remarked, this test shows a wide inter-individual variability in the relative weight of visual context cues.

Statistically optimal Bayesian models also predict SVV responses (Clemens et al., 2011; De Vrijer et al., 2008; MacNeilage, Banks, Berger, & Bulthoff, 2007). The inputs are represented by the head orientation in space and the visual orientation of the luminous bar relative to the retina. The visual signals are precise, whereas the head tilt signals are noisy. However, head tilt signals are combined with a prior assumption that the head and the body are typically oriented near the vertical, resulting in a statistically optimal estimate. Both the vectorial model and the Bayesian model account for the accurate perception of verticality in the default upright posture, as well as for the spatial biases observed in less ecological tilted postures.

A probabilistic model has also been proposed by Battaglia et al. (2013) to account for subjective judgments of stability under the perturbing influence of gravity of virtual towers of stacked blocks of bricks (see discussion that follows). The conceptual architecture of their model is very similar to that of computer graphics engines that are now used in realistic videogames.

Spatial Disorientation (SpDi)

Spatial disorientation corresponds to a strongly biased perception or complete inability to determine one's true position and orientation in space. Here we are concerned only with SpDi relative to gravity. SpDi may result from underwater immersion (Cheung, Hofer, Brooks, & Gibbs, 2000), lack of gravity effects during space flight (Lackner & DiZio, 2000; Oman, 2007), or neurological disorders (Monacelli et al., 2003). For brevity and because of its relevance to the problems of multisensory fusion involved in the gravity reference, I will consider only SpDi in aviation. This is defined by Benson (1988, p. 277) as the situation in which "the pilot fails to sense correctly the position, motion, or attitude of his aircraft or of himself within the fixed coordinate system provided by the surface of the Earth and the gravitational vertical." According to Gibb, Ercoline, & Scharff (2011), SpDi contributes to at least 30% of all aircraft accidents and causes the highest number of fatalities.

It is important to note the multisensory contributions to SpDi, vestibular misperceptions being the most studied though not necessarily the most frequent causes of SpDi. Vestibular misperceptions are worsened by low visibility and lack of reliance on instruments. During takeoff, pilots may overestimate their upward pitch, because the resultant of the vector sum of gravity and backward inertial acceleration is misperceived as the actual orientation relative to the vertical. To compensate, pilots may lower the plane nose and crash on the ground. Pilots in a constant bank-angle turn feel that their body posture is upright instead of being tilted relative to gravity. When turning gradually, they may feel as though they are flying straight but ascending; when the turn is corrected, the sensation is that of descending. The so-called "graveyard spin" is an illusion occurring while entering a spin. Pilots may feel stationary if the spin is continued long enough; when they correct the spin, they feel spinning in the opposite direction, and start countering their own corrective actions, thus returning in the original spinning pattern. If enough altitude is lost before the illusion is recognized and corrective actions are not taken, the plane will crash on the ground. The "graveyard spiral" occurs when the sensation of turning is lost in a banked turn. Because the pilots' instruments show that they are descending, they may add power entering into spiral motion. In the oculogyral illusion, a turning target seen by pilots while they are turning appears to move faster than it is actually moving. In the oculogravic illusion, a watched target appears to rise if the pilot is momentarily in weightlessness, and appears to fall if the pilot is in hypergravity.

Memory of Moving Targets

When observers are asked to indicate the last seen position of a moving target that has just disappeared, they make small but systematic errors forward in the direction of motion. The spatial bias is consistent with the idea that the brain has internalized the basic property of physical momentum (representational momentum), namely that an object in motion will continue to move unless halted by an external force (Freyd & Finke, 1984). This idea has been extended to include the effects of internalized gravity (representational gravity) and friction (representational friction; Hubbard, 1995b, 2005b). Here we are concerned with representational gravity (for the general issue, see Chapter 8 in this volume).

Hubbard and Bharucha (1988) showed that a vanishing target that previously moved horizontally results in a forward and downward memory displacement. Such spatial bias is consistent with the fact that objects traveling horizontally on a support, once without support, will fall under gravity forward and downward along a parabola. De sá Teixeira et al. (2013) reassessed this issue by introducing variable temporal delays between the target disappearance and the observer's response in order to track the changes in the remembered location over time. They found that the remembered vanishing positions evolved with time in a pattern that indeed roughly mimics a parabolic trajectory. However, this representational trajectory does not correspond to a full extrapolation of actual target motion over the delay interval, but to a much slower drift of the last seen location. Moreover, the extrapolated trajectories are highly distorted copies of actual target motion.

Several other findings argue for representational gravity. Thus, observers in an upright posture typically exhibit a larger memory displacement for a downward motion than for an upward motion of a stimulus (Hubbard & Bharucha, 1988), and memory for the location of a stationary object implying motion is shifted downward (Bertamini, 1993; Freyd, Pantzer, & Cheng, 1988; Hubbard & Ruppel, 2000). Hubbard (2001) found that forward displacement was larger for ascending or descending targets that were lower in the picture plane than for ascending or descending targets that were higher in the picture plane. This is consistent with the belief that descending objects accelerate due to implied gravity and ascending objects decelerate due to implied gravity. Observers lying on their side show memory displacements that reflect both gravicentric and bodycentric effects (de sá Teixeira & Hecht, 2014). Accordingly, the latter authors proposed that the dynamic representation of gravity effects for the task of locating memorized targets involves a weighted combination of cues related to the direction of physical gravity and cues related to the direction of the long body axis, much in the same way as it occurs for SVV (but see Nagai, Kazai, & Yagi, 2002 for a different conclusion).

It should be noticed that representational gravity does not reflect Newtonian physics faithfully. As mentioned earlier, targets moving downward produce a larger memory displacement than do targets moving upward (Hubbard & Bharucha, 1988). Moreover, Hubbard (1997) found that larger targets (perceived as more massive) induce a greater downward displacement than smaller (lighter) targets in upright observers. Kozhevnikov and Hegarty (2001) reported that for ascending motion, larger targets are perceived as slower than smaller ones. Altogether, these findings are more consistent with naïve physics than Newtonian physics. In particular, according to the impetus beliefs of naïve physics, falling objects that accelerate under gravity should move faster than the same objects that decelerate while ascending. Moreover, heavier objects should move faster on the way down and slower on the way up than lighter objects. In contrast, Newtonian physics predicts that all objects, regardless of mass and shape, fall and rise at the same rate in the absence of air drag. In the presence of significant air drag (as for light, large objects), the descent phase takes longer than the ascent phase of the same length; also, more massive objects ascend faster than less massive objects of the same shape. On the whole, representational gravity – just as representational momentum – is a complex phenomenon, with several contributing factors, ranging from sensorimotor cues (e.g., eye movements, visual and vestibular signals) to higher-level cognitive cues (e.g., naïve physics, attention, psychological traits).

Target Interception

Considerable evidence has been accumulated that the interception of a descending target includes prior knowledge of gravity effects (de la Malla & López-Moliner, 2015; Lacquaniti & Maioli, 1989; McIntyre, Zago, Berthoz, & Lacquaniti, 2001; Vishton, Reardon, & Stevens,

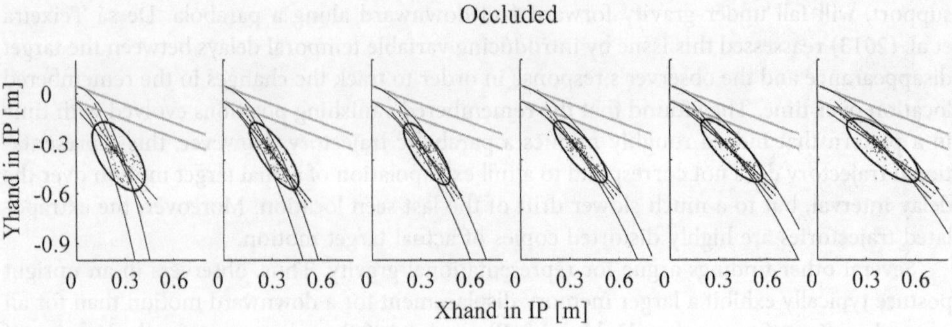

Figure 10.2 Intersection points, for all targets, between the ball surface and the tip of the hand-held rod. Ellipses representing the frontal projections of the tolerance ellipsoids including 95% of the intersection points, centered on the mean intersection point, are drawn along with the first eigenvector (and its 95% confidence cone). Gray continuous lines represent the envelope of ball trajectory, whereas gray dotted lines represent the trajectory that would be followed by the ball if it continued moving on an extended inclined plane. (modified from La Scaleia et al., 2015)

2010; Zago & Lacquaniti, 2005; Zago et al., 2004, 2005; Zago et al., 2009). Under ordinary conditions, participants can accurately place their hand at the right time and the right position to catch or hit a ball falling under gravity along either the vertical (Lacquaniti & Maioli, 1989; Zago et al., 2004, 2005) or a parabolic trajectory (Bosco et al., 2012; Cesqui, d'Avella, Portone, & Lacquaniti, 2012; La Scaleia, Lacquaniti, & Zago, 2014; La Scaleia, Zago, & Lacquaniti, 2015, 2015).

In a recent study (La Scaleia et al., 2015), participants were asked to hit a ball that rolled down an incline at 0.2 g, and then fell in air at 1 g along a parabola. By varying starting position, ball velocity and trajectory differed between trials. Motion on the incline was always visible, while parabolic motion was either visible or occluded. Strikingly, participants were equally successful at hitting the falling ball in both visible and occluded conditions. Moreover, in different trials, the intersection points were distributed along the parabolic trajectories of the ball, indicating that participants were able to extrapolate an extended segment of the target trajectory (Figure 10.2). Remarkably, this trend was observed even in the very first trials. The only way to extrapolate ball motion correctly during the occlusion was to assume that the ball would fall under gravity and air drag when hidden from view. Such an assumption had to be derived from prior experience.

In contrast with the accurate performance observed under ecological conditions, systematic errors in both time and space are made when the target descends with a kinematics violating gravity under unusual or nonecological conditions (Bosco et al., 2012; Indovina et al., 2005; McIntyre et al., 2001; Zago et al., 2004, 2005). Thus, astronauts continue to anticipate the effects of Earth gravity on a ball that is projected at constant speed from the ceiling of the space shuttle, with the result that they move their arm too early, then stop and reverse direction (McIntyre et al., 2001). Artificial gravity conditions can be simulated in the laboratory. Thus, in a baseball videogame, participants shifted a cursor to intercept a landing fly ball (Bosco et al., 2012). Ball trajectories were either fully visible or occluded for variable epochs before landing. During descent, the ball could move under natural gravity (1 g), zero gravity (0 g), or hypergravity (2 g) in different, randomized trials. In occluded trials, 500 msec of perturbed motion was always visible before ball disappearance. The effect of air drag was included in all video animations. It was found that participants placed the

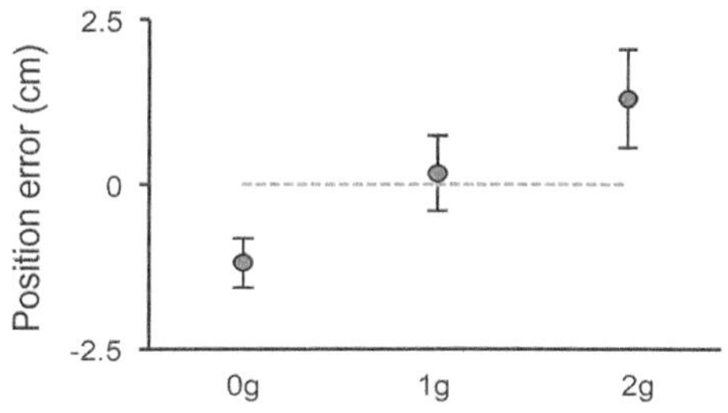

Figure 10.3 Distributions of positional errors in the occluded trials (average values among subjects ± SEM) for each ball acceleration. Positional error was the difference in centimeters between the horizontal position of the cursor at the time of the button press and the landing ball position. Negative values indicated horizontal underestimate of the landing position of the ball, while positive values indicated overestimate. (modified from Bosco et al., 2015)

cursor at the ball landing position for all gravity levels in the fully visible trials but only for natural gravity in the occluded trials. Instead, the landing position was systematically undershot with 0 g balls and overshot with 2 g balls in the occluded trials (Figure 10.3). These spatial biases are explained by the hypothesis that observers use a priori expectation that a fly ball will always fall under natural gravity. Applying this a priori by default results in undershooting an artificial 0 g ball and overshooting a 2 g ball. The same systematic errors occur in the time domain: participants try to intercept 0 g balls too early and 2 g balls too late (Bosco et al., 2012). Algorithms combining online visual information about target position and speed with a prior model of 1 g acceleration are able to explain both the accurate performance under normal gravity as well as the errors under unusual gravity conditions (Bosco et al., 2015; de la Malla & López-Moliner, 2015; McIntyre et al., 2001; Zago & Lacquaniti, 2005; Zago et al., 2004, 2005).

As in the case of representational gravity for memorized targets, also the internal model of gravity effects for target interception involves an approximation of physical laws under unusual nonecological conditions (Zago et al., 2008). Evidence for a naïve physics bias was provided by asking participants to intercept an approaching ball in a virtual reality setup (Senot, Zago, Lacquaniti, & McIntyre, 2005). Sitting participants either pitched their head backward so as to look up toward the ball falling from a ceiling, or they pitched their head forward so as to look down toward the ball rising from a floor. The ball could move at −1 g, 0 g, or 1 g in different, randomized trials, for both downward and upward trajectories. Participants responded earlier when the ball was launched downward from above than when it was launched upward from below, irrespective of ball acceleration profile. This finding is reminiscent of the observation in representational gravity studies that targets moving downward produce a larger memory displacement than do targets moving upward (Hubbard & Bharucha, 1988), and is consistent with the naïve expectation that objects move faster when they fall than when they rise.

Senot et al. (2005) further showed that the response bias related to whether the target was projected from the visual ceiling or from the visual floor disappeared when participants looked straight ahead, indicating that the reference frame used to plan the interceptive responses is anchored to physical gravity. Evidence for a direct contribution of otolith signals in these interception responses was provided by a parabolic flight study involving the same virtual reality setup as above (Senot et al., 2012). The downward/upward bias of the responses reversed sign during the weightless phases compared with the responses on the ground. This reversal can be attributed to a corresponding reversal of the otolith responses during the transition from hypergravity to hypogravity in parabolic flight.

In sum, the interception of moving targets potentially affected by gravity is driven by a complex process of integration of visual, vestibular, and somesthesic cues with an internal model of gravity effects. Although the internal model of gravity revealed by the interception studies shares several features with the representational gravity of spatial memory task reviewed earlier, they are not necessarily the same thing. In general, motor actions directed to a target accelerated by gravity under normal, ecological conditions are quite accurate and reflect true physics, and only under unusual or nonecological circumstances (e.g., weightlessness conditions of space flight or virtual reality simulations of motions violating gravity constraints) do motor actions become inaccurate, the inaccuracy reflecting the presence of ecological biases. Thus, the anticipatory timing of interception of different balls falling from a given height is constant, independently of ball mass (Lacquaniti & Maioli, 1989). By contrast, in spatial memory tasks, larger targets perceived as more massive induce a greater downward displacement than smaller, lighter targets, a result inconsistent with physics (Hubbard, 1997). Also, balls rolling down an incline and then falling along a parabolic trajectory are always intercepted correctly from the very first trials, whether the parabola is visible or completely obscured (La Scaleia et al., 2015). Moreover, in different trials the intersection points are distributed along the parabolic trajectories, indicating that participants are able to extrapolate the trajectory quite accurately. By contrast, target extrapolation in spatial memory tasks derived by introducing variable temporal delays between the target disappearance and the observer's response are highly distorted copies of actual target motion (de sá Teixeira et al., 2013).

Conclusions

I reviewed several lines of research indicating that both the accurate estimates as well as the presence of significant biases in the spatiotemporal localization of targets relative to gravity result from the integration of multisensory information with prior assumptions about the environment derived from past experience. Prior assumptions are necessary in the multisensory fusion because sensory signals are noisy and ambiguous. Moreover, the process of multisensory fusion appears to be governed by probabilistic rules. There is still no consensus on the nature of the prior assumptions underlying the spatiotemporal estimates. According to hard versions of constructivism, internal representations mirror the actual physical principles that govern the environment (Shepard, 1994). Thus, neural anticipation of a gravitational motion would be based on exact Newton's laws. Indeed, when people are tested under ecological contexts, their automatic motor responses are often in accord with physics. On the other hand, cognitive judgments often are inconsistent with Newtonian mechanics, and people's intuitive physics may be based on heuristics rather than realistic models of physics (McCloskey, 1983a).

Acknowledgments

The work was supported by the Italian Space Agency (SLINK grant ASI-2013-038-I.0, and ARIANNA grant ASI-2014-008-R.0). I thank Francesco Lacquaniti for helpful input, comments and suggestions.

Chapter 11

Auditory Biases in Visual Motion Perception

Wataru Teramoto, Souta Hidaka, and Yoichi Sugita

Introduction

Detecting and localizing moving objects is essential to avoiding obstacles and acting appropriately in response to these objects. To do this efficiently, humans use multiple sensory signals from the surrounding environment. Let us take moving cars as an example. Not only do they provide visual information (i.e., moving car-bodies), but also generate dynamically changing sounds. Sometimes one or both of these signals may be embedded in noise originating from the surrounding environment, such as someone's voice or other objects in the proximity. In such cases, using information from multiple sensory modalities is a useful and effective way to compensate for the signal reduction in either modality, and to disambiguate the signals of a moving car.

Studies in the last few decades have increasingly demonstrated that the brain integrates information from different modalities in various aspects of perception, cognition, and action. While classic studies have demonstrated visual dominance in a variety of perceptual phenomena, recent findings have consistently shown that visual dominance does not always occur, even in spatial and spatiotemporal perception (see also other reviews, e.g., Alais, Newell, & Mamassian, 2010; Chen & Vroomen, 2013; Hidaka, Teramoto, & Sugita, 2015; Shams & Kim, 2010; Shimojo & Shams, 2001). For example, Sekuler, Sekuler, and Lau (1997) reported a "stream-bounce illusion." In this illusion, two identical objects move toward each other, coincide, and move apart. The visual stimuli are equally likely to be perceived as overlapping and then continuing to move along their trajectory (i.e., streaming) or as bouncing off each other and then reversing direction (i.e., bouncing). However, playing a transient sound at the time of overlap induces perception of bouncing more frequently than perception of streaming. This suggests that auditory information can affect spatial aspects of visual motion processing. In this chapter, we discuss crossmodal, spatiotemporal biases observed in the large literature of multisensory interaction/integration, with a particular focus on the effects of auditory cues on visual motion perception.

Effects of Auditory Temporal Cues on Visual Motion Perception

Visual dominance over other sensory modalities in spatial tasks has been well known. A sound is often perceived as occurring at the position of a visual event, even when the sound and visual events actually occur at discrepant positions (ventriloquism effect; Howard & Templeton, 1966). The sound /ba/ can be perceived as /da/ or /ga/ when it is presented with a video of lips articulating /ga/ (McGurk effect; McGurk & MacDonald, 1976). In contrast, in temporal processing and tasks, auditory information is often dominant over visual information. While perceived flicker rate of a light while listening to fluttering sounds can be altered by the rate of fluttering sounds, visual flicker has little effect on the perceived rate of sound fluttering (auditory driving; Gebhard & Mowbray, 1959; Shipley, 1964; Welch, DutionHurt, & Warren, 1986). A single flash can be perceived as two flashes when two sounds are presented simultaneously with the flash. This is known as the double flash illusion (Shams, Kamitani, & Shimojo, 2000; see also Courtney, Motes, & Hubbard, 2007). In another example, temporal order judgments of two visual flashes improve when two sounds are presented before and after the flashes, but are impaired when the sounds are interspersed between the flashes (Morein-Zamir, Soto-Faraco, & Kingstone, 2003; Scheier, Nijhawan, & Shimojo, 1999; see also Fendrich & Corballis, 2001). These findings indicate that the perceived temporal position of visual stimuli is captured by sounds.

This superiority of audition over vision in temporal processing has also been observed in phenomena associated with visual motion perception. For instance, Vroomen and de Gelder (2004) demonstrated an effect of audition on visual motion perception using a flash-lag display, where a flash is perceived to be misaligned with respect to a moving visual object, even if the flash and object are physically presented at the same location (flash-lag effect, FLE; Nijhawan, 1994; see also Hubbard, 2014b, Chapter 9 in this volume). In their experiment, a brief tone (16.7 msec) was presented in conjunction with a flash at various lags. The results showed that the magnitude of the FLE was reduced when the sound was presented before the flash, and increased when the sound was presented after the flash. Heron, Whitaker, & McGraw (2004) demonstrated that the timing at which a brief sound was delivered was able to influence the bounce position of a horizontally moving visual object against a vertical virtual surface. Compared with a no-sound condition, when the sound occurred prior to the actual bounce, the bounce position appeared to be displaced in the direction opposite to the preceding visual motion (i.e., before the actual bounce position). These findings suggest that sounds can attract (and sharpen) visual event onset timing (flash or bounce) and, consequently, modulate perceived spatial position, measured as the flash-lag effect magnitude (Vroomen & de Gelder, 2004) or the apparent visual stimulus bounce position (Heron et al., 2004).

The effects of auditory temporal cues on visual motion perception have also been investigated using apparent motion displays. When two spatially discrete visual stimuli are presented in alternation, observers perceive motion between the visual stimuli. Crucial factors for the occurrence of apparent motion are spatiotemporal distance between the visual stimuli and stimulus duration. When spatial distance and duration are held constant, temporal distance is an important factor that affects motion perception. Getzmann (2007) investigated whether short sounds presented proximally in time to visual stimuli influenced the quality of apparent motion perception. Two visual stimuli were presented in rapid succession with various onset intervals. Short sounds were presented either 75 msec before and after the first and second visual stimuli, at the same time as each visual stimulus, or in the interval between them. Participants were asked to classify their impression of motion. Presenting sounds in the interval between the visual stimuli facilitated the impression of continuous motion. In contrast, presenting sounds before and after or at the same time as each visual stimulus disrupted continuous motion perception. Freeman and Driver (2008) presented two visual stimuli separated by 14° in alternation for several seconds to produce apparent visual horizontal motion. The stimulus onset asynchrony (SOA) and spatial position of the visual stimuli was kept constant, while the timing of concurrently presented auditory beeps varied. Total duration of right-left and left-right presentation was kept

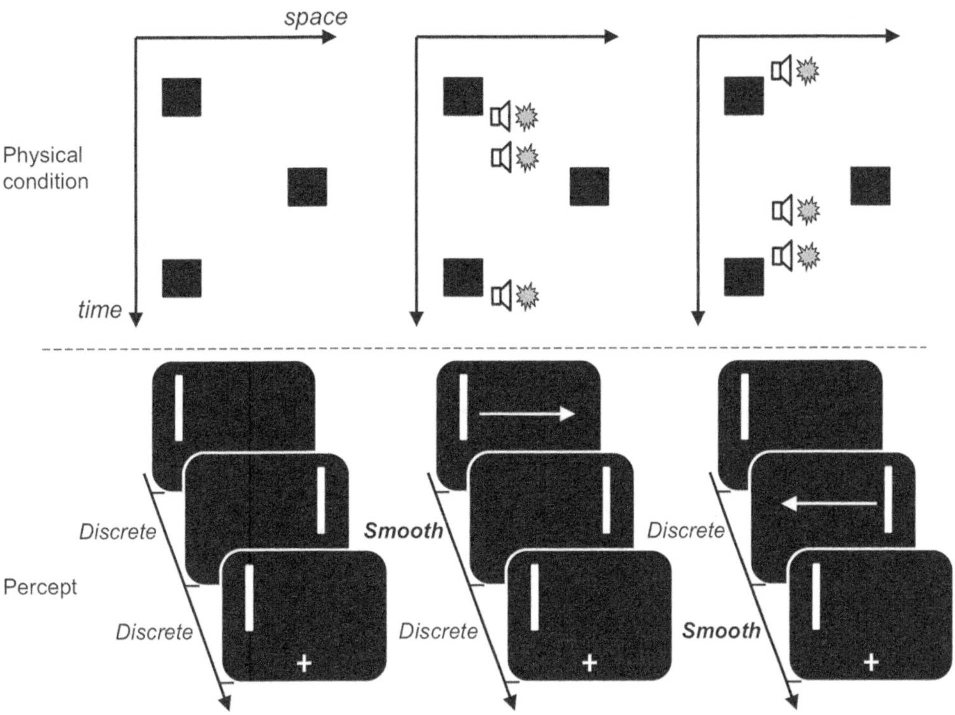

Figure 11.1 Schematic illustration of spatiotemporal positions of audiovisual stimuli in Freeman and Driver (2008). A bar was presented in alternation in the upper visual field. In the absence of any sounds, two discrete visual stimuli were perceived as "discrete," consistent with the physical condition (left panel). In contrast, accompanying sounds changed the percept of these visual stimuli. When tone bursts were presented immediately after the left and before the right visual stimulus, smooth rightward apparent motion was usually perceived (middle panel). In contrast, when tone bursts were presented immediately after the right and before the left visual stimulus, smooth leftward apparent motion was usually perceived (right panel).

constant at 1.0 sec. Leftward motion perception should be dominant over rightward motion perception when the SOA is shorter for right-left than left-right stimulus presentation. Conversely, rightward motion perception should be dominant over leftward motion perception when the SOA is shorter for left-right than right-left presentation. The results showed that leftward motion perception was dominant when the first and second beeps lagged behind the right stimulus and preceded the left stimulus, respectively. Similarly, rightward motion was dominant over leftward motion when the SOA relationship was reversed (Figure 11.1). Moreover, prolonged presentation of such audiovisual stimuli induced motion aftereffects in the opposite direction of the previously perceived visual motion. Thus, these studies demonstrate that auditory information can change the perceived temporal position of visual stimuli in an apparent motion display, altering the perceptual quality and direction of visual motion.

While the aforementioned studies focused on perceptual changes in visual onset timing mediated by auditory stimuli, there are several studies demonstrating that auditory information can also alter the perceived offset timing (and offset position) of visual motion stimuli. In the vision science literature, it is well known that the final position of a moving visual stimulus is mislocalized in the direction of motion (see Hubbard, 1995a, 2005b,

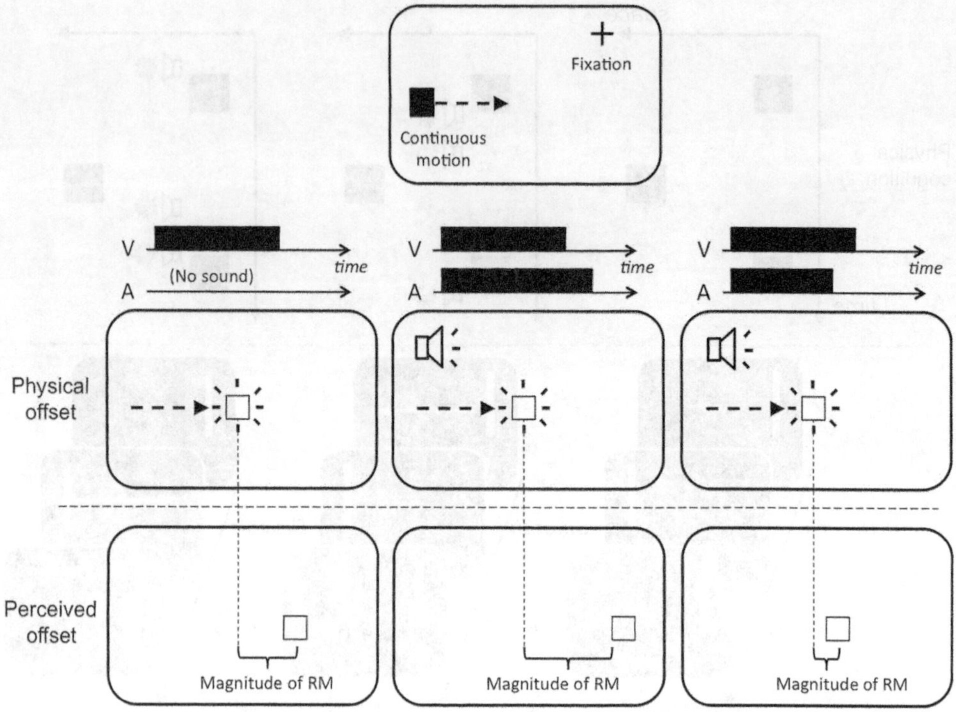

Figure 11.2 Example of the visual display and the presentation sequence used in Teramoto, Hidaka, Gyoba, and Suzuki (2010). A white square (target) moved laterally from either side of the CRT display to the center at a speed of 7.5°/s for 1,200 msec, and disappeared at an unpredictable position. A probe was presented 500 msec after the target disappeared. The participants' task was to judge on which side (left or right) the probe was located relative to target offset position. On some trials, a sound was presented via headphones from the beginning of the target motion and stopped before, at, or after target offset. The magnitude of representational momentum (RM) decreased when the sound stopped before visual offset, and increased when the sound stopped after visual offset.

2015a, Chapter 8 in this volume for a review). Freyd and Finke (1984) termed this mislocalization "representational momentum (RM)" because the motion of a moving stimulus is prolonged in the direction of the object's implied path of motion, similar to the manner in which physical object motion is prolonged. Although the mechanisms underlying RM remain unclear, it is generally accepted that some visual system extrapolation processes may maintain a mental representation of the moving visual stimulus for a brief period after the visual stimulus physically vanishes. Teramoto, Hidaka, Gyoba, and Suzuki (2010) used this phenomenon to investigate the effect of sounds on visual motion perception. In their experiments, a visual stimulus moved from either side of a monitor and disappeared at unpredictable positions around the center of the monitor. A complex tone without any motion cues was presented with the moving visual stimulus and vanished at various times relative to the offset of visual motion (Figure 11.2). Visual RM was enhanced when the sound was continuously presented from the beginning of the visual motion, and lasted longer than the visual motion. The magnitude of RM increased as the auditory offset lag relative to visual motion increased, and reached asymptote around 150 msec. In contrast, visual RM was reduced when the sound terminated shortly before the visual motion. The effect of sound on RM reduction was greatest when the auditory stimulus disappeared 100 msec before the visual motion, and was almost zero when it disappeared 150 msec before the visual motion. However, RM magnitude was not modulated by a brief tone burst (20 msec) around the visual motion offset (but see Chien, Ono, & Watanabe, 2013), nor by a long-lasting sound that was not synchronized with the visual motion. These findings suggest that when an association between auditory and visual signals is sufficiently formed, auditory information can alter perceived visual motion offset.

Taken together, auditory information can modulate temporal aspects (onset and offset) of visual motion events, resulting in an alternation of the spatial perception of these events. However, it should be noted that temporal auditory cues do not always influence visual motion events. A crucial factor for the modulation of visual motion events is relative temporal certainty between the auditory and visual signals. Heron et al. (2004) investigated the effect of temporal certainty of auditory information on visual events by manipulating sound duration (2.33 msec [high certainty] vs. 74.56 msec [low certainty]). Effects were only found in the high temporal certainty condition. This is inconsistent with the modality appropriateness hypothesis (Welch & Warren, 1980, 1986), which states that audition dominates vision for temporal tasks, while vision dominates audition for spatial tasks. This hypothesis predicts that auditory information *always* influences visual perception, as long as tasks require a temporal domain. Thus, the results of Heron et al. (2004) indicate that, in the temporal domain, the perceptual system uses auditory information, not because it originates from the auditory system per se, but because it is more reliable than other simultaneous sensory inputs.

Another important factor is the temporal distance between auditory and visual events. This is known as the "temporal window" in which multisensory effects occur. The double flash illusion becomes less likely to be observed when an asynchrony between auditory and visual stimuli exceeds ±70 msec (Shams, Kamitani, & Shimojo, 2002). Similarly, auditory effects on visual motion events reach asymptote around 100~150 msec and then decrease (Heron et al., 2004; Teramoto, Hidaka, Gyoba, and Suzuki (2010). Additionally, Teramoto, Hidaka, Gyoba, and Suzuki (2010) showed that sounds did not affect visual RM when the offset lag was kept within ±100 msec, but rather when large onset asynchronies (>400 msec) were introduced between a visual motion stimulus and an accompanying long-lasting sound. Thus, there is a temporal limit within which the perceptual system considers visual and auditory signals as originating from the same event.

Effects of Auditory Spatial Cues on Visual Motion Perception

Several studies have investigated spatial audiovisual interactions in motion perception, and most of these have found evidence for visual dominance (e.g., Allen & Koelers, 1981; Kitagawa & Ichihara, 2002; Kitajima & Yamashita, 1999; Mateeff, Hohnsbein, & Noack, 1985; Sanabria, Spence, Soto-Faraco, 2007; Soto-Faraco, Lyons, Gazzaniga, Spence, & Kingstone, 2002; Soto-Faraco, Spence, & Kingstone, 2004, 2005; Soto-Faraco, Kingstone, & Spence, 2003; Strybel & Vatakis, 2004). For example, Matteeff et al. (1985) demonstrated that stationary auditory stimuli tended to be perceived as moving in the same direction as simultaneously presented moving visual stimuli, a phenomenon termed "dynamic visual capture."

Kitajima and Yamashita (1999) and Soto-Faraco and colleagues (Sanabria et al., 2007; Soto-Faraco et al., 2002, 2003, 2004, 2005; Strybel & Vatakis, 2004) extended the findings of Matteeff et al. (1985) to moving auditory stimuli. Kitajima and Yamashita (1999) presented auditory stimuli that moved smoothly horizontally, vertically, or in depth in conjunction with visual stimuli moving in the same or opposite direction as the auditory motion. Perceived direction of the moving auditory stimulus was strongly influenced by visual motion in all directions, especially vertically and in depth. In the study presented by Soto-Faraco et al. (2002), auditory stimuli were presented via loudspeakers placed at horizontally separated positions, so that auditory apparent motion was perceived. Visual stimuli were concurrently presented from LEDs attached to the speakers to produce visual apparent motion in the same or opposite direction to the auditory motion. Participants correctly reported the motion direction (left/right) of auditory apparent motion when its direction was consistent with that of visual apparent motion. However, when auditory and visual motion directions were inconsistent, participants often misperceived the direction of auditory motion as consistent with the visual direction. Interestingly, auditory motion had little or no influence on the perceived direction of visual motion (Soto-Faraco et al., 2004; Strybel & Vatakis, 2004). Kitagawa and Ichihara (2002) also demonstrated asymmetrical effects in motion perception between auditory and visual modalities. In their experiment, participants adapted to visual size change (increasing or decreasing visual image size), auditory loudness change (increasing or decreasing sound pressure level), both visual and auditory stimuli in the same (congruent) direction, or both in the opposite (incongruent) direction. These changes in physical attributes occurred as the objects moved in the depth direction. Auditory loudness aftereffects that occurred after adaptation to visual size change alone were enhanced after adaptation to congruent audiovisual stimuli, and disappeared after adaptation to incongruent audiovisual stimuli. In contrast, adaptation to auditory loudness change alone did not influence visual size change aftereffects. These findings seem consistent with the notion of visual dominance in spatial perception (see also Alais & Burr, 2004a; Meyer & Wuerger, 2001; Wuerger, Hofbauer, & Meyer, 2003).

In the literature of ventriloquism effects, Alais and Burr (2004b) demonstrated that visual dominance in spatial tasks is not always the case when the visibility or reliability of visual inputs is degraded. In this study, blurred visual stimuli were used to increase or decrease position certainty. Participants performed a localization task with visual, auditory, and audiovisual targets. In the audiovisual target condition, the perceived position of auditory stimuli shifted toward concurrent visual stimuli when positional uncertainty of the visual stimuli was low. In contrast, the perceived position of visual stimuli with high positional uncertainty was often captured by auditory stimuli. Similar to the temporal domain described in the previous section (Heron et al., 2004), it seems likely that the perceptual system often uses visual information in spatial processing, not because it originates from the visual system per se, but because it is more reliable compared to other incoming sensory inputs. This indicates a possibility that auditory information influences visual *motion* perception when visual information is degraded.

Hidaka et al. (2009) investigated whether auditory motion information affected the perception of visual stimuli presented in the peripheral visual field. In this paper, the authors noted that previous studies reporting visual dominance effects in audiovisual motion perception presented visual stimuli in the central or paracentral retinal regions, where the spatial resolution of the visual system is excellent. Perrott, Costantino, and Cisneros (1993) reported that location discrimination performance, at azimuth angles of 20° or larger, was better suited for the auditory versus visual modality. Therefore, there is a possibility that auditory spatial information can modulate visual perception when visual stimuli are presented in the peripheral visual field. In the experiment conducted by Hidaka, Kawachi, and Gyoba (2009), apparent auditory motion was presented through headphones,

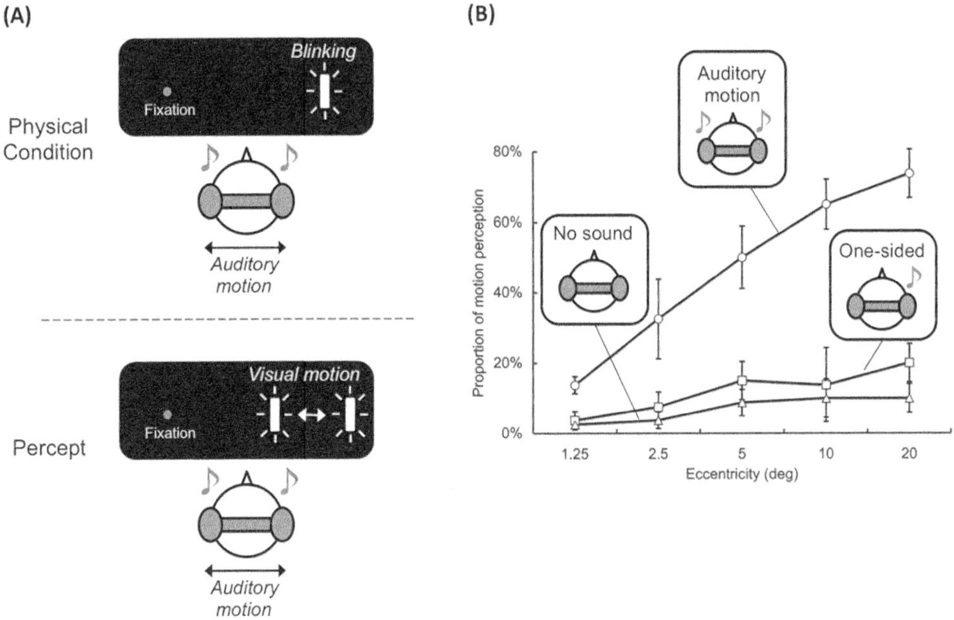

Figure 11.3 Schematic illustration of the sound-induced visual motion illusion (Hidaka et al., 2009). (A) A vertical bar blinking at a fixed position was presented, and its onset was synchronized with a tone burst alternating between the left and right ears. The visual stimuli in such an auditory condition could be perceived as moving laterally. This is the first demonstration of an auditory driving effect on visual motion perception. (B) The results of Hidaka et al. (2009). The auditory effect increased as retinal eccentricity of the visual stimuli increased.

and a blinking visual target was presented at various retinal eccentricities (Figure 11.3). Results showed that the auditory motion signal drove motion perception of a static visual target. The frequency of visual motion perception increased as retinal eccentricity increased. This auditory driving effect on visual motion perception was reported for both horizontal and vertical auditory motion (Teramoto et al., 2010).

In a different study, Hidaka et al. (2011) found that continuous auditory motion signals, generated by cross-fading two white noises between the left and right ears, could also induce motion perception for a static visual target when it was presented in the peripheral visual field. In addition, continuous auditory motion determined the perceived direction of an ambiguous visual global motion display, in which motion information is extracted from the integration of multiple local motion signals (Williams & Sekuler, 1984). Furthermore, Teramoto et al. (2012) demonstrated that perceived direction of visual motion could be altered by auditory motion information. In these experiments, apparent visual and auditory motion stimuli were simultaneously presented in two orthogonal directions (i.e., one moved vertically while the other moved horizontally). The perceived direction of visual motion could be consistent with the direction of auditory motion, or was between this direction and that of actual visual motion. Deviation of perceived direction of motion from actual direction was more likely at larger retinal eccentricities. In addition, a separate study reported that the auditory effects on visual motion offset localization were also more salient in the peripheral visual field (Schmiedchen, Freigang, Nitsche, & Rübsamen, 2012). These studies suggest that auditory information can influence visual motion events when visual information is degraded.

One study increased the spatial precision of auditory stimuli instead of decreasing the reliability of visual stimuli. Meyer, Wuerger, Rührbein, and Zetzsche (2005) created auditory motion by switching physical sound sources instead of manipulating differences in the interaural level (Meyer & Wuerger, 2001; Wuerger et al., 2003) or time (Alais & Burr, 2004a). Motion detection thresholds were measured when either or both auditory and visual motion information were presented. Motion detection thresholds improved when auditory and visual motion signals were spatiotemporally consistent. This study highlights the importance of the quality of auditory spatial information (but see also Kim, Peters, & Shams, 2012).

To summarize, in contrast to the traditional view of visual dominance for motion processing, recent studies have demonstrated that auditory spatial cues can exert a strong influence on visual motion perception. It should be noted that this is not always the case, however. Necessary conditions include high positional uncertainty of visual, relative to auditory stimuli (Hidaka et al., 2009; Hidaka, Teramoto, Sugita, Manaka, Sakamoto, & Suzuki, 201; Teramoto, Manaka, Hidaka, Sugita, Miyauchi, Sakamoto, Iwaya, & Suzuki, 2010; Teramoto et al., 2012; Schmiedchen et al., 2012), and/or auditory signals that include high-quality spatial information (Meyer et al., 2005). In other words, the relative reliability of auditory and visual information is crucial. These findings are well consistent with the predictions made by the maximum likelihood estimation (MLE) model of multisensory interaction/integration (Alais & Burr, 2004b; Ernst & Banks, 2002; Ernst & Bülthoff, 2004). Signals from each sensory modality are weighted by their reliability, such that reliable and unreliable sensory signals are assigned a high and low weight, respectively. Then, these weighted signals are combined (weighted average) so that the brain ultimately receives a more reliable (or less variable) estimate of the event. Reliability is typically defined as the inverse variance of the probability distribution representing the likelihood of the occurrence of an event in the given sensory modality. Thus, this model assumes that how sensory modalities interact can be flexibly changed depending on the relative reliabilities of sensory signals. It seems likely that the MLE model can capture most of the previous data on audio-visual motion perception.

Effects of Auditory Synesthetic or Semantic Cues on Visual Motion Perception

In addition to spatial and temporal information in multisensory signals, it is well known that semantic congruency plays an important role in multisensory interaction/integration. Semantic congruency is defined by whether auditory and visual signals are matched in terms of identity and/or meaning (Spence, 2011). For example, a visual image of a dog barking is semantically congruent with the sound "bow-wow," but not "meow" (e.g., Amedi, von Kriegstein, van Atteveldt, Beauchamp, & Naumer, 2004; Laurienti, Kraft, Maldjian, Burdert, & Wallace, 2004). According to Spence (2011), semantic congruency is different from synesthetic congruency, which refers to "correspondences between putatively non-redundant stimulus attributes or dimensions that happen to be shared by many people" (Spence, 2011, p. 927). For example, sound pitch can provide not only a high-low sensation of pitch, but also an up-down (or high-low) impression in space (Bernstein & Edelstein, 1971; Evans & Treisman, 2010; Mudd, 1963; Pratt, 1930; Roffler & Butler, 1968; Rusconi, Kwan, Giordano, Umiltà, & Butterworth, 2006). The effects of semantic/synesthetic acoustic cues on visual motion perception have also been investigated in the audiovisual motion perception literature. Maeda, Kanai, and Shimojo (2004) investigated whether ascending or descending pitch can bias visual motion perception. Ascending pitch tends to be associated with upward motion, while descending pitch tends to be associated with downward motion. They presented ambiguous visual motion stimuli that included both upward and downward motion information in conjunction with either ascending or descending pure tone pitches. Results showed that ascending pitch induced more upward than downward motion responses, and vice versa. Interestingly, sounds that were not synchronized with visual motion onset or actual words representing upward or downward motion had little or no effect on visual motion perception. Furthermore, the effect decreased as orientation disparity increased. These findings suggest that synesthetically congruent sounds can modulate visual motion perception.

Recently, Hidaka, Teramoto, Keetels, and Vroomen (2013) investigated whether this pitch-space correspondence could induce motion perception to a static visual stimulus, using a method adapted from Hidaka et al. (2009). In this experiment, two loudspeakers were set at 50 cm above and below the center of the visual display where a blinking visual stimulus was presented. In one condition, two discrete tones with high- and low-frequency bands were presented in alternation from both loudspeakers. In the other condition, high- and low-pitched tones were presented alternately from the upper and lower speakers, respectively. Alternation of pitch information (high-low or low-high tone sequences) did not induce vertical visual apparent motion perception, while alternation of position information induced motion perception in a manner similar to that observed by Hidaka et al. (2009). Hubbard and Courtney (2010) investigated the effects of ascending/descending tone frequencies on visual RM. They found that changing frequency had little effect on visual RM when vertical visual motion was presented, but induced slight mislocalization in the direction of implied gravity (i.e., visual representational gravity; Hubbard, 1990, 1995b, 1997; Hubbard & Bharucha, 1988) when horizontal visual motion was presented. Thus, the results of these three studies are not consistent, possibly due to a difference in auditory stimuli, as suggested by Hidaka et al. (2013). While both Hidaka et al. (2013) and Hubbard and Courtney (2010) used sequences of discrete sounds, Maeda et al. (2004) used continuous frequency changes (glides), which can elicit a greater impression of motion (Walker, 1987). However, there are several other differences in addition to auditory stimuli, such as the parameters of visual stimuli and experimental procedures. Thus, further studies are needed to clarify the effects of pitch-space correspondence on audiovisual motion processing.

Auditory effects on biological motion perception have also been investigated (Arrighi, Marini, & Burr, 2009; Brooks, van der Zwan, Billard, Petreska, Clarke, & Blanke, 2007; Saygin, Driver, & de Sa, 2008; Schouten, Troje, Vroomen, & Verfaillie, 2011). According to this literature, semantically congruent sounds can improve the perception of visual biological motion. Arrighi et al. (2009) had participants detect point-light tap dance sequences embedded in visual noise. In some trials, synchronous or asynchronous tap dance sounds were added to the visual sequence. In other trials, no sound was added. Synchronous tap dance sounds improved detection of visual biological motion, irrespective of whether the sounds were task-relevant or not. Because the improvement was better than predicted by the MLE model, the authors suggested that there was physiological summation of audiovisual signals.

Some studies have shown that synesthetically or semantically congruent sounds can facilitate visual motion perception, while other studies have not. Overall, relatively little is known about the influence of this type of auditory cue on visual motion perception. Spence (2011) classified this type of multisensory interaction/integration into three categories based on the assumed origin: structural, statistical, and semantically mediated correspondences. Structural correspondence refers to physiological correspondence, that is, the proximity/similarity of brain areas/activity when processing signals from different sensory modalities. Statistical correspondence refers to the learned natural correlation between attributes of different sensory modalities. Semantically mediated correspondences refer to links mediated by common linguistic terms. Pitch-space correspondence would be classified as semantically mediated, while the tap dance scenario is an example of statistical correspondence. These differences may modulate the influence of auditory cues on visual motion perception. Statistical correspondence is discussed in more detail in the next section.

Effects of Audiovisual Associative Learning on Visual Motion Perception

Although the MLE model specifies sensory weighting mechanisms, it does not describe how the brain knows which signals, from different sensory modalities, should be integrated or segregated. This is important for establishing crossmodal correspondence relationships. Spatiotemporal consistency (e.g., Calvert, Spence, & Stein, 2004) and temporal correlation (Parise, Harrar, Ernst, & Spence, 2013; Parise, Spence, & Ernst, 2012) of signals provide cues about correspondences. However, because each sensory modality receives many inputs each second, it seems implausible that the brain relies only on such spatiotemporal information. A strategy in which preliminary associations between signals are formed would likely help the brain bind signals efficiently. Recent research has approached this question using Bayesian estimation theory in multisensory integration (e.g., Bresciani, Dammeier, & Ernst, 2006; Ernst, 2005, 2007, 2012; Körding, Beierholm, Ma, Quartz, Tenenbaum, & Shams, 2007; Roach, Heron, & McGraw, 2006; Shams, Ma, & Beierholm, 2005). According to this theory, sensory inputs are combined with prior knowledge about whether stimuli from one sensory modality and those of another modality should be integrated to determine perception. Ernst (2007) demonstrated that prior knowledge could be learned across arbitrary crossmodal inputs, even by adults. In training sessions, arbitrary but correlated signals from vision and touch, such as object luminance and stiffness (i.e., the brighter the object, the stiffer it was), were presented. Before and after training, participants performed an object discrimination task. Ernst (2007) predicted that if these sensory signals were processed independently, visual luminance should not influence tactile discrimination performance, and vice versa. Although no effect was found before training, discrimination performance was worse for stimuli where luminance-stiffness was uncorrelated than correlated after training. Seitz, Kim, van Wassenhove, and Shams (2007) also demonstrated that this type of associative learning effect could occur in audiovisual object identification. These findings suggest that newly learned relationships (i.e., coupling prior) can contribute to subsequent perception, as predicted by a Bayesian model.

Recently, Teramoto, Hidaka, and Sugita (2010) demonstrated that arbitrary crossmodal associations could also be established in audiovisual motion perception. In these experiments, two side-by-side visual stimuli were presented in alternation, producing horizontal apparent motion. The onsets of the two visual stimuli were synchronized with high (H:

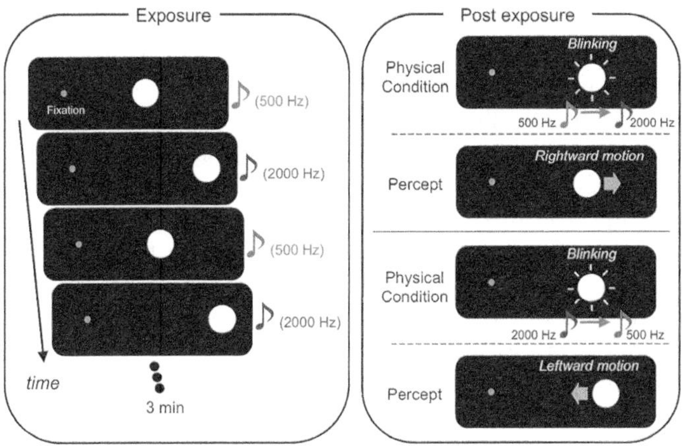

Figure 11.4 Schematic illustration of perceptual associative learning effects between auditory and visual stimuli reported in Teramoto, Hidaka, and Sugita (2010). Two visual stimuli were presented in alternation. Each stimulus was paired with a tone of unique frequency (500 Hz or 2,000 Hz). After 3 min exposure to such audiovisual stimuli, the tone sequence induced lateral motion perception to a visual stimulus blinking at a fixed location, in the same manner as the audiovisual relationship at exposure.

2,000 Hz) and low (L: 500 Hz) frequency tone bursts (Figure 11.4). Participants were exposed to these audiovisual stimuli for 3 min. Prior to exposure, the sound sequence had no influence on visual motion perception. However, after exposure, the sound sequence induced visual motion perception of a visual stimulus blinking at a fixed location. For example, when the left and right visual stimuli were presented with L and H tones, respectively, in the exposure phase, the LH and HL tone sequences induced rightward and leftward visual motion perception, respectively. The association effect disappeared when the inter-stimulus interval of the visual stimuli was too long to be perceived as apparent motion during exposure. In addition, more recent studies have demonstrated that this effect generalizes to other types of visual motion, such as global motion (Hidaka, Teramoto, Kobayashi, & Sugita, 2011) and higher-order visual motion (Kafaligonul & Oluk, 2015). These findings suggest that an audiovisual association in motion perception can be formed at several levels of visual motion processing. Furthermore, Kobayashi, Teramoto, Hidaka, and Sugita (2012a) tested the association effect using two tone bursts, instead of pure tones, that were physically different but perceptually indiscriminable. In their first experiment, two complex tones that consisted of eight frequencies (seven were shared and one was unique) were used and associated with visual motion. Even these sounds influenced visual motion perception. In the second experiment, another type of indiscriminable tone bursts, which were created by applying peak and notch filters centered at 500 Hz and 2,000 Hz, respectively (or vice versa), to white noise, were tested. The results were replicated with these sounds, suggesting that explicit knowledge of the relationship between visual motion and sound sequence is not necessary to form an association.

Interestingly, Teramoto, Hidaka, and Sugita (2010) reported that the effect was not observed when the retinal position of the visual stimuli differed between the exposure and test phase (by at least 5°) (see also Hidaka, Teramoto, Kobayashi, & Sugita, 2011). Since visual receptive fields typically get larger with progressing visual processing, this finding suggests that relatively lower stages of visual processing are involved. Furthermore,

Kobayashi et al. (2012a) showed that the effect did not occur when the stimulated eye differed between exposure and test phases, suggesting visual processing stages before integrating information from both eyes are involved in this association.[1] As for the auditory processing stages involved in this association effect, Kobayashi, Teramoto, Hidaka, and Sugita (2012b) reported that the association effect had sharp frequency selectivity. After exposure to visual apparent motion and specific tone frequency changes (e.g., 400 and 2,100 Hz), tones with frequencies that differed by 0.25 octave (e.g., 476–2,496 Hz or 566–2,970 Hz) did not have any effect on visual motion perception in the subsequent test session. Similarly, the observed effect was specific to the exposed ear. These findings suggest that auditory processing stages that have relatively sharp frequency selectivity (i.e., relatively lower stages of auditory processing) are involved.

There are two points that should be noted here. First, the association effect is reciprocal. Teramoto, Kobayashi, Hidaka, and Sugita (2013) demonstrated that pitch perception of auditory stimuli can be contingent on visual motion direction, after prolonged exposure to audiovisual stimuli. When leftward apparent visual motion was paired with high–low-frequency sequences during exposure, a test tone sequence was more frequently perceived as a high–low-pitch sequence when leftward apparent visual motion was presented. Thus, once an association is formed, related signals from different sensory modalities influence each other. Second, crossmodal associative learning in spatiotemporal processing is not limited to audiovisual domains. Recently, Kuang and Zhang (2014) presented changes in smells (banana and fennel) with a visual global motion display. After exposure, the smells affected perceived direction of visual global motion. This suggests that crossmodal associative learning is a general phenomenon that occurs between a variety of sensory modality pairs.

In summary, recent studies have clearly demonstrated that new perceptual associations can be easily established between signals from different modalities. After associations are formed, inputs from one sensory modality influence perception of the other sensory modality. This fits well with the role of prior knowledge in the Bayesian estimation theory of multisensory interaction/integration. These findings suggest that perceptual associative learning is one of the most plausible underlying mechanisms for establishing common perceptual and neural mechanisms, not only in audiovisual motion processing, but also in crossmodal processing more generally.

Concluding Remarks

In this chapter, we summarized recent studies on the influence of auditory cues on visual motion processing. Traditionally, crossmodal studies have demonstrated that vision is dominant over other sensory modalities (i.e., visual capture) in *spatial* processing. Spatial ventriloquism and McGurk effects are typical examples supporting this notion. This had been considered to be the case with motion perception. Until recently, it was believed that

[1] The Spatial-Musical Association of Response Codes (SMARC) effects (Rusconi, Kwan, Giordano, Umiltà, & Butterworth, 2006) described in Chapter 4 of this book may appear similar to sound-contingent visual motion perception reported by Teramoto, Hidaka, and Sugita (2010), but differ in several points, although statistical learning may contribute to the formation of both association effects. While the former is one of the stimulus-response compatibility effects (left and right manual responses to low- and high-pitched tones, respectively, are facilitated as compared with when the reversed stimulus-response mapping is used), the latter occurs at the perceptual level. Furthermore, the high-right/low-left mapping is fixed in the former but not in the latter.

visual information exerts a critical influence on auditory motion perception, while auditory information has little or no influence on visual motion perception. However, recent findings have clearly demonstrated that various auditory cues can influence visual motion perception. First, temporal auditory cues can change the perceived timing of concurrently presented visual stimuli, resulting in improvement/impairment of apparent motion perception (Freeman & Driver, 2008; Getzmann, 2007), as well as the mislocalization of visual motion (Chien et al., 2013; Heron et al., 2004; Teramoto, Hidaka, Gyoba, & Suzuki, 2010) and visual motion events (Vroomen & de Gelder, 2004). These results are consistent with a traditional view that audition is more dominant than vision in *temporal* processing. However, in contrast to the traditional view of visual dominance in motion processing, auditory *spatial* cues also can exert a strong influence on visual motion perception. Notably, auditory dominance occurs in conditions where positional uncertainty of visual, relative to auditory, stimuli is high, for example, when visual stimuli are presented in the peripheral visual field (Hidaka et al., 2009; Hidaka, Teramoto, Sugita, Manaka, Sakamoto, & Suzuki, 2011; Teramoto, Manaka, Hidaka, Sugita, Miyauchi, Sakamoto, Iwaya, & Suzuki, 2010; Teramoto et al., 2012; Schmiedchen et al., 2012), and/or there is high-quality spatial information in the auditory signals (Meyer et al., 2005). This is clearly consistent with the MLE model of multisensory interaction/integration (e.g., Ernst & Banks, 2002). Specifically, audiovisual integration/interaction in motion perception is flexible and based on the reliability and saliency of spatiotemporal information.

In addition to auditory spatiotemporal cues, synesthetic or semantic auditory information can also influence visual motion perception (e.g., Arrighi et al., 2009; Hidaka et al., 2013; Hubbard & Courtney, 2010; Maeda et al., 2004; Schouten et al., 2011). It is likely that associative learning between auditory and visual information influences this type of auditory effect on visual motion perception (Teramoto, Hidaka, & Sugita, 2010). These findings clearly suggest that auditory and visual information closely interact in motion perception, and that common neural mechanisms for audiovisual motion processing exist at multiple stages, from very early to higher-level cognitive processing. Similarly, recent neuroimaging studies have shown that several regions are commonly activated by auditory and visual motion stimuli (Alink, Singer, & Muckli, 2008; Baumann & Greenlee, 2007; Hidaka, Higuchi, Teramoto, & Sugita, 2017; Lewis, Beauchamp, & DeYoe, 2000; Scheef et al., 2009). However, relatively little is known about the neural mechanisms of audiovisual motion processing. Further investigation on this topic could contribute to a comprehensive understanding of the influence of crossmodal interactions on spatiotemporal, spatial, and temporal processing.

Chapter 12

Adaptive Biases in Visual and Auditory Looming Perception

John G. Neuhoff

In the scientific community, it is almost universally accepted that the remarkable features of our auditory and visual systems have evolved over millions of years. Adaptive perceptual mechanisms that aid survival and reproduction are passed on to subsequent generations. Yet many in the scientific community also hold the belief that our perceptual systems have evolved to give us an "accurate" representation of our environment. Any observed inaccuracies in our perceptual abilities are often considered imperfections in the evolutionary process. However, from an evolutionary perspective, perceptual accuracy is far less important than *utility* in terms of survival and reproduction. While it is true that our perception of the world must be accurate enough for us to perform all of the remarkable feats that we do, it is also true that evolution will always favor perceptual distortions and cognitive biases over veridical representation of the environment when the former provide a greater selective advantage. Perceptual systems have not evolved because they provide an accurate representation of the environment. They have evolved because they bestow specific advantages in survival and reproduction (Popper & Fay, 1997). If a perceptual bias provides a greater evolutionary advantage than an accurate representation, it will be passed on to subsequent generations at a higher rate than veridical perceptual abilities. The biases that occur when perceiving looming objects are examples of this phenomenon.

However, providing evidence to support this assertion presents a unique challenge. Although few doubt that perceptual abilities in some nebulous sense are a product of evolution, the burden of evidence crystalizes considerably when one begins to make a case for the evolution of a specific perceptual trait. Ethical considerations and the time course of human evolution dictate that the traditional experimental method used to test behavioral hypotheses must be modified when our hypotheses regard the evolution of a specific perceptual trait. We can't simply apply an independent variable to a sample and evaluate how it affects human evolution. As such, making a convincing case that a perceptual bias (or any other behavioral trait) has been shaped by evolution requires converging empirical evidence from a wide array of disciplines and methods including physiological, comparative, behavioral, theoretical, cross-cultural, anthropological, medical, and even genetic areas of investigation (Schmitt & Pilcher, 2004). Not every perceptual bias is an adaptation. The dim visual afterimage that occurs after a bright camera flash is a perceptual bias. However, this bias provides no ostensible selective advantage. It is simply a by-product of how the visual system functions. Thus, the degree to which we can be confident that a given bias is an evolutionary adaptation is dependent on the quality and quantity of converging evidence that can be gathered from these diverse areas.

In this chapter, I examine auditory and visual perception of looming objects – objects in motion that approach an observer. Looming objects are a very special class of stimuli that are treated with priority by both the auditory and visual systems because of their importance in ecological and evolutionary terms. However, the auditory and visual systems have evolved to deal with looming objects in different ways and have different strengths and weaknesses. The visual system provides estimates of arrival time that are relatively accurate and precise under good viewing conditions. The auditory system is less accurate and less precise but can be characterized as an "advanced warning" system that provides input into a categorical decision about whether there is time to direct the eyes toward the looming object, or whether evasive actions need to be initiated immediately (Guski, 1992; Seifritz et al., 2002). Unlike vision, the auditory system functions well when visibility is poor and when objects are occluded or are out of the line of sight. Together the two systems provide a perceptual representation that, while not always accurate, enables highly successful interaction with looming objects. We begin this chapter with a review of the literature on unimodal looming perception in both audition and vision. We then examine the smaller body of research that has examined multisensory integration of looming perception.

Auditory Looming

The Auditory Looming Bias

The auditory looming bias is the strong tendency for listeners to underestimate the arrival time of an approaching sound source. Essentially listeners perceive that the source has arrived when it is still some distance away (Neuhoff, 2001). A looming sound source creates several important sources of dynamic acoustic information that listeners can use to make judgments about the arrival time. Perhaps the most studied is the change in intensity that occurs as a source approaches. Simple changes in intensity are sufficient for activating motion-sensitive areas of the brain (Seifritz et al., 2002), and the pattern of rising intensity that occurs as a source approaches can physically specify the arrival time, a variable termed *acoustic tau* (Guski, 1992; Shaw, McGowan, & Turvey, 1991). For a sound source approaching a listener on a close bypass trajectory, the Doppler shift specifies that the frequency observed at the listening point is slightly higher than that emitted by the source and drops gradually as the source draws near. It then drops dramatically as the source passes the listener and continues to drop gradually as the source recedes. Despite this continual drop in frequency, listeners report a rise in pitch as the source approaches, a phenomenon referred to as the Doppler Illusion (Neuhoff & McBeath, 1996). The illusion is likely due to the rising intensity that occurs as the source approaches and the integral processing of pitch and loudness (McBeath & Neuhoff, 2002; Neuhoff & McBeath, 1996; Neuhoff, McBeath, & Wanzie, 1999).

Rosenblum, Carello, and Pastore (1987) examined the relative effectiveness of Doppler shift, interaural time differences, and overall intensity change in cuing listeners to the arrival time of a looming sound.[1] Intensity change was found to be the dominant cue to the point of closest approach for a passing sound source, followed by interaural temporal differences, then the Doppler effect. Performance was best under full cue conditions. However, even under full cue conditions, listeners still exhibited a systematic bias to hear the sound arrive before it actually did. Subsequent work showed that providing explicit

[1] The term "looming sound" is used in this chapter to refer to the approach of a sounding object and should not be confused with the acoustic pressure wave that approaches at the speed of sound.

feedback on the accuracy of the judgments after each trial diminished but did not eliminate the anticipatory bias (Rosenblum, Gordon, & Wuestefeld, 2000).

There are other potential sources of acoustic information that listeners may use in judging the arrival time of a looming sound, which have yet to be systematically investigated. For example, the ratio of direct to reverberant sound increases as the source draws closer to the listener. The spectral profile of the sound at the listening point also changes due to the decrease in atmospheric damping of high frequencies as the source approaches. Some work has shown greater physiological effects of looming sounds when multiple cues are employed over conditions in which just amplitude changes (Bach, Neuhoff, Perrig, & Seifritz, 2009). This implies that listeners are sensitive to these additional cues and may in fact use them in judging arrival time. Other work has shown that the perceived urgency of the sound source can influence judged arrival time, with more urgent sounds being perceived to arrive sooner (Gordon, Russo, & MacDonald, 2013; Neuhoff, Hamilton, Gittleson, & Mejia, 2014).

The auditory looming bias is closely linked to the bias to hear rising intensity as changing in loudness more than equivalent falling intensity. If listeners overestimate the approach of a looming sound and loudness change is the dominant cue to approach, then the overestimation of rising loudness may be at the heart of the auditory looming bias. However, if we are to argue that this perceptual bias is an adaptation shaped by evolution, then it should be specific to the conditions under which it would provide the greatest advantage in the natural listening environment over our evolutionary history. To test this prediction, Neuhoff (1998) presented listeners with rising and falling intensity tones and asked them to use a slider to indicate the amount of loudness change they heard in each sound. Rising intensity was consistently heard to change more than equivalent falling intensity, and the louder the range of intensity change presented, the greater the disparity between rising and falling loudness change. In a natural listening environment where closer sounds are louder than equivalent distant sounds, these findings suggest a perceptual priority for looming sounds that are close over those that are more distant. Although the effect occurred with harmonic tones, it did not occur when listeners were presented with broadband noise, a finding that is also consistent with the priorities of localizing moving sound sources in a natural environment. Periodic sounds that have a tonal quality are produced by a wide variety of biological organisms and can act as a reliable marker for the identity of a single source. Although broadband noise can be produced by biological organisms, it can also be produced by widely dispersed nonbiological sources (e.g., wind and rain) for which localization may be less important. Thus, the looming bias is most robust under conditions in which it would be most advantageous to survival (see also McCarthy & Olsen, 2017). Recent magnetoencephalography work supports this hypothesis by showing that strength of sustained magnetic fields over bilateral temporal sensors linearly tracks intensity change in a looming harmonic tone but not in looming broadband noise (Bach, Furl, Barnes, & Dolan, 2015)

The argument for the looming bias as an adaptation was initially challenged because of the limited number of conditions under which the effect was first demonstrated (Canevet, Scharf, Schlauch, Teghtsoonian, & Teghtsoonian, 1999; Neuhoff, 1999). However, it has now been replicated under a wide variety of experimental conditions and settings (DiGiovanni & Schlauch, 2007; Grassi, 2010; Grassi & Darwin, 2006; Olsen & Stevens, 2010; Olsen, Stevens, & Tardieu, 2010; Ponsot, Meunier, Kacem, Chatron, & Susini, 2015; Ponsot, Susini, & Meunier, 2015; Teghtsoonian, Teghtsoonian, & Canevet, 2005). One of the first studies to replicate the effect used a moving loudspeaker in an outdoor environment (Neuhoff, 2001). Blindfolded listeners made terminal egocentric distance estimates of a moving loudspeaker that either approached or receded. The sounding loudspeaker approached from a distance of 12.2 meters and came to rest 6.1 meters from the listener. The receding loudspeaker began directly in front of the listener and stopped at the same distance, 6.1 meters from the listener. Analogous to the rising and falling loudness sounds presented over headphones, the results showed that approaching sounds were perceived to stop closer to the listener than receding sounds, and tones were perceived as stopping closer than noise despite an equal stopping distance from the observation point in all conditions.

The Logic Behind the Adaptation: Error Management Theory

Error Management Theory (EMT) predicts that a wide range of social, cognitive, and perceptual biases have evolved because they increase the likelihood of survival and reproduction. EMT proposes that biases will evolve when judgments are made under conditions of uncertainty, when the decisions have historically had an impact on evolutionary fitness, and when there is an asymmetric cost of making false-positive and false-negative errors (Haselton & Buss, 2000; Haselton & Nettle, 2006; Haselton et al., 2009). All of these conditions are met when a sound source approaches a listener. All perceptual judgments are made under a degree of uncertainty (Mathys et al., 2014), and predicting the arrival time of a looming sound source may have more uncertainty than many other perceptual judgments. Such decisions could also have life or death consequences and thus undoubtedly have had an impact on evolutionary fitness. Finally, the cost of a false positive (responding too early to a looming sound source by escape or avoidance behaviors) pales in comparison to the potentially deadly cost of a false negative (responding too late). Thus, an organism whose neural architecture represented looming sounds as closer than they actually were would have a selective advantage.

In the context of this discussion, the question is often posed: *Wouldn't it be more advantageous to have veridical perception of a looming sound source and let the listener cognitively decide what appropriate actions should be taken?* There are several critical reasons why the answer to this question is "no." First, let us contrast a hypothetical listener with veridical perception to a listener with a bias to hear looming sounds as closer than they are. Each is tasked with predicting the arrival time of a dangerous looming source. On average, the listener with veridical perception predicts perfectly the arrival time of the source, a point we will call 0 seconds to contact. The listener with the looming bias responds consistently early, for example at an average of 500 msec before contact. However, each listener also has a degree of variability associated with their arrival time judgments, each sometimes responding slightly earlier or later than their respective means. Early judgments are not problematic for either listener, as they provide slightly more time than expected to prepare for the arrival of the source. However, a listener with veridical perception who responds just a half-second late is responding after the source has already arrived. Even if the listener with veridical perception makes the cognitive decision to initiate motor behaviors sooner rather than later in order to prepare for the arrival of the source, their responses are still based on a "veridical" judgment of arrival time that in this instance is a half-second too late and pushes the motor response back by a potentially perilous half-second as well. A late response by the listener with the looming bias still leaves enough time to respond safely.

It is also the case that cognitive resources are limited. If the decision to engage the motor system in the face of a looming sound source were entirely under conscious control, then anyone engaged with a high cognitive load at the time would be disadvantaged in that there would be fewer cognitive resources to devote to the approaching danger. However, a recent study that manipulated cognitive load while participants judged the arrival of a looming sound found just the opposite. McGuire, Gillath, and Vitevitch (2015) asked listeners to judge when a looming sound would reach them while under high cognitive load (memorizing a seven-digit number) or low cognitive load (memorizing a two digit number). They found that the looming bias was significantly larger under high cognitive load. That listeners respond sooner rather than later under high cognitive load suggests that the bias to hear sounds as closer than actual is an automatic process that requires little effortful cognitive processing. Rather than (or at least in addition to) a "decision to respond early," the looming bias appears to be a perceptual phenomenon that has evolved to keep organisms safe. This finding is consistent with work in representational momentum and boundary extension that also shows an increase in the magnitude of perceptual bias under high cognitive load (Hayes & Freyd, 2002; Hubbard, Hutchison, & Courtney, 2010; Intraub, Daniels, Horowitz, & Wolfe, 2008).

Converging Evidence for the Looming Bias Adaptation

Behavioral evidence. A wide range of behavioral research has now demonstrated that looming sounds are treated with priority by the auditory system and that listeners demonstrate a consistent underestimation of distance and arrival time when faced with a looming sound (Neuhoff, 1998, 2001, 2016; Neuhoff et al., 2014; Neuhoff, Long, & Worthington, 2012; Riskind, Kleiman, Seifritz, & Neuhoff, 2014; Rosenblum, Wuestefeld, & Saldana, 1993; Rosenblum et al., 1987; Rosenblum et al., 2000). The prioritization of looming sounds is present in infancy. Infants as young as four months old show significant differential responding to looming versus receding sounds by exhibiting defensive avoidance responses to looming sounds that do not occur with equivalent receding sounds (Freiberg, Tually, & Crassini, 2001). By six months, infants have better discrimination abilities for looming versus receding sounds (Morrongiello, Hewitt, & Gotowiec, 1991).

Sex differences have also been demonstrated in the looming bias, with women tending to perceive auditory arrival time as occurring sooner than do men (Neuhoff, Planisek, & Seifritz, 2009; Schiff & Oldak, 1990). In the study of the evolution of behavior, sex differences can be a key piece of evidence for behavioral adaptations. To the extent that men and women have faced different challenges to survival and reproduction over our evolutionary history, we should expect slightly different behavioral adaptations to have evolved to deal with these challenges (Buss, 1995; Byrd-Craven & Geary, 2007).

Although judging the arrival time of a looming sound is essentially a spatial transformation task, the sex difference in the perception of looming sounds is likely not the product of the well-known male advantage in spatial transformation (Kimura, 1999; Silverman, Choi, & Peters, 2007; Voyer, Voyer, & Bryden, 1995). If it were, we would expect sex differences for the perception of sounds that move both toward and away from the listener. However, Neuhoff et al. (2009) presented male and female listeners with both approaching and receding sounds that stopped equidistant from the listening point. Women perceived the looming sounds to be significantly closer than did men. However, there was no difference between the male and female judgments for the receding sounds. If sex differences in the perception of looming sounds were simply the result of more accurate spatial transformation abilities by men, then we should expect those same differences to occur for receding sounds.

That there were no sex differences in the perceived distance of receding sounds may highlight a particular differential environmental challenge that men and women have faced over our evolutionary history – dealing with an approaching threat. Some of the most reliable differences between males and females are in physical strength, running speed (Nicolay & Walker, 2005; Whipp & Ward, 1992). Both of these characteristics could be crucial in dealing with a dangerous looming sound source. Thus, women would stand to benefit from a larger margin of safety in anticipating the arrival of a looming source. Essentially, those least well prepared to engage a dangerous looming sound source should have the greatest auditory looming bias.

Evidence to support this hypothesis comes from work showing that the magnitude of the looming bias is negatively correlated with strength and physical fitness (Neuhoff et al., 2012). Individual within-sex correlations were also significant, suggesting that the sex differences in the looming bias may be largely due to differences in strength and fitness rather than biological sex per se. The magnitude of the looming bias is also modulated by factors such as affect, anxiety, depression, and even schizophrenia (Bach, Buxtorf, Strik, Neuhoff, & Seifritz, 2011; Ferri, Tajadura-Jimenez, Valjamae, Vastano, & Costantini, 2015; Neuhoff et al., 2014; Tajadura-Jimenez, Valjamae, Asutay, & Vastfjall, 2010)

Comparative Evidence. The argument that a given human behavior is an evolutionary adaptation can be made stronger if the behavior is also found in a closely related species. Thus, if the human bias to perceive looming sounds as closer than actual is an adaptation, then we might expect to observe the bias in related species that have faced similar evolutionary challenges. Ghazanfar, Neuhoff, and Logothetis (2002) presented rhesus monkeys with the same simulated looming and receding sounds that were presented to human listeners in the work by Neuhoff (1998). They found a preferential orienting response to looming sounds over receding sounds that matched the pattern of results found in humans. As with humans, the bias occurred for harmonic tones but not for broadband noise. The concomitant neural activity that underlies the perceptual priority for looming sounds shows the same directional asymmetry (Hall & Moore, 2003). Activity in the lateral belt auditory cortex of monkeys is stronger with looming than with receding sounds, and the processing of looming stimuli appears to involve an interaction between the auditory cortex and the superior temporal sulcus that is not apparent with equivalent receding stimuli (Maier, Chandrasekaran, & Ghazanfar, 2008; Maier & Ghazanfar, 2007).

Physiological Evidence. Support for the hypothesis that any perceptual trait is an evolutionary adaptation can be strengthened by identifying specific physiological mechanisms that support the trait. Seifritz et al. (2002) used neuroimaging to examine the neural processing of looming sounds in comparison with receding sounds. They found that looming sounds preferentially activate a distributed neural network that is known to subserve auditory motion perception, attention, and motor planning. These findings support an earlier hypothesis by Guski (1992) that suggested that the function of the auditory system in the face of a looming sound is to provide input into a decision about the appropriate motor behaviors in which to engage. Faced with a looming sound, the auditory system provides advanced warning as evidenced by enhanced autonomic responses. Looming sounds when compared with equivalent receding sounds produce stronger amygdala activation, greater skin conductance, more robust pupil dilation, enhanced phasic alertness, and greater emotional arousal (Bach et al., 2008; Bach et al., 2009; Ferri et al., 2015; Fletcher et al., 2015; Tajadura-Jimenez et al., 2010).

Visual Looming

If the auditory looming bias provides advanced warning of looming objects, the visual system provides a more accurate means of dealing with the source on approach. The anticipatory looming bias that occurs in audition is significantly diminished in vision, and observers are significantly more accurate in making visual time-to-arrival judgments than they are when making equivalent auditory judgments (DeLucia, Preddy, & Oberfeld, 2015; Schiff & Oldak, 1990). However, the evolutionary implications of impending collision are still present in vision. It should also be noted that many of the studies that show an underestimation of visual arrival time introduce uncertainty by occluding the final approach of the object. In these experiments, the object first approaches then disappears before reaching the observer, and observers are asked to judge when the object would have reached them had it not disappeared. Performance increases significantly when viewers can see the entire approach of the object as is evidenced by our ability to interact with objects in the real world successfully (e.g., catching a ball; McBeath, Shaffer, & Kaiser, 1995; Shaffer & McBeath, 2002; Wang, McBeath, & Sugar, 2015b). In audition, the anticipatory looming bias remains even when listeners can hear the full approach of the sound source (Neuhoff et al., 2009; Neuhoff et al., 2012; Neuhoff et al., 2014).

Under some conditions, there are also sex differences in judgments of visual arrival time that mirror those in audition. Females tend to show greater underestimation of arrival time than males (Hancock & Manser, 1997; Manser & Hancock, 1996; McLeod & Ross, 1983; Montgomery, Kusano, & Gabler, 2014; Schiff & Oldak, 1990). Moreover, an incredibly wide variety of species ranging from insects to humans exhibit defensive responses when presented with visually looming objects (W. Ball & Tronick, 1971; Bower, Broughto, & Moore, 1971; King, Dykeman, Redgrave, & Dean, 1992; Lima, Blackwell, DeVault, & Fernandez-Juricic, 2015; Sato & Yamawaki, 2014; Schiff, 1965; Schiff, Caviness, & Gibson, 1962; Yilmaz & Meister, 2013; Zurek, Perkins, & Gilbert, 2014). Like looming sounds, visually looming objects also prime the human motor system for action and activate the autonomic nervous system (Low, Lang, Smith, & Bradley, 2008; Skarratt, Cole, & Gellatly, 2009; Skarratt, Gellatly, Cole, Pilling, & Hulleman, 2014). All of these factors, including the more accurate judgments of visual time-to-arrival, provide evidence for an evolutionary role in shaping our responses to looming visual stimuli.

Cues to Visual Looming

As an object approaches an observer,. the image cast on the retina dilates. This optical dilation was one of the first investigated sources of information in judging the arrival time of a looming visual object. The inverse of the rate of dilation (called optical tau) can specify time-to-arrival under some conditions, and observers have been shown to be sensitive to this information (Kaiser & Mowafy, 1993; Lee, 1976; Lee & Reddish, 1981; Regan & Hamstra, 1993; Todd, 1981). Tau was initially proposed as a candidate for completely explaining how observers judge visual time-to-arrival (Lee, 1976; Savelsbergh, Whiting, & Bootsma, 1991; Turvey & Carello, 1986). However, this version of tau is limited in that it breaks down when looming objects do not have a constant acceleration, are not symmetrical, and are not on a collision course with the observer (Tresilian, 1999). Optical tau alone also fails to account for how observers can predict the arrival of very small objects or objects that undergo short falls from gravity (Gray & Regan, 1998; Tresilian, 1993). Subsequent work showed that optical tau is just one of a number of available cues. Several other dynamic tau variables, heuristics, and even pictorial cues such as image size can be used to estimate arrival time (DeLucia, 1991, 2004; Kaiser & Mowafy, 1993; Tresilian, 1993, 1994). For a review, see Tresilian (1999).

Bias for Looming Visual Motion

Despite the relatively accurate perception of visual arrival time under full viewing conditions, there are nonetheless well-documented asymmetries in the perceptual processing of looming versus receding motion. For example, it has been known more than 100 years ago that the motion aftereffect for looming motion persists significantly longer than that for equivalent receding motion (Scott, Lavender, McWhirt, & Powell, 1966; Wohlgemuth, 1911). More recent work has shown that the perceptual asymmetry in perceiving looming and receding develops within the first few months of life (Shirai et al., 2009; Shirai, Kanazawa, & Yamaguchi, 2004, 2006). Observers also perceive looming motion as faster than receding motion, detect the onset of looming motion sooner, and show better detection of changes in looming stimuli than in receding ones (K. Ball & Sekuler, 1980; Bex & Makous, 1997; Geesaman & Qian, 1996; Petersik & Thiel, 2010). Under conditions of uncertainty, viewers show a predisposition to perceive objects as looming rather than receding. For example, C. F. Lewis and McBeath (2004) presented viewers with a 3D bistable apparent motion display that could be perceived as either looming or receding. They found that viewers showed a significant bias to perceive the display as looming.

The emotional characteristics of a looming visual stimulus can influence perceived arrival time. Looming threatening stimuli (spiders and snakes) are perceived to arrive more quickly than nonthreatening stimuli (bunnies and butterflies), and the ratings of fear of the stimulus are correlated with perceived arrival time (Vagnoni, Lourenco, & Longo, 2012, 2015). The more fearful the viewer, the sooner the object is perceived to have arrived. Other threatening stimuli (e.g., threatening faces) show similar effects (Brendel, DeLucia, Hecht, Stacy, & Larsen, 2012). However, the notion that "threat" alone is the critical factor at work here is likely an oversimplification. More recent work has also implicated the role of arousal and stimulus complexity. For example, highly arousing positive stimuli (erotica and money) are perceived to arrive as soon as fearful stimuli, and pictorial stimuli, in general, are perceived to arrive sooner that simple colored rectangles (Brendel, Hecht, DeLucia, & Gamer, 2014).

Looming and receding visual stimuli also have asymmetrical effects on attention. Objects that approach an observer attract attention and are treated with priority. For example, in a visual search task, increasing the number of distractors in a search for a receding target significantly increases search time. However, the same increase in the number of distractors has no effect on search time for looming targets (Takeuchi, 1997). This suggests that detecting looming stimuli occurs automatically. Subsequent work has shown that looming, but not receding, objects capture attention even when they are not relevant to the task at hand (Franconeri & Simons, 2003). The attentional capture effects of looming stimuli appear to be subserved by covert attentional mechanisms that do not require effortful cognitive processing (Kahan, Colligan, & Wiedman, 2011; J. E. Lewis & Neider, 2015). The evolutionary implications of the priority for looming objects are underscored by the finding that the effects of looming stimuli are magnified when the stimuli are on a collision path with the observer and are not dependent on the sudden onset of motion (Lin, Franconeri, & Enns, 2008; von Muhlenen & Lleras, 2007).

Bias for Receding Visual Motion?

Despite the wealth of data demonstrating a perceptual bias for looming visual motion, there are a number of apparently conflicting studies that find a bias for receding motion. For example, observers show better sensitivity as measured by a discrimination task for a contracting pattern of dots than an expanding pattern of dots (Edwards & Badcock, 1993; Edwards & Ibbotson, 2007). Viewers have also demonstrated better sensitivity to acceleration when viewing receding versus looming dot patterns (Mueller & Timney, 2014). Some researchers have suggested that the better discrimination for these receding optic flow fields is related to postural stability, where falling backward (which creates receding optic flow) is a greater danger than falling forward (Edwards & Ibbotson, 2007; Mueller & Timney, 2014; but see Holten, Donker, Stuit, Verstraten, & van der Smagt, 2015). Other work suggests that the direction of asymmetry is also dependent on the speed of the optic flow pattern, which varies across studies (Naito, Sato, & Osaka, 2010).

Variability in the nature of the tasks employed may also in part be responsible for the conflicting results. The majority of the studies that find an advantage in the perception of receding stimuli use optic flow fields that are composed of dots. The analogous real-world looming condition would be moving through the world without any indication that a collision is imminent. On the other hand, many of the studies that show a perceptual advantage for looming present a target stimulus that appears to approach the observer on a collision course. Given these differences, both the "postural stability" and the "looming threat" hypotheses may have merit depending on the specific circumstances. Another consideration is that the discrimination tasks typically used in studies that find an advantage for receding stimuli may require more effortful cognitive processing than, for example, visual search or time-to-arrival where attention is captured automatically or where a defensive motor response is appropriate (e.g., Franconeri & Simons, 2003; King et al., 1992; J. E. Lewis & Neider, 2015; Lin et al., 2008). Because the visual features of a looming object are thought to be processed without depleting effortful cognitive resources (Kahan et al., 2011; J. E. Lewis & Neider, 2015), looming objects may impact discrimination and visual search differentially. Looming stimuli activate motor defense mechanisms and responses of the autonomic nervous system (King et al., 1992; Low et al., 2008; Skarratt et al., 2009; Skarratt et al., 2014). This may *facilitate* the automatic responses to looming objects that do not require effortful processing but *interfere* with effortful cognitive judgments such as discrimination. Essentially, the different tasks may differentially invoke either visuo-motor or visuo-cognitive systems and account for the conflicting findings (Goodale & Milner, 1992; Tresilian, 1995).

Multisensory Integration of Looming Stimuli

Despite the fact that most of the research conducted on the perception of looming objects has been unimodal in nature, many looming objects simultaneously produce both visual and auditory cues as they approach. Some work has examined how the presence of both visual and auditory information influences the perception of looming objects and how this multimodal information is integrated. Other work has hypothesized that the information contained in a looming object is "modality-neutral." In other words, regardless of modality, the tau variable information for an approaching source can specify time-to-arrival. If the system is sensitive to this underlying change in information over time, then in theory, the modality with which the approach is perceived should have little effect on the perceived time-to-arrival (Gordon & Rosenblum, 2005).

However, despite the potential equivalence of information in the two modalities, observers show stark differences in how they use auditory versus visual information in judging arrival time (DeLucia et al., 2015; Schiff & Oldak, 1990). The visual and auditory systems have evolved to solve quite different evolutionary problems in dealing with looming objects. Guski (1992) suggested a kind of "handshaking" that occurs between the two systems, in which audition acts as an advanced warning system that provides information for a categorical decision about engaging appropriate motor behaviors (e.g., either evasive or engaging in visual tracking). The visual system can then provide the more accurate estimates of arrival time that can be used to deal appropriately with the looming object. The "warning system" hypothesis for audition is supported by work that examines time-to-arrival judgments under auditory only, visual only, and audiovisual conditions. A number of these studies have shown that performance in predicting the arrival time of a looming object with both visual and auditory information does not significantly differ from performance when only visual information is available (DeLucia et al., 2015; Hofbauer et al., 2004; Schiff & Oldak, 1990; Zhou, Yan, Liu, Li, & Xie, 2007).

The integration of auditory and visual information is present in infants as early as five months. When presented with matched and mismatched auditory and visual looming and receding signals, infants show greater attention to the stimuli that are matched in direction of travel (Walker-Andrews & Lennon, 1985). As might be expected, they also show greater attention to looming versus receding stimuli. Work with rhesus monkeys shows a similar strong orienting preference for coincident visual and auditory looming stimuli but no analogous response for coincident stimuli that were receding. Consistent with previous work in auditory looming, the orienting preference effect occurred only with tonal auditory stimuli and not with broadband noise (Maier, Neuhoff, Logothetis, & Ghazanfar, 2004; Maier et al., 2008). Preferential processing and integration of looming versus receding multisensory stimuli also occurs in adult humans and is supported by recent behavioral, EEG, and fMRI experiments (Cappe, Thelen, Romei, Thut, & Murray, 2012; Cappe, Thut, Romei, & Murraya, 2009; Maier et al., 2008; Ogawa & Macaluso, 2013; Tyll et al., 2013).

The processing advantage and environmental importance of looming stimuli are further evidenced when the perception of multimodal looming and receding stimuli are examined under congruent and incongruent conditions. Under congruent conditions, both the auditory and visual stimuli move in the same direction, either toward or away from the observer. Under incongruent conditions, the auditory and visual stimuli move in opposite directions. If listeners are asked to judge the direction of the moving sound and ignore the visual stimulus, performance is better in the presence of receding stimuli. In essence, observers can more easily ignore the incongruent receding visual stimulus. This indicates that looming visual stimuli show stronger visual capture effects than receding visual stimuli (Harrison, 2012). Similar work has shown that varying the congruency of a visual stimulus has little effect on detection rates for looming sounds (where accuracy is generally high). However, a congruent versus an incongruent visual stimulus does increase the detection of receding auditory stimuli (Liu, Mercado, & Church, 2011).

Looming auditory and visual objects have also been shown to influence tactile perception. For example, a looming visual object that approaches the face increases tactile sensitivity at the predicted time and location that the object would contact the face (Clery, Guipponi, Odouard, Wardak, & Ben Hamed, 2015). Similar tactile effects are obtained with looming sounds (Teneggi, Canzoneri, di Pellegrino, & Serino, 2013). Canzoneri, Magosso, and Serino (2012) found that a sound moving toward an observer's hand speeded up the processing of a tactile stimulus at the hand when the sound was within the boundaries of peripersonal space representation. The effect was significantly stronger for looming sounds than for receding ones (but see Finisguerra, Canzoneri, Serino, Pozzo, & Bassolino, 2015).

Conclusions

Looming objects are salient and behaviorally relevant stimuli that are perceptually and neurally processed with priority when compared with equivalent receding objects. Looming stimuli also preferentially activate the autonomic nervous system and the motor system. They bring about differential emotional, cognitive, and defensive responses. Although this perceptual anisotropy can sometimes create distortions in perceived egocentric spatial relations, they are distortions that over time have enhanced the probability of survival and reproduction. Thus, these perceptual biases have been favored by evolution.

Part III

Dynamics in Scenes

Chapter 13

Expanding Space

Does Imagination Affect Boundary Extension for Visual Scenes?

Helene Intraub and Kristin M. Gagnier

Memory errors are known to occur in the face of misinformation, divided attention, and memory overload (Koriat, Goldsmith, & Pansky, 2000). A memory error that requires none of these is boundary extension, false memory for the continuation of a scene beyond the physical bounds of a view (see Hubbard, Hutchison, & Courtney, 2010; Intraub, 2010). This anticipatory error is difficult to explain in the context of a traditional modality-centric model of scene representation (i.e., a "visual representation", e.g., Henderson & Hollingworth, 1999), because memory beyond the view has no visual-sensory correlate. We discuss an alternative spatio-centric view in which the bottom-up source (the visual input), in combination with multiple top-down sources that predict the likely continuation of the scene, are organized within the observer's framework of surrounding space (Intraub, 2010, 2012). The anticipatory component of this construction may facilitate integration of successive views during visual scanning and navigation. In this chapter, we describe a model of scene representation that takes this spatio-centric perspective, discuss a variety of factors known to affect boundary extension, and explore the potential effects of imagination on the size of the boundary error. We suggest that determining the factors that do and do not affect boundary extension will help us to refine our understanding of scene representation, and that at this point, the constraints we report appear to underscore the adaptive value of this anticipatory error.

Boundary Extension: A Brief Overview

Originally discovered in the context of long-term memory (Intraub & Richardson, 1989; Intraub, Bender, & Mangels, 1992), it is now known that boundary extension occurs for briefly presented pictures (e.g., 250 msec) across masked retention intervals ranging from seconds (Intraub, Gottesman, Willey, & Zuk, 1996; Intraub, Hoffman, Wetherhold, & Stoehs, 2006) to fractions of a second brief enough to be commensurate with the duration of a saccadic eye movement (e.g., 42 msec; Dickinson & Intraub, 2008; Intraub & Dickinson, 2008). In saccade-contingent eye-tracking experiments, boundary extension was also observed across an actual saccade, when a briefly presented stimulus was presented on one side of the screen and a saccade-contingent test picture was presented on the other, requiring an eye movement (Dickinson & Intraub, 2008; Intraub & Dickinson, 2008). Although most research on boundary extension tests visual memory, other studies have demonstrated boundary extension following haptic exploration of scene-regions (bounded by wooden frame, like a window frame) when participants were blindfolded (Intraub, 2004; Intraub, Morelli, & Gagnier, 2015; Mullally, Intraub, & Maguire, 2012) and when a woman who had been deaf and blind since early life explored the scene-regions (Intraub, 2004).

The fundamental nature of boundary extension is supported by developmental research. When infants as young as 3–4 months of age were presented with a photograph and then closer and wider views of the same scene were shown, preferential looking was consistent with boundary extension (Quinn & Intraub, 2007). Boundary Extension has also been observed in a variety of memory tasks with children (4–16 years old; Candel, Merckelbach, Houben, & Vandyck, 2004; Chapman, Ropar, Mitchell, & Ackroyd, 2005; Kreindel & Intraub, 2017; Seamon, Schlegel, Hiester, Landau, & Blumenthal, 2002), as well as in studies with older adults (Multhaup, Munger, & Smith, 2016; Seamons et al., 2002). Neuropsychological and neuroimaging studies have suggested that scene construction may be associated with hippocampal activity (Chadwick, Mullally, & Maguire, 2013; Maguire, Intraub, & Mullally, 2015; Mullally et al, 2012), and that neural activation in brain regions associated with perception of scene layout and with understanding a view within its broader geographic context (parahippocampal cortex and retrosplenical complex, respectively; Epstein, 2008, 2011) is associated with boundary extension (Chadwick et al., 2013; Park, Intraub, Yi, Widders, & Chun, 2007).

Multisource Scene Representation

Given that boundary extension can occur after only a brief presentation and a 42 msec retention interval, Intraub and Dickinson (2008) proposed that scene construction may begin during scene perception, while the stimulus is still in view. They proposed the *multisource model of scene representation* in which rapid *scene construction* beyond the bounds of the view draws on multiple top-down sources of information, such as: (1) amodal perception of objects (Kanizsa, 1979) and surfaces (Fantoni, Hilger, Gerbino, & Kellman, 2008; McDunn, Siddiqui, & Brown, 2014) cropped by the view-boundaries, (2) rapid perception of layout and context (e.g., Greene & Oliva, 2009) that support expectations about continuity, and (3) object-to-context associations (Bar, 2004). These sources, in combination with the bottom-up source (in this case, vision), are organized within a surrounding spatial framework centered on the given view. Thus the ensuing representation is a multisource representation set within a spatial framework that captures the continuous nature of scenes. To explain this further we offer the following illustration.

As recounted in Intraub (2010), the author had always understood the photograph in Figure 13.1 (panel A) as showing curbside garbage receptacles outside a house, awaiting pickup, taken by a photographer who stood in the street with the house across the street behind him. In fact, she admonished subsequent student photographers not to set the tripod in the street for safety reasons. When asked, another researcher disagreed, saying he had always understood the receptacles as being in the owner's backyard, with the owner's house behind the photographer. To settle the "dispute," a third researcher (the coauthor of the current chapter) was asked, and reported that the receptacles and fence were in an alley, and that the photographer was in the center of the alley with another fence behind him, adding, "Where else would it be?" Six other researchers were queried (reported in Intraub, 2014), and they provided similar descriptions, except for one respondent who reported only that the garbage receptacles were "outdoors." This respondent resided outside the United States, and reported having seen such garbage receptacles only in cartoons.

These accounts suggest that the photograph was not remembered simply as a photograph, but that respondents had interpreted the photograph within a broader scene context, the most generic being "outdoors," but with the others (which were offered by respondents

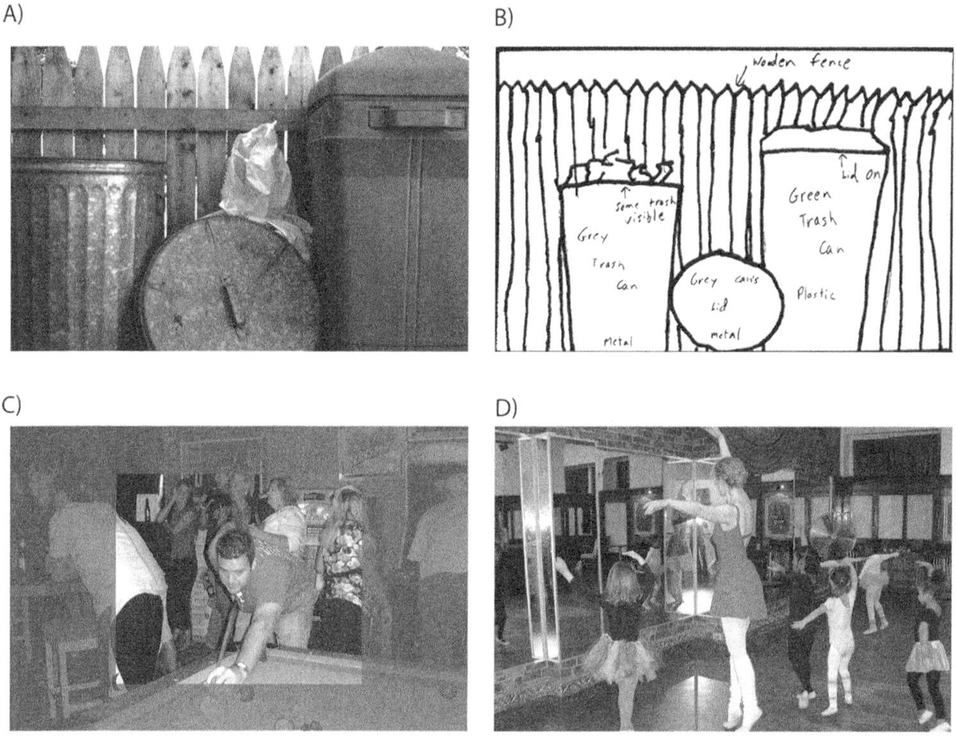

Figure 13.1 The figure shows a stimulus view (panel A) and a drawing made by a participant demonstrating boundary extension (panel b; images are taken from Figure 1 in Intraub & Richardson, 1989). Panel C is a frame of the animation shown to participants to illustrate imagination "filling out" the scene; and panel D shows an example of the stimuli presented in Gagnier and Intraub (2012) and in Gagnier (2010). Note that all photographs were presented in color in their associated experiments.

who were familiar with these types of receptacles and their associated context), being quite specific. According to the multisource model, most of the construction beyond the boundaries would not be misattributed to vision. Misattribution would include a region just beyond the boundaries – one that would be heavily supported by amodal perception. The notion is that amodally perceived information just *outside* a boundary is so highly constrained by the visual information just *inside* that boundary that it is sufficiently similar to be accepted as having been seen before. Intraub and Dickinson (2008; also Intraub, 2010, 2012) proposed that the decision processes used to determine how much one saw before are the same as those described in the general source monitoring framework provided by Johnson, Hashtroudi, and Lindsay (1993; also Lindsay, 2008) – a model that was developed to explain source monitoring errors in long-term memory, but which also provides a good explanation of boundary extension.

The key concept in the source-monitoring framework is that memories are not tagged with respect to their source. Rather, memories associated with different sources have different "profiles" that vary across a set of characteristics (such as sensory detail, coherence, co-temporal events, and awareness of cognitive operations). Misattribution occurs when the profile associated with a specific memory matches less with the expected profile of its source than with the profile of an alternative source. For example, if a dream is unusually vivid and percept-like, highly coherent, and encompasses a plausible flow of changes over time (a profile more similar to memory for a perceived event, than to memory for a typical dream), one might misattribute the remembered dream, at least without additional, more detailed monitoring, to perception (i.e., to having actually occurred).

In terms of boundary extension, according to the multisource model, the scene representation is inhomogeneous (similar to the visual field) with quality and detail decreasing from memory for visual information, to memory for amodally generated information, to memory for anticipated information based on scene layout and object-to-context associations. In this view, the constructed scene content just beyond the boundaries is indistinguishable from memory for the visually presented information (inside the boundaries), and is therefore misattributed to vision. In other words, the farther from the boundaries, the greater the disparity between the visual representation and the graded scene construction.

In considering the respondents' descriptions of the photograph in Figure 13.1A (described earlier), for all respondents, the fence and the receptacles were outdoors, with the sky above. However, depending on familiarity with the objects in the scene, respondents either placed the location within a specific suburban U.S setting or, without this experience, represented simply as "outdoors". In no case did any of the respondents believe that they had actually viewed the more distant surrounding environment (which included what they "always" thought existed *behind* the photographer). We have no measure of their boundary extension, but we would expect it to be no different than the scores of participants who have drawn this photograph from memory. That is, the respondents' drawings would likely have included the completed receptacles, more of the fence to the sides, more of the sky above, and object completion at the bottom boundary. Similar to the way that scenes surround us in the world, the respondents had remembered the view, not so much as a bounded picture, but as receptacles and a fence in a location that continued well beyond the bounds of the photograph. In this view, boundary extension is just the "tip of the iceberg" – the leading edge of a more expansive, multisource scene representation.

So what determines how much anticipated space will be misattributed to vision (i.e., how large a boundary extension error will occur)? There are several *stimulus* factors that have been shown to affect boundary extension: for example, the scope of the view (i.e., how close-up or wide-angle it is: e.g., Intraub et al., 1992), the proximity of objects to the boundaries of the picture (Bertamini, Jones, Spooner, & Hecht, 2005), and when the distance between the object and boundaries is held constant, the amount of the background that is visible as the angle of view changes (Gagnier, Intraub, Oliva, & Wolfe, 2011). In the next section we will consider other kinds of factors that have been shown to modulate boundary extension when stimulus factors are held constant.

Cognitive Factors That Affect Boundary Extension

According to the multisource model, factors that enhance the difference between the constructed continuation of the scene and memory for the visually perceived information (within the boundaries) would be expected to decrease boundary extension (reduce the misattribution error), whereas factors that minimize the difference would be expected to increase boundary extension (increase the misattribution error). In this section we will discuss five factors that have been found to modulate boundary extension: (1) *dynamic test effects*, (2) *attention effects*, (3) effects of the *medium of presentation* (line drawings vs. photographs), (4) *neuropsychological effects* (in the case of bilateral hippocampal lesions), and (5) *developmental effects* (when preschoolers and adults study the same pictures).

Dynamic Test Effects. In dynamic tests, participants were asked to adjust the view boundaries of a test image on the computer, to match memory. When the initial test view showed a more expansive view than the stimulus, participants exhibited greater boundary extension than when the initial test view showed a less expansive view (Chapmanet al., 2005; Dickinson & Intraub, 2009; Intraub et al., 2006; Gagnier, Dickinson et al., 2014; Gagnier et al., 2011). It is tempting to think of this as a simple anchoring effect, in which a boundary's initial position affects how far participants will move that boundary as they recreate the remembered view (Chapman et al., 2005; Dickinson & Intraub, 2009). However, when adjusting boundaries at test, participants do not tend to move the boundaries in one direction and stop (i.e., inward when the initial test view is more expansive than the stimulus, and outward when the initial test view is less expansive that the stimulus). Instead, they tend to shift the boundaries inward and outward as they hone their reconstruction, irrespective of the scope of the initial test view, confounding a simple anchoring account. Greater boundary extension when the initial test view was more expansive than the stimulus view is a robust finding, occurring even when image size and image placement on the screen were changed between presentation and test, thus eliminating any benefit of using the screen borders as reference points (Intraub et al., 2006). The multisource model provides a different explanation of the effects of the scope of the initial test view on boundary extension.

According to this account (Intraub et al., 2006), when the test view initially shows more of the scene than before, this information (now seen) strengthens the constructed representation just beyond the original boundaries causing participants to subsequently accept a greater swath of constructed space as having been seen than when the initial view showed less of the scene than the stimulus. When the initial test view shows less of the scene than was originally shown, boundary extension is not eliminated, but is reduced in size as a smaller swath of the constructed space is misattributed to having been seen before. When the initial test view is identical to the stimulus view (Gagnier et al., 2011; Gagnier et al., 2014), boundary extension also occurs, but is smaller than when the initial test view is more expansive than the stimulus. In this account the expanse of the initial test view can impact the participant's source-monitoring decision.

Divided Attention. In research on boundary extension, participants are typically instructed to try to remember each picture in as much detail as possible. One possibility is that this intention to remember might be instrumental in eliciting scene construction. Intraub, Daniels, Horowitz, and Wolfe (2008) set out to determine if engaging in a very demanding visual search task (primary focus) while viewing a briefly presented photograph might eliminate (or minimize) boundary extension. Superimposed on close-up photographs were block 2s and 5s (stimuli that are very difficult to discriminate). On each trial the composite display was presented for 750 msec. There were three independent conditions: (1) *search-alone* where participants reported the number of fives and were told not to let the picture distract them; (2) *dual-task* where participants' primary task was to report the number of fives and that scene memory would also be tested; or (3) *memory-alone* where participants were told to ignore the numbers and that scene memory would be tested. Critically, search-alone performance verified the difficulty of the task, and search performance did not differ between the search-alone and the dual-task conditions. Results showed that dividing attention did affect boundary extension. However, rather than eliminating (or reducing) it, divided attention led to *greater* boundary extension.

The authors concluded that divided attention did not prevent scene construction, but did reduce the quality of visual memory for the briefly presented scenes. With lower visual quality, the similarity between the visual information and the scene construction beyond is greater than when participants can focus their full attention on the briefly presented picture. As a result, divided attention led to greater boundary extension. Following a similar logic, in other research, when instructions drew attention (and oculomotor activity) to the boundaries during presentation, a condition that is expected to enhance memory for the boundary region, boundary extension was smaller than when participants were simply told to remember the pictures without mentioning the boundaries (Gagnier, Dickinson, & Intraub, 2013; Intraub & Bodamer, 1993).

Presentation Medium (Line Drawings vs. Color Photographs). Gagnier and Intraub (2012) compared boundary memory for photographs and for line drawings that were traced from the photographs. This allowed them to create versions of the same scenes in which the layout of the objects and the general meaning ("gist") of the scenes were preserved but the medium of presentation differed. After the participants studied either line drawings or photographs for 15 seconds each, boundary extension was assessed by presenting the same pictures again and asking participants to rate each one as being "the same," "closer," or "farther away" than before, on a 5-point scale. Line drawings led to greater boundary extension than photographs, regardless of whether the scenes were complex (e.g., the ballet class scene shown in Figure 13.1D: Experiment 1) or very simple (a single object on a background surface; e.g., a man sitting in front of a wall: Experiment 2).

To explain the results, Gagnier and Intraub (2012) appealed to the same "similarity" mechanism discussed in the previous section. If we assume that the top-down continuation of the simple lines in the line drawings led to greater similarity between the remembered visual content and the constructed continuation of the view, than for the photographs (which, in addition to lines, include textures, colors, luminance changes, and so forth), then greater boundary extension for the line drawings can be explained.

Neuropsychological Studies: Scene Construction Deficit. Hassabis, Kumaran, Vann, and Maguire (2007) reported a previously unknown disability in a group of bilateral hippocampal patients. In addition to the expected memory deficits associated with these lesions, patients exhibited a profound inability to construct a spatially coherent image of surrounding space. If asked to imagine themselves on a sandy beach, for example, they could list items likely found on a beach (semantic associations) but were unable to create a spatial image. Their subjective reports and analysis of their descriptions suggested that they were unable to experience a coherent spatial representation of a novel scene. Mullally et al. (2012) sought to determine if this disability might impact scene construction beyond the boundaries of the view, causing the patients (who have very impaired memory) to actually exhibit *less* boundary extension (i.e., a more veridical memory for the boundaries) than healthy control participants.

Mullally et al. (2012) demonstrated that their seven patients (six new participants and one prior participant who had no memory of the earlier tasks) exhibited the same scene construction deficits as in Hassabis et al. (2009). For example, one participant's response to imagining a novel scene was, "There is no scene in front of me here. It's frustrating because I feel like there should be. I feel like I'm listening to the radio instead of watching it on the TV. I'm imagining different things happening, but there's no visual scene opening out in front of me" (Mullally et al., 2012, p. 266). Participants also took part in a "step-back task" in which they viewed a photograph and described what the scene might look if they could "step back" and see beyond the photograph's boundaries. They made semantically relevant predictions, but again claimed to not be able to "visualize" this, and indeed analysis of their verbal descriptions differed from control participants *only* in terms of their relatively limited use of spatial terms in describing what they would likely see.

To test boundary extension, the patients and their controls participated in three different boundary extension tasks: a drawing task, an immediate recognition task (across a 250-msec masked interval; as in Intraub & Dickinson, 2008), and a haptic exploration task. On the two visual scene tasks, patients exhibited less boundary extension than did control participants. Patients' drawings did not differ in terms of quality or detail (judges were blind to group membership, and a sorting task showed that they could not discriminate patients' drawings from those made by control participants). Patients simply made a smaller boundary extension error. In the brief boundary extension task, again, patients exhibited less boundary extension, correctly recognizing that the test pictures showed the same view as the stimuli more frequently than did the control participants. Whereas control participants were more confident on trials on which they made a boundary extension error, patients were more confident on trials on which they correctly identified the test picture as being identical to the stimulus. In the haptic task, on average, control participants exhibited boundary extension whereas patients did not.

It is important to note that attenuated boundary extension for visual scenes was associated with patient's *scene construction* disability. When another group of bilateral hippocampal patients, who *did not* exhibit this disability, participated in a drawing task and a brief recognition memory task to assess boundary extension, their performance did *not* differ from control participants' performance (Kim, Dede, Hopkins, & Squire, 2015; for a discussion of the two studies see Maguire et al., 2015). Results suggest that a disability affecting scene construction (the first stage in the multisource model) reduced boundary extension.

Developmental Differences. Although early research with children (6–11 years of age; Candel, 2004; Seamon et al., 2002) provided evidence for boundary extension in drawing tasks, developmental differences in drawing ability prevented researchers from drawing conclusions about changes across age groups. The source monitoring literature suggests a developmental trajectory (from 3 to at least 8 years of age) in the ability to distinguish self-generated from perceived information (e.g., Foley & Johnson, 1985; Lindsay, 2008; Sluzenski, Newcombe, & Ottinger, 2004; Sussman, 2001). If, as proposed in the multisource model, boundary extension is a source monitoring error, then young children might be expected to accept more of their constructed scene representation as having been seen before than would adults, resulting in a greater tendency for boundary extension.

Kreindel and Intraub (2017) developed a two-alternative forced-choice recognition test appropriate for young children (4–5 years of age) and adults (undergraduates) that eliminated the problems inherent in the drawing task. Beginning with a set of 40 scenes, they created a close-up and a wide-angle version of each scene. On each trial, either the close-up or the wide-angle view was presented for study (15 seconds), followed by both members of the pair. Participants indicated which member of the pair matched the study picture. To increase the sensitivity of the test (because one does not know a priori how large boundary extension might be), they included two types of trials: high-similarity trials in which the wider view revealed 13% more of the scene than the close-up view did, and low-similarity trials in which the wider view revealed 30% more. If boundary extension occurs, one would expect to see more errors on trials where the study view was the close-up (and participants erroneously selected the wider view) than when the study view was the wide-angle view (and participants erroneously selected the close-up).

Children and adults both exhibited boundary extension with children showing a greater effect. A guessing-game experiment demonstrated that this outcome did not reflect a bias to select the wider view in the test pairs; in fact, when guessing, children tended to select the closer view (a bias that in the memory task would result in boundary restriction). More importantly, in another experiment, when a new set of 40 scenes that were relatively wide-angled were used to create the stimuli (a stimulus factor known to reduce or eliminate boundary extension in adults; Intraub et al., 1992; see Hubbard et al., 2010 and Intraub 2002), children again exhibited robust boundary extension, whereas adults rarely made any errors at all. Only when stimulus duration was reduced to 5 seconds (from 15 seconds), making the task more difficult, did adults exhibit boundary extension for the more wide-angle set. This, along with subsequent research on younger and older children (Spanò, Intraub, & Edgin, 2017), suggests that young children's relatively immature source-monitoring ability leads to greater boundary extension.

Is Imagination One of the Cognitive Factors That Affects Boundary Extension?

In the last section, we presented a number of cognitive factors that have been found to increase or decrease boundary extension in memory for visually presented scenes. Next we ask if visual imagination during encoding is another cognitive factor that might impact the size of the boundary error. The premise here is that visual imagination might add details to the constructed continuation of the scene beyond the boundary, thus increasing its similarity to visual memory and enhancing boundary extension.

Turning to the traditional source monitoring literature, which focuses on the ability of participants to discern the source of different events, there is evidence to support the plausibility of this hypothesis. Johnson and colleagues (Johnson, Taylor, & Raye, 1977; Johnson, Raye, Wang, & Taylor, 1979) used a frequency estimation task to evaluate source-monitoring errors. In one experiment, participants were presented with repetitions of line-drawn objects (2, 5, or 8 times) and, on other intermixed trials, were required to generate images of those objects, in response to verbal labels (Johnson et al., 1979). Participants' estimates of how often they had seen each object increased as a function of how often that object had been imagined. Furthermore, they found that this effect was enhanced for "good imagers" as compared to "poor imagers" (participants who had scored at the top 35% or bottom 35%, respectively, on an imagery test). Results suggested that self-generated images of objects could be confused with perception of line drawings of those objects, and that the more "percept-like" the imagery, the greater the likelihood of the type this misattribution.

Whereas imagination can increase source misattributions by enhancing perceptual detail, there is evidence that explicit imagination instructions can also reduce the incidence of source monitoring errors, under conditions that heighten participant's awareness of having recruited imagination (*awareness of cognitive operations*; Johnson & Raye, 1981; Johnson et al., 1993). For example, in Durso and Johnson (1980), participants viewed words and pictures of the same concepts. During presentation, participants either (1) made judgments about the function of each item, (2) identified a salient feature of each item, or (3) created an image of the item and rated its quality on a 10-point scale. The first two questions were intended to lead to incidental imagery, whereas the last question (which explicitly instructed participants to imagine) was intended as a means of raising participants' awareness of cognitive operations. At test, participants heard a list of concepts, and indicated whether each had been presented, and, if it had, then indicated whether it had been presented as a picture or a word. In line with their prediction, results indicated that source-monitoring errors were least likely in the "create an image" condition where awareness of cognitive operation was expected to be relatively high.

Intraub and Hoffman (1992) provided additional support for this proposal. First they asked participants to take part in a photograph memory experiment. Participants were asked to remember a long sequence of photographs (followed by two recall questions to give plausibility to this cover story). Next, they participated in a reading comprehension experiment. On each trial, they read a brief paragraph describing a scene and answered a question about it (intended to encourage them to think about the scene). A small subset of paragraphs actually described photographs from the earlier "experiment." After "both experiments" were completed, an unexpected paragraph recognition memory and source-monitoring task required participants to indicate which paragraphs they had seen before, and then, if they recognized a paragraph, to indicate whether or not they had seen a photograph corresponding to that paragraph in the earlier "photograph memory experiment." Participants made many source misattribution errors, reporting seeing photographs that had never been presented. In one experiment, however, to determine if awareness of image creation during reading would affect the number of misattribution errors, a similar experiment was run, but a second group was added. These participants did not answer a question about each paragraph during the reading phase of the study. Instead, they rated how easy or difficult it was to *visualize* what they read. At test, these participants made fewer source-monitoring errors than did participants who answered a question about each paragraph, although they were just as good at identifying paragraphs that actually did correspond to seen photographs. This suggests that awareness of imagining helped participants to distinguish the source of their image (self-generated vs. seen).

These findings from the source monitoring literature motivate two hypotheses about the relationship between explicit vs. implicit imagination to boundary extension. Explicit imagination beyond the bounds of a view may increase the probability of source misattributions, although it is possible that awareness of cognitive operations could work against that. In contrast, tasks that elicit implicit imagination beyond the bounds of a view may be more likely to increase boundary extension. Implicit imagination might enhance scene construction beyond the view without increasing the participant's awareness of cognitive operations.

Explicit Imagination. In this section we will discuss the effects of explicit imagination on boundary extension for non-scene displays in which the background was not visually presented, but was imagined, and for visual scenes (color photographs) in which visual information about the scene's background was present and participants were asked to augment their experience by engaging in a specified imagination task.

Explicit visual imagination beyond the boundaries of non-scene displays. Intraub, Gottesman, and Bills (1998) examined whether explicit imagination of a scene background would be sufficient to produce boundary extension. They showed participants pictures of line-drawn objects on blank backgrounds (Intraub et al., 1998; also see Gottesman & Intraub, 2002). Participants studied the line drawings for 15 sec each after being instructed to remember each one in as much detail as possible, including the size of the object. In the *control condition*, participants received only that instruction. In the *imagine-object-color condition*, they were provided with the colors of the object and were asked to "project them onto the picture to help memory." In neither condition was there any suggestion of a scene, and boundary extension did not occur, instead memory errors involved regression to the mean object size. However, when other participants were asked to imagine a background while remembering the same stimuli, projecting it onto the picture, boundary extension occurred.

These results demonstrated that recruiting scene construction through explicit imagination of a background was sufficient to elicit boundary extension for displays that otherwise did not yield the boundary extension error. This raises the question of whether explicit imagination instructions would also affect boundary extension under conditions in which the background in the stimulus actually contains detailed visual information (a photograph of a scene).

Explicit visual imagination beyond the boundaries of a view. In an experiment we conducted in preparation for Gagnier's (2010) dissertation research, we tested the effect of imagination instructions on memory for visually presented scenes. In both the control group and the explicit-imagination group, participants were instructed to remember the scenes in as much detail as possible, and were warned that the background was as important to remember as the objects. In the *explicit-imagination* group an additional instruction was presented. Participants were instructed to imagine the continuation of each scene and to mentally "project" it onto the screen while viewing the photograph. To illustrate the instruction, participants were presented with an animation in which the peripheral region of a scene slowly faded into view – an illustration of a time slice from the animation is presented in Figure 13.1 (panel C). Participants (N = 32 in each group) viewed 12 close-up complex, multi-object photographs such as is shown in Figure 13.1 (panel D) for 14 seconds each. After the 12-item sequence was presented, the same close-up pictures were presented at test, and participants rated each test picture, indicating on a 5-point scale if the camera's position was "the same," "closer up," or "farther away" than before (as in Intraub & Richardson, 1989). The mean rating for each condition is shown in Figure 13.2 (panel A). As may be seen in the figure, significant boundary extension occurred in both conditions, but explicit imagination had no affect boundary extension, $t(62) = .49$, *ns*.

More recently, in a series of experiments, Munger and Multhaup (2016) took a different approach to studying the effects of explicit imagination instructions on boundary extension, and also observed null effects. In one experiment, participants viewed complex scenes (e.g., city markets, deserts, animals, and natural landscapes) and were instructed either to visually imagine the scene beyond the left or right boundary of each photograph (Experiment 3A) or to mentally "zoom in or zoom out" of the picture the way a photographer can do with a camera (Experiment 3B). Following Intraub and Richardson's (1989) original design, they created a close-up and a more wide-angle view of each scene and presented half of the photographs in their close-up version and half in their more wide-angle version (counterbalanced across participants). Test pictures showed either the same version of the scene or the alternate version. In both experiments, participants exhibited boundary extension when the identical picture was presented at test and the asymmetric response pattern that is diagnostic of boundary extension when the alternate picture was presented at test (when the test picture was a wider view, participants rated it as more similar to the original than when the test picture was a closer view).

Explicit multimodal imagination beyond the boundaries of the view. In other experiments, Munger and Multhaup (2016) used different explicit imagination instructions in an

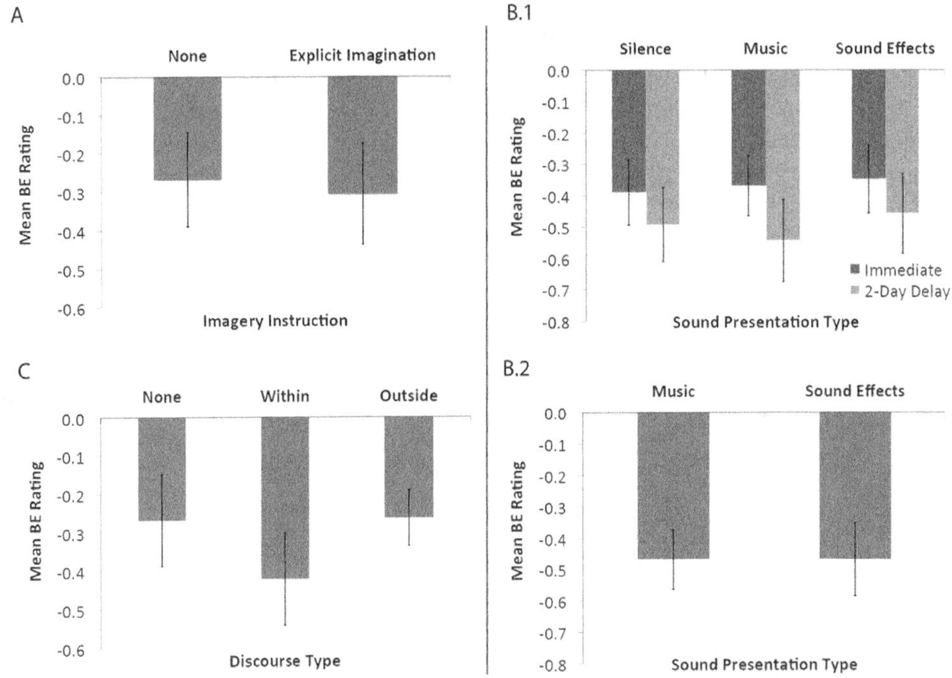

Figure 13.2 The figure shows boundary scores in the explicit imagination experiment conducted by the authors (panel A); the sound effects experiment (Gagnier 2010; Section 3 experiments) (panels B.1 and B.2); and the discourse experiment (Gagnier 2010; Section 2 experiments) (panel C). Error bars designate the .95 confidence interval. Negative scores that differ from 0, "same," indicate boundary extension. None of these factors affected boundary extension. Note that the only factor that did affect boundary extension was an increase in retention interval between the study and test (from immediate to a two-day delay).

attempt to increase boundary extension. They instructed participants to "enrich the scene representation" by imagining the sounds and smells associated with each scene. The rationale for why this type of imagination (focused on the content of the view rather than the sounds or smells from beyond the boundaries) would be expected to enhance imagination beyond the view was not clear, but the implication was that enriching the experience of viewing the scene might also enhance the constructed continuation of the view. They conducted the same experiment with complex views (Experiments 1A and 1B) and with single-object scenes of the type of that have most often been used in boundary extension studies (Experiments 2A and 2B). In all conditions, all the patterns associated with boundary extension were obtained, but in no case was any difference observed between the multimodal imagination condition and the control condition. In sum, neither in our experiment nor in those conducted by Munger and Multhaup (2016) was boundary extension affected by explicit imagination instructions.

Implicit Imagination. As part of Gagnier's (2010) dissertation research, she conducted a series of experiments in which she attempted to induce imagination in the absence of any imagination instructions. She reasoned that awareness of cognitive operations might have prevented an effect of explicit imagination in the imagination study we conducted (described in the preceding section). In one series of experiments she attempted to induce imagination beyond the bounds of the view by pairing each picture with related sound effects, and in a second series she sought to achieve the same goal by pairing each view with recorded discourse that referred to people or objects falling outside the bounds of the view. We describe these experiments in the next sections.

Implicit multimodal imagination of scene content beyond the boundaries of a view. Henkel, Franklin, and Johnson (2000) demonstrated that sound effects in combination with imagination (e.g., the sound of a bouncing basketball while visually imaging the bouncing ball) increased the misattribution of imagined events to having been seen. They had participants experience an event, such as a person bouncing a basketball, via seeing, hearing, imagined seeing, imagined hearing, or a combination of imagined seeing plus hearing and asked whether hearing along with imagining seeing would lead participants to confuse the source of the two experiences. Their results indicated that participants were more likely to say they had seen a picture when they had imagined it and were provided with auditory information from the picture. This finding suggests that imagining events for which there is also perceptual evidence increases the likelihood of source-monitoring errors.

Drawing on this finding, Gagnier (2010; Section 3, Experiment 1) sought to determine if sound effects might also serve to strengthen scene construction beyond the boundaries of the view. To select sound effects that evoked a sense of space beyond the view versus sound effects that emanated from the visible portion of the photograph, she pretested combinations of sound effects and photographs. Participants rated each of 24 photograph-sound combinations as: (a) evoking a sense that the scene continues beyond what you can see in the picture, (b) being limited to the picture (not evoking a sense of the scene continuing), or (c) a mismatch. The pictures consisted of close-up views of complex, multi-object scenes, and each photograph-sound combination was presented for 14 seconds. Gagnier selected picture-sound combinations that were rated as eliciting a sense of surrounding space for at least 70% of the participants; 16 photograph-sound combinations met this criterion. For example, all participants agreed that the close-up photograph of a batter in a stadium hitting a baseball + the sound of a ball being hit and the crowd cheering evoked a sense of the scene's surroundings, whereas only 18% rated a picture of a sailboat + the sounds of seagulls and water splashing as evoking this sense (82% reported that the sound seemed to emanate from the picture).

All participants viewed the 16 photographs for 14 seconds each and were instructed to remember the pictures in as much detail as possible, including the background. In the *silent-control* group the pictures were shown without any sound. In the *sound effects* group, the photograph-sound combinations were presented, and in the *music* group (included to control for the effect of any meaningful sound during presentation) each picture was paired with a 14-second segment of instrumental guitar music (from "Birdsong" by Phil Campbell). In all conditions, memory was tested for the pictures alone using the same 5-point scale as in the previously described studies. The mean boundary rating for each condition is shown in Figure 13.2 (panel B.1 top; dark-gray bars); as may be seen in the figure, boundary extension occurred in all three groups. Planned comparisons verified no difference in boundary ratings between the silent-control group and either the sound effects group, $t(93) = .56$, $p = .58$, or the music group, $t(93) = .29$, $p = .77$. Focusing on the two sound conditions, a t-test showed no effect, $t(62) = .29$, $p = .77$. Although the sound effects were chosen because other participants had rated them as evoking a sense of the surrounding scene, these sounds did not cause participants to misattribute a greater amount of surrounding space to the visual modality when boundary extension was tested within a few minutes.

Given that source-monitoring errors can increase with delay (Intraub & Hoffman, 1992; Jacoby, Kelley, Brown, & Jasechko, 1989), Gagnier (2010, Section 3, Experiment 2) conducted the same experiment, but this time included a two-day retention interval between presentation and test. As shown in Figure 13.2 (panel B.1 top; light-gray bars), again, boundary extension occurred in all three conditions, but as before, no difference between the groups was observed. When performance was contrasted with that in the silent-control group, planned comparison revealed no effect of sound effects, $t(93) = .39, p = .70$, or music, $t(93) = .60, p = .55$. However, a comparison across the two experiments demonstrated that boundary extension for these stimuli was not "fixed" – boundary extension was affected by the two-day delay. A 3 (sound condition) × 2 (retention interval) ANOVA showed that boundary extension was significantly greater after a two-day delay, than when memory was tested immediately, $F(1,186) = 8.03, p<.01$. Again there was no effect of sound condition, and there were no interactions ($F<1$ in both cases). The effect of time was similar to that observed in Intraub et al. (1992: Experiment 2). In terms of source monitoring, the delay may have reduced the quality of visual memory, decreasing the difference between visual memory for scene content just inside the boundaries and the constructed continuation just beyond the view, but if the sound effects evoked imagination of surrounding space, it clearly did not cause more constructed space to be misattributed to vision.

Finally, Gagnier (2010; Section 3, Experiment 3) sought to determine if sound effects might impact spatial memory if participants were not informed that memory for the pictures would be tested. Participants were randomly assigned to the two sound conditions (sound effects and the music control), using the same presentation sequences as in Experiment 1 (where participants viewed each picture for 14 seconds), but instead of instructing participants to remember the pictures, she asked them to rate how well the picture-sound combinations fit together after viewing the presentation sequence. Participants were told they would see a slide show with a soundtrack and that for each slide they would rate (1) how well this particular sight and sound combination meshed on a 5-point scale ranging from very well to not at all, and (2) whether the integration of the two was the same across the 14 seconds, improved with time, decreased with time, or was the best in the middle. After viewing each picture-sound pair, participants were given 10.5 seconds to make their two ratings about the fit before the next combination was presented. After viewing all picture-sound combinations, participants received an incidental boundary memory test, identical to the test in Experiment 1. The mean boundary ratings in the sound effects and music conditions are shown in Figure 13.2 (panel B.2 bottom). Boundary extension occurred in both conditions, but there was no difference between them, $t(62) = .01, p = .99$.

In sum, source monitoring in scene memory is clearly not so malleable as to cause confusion *across* modalities in these experiments. In other research, Intraub et al. (2015) observed a similar lack of a crossmodal source-monitoring error – in this case between vision and haptic perception. Although visual perception of 3D scene-regions led to greater boundary extension than did haptic perception of the same regions, for participants who viewed the scene while they haptically explored them (i.e., bimodal exploration), boundary extension was the same as with haptic exploration alone. Boundary extension was unaffected by the addition of vision.

Implicit imagination beyond the boundaries of a view guided by language. Gagnier (2010; Section 2, Experiments 1–4B) took a different approach to implicit imagination in the next series of experiments. She presented discourse during picture-viewing to evoke implicit visual imagination of the scene just beyond the photograph's boundaries.[1] Pictures such as the one in Figure 13.1 (panel D) were paired with a sound recording made by a professional voice-over artist.[2] Across a series of experiments, in the *outside* condition, the discourse referred to information that was not visible in the scene but provided plausible content for the continuation of the scene *outside* the view boundaries. For example, the *outside* discourse for the picture in Figure 13.1 (panel D) was, "Sitting in the back of the ballet class, the parents are clapping approvingly." In the *within* condition, the discourse referred to visible information within the stimulus boundaries and thus would not be expected to elicit anything about the content outside the boundaries. For example, the *within* discourse for the same photograph was, "The ballet teacher is showing the ballet class how to hold up their arms." A silent-control condition was also included to determine if discourse (irrespective of content) would have any effect on boundary memory.

Gagnier (2010, Section 2, Experiment 1) presented a sequence of 12 photographs for 14 seconds each. All participants were instructed to remember the pictures in as much detail as possible, and that the background was as important as the objects. Participants in the *within* and *outside* conditions were told that to help them remember each picture, it would be presented along with a voice-over about the scene. One second following the onset of each picture, the associated discourse began. The same views were repeated at test, and participants rated them on the same 5-point scale as was used in the earlier studies. Mean boundary ratings are shown in Figure 13.2 (panel C). As shown in the figure, boundary extension occurred in all conditions. However, discourse from outside the scene did not influence boundary extension. In fact, planned comparisons between the silent-control group and the other two groups showed greater boundary extension in the *within* discourse condition, $t(93) = 2.02$, $p < .05$, but not in the *outside* discourse condition, $t(93) = .36$, $p = .70$. Further analysis revealed that this effect was driven by a single picture, and with this picture removed, planned comparisons showed no difference between the silent-control group and the *outside* discourse condition, $t(93) = .10$, $p = .92$, or the *within* discourse, $t(93) = 1.55$, $p = .13$. Focusing only on the sound conditions, with that picture removed, a t-test also revealed no difference in boundary extension $t(62) = 1.55$, $p = .13$

In a series of four follow-up experiments, the same pattern of results occurred across a variety of different manipulations (Gagnier, 2010, Section 2, Experiments 2–4b). Gagnier reasoned that perhaps because the statements were presented 1 second into picture viewing, this created a conflict between the information provided in the statement and participants' natural and automatic expectations beyond the boundaries. To test this, she presented both the "within" and "outside" sentences *prior* to the onset of each picture (Gagnier, 2010, Section 2, Experiment 2). Again, boundary extension occurred, but was unaffected by the content of the discourse. To determine if simple scenes might be more susceptible to the effects of discourse, she conducted the same experiment with tight close-ups of single-object scenes (e.g., a traffic cone on an asphalt background), and the same outcome was observed (Gagnier, 2010, Section 2 Experiment 4a). To determine if discourse would affect boundary

[1] Note that memory for prose (without a pictorial context) does not yield a boundary-extension-like error in which participants remember hearing anticipated prose beyond the bounds of the stimulus; the same was true for segments of popular music (Hutchison, Hubbard, Ferrandino, Brigante, Wright, & Rypma, 2012).

[2] The authors thank voice-over artist, Neil Intraub, for recording the discourse for the "within" and "outside conditions" in the experiment.

extension if the picture duration were reduced, Gagnier decreased the duration of each picture to 1 second followed by a 14-second masked interval in which the sentence was presented. Again, boundary extension occurred, but there was no effect of the discourse. This was true both for complex, multi-object scenes (Gagnier, 2010, Section 2 Experiment 3) and for simple, single-object scenes (Gagnier, 2010, Section 2 Experiment 4b). In sum, across five experiments, discourse about aspects of the unseen surrounding world did not affect boundary extension. The same-sized error occurred when the discourse focused only on describing the visible content of the scenes and when the scenes were paired with silence.

Discussion: Imagination Studies. Boundary extension in memory for photographs was *unaffected* by several explicit imagination tasks (Gagnier & Intraub, reported earlier; Munger & Multhaup, 2016) and was also *unaffected* when sound effects or descriptive discourse was presented along with photographs to encourage imagination beyond the boundaries of each view (Gagnier, 2010). Repeatedly, across experiments, performance in these conditions was no different than in the control conditions. These observations suggest an important qualification about the nature of scene representation and also suggest an important constraint on boundary extension that may in itself have adaptive value. We will discuss each of these points in turn.

As described in the multisource model (Intraub, 2010, 2012; Intraub & Dickinson, 2008), scene construction beyond a view occurs early in processing (e.g., Intraub & Dickinson, 2008). It appears to be automatic, occurring, for example, even when attention was diverted by a demanding primary visual attention task during the time a briefly presented photograph was on screen (Intraub et al., 2008). The lack of an imagination effect across these experiments suggests that the rapidly available scene construction, when grounded in amodal perception (e.g., Fantoni et al., 2008; Kanisza, 1979) and supported by rapidly available expectations about layout and context (e.g., Greene & Oliva, 2009), is fundamentally different from imagination. The two are not readily confused. The addition of sound effects may enrich the sense of surrounding space (as claimed by participants who rated the photograph-sound pairs), but what it did *not* do was cause participants in the memory experiment to believe they have *visually observed* more of the scene, whether memory was tested immediately or after a two-day delay (Gagnier, 2010).

The only known effect of visual imagination of a scene on boundary extension has been reported in the case of non-scene displays, in which line-drawn objects were presented on blank backgrounds and participants imagined scenic backgrounds described by the experimenter (Intraub et al., 1998). In contrast to the control condition which yielded normalization in memory for the objects (regression to the mean object size), the imagery condition led to boundary extension, causing a different error pattern in memory for the same outline objects. In contrast, when visual scenes were presented (Gagnier, 2010; Munger & Multhaup, 2016), neither explicit nor implicit imagination instructions affected boundary extension. We suggest that in these cases, scene construction occurred rapidly and automatically in response to the scene stimulus. The imagination tasks were apparently superfluous to rapid scene construction.

Overview and Conclusions

Boundary extension is an anticipatory error in which the observer remembers seeing beyond the boundaries of a view. According the multisource model of scene representation (Intraub & Dickinson, 2008; Intraub, 2010, 2012) bottom-up visual content and top-down expectations provide a sense of the surrounding world that a single view only partially reveals. We summarized several stimulus factors that have been shown to affect the size of the boundary error (e.g., the scope of a view), and several cognitive factors that also have an effect (e.g., dividing attention), but the key purpose of our paper was to consider the possible effects of imagination on boundary extension. Across a large number of experiments conducted by Munger & Multhaup (2016), and in our own lab (including Gagnier, 2010), we found no evidence that either explicit or implicit imagination instructions affect boundary extension for visual scenes. Scene construction may be a rapid automatic process that is not the same as simply imagining the world.

We suggest that anticipatory representation is adaptive, and that boundary extension may play a role in facilitating view integration during visual scanning and navigation. As such, it may be beneficial that there are constraints on the cognitive factors that can affect the size of the boundary extension error. Boundary extension that projects too far into the distance in response to flights of imagination or ongoing conversations while navigating, could create misconceptions about the environment that are maladaptive rather than beneficial. For the observer, embedded within a surrounding world, these constraints may help strike a balance between seen and anticipated scene content, thus providing a well-grounded anticipatory construction of the likely content of an upcoming view.

Chapter 14

Spatial Contraries and Mirrors

Ivana Bianchi and Ugo Savardi

Introduction

Every so often experimental psychologists attempt to approach experimentally and operationalize constructs that have already been the subject of philosophical inquiries. For example, they have delved into the origins and nature of human intuitions about natural numbers (Carey, 2009), the geometrical properties of points, lines, and figures (or Euclinean intuitions, e.g., Spelke, Lee, & Izard, 2010), causality (e.g., Hubbard, 2013c, 2013d, 2013e; Michotte, 1963), force and resistance (e.g., Hubbard & Ruppel, 2017b; White, 2011, 2012), similarity (Medin, Goldston, & Gentner, 1990; Tversky, 1977), and opposition (e.g., Kelso & Engstrøm, 2006). All of these attempts have provided evidence that logical-physical-mathematical descriptions of a "stimulus" are independent from the "corresponding" psychological correlates that characterize human experience.

Just as experimental studies of similarity have approached the subject from different perspectives – i.e., from the perspective of a perceptual relationship (e.g., Goldmeier, 1936/1972), a lexical relationship (e.g., Miller & Fellbaum, 1991; Murphy, 2003), a structural relationship (e.g., Gentner & Markman, 1994; Markman & Gentner, 1993), and a category formation process (e.g., Goldston, 1994; Sloutsky, 2003) – the experimental investigation of contrariety, opposition, and contrast has also been approached from various points of view and within different domains, i.e., cognitive linguistics (e.g., Croft & Cruse, 2004; Mettinger, 1994; Paradis & Willners, 2011), the psychology of perception (e.g., Bianchi & Savardi, 2008a), inductive reasoning (e.g., Gale & Ball, 2012), insight problem solving (e.g., Branchini, Bianchi, Burro, Capitani, & Savardi, 2016; Branchini, Burro, Bianchi, & Savardi, 2015), counterfactual thinking (e.g., Byrne, 2005; Roese, 1997), and relational reasoning (e.g., Dumas & Alexander, 2016; Goodwin & Johnson-Laird, 2006). The oppositional basis of the human direct experience of space has been demonstrated in a series of studies (Bianchi, Savardi, & Kubovy, 2011), which have shown, firstly, that the structure of space that people experience is based on a number of oppositional dimensions, each of which is anchored to two contrary properties; and secondly, that these oppositional dimensions have a precise structure that is defined by the nature of their components that are either gradable or defined by specific points (i.e., the two opposite pole properties and the intermediate region). For instance, *closed*, *still*, *full*, *parallel*, and *neither above nor below* are all point (single) properties, while *open*, *divergent*, *large*, and *neither full nor empty* are all gradable properties.

There are a number of interesting questions that arise when we start to analyze the human experience of space in terms of these (perceptually grounded) oppositional dimensions. For example, are observers consistent in defining the boundaries of one property and its corresponding opposite – that is, when does a person's experience of *near* end and their experience of *far* begin? Is the boundary between two opposite properties sharply defined? Are the experiences that are perceived as neither one pole nor the other really equidistant from the two poles or are they perceived as more similar to one pole? These are all empirical questions, and some empirical answers have already been provided (Bianchi, Burro, Torquati, & Savardi, 2013; Bianchi et al., 2017; Bianchi & Savardi, 2008a; Bianchi, Savardi, & Burro, 2011; Bianchi, Savardi, & Kubovy, 2011; Savardi & Bianchi, 2009).

Given this premise, we will now zoom in on the role of opposites in the perception and cognition of space by looking at mirror reflections (starting from how people directly perceive them to how they think about them) and showing how the oppositional nature of space is the bedrock on which the relationship between an object or body and its reflection rests.

Bedrock Architecture

Before looking specifically at mirrors, let's put the general claim we are making in this chapter concerning modeling space perception in terms of oppositional dimensions within a theoretical context that has been proven to be capable of accounting for the complexity of ecological experience of space. A reading of Gibson's (1979) ecological optics from this perspective provides evidence in support of the claim that opposites constitute the bedrock architecture of human spatial experiences.

To start with, a salient experiential contrast is that perceived between a region of space that observers recognize as pertaining to the observer *himself* or *herself* and an *outside world* or *surrounding environment* (Koffka, 1935, p. 322; Gibson, 1979, p. 317; see Figure 14.1, top left image).

Another basic opposition that is evident in the organization of the visual field relates to *viewpoint* and *vanishing point*. Everything that falls within an observer's visual field is necessarily seen from one viewpoint (i.e., that of the observer). But it is also true that every single viewpoint necessarily corresponds to its own individual vanishing point.[1] This basic opposition is manifested not only in terms of the location in space of these two points (they are the two opposite *endpoints* in the line of sight) but also in terms of the *divergence* vs. *convergence* of the perspective lines that are clearly evident in the organization of the visual field. These are the lines to the left and right of the observer that converge toward the vanishing point and diverge with respect to the viewpoint; the same goes for the perspective lines above and below the observer.

The perception of distance phenomenally consists of another fundamental oppositional structure: *near–far*. Even in a minimal condition of stimulation – that is, in front of an isoluminant surface (*Ganzfeld*) – the perceptual experience of space already contains this opposition (Avant, 1965; Bianchi & Savardi, 2008a, pp. 49–54; Cohen, 1957; Metzger, 1930). The three spatial regions identified by Cutting and Vishton (1995) by means of an

[1] The geometry that relates the viewpoint to the vanishing point has been extensively studied, starting with Euclid's *Optics* (325–265 BC) and continuing on through pictorial studies on perspective (e.g., Alberti, 1404–1472; Brunelleschi, 1377–1446; Piero della Francesca, 1415–1492) right up to the analyses of perceptual fundaments developed in the field of cognitive science (Cutting, 1986; Deregowski & Parker, 1988; Epstein & Rogers, 1995; Hagen, 1980; Kemp, 1992; Kubovy, 1986; Pirenne, 1970). However, nobody has ever emphasized the oppositional elements characterizing it, which is instead the aspect emphasized in our analysis.

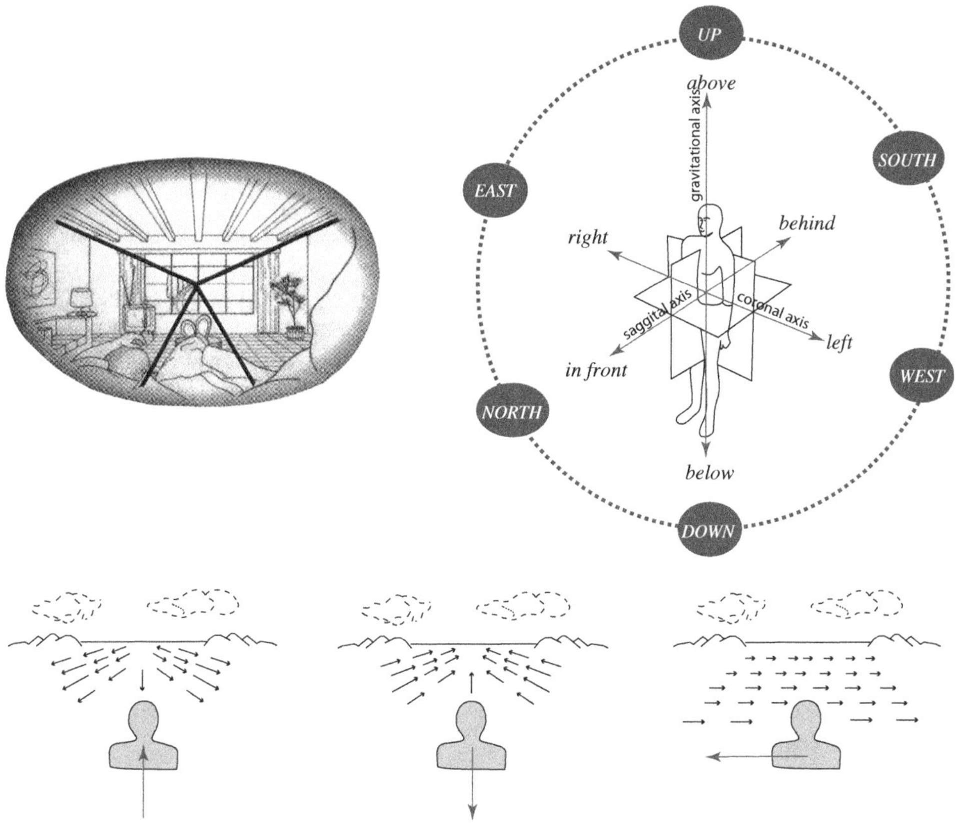

Figure 14.1 Basic spatial oppositions. Top left: Visual field of an observer looking from his/her left eye (adapted from Gibson, 1979). Top right: The egocentric oppositions relating to *front–back, above–below, left–right* based on the persons' body coordinates (adapted from Richards, 1975) in relation to the basic environmental (or allocentric) oppositions. Bottom figures: The opposite direction of the optical flow with respect to three directions of motion of the observer. (adapted from Gibson, 1979)

integrated psychophysical analysis of the relative potency of the various visual cues for distance (personal space, action space, and vista space) pertain to three consecutive regions delineated along the *near–far* dimension.

If we consider all of the visual cues that relate to the *near–far* dimension (i.e., relative size, the height on the visual field, occlusion, etc.) from the point of view of the visual features that make them evident, we realize that they are all manifested by oppositional properties: objects (the projected size of which depends on their distance from the observer) appear *bigger* nearer to the observer and *smaller* farther away, and stimuli that are below the horizon appear to be *low* in the visual field when they are near and *high* when they are far away. Likewise, occlusion depends on whether an object is *in front of* or *behind* another object.

The dynamics pertaining to optical flow are also oppositional. The optical flow always moves in the opposite direction with respect to that of the observer, whatever his/her direction of motion: when an observer moves forward, the environment moves toward him/her in the opposite direction (Figure 14.1, bottom row, diagram on the left); when the observer moves backward (Figure 14.1, bottom row, center diagram), the environment moves away from her/him in the opposite direction; and when the observer moves to the side (Figure 14.1, bottom row, diagram on the right), the environment moves in the opposite direction. This is maximally evident when the motion of the viewpoint is faster than walking speed (for instance, a moving vehicle). The rule also holds when it is applied to an immobile observer and a cinematic simulation (i.e., an induced perception of motion).

If we wish to describe two objects as moving (or pointing) in "opposite directions," we need a system of coordinates to refer to. The fact that "the mind is locked in a body that, at any time, occupies a specific place and faces a specific direction" (Tversky & Hard, 2009, p. 124) has often been cited in support of the primacy of the egocentric frame of reference over the allocentric (or exocentric or environmental) frame of reference, both in terms of the system of coordinates relating to the observer's own body (e.g., Golledge, 1992; Howard & Templeton, 1966; Levelt, 1989; Shelton & McNamara, 1997) and that of another person's body (e.g., Mainwaring, Tversky, Ohgishi & Schiano, 2003; Shober, 1995; Tversky & Hard, 2009). However, experimental evidence arising from other studies has shown the primacy of allocentric representations of space in scene recognition, locomotion and reorientation in space (e.g. Holmes & Sholl, 2005; Klatzky, 1998; Mou, McNamara, Rump, & Xiao, 2006; Mou, Fan, McNamara, & Owen, 2008; Mou, McNamara, Valiquette, & Rump, 2004; Tversky, Lee, & Mainwaring, 1999). Whatever the relative importance of the two systems of coordinates (egocentric and allocentric), both are grounded in three basic oppositions: *front–back* (sagittal axis), *left–right* (coronal axis), and *above–below* (gravitational axis) for the egocentric system and *north–south*, *east–west*, and *up–down* (Figure 14.1, top right diagram) for the allocentric system.

Opposites constitute the pattern of the human perceptual organization of space also beyond the structures mentioned thus far, and they are already present in preverbal stages of human development (Casasola, 2008; Casasola, Cohen, & Chiarello, 2003; Hespos & Spelke, 2004; McDonough, Choi, & Mandler, 2003; Quinn, Cummins, Kase, Martin, & Weisman, 1996). From a phenomenological perspective, all characteristics of space involving extension, shape, orientation, or location lay on oppositional dimensions, and the psychophysical structure of these dimensions is consistent among individuals (Bianchi et al., 2013; Bianchi et al., 2017; Bianchi & Savardi, 2008a; Bianchi, Savardi, & Burro, 2011; Bianchi, Savardi, & Kubovy, 2011; Savardi & Bianchi, 2009). Not only are things perceived as near or far away (or neither near nor far away) but they are also perceived as long or short (or neither long nor short) and high or low (or neither high nor low). They might appear to be regular or irregular, symmetrical or asymmetrical, rounded or angular, straight or bent, or open or closed. A road is uphill or downhill (at various inclinations) or it is flat. An object can be at the top of a mountain or at the bottom of it, or in many intermediate locations that are neither at the top nor at the bottom. It may be in front of us or behind us, or neither in front nor behind if it is to one side. And, if the latter is the case, it can be to our left or to our right. An organized space such as this is the framework not only for our field of vision but also for any movements we make or actions we perform. Within this space gestures are made and people interact with objects and with other humans (and also with mirror reflections). Therefore, it is not surprising that people are immediately intuitive about what doing the opposite means (Bianchi, Savardi, Burro, & Martelli, 2014), in addition to imitating (e.g., Heyes, 2011; Iacoboni et al., 1999; Prinz, 2002).

One might wonder whether the fact that opposites are central to conceptual spaces (Gärdenfors, 2000, 2014; Paradis, Hudson, & Magnusson, 2013) might not simply reflect that they are central to spatial perception. We might also ask whether the special status of opposites in all natural languages (Croft & Cruse, 2004; Cruse & Pagona, 1995; Jones, 2002; Paradis & Willners, 2011) is a result of this (Bianchi & Savardi, 2008a, 2012b).

Spatial Opposites and Mirrors

Opposition and identity are the two key structures in the spatial perception and cognition relating to mirrors. What naive observers looking at their image in a plane mirror perceive (and report) is the opposite orientation of the reflection with respect to their material body, in addition to the opposite localization of the reflection with respect to the mirror surface (Bianchi & Savardi, 2008b, 2009). “My left arm is my right arm in the mirror”: this egocentric opposition is what is genuinely intriguing about mirror images (Bianchi & Savardi, 2009; Corballis, 2000; Ittelson, 1993; Takano, 1998). It is not a pseudo-problem or a linguistic problem, as some have tried to insinuate (Gregory, 1996; Haig, 1993; Tabata & Okuda 2000). The left-right reversal is a visual characteristic that is visible in reflections of human bodies in plane mirrors, which observers notice especially when they are encouraged to focus on the lateralization of the reflection. In addition to this egocentric opposition, they also immediately notice various allocentric oppositions. In a mirror set vertically on a wall in front of them (Figure 14.2, diagram on the left), they immediately notice that they are facing one direction (e.g., north) while the reflection is facing the opposite direction (i.e., south), and when they walk toward the mirror, the person in the reflection moves in the opposite direction (opposition in the sagittal axis). When the mirror is on the wall beside the observer (Figure 14.2, diagram on the right), the reflection still appears to be opposite: it has an identical sagittal and gravitational orientation but is opposite in terms of the coronal axis, and if the person moves laterally to his/her left (e.g., westward), the reflection moves in the opposite direction (i.e., to the right and eastward). If the observer then positions him/herself on top of a mirror lying on the floor (Figure 14.2, center diagram), he/she again perceives the reflection as being opposite. This time the opposition is along the gravitational axes, as the reflection is upside-down.

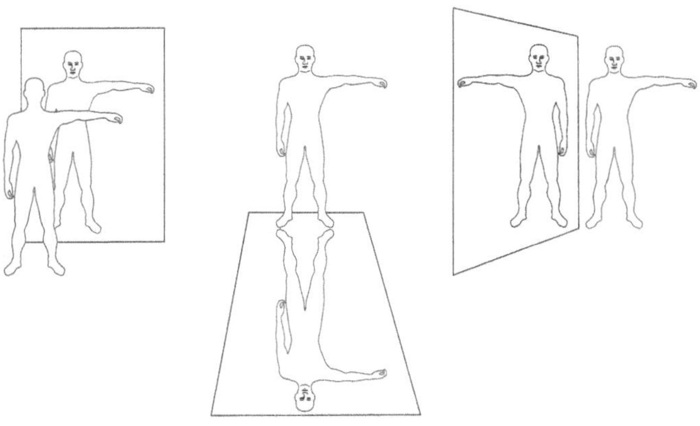

Figure 14.2 Diagrams showing reflections in mirrors set in different positions: vertically in front the "real" body (left-hand diagram), below the "real" body (central diagram), or beside the "real" body (right-hand diagram).

These elements of opposition are prominent in people's perception of the spatial structure of their reflections. What they notice first when they are asked how they perceive the relationship between their body and its reflection is the allocentrically based opposition rather than the egocentric left–right inversion. Bianchi and Savardi (2008b, p. 674) report that "opposite" was the relationship chosen in 87% of the cases when the participants in their experiment were asked to describe the relationship between their orientation and the reflection of themselves in a mirror. When requested to specify which aspect of the scene they were focusing on, they never referred to the egocentric left–right reversal in the gravitational condition (Figure 14.2, center diagram) and only referred to it in a very few instances in the frontal condition (Figure 14.2, diagram on the left). They usually explained their choice in terms of allocentric opposition. The same thing happened when participants in another study were asked to describe the reflections of simple objects and movements in various positions with respect to the mirror surface (Savardi, Bianchi, & Bertamini, 2010, Exps. 4–5).

It is worth noting that many of the errors that people make when asked to predict the behavior of reflections are based on this perception of opposition. Research into "naive optics" (Croucher, Bertamini, & Hecht, 2002), which can be conceived of as a branch of "naive physics" (e.g., McCloskey, 1983a, 1983b; McCloskey, Caramazza, & Green, 1980; McCloskey, Washburn, & Felch, 1983), has clearly demonstrated that when people make predictions about the orientation, speed, or trajectory of moving objects in reflections, they do not necessarily reason in terms of the physical or optical laws they were exposed to at school (e.g. Bertamini & Parks, 2005; Bertamini, Spooner, & Hecht, 2003; Bianchi, Bertamini, & Savardi, 2015; Bianchi & Savardi, 2012a; Croucher et al., 2002; Hecht, Bertamini, & Gamer, 2005; Lawson, 2010, 2012; Lawson & Bertamini, 2006; Lawson, Bertamini, & Liu, 2007). Rather, they use their imagination, but their prediction is then strictly dependent on the scenario they have in mind, and therefore imagining different scenarios can lead to different predictions. The systematic errors made by participants in prediction tasks are interesting because they tell cognitive scientists something about the various different processes underlying predictions. This is what has stimulated 30 years of research on naive physics and more than a decade of research on naive optics – research that is still ongoing.

For instance, 20–40% of people expect a person entering a room and making a parallel approach to a plane mirror hanging on the wall to see their reflection appearing at the mirror's farther edge (Figure 14.3a, in the position indicated by the letter A) rather than the nearer edge (Figure 14.3a, in the position indicated by the letter D) (Bertamini et al., 2003; Croucher et al., 2002). This is a considerable percentage for such a blatant error: nobody approaching a mirror walking parallel to it has ever seen his/her reflection appearing at the farther edge of the mirror. Moreover, a significant percentage do not expect the reflection of an object moving along a straight line at various angles with respect to a mirror to follow a symmetrical pattern (Figure 14.3c). They expect it to move along the same line either in the same allocentric direction (i.e., if the object is moving toward the mirror, in the reflection it will move away from it; Figure 14.3d) or in the opposite direction (i.e., the object and its reflection move toward each other; Figure 14.3e). In paper-and-pencil tasks, these errors occur in up to 40% of responses for movements with an angle of inclination and in 20% of responses for movements that are parallel or orthogonal to the surface of the mirror (Savardi et al., 2010, Exps. 1–2). The pattern shown in Figure 14.3e (where the reflection moves along the same line but in the opposite direction to the trajectory of

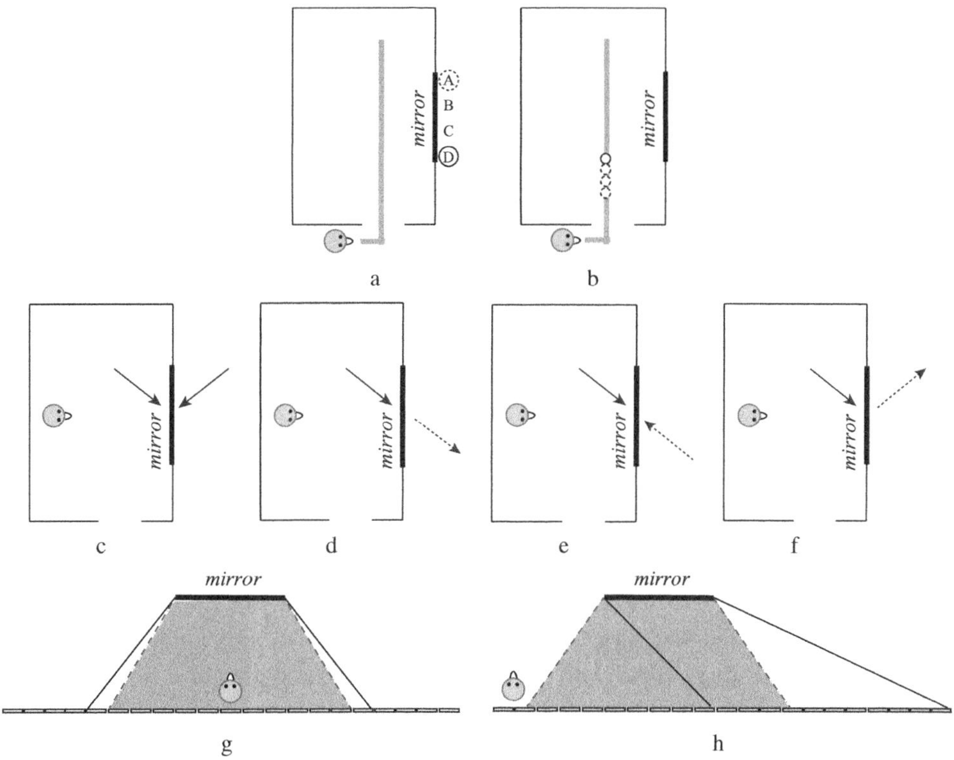

Figure 14.3 Some false beliefs concerning naive optics; all are represented using aerial views.

Top row: Many people expect a person entering a room and walking along the light-gray path to see his/her reflection appear at the farther edge of the mirror in the position indicated by A rather than where it would in reality appear, D (see diagram 3a) or that he/she will see their reflection before reaching the near edge of the mirror (b).

Second row: A good number of people do not expect a person looking at an object moving toward a mirror along the path indicated by the solid arrow (c–f) to see its reflection moving along the correct path (c) but rather to see it moving along the same trajectory with an identical orientation (as indicated in d) or along the same trajectory but with an opposite orientation (e); a smaller percentage of people predict that the movement will follow the path shown in f.

Third row: A person looking in a mirror from a central (g) or eccentric (h) position expects to see more or less the same area reflected (i.e., the gray area delimited by dashed lines) while the correct cone of view is very different (as delimited by the continuous lines).

the material object) is particularly robust, since it was manifested also when a real-life setting rather than a paper-and-pencil task was used (Bianchi et al., 2015, Exp. 2; Savardi et al., 2010, Exp. 3).

Some of these errors are compatible with the hypothesis that people might think of the virtual world in a mirror in terms of a rotation of the actual world through the surface of the mirror (Hecht et al., 2005). However, when asked in an experiment to identify the correct reflection from a series of pictures showing a room and its reflection in a mirror hung on the wall, participants recognized that the picture showing a 180° rotation of the room was incorrect (Croucher et al., 2002). Likewise, when people are asked to look at reflections of simple objects and to describe the relationship perceived between the reflection and the object, they almost never describe the reflection as being rotated

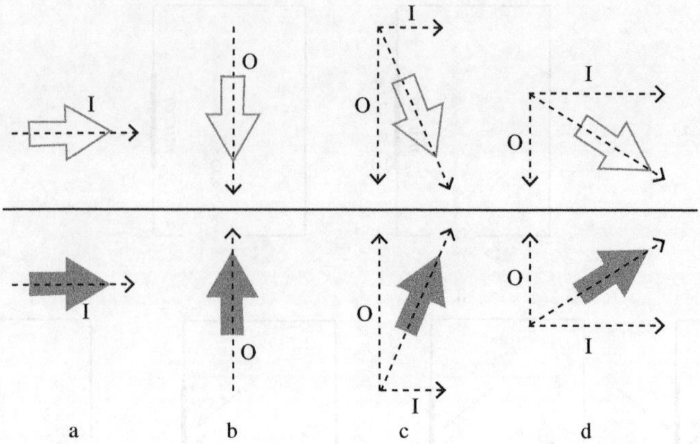

Figure 14.4 Diagrams showing the relationship between the orientation of a material object or movement (indicated by the thick grey arrows) and that of its reflection (indicated by the arrows with grey outlines) in terms of two vector components: one parallel to the mirror (with the material object/movement and the reflection identically oriented, I) and the other orthogonal to the mirror (with the material object/movement and the reflection oriented opposite each other, O). When the object is oriented (or moving) parallel to the mirror (a) or orthogonal to the mirror (b), one of the two components is null. For more details, see the main text.

with respect to the object (Savardi et al., 2010, exps. 4–5). These findings are inconsistent with the hypothesis that people think of the reflected world in terms of a rotation of the real world.[2]

The geometry of the errors shown in Figure 14.3 is compatible with a generalization of the simplified heuristics stating that "the mirror does the same" and "the mirror does the opposite" (Bianchi et al., 2015; Bianchi & Savardi, 2012a; Savardi et al., 2010, Exps. 1–3). But why do we say that these heuristics are *simplified*? Let's suppose that the geometry that naive observers have in their minds regarding reflections in plane mirrors is a lawful combination of two components (see Figure 14.4): one that is parallel to the mirror (relating to the identicalness component, I), and another that is orthogonal to the mirror (relating to the oppositional component, O).

When an object is positioned parallel to a mirror or orthogonal to it, one of the two components is null and the relationship between the real object and its reflection is defined by the other component: i.e., the orientation is identical in the first case (Figure 14.4a – "the reflection does the same") and opposite in the second case (Figure 14.4b – "the reflection does the opposite"). From 75% to 85% of people are able to predict the behavior of the reflections correctly in these two positions (Bianchi et al., 2015; Savardi et al., 2010). But when an object is positioned at an angle (Figures 14.4c, 14.4d), the standard of performance drops and 32% to 51% of responses are incorrect. This is because when an object is positioned at an angle, the two components interact and the result is a combination of the two: the trajectory of the reflection is opposite to that of the material object with respect to the orthogonal component but identical with respect to the parallel component.

[2] The rotational hypothesis has been re-proposed in a less radical version by Muelenz, Hecht, and Gamer (2010), not to explain the errors mentioned above, but to account for the small quantitative localization bias that they found indicating that reconstructions of the virtual world are systematically rotated counterclockwise by an average angle of two degrees.

The movement is identical *and* opposite. Which of the two components is more evident depends on the angle of inclination: the identity component is more evident for movements that are angled more closely to the parallel line, and the opposite component is more evident for movements angled more closely to the orthogonal line. If observers simplify this more complex condition and apply the rules saying that "the reflection does the same" or "the reflection does the opposite" to cover all angles of incidence, the patterns of error described in Figure 14.3 occur: the reflection is expected to appear at the farther edge of the mirror and move contrary to the real person even when he/she is walking parallel to the mirror ("the reflection does the opposite" to the real person, Figure 14.3a). Likewise, objects are expected to move along the two patterns shown in Figures 14.3d and 14.3e, which are generalizations, respectively, of the rules saying that the mirror does the same (Figure 14.3d) and the mirror does the opposite (Figure 14.3e). Evidence that people *simplify* the geometry of reflections and do not simply *confuse* the two components is reinforced by the fact that errors manifesting the confusion of the two components (Figure 14.3f) are rare – less than 7% of the total number of responses.

The two heuristics ("the reflection does the same" and "the reflection does the opposite") seem to have different influences, relatively speaking, depending on whether the task of the observer is to predict a reflection based on what he/she sees in the real world or vice versa, predict the real world starting from the reflection. The expectation that something will be opposite in a mirror is more salient when people imagine how a reflection will relate to the corresponding real world as compared to when they imagine how the material world will be with respect to the corresponding reflection. An asymmetry in terms of the number of errors (i.e., the difficulty of the task) also emerged: people perform better when they are asked to predict a real movement starting from the reflection than vice versa. These two asymmetries (from the point of view of the salience of the two heuristics and the easiness of the task) are related to the geometry underlying the processes involved when people mentally derive the virtual world from the material one and vice versa (Bianchi, Bertamini, & Savardi, 2015). Whether they might also tell us something about a naïve ontology of reflections – and the different roles played by the material and virtual worlds – is not yet possible based on these results. An explanation of this sort has been put forward by Sareen, Ehinger, and Wolfe (2014) to interpret the asymmetry that they found. Observers presented with indoor scenes including mirrors that reflected several objects in the room were asked to label "everything they saw." They labeled reflected objects less frequently than their material counterparts. Moreover, in change detection tasks, the disappearance of a reflected object was less easily detected than the disappearance of its material counterpart. These findings were interpreted by the authors as evidence that people treat the parts of images that represent reflections as somewhat less "real" than the parts that represent material objects/spaces. After all, the reflection is a copy of the visual information that must be available somewhere else in the scene (therefore redundant). And when the material object is not visible (i.e., the reflection is not redundant), the reflection is in any case discounted because mirror information misinforms about the "real" visual aspects of the objects. Whether the idea of reflections as copies (with emphasis on what we call the *identity* element) is more relevant when people move mentally from the virtual world to the material world, while the idea that reflections misinform (with emphasis on what we called the *opposition* element) is more relevant when people move mentally from the real world to the virtual world, would be an intriguing hypothesis to investigate.

Further Simplification Biases

Other errors that have emerged in the literature on naive optics manifest a simplification in the mental models of mirrors that people have in mind. For example, a surprising percentage of observers (62% in Bertamini et al., 2003) make what has been called the "early error" (Figure 14.3b): When asked to predict where a person approaching a mirror from the side and walking parallel to it would start to see herself/himself reflected in the mirror, they predict that they will start seeing the reflection before they actually do, i.e., before reaching the near edge of the mirror (see also Croucher et al., 2002; Hecht et al., 2005; Lawson & Bertamini, 2006).

This error is compatible with another class of findings which came out in various different experimental conditions, from paper and pencil tasks (Bertamini, Lawson, Jones, & Winters, 2010, exp. 5; Bianchi & Savardi, 2012a, exps. 1, 3, 4), to real-life settings (Bianchi & Savardi, 2012a, exp. 2; Lawson, 2012), to photographs (Bertamini et al., 2010, exps. 1–4) and paintings (Bertamini, Latto, & Spooner, 2003). When asked to predict what part of the surrounding space an observer would see reflected in a mirror on a wall, given various viewpoints, two biases emerged. First, when the observer is central to the mirror (i.e., not displaced laterally), adults expect the field of view to be expanded at both of the mirror's lateral edges, but with a smaller angle than it would be in reality (Figure 14.3g). This is an attenuated version of the frontal error found in children who expect the mirror to show what is just in front of it (Bertamini & Wynne, 2009). Secondly, people expect the field of view to be basically invariant whatever the eccentricity of the observer's viewing position: approximately 50% of people in the case of close viewpoints and up to 80% for greater distances expect the observer to see *more or less the same field* for both central and eccentric viewpoints, i.e., the space directly in front of the mirror plus a certain area beyond *both* its edges. The area is represented in Figures 14.3g and 14.3h by the grey areas delimited by dashed lines (the diagrams refer to the study by Bianchi & Savardi, 2012a, but similar results were found in Bertamini et al., 2010, exp. 5). This trapezoidal shape is true for central viewpoints (in Figure 14.3g the straight lines delimitate the correct area of visibility), whereas for eccentric viewpoints the shape of the visual field is different: one sees an expansion at the farther edge while part of the space in front of the mirror at the nearest edge is not visible (Figure 14.3h – "parallelogram" area between the straight lines).

This widespread inability to predict what in effect an observer sees in a mirror (given various positions of the observer and various positions and inclinations of one or more mirrors) is at the heart of what has been called the "Venus effect" (Bertamini et al., 2003; Bertamini et al., 2010). In a number of famous paintings involving a person (usually Venus or a lady) gazing into a mirror, the face in the mirror is located incorrectly, but observers seem to be happy with this artistic license (e.g., Hans von Aachen, *Couple with Mirror*, 1596; Simon Vouet, *Toilet of Venus*, c1628; Titian, *Venus with a Mirror*, 1555; Velázquez, *The Toilet of Venus*, 1647–1651). This underestimation, which characterizes mental models of reflections with regard to the role of the point of view, has also emerged in tasks involving estimates of the size of a reflection and how it would vary as the distance of the observer or the distance of the target object increases or decreases (Bertamini, Lawson, & Liu 2008; Bertamini & Parks, 2005; Lawson & Bertamini, 2006).

It has already been suggested that the fact that people seem to be incapable of taking into account the role of the viewpoint in naive optics can be explained by the fact that, in general, the visual system works to make the outside world constant despite continuous changes in observer position (Bertamini & Parks, 2005; Croucher, Bertamini & Hecht, 2002). It would also reinforce the hypothesis put forward by Bertamini and Parks (2005) and also discussed in Bertamini et al. (2010) that mirrors are treated as devices that capture an image (or a space), and therefore, what fits into a mirror depends mainly on what is contiguous to it rather than on the location of the observer, which is, to a great extent, disregarded.

What is important to note is that (1) a *simplification bias* is invariant to all the errors mentioned in this chapter, and (2) what is generalized beyond the conditions of validity is a *salient perceptual aspect* of direct visual experience. This last statement might sound odd, and one might object that nobody, when walking parallel to a plane mirror, has ever seen their reflection before reaching the nearer edge (Figure 14.3b), or seen it appear at the farther edge of the mirror (Figure 14.3a), or seen it move as represented in Figures 14.3d–14.3f. It is also a fact that nobody has ever seen a mirror reflect only the portion of the room directly in front of it independently of whether the observer is positioned centrally or laterally with respect to the mirror (Figure 14.3g–14.3h). These are, in Lawson and Bertamini's (2006) terms, "conceptual errors" and not "perceptual errors": the latter do not disappear when the experiment is repeated and the participants are asked to look at reflections in a mirror (i.e., visual feedback is provided), since what they *see* is exactly what they had predicted. Conversely "conceptual errors" do in fact disappear when visual feedback is provided, because what people then see does not conform to what they had predicted. In other words, the origin of these errors has nothing to do with what people strictly speaking *see*.

The distinction between these two kinds of errors is of course tenable, if only for the fact that the reactions to visual feedback are different in each case. What we are asserting here (see also Bianchi & Savardi, 2014) is that these conceptual errors are in any case shaped by what people see. They are generalizations of some *salient* aspects of perceptual experience, which are widely applied not only to correct conditions but also to incorrect ones. We have already explained this with reference to errors based on people recalling the rule that "a reflection does the same" and/or the rule that "a reflection does the opposite" as their visual experiences have led them to believe. However, this explanation works equally well for the other errors discussed in this chapter. Consider, for example, the error that people make when they expect to see the area beyond the left and right edges of a mirror reflected whatever the position of the observer (Figures 14.3b, 14.3g, 14.3h). When people look in a mirror either intentionally or by chance in everyday life, the incidence of their line of vision is at different angles. Portions of the real world beyond the edges of the mirror are always visible: when an observer is positioned "inside" the mirror edges, this holds for all four edges; when he/she is displaced laterally, this holds for all the edges except the nearest. In other words, people normally perceive an expansion. If they do not carefully analyze what happens at the nearer as opposed to the farther edge of a mirror, they will easily generalize the experience, concluding that mirrors reflect the part of the world that lies beyond their boundaries.[3] It is also very likely that in everyday life people do not carefully analyze what

[3] This is not related to the boundary extension phenomenon, i.e., the fact that people tend to remember that a view contains information beyond its actual boundaries (for a review, see Hubbard, Hutchison, & Courtney, 2010, and also Intraub & Gagnier, Chapter 13 in this volume). Given various eccentric or central viewpoints with respect to a mirror, the reflected scene always shows a field of view that is wider overall than that which is within the orthogonal projections relative to the

happens at the edges of a mirror, since they usually only pay attention to the small area that interests them, i.e., their face if they are shaving or putting on makeup, and tend to ignore the rest of the reflection. Similarly, they will generalize their experience that mirrors show what is positioned directly in front of them (or nearby), and thus every object or observer in this area (Venus included) is expected to have its reflected counterpart visible in the mirror.

Two Final Notes

In this chapter we have focused on some of the many interesting questions about mirrors, which have been raised by cognitive scientists. We have looked at how people perceive reflections or "reason" about them when predicting their behavior (based on heuristics, intuition, mental models, etc.). We have shown that, in spite of different perspectives and the fact that this has been explicitly or implicitly addressed, there is recurring experimental evidence that people recognize an oppositional relationship between a material object/body and its reflection. Beyond presenting various different errors and biases analytically, there are two considerations that we hope the reader has been stimulated to reflect on. These considerations are related to each other.

1. First is the idea that opposition (as well as identity) is a key feature of direct (naive) human experiences of reflections. This has been revealed in perceptual tasks and confirmed by the types of errors that people make when asked to predict the behavior of reflections. If today the "mirror question" (i.e., why does a mirror invert left/right but not up/down?) was to be reconsidered in an empirical context and framed exclusively in terms of directly experienced variables (i.e., the features and relationships perceived by an observer), we believe it should be discussed within the mind-set of the perceptually evident and cognitively salient relationships of identity and opposition. Contrary to what Tabata and Okuda (2000) maintained, this means that we must get rid of all considerations concerning the physics of reflections, since they do not make the problem disappear for cognitive scientists (Bianchi & Savardi, 2008b; Takano, 1998). We should instead focus on the whole set of perceptions of identity and opposition experienced by observers when they look in a mirror. These are anchored not only to the material object's or body's egocentric coordinate system (i.e., on the observer's right, on the left in the reflection) but also to an allocentric spatial frame of reference (i.e., the reflection is oriented and moving in an opposite direction with respect to the orientation and movement in the material world).
2. Second, a surprising number and variety of errors are made in very simple prediction tasks (i.e., those that do not make use of convex or concave mirrors or involve complex movements in complicated scenarios). This is particularly surprising if one considers that people are daily exposed to mirrors and reflecting surfaces. The correspondence between the real world and the reflected world is easy to see. We see it every time we recognize something *as* a reflection. Mirrors look like windows in that they "show something on the other side." But there is a complexity to the apparently simple geometry of mirrors that naive observers are not able to fully come to terms with, independently of whether or not they have explicit knowledge of the physical

mirror edges. There are differences between eccentric and central viewpoint positions, and generally speaking, people tend to predict an expansion that is smaller than what it would be in reality (Bianchi & Savardi, 2012a). Therefore, it is a boundary reduction rather than an expansion.

rules regarding reflections. We would suggest that what makes the geometry of mirrors difficult is specifically related to the oppositional structure inherent to reflections that cannot be eliminated. Reflected objects and spaces are oriented and located oppositely in relation to their real counterparts. This means that they look identical, but in reality they are not the same: they "match" in the same way as your right hand matches your left hand – i.e., they are enantiomorphic, not "the same." Moreover, reflections are spatially perceived *beyond* the frame of the mirror (just like windows), but we understand that what we see is not physically located there but is on the same side of the mirror as we are, along with all the counterparts of everything contained in the reflection. So in order to make the two worlds match, we have to "fold back" the reflected world visible beyond the mirror, at the same time mentally getting rid of the internal enantiomorphic opposition. It is these oppositional elements embedded in the impression of identity that characterize direct experiences of reflections that make the mental transformation complicated. This is the reason why people make mistakes, but it is also the reason why we will always be fascinated by the world in the looking glass.

Chapter 15

Aesthetics and Preferences in Spatial and Scene Composition

Timothy L. Hubbard

A fundamental issue in visual art and aesthetics is where to place elements within a scene or frame in order to maximize aesthetic response of the viewer, and this is referred to as *spatial composition* or *scene composition*. Even though studies of aesthetics often focus on how individuals respond to art objects, aesthetics is a broader topic involving evaluative responses (e.g., "I love it!", "I hate it!") to any stimulus and is not limited to responses to art objects (e.g., Leder, Belke, Oeberst, & Augustin, 2004; Sammartino & Palmer, 2012a). Studies of aesthetics often involve judgment of preferences, and when these preferences involve placement, orientation, or configuration of objects, such preferences are considered to be spatial biases. Although many spatial biases have been investigated within domains of visual art and aesthetics, other spatial biases are not clearly related to visual art or aesthetics. Thus, biases in spatial composition and scene composition are considered to reflect general aspects of perceptual and cognitive functioning based (at least in part) on experience and expectations (and how a stimulus conforms with those experiences and expectations). In this chapter, spatial biases involving object placement, facing direction, contour, ratio, and Gestalt effects are described; accounts for some of these spatial biases are briefly considered, and some general properties of spatial biases are suggested.

There are several methods for studying biases in spatial or scene composition. Three methods adapted from studies of aesthetics and initially described by Fechner (1876) involve choice (i.e., stimuli created by an experimenter are rated or compared by participants), production (i.e., participants create objects when asked to make pleasing stimuli), and use (i.e., documenting preferred patterns in the world of everyday life). Additional methods based on cognitive and information-processing approaches include examination of perception or memory for qualities of a stimulus (e.g., object location, Hubbard, 2005b; distance, Thorndyke, 1981; boundary location, Intraub, 2012). A variety of stimuli have been used, with some studies presenting simple geometric stimuli (e.g., Chen & Scholl, 2014; Palmer & Guidi, 2011), and other studies presenting reproductions of art objects (e.g., Mather, 2012; McManus & Kitson, 1995) or other pictures (Gardner, Fowlkes, Nothelfer, & Palmer, 2008; Thomas, Burkitt, & Saucier, 2006). A complete review of each spatial bias noted here is beyond the scope of this chapter; however, selected studies related to each bias are mentioned. A catalog of biases in spatial composition and scene composition and spanning aesthetic, perceptual, and cognitive literatures has not previously been presented, and one purpose of this chapter to present such a catalog.

Biases in Placement

Biases in placement involve preferences regarding where objects are located within a scene. Eight such biases are considered and include (a) influences of the skeletal structure, (b) balance across the scene, (c) preference for objects to be nearer to the center, (d) preference for more salient objects to be on the right, (e) preferences for different types of objects at different heights within the picture plane, (f) the clutter effect, (g) preferences based on identity and semantic relatedness of objects, and (h) effects of a title or other descriptive or explanatory information.

Skeletal Structure

Arnheim (1974, 1988) proposed pictures contained a skeletal structure that arose from interactions of the object and surrounding frame. This structure generally involves axes of the picture, and intersections of the axes (e.g., the center where horizontal, vertical, and diagonal axes cross) are the most important and stable locations within the picture. Palmer and Guidi (2011; Palmer, 1991) tested Arnheim's claims regarding skeletal structure by presenting participants with a frame containing a dot, and location of the dot varied across trials. Participants rated "goodness of fit" of the dot within the frame, and the highest scores occurred for dots at center, followed by dots on the vertical axis, then by dots on the horizontal axis, and lastly by dots located elsewhere. When stimuli were near a vertex of the frame or within a subframe (division) of the larger frame, ratings of fit were related more to local symmetry axes than to global axes. These patterns are consistent with Arnheim's notions, but Palmer and Guidi suggested their results reflected figural goodness rather than position of the dot relative to skeletal structure per se. Even so, it is not entirely clear that figural goodness might not depend at least in part on the form of the skeletal structure (cf. Pinna & Albertazzi, 2011). Relatedly, Bryant and Subbiah (1994) presented dots within a frame, and subsequent reproduction of dot locations by participants was more accurate when dots had been located on intersections of axes defined by markers on the edge of the space.

McManus and Kitson (1995) had participants mark on reproductions of artworks the locations of "important points." They hypothesized that if geometric composition played a role in picture construction, then major axes (and the intersections of those axes) should underlie important points across multiple pictures. Judged locations of the important points were not homogenous over pictorial space; rather, important points were more common in the center than the periphery and in the bottom half than the top half (cf. dot placement in Palmer & Guidi, 2011; up-down principle in Winner, Dion, Rosenblatt, & Gardner, 1987). In follow-up studies, pictures with material removed from top, bottom, left, or right sides were compared with intact pictures, and there was a slight preference for intact pictures, suggesting that skeletal structure and balance were related to an entire composition rather than to specific objects. McManus and Kitson suggested there was relatively little evidence participants exhibited preference for geometrical patterns or used detection of such patterns in aesthetic judgments. McManus, Stöver, and Kim (2011) had participants move a "crop box" over an image until those participants determined which cropping "looked best"; the chosen croppings produced images in which the center of mass was closer to a principle axis of the skeletal structure (cf. Arnheim, 1974) than would be the case for random cropping. Relatedly, in portraiture one of the two eyes is usually placed along the vertical midline (Tyler, 1998; but see McManus & Thomas, 2007).

Balance

Balance in spatial composition is often phrased in terms of consequences of a gravitational field (e.g., weight, stress, tension, force, stability; Gershoni & Hochstein, 2011) and has been thought to be a prime determinant of aesthetic preference (e.g., Arnheim, 1974, 1988). Balance involves visual weight and is influenced by variables that influence visual weight such as size (e.g. size-weight illusion; Jones, 1988), color (e.g., Pinkerton & Humphrey, 1974), and directionality and dynamics (e.g., Locher & Stappers, 2002). Cropping the left or right side of a photograph moves the perceived center of balance of the remaining scene, and this suggests balance involves integration over the entire visible scene (McManus, Edmondson, & Rodger, 1985; McManus & Kitson, 1995). In a study of balance in Japanese kanji calligraphy, Gershoni and Hochstein (2011) found ratings of balance were based on the presence of approximately horizontal or vertical lines, vertical mirror asymmetry, and grouping of elements into meaningful objects; perhaps more critically, they found judgment of balance (and aesthetic preference more generally) depended on contextual meaning (cf. Sammartino & Palmer, 2102b) and not just formal attributes such as size and color. Detection of balance appears to be a rapid and low-level response formed within a single fixation (Gershoni & Hochstein, 2011; Locher & Stappers, 2002). Interestingly, balance on its own is not sufficient to distinguish between works of different artistic quality (Vartanian, Martindale, Podsiadlo, Overbay, & Borkum, 2005).

Arnheim (1974) suggested a picture is balanced if the center of attention is on one of the major axes of the frame. Several other authors suggested geometrical relationships among elements in a scene might be the basis of judgment of balance (e.g., Bouleau, 2014; Thomas, 1971), although such a basis has been disputed on statistical grounds (McManus & Kitson, 1995). Indeed, participants instructed to create designs using different and multiple shapes often create designs in which the center of mass is in or near the geometric center of the design, and structural weight of the distribution of composition elements is evenly distributed (balanced) around the center of mass (Locher, Cornelis, Wagemans, & Stappers, 2001; Locher, Stappers, & Overbeeke, 1998). Balance is related to symmetry but is not the same as symmetry (e.g., a large weight close to a fulcrum might be balanced by a small weight far from the fulcrum, but such a stimulus would not be symmetrical). McManus et al. (1985) presented observers with postcard-sized renditions of famous paintings, and observers adjusted the horizontal location of an equilateral triangle fulcrum beneath the reproduction so that the depicted scene would balance. The indicated locations were clustered around the center, although there was considerable variance, and judgments were not influenced by sex, handedness, or eye dominance. A similar pattern was found in a second experiment in which square stimuli of various colors were presented, with position being more important for judgment of balance than was size or color.

Center

Arnheim (1982, 1988) suggested the center was an important part of spatial composition and that whatever was at the center (a single object or a group of objects) of a picture was of greater visual importance. Palmer, Gardner, and Wickens (2008) presented pictures of single objects within a larger horizontal frame, and participants preferred symmetrical objects to be located at the center of the frame. Identification of reversible figures in the periphery is biased toward the interpretation consistent with that figure facing the center (Chen & Scholl, 2014). Sammartino and Palmer (2012a) found that when pictures of objects showed a view from directly above or below those objects, participants preferred objects to be located at the center of the frame (although asymmetric side views were preferred to be located closer to the bottom or top of the frame). Leyssen, Linsen, Sammartino, and Palmer (2012) found a center bias in multi-object configurations. The importance of the center is not limited to the global center for an entire picture, but can include multiple local centers. Huttenlocher, Hedges, and Duncan (1991) had participants report position of a dot placed within a circle. Dots were mislocalized toward the center of each quadrant of the circle, and Huttenlocher and colleagues suggested participants superimposed vertical and horizontal axes on the circle.

Locher, Stappers, and Overbeeke (1998; Locher, Cornelis, Wagemans, & Stappers, 2001) videotaped adults instructed to make designs using various geometric shapes that were "interesting and pleasant." The participants appeared to use the center of the picture space as an anchor and distributed structural weight evenly about the center. Consistent with this, art objects tend to have their center of mass nearer a major axis than do random control images; however, this difference is diminished if contrast is ramped across the image (thus shifting center of mass but not changing content; McManus, Stöver, & Kim, 2011). However, when participants are asked to indicate the center of a stimulus, their judgments are often biased away from the center. In line bisection tasks (in which participants indicate the center of a line) participants typically mark to the left of center (referred to as *pseudo-neglect*, e.g., Jewell & McCourt, 2000, but see Mohr & Leonards, 2007). This bias is stronger for explicit than for implicit judgments (Noël, van der Kamp, Weigelt, & Memmert, 2015) and might be reversed when judgments involve a line beyond peripersonal space (e.g., Halligan & Marshall, 1991; but see Longo & Lourenco, 2006). Judgment of the location of the center is influenced by the length and type of stimulus, as lines are bisected to the left of center regardless of length, but orthographic strings are bisected to the right of center when such lines are short and to the left of center when such lines are longer (Arduino, Previtali, & Girelli, 2010).

Right Side

Contrary to the importance of the center or of symmetry (see discussion below), there is evidence of preference for the most important stimuli in a scene to be located on the right side. Levy (1976) found right-handed participants prefer pictures in which the more salient object is located on the right side, and Beaumont (1985) found a preference to put the more salient of two objects on the right. McManus and Kitson (1995) found that participants marked more important points on the right side than on the left side. Chatterjee (2002) suggested left and right sides were biased toward different roles in expressing action, power, or motion, with agents located on the left and recipients located on the right. Consistent with this, a left-to-right sequence is often preferred and judged as more balanced and more dynamic than is a right-to-left sequence (Friemuth & Wapner, 1979; Walker, 2015). Elements in a painting portrayed as moving left-to-right are more likely to be perceived as accelerating, whereas elements portrayed as moving right-to-left are more likely to be perceived as decelerating (Gaffron, 1950). Banich, Heller, and Levy (1989) found a preference for content to be on the right side but a preference for right-to-left motion with right-handed participants, the latter of which they proposed helped draw attention back to the center of the picture. However, not all studies support claims of a difference between left and right sides of a picture (e.g., Winner et al., 1987). A right-side preference might arise from typical reading directions of participants in most studies, as participants usually exhibit a preference for stimuli to have the same directionality as their reading (Chokron & de Agnostini, 2000; Nachson, Argaman, & Luria, 1999).

Vertical Location

The majority of research on aesthetic preference in object placement focused on horizontal placement. Sammartino and Palmer (2012a) varied location of a single stimulus along the vertical axis. Participants preferred objects viewed from directly above or directly below to be located near the vertical center of the frame. However, participants preferred objects typically located below an observer's viewpoint (e.g., bowl, swimming stingray) to be located lower in the frame and objects typically located above an observer's viewpoint (e.g., light bulb, flying eagle) to be located higher in the frame. This latter finding is consistent with an inward bias (see below), in that effects or affordances of objects were directed toward the center of the frame. However, a potentially simpler notion is that objects viewed from the side imply a different perspective than do objects viewed from directly above or directly below. Winner et al. (1987) reported evidence consistent with an "up-down principle," in which an object appears to weigh more when it is in the upper part of a picture than when it is in the lower part of a picture (and so, if more mass is in the lower part of the picture, then that picture looks more balanced than if there is more mass in the upper part of the picture). The emotional valence of an object influences remembered vertical location, such that positively valenced items are remembered as slightly higher in the visual field than are negatively valenced items (Crawford, Margolies, Drake, & Murphy, 2006).

Clutter Effect

There are large literatures regarding how visual clutter and crowding influence visual target detection (e.g., Asher, Tolhurst, Troscianko, & Gilchrist, 2013; Rosenholtz, Li, & Nakano, 2007) and map-reading (e.g., Phillips & Noyes, 1982). What is less well known is that visual clutter also leads to a specific spatial bias; more specifically, an increase in clutter (i.e., an increase in the number of intervening items) between two locations leads to an increase in judged distance between those two locations.[1] A typical explanation is that in estimating distance, an individual scans between two locations. Each item encountered along the scan path presumably requires additional time to represent or process, and so the more intervening items there are, the longer the time required to scan between the two locations; the longer the time required to scan between the two locations, the greater the judged distance. In an early study, Oppel (1855) found a row of several dots appeared to be longer than if only the two endpoint dots were presented. Kosslyn, Pick, and Fariello (1974) found judgment of the distance between two objects in a room increased when barriers were placed between those objects, and Hubbard and

[1] An analogous bias in audition is referred to as the *filled duration illusion* and has attracted the attention of researchers (e.g., Wearden, Norton, Martin, & Montford-Bebb, 2007).

Baird (1993) found adding objects to an imaged scene increased distance estimates in that scene. Thorndyke (1981) had participants learn maps of fictitious locales containing many cities, and for a given distance, judgments were larger when there was more clutter along the route. Similar results were found for memory and perception. Such an effect is widespread and occurs in many studies of cognitive mapping (see Tversky, Chapter 16 in this volume).

Semantic Relatedness

Semantic information involving object identity and relationships between categories of objects in a scene influence preferred placement of those objects. Leyssen et al. (2012) demonstrated semantic relatedness had strong influence on object placement within a frame. The location of one object (e.g., a bottle of liquid soap) was fixed, and participants placed a second object (e.g., a cake or a sponge) at the location most preferred. The most preferred location was closer to the fixed object when the second object was semantically related (e.g., sponge) than when the second object was not semantically related (e.g., cake), even when overall shapes and sizes of objects were preserved. An analogous pattern was observed when participants viewed pairs of objects. Interestingly, preference for locating semantically related objects closer than unrelated objects appeared more important than preference for balance. Participants also preferred configurations in more central locations within the frame and in which the objects faced or pointed into (i.e., toward the center of) the frame, and this inward bias is consistent with reports for single object placement (see below). As discussed below, semantic effects on judgments of object location are also found in spatial biases for moving objects (e.g., representational momentum, Hubbard, 2005b, 2014a, Chapter 8 in this volume).

Title or Description

Semantic information involving a title or description influences spatial preference regarding object placement within a picture. Sammartino and Palmer (2012b) presented pictures in which a left-facing or right-facing object was at different horizontal locations within the frame, and three potential titles were given. One title was a neutral description, and the other titles were elaborative and referred to unseen objects or events (e.g., a picture of a galloping racehorse could have a neutral title of "racehorse" or a elaborative title of "frontrunner" or "dead last"). Participants judged which title best fit the picture; they preferred neutral titles when no other information was implied, but preferred elaborative titles when they "fit" the placement of the object in the picture (e.g., for a right-facing racehorse, "frontrunner" fit a horse to the right of center, and "dead last" fit a horse to the left of center). Preferences based on perspective were also found (e.g., if a person was walking toward the right, a title of "arriving" was chosen if that person was near the left edge of the frame, but a title of "departing" was chosen if that person was near the right edge of the frame). For multi-object pictures, objects on the left have a greater influence on judged suitability of the title than do objects on the right (Nelson & MacDonald, 1971), and this is consistent with Chatterjee's (2002) notion of preference for an agent to be on the left side. Relatedly, participants report greater aesthetic experience when a work of art is given an elaborative title or description (Leder, Carbon, & Ripsas, 2006; Millis, 2001).

Biases in Facing Direction

Biases in facing direction involve preferences regarding orientation of objects within a scene. Five such biases are considered and include (a) differences between the front and back of an object, (b) inward bias, (c) rightward bias, (d) preferences for horizontal lines or vertical lines over oblique lines, and (e) preference for suggested or apparent motion to be in the facing direction.

Front and Back

Information about locations in front of one's self is generally easier to access and more detailed than is information about locations behind one's self (e.g., Franklin, Henkel, & Zangas, 1995; Sholl, 1987; Werner & Schmidt, 2000). Such differences suggest there might be a corresponding bias regarding the preferred direction an object in a scene faces or points (e.g., perhaps maximizing affordance space in front of the object; see discussion below). Most objects have a perceptually distinct front and back; indeed, the term "facing direction" draws upon differences between the front and back of an object. Preferred placement of a single object within a frame is influenced by the direction the object is facing (i.e., the front side), such that facing side is directed toward the center (e.g., Palmer et al., 2008). Bertamini, Bennett, and Bode (2011) examined facing direction of animals in paintings by artists (Bewick, 1753–1828; Stubbs, 1724–1806) in a medieval bestiary. Bertamini and colleagues found more open space in front of the animal (i.e., in the direction the animal faced) than behind, and they referred to this as an *anterior bias* (consistent with an inward bias, but relative sizes of animals and context did not allow animals to actually face the center). The perceived direction of ambiguous apparent motion is influenced by the direction in which the front of the object undergoing apparent motion appears to face (McBeath, Morikawa, & Kaiser, 1992).

Inward Bias

In the absence of an elaborative title or other semantic information, if a person or object is facing rightward, then the preferred position of that person or object is on the left side of the picture, whereas if a person or object is facing leftward, then the preferred position of that person or object is on the right side of the picture (Palmer et al., 2008). In other words, there is a preference to have more pictorial space in front of the person or object (the direction in which that person or object is facing) than behind the person or object (cf. anterior bias). Chen and Scholl (2014) presented ambiguous figures (e.g., duck/rabbit) or geometric shapes (e.g., triangles) within a framed boundary. When those figures and shapes were placed near the frame, participants were more likely to interpret figures as whichever identity or pointing direction was consistent with facing inward toward the center of the frame. Presence of an inward bias in pictures more generally has been confirmed by Gardner et al.'s (2008) analysis of stock photography. The inward bias might reflect a combination of frame structure and differences in relative importance of the front and back of an object (Palmer, Schloss, & Sammertino, 2012). As noted earlier, the preference reported by Sammartino and Palmer (2012a) for object location along the vertical axis is consistent with inward bias, as the preferred location for an object facing upward (e.g., bowl) is below vertical center and the preferred location for an object facing downward (e.g., light bulb) is above vertical center.

Rightward Bias

In a study of Italian renaissance portraiture, McManus (2005) found portraits were more likely to show the left than right cheek of the individual, and this was especially prevalent in portraits of women. However, self-portraits tended to show the right rather than left cheek. Such a difference might reflect differences in emotional expressivity (e.g., Nicholls, Wolfgang, Clode, & Lindell, 2002; Schirillo, 2000), dominance (Keating, 1985), and "like me" or "not like me" (McManus, 2005). Thomas et al. (2006) found pictures of human infants and of nonhuman animals were more likely to show the left than right cheek. Chatterjee (2002) suggested a preference to display more of the left or right side of the face in portraiture is related to findings that right-handed individuals conceptualize agents of actions to the left and recipients of actions to the right. Palmer et al. (2008) found a weak preference for objects to face rightward rather than leftward, regardless of the horizontal location of the object within the frame. Bertamini, Bennett, and Bode (2011) reported a nonsignificant trend for illustrations of animals in a medieval bestiary to face rightward rather than leftward. Bertamini and colleagues also found a larger anterior bias for animals facing toward the left, and this is consistent with rightward bias for placement. Suitner and Maass (2007) found the left cheek preference in portraits was larger in paintings of infants or elderly individuals, and that female painters were less likely to exhibit this than were male painters.

Orientation

Orientation of the skeletal structure in a painting involving representational art or in a photograph is often clear, as objects depicted in that painting or photograph have canonical orientations in everyday experience, and orientation of the depicted objects can specify orientation of the artwork (i.e., which side is "up"). Indeed, recognition is facilitated for objects at canonical orientations relative to recognition of objects at noncanonical orientations (e.g., Palmer, Rosch, & Chase, 1981). Also, the vertical axis seems to be preferred, as vertical symmetry is detected faster than is horizontal symmetry (Palmer & Hemenway, 1978; Wagemans, 1995; Wenderoth, 1994), and it is easier to imaginally complete half of a symmetrical form if the figure is symmetrical about the vertical axis rather than the horizontal axis (Finke, Johnson, & Shyi, 1988). Triangles pointing upward (i.e., with a base parallel to the ground and a vertical axis of symmetry) are rated as more attractive (Friedenberg, 2012). Triangles pointing rightward had the second highest rating, and this is consistent with a rightward bias. There is a robust sex effect on orientation bias of a diagonal stimulus, as females prefer or perform better with diagonals from top-right to bottom-left, and males prefer or perform better with diagonals from top-left to bottom-right; this difference was found in tasks involving aesthetic ratings of paintings, searching for differences between paintings, and visual search for a tilted line (Avrahami, Argaman, & Weiss-Chasum, 2004).

Orientation of the skeletal structure in a painting involving abstract art is often unclear, as there are typically no objects that define orientation of the artwork. Mather (2012) presented images of abstract paintings in different orientations, and participants chose which orientation was "most pleasing or meaningful." Participants preferred the artist's orientation in approximately half of the trials (similar to Lindauer, 1969); however, a follow-up study suggested meaningful content could be recognized in some pictures, and this might have influenced participants' ratings. Latto, Brain, and Kelly (2000) presented participants with pictures of Mondrian paintings, and each painting was shown at eight different orientations (aligned with the cardinal and diagonal axes). Orientations containing horizontal or vertical lines were rated as more pleasing than were orientations containing oblique lines, and this was referred to as the *aesthetic oblique effect*. Participants gave higher ratings to paintings in their original orientation, even though they were not familiar with the original paintings (see also Plumhoff & Schirillo, 2009). However, Swartz and Hewitt (1970) did not find an effect of rotation or mirror reversal on aesthetic ratings of abstract art. Latto and Russell-Duff (2002) found more horizontal and vertical lines than oblique lines in a large sample of representational art paintings. Miller (2007) replicated this result but also found no differences in frequency of horizontal or vertical lines and oblique lines in abstract art generated by participants.

Motion

Arnheim (2006) suggested spatial composition is important for moving images as well as static images. Bertamini, Bode, and Bruno (2011) showed participants normal or mirror-reversed versions of Kurosawa films, and surprisingly, nearly all participants who viewed a reversed version did not notice the reversal (despite a majority of characters appearing left-handed in the reversals). As noted earlier, direction in which a set of objects appears to face or point biases perceived direction of an otherwise ambiguous apparent motion (McBeath et al., 1992), and such an effect is largest with geometrical stimuli (e.g., pointing triangles; see also Freyd & Panzer's, 1995, findings regarding memory for pointing shapes), next largest with drawings of stick figure drawings and mouse faces, and smallest with letters. Dynamic information implied by an object biases spatial judgments in ways consistent with that information. For example, if participants view a "frozen-action" photograph from the middle of a motion sequence (e.g., a jumper in mid-leap), memory for location of the pictured moving object is displaced in the direction implied by that object's momentum (i.e., participants are more likely to accept a photograph from slightly later than from slightly earlier in the motion sequence as the same as the target; e.g., Futterweit & Beilin, 1994). Relatedly, Freyd (1993) and Hubbard and Ruppel (2017a) speculated that mismatch in represented location between a recent (displaced) view and a current view of an aesthetic object or scene might result in an increase in arousal consistent with increased aesthetic response.[2]

Biases in Contour

Biases in contour involve preferences regarding the shape of an object. Three such biases are considered and include preferences for (a) symmetrical shapes over asymmetrical shapes, (b) curved contours over angular contours, and (c) longer contour lengths over shorter contour lengths.

[2] Original artworks result in higher ratings of hedonic value than do reproductions, although originals and reproductions are rated similarly in pictorial qualities (Locher, Smith, & Smith, 2001), and this might reflect dynamic information inherent in original artworks but not captured by reproductions (e.g., brushstrokes, Freyd, 1993).

Symmetry

Symmetry is detected early in visual processing (Gershoni & Hochstein, 2011; Locher & Nodine, 1987), and an axis of symmetry often acts as a perceptual landmark for visual exploration (Locher & Nodine, 1989). Detection of symmetry is a critical aspect of figure-ground discrimination (for review, see Wagemans et al., 2012). Symmetrical figures are preferred over asymmetrical figures (Frith & Nias, 1974). Symmetry has been linked with greater beauty (for review, see McManus, 2005; Weyl, 1952), although some authors suggested beauty is greater between symmetry and asymmetry (e.g., Gombrich, 1979). Asymmetry is linked with tension, and this is consistent with Leyton's (1989, 1992) notion that deviations from symmetry reflect the history of forces that acted on an object and are encoded in the shape of that object. Arnheim (1986) suggested a simple-complex dimension underlies symmetry-asymmetry, and so a preference for intermediate complexity would suggest preference for the slightly asymmetrical. This is consistent with Berlyne's (1971) theory linking aesthetic response and physiological arousal. However, Gartus and Leder (2013) presented participants with patterns exhibiting differing amounts of asymmetry, and patterns with slight asymmetries were liked less than were symmetric patterns.

Angularity

In early studies, angular lines were associated with agitation, fury, and seriousness, whereas curved lines were associated with gentleness, sadness, and quietness (e.g., Hevner, 1935; Lundholm, 1921; Poffenberger & Barrows, 1924). More recently, Bar and Neta (2006) presented participants with pictures of objects with similar semantic meaning and general appearance, but some objects had curved contours and other objects had angular contours. Participants preferred objects with curved contours more than objects with angular contours. Bar and Neta hypothesized this occurred because objects with sharp contours were perceived as more threatening or dangerous. Consistent with this, Bar and Neta (2007) reported angular objects evoked greater activation of the amygdala than did curved objects. Carbon (2010) reported a high correlation between curvedness of automobile exteriors and liking; however, this correlation weakened if information regarding history and innovation of automobiles was provided, and Carbon suggested preference for curved or angular contours is flexible and changes over time. Silvia and Barona (2009) presented participants with arrays of circles or hexagons, and participants rated how pleasing the array was. Participants with artistic training did not have a preference, but participants with less artistic training preferred displays containing circles. In a follow-up experiment, participants viewed random polygons exhibiting sharp angular corners or smoothed rounded corners. Rounded polygons were preferred, and there was a greater difference in ratings for artistic novices than for artistic experts.

Length

Across a wide range of stimuli typically encountered, contour (perimeter) length of an object is confounded with size of that object, that is, an object with larger surface area (or volume) would typically have a longer contour. One study assessing preferences for different triangles found preferences were highest when sides of the triangle were approximately equal in length (Friedenberg, 2012), and this was hypothesized to reflect preference for compactness. Such preference was speculated to occur because a more compact object is more perceptually stable (e.g., less likely to fall or less likely to break). In compact shapes, ratio of contour length to area (volume) is minimized, and so if there is preference for compact shapes, then (if area is held constant) participants should exhibit preference for shapes with shorter contour lengths. The relationship between contour length and preferences in irregular polygons was tested by Friedenberg and Bertamini (2015), who created irregular polygons with varying contour lengths by superimposing a radial grid on a regular polygon and moving the contour at each grid position toward or away from the center in a way that preserved the approximate total area of the polygon. Surprisingly, preference increased as length of the contour increased. Also, contours with more alterations (i.e., changes between concave and convex) were more preferred.

Biases in Ratio

Biases in ratio involve preferences regarding relative location of a single object or of multiple objects within a bounded space (and so are relevant to biases in placement) or preferences regarding proportions within an object or between objects (and so are relevant to biases in contour). Two such biases are considered and include (a) the rule-of-thirds and (b) the golden ratio.

Rule-of-Thirds

One of the best-known composition rules in photography and painting is the rule-of-thirds, and this rule states that the most salient object in a scene should be placed at a location one-third the distance between the margins both horizontally and vertically (e.g., Gooch, Reinhard, Moulding, & Shirley, 2001; Kragen, 2005; Mai, Le, Niu, & Liu, 2011). Indeed, the rule-of-thirds is often used as a metric in assessing the aesthetic quality of photographs and paintings (e.g., Datta, Joshi, Li, & Wang, 2006; Li & Chen, 2009). In a test of the rule-of-thirds, Amirshahi, Hayn-Leichsenring, Denzler, and Redies (2014) presented observers with photographs and reproductions of paintings, and participants rated whether stimuli were aesthetic by positioning a cursor along a line with the anchor terms "not aesthetic" and "aesthetic." There was no correlation between the extent to which a stimulus corresponded to the rule-of-thirds and aesthetic ratings, and Amirshahi and colleagues suggested the rule-of-thirds played only a minor, if any, role in aesthetics of paintings and photographs. Furthermore, studies of preferred placement location of stimuli along the horizontal (e.g., Palmer & Guidi, 2011; Palmer et al., 2008) or vertical (Sammartino & Palmer, 2012a) axis have not supported the rule-of-thirds. The rule-of-thirds is incompatible with Arnheim's (1974, 1982) emphasis on the center as the most important or aesthetic location, but it is more compatible with placement based on a golden ratio.

Golden Ratio

Perhaps the most famous ratio in aesthetics is the "golden ratio" or "golden section" (for reviews, see Benjafield, 2010; Green, 1995; Herz-Fischler, 1998; Höge, 1995; McManus, 1980). The golden ratio is a division of a line at the point at which the ratio of the smaller part to the larger part is the same as the ratio of the larger part to the original line. In other words, if c is the length of the original long line, and a and b are the lengths of the short and intermediate lines created by dividing c, then $a/b = b/c$. This ratio is equal to approximately 0.618, and it is widespread in painting, sculpture, and architecture, as well as in the natural world (e.g., Konečni, 2001; Livio, 2003; McWhinnie, 1986). Fechner (1876) performed some of the first experiments on the golden ratio by presenting observers with rectangles differing in the ratios of lengths of horizontal sides and vertical sides, and he reported the rectangle his observers were most likely to prefer corresponded to a golden ratio (i.e., lengths of the shorter and longer sides were in the golden ratio). McManus, Cook, and Hunt (2010; McManus & Weatherby, 1997) pointed out research on preference for the golden ratio usually focuses on population trends, and they called for an increase in research on individual differences. However, McManus and Wu (2013) found preference for a specific ratio in rectangles does not correlate with any of several personality or cognitive measures.

Interest in the golden ratio has waxed and waned since Fechner's time. Boselie (1984), McManus (1980), and Russell (2000) rejected the hypothesis the golden ratio is preferred by human observers, but McManus and Weatherby (1997) found a population preference for object placement at a golden ratio along the horizontal axis and between the golden ratios along the vertical axis. Konečni (2005) pointed out effects of the golden ratio are often modulated by context. Consistent with this, Westpahl-Fitch, Oh, and Fitch (2013) had participants generate rectangles of varying proportions within framing contexts that were square, longer horizontally than vertically, or longer vertically than horizontally; a preference near the golden ratio was found, but only if a surrounding rectangular frame was present. A preference for a square was also reported. Wölfflin (1994 [1886]) suggested preference for the golden ratio resulted from empathic, embodied responses to gravity (see discussion in McManus & Wu, 2013), and Stieger and Swami (2015) suggested preference for the golden ratio resulted from art expertise. Although the majority of research on the golden ratio involves lines or rectangles, the golden ratio occurs in a variety of other shapes, although it isn't always linked to preferences in those shapes (e.g., Boselie, 1997; Davis & Jahnke, 1991; Friedenberg, 2012; McManus, 1980). Green (1995) and Konečni (2005) suggest a preference for the golden ratio exists, but it is weak and easily overwhelmed by contextual elements or poor experimental methodology.

Biases Due to Gestalt Effects

Arnheim's (1974, 1988) approach to aesthetics was firmly grounded in Gestalt psychology, and as is evident in the earlier discussion, many of his ideas motivated empirical studies of spatial biases related to aesthetic preferences. Although not usually interpreted as such, a case can be made that other Gestalt notions are also related to spatial bias. Two such notions are considered and include (a) Gestalt principles of perceptual grouping and (b) Gestalt illusions.

Gestalt Principles of Perceptual Grouping

Most textbooks on Introductory Psychology or on Sensation and Perception discuss Gestalt principles of perceptual grouping (see also Palmer, 1999; Wagemans et al., 2012). As illustrated in Figure 15.1, there are several well-known principles. Traditional Gestalt principles include proximity (elements closer together are more likely to be grouped together), similarity (more similar elements are more likely to be grouped together), good continuation (elements that are aligned are more likely to be grouped together), closure (gaps between elements are filled in so an incomplete figure is perceived as a complete

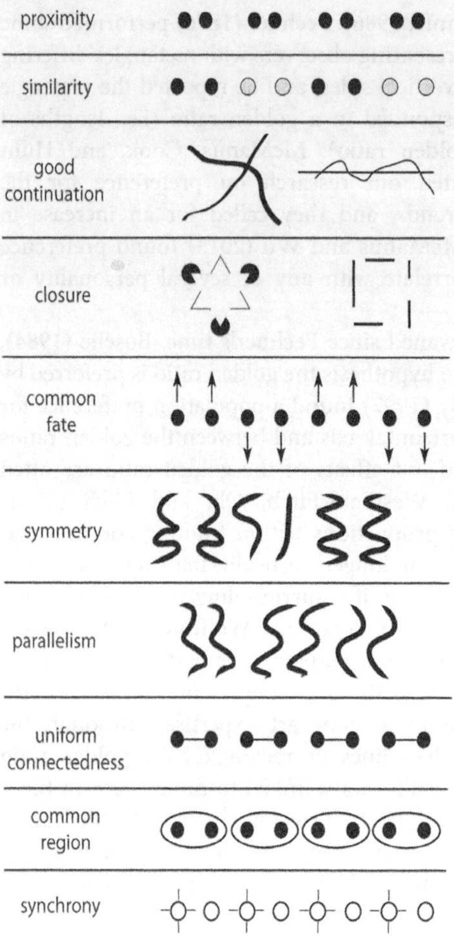

Figure 15.1 Illustrations of the Gestalt principles of perceptual grouping including proximity (dots that are closer together are grouped together), similarity (dots that are more similar are grouped together), good continuation (line segments that are aligned are grouped together), closure (incomplete figures are completed), common fate (dots that move in the same direction are grouped together), symmetry (lines that form a symmetrical pattern are grouped together), parallelism (lines that are parallel to each other are grouped together), uniform connectedness (dots that are connected are more likely to be grouped together), common region (dots enclosed within the same ellipse are grouped together), and synchrony (lights that flash at the same time are grouped together).

figure), and common fate (elements moving in the same direction are more likely to be grouped together). Several other Gestalt principles have been documented recently, and these include the law of symmetry (elements that form a symmetrical pattern are more likely to be grouped together), parallelism (parallel elements are more likely to be grouped together), uniform connectedness (visually connected elements are more likely to be grouped together), common region (elements are more likely to be connected if they are in a common closed region), and synchrony (simultaneously changing elements are more likely to be grouped together). Gestalt principles of perceptual grouping bias spatial perception in highly specific and consistent ways but, perhaps surprisingly, are rarely considered in discussions of spatial bias.

Gestalt Illusions

Several biases are related to illusions that arise from operation of Gestalt principles. Three such types of biases are considered here, and these include biases based on (a) perceptual grouping, (b) boundary extension, and (c) environmentally invariant physical principles.

Perceptual Grouping. In addition to biasing which elements are grouped together, Gestalt principles of perceptual grouping are linked to spatial biases in judgment of length or distance. Coren and Girgus (1980) presented stimuli designed to evoke specific principles of perceptual grouping, and they incorporated two test lines within each stimulus; one line was within a set of elements that were grouped together (e.g., two dots grouped by proximity), and the other line was between two elements that were not grouped together (e.g., two dots not grouped with each other, but were each grouped with a different other dot). Importantly, test lines within a given stimulus were always the same length. Participants' reproductions of lines between elements that were grouped together were shorter than reproductions of lines between elements that were not grouped together. This effect was found for lines involving dots grouped by proximity, similarity, closure, and good continuation. Enns and Girgus (1985) replicated these findings and found that such biases were larger in younger children. Pinna and Albertazzi (2011) presented numerous examples of how Gestalt principles of perceptual grouping related to global shape of an object distorted the shape of individual elements composing that object (e.g., square elements in a larger stimulus array parsed as rows or columns as a function of proximity or similarity appear longer along their horizontal or vertical sides than along their vertical or horizontal sides, respectively).

Boundary Extension. Like the clutter effect, boundary extension reflects bias regarding the space (scene) in which objects are located. Memory for what was contained within a previously viewed scene is often extended beyond the edges of the actual view, and this is referred to as *boundary extension* (see Hubbard, Hutchison, & Courtney, 2010; Intraub, 2012; Intraub & Gagnier, Chapter 13 in this volume). Boundary extension is not simply based on closure or good continuation, as boundary extension occurs whether or not stimuli at the edges of the view are truncated (e.g., Gagnier, Dickinson, & Intraub, 2013); rather, boundary extension has been hypothesized to reflect continuity of spatial layout (e.g., Gottesman & Intraub, 2002; Intraub, Gottesman, & Bills, 1998). Boundary extension exhibits several properties in common with Gestalt principles of perceptual grouping (e.g., displacement in remembered location, reflection of environmental regularities, a basis in isomorphism, decreases in information processed, dependence on surrounding context, interpretation as a laboratory-based illusion or as a successful adaptation, reflection of general rather than domain-specific principles, and automatic application), and given these commonalities, Hubbard (2011a) suggested boundary extension could be considered a Gestalt principle. Indeed, boundary extension, like Gestalt principles of perceptual grouping, is an emergent process or property in which spatial properties of the perceived whole are greater than the sum of spatial properties of the perceived parts.

Environmentally Invariant Physical Principles. Studies on memory for location revealed several spatial biases consistent with implied influence of environmentally invariant physical principles (Hubbard, 1999). The most well-known of these is referred to as *representational momentum*, in which memory for the location of a target is displaced in the direction of anticipated motion (see Hubbard, 2005b, 2014a, Chapter 8 in this volume). These displacements are modulated by semantic knowledge regarding expected target behavior (e.g., if a target is expected to change direction, displacement is in the direction of expected motion, Hubbard & Bharucha, 1988; if an object is labeled "rocket," forward displacement with upward motion is greater than if the object is labeled "steeple," Reed & Vinson, 1996), and this is consistent with effects of semantic knowledge and titles, respectively, on placement preferences. Representational momentum exhibits several properties in common with Gestalt principles of perceptual grouping (e.g., displacement in remembered location, reflection of environmental regularities, a basis in isomorphism, decreases in information processed, dependence on surrounding context, interpretation as a laboratory-based illusion or a successful adaptation, automatic application), and given these commonalities, Hubbard (2011b) suggested representational momentum could be considered a Gestalt principle; indeed, representational momentum is similar to the principle of good continuation.

In addition to a spatial bias based on implied momentum (i.e., representational momentum), spatial biases based on implied gravitational attraction, friction, and centripetal force have been reported. Memory for locations of static objects is displaced downward (e.g., Freyd, Pantzer, & Cheng, 1988); memory for locations of horizontally moving objects is displaced downward, and forward displacement for descending objects is larger than forward displacement for ascending objects (e.g., Hubbard, 1990). Such patterns are consistent with effects of implied gravitational attraction (e.g., see also de sá Teixeira, 2014; de sá Teixeira & Hecht, 2014; Hubbard, 1997; Motes, Hubbard, Courtney, & Rypma, 2008; Zago, Chapter 10 in this volume), and this is referred to as *representational gravity* (cf. visual weight). Increases in implied friction on a moving target decreases forward displacement in judged location of that target (*representational friction*, Hubbard, 1995b, 1998), and judgment of the location of a target moving along a circular path is displaced inward consistent with centripetal force (*representational centripetal force*, Hubbard, 1996b). Importantly, effects of implied momentum, gravitational attraction, friction, and centripetal force occur with computer-generated displays in which animated targets do not possess mass and do not experience effects of physical principles, and so any influence of implied physical principles on judged location demonstrates the robustness of such spatial biases (Hubbard, 1999, 2006b).

Why Do Biases in Spatial and Scene Composition Occur?

There are several possible accounts for why biases in spatial and scene composition occur. It is unlikely a single explanation will account for all such biases, and it is more likely that different biases occur for different reasons or arise from different mechanisms. Five theories with wide potential applicability are considered and include (a) neurological functioning, (b) perceptual fluency, (c) ecological expectations, (d) representational fit, and (e) unintended consequences.

Neurological Functioning

Early investigations of spatial biases were generally not concerned with neurological bases for such biases. When speculation regarding such biases occurred, it was often inconsistent with subsequent neurological findings (e.g., electromagnetic fields in the brain accounting for Gestalt principles of perceptual grouping; Kohler, 1938 [1920]). Some spatial biases considered in this chapter (e.g., preference for the right side of a picture) might be based on hemispheric asymmetries (e.g., Banich et al., 1989; Bromberger, Sternschein, Widick, Smith, & Chatterjee, 2011; Corbetta, Patel, & Shulman, 2008; Schirillo, 2000). Given that many spatial biases discussed in this chapter are related to aesthetics, it is possible research on "neuroaesthetics" (i.e., neural underpinnings of aesthetic experience; see Chatterjee, 2012, 2014; Ramachandran, 2011; Ramachandran & Hirsten, 1999; Zeki, 1999) might shed light on neurological variables involved in those biases. Models in neuroaesthetics are beginning to address specific processes and processing stages (e.g., Leder et al., 2004; Vartanian & Nadal, 2007). Although it might be possible at some point to account for many spatial biases discussed here in terms of underlying neurology, it is not yet possible to do so. However, a potential danger in a neuroreductive approach is the lack of a larger framework or principle addressing why such biases occur (i.e., lack of computational theory [cf. Marr, 1982] of spatial bias). More behaviorally based accounts described below generally posit such larger frameworks or principles.

Perceptual Fluency

Perceptual fluency refers to the ease with which a stimulus is perceived, and stimuli perceived more easily (quickly, etc.) are said to be more fluent (e.g., Reber, 2012). People generally prefer scenes in which the content is easier to perceive or process (Reber, 2012; Reber, Schwartz, & Winkielman, 2004). A flying eagle or a light bulb is usually perceived above eye level, and so those objects are perceived more easily or more readily if they are presented higher in the frame. Similarly, a center bias exists because resolution is best at the center, and inward bias puts the most important (actionable) side of a stimulus toward the center. Perceptual fluency theory can account for effects of symmetry, typicality, and orientation, as symmetrical objects, prototypical objects, and objects aligned with vertical or horizontal axes are easier to process. However, it is not clear how perceptual fluency might account for effects on preferences in location and facing direction based on semantic information in a title or description (Sammartino & Palmer, 2012b), nor is it clear how perceptual fluency theory could account for the inverted-U shaped function of aesthetics as a function of complexity (e.g., Berlyne, 1971). Importantly, violation of expectancies might be a critical aspect of aesthetic response (e.g., Huron, 2006; Meyer, 1956), but the stimulus processed with the most fluency would be the stimulus that least violated expectations (Palmer, Schloss, & Sammartino, 2013), and more frequently encountered (i.e., more fluent) stimuli are not always judged as more preferable.

Ecological Experience

The extent to which a stimulus accords with everyday experience might influence spatial bias. Individuals usually prefer objects typically viewed from above (bowl on a table, swimming stingray) to be lower in the frame and objects typically viewed from below (e.g., light bulb, flying eagle) to be higher in the frame (Sammartino & Palmer, 2012a). Such a pattern is consistent with findings that objects are perceived more quickly and easily if viewed from a canonical perspective (e.g., Palmer et al., 1981). Ecological experience might lead to preferences that maximize available affordance space (i.e., the space around an object in which activities regarding that object occur, Sammartino & Palmer, 2012a) by locating objects so that the direction of any potential effects or influences is toward the center of the picture space. Such a notion is consistent with an inward bias. The ecological experience theory is consistent with spatial biases that arise from Gestalt principles; principles of perceptual grouping facilitate detection of objects, and boundary extension and biases arising from environmentally invariant physical principles facilitate interactions with stimuli that might be encountered in the next fixation or few hundred milliseconds. Ecological experience theory is similar to perceptual fluency theory, but it is broader in that biases and preferences would also be exhibited for stimuli that have not been previously encountered and for which no opportunities to acquire fluency have occurred (e.g., representational momentum occurs with the first viewing of a specific target).

Representational Fit

The idea of representational fit suggests aesthetic preference for a stimulus is related to how well that stimulus conveys a meaningful message to the observer (Sammartino & Palmer, 2012b). Such a message can be sensory, cognitive, or emotional, and the message observers perceive might or might not be the message the artist intended. The location and orientation for an object that would maximize available information is the location and orientation at which the object is most easily recognizable (cf. recognition of objects in nontypical locations, Beiderman, Mezzanotte, & Rabinowitz, 1982; Henderson & Hollingworth, 1999). When meaning is added by a title or other description, object placements that are otherwise nonpreferred might become more preferred (e.g., titles such as "frontrunner" or "dead last" in Sammartino & Palmer, 2012b). Palmer, Schloss, and Sammartino (2013) suggest representational fit and fluency theory can be viewed as compatible accounts that differ in emphasis: representational fit focuses on why processing of a given stimulus is easier, and perceptual fluency focuses on the ease of processing. However, it could be suggested that ecological experience might offer a better account for why processing of a given stimulus might be easier, and that representational fit describes the match between the expected stimulus information and the actual stimulus information.

Unintended Consequences

A case can be made that many spatial biases are adaptive (e.g., spatial biases due to Gestalt effects, and preferred placement or facing direction, facilitate perception of stimuli). Along these lines, Ramachandran (2011; Ramachandran & Seckel, 2012) suggested evolution might shape aesthetic principles and aesthetic principles might drive evolution (e.g., bowers of bowerbirds, pecking behavior of seagull chicks). However, it need not be that all spatial biases were necessarily selected for or necessarily adaptive. It might be that some spatial biases (e.g., asymmetries between left and right sides of space) are an unintended consequence of some other trait that was selected for, and that the unintended spatial bias was not destructive or dangerous enough to be selected against (e.g., Corbetta et al., 2008, suggest that preference for the right side of space might reflect differences in functioning of left and right parietal lobes). In this case, such spatial biases would be the perceptual equivalent of spandrels (i.e., a by-product of evolution of some other characteristics and not a direct product of adaptation; Gould & Lewontin, 1979). Consistent with this, Ramachandran and Seckel (2012) note some aesthetic preferences might be adventitious by-products of visual principles that were evolved for some other purpose. One caveat is that such a theory cannot be conclusively demonstrated, as it is possible in principle that adaptive effects might be discovered in the future for those spatial biases not currently thought to be adaptive.

Summary and Conclusions

Biases in spatial and scene composition involve preferences regarding location of object placement within a frame, facing direction of an object, contour of an object, ratio between an object and the frame or between parts of an object, perceptual grouping, and invariant physical principles. The preferred placements of objects within a scene are influenced by horizontal, vertical, and oblique axes, with preferences for objects to be located near the center or near a primary axis and for distribution of mass to be balanced about and near the geometric center. Horizontal and vertical orientations are more preferred than oblique orientations. Some studies found preference for symmetry, but other studies found preference for asymmetry. There is a slight preference for objects to be located on the right side of a scene, although this interacts with facing direction, such that the preferred location for objects that face rightward or leftward is near the left edge or right edge, respectively, of a scene. Similarly, there is a preference for objects that face downward or upward to be located near the top or bottom edge, respectively, of a scene. Semantic information regarding object type, relationships between categories of objects, and title influences preferences. Although there have been suggestions for location preferences one-third of the distance across the frame or at the golden ratio, evidence for such preferences is weak. Judgment of distance and location is biased by perceptual grouping and spatial continuity and by implied physical principles such momentum, gravitational attraction, friction, and centripetal force.

There are several possible theories to account for spatial biases, and these include neurological functioning, perceptual fluency, ecological experience, and representational fit. These theories are not mutually exclusive; indeed, it is likely that many biases reflect a combination of influences. Although some spatial biases might have positive consequences and were selected for, other spatial biases might reflect unselected consequences of other traits that were selected for. Neurological functioning involving factors such as asymmetries in the cerebral hemispheres or relatively greater activation of specific neural structures might also contribute to some spatial biases. However, research on the neural basis of spatial biases is still in its infancy, and even if neurological bases are found, such preferences will also need to be accounted for within a more cognitive framework. More cognitively oriented theories involving perceptual fluency, ecological experience, and representational fit all stress the role of experience and learning in development of spatial preferences and biases, and this is consistent with Carbon's (2011) suggestion that mechanisms underlying aesthetic judgment are probably flexible and not hard-wired. In addition, perceptual fluency theory stresses the relative ease with which a stimulus can be processed, ecological experience theory stresses adaptation (based on the history of interactions with similar stimuli in everyday life) for types of stimuli likely to be encountered, and representation fit stresses how easily an observer can access the information in the stimulus and the match between the information in the stimulus and the information possessed by the observer.

Spatial biases in spatial and scene composition have several properties. First, spatial biases are not limited to art objects or other aesthetic stimuli, but involve a wide range of stimuli from everyday experience (e.g., Carbon, 2010; Intraub, 2012). Second, spatial biases appear to be relatively automatic and emerge very quickly (e.g., Gershoni & Hochstein, 2011; Locher & Nodine, 1987; Locher & Stappers, 2002). Third, spatial biases are influenced by additional information (e.g., titles given to works of art, Sammartino & Palmer, 2012b; knowledge of object type in representational momentum, Reed & Vinson, 1996). Fourth, spatial biases operate individually or in combination (e.g., Palmer et al., 2008, suggested inward bias results from a combination of effects of the skeletal structure and differences between the front and back of a stimulus; Hubbard, 2006a, suggested remembered location of a moving target resulted from a combination of information involving implied momentum, gravitational attraction, and friction, as well as semantic information regarding expected target behavior). Fifth, spatial biases are often related to previous experience and future expectations, and are tuned to facilitate processing of stimuli in the relative locations in which those stimuli are likely to occur in the future. In this view, at least some spatial biases reflect successful adaptations. Further investigation will likely discover more properties of spatial biases in spatial and scene composition, and additional investigation could potentially reveal additional aspects regarding how our perceptual and cognitive processes have adapted to our environment.

Chapter 16

Spatial Biases in Thought and Judgment

Reference Theory

Barbara Tversky

We can't escape space, even in our imaginations with our eyes shut tight; we're stuck there. Everything we do we do in space. Knowledge of space and of how our bodies act in space is essential for survival. Until there are robots that can read our minds, know where we need to go and how to get there without bumping into anything, how to bring us what we need when we need it, and how to feed us, bathe us, and dress us, we must do those things ourselves. But even if that frightening future comes to pass, we need to act and move in space in order to sustain our bodies, our brains, and our minds. Or, until we exist as placeless, bodiless ethereal beings, fulfilling those essentials of life depends on our knowledge of space, how our bodies act in space, how the space of the world constrains our actions, how to orient ourselves in the space that surrounds us, how to navigate to the places we need to go. Learning about the space of our bodies, the space around our bodies, and the space of navigation occurs from the beginning of life and changes constantly as our bodies and the spaces around them change. That knowledge is created from the interactions of our bodies and the surrounding spaces. As such, it depends on our perceptions and our actions in space; those perceptions and actions differ for the different spaces central to our lives, the space of the body, the space around the body, and the space of navigation (e.g., Tversky, 1998, 2000, 2005a, 2005b).

Our knowledge of space, then, is inextricably bound to our perception of space and our actions in it. As such, it derives from experiences that are inherently embodied. That knowledge is called upon nearly every moment of our lives as we adjust our posture, as we walk around, as we reach for the things we need, as we navigate a city, as we plan travel, and as we use space to make a multitude of inferences: distances, directions, travel time, voting patterns, or disease spread.

Here the focus is on the larger spaces that cannot be seen from a glance, spaces we might actually navigate or think about in a range of contexts such as planning travel, accounting for housing prices, or crime statistics. We come to know these spaces from a multitude of sources, from navigation, from descriptions, and from a variety of maps. Some of the biases in judgment and memory for small spaces that we can see from a single viewpoint are similar to those for larger ones, for example, the bias to perceive nearly symmetric objects as more symmetric than they are (e.g., Freyd & Tversky, 1984). Conceptual biases also affect both diagrammatic and real spaces. For example, when a line embedded in parallel rudimentary axes is interpreted as a graph line, it is remembered as closer to 45 degrees, but when it is interpreted as a path in a map, memory is not biased (Schiano & Tversky, 1989; Tversky & Schiano, 1997).

The current focus is on armchair judgments, not on actions in the world. Actions in the world have additional sources of information from the body and the world, visual, haptic, auditory, proprioceptive, temporal, social, and more that can serve both to correct errors of the mind and to provide additional sources of error. Research on action in space is nicely covered in an entire section, especially in the chapters by McBeath (Chapter 19), Witt (Chapter 21), and Wiener and Meilinger (Chapter 22). Armchair judgments are disembodied; nevertheless, they often derive from embodied experience. Armchair judgments pull together whatever information seems relevant to the question. That information is often partial and incomplete. Pulling together the information that seems relevant might also bring together information that is not reconcilable or consistent. For this reason, we long ago noted that cognitive maps are impossible figures, like Escher's etchings (Tversky, 1981). Nevertheless, for spatial judgments as for other judgments, more information is better, as errors are typically independent and can correct each other (e.g., Baird, Merrill, & Tannenbaum, 1979).

People don't come with built-in devices that can directly measure distance, direction, or time, much less make more complex inferences. Instead, we estimate, drawing on whatever knowledge from the mind, from the body, and from the world seems to be relevant. Those mechanisms substitute for measurement gauges that we don't have and for information that we lack. Like measurements that surveyors of the world make, these mechanisms are relative, not to physical rulers or gauges, but to conceptual ones. Many of the errors are systematic, because the conceptual rulers and gauges culled and used to predict or estimate are themselves approximate and biased. In addition, there may be no clear way to reconcile the disparate sources of information, for example, integrating navigation experiences, memories of maps, and descriptions of spaces. Thus, bias is inevitable.

Uncovering systematic bias is not an easy and straightforward matter. You have to intuit the sources of bias ahead of time to know how to design research to reveal them. Otherwise, in more ordinary research, errors enter averages as variance, and are swamped by other sources of error and aren't even noticed. Studying bias and error has a long tradition in cognition, perception, action, and related fields for its role in discovering the mechanisms that underlie both successful performance and failures (e.g., Coren & Girgus, 1980; Poulton, 1989).

Here we present a theory, the Spatial Reference Theory, that accounts for many of them. A scattered set of fascinating biases has been uncovered by various researchers, but they have not been theoretically organized. The referential points and frames that have been shown to create bias include grouping, reference frame, reference points, paths, and schemas. But because this review is meant to be useful to many communities that are interested in spatial thinking, we will also organize by type of judgment, distance, direction, and route planning.

Spatial Reference Theory: Objects and Frames

Conceptually, space is empty. Of course the spaces important to us, the spaces we think about have many thing in them, people, objects, buildings, streets, parks, cities, oceans, and countless more. We can and do think about the things qua things, but thinking about the things in space is not yet spatial thinking. Spatial thought is inherently relational. We think of the things in space relative to other things or relative to frames of reference that encompass the things. The "relative to" can involve things at the same level of analysis or things or frames at a larger level of analysis. The "relative to" can be a sources of bias. Those that have been studied include grouping, reference frames, reference points, paths, and schemas.

Manner of thought is often revealed in manner of talk. Importantly, these relative to's are also the way we talk about spatial relations (cf. Levelt, 1982a, 1982b; Levinson, 1996; Talmy, 1983, 2000; Tversky, 1996). I might say "your bike is by the gate," using a reference object at the same level of analysis, or "your bike is behind you," using a privileged reference object, the perspective of an ego. I could say "your bike is to the right of the gate," using both a reference object and the perspective of an ego. I might say "your bike is in the yard," using a reference frame at an encompassing level of analysis, or "your bike is north of the gate," using a reference object and an encompassing frame of reference. Similarly for thought. We can think about one thing with respect to another at the same level, for example, with respect to a more salient or familiar or stable entity, such as a landmark. We can think of things from a specific viewpoint within a space real or imagined. We can think about one thing as part of a larger set that contains it, and then think of that larger set with reference to some other set or sets. Cities embedded in countries and countries in continents. This is what is meant by grouping and by hierarchical organization. We can think about one thing with respect to a larger frame of reference that serves to locate it more or less precisely. The canonical frame of reference for things in allocentric space is the NSEW axes, more familiar now as GPS coordinates.

These are but some of the many comparisons, perspectives, and frames of reference that can be taken on entities in space. Of special significance to our everyday lives are the personal perspectives, yours and mine. Although our experience in space is necessarily from a personal perspective, we, along with rats (e.g., Moser, Kropff, & Moser, 2008; O'Keefe & Nadel, 1978), bees (Menzel et al., 1996), and other creatures, rapidly create allocentric representations of space, representations of space that may have been created from an egocentric perspective but become independent of it by anchoring the things in space to some external frame of reference. Creating allocentric representations is crucial to navigation, as rats, bees, and people enter spaces from many different places. It is the organization of both egocentric and allocentric representations that seem responsible for the biases and errors we will review. These two perspectives, inside and outside, egocentric and allocentric, the former regarding a space from within and the latter regarding a space from outside, have deep consequences for spatial thought, social thought, and abstract thought. Finally, we develop general schemas for or representations of the designs of the various spaces we inhabit, such as grid plans for cities or layouts of houses or supermarkets; these, like schemas for objects and events, guide our expectations but also bias them.

Grouping and Hierarchical Organization

When I was eight, I was taught to write my return address on letters to my grandparents. That simple experience was eye-opening: it anchored me to a specific spot in space and at the same time connected me to everything in the universe, opening me to the wonder of the infinity of space. For a time, I delighted in writing my return address as my name, my number, my street, my town, my state, Earth, the Milky Way, the Universe. Sets within sets within sets. Similarly, I could think of myself as part of a family, the family as part of a neighborhood, the neighborhood as part of a town, the town as part of state, and so on, each superset bringing in new connections and new meanings. Space has many kinds and levels of entities. An important way to think about entities, to organize them and make sense of them, is through the larger structures that encompass them. Space is virtually flat, as in maps, yet people think of smaller entities as included in larger ones, cities inside states inside countries inside continents (e.g., Stevens & Coupe, 1978). In fact, people seem to subdivide any large set of things into subsets and subdivide large subsets yet again (e.g., Mandler, 1967). For categories, people regard entities in the same category as more similar to each other than entities in different categories (e.g., Rosch, 1978). Similarity in conceptual spaces translates to distance in actual space. People judge two cities that are in the same state or country to be closer to each other than two cities in different states or countries even when the pairs of cities in different states or countries are actually closer than those within the same geographic entity (e.g., Allen & Kirasic, 1985; Hirtle & Jonides, 1985). Spatial distances within groups are judged smaller than spatial distances between groups for conceptual groupings as well as spatial ones. Residents of Ann Arbor, Michigan, judged distances among academic buildings and among commercial buildings to be closer than the same or smaller distances across those categories (Hirtle & Jonides, 1985). Israelis judge distances between Jewish settlements or between Arab settlements to be smaller than distances between Arab and Jewish settlements that are actually equivalent or larger (Portugali, 1993). In short, conceptual categories as well as geographic categories bias judgments of spatial distance.

Groupings bias judgments of direction as well as judgments of distance. Famously, a majority of students in San Diego, California, erroneously judged San Diego to be west of Reno, Nevada (Stevens & Coupe, 1978). The error seems to have come from using the overall direction of the states that contain those cities to make inferences about the directions of the cities. Since Nevada is east of California, then Reno must be east of San Diego. But Reno is in fact west of San Diego. The same error was obtained for other pairs of cities and for artificial maps.

Higher-level categories also affect speed of judgments of direction. People are faster to judge whether one city is east or north of another when the cities belong to separate states or countries than when they are parts of the same geographic entity (Maki, 1981; Wilton, 1979).

Categorization biases similarity judgments of people just as categorization biases distance judgments of places. That is, people treat members of the same social category as more similar than members of different social categories even when the assignment to categories is random (e.g., Billig & Tajfel, 1973). This phenomenon will be familiar to readers of Dr. Seuss's *Sneetches* (1953).

Reference Frame: Canonical Axes

Reference frames form a special kind of encompassing category that can involve two (or perhaps three or even four) dimensions. The standard reference frame for geographic entities are the canonical axes, north–south, east–west. People tend to remember geographic entities as more parallel or perpendicular to the canonical axes than they actually are (Tversky, 1981). For example, when students were given cutouts of South America and asked to place them in a NSEW reference frame, they oriented South America more upright than it actually is. Similarly, residents of the Bay Area in California think of the San Francisco Bay as oriented closer to NSEW than it actually is. The same effects have been found for direction estimates of pairs of cities, so people judge Berkeley to be east of Stanford (both in California), when it is actually west of Stanford. The distortions appear for other regions of the world (Glicksohn, 1994; Lloyd & Heivly, 1987) as well as for artificial maps and meaningless bloblike shapes (Tversky, 1981).

Reference Point Asymmetries

Perhaps the most common way to locate a place is with reference to some other place. Your keys are on your desk. Usually the reference point is a better-known or more salient place or landmark. A not uncommon experiment on college campuses is to collect lists of landmarks from one set of students and then ask another set to estimate the distances of ordinary buildings to landmarks, or vice versa, the distances of landmarks to ordinary buildings. The surprising finding is that the distance estimates to/from landmarks are not the same! Distances from undistinguished structures to landmarks are estimated to be smaller than distances from landmarks to undistinguished structures (e.g., McNamara & Diwadker, 1997; Sadalla et al., 1980).

What makes this effect surprising – and has led to resistance – is that the distance asymmetry violates any metric model of space. A metric model cannot tolerate the distance from A to B to be greater than the distance from B to A. There are explanations for this phenomenon on several levels. Landmarks serve to define and refer to neighborhoods, not just to themselves, and so are thought to include more territory than they themselves occupy. Indeed, landmarks serve as reference points for many less distinguished elements (e.g., Couclelis, Golledge, Gale, & Tobler, 1987; Shanon, 1983). A Parisian might say, "I live near the Eiffel Tower," or a New Yorker, "I work in the Flatiron District." Not so for ordinary buildings. However, there are parallel asymmetries in other kinds of judgments, notably similarity judgments. People judge magenta to be more similar to red than red to magenta; that is, they regard atypical colors to be closer to prototypic colors than vice versa (Rosch, 1975). *Red* can be regarded as a large category that includes many shades of red. In the 1970s, people judged North Korea as more similar to Communist China than vice versa, and sons as more similar to their fathers than fathers to sons (Tversky & Gati, 1978). A prototype, whether a color or an institution, is regarded to represent a larger set of progeny or instances than a more ordinary example.

Perspective

When we are at a vantage point, a mountain top or top of a skyscraper, the distances between objects in front of us that are near our viewpoint seem larger than the distances between objects far from our viewpoint. The distance objects are telescoped. The same occurs for judgments for imagined viewpoints. Students in Ann Arbor were randomly asked to imagine themselves in New York or San Francisco (Holyoak & Mah, 1982). Then they were asked to judge the relative distances between pairs of cities distributed fairly evenly along an east–west line across the United States. Those with a west coast perspective judged the distance from San Francisco to Salt Lake City, Utah, to be larger than those with an east coast perspective. Similarly, those with an east coast perspective judged the distance from New York to Pittsburgh to be greater than those with a west coast perspective. Of course this effect is an instantiation of the Weber-Fechner Law. The exaggeration of distances close to a viewpoint relative to those farther from a viewpoint was noted much earlier by cartoonists, notably Steinberg's *New Yorker* covers, telescoped views of the world from his window. The psychological research seems to have added that the viewpoint can be determined by suggestion and that the results are quantifiable.

As for many of the spatial biases, this phenomenon has parallels in abstract thought. Telescoping occurs for temporal judgments as well as spatial ones (Bradburn, Rips, & Shevell, 1987; Brown, Rips, & Shevell, 1985; Loftus & Marberger, 1983) and in temporal inferences and planning (Liberman, Trope, & Stephan, 2007; Trope & Liberman, 2003). It also occurs in social judgments: People in socially distant categories are judged more similar to each other than those close to us, where the differences among people close to us loom larger (Quattrone, 1986).

Paths

When people explore environments, their exploration is typically along a virtual line, a road, a sidewalk, a bike path. This is one of the ways we acquire knowledge of how environments are structured. That experience biases judgments, especially of distance. Put simply, when there are more things – sometimes called "clutter" – along a traversed path, the path is judged to be longer (Thorndyke, 1981). Barriers along a route make the route seem longer to both children and adults (e.g., Kosslyn, Pick, & Fariello, 1974; Newcombe & Liben, 1982). Similarly, more turns (Sadalla & Magel, 1980) or nodes (Sadalla & Staplin, 1980b) or greater remembered information (Sadalla & Staplin, 1980a) or indirectness of path (Klippel, Knuf, Hommel, & Freksa, 2004) all increase judgments of distance. In short, increasing the amount of information that receives attention increases estimates of distance, undoubtedly one of the reasons why journeys back home seem shorter than journeys away from home.

Schemas

The spaces we inhabit have structures. Some, like forests and beaches, have been designed by nature, though humans frequently tinker with their design, adding parking lots and trails and concession stands. Others, such as homes and schools and cities, have been designed by people to enable and guide their behavior (Tversky, 2011b, 2016). People form representations of these varied environments, representations that include but also bias the way they appear and the activities associated with them (Tversky & Hemenway, 1983). A powerful expectation for navigation in and judgments of cities is a grid plan, that streets run parallel and perpendicular to each other, intersecting at right angles. This general schema biases judgments so that people incorrectly report that well-known streets are more parallel than they actually are and familiar intersections closer to 90 degrees than they actually are (e.g., Tversky, 1981). Even experienced taxi drivers show those biases (Chase, 1983). The biases extend to rivers designed (primarily) by nature in addition to roads designed by people; Parisians think the Seine is straighter than it actually is (Milgram & Jodelet, 1976). Now we switch to anecdotes, and a friend is fond of saying that the plural of anecdote is not data. Nevertheless, just as the Seine is mentally straightened by Parisians, the US–Canadian border seems to be mentally straightened as well, so that many newcomers to Michigan are surprised to learn that the Canadian city of Windsor is south of the US city of Detroit.

Route Planning

Route planning is a complex and fascinating topic in itself, and for the most part beyond the scope of this chapter. Here we present only a single finding, that people are biased toward selecting south-going routes over north-going routes (Brunyé et al., 2012; Brunyé, Mahoney, Gardony, & Taylor, 2010). This is yet another example of "contamination" of irrelevant information on spatial judgments. As noted, distance judgments within conceptual categories, university in contrast to commercial, Palestinian in contrast to Jewish, are judged smaller than distances between conceptual categories. Here, the preference for south-going routes seems to be based on a confusion of northward and upward – it's harder to go uphill than down.

Almost all the spatial biases have parallel conceptual biases, notably in temporal and social judgments. This could be because spatial thinking serves as a foundation for all thought. Or it could be because there is a common basis for both spatial and abstract thought. Or it could simply be chance – but that doesn't seem likely.

Kinds of Judgments

Distance Judgments

Distance judgments are shorter for paths that have less information to attend to: fewer turns or landmarks (Sadalla, Burroughs, & Staplin, 1980; Sadalla & Magel, 1980; Thorndyke 1981), number of nodes (Sadalla & Staplin, 1980b), information remembered (Sadalla & Staplin, 1980a), fewer barriers on a route (Kosslyn, Pick, & Fariello, 1974; Newcombe & Liben, 1982), or directly connected on a map (Klippel, Knuf, Hommel, & Freksa, 2004). Distance to a landmark from an ordinary building is judged less than distance from a landmark to an ordinary building (McNamara & Didwadker, 1997; Sadalla & Staplin, 1980b). Pairs of distances farther from a point of view are judged less than pairs of distances closer to a viewpoint (Holyoak & Mah, 1982). Distances within geographic (Allen & Kirasic, 1985; Hirtle & Jonides, 1985; Stevens & Coupe, 1978) or conceptual categories (Hirtle & Jonides, 1985; Portugali, 1993) are judged smaller than distances between categories. Smaller distances are overestimated relative to larger ones (e.g., Lloyd, 1989), a general finding in estimates of quantity (e.g., Poulton, 1989). In another example of "contamination" of irrelevant information on spatial judgments, positive affect reduces distance estimates (Briggs, 1973; Golledge & Zannaras, 1973), similar to the reduction of slope judgments from good company (Schnall, Harber, Stefanaucci, & Proffitt, 2008) or good nutrition (Schnall, Zadra, & Proffitt, 2010).

Direction Judgments

Direction, like distance, is biased by grouping; the overall directions of the encompassing group biases the direction of entities within the group (Stevens & Coupe, 1978; Tversky, 1981). Direction is also biased in the direction of the canonical axes (Tversky, 1981).

Some Implications

We have presented the Spatial Reference Theory, a theory that integrates and accounts for many of the armchair biases and errors that people make when thinking about space and when making judgments about space, primarily distance and direction, but other judgments as well. Judgments of space are inherently relative, relative to other entities, to frames of reference, to perspectives, and more. Because we do not carry detailed and accurate maps in our heads from which we can make accurate judgments, because we do not come equipped with the precise measuring devices it took generations to invent (and even they make errors), we pull together whatever information seems relevant to make a range of spatial judgments and inferences. That information is typically incomplete and approximate, prone to error in itself.

You might be thinking: If we are so prone to bias and error, how on earth do we get around (or how do we get around on earth)? Sometimes those errors make a difference;

when we rely only on our minds, we do get lost on occasion, and we are wrong about distances or directions. Once, an Italian travel agent (!) booked me a ticket to Rome leaving from Los Angeles instead of San Francisco, a distance of 450 miles, far greater than the distance from Milan to Rome. But in many cases those errors go undetected; they have no consequences, they make no difference in our lives, or they are corrected by the situation on the ground (e.g., Tversky, 2003). If we think a turn is 90 degrees and it turns out to be less or more, we go with the road; in fact, we treat the expectation of 90 degrees as an approximation, knowing that the world will fill in our approximations. As before, our talk reflects our thought: we say, "turn right," without specifying how much; we say, "drive about an hour," distance unspecified. So often approximations are sufficient.

The Future

What about the future? Who knows, maybe someday we will all be disembodied minds that take up no place in space, so moving in space and knowledge of space will be irrelevant. But then one wonders: How we would think, how could we think? So much of our abstract thought and abstract talk is rooted in spatial thought (e.g., Gattis, 2001; Lakoff & Johnson, 1980; Schubert & Maass, 2011; Tversky, 2001, 2011b) and expressed by our bodies, hands, and faces. We have drawn many parallels between the biases of spatial thought and biases in abstract thought. It is easier to imagine a path to the future continuous with the path that led us to today. Spatial thinking and acting in space have long been aided by technology. Technology is not a 20th-century invention. Eons ago humans gestured and spoke. They are core technologies for learning about space and communicating about space: remember Homer, Herodotus, Song Lines, and Marco Polo? Maps and trail markers are ancient technologies (e.g., Brown, 1979), more and more in use today as every building and street is diagrammed (Tversky, 2011b). Digital technologies, personal as well as public ones, are already shaping our actions and our thoughts, and those will continue to impact and enrich our lives. Alternative realities, augmented and virtual, will bring us to places and spaces we can barely imagine. So rather than becoming disembodied, it seems that we will be more and more embodied in richer and richer spaces. Stay tuned.

Author Note

Preparation of the manuscript was aided in part by funding from the Varieties of Understanding program of the John Templeton Foundation and grants NSF CHS-1513841, NSF HHC 090541, NSF IIS-0725223, and NSF IIS-0855995. The opinions expressed in this publication are those of the author and do not necessarily reflect the views of the funders.

Chapter 17

Categorical Influences on Spatial Bias

Nora S. Newcombe

Navigation is an essential part of adaptation for any mobile species, humans included. We need to search for food and water, return to shelter for sleep, and avoid dangers that include both stationary and moving threats. Because these functions are vital for survival and reproduction, it would be natural to expect that we would have a cognitive and neural system that can accurately encode the spatial environment and that might even operate automatically. Instead, human spatial judgments show odd biases, and even outright incoherencies. Such phenomena are surprising. But we need to face facts: There is now a large literature documenting spatial biases and oddities, usefully gathered together in this volume.

Note, however, that not all of these biases are necessarily the target of evolutionary pressure, and some do not constitute likely threats to survival. Simplifying heuristics for spatial relations at geographic scale (e.g., lining up geographic areas along north–south or east–west axes, Tversky, 1981; judging the relative locations of cities in terms of the relative locations of their states, Stevens & Coupe, 1978) may reflect how we deal with knowledge derived from maps rather than actual travel, a task probably not subject to evolutionary pressure. Errors in judgments of local environments seem a bit more serious, as when Stanford undergraduates draw El Camino Real running north to south rather than from northwest to southeast (Tversky, 1981), when students at the University of Michigan or Northwestern represent their campus as locations clustered into regions with inter-region distances exaggerated with respect to intra-region ones (Hirtle & Jonides, 1985; Uttal, Friedman, Hand, & Warren, 2010) or when people prefer to start journeys by going as directly as possible toward their goal, leading to differences between return journeys and initial routes and longer routes than necessary (Bailenson, Shum, & Uttal, 2000). But these errors may reflect rational reductions in cognitive load, given they are relatively low cost, at worst probably arriving at a goal a few minutes later than necessary (Bailenson et al., 2000).

Spatial judgments that seem downright illogical pose a more troubling challenge to the idea that evolution should have ensured adequate spatial functioning. One well-studied phenomenon of this kind is asymmetries, i.e., when people judge a distance from point A to point B as different from the distance from point B to point A (Baird, Wagner, & Noma, 1982; Holyoak & Mah, 1982; McNamara & Diwadkar, 1997; Sadalla, Burroughs, & Staplin, 1980). Another example occurs when participants in virtual reality experiments do not realize that they are in impossible environments (Kluss, Marsh, Zetzsche, & Schill, 2015; Warren, Rothman, Schnapp, & Ericson, 2017; Zetzsche, Wolter, Galbraith, & Schill, 2009). Impressed by these incoherencies, but also arguing from the overall literature on spatial bias, some investigators have suggested that our spatial representations are nonmetric or

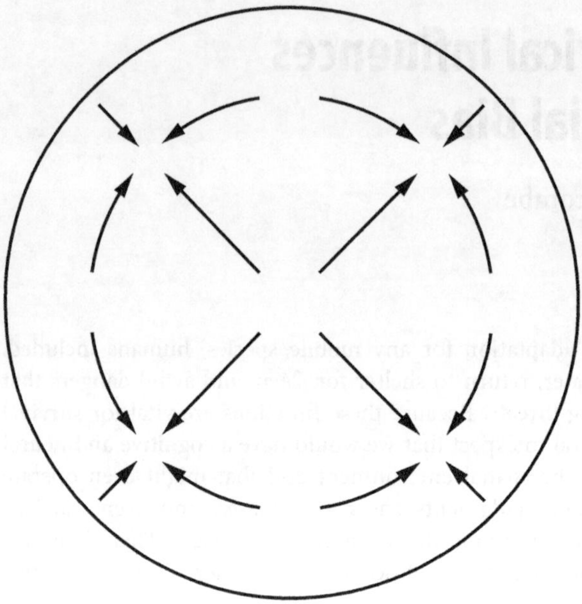

Figure 17.1 Idealized pattern of bias found in experiments on memory for the location of dots in a circle. From Huttenlocher, Newcombe and Sandberg (1994).

even associative (Foo, Warren, Duchon, & Tarr, 2005; McNamara, 1991; Tversky, 1981). As Tversky (1981, p. 432) put it, "Cognitive maps may be impossible figures."

So, how has our species survived? One answer might be that associative spatial representations are "good enough" to get around the environment. However, there is an alternative. Huttenlocher, Hedges, and Duncan (1991) proposed that spatial representations are fundamentally accurate and coherent, even while exhibiting oddities and biases. In their Category Adjustment Model (CAM), fine-grained representations of location are nested within categories. When the fine-grained information is relatively uncertain, categorical information is more heavily weighted, and the result is biased toward the location of the category prototype. The distinction between categorical and fine-grained information maps well onto analogous distinctions between categorical and coordinate information (Kosslyn, 1987) and between qualitative and quantitative information (Forbus, 2011; Klippel, 2012). CAM adds the idea of an adaptive combinatorial process.

This combinatorial process has been mathematically modeled in Huttenlocher et al. (1991) and in subsequent papers. It can be intuitively understood using the diagram in Figure 17.1. The arrows show the idealized results of multiple experiments in which participants saw dots located in circles, and then located those dots after brief delays. Point estimates move toward the centroid of the four quarters of the circle, as divided by horizontal and vertical lines. Points move farther when they are farther from the centroid, resulting in increased bias for those points. Boundaries also exert effects by truncating the distributions of locational uncertainty. For example, a point toward the circumference of the circle cannot move outward without crossing the boundary of the circle, an unlikely violation (Huttenlocher, Hedges, Lourenco, Crawford, & Corrigan, 2007). Overall, the effect of combining fine-grained and category information is to increase accuracy by constraining location given uncertainty, even at the price of introducing bias.

A somewhat more formal explanation of CAM can be derived from Figure 17.2, taken from Duffy, Huttenlocher, Hedges, and Crawford (2010). The larger normal distribution

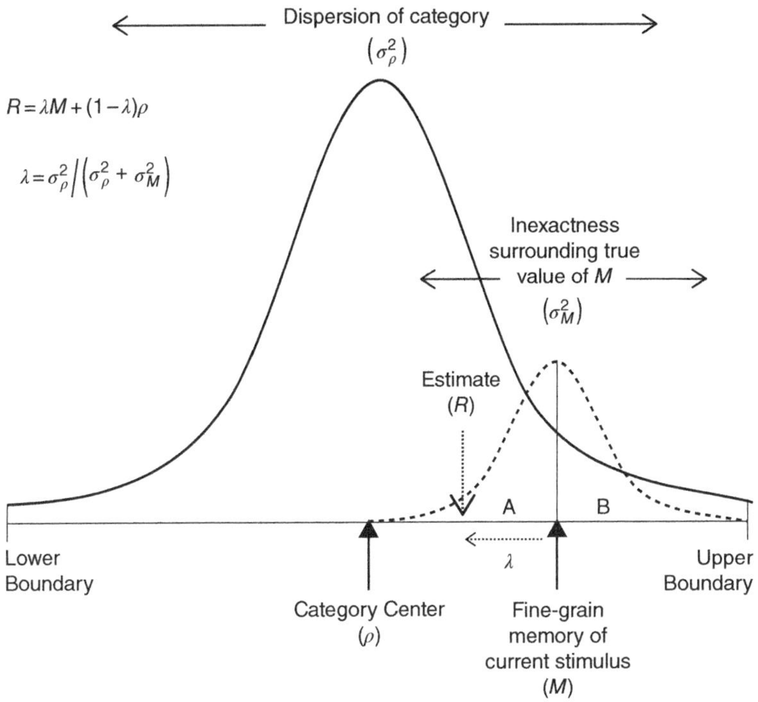

Figure 17.2 Model of the Category Adjustment Model (CAM). From Duffy et al. (2010).

shows the likelihood of any particular point's location, given that we know only that it is within the category, centered at the category prototype, and truncated at the boundaries. The smaller normal distribution shows the likelihood of a point's location as determined by fine-grained memory. This distribution is much more constrained, but also has inexactness. It could be centered at any point within the overall category distribution.

CAM applies to various kinds of judgments, not just spatial ones. In fact, the earliest publications concerning the model involved time estimation (Huttenlocher, Hedges, & Bradburn, 1990; Huttenlocher, Hedges, & Prohaska, 1988, 1992). Subsequent applications of the model concentrated on category judgments (Huttenlocher, Hedges, & Vevea, 2000), and much subsequent work on the model has focused on forming categories for stimuli such as fat and thin fish or long and short lines.

CAM is a Bayesian model in that it weights categorical information more heavily as the variability, and hence uncertainty of fine-grained information increases. In this sense, it is quite similar to Bayesian models of sensory combination and its development (e.g., Ernst & Banks, 2002; Nardini, Begus, & Mareschal, 2013) and also quite similar to Bayesian models of combination of spatial information from path integration and from landmarks and its development (e.g., Nardini, Jones, Bedford, & Braddick, 2008; Zhao & Warren, 2015a, 2015b). An overview of CAM and its relation to other Bayesian approaches to spatial behavior explains this family of approaches in more detail (Cheng, Shettleworth, Huttenlocher, & Rieser, 2007). CAM is similar, but less so, to Bayesian models of reasoning and its development, which utilize probabilities of one event given another, with the probabilities based on priors either gathered by prior experience or built in, or both (e.g., Oaksford & Chater, 2001; Perfors, Tenenbaum, Griffiths, & Xu, 2011).

Crucially, CAM can explain asymmetries in spatial judgment. Newcombe, Huttenlocher, Sandberg, Lie, and Johnson (1999) demonstrated this aspect of the model. Their Experiment 1 again used the paradigm of estimating the location of dots in circles. People saw two dots sequentially. One dot was closer to the category prototype (i.e., the centroid of one of the quadrants of the circle shown in Figure 17.1), and the other dot was farther away. They were then shown the location of one of the dots (either the first or the second presented) and asked to locate the other dot. When the to-be-estimated dot was between the fixed dot and the centroid, overestimation of the distance was predicted, and found. When the centroid was between the to-be-estimated dot and the fixed dot, underestimation was predicted, and found. An additional prediction was also confirmed, namely that this pattern would be stronger when the to-be-estimated dot had been presented first, because fine-grained information about its location would have decayed, and hence the category information would be more heavily weighted. The bottom line is that asymmetries of various kinds and magnitudes are expected, and their nature and size can be predicted, from a model in which people generate spatial estimates from underlying information without bias.

Two subsequent experiments used a different paradigm, examining asymmetries more directly. Participants were asked to learn a simple map showing locations (such as a church) in a hypothetical town. They learned the map either as a whole (in Experiment 2) or one quadrant at a time (in Experiment 3). In both cases, locations were grouped into quadrants by clustering around a centrally located spatial prototype, such as the HOSPITAL in the upper left, further highlighted by an asterisk rather than a point, designated with a star and labeled in capital letters. After the participants reached a learning criterion, their spatial memory was tested in a series of pair-wise reproduction tests. One building (e.g., hospital) was shown and a second point (e.g., church) had to be placed relative to it. CAM accurately predicted when asymmetries would and would not be found.

We see then, overall, that the delineation of CAM and the fact that it explains asymmetries offer some hope of reconciling reflections on the evolutionary value of accurate spatial representations with the facts of spatial bias. In the context of this promise, the main purpose of this chapter is to review the various directions work on CAM has taken since its initial formulation, covering six topics: (1) creating and testing a process model of CAM, (2) examining how CAM might be used in the real world, (3) addressing the basis of natural spatial categories, (4) evaluating whether the categories used in spatial estimation align with linguistically marked spatial categories, (5) exploring development and individual differences, and (6) criticisms of CAM.

Testing CAM through Formulating a Process Model

Although CAM has much in common with research on a Bayesian approach to sensory combination, there is a crucial difference, namely the difficulty of separately measuring the reliability of categorical and fine-grained information and then determining if the combination process uses that information to increase reliability. Separate measurement is challenging, given that one kind of information is hierarchically nested within the other. Huttenlocher, Hedges, and their collaborators have instead relied on mathematical models showing that the use of categorical models improves the reliability of fine-grained information considered alone and is hence a Bayesian process (e.g., Huttenlocher et al., 2007).

Empirical research to deal with the issue of separate measurement of component processes and combined judgments is also potentially possible. One promising approach was charted by Friedman, Ludvig, Legge, and Vuong (2013), who presented a model of combining two dimensions. They worked with x and y coordinates given their use of geographic stimuli. Using spatial judgment data gathered either in a perceptual or in a semantic context, they performed analyses separately for each single dimension (i.e., x-axis or y-axis) and then evaluated the two-dimensional (or combined) situation. More work of this kind is needed.

Another route forward is to test an implementation or process model of how CAM operates, rather than staying at the computational or algorithmic levels (Marr, 1982), and hence to test whether CAM makes predictions that are empirically confirmed. Work using this strategy began with examining the effect of delay, which should increase category bias by decreasing the accuracy of fine-grained information. Along similar lines, increased cognitive load might be expected to increase bias by making fine-grained information harder to encode and maintain. A second process question is whether category adjustment in fact occurs during retrieval, reflecting weighting of two underlying sources of information, neither of which contains bias, but which vary in certainty. A contrasting possibility is that the two kinds of information blend with each other as an ongoing process during retention (e.g., Spencer & Hund, 2002).

Does Decreasing Reliability of Fine-Grained Information Increase Bias?

The passage of time and the addition of interference should degrade the accuracy and reliability of fine-grained information quite quickly in short-term dot-location experiments, hence increasing the weighting of category information and leading to increases in bias. One set of findings has already been discussed; Newcombe et al. (1999) found the over- and underestimations that would lead to asymmetries only with first-presented dots, which had been subject to longer delays than second-presented dots. The effects of delay have also been confirmed in other studies, including in natural environments (Holden, Newcombe, & Shipley, 2013), with children (Hund & Plumert, 2002, 2005), and in angle estimation (Crawford, Huttenlocher, & Engebretson, 2000), although there were no effects of a 15-second delay in another study (Haun, Allen, & Wedell, 2005).

Cognitive load might also sap the ability to encode and/or maintain fine-grained estimations, and increased cognitive load has recently been found to increase categorical bias in the estimation of line lengths (Allred, Crawford, Duffy, & Smith, 2016). However, the number of locations to be remembered and the presence or absence of a concurrent task did not increase categorical bias in the same spatial estimation study that failed to find an effect of delay (Haun et al., 2005). One explanation for these null effects may be that participants in Haun et al. (2005) showed categorical effects that were very pronounced even though they only had to encode one location and had no concurrent interference task. This pattern suggests poor fine-grained coding even in the baseline condition, making it difficult to observe any increases in the effects of categories.

Fine-grained information does not always decay over time. It is acquired and retained in natural environments that are repeatedly experienced, as when university students get to know their campus (Uttal et al., 2010). In fact, we would hope that such information would be robustly encoded, because without it, the kind of adaptive navigation that we imagine evolution should ensure would be challenging. However, there is evidence that cognitive load may be a factor in acquiring fine-grained information for natural spaces; working memory is significantly correlated with more accurate spatial estimates in learning a virtual environment (Weisberg & Newcombe, 2016).

Does Combination Occur at Retrieval?

Using the dots-in-a-circle task, Sampaio and Wang (2009) found that people prefer the correct location when it is pitted against their own (biased) estimate in a recognition task, suggesting maintenance of unbiased information and supporting a combination-at-retrieval model. Using natural scenes, Holden, Newcombe, and Shipley (2015) replicated this key result, and also performed a further manipulation, in which participants were allowed to skip trials if unsure. However, in 25% of the cases, people were not allowed to skip the trial and instead compelled to respond. Recognition trials could thus be sorted into the cases where people were sure enough about location to want to respond versus trials on which they had considerable uncertainty. The key prediction was that categorical information would be most heavily weighted when participants were unsure and were forced to respond anyway. Indeed, such a pattern was observed to a truly striking extent. When people were relatively sure, they preferred the correct location in a recognition test, as shown in the panel at the top right. When they were unsure, they preferred the category prototype, even in recognition, as shown in the panel at the lower right.

Categories and Combination in the Real World

Remembering the locations of dots in circles for very brief periods of time is clearly not the same kind of task as navigating in an environment. Learning the highly schematic maps used by Newcombe et al. (1999) does not take us much closer to spatial reality. Thus, a crucial challenge for CAM is whether it applies to the real world. Posing this question also involves us in asking another vital question, namely what determines spatial categories in the real world. To move towards real-world contexts and to begin to explore the nature of natural spatial categories, Holden, Curby, Newcombe, and Shipley (2010) studied memory for dot locations on photographs of natural scenes such as sand dunes, mountains, or lakes. To define the categories in order to investigate the predictions of CAM, there were both conceptual and perceptual categories to be considered. For example, people have a concept of "lake," but they also see the shadows and reflections on the lake's surface as different in color and other visual characteristics from the clear blue color of the lake elsewhere. To evaluate CAM, we began by defining categories in the scenes in two ways: using a perceptual clustering machine algorithm, and by obtaining judgments by human observers. Interestingly, these two methods converged quite well. This agreement may suggest that people use perceptual variations to subcategorize concepts such as lake, or may suggest that people also have concepts such as "shadow" and "reflection." In any case, bias was found toward the center of scene categories, just as bias is exhibited toward the prototype of a quadrant of a circle. Holden et al. (2010) collected further data using inverted versions of the photos, which we thought would primarily weaken access to the conceptual content, and color-negative versions of the photos, which might interfere more with perceptual than with conceptual processing. People showed categorical bias effects in both cases, although the categories were somewhat different, both from each other and from the natural scenes.

Natural scenes move us toward the real world, but not all the way into it. If we would like to use CAM in thinking about spatial navigation in the actual three-dimensional world, we need to conduct experiments in that world. There have started to be efforts to do so. Holden, Newcombe, and Shipley (2013) asked participants to remember the location of dots flashed with a laser pointer on the three-dimensional world, e.g., a bench. We again found bias toward the center of spatial regions. Pyoun, Sargent, Dopkins, and Philbeck (2013) also found category effects, in an analogue of the work on the circle in which participants were placed in a real circular arena. People seemed to divide a surrounding circle into halves, not quarters, considering there to be a front half (ahead of them) and a rear half (behind them). Furthermore, the category prototypes were not at the center of those two halves, but rather "pulled forward" so that they were within the field of vision, on the left and right sides of the front half. Even after people rotated, they preserved this organization when offered cues as to where the front and rear had been (either doors or objects such as a lamp).

Spatial Categories in the Real World

In the asymmetry experiments involving maps conducted by Newcombe et al. (1999), the quadrant organization was suggested by the stimuli (clustering locations and adding stars and capitalized labels). In the circle experiments, the gravitational vertical and the corresponding vertical are very powerful categorization cues, which can also organize other geometric shapes (Wedell, Fitting, & Allen, 2007). But the natural environment is rarely structured completely by convenient external organization or by a gravitation-imposed axis structure. Further, the power of the gravitation-defined axes for the circle actually poses a challenge to the Bayesianism of CAM. People are resistant to organizing their spatial memory in other ways even when evidence accumulates that the frequencies of dot locations are not uniform (Huttenlocher, Hedges, Corrigan, & Crawford, 2004). They can be pushed into using stable external cues when the circle rotates (Fitting, Wedell, & Allen, 2007), as they should when horizontal and vertical bisection becomes an unreliable categorical cue. But the resistance to using frequency data remains troubling. From a Bayesian point of view, it is important to be flexible in using categories and to rely on categories that are stable, e.g., shadowed areas are poor categories for anything but short-term spatial tasks, because shadows shift with time of the day and whether the sun is shining. Interestingly, after repeated trials, people utilize information about the density of locational distributions even in the circle (Lipinski, Simmering, Johnson, & Spencer, 2010), acting in a Bayesian way by taking into account the probability that a point falls into a particular local area.

Most of the research on the formation of category prototypes in CAM has focused on attributes of objects, i.e., fatness of fish, length of lines. In such cases, prototypes might be running averages, but they might also weight recent examples more heavily or early examples more heavily. Sailor and Antoine (2005) found evidence for greater influence from recent examples, but Duffy, Huttenlocher, Hedges, and Crawford (2010) found instead that people use running averages as their central tendencies or categories. Prototypic line lengths increased if stimuli were getting progressively larger and decreased if they were getting progressively smaller. There was no greater influence from recent items. Duffy et al. (2010) did not evaluate possible primacy effects, but Duffy and Crawford (2008) had done so, and found them. Is this adaptive? Duffy and Crawford argue that it may be, because forming a category prototype quickly provides support for subsequent memory, and the likelihood that it is erroneous may not usually be high.

There has been less attention to how spatial prototypes are formed, for various reasons. However, Crawford and Duffy (2010) report a primacy effect: Participants form a category prototype that is too far to the left when they have seen a sequence of dots in a rectangle presented from left to right and too far to the right when they have seen a dot sequence progressing from right to left. Similarly, for children as old as 9 and 11 years, though not for adults, category effects are seen only when locations are learned in a temporally grouped way and/or placed in an order that emphasizes the categorical grouping (Hund, Plumert, & Benney, 2002). People group categorically by the identities of the objects in experiments using objects with varying identities instead of plain dots (Crawford & Jones, 2011; Hund & Plumert, 2003; Sampaio & Wang, 2010), possibly forming categories of the "fruit section" and so forth.

From a Bayesian point of view, it is important to use the smallest available categories, because location is then more effectively constrained. Holden, Newcombe, Resnick, and Shipley (2015) were able to evaluate the interacting use of perceptual and conceptual information and the ideas of flexibility and constraint by contrasting categorical bias in experts and novices. Expert geologists and participants with little geological knowledge viewed pictures in which categories were available to the experts that novices would be likely to miss. In some cases, the expert categories were smaller than the perceptual categories that were the only spatial categories available to novices. In other cases the expert categories were larger. We found that novices showed categorical bias toward the perceptual categories. The expert geologists also showed such bias when the perceptual category was smaller and hence constrained location more than their expert category would have done. They used their expert categories when they offered further constraint. These data strikingly confirm that spatial categories have both perceptual and conceptual bases, that they are flexible, and that they are adaptive in that they utilize the smallest available category to provide more accurate spatial estimation.

Becoming a geologist is a lengthy process, so it may be that the formation of conceptual spatial categories takes some time. Indeed, the Uttal et al.'s (2010) study of Northwestern University undergraduates' knowledge of their campus did not find categorical clustering of campus locations in first-year students, but such grouping was present in the sophomores, juniors, and seniors. Furthermore, Newcombe and Chiang (2007) did not find that participants easily formed strong categories of countries on hypothetical maps, even though a series of studies by Friedman and colleagues have found pronounced categorical effects for geographic knowledge in the real world (Friedman, 2009; Friedman & Brown, 2000a, 2000b; Friedman, Brown, & McGaffey, 2002; Friedman, Kerkman, Brown, Stea, & Cappello, 2005; Friedman & Montello, 2006). Such categorical effects begin to appear between 9 and 11 years of age, perhaps reflecting the growth of awareness of countries and their distinctive characteristics (Kerkman, Friedman, Brown, Stea, & Carmichael, 2003).

Spatial Categories and Linguistic Categories

Contrasting views concerning the relations between language and thought are one of the longest-running controversies in cognitive science. Recently, in the spatial domain, a great deal of attention has been focused on relative and absolute frames of reference. Languages vary in which they prefer, and strong positions have been taken on whether such variation transforms spatial thought (Gleitman & Papafragou, 2012; Levinson, 2003). A second language-thought issue in the spatial domain, however, has been whether the categories that language encodes with single words correspond to the nonlinguistic categories used in thought. Research on this topic has typically involved small closed sets of spatial terms such

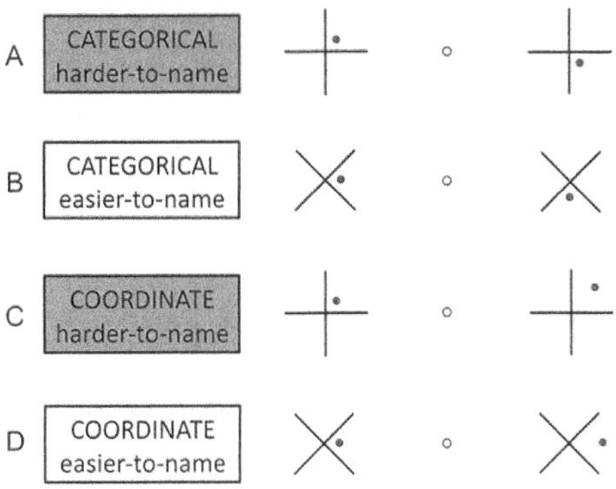

Figure 17.3 Experimental procedure in Kranjec, et al. (2014).

as front/back/left/right or above/below/left/right. Hayward and Tarr (1995) claimed to find close correspondence between linguistic terms and nonlinguistic thought for the latter foursome, with categories centered on the vertical and horizontal dimensions. This finding is surprising in the light of the dot-in-circle experiments, in which the category prototypes were in the top right, bottom left, and so forth, with the horizontal and vertical axes constituting the quadrant boundaries. Using a common paradigm to examine the contrasting findings, Crawford, Regier, and Huttenlocher (2000) showed that there was indeed a contrast between nonlinguistic category prototypes and spatial language, and suggested that language refers to the boundaries of nonlinguistic categories rather than to the prototypes used in estimating spatial location. Thus, they suggest that there is a mismatch, albeit an understandable and functional one, between linguistic spatial categories and nonlinguistic spatial categories.

Kranjec, Lupyan, and Chatterjee (2014) worked with categorical and coordinate (i.e., fine-grained) tasks. Dots were located within quadrants defined either by horizontal and vertical axes, or by diagonal axes, as shown in Figure 17.3. They label the former as "harder-to-name" because the quadrants in this case require two words, not one, e.g., *top right*, not simply *top*. However, this case could also be called "easier-to-encode" in the sense that these axes provide the quadrants typically used for improving retrieval of coordinate information using categorical adjustment. Kranjec and colleagues found that the accuracy of coordinate information was higher in the "easier-to-encode" case. This pattern again suggests a contrast between linguistic and nonlinguistic categories.

The categorical and coordinate terminology used by Kranjec and colleagues was taken from Kosslyn (1987). As we have mentioned, Kosslyn's approach did not tackle the problem of how these two kinds of information may be combined and used together. But it did tackle a different problem that Huttenlocher and her collaborators never addressed, namely the neural substrates of the two kinds of information. Kosslyn proposed that encoding and maintaining coordinate information was a right-hemisphere function, and encoding and maintaining categorical information was a left-hemisphere function, convenient enough if language picks out categories. Further research has provided considerable support for this hypothesis (e.g., Amorapanth, Widick, & Chatterjee, 2010; Kosslyn, Koenig, Barrett, Cave, Tang, & Gabrieli, 1989).

Thinking about these neural facts suggests an approach to the problem of the relation between spatial language and spatial thought. In the domain of color, Gilbert, Regier, Kay, and Ivry (2006) found that perception is congruent with linguistic terms in the left hemisphere (as suggested by experiments lateralizing input to the right visual field) but not congruent in the right hemisphere. Similar patterns were found for categorization of profiles of cats and dogs (Gilbert, Regier, Kay, & Ivry, 2006, 2008) and for novel figures that participants learned to categorize (Holmes & Wolff, 2012). As Regier and Kay (2009) put it, "Whorf was half right." Perhaps this approach can be extended to spatial categories. The right hemisphere, which is primarily concerned with coordinate or fine-grained information, might use the kind of categories that best address unreliability in that kind of information, in a combinatorial process. By contrast, without a strong concern with the reliability of fine-grained information, the left hemisphere might aim to maximize accuracy and precision of the usage of linguistic terms by centering categories on the easily encoded axes. Evaluation of this hypothesis is needed; one method might be to use Kranjec et al.'s paradigm in a design that would allow for evaluation of hemispheric differences.

In pursuing the question of spatial categories and linguistic encoding, we also need to move beyond the small, closed fields of spatial terminology, such as above/below, because the spatial categories used in natural scenes, natural environments, and geological categories involve irregular shapes, many of which have names, although not all do. The names, however, suggest shapes (e.g., *crater* suggests a concave depression), but the particular shape is not strongly constrained. It is this kind of spatial meaning that Landau and Jackendoff (1993) had in mind when they discussed nouns as encoding spatial information. Whether or not names for these categories matter for their use in a combinatorial process has not been investigated. But if the names matter, they can't matter too much. For instance, consider that the geologists in Holden et al. (2015) did not use the expert categories for which they surely have names if a smaller novice category more narrowly constrained a location.

Development, Aging, and Individual Differences

Considerable research has been done on children's ability to estimate spatial location, converging on the view that there is evidence of early competence in both fine-grained and categorical estimation, as well as evidence of combinatorial processes (see review by Holden & Newcombe, 2013). At the same time, we also see developmental change, of a variety of kinds and with a variety of probable causes. First, the exactness of fine-grained estimation increases; increases in spatial precision have been used to model development in many tasks (Simmering, Schutte, & Spencer, 2008). Second, categories may change, becoming richer as well as finer with development as more categories are acquired, especially in the natural environment, where categories have semantic content. Third, increases in working memory capacity may allow for combinatorial processes of wider scope (Sandberg, Huttenlocher, & Newcombe, 1996). Fourth, feedback from the success of way-finding and searches for lost objects may lead to progressive refinement in Bayesian combination. Such change has yet to be evaluated for CAM in the natural environment, but has been found for sensory combination (Nardini et al., 2013) and for combination of path integration and landmark use (Nardini et al., 2008).

There is not yet much research on individual differences in spatial location memory, other than age differences. A few papers have started to appear, however. Holden, Duff-Canning, and Hampson (2015) found that women emphasize categorical information more than men do in encoding both dots-in-a-circle and natural scenes, with this emphasis likely occurring during encoding, perhaps due to differential attention (see also Holden & Hampson, 2014 on sex differences in angle estimation). Crawford, Landy, and Salthouse (2016) reanalyzed data from a large study of cognitive aging. They found that spatial working memory capacity was related to spatial bias for a task involving memory for the location of a dot on a computer screen. Lower-capacity individuals showed higher bias, as predicted from data already reviewed, suggesting that working memory is relevant to the acquisition of fine-grained information. If such information is less reliable, we expect greater bias. In addition, in an important methodological development, Crawford and colleagues fitted separate models to individual data sets, which has never been done previously. They found that only some people showed patterns that were consistent with CAM. A good proportion of participants showed different patterns, which will need further exploration.

Criticisms of CAM

Over the years, CAM has been the target of a variety of criticisms. Some of them are a matter of theoretical taste. For example, Barth, Lesser, Taggart, and Slusser (2015) argued that a Bayesian approach is not needed for spatial estimation tasks, because a simpler psychophysical model is sufficient to explain the data. While that is probably true, Bayesianism may be considered the more attractive interpretive framework because it explains a wider variety of phenomena, generates more predictions about behavior in other situations, and provides a psychological model of what is occurring when. Indeed, much of this chapter has reviewed tests of this model and overviewed its scope.

Other investigators have focused on the issue possibly implicit in Barth et al.'s criticism, namely whether perceptual factors are sufficient to account for some of the phenomena discussed in this chapter without invoking cognitive and conceptual issues. An early exchange involved research on the perceived and remembered tilt of lines in right-angled frames. Tversky and Schiano (1989) had explored the tilt of lines in L-shaped frames in the context of research on people's use of graphs. Responses varied depending on whether people thought they were looking at graphs or not, and Tversky and Schiano argued that there is a perceptual bias even when there isn't a conceptual one. Engebretson and Huttenlocher (1996) used L-shaped and also V-shaped frames without interpretive contexts, and found varying bias patterns that depended on the certainty with which vertical and horizontal lines can be remembered, which fit the CAM framework. This exchange did not develop into controversy, however, as Tversky and Schiano (1997) simply agreed that the phenomena were interesting, although seeing them as not inconsistent with their interpretation or with prior research on such displays.

The issue of the relevance of perceptual and conceptual phenomena to bias effects has continued to generate varying views. In a series of articles, Spencer and his colleagues (e.g., Simmering et al., 2008) have argued for the adequacy of a fundamentally perceptual account in their Dynamic Field Theory (DFT) approach. Furthermore, Friedman, Montello, and Burte (2012) found that category effects in experiments on placements of geographic locations were similar even when participants saw polygons shaped like Alberta or California and had to remember the location of dots, suggesting that perception alone could account for the phenomena. However, these criticisms do not explain the importance of the expert categories used in location estimations by geology experts (Holden et al., 2015). Furthermore, Crawford, Huttenlocher and

Engebretson (2000) showed that line estimation in the context of the Muller-Lyer illusion has both perceptual and conceptual components.

Conclusion

We have reviewed a variety of evidence concerning the success of CAM in explaining why and when people show biases in spatial memory. This debate is part of a more general debate about the existence of cognitive maps. After all, if spatial location judgments are subject to biases and, worse, illogical judgment patterns, it is hard to see how they could be map-like. Recent models of cognitive maps are, however, starting to suggest rapprochements. There may be locally metric representations with broad directional relations among them (Chrastil & Warren, 2013, 2014; Jacobs & Schenk, 2003; Kuipers & Byun, 1991), along with individual differences in the ability for formation of both the component representations and the directional relations (Schinazi, Nardi, Newcombe, Shipley, & Epstein, 2013; Weisberg & Newcombe, 2016; Weisberg, Schinazi, Newcombe, Shipley, & Epstein, 2014).

This view, derived from cognitive, behavioral, and modeling approaches, is beginning to converge with data at the neural level. The discovery of grid cells (Hafting, Fyhn, Molden, Moser, & Moser, 2005), and their gradation from finer to broader tunings at various levels (Giocomo, Zilli, Fransén, & Hasselmo, 2007), has combined with decades of work on place cells and head direction cells to begin to create a sophisticated model of way-finding and spatial representation. Interestingly, grid cells and place cells seem to encode spatial relations in particular enclosed spaces, raising the question of how the spaces are identified (e.g., Julian, Keinath, Muzzio, & Epstein, 2015) and interrelated (Grieves, Jenkins, Harland, Wood, & Dudchenko, 2016).

In turn, debates about both biases in spatial representation and cognitive maps are part of a wider effort to determine the nature and limits of human rationality, to understand the role of evolution in shaping "good enough" rationality to ensure adaptation, survival, and reproduction, and to trace the origins and development of this "good enough" rationality while avoiding the extremes of nativism and empiricism.

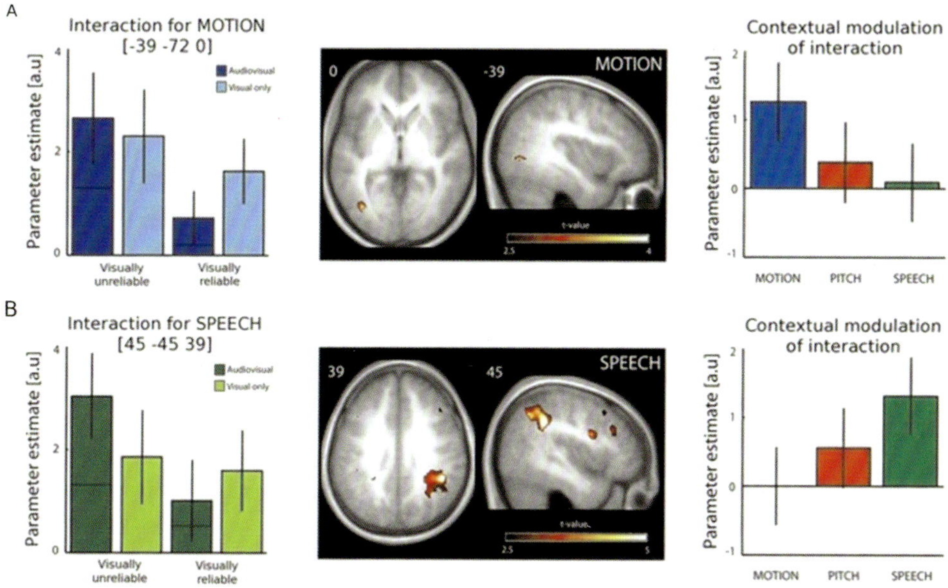

Figure 5.3 Results of Sadaghiani et al.'s (2009) neuroimaging study looking at the neural substrates of crossmodal effects of auditory stimulation on visual motion perception. Up or down could be conveyed by speech (the words "up" or "down"), or by physical motion, or by ascending or descending frequency sweep. (Reprinted with permission from Sadaghiani et al. [2009].)

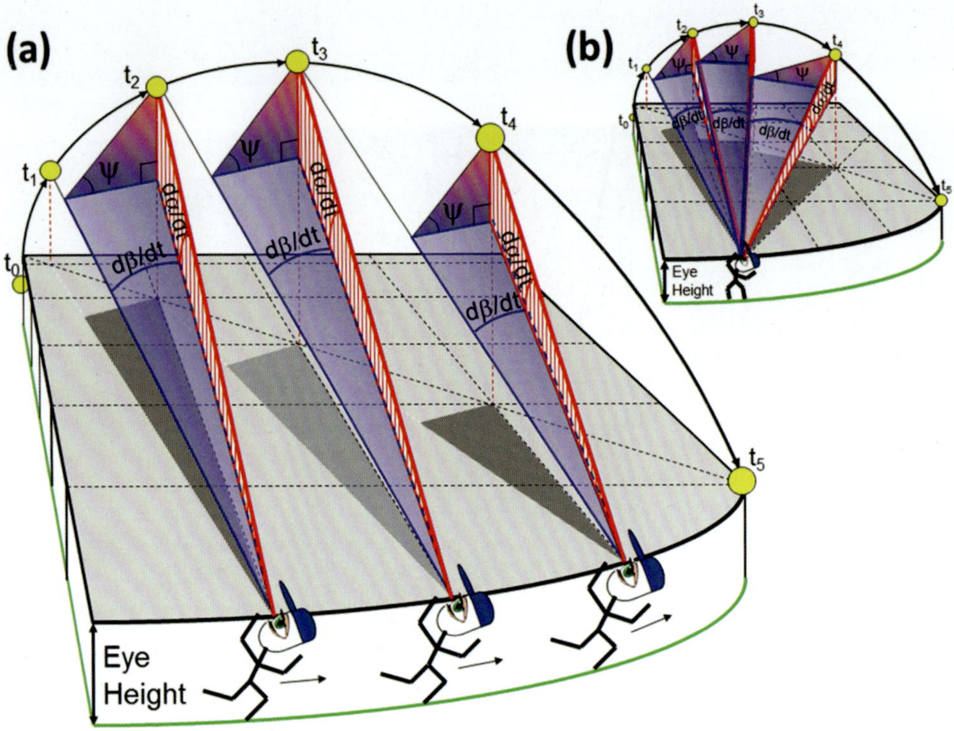

Figure 19.3 The linear optical trajectory (LOT) interception control strategy: An example of the invariant property of maintaining angular constancy. (a) The fielder runs along a path to keep the ball image monotonically rising along a constant projected optical angle, Ψ. (b) If the fielder remains stationary or selects a path that does not converge on the ball (as shown in the inset figure), Ψ will change, and the optical trajectory will curve toward the image of the ground. Continuously cancelling out optical curvature to keep the image of the ball rising along a straight optical line insures interception, because the ball image will never approach that of the ground.

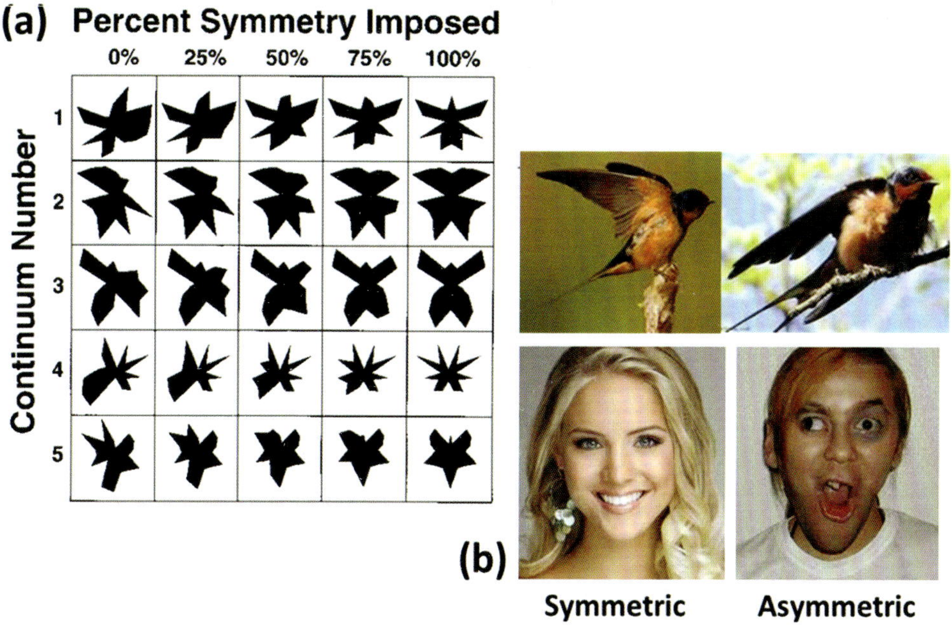

Figure 19.5 The natural regularity of object symmetry. (a) The five 18-sided random polygons in the left column were created and then either the left or right half was folded over to create the five symmetric figures in the right column. The figures in between were made by successively morphing to create a set of random shapes that varied in degree of outline symmetry. Participants were asked to indicate what the figures look like and which direction they faced. Over 95% of selections were symmetric objects such as animals or planes, so degree of outline symmetry could be used to specify their orientation. Virtually all of left-column figures were described as animals or objects at skewed or profile views, and virtually all of right-column figures were described as ones at straight-on views. (b) Symmetry is naturally selected because it is more stable and balanced in actions such as locomotion, and because of sexual selection, with symmetric individuals viewed as more attractive in a number of species including swallows and humans.

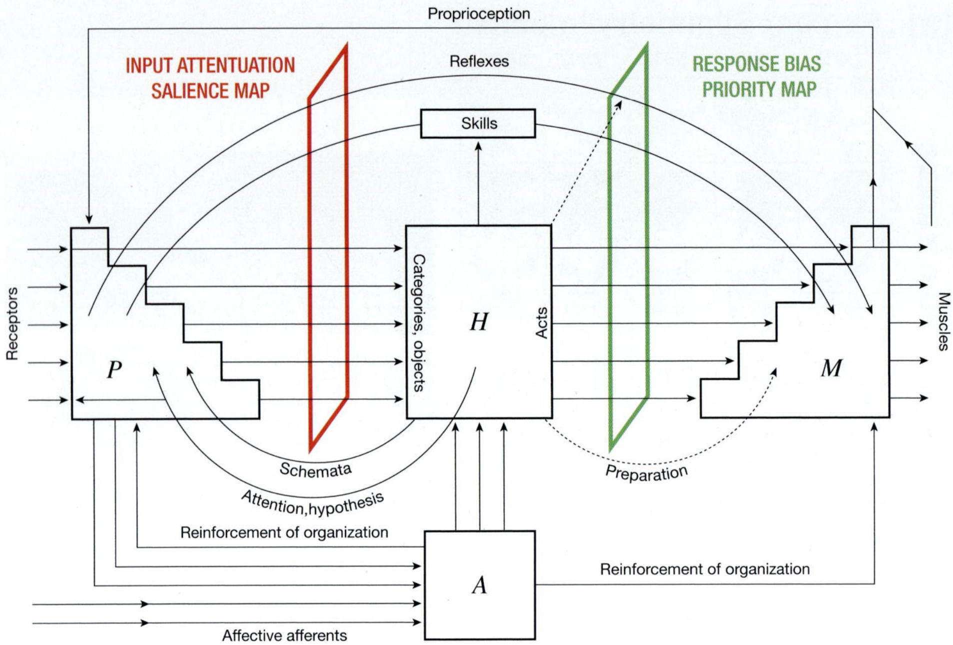

Figure 20.1 A figure illustrating the flow of information between cognitive modules from perceptual (P), input-related processing to motoric (M), output-related processing. A central stage of processing (H), which can be bypassed, is the site of awareness of perceptually derived categories and objects and of the initiation of some actions. The two dashed arrows have been added by RMK and are not in the original figure, which has been redrawn from Attneave (1961). We have inserted into this framework (red and green surfaces) where the two forms of inhibition described in this chapter might be operating.

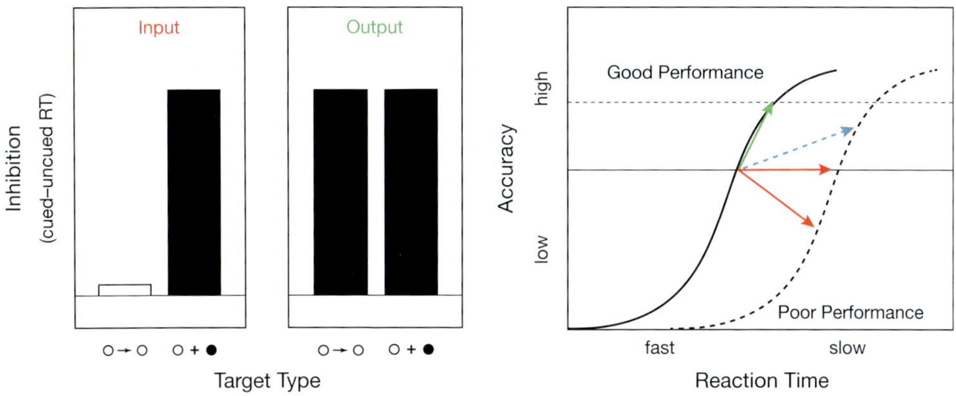

Figure 20.2 Diagnostics for determining whether an effect is early (input) or late (decision/output) in processing. (A) Target-type diagnostic. Two hypothetical patterns of results when the different target types (central arrow or peripheral target, shown on the x-axis) are calling for a localization response are presented about 1 second after an uninformative peripheral cue. (B) Speed–accuracy diagnostic. A nonspatial, two-alternative, forced-choice response is required to a continuous target, and performance is examined in speed–accuracy space. Two hypothetical information accrual functions (solid and dashed curves) and two hypothetical criteria for responding (horizontal lines) are illustrated. See text for explanation.

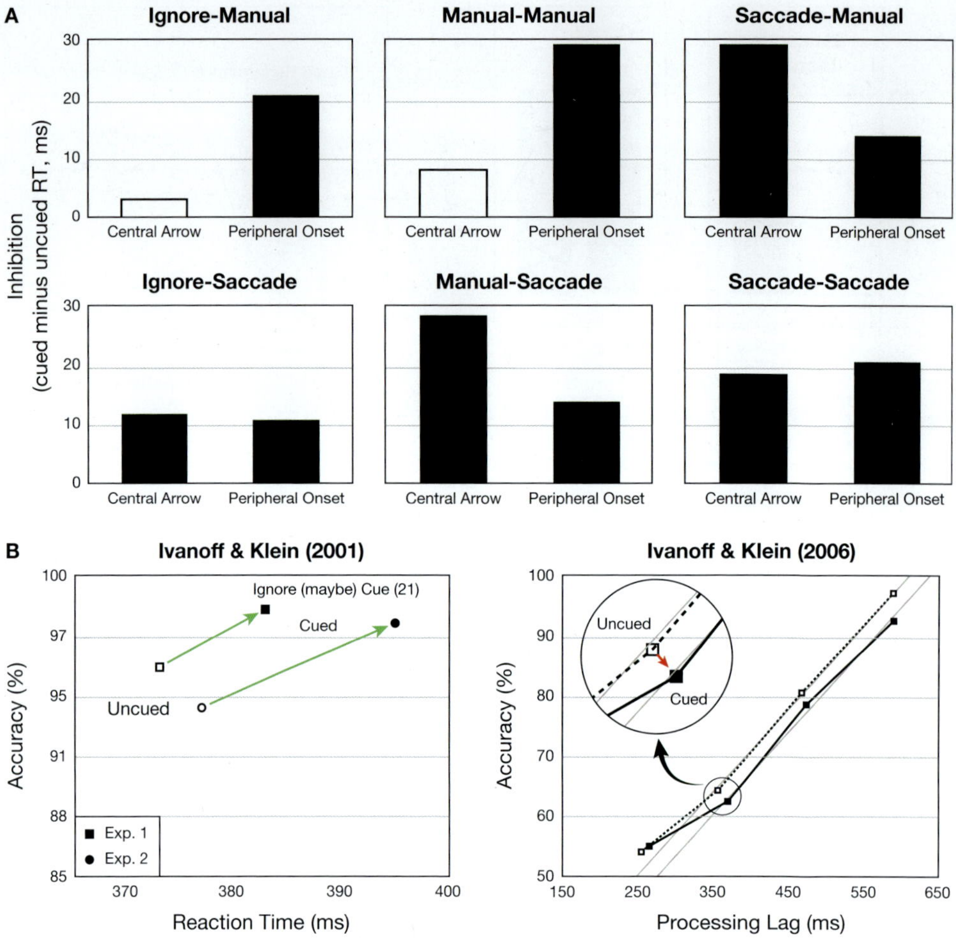

Figure 20.3 (A) Results from the target-type diagnostic as employed in Taylor's dissertation (Taylor & Klein, 2000). Significant IOR is represented by black bars; non-significant IOR is represented by white bars. Each panel represents a different combination of tasks the participant performed in response to S1 and S2 (e.g., Ignore-Manual denotes ignoring S1 and making a manual response to S2). (B) Results from the speed–accuracy diagnostic as employed in Ivanoff's dissertation (Ivanoff & Klein, 2001, 2006). RT and accuracy are plotted on the x- and y-axes, respectively, for targets presented at the cued (filled symbols) and uncued (unfilled symbols) conditions. In the left panel participants performed a go/no-go task with typical reaction time methodology. In the right panel a response-window technique was used to generate full SAT functions. The green arrows signify that IOR is characterized by a speed–accuracy trade-off and the possibility of an output effect. The red arrow signifies a genuine decrease in information processing efficiency.

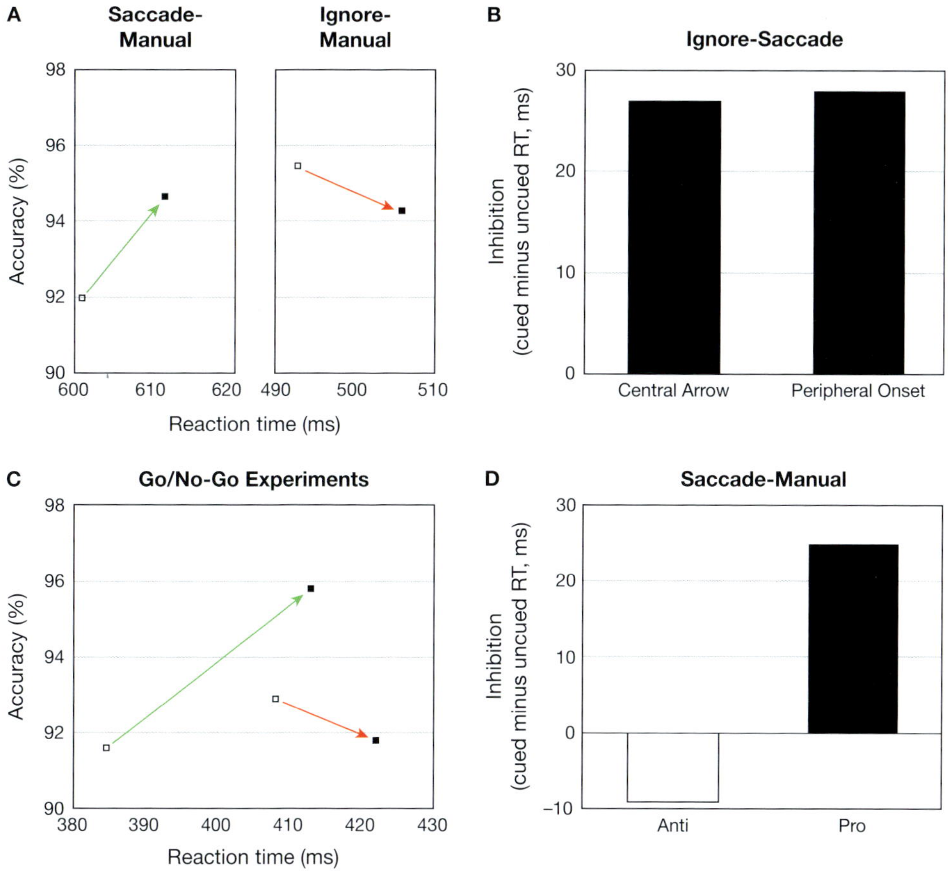

Figure 20.4 Patterns of findings when the two diagnostics illustrated in Figure 20.2 are applied to our studies and to the literature. In A, C, E, and F, data from cued targets are plotted with filled symbols; data from uncued targets with unfilled symbols. (A) Results from Chica et al. (2010). Data connected by the green arrow are from Experiments 1B and 3B; data connected by the red arrow are from 3A. (B) Data from Hilchey, Klein & Ivanoff (2012). (C) Data are from a go/no-go tasks (errors are false alarms). Data connected by the red arrow are from Hilchey, Hashish et al. (2014) in which there was careful monitoring of eye position and participants were given feedback whenever an untoward eye movement occurred. Data connected by the green arrow are collapsed from six studies in the literature (Ivanoff & Klein, 2001, 2003, 2004, Exp. 1; Taylor & Ivanoff, 2003; Prime & Ward, 2008, Exp. 3; Prime & Jolicoeur, 2009, Exp. 1). (D) Data are from Hilchey, Dohmen, Crowder, & Klein (2016). In separate blocks participants made pro- or anti-saccades in response to a peripheral luminance increment, and ~1 second after the onset of this stimulus they made a manual response in the direction of a centrally presented arrow. (E) Data are from Redden, Hilchey, & Klein (2016). Different participants made a pro- or anti-saccade in response to a peripheral luminance increment, and ~1 second after the onset of this stimulus they made a nonspatial 2-AFC. (F) From Redden, Hilchey, and Klein (2015a). Participants made a pro-saccade in response to a peripheral luminance increment, and ~1 second after the onset of this stimulus a target was presented calling for a nonspatial 2-AFC response. Importantly, speed–accuracy trade-off functions were obtained for cued and uncued targets using a window method.

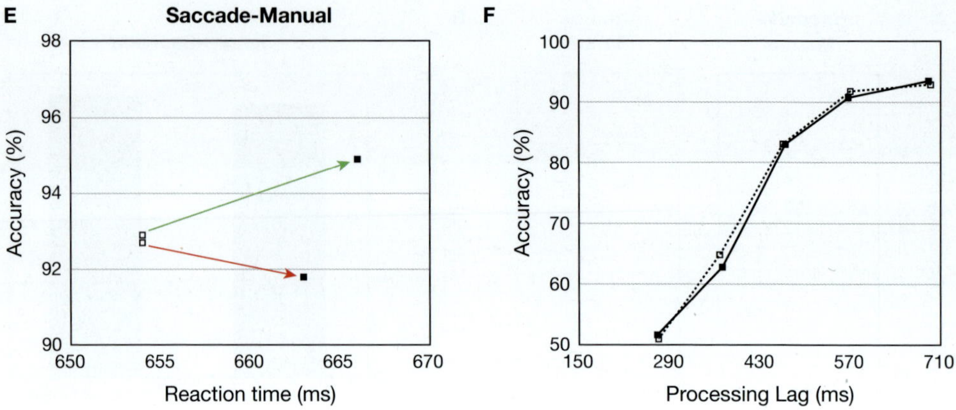

Figure 20.4 (*cont.*)

Part

IV Perception and Action

Chapter 18

Spatial Bias after Brain Damage

The Case of Visual Neglect

Pom Charras, Juan Lupianez, and Paolo Bartolomeo

Visual Neglect

Unilateral lesions of the right hemisphere (most often of vascular origin) frequently give rise to signs of left unilateral neglect (about 85% of stroke patients in the subacute phase).[1] Neglect patients behave as if the left part of the world does not exist anymore: they bump into left-sided obstacles, do not eat from the left side of their dish, and may not answer people on their left (Bartolomeo, 2014). Especially in the acute phase, their head and eyes may be turned toward the right side, even at rest (De Renzi et al., 1982), or whenever a stimulus appears in their visual fields (Bourgeois et al., 2015; Gainotti, D'Erme, & Bartolomeo, 1991). The presence of sensorimotor deficits (respectively hemianopia – visual field loss, and hemiplegia or hemiparesis – paralysis or weakness of the left side of the body) may accompany these cognitive disabilities. Although independent, these sensory-motor deficits accentuate the severity of neglect. Neglect patients often spontaneously recover with time (but see Lunven et al., 2015); however, residual signs of spatial bias, such as slowed response times (RTs) to left-sided targets (Bartolomeo, 1997), may persist and put the patient at risk in rapidly changing situations (e.g., during driving).

Visual neglect is not only characterized by a lateralized bias in information processing but also encompasses additional nonspatial aspects such as reduced alertness and sustained attention (Husain & Rorden, 2003; Robertson, 2001) and impaired spatial working memory (Wansard et al., 2014, 2015; Wojciulik et al., 2001) or temporal processing (Husain & Rorden, 2003). The amount of nonspatial deficits correlates with the degree and severity of neglect (for reviews, see Bonato, 2012; Husain & Rorden, 2003).

In the present chapter, we will focus on asymmetries of performance in visual neglect. First, we will review the literature that gathered evidence for a left–right imbalance in visuospatial and representational processes; then we will provide support for attentional accounts of neglect by discussing how the anatomical and functional organization of the attentional networks may underlie normal visuospatial processing. We will finally review evidence that pre-attentive processing can be relatively spared in neglect patients, thus accounting for the absence of spatial bias in specific contexts.

---

[1] Signs of contralesional (right) neglect may also occur after left hemisphere damage (Beis et al., 2004), but they are less frequent, severe, and durable.

Left–Right Imbalance in Neglect

Spatial Bias Associated with Visual Neglect

Patients with left visual neglect subsequent to right hemisphere damage show a dramatic left–right imbalance in visual processing (Bartolomeo, 2007; Mort et al., 2003; Vallar, 2001), as attested by impaired performance on several tasks. A typical neuropsychological test assessing neglect is the line bisection task. Participants are presented with a horizontal line centered on their body and are required to mark its midpoint. Since the left side of space and/or objects is ignored, patients usually show a strong rightward deviation at fixation (Marshall & Halligan, 1990; Schenkenberg, Bradford, & Ajax, 1980). A similar left-right imbalance may occur in tasks requiring to mark the subjective endpoints of a line (endpoint task; Bisiach, Rusconi, Peretti, & Vallar, 1994; Bisiach, Pizzamiglio, Nico, & Antonucci, 1996, 1996; Urbanski & Bartolomeo, 2008), or to extend a line leftward or rightward (line extension; Geminiani, Camaschella, Mariani, Alberoni, & Farina, 2002; but see Ishiai et al., 1994a, 1994b; Perri, Bartolomeo, & Gainotti, 2000). Deviations are often attributed to left inattention, but other deficits might contribute, such as directional hypokinesia (Bartolomeo, Chokron, & Gainotti, 2001; Coslett et al., 1990), i.e., the failure in executing movements fully in or toward the contralateral hemispace. Using analogous tasks that minimize or even do not require movements provided evidence for left–right imbalance even in the absence of motor actions, thus suggesting a possible dissociation between perceptual and motor deficits (Chokron et al., 1998). Moreover, when patients have to verbally estimate the length of horizontal bisected lines, such as in the landmark task, a strong rightward bias alters performance (Milner, Brechmann, & Pagliarini, 1992; Milner et al., 1993). Even in purely perceptual tasks, then, the left portion of a horizontal line seems to be ignored, the right portion overemphasized, or both.

As salience ascribed to a stimulus across the visual field is not uniform, spatial imbalance in neglect is not restricted to misperception of the position of the center of the line, but also extends to any processing of left-lateralized events. In tasks where visual targets are briefly displayed either to the left or right of a centered fixation point, neglect patients typically fail to detect target appearance or show abnormally delayed responses for left-lateralized events (Bartolomeo, 1997; Rastelli et al., 2013). Moreover, when two targets are simultaneously presented on both sides, patients often report the presence of the ipsilesional target alone. This phenomenon, known as visual extinction, emphasizes the biased competition between left and right visual stimuli. Again, these instances of imbalance in lateralized target processing might result from processing impairment of left events, over-prioritization of right events, or both.

Visual search tasks, such as the cancellation tasks commonly used in clinical neuropsychology, can disclose abnormalities in the processing of targets appearing on the left, neglected side, but sometimes also targets situated on the right, allegedly "normal" side. On these tasks, patients are presented with a sheet of paper with many objects, equally distributed on both sides. The task is to detect all the targets, for example to circle all the silhouettes of a bell among other objects (Gauthier et al., 1989). The results are clear-cut and twofold: objects located on the left part of the sheet are mostly ignored, and sometimes, patients perseverate in re-marking right-sided objects several times (Husain et al., 2001; Wansard et al., 2014). Indeed, when the task is implemented on a computer and eye movements are recorded, the amount of right-sided saccades is exaggerated, with the very same stimulus being explored several times (Driver & Husain, 2002; Mannan et al., 2005;

Nys et al., 2006; Wojciulik et al., 2004). Variants of the cancellation task also provided evidence of over-prioritization of right-sided items. For example, when targets have to be erased, and not canceled by a pencil stroke, left-sided omissions decrease, presumably because erased right-sided items are obviously unable to capture attention ipsilesionally (Mark et al., 1988; see also Bartolomeo et al., 2004).

A large body of evidence pointed out that space or the external world is not uniformly coded in the healthy brain. Since interactions depend on whether objects or events occur in a potentially reachable space (Colby & Duhamel, 1996; Joanette et al., 1986; Rizzolatti, Riggio, & Sheliga, 1994), a crucial determinant of spatial coding lies on the location within or outside the reach (Colby, 1998 for monkey models). Thus, in monkeys and humans, reaching distance determines spatial coding and gives rise to three frames of reference labeled personal, peri-personal, and extra-personal space. Peri-personal space (within arm reaching) and extra-personal space (beyond reaching) define the external world, in opposition to personal space, which refers to the space occupied by our body. Peri-personal space refers to the immediate surroundings of the body whereas extra-personal space implies that locomotion is necessary to physically interact with the relevant object or event. Neglect syndrome has been widely investigated to further understand how space is coded in regards to the frames of reference. Double dissociations between far and near extra-personal space neglect (Cowey et al., 1994; Halligan & Marshall, 1991; Vuilleumier et al., 1998) corroborated evidence in monkeys and support the view of different ways of coding with respect to body position. Nonetheless, spatial coding is not fixed, but flexible, as demonstrated by functional remapping from far space to near space during tool use (see Berti & Frassinetti, 2000). Dissociations between personal and extra-personal have also been reported in neglect (Bisiach et al., 1986; Guariglia & Antonucci, 1992; Marangolo et al., 2003; Peru & Pinna, 1997). Perhaps as a consequence of the multiplicity and flexibility of spatial processing mechanisms, the nature of the neglected space may vary across patients, or even within patients on different experimental conditions. For example, some patients experience egocentric spatial neglect, where stimuli falling on the contralesional side of the body are ignored (Doricchi & Galati, 2000; Riddoch & Humphreys, 1983), while others ignore parts that fall on the contralesional side of objects regardless of the objects' egocentric position (allocentric neglect; Doricchi & Galati, 2000; Kleinman et al., 2007; Olson, 2003; Walker, Findlay, Young, & Lincoln, 1996; Walker & Young, 1996). Egocentric and allocentric neglect can co-occur but also dissociate across patients. Traditionally, the two phenomena are considered as behaviorally and anatomically dissociable (Marsh & Hillis, 2008); however, recent work suggests a unitary underlying disorder (Rorden et al., 2012).

Taken together, evidence from neglect patients' performance on a variety of tasks such as line bisection, lateralized target detection and target cancellation, and tapping several spatial frames of reference pinpoints the same conclusions: on the one hand, left-sided events are ignored, on the other hand, right-sided events are not normally processed either because they can be abnormally over-prioritized (see Bartolomeo et al., 1999; Bourgeois et al., 2012; Snow & Mattingley, 2006), albeit in a context of defective attention (Bartolomeo & Chokron, 1999). However, important differences remain between patterns of performance in different patients. This raises major theoretical issues on the very status of visual neglect as a homogenous entity. One possibility is that neglect signs result from the interaction of multiple component deficits, which can be spatial and nonspatial and whose severity can vary across patients (Bartolomeo, 2007). The next section describes a particularly intriguing form of neglect, that concerning visual mental images.

Spatial Bias in Representational Neglect

Spatial neglect is prominent in the visual domain, but it is not limited to visual processing. Some neglect patients show a deficit in exploring and manipulating visual mental representations. In a seminal study, Bisiach and Luzzatti (1978) asked two Milanese patients to describe from memory a very familiar place (the Piazza del Duomo in Milan). The amount of information verbally reported was clearly spatially biased with respect to the imagined viewing point. Left-sided elements of the mentally imagined scene were ignored and right-sided elements perfectly remembered, respective to the viewing point. The spatial deficit defined by a left–right imbalance in processing mental representation is referred to as representational or imaginal neglect (Bartolomeo et al., 1994, 2005). The inability to explore the contralesional side of visual mental representation often coexists with visual neglect (see Bartolomeo, 2002, for review), but in its pure form, representational neglect emerges while visual perception is preserved, thus suggesting the possibility of a dissociation (Beschin et al., 1997; Coslett, 1997; Guariglia et al., 1993). However, pure representational neglect is comparatively rare with respect to visual neglect and also less severe (Bartolomeo, 1994; Bourlon et al., 2008, 2011).

Another form of representational neglect comes from studies based on numerical cognition. Adapting the physical line bisection task to numerical cognition, participants are asked to mentally bisect a numerical interval, without calculating, by giving the number they think to be halfway between the two presented numbers (Fischer, 2001; Zorzi et al., 2002). A rightward bias was reported for neglect patients, since the halfway number, supposed to be perceived as the center, was larger than the actual number. This deviation toward large numbers has been accounted for by the assumption of a mental number line on which numbers are spatially organized. Small numbers are therefore mapped on the left side of the line and large numbers on the right (Dehaene et al., 2003; Gevers et al., 2010). As a consequence, when estimating the middle between 3 and 9, neglect patients tend to deviate toward numbers larger than 6 (the true middle), and in an RT task, responses to small numbers on the left side of the mental number line are faster (Vuilleumier et al., 2004). The spatial-numerical correspondence results in deviation toward larger numbers for neglect patients. Note, however, that the severity of neglect on bisecting physical lines does not correlate with performance for bisecting numerical lines (Doricchi et al., 2005, 2009; Loetscher & Brugger, 2009; Loetscher et al., 2010; Pia et al., 2012; Rossetti et al., 2004, 2011; Tian et al., 2011; van Dijck et al., 2011). It has thus been suggested that this phenomenon is not really spatial in nature, but rather results from a deficit in representing small magnitudes (Aiello et al., 2012). See Cipora, Patro, and Nürk (Chapter 3 in this volume) for further details on spatial components of the SNARC effect.

Another dimension that is hypothesized to rely on spatial attributes to foster its representation is time. Time is an abstract dimension that can be encoded in terms of spatial analogies or metaphors. Extensive investigation of the spatial organization of time has suggests that time, when referring to the past and future, can map onto a horizontally oriented mental time line (Fuhrman & Boroditsky, 2010; Ouellet et al., 2010). Past elements would thus be associated to the left and future ones to the right. Following this theoretical assumption, Saj et al. (2014) tested whether neglect patients would show a selective deficit in representing events as belonging to the past, that is, left-sided items, in contrast to right-sided items, such as events referring to the future. As expected, the inability to accurately represent space had a direct impact on the mental representation of time and led to temporal distortion.

Past events were mislabeled as future events more often in neglect patients, suggesting a representational crowding effect in the right portion of the time mental line.

A Spatial Bias in Normal Spatial Cognition: Visual and Representational Pseudoneglect

The spatial bias in visual and representational neglect underlines a robust and disabling left–right imbalance in information processing for neglect patients. Surprisingly, however, neurologically healthy individuals also present signs of spatial bias, although in the opposite direction and to a much lesser degree of magnitude. When asked to bisect the middle of a horizontal line, they slightly deviate the bisection mark toward the left side of the line (Bowers & Heilman, 1980). This bias – referred to as pseudoneglect – is subtle but highly reliable (for a review, see Jewell & McCourt, 2000). Recent studies even suggest that pseudoneglect is not limited to meticulous research tasks in labs, but also accounts for a general and observable preference for the left side of space (Nicholls et al., 2007; but see Hatin et al., 2012; see also Hubbard, Chapter 15 in this volume, for a discussion on the left side preference in spatial and scene composition). Evidence from behavioral (Toba et al., 2011) and anatomical (Thiebaut de Schotten et al., 2011) studies suggests an important role for spatial attention in pseudoneglect. As observed for neglect patients, healthy individuals can also show a representational form of pseudoneglect, in the sense that they tend to verbally report more left-sided events of an imagined scene (Bourlon et al., 2011 – non-significant trend – and Della Sala et al., 2010; Dickinson & Intraub, 2009; McGeorge et al., 2007). Similarly, pseudoneglect modulated performance on a numerical bisection task, in which participants had to decide which flanker number in a triplet was further away from the middle number (i.e., 15–22–47). Regardless of whether triplets were presented in ascending or descending order, the bisection point was consistently biased in the direction of the lower numerical value, i.e., leftward (Loftus et al., 2009).

As already mentioned, research over the past last decades has convincingly shown that, far from being a unitary deficit, neglect is a multicomponent disorder. Obviously, no theoretical account can give full account for all the deficits and dissociations reported in the neuropsychological literature. In the remainder of this chapter, we will focus on the possible origins of the spatial bias, characterized by left–right imbalance in stimulus processing in neglect. Because spatial bias in neglect appears to mimic and mirror normal pseudoneglect, we will discuss how attentional mechanisms can account for these spatial processing features in normal spatial cognition and also how attentional deficits can contribute to major spatial impairments in information processing, as well as in the nonspatial aspect of neglect.

Attentional Accounts of Neglect

Anatomical and Functional Organization of the Visuospatial Attentional Networks

Visuospatial attention allows us to efficiently interact with objects in the environment, by selecting important items based on their perceptual salience but also on our goals (Desimone & Duncan, 1995). It reflects flexible mechanisms that allow for processing of unexpected but potentially relevant events (exogenous orienting; Yantis, 1995), while ensuring the maintenance of the task at hand (endogenous orienting; Laberge et al., 2000).

Anatomically, visuospatial attention relies on two large fronto-parietal networks (Bartolomeo, Thiebaut de Schotten, & Chica, 2012; Corbetta & Shulman, 2002, 2011), linked by three branches of the superior longitudinal fasciculus (SLF) (Thiebaut de Schotten et al., 2011). A dorsal attentional network includes the intraparietal sulcus/superior parietal lobule and the frontal eye field/dorsolateral prefrontal cortex (Corbetta et al., 2000; Hopfinger, Buonocore, & Mangun, 2000; Kastner et al., 1999). Neuroimaging studies suggest that in healthy subjects, the dorsal network is recruited for the control of spatial orienting of attention (Corbetta & Shulman, 2002; Macaluso, Frith, & Driver, 2002; Sereno, Pitzalis, & Martinez, 2001; Silver, Ress, & Heeger, 2005). Additionally, a more ventral attentional network that includes the temporo-parietal junction and the ventral frontal cortex (inferior and middle frontal gyri) seems to be crucial for several non-lateralized attentional functions, such as visual and temporal capacity (Husain & Rorden, 2003; Peers et al., 2005; Shapiro, Hillstrom, & Husain, 2002), arousal and vigilance (Coull, Frackowiak, & Frith, 1998; Pardo, Fox, & Raichle, 1991; Rueckert & Grafman, 1996; Wilkins, Shallice, & McCarthy, 1987), saliency detection (Downar, Crawley, Mikulis, & Davis, 2000; Serences, Yantis, Culberson, & Awh, 2004; Shulman et al., 2010), and the reorienting of attention (Arrington, Carr, Mayer, & Rao, 2000; Corbetta et al., 2000; He et al., 2007; Macaluso et al., 2002; Rengachary, He, Shulman, & Corbetta, 2011); it may also be important in the case of late consequences of exogenous orienting of attention, such as inhibition of return (Chica, Bartolomeo, & Valero-Cabre, 2011). Of utmost importance, the ventral attentional network appears to be largely lateralized to the right hemisphere (Corbetta & Shulman, 2002), and consequently, damage to right hemisphere ventral network is assumed to cause a functional imbalance between the left and right dorsal networks. A hyperactivity of the left dorsal fronto-parietal network would provoke an attentional bias toward right-sided objects and neglect of left-sided items (He et al., 2007). Indeed, the ventral attentional network is not only dedicated to salience detection in a stimulus-driven way but is also responsible for maintaining attention on goals or task demands (endogenous control). Thus, deficits in these nonspatial aspects of attention may lead to an exacerbation of the spatial bias in neglect patients (Husain & Nachev, 2007).

A recent study provided evidence that in humans, as in monkeys, three distinct branches of the SLF link frontal to parietal regions (Thiebaut de Schotten et al., 2011). Interestingly, the intermediate branch (SLF II) seems to provide direct communication between the ventral and dorsal attentional networks by connecting the parietal component of the ventral network to the prefrontal component of the dorsal network. Anatomical asymmetries of the right and left SLF II correlate with behavioral signs of right hemisphere specialization for visuospatial attention, such as pseudoneglect in line bisection, and with asymmetries in the speed of detection between the right and the left hemi-fields (Thiebaut de Schotten et al., 2011). Moreover, temporary inactivation of the SLF II in the human right hemisphere impairs the symmetrical distribution of visual attention, by determining dramatic rightward shifts on line bisection (Thiebaut de Schotten et al., 2005). Damage to SLF networks in the right hemisphere is frequently associated with neglect (Bartolomeo, 2006, 2007; Doricchi et al., 2008; Lunven et al., 2015; Thiebaut de Schotten et al., 2008).

Altogether, this evidence suggests a causal link between attentional dysfunctions in neglect and the typical signs of spatial bias observed in this syndrome. However, the specific mechanisms are as yet uncertain. For example, how can attentional disorders determine under- or overestimation of the length of portions of horizontal lines? Mislocalizing the midpoint of a horizontal line could possibly reflect two dissociable phenomena: an underestimation of the left portion, and also an overestimation of the right portion. In the following section, we will review evidence that attentional orienting can influence length perception, and also that, depending on the context, line length distortion can result from both left underestimation and/or right overestimation.

Exogenous Attention Can Distort Perceived Length in Normal Participants

Whether exogenous orienting of attention could modulate perceived horizontal line length in healthy observers was examined by Toba et al. (2011). Normal observers were tested with a modified version of the Attentional Repulsion test (Suzuki & Cavanagh, 1997), whereby briefly presented probes appear displaced away from the focus of attention. When presented with horizontal lines pre-bisected with a vertical mark near the center, observers misjudged the middle as function of whether the line was preceded by a cue lateralized to the left or to the right, or by no cue. While no cue favored the emergence of pseudoneglect (i.e., a slight but reliable leftward deviation), right-sided cues triggered leftward displacement of the true center, suggesting a transient neglect-like effect (Toba et al., 2011). Overall, Toba et al.'s results suggest a basic mechanism linking exogenous attention and perceived length.

Line Length Perception: Left Underestimation or Right Overestimation?

The spatial bias in unilateral neglect is mainly characterized by a left–right imbalance in stimulus processing. In order to better understand spatial cognition in humans, it is thus crucial to determine whether this imbalance only emerges when left- and right-sided events are competing for attention or whether these two biases coexist independently. To dissociate the possible confounds, Charras, Lupianez, and Bartolomeo (2010) proposed a new paradigm for investigating the left–right imbalance by directly comparing the lengths of horizontal to vertical lines. In a line extension task, participants were either presented with a horizontal or vertical line and were to draw the missing line (vertical vs. horizontal) to form an L shape, either oriented leftward or rightward (Figure 18.1). The L figure is commonly used to study the vertical–horizontal spatial bias, whereby the length of vertical lines is overestimated as compared to horizontal lines (Avery & Day, 1969; Finger & Spelt, 1947;

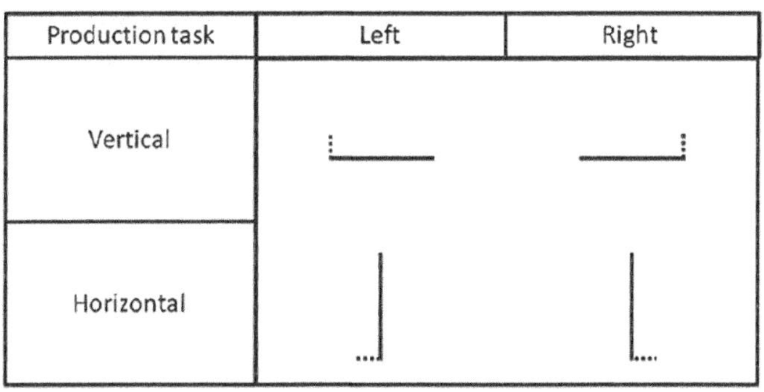

Figure 18.1 The four configurations used in the study by Charras et al. (2010).

Howe & Purves, 2002; Künnapas 1955; Miller & Al-Attar, 2000; Raudsepp, 2001; Raudsepp & Djupsjobacka, 2005; Yang, Dixon, & Proffitt, 1999). The objective for using the L figure was to avoid the left–right competition observed in a typical horizontal line bisection task. Here, instead, the presented line was always centered, while the "to-be-drawn" line was either left or right-sided (see Figure 18.1).

Therefore, in the case of drawing a horizontal line, a centered vertical line could be compared to either a left-sided horizontal line to measure a possible left bias (overextension, see also Geminiani et al., 2002; Ishiai et al., 1994a, 1994b; Perri et al., 2000), or to a right-sided horizontal line to test right bias. By comparing patients' performance for left- and right-sided segments to that of control participants, Charras and colleagues separately assessed the extent of left underestimation and the role of right overestimation in the context of no left–right competition. Healthy participants showed a slight but reliable imbalance between leftward and rightward horizontal lines, consistent with pseudoneglect: left-sided horizontal lines were under-extended as compared to right-sided segments. As expected, neglect patients showed the opposite pattern of left-right imbalance, with left-sided segments being overextended. Interestingly, patients' performance for right-sided lines was similar to that of neurologically healthy participants; however, leftward extension was disrupted. These results invited the conclusion that in the absence of left–right competition, left–right imbalance results more from a left impairment in stimulus processing than from right hyper-attraction.

In apparent contradiction with this conclusion, Urbanski and Bartolomeo (2008) underlined the importance of the right portion of visual stimuli in the severity of the left–right imbalance by using an endpoint task. In this task, participants are usually instructed to mark the subjective endpoints of a line (Bisiach et al., 1994, 1996). Urbanski and Bartolomeo (2008) asked participants to set the endpoints and the center of perceived or imaginary lines. In left-to-right trials, the absence of stimuli attracting attention on the right highly reduced the imbalance, thus leading to the claim that left–right competition might be an important clue in determining the imbalance. Critically, however, in the study of Charras et al. (2010), there was no right-sided horizontal line whose extremity could capture patients' attention (see Gainotti et al., 1991), which presumably induced the absence of right overestimation. Taken together, Charras et al.'s results provide evidence that in the absence of left–right competition, the imbalance in stimulus processing is caused by a left defective bias. With left–right competition, the right portion of the line might instead play a crucial role.

This conclusion has been further supported by a study in which the lengths of horizontal and vertical lines were compared in a perceptual task (Charras, Lupianez, Migliaccio, Toba, Pradat-Diehl, Duret, & Bartolomeo 2012). As mentioned previously, horizontal–vertical comparisons are commonly used to investigate the vertical spatial bias. In an inverted T figure, the length of vertical lines is strongly and reliably overestimated as compared to horizontals (referred to as the *horizontal–vertical illusion* in the literature). This bias reflects additive effects of the vertical bias and bisection (Charras & Lupianez, 2009, 2010; Künnapas, 1955; Wolfe, Maloney, & Tam, 2005). Bisection, more specifically the bisection a/symmetry, plays a major role in length perception such that lines symmetrically bisected were underestimated but lines asymmetrically bisected overestimated (Charras & Lupianez, 2009, 2010). We refer to this effect as the *symmetry law*. In the study of Charras, Lupianez, Migliaccio, Toba, Pradat-Diehl, Duret, and Bartolomeo (2012), patients were asked to compare the length of horizontal and vertical segments to assess the potential left–right imbalance. The two segments were displayed in L-, T-, or X-shaped

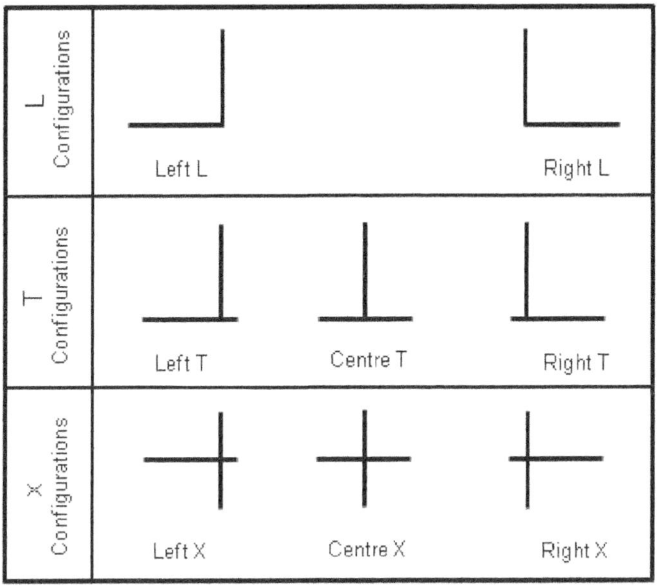

Figure 18.2 The eight configurations used in the study by Charras, Lupianez, Migliaccio, Toba, Pradat-Diehl, Duret, and Bartolomeo (2012a).

configurations, in which horizontal lines were either centered, left-sided, or right-sided (Figure 18.2). The biases of under- or overestimation stemming from bisection a/symmetry were manipulated to investigate the left–right imbalance in neglect patients.

In the L configurations, neglect patients exhibited a left–right imbalance clearly resulting from left underestimation. The lack of right overestimation replicates the results obtained in the line extension task (Charras et al., 2010) and confirms the previous conclusion according to which, in the absence of competition between left and right horizontal segments, the spatial bias is consecutive to left-sided line neglect.

For the T configurations, the results also suggested that the spatial bias is consecutive to left-sided line neglect. Performance for participants and neglect patients was similar for right-sided lines, but clearly differed in the left T configuration, since left-sided segments were largely underestimated (see Figure 18.3). When the emphasis was put on the vertical dimension of the stimulus (as demonstrated by participants' strong tendency to overestimate the vertical portion of the stimulus), thus preventing left–right integration of the horizontal segments, patients mainly neglected left-sided segments.

Finally, in the X configurations, performance between controls and neglect patients differed not only for left-sided lines but also for right-sided lines. Indeed, neglect patients overestimated right-sided lines (right X, see Figure 18.3). The authors concluded that when left-sided and right-sided segments competed to be integrated in a single percept, the processing of the right portion contributed to worsen the left–right imbalance (see also Urbanski & Bartolomeo, 2008).

The overall results suggested that the spatial bias in line length processing is mainly caused by left neglect (in the absence of left–right competition for stimulus processing) but is accompanied by a right over-processing when left and right compete to form a single percept. Interestingly, the presence of left homonymous hemianopia worsened left

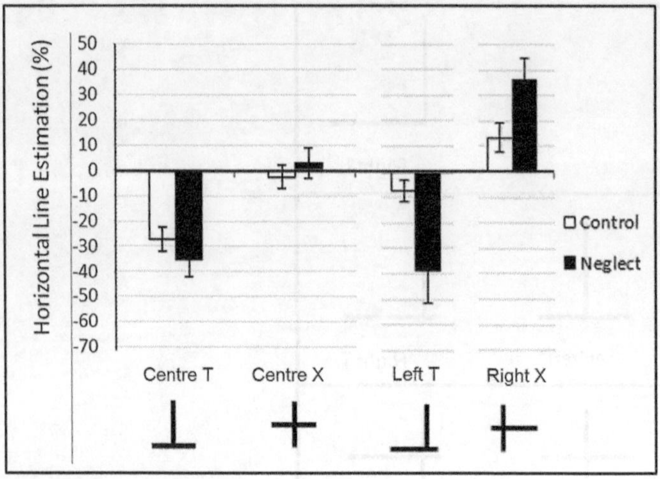

Figure 18.3 The mean percentage and SDs of horizontal underestimation (negative values) and horizontal overestimation (positive values) as a function of Configuration (Center T, Center X, Left T, and Right X) for Healthy controls and Neglect patients.

underestimation but did not modulate right overestimation. It is tempting to relate left underestimation to the defective functioning of right-hemisphere attention networks typical of neglect (Bartolomeo, Thiebaut de Schotten, & Chica, 2012), and right overestimation to inappropriate functioning of left-hemisphere attention networks (Corbetta, Kincade, Lewis, Snyder, & Sapir, 2005), which may be sustained over time by a concurrent deficit of posterior callosal communication between the hemispheres (Lunven et al., 2015).

Pre-Attentive Processing in Neglect

Left–Right Imbalance and Pre-Attentive Mechanisms in Neglect

Considering that attentional dysfunctions are important for the rightward deviation in line bisection, and therefore the spatial bias in neglect, may imply that pre-attentive mechanisms are spared and still functional in visual neglect. Basic perceptual processing, which is supposed to take place pre-attentively, might operate normally, despite a lack of conscious awareness in the defective visual hemi-field (Driver & Vuilleumier, 2001). The preservation of low-level, perhaps pre-attentive, perceptual mechanisms in visual neglect was suggested by a study using line bisection with illusory figures. Illusory effects were elicited by adding unilateral or bilateral fins to horizontal lines, thus reproducing a Müller–Lyer illusion. While rightward deviation was observed for neglect patients in the control condition (no fins triggering line length distortion), this bias was modulated by the presence of contralateral fins (Mattingley, Bradshaw, & Bradshaw, 1995; see also Vallar, Daini, & Antonucci, 2000). In a similar study, Daini, Angelelli, Antonucci, Cappa, and Vallar (2002) replicated the effect of illusory figures on the spatial bias, but also provided evidence for a dissociation between left neglect and left homonymous hemianopia. While neglect patients were sensitive to the contralateral elements triggering a length illusion, left homonymous hemianopia patients were totally immune (Daini et al., 2002). Moreover, performance on a line bisection task can also be biased by implicit grouping, such as when the inducers are not consciously perceived. In accordance with the dissociation between neglect and hemianopia, this ability to process contralateral inducers might rely on the sparing of lateral occipital regions at early stages of visual processing (Vuilleumier & Rafal, 2000).

More evidence for a preservation of pre-attentive processes in visual neglect comes from studies using brief presentation of lateralized targets. In a simple detection task, right brain–damaged patients who showed extinction tended to omit left-sided targets during bilateral presentation. The authors provided evidence that the number of left omissions was reduced when the two targets could be grouped, following Gestalt factors such as symmetry, similarity, or familiar configuration (Ward, Goodrich, & Driver, 1994).

Studies investigating summary statistics corroborated the idea of integrity of pre-attentive mechanisms. When confronted with a bunch of similar objects or events in the visual scene, the visual system is able to rapidly and effortlessly extract information and calculate summary statistics for sets of similar objects (Ariely, 2001; see Alvarez, 2011 for a review). Participants can therefore judge the average size of circles (Ariely, 2001; Chong & Treisman, 2003), the average orientation of items (Parkes, Lund, Angelluci, Solomon, & Morgan, 2001), or average facial expressions (Haberman & Whitney, 2007, 2009) in a visual display. Interestingly, patients with unilateral attentional deficits can successfully build statistical summary representations and use this information (Pavlovskaya, Bonneh, Soroker, & Hochstein, 2010; Pavlovskaya, Bonneh, Soroker, & Hochstein, 2011; Yamanashi Leib et al., 2012). It had already been demonstrated in healthy participants that this ability requires few attentional resources (Alvarez & Oliva, 2009; Haberman & Whitney, 2011; Joo, Shin, Chong, & Blake, 2009).

Some sparing of pre-attentive processes has also been highlighted in the previously mentioned study of Charras, Lupianez, Migliaccio, Toba, Pradat-Diehl, Duret, and Bartolomeo (2012). In this study, both control participants and neglect patients overestimated a vertical line implemented in an inverted T configuration (see Figure 18.2), consistent with the horizontal–vertical illusion (Avery & Day, 1969; Fick, 1851; Finger & Spelt, 1947; Künnapas, 1955, 1957). Indeed, the length of lines symmetrically bisected is typically underestimated (symmetry law; see Charras & Lupiáñez, 2009, 2010; Wolfe et al., 2005). Neglect patients were affected to the same extent as controls by the symmetry law: the left-sided portion of the horizontal segment was processed and taken into account for the line length estimation. Unequivocal evidence for a preserved symmetry law in neglect patients was provided by the center X configuration, in which lines were appropriately perceived by both neurologically healthy individuals and neglect patients (see Figure 18.3). Overall, when the horizontal line was symmetrically bisected, neglect patients were able to perceive both the left and the right segment parts. Ultimately, despite normal performance for symmetric bisections (center T and center X, Figure 18.3), neglect patients showed strong spatial biases for the left T (underestimation) and right X (overestimation) configurations (see Figure 18.3; Charras, Lupianez, Migliaccio, Toba, Pradat-Diehl, Duret, & Bartolomeo, 2012).

This preservation raises the issue of whether the symmetry law is driven by perceptual mechanisms at a different level of processing with respect to attentional bias in neglect. The finding that a/symmetry influenced both control and neglect participants reveals that the detection of a/symmetry might be driven by low perceptual level or pre-attentive mechanisms. As mentioned earlier, behaviors controlled by pre-attentive mechanisms may be intact after right parietal damage and neglect or extinction (Halligan & Marshall, 1995; Mattingley, David, & Driver, 1997). Gestalt laws such as grouping or connectedness may remain operative after right parietal damage (Esterman, McGlinchey-Berroth, & Milberg,

2000; Mattingley et al., 1997; Roberston, Mirjam, & Knight, 2003). For instance, perceptual grouping can help neglect patients detect left-sided targets (Robertson et al., 2003), perhaps because left and right were integrated into a "perceptual whole." The authors concluded that pre-attentive mechanisms responsible for grouping enable the conceptualization of left and right as a unique object, thus preventing or at least decreasing left neglect. (Note, however, that these phenomena are variable and do not always permit compensation of neglect. Even within single perceptual objects, attention appears to operate with a right-to-left spatial gradient, such that the left part of the object may still be neglected; see Driver & Pouget, 2000.) The symmetry law explored here might rely on similar mechanisms. The detection of a/symmetry may shape the attentional distribution all around the locus of bisection, in order to map the whole configuration. Consequently, the distribution of attention might encourage the orienting of attention to the left, and thus reduce the left underestimation.

A Possible Account for the Symmetry Law in Perceived Length

To deal with the overabundant amount of information present in the environment, the visual system organizes inputs by selecting and perceiving objects. Visual organization is determinant as selected objects are attended and processed in a particular way. Research on object perception demonstrated that visual short-term memory is characterized by a fixed number of items (up to four) based on spatial location (Luck & Vogel, 1997; but see Bays & Husain, 2008). However, the ability to detect the presence of objects does not always lead to efficient identification. Sagi and Julesz (1985) first reported that the visual system allows us to rapidly detect a target, but that identification is a much slower process. Notably, the capacity to detect multiple targets seems to be independent of object features (Luck & Vogel, 1997), while identification is dependent on attentional resources and influenced by object complexity. In congruence with this dissociation between object detection and identification, Kahneman, Treisman, and Gibbs (1992) proposed an "object file" theory, whereby target detection results from object individuation that leads to "open" an object file, whereas target identification entails processing of object features. As a consequence, in opposition to object complexity, which impairs target identification, grouping and object repetition clearly enhance this process but do not affect target detection (Xu & Chun 2007, 2009). In their neural object-file theory, Xu and Chun questioned the traditional dichotomy between ventral and dorsal pathways (Ungerleider & Mishkin, 1982). Xu and Chun (Xu, 2007, 2009; Xu & Chun, 2007, 2009) proposed that the parietal cortex is involved in both "where" and "what" visual processing (see also Konen & Kastner, 2008; Sereno & Maunsell, 1998). More specifically, neuroimaging evidence suggests that distinct cerebral regions underlie the individuation and identification phases. While activity in the inferior intraparietal sulcus (IPS) reflects the individuation process, identification is implemented in the superior IPS and lateral occipital cortex (LOC) (Xu, 2009; Xu & Chun, 2009, but for a controversy, see Naughtin, Mattingly, & Dux, 2016; Naughtin, Tamber-Rosenau, & Dux, 2014).

Drawing on these hypotheses, we attempt here to provide an account for the underestimation of lines symmetrically bisected. In symmetric bisection, the relevant property – length in this task – is identical for the two line segments (left and right parts). In the light of the neural object-file theory (Xu, 2009; Xu & Chun, 2009), we propose that the processing of equal-length segments requires less effort than that required for two different lengths. Since two equal segments could be processed as a single repeated stimulus, less effortful processing would result in underestimation, which would emerge from a heuristic linking cognitive cost and magnitude estimation. Symmetric parts of a segment are underestimated because they are processed as a single, repeated stimulus. The underestimation might result from lack of increased activity in the superior IPS and LOC for repeated objects, as compared to different objects. Indeed, activity in these regions is similar for several identical objects and for a single object (Xu, 2009). In other words, repeated stimuli are processed as distinct objects in the object individuation phase – revealed by an increased involvement of inferior IPS for multiple object displays, but as a single object in the object identification phase (Xu, 2009; for a review, see Xu & Chun, 2009).

Following this assumption, we predict that the magnitude underestimation for repeated stimuli might influence any kind of magnitude (Cantlon et al., 2009; Walsh, 2003). Evidence for such an effect has been gathered in the field of numerical cognition, in which stimulus repetition leads to bias magnitude estimation toward an underestimation (Charras, Brod, & Lupianez, 2012; Charras et al., 2014). Consistent results have also been reported for interval timing estimation and support the idea that neural energy required to process a stimulus is correlated with the subjective perceived duration (Eagleman, 2008; Eagleman & Pariyadath, 2009; for a review; Pariyadath & Eagleman, 2007, 2008). Although further research is needed to better understand the effect of stimulus repetition on magnitude estimation, it seems already uncontroversial that repeated stimuli are not processed in the same way as different stimuli.

Interestingly, this account based on the individuation–identification distinction also accounts for the preservation of the symmetry law in neglect, as well as the left–right imbalance in line length perception. The ease in perceiving two identical objects would favor the two-line binding, thus accounting for similar performance between healthy participants and neglect patients for symmetrically divided lines. However, binding two different-length segments would have an additional cost, which generates a global length overestimation for healthy participants but a left–right imbalance for neglect patients. As a consequence, when binding two segments is resources consuming (asymmetric bisection), neglect patients show the typical left–right imbalance in stimulus processing, resulting in a left underestimation or/and a right overestimation. This idea is congruent with a large amount of evidence indicating that nonspatial task demands can boost left–right imbalances in neglect patients (Bartolomeo, 2000; Bonato, 2012) and also in healthy participants (Dodds et al., 2008; O'Connell, Schneider, Hester, Mattingley, & Bellgrove, 2011; Peers, Cusack, & Duncan, 2006).

As a conclusion, the present chapter gathered and reviewed recent evidence that although neglect is a multicomponent disorder, it is mainly characterized by unbalanced spatial processing across the visual field. Moreover, the spatial bias in neglect mimics a subtle phenomenon occurring in normal spatial processing known as pseudoneglect. Consequently, spatial bias in visual neglect can highlight important mechanisms of normal spatial processing in visual perception in terms of behavioral and anatomo-functional perspectives.

Chapter 19

Natural Regularities and Coupled Predictive Perceptual and Cognitive Biases

Why We Evolved to Systematically Experience Spatial Illusions

Michael K. McBeath

Historically, spatial illusions are often framed merely as curious perceptual or cognitive processing errors. Orthodox ecological theorists such as J. J. Gibson note that spatial illusions are rare phenomena, largely absent in fully realistic perception-action environments, and therefore of little interest (Gibson, 1979). People typically perform remarkably well in the vast majority of common, daily perception-action activities such as reaching, grasping, and navigating. In contrast, more progressive ecological theorists such as Roger Shepard and myself suggest that active, observant organisms evolved to anticipate and predict likely environmental patterns, including patterns that are only somewhat reliable (McBeath & Shepard, 1989; Shepard, 1984, 1994, 2001). Such predictive expectations could systematically bias perceptions and cognition in naturally selected optimizing directions, at the expense of occasional perceptual and cognitive errors when those expectations are not met (McBeath & Sugar, 2005). While the orthodox approach emphasizes that active observers couple perception-actions to the array of invariant geometric properties in the world, the progressive approach emphasizes that these perception-action relationships with the environment can be extended to also include looser couplings to an array of natural regularities that vary in reliability. Such natural regularities can span from universal, invariant geometric properties (such as the systematic increase in the optical angle of an approaching, looming object) to weak socially learned tendencies (such as the tendency for approaching American pedestrians to veer off to their respective right sides). The principal theme of this chapter is to outline this progressive approach and support that observer-actors possess an array of perceptual or cognitive spatial biases that vary in robustness as a function of the reliability (or predictability) of the driving natural regularities. Examples are provided of natural regularities with various levels of reliability, along with the biased perceptual or cognitive tendencies that are coupled to the natural regularities. The chapter concludes with a discussion of different types of natural selection mechanisms that may produce and regulate these biases.

Invariant Universals

The Universal of Breathable Air

On March 19, 1981, tragedy struck NASA's Cape Canaveral launch complex for the space shuttle *Columbia*. NASA-Rockwell technicians, John Bjornstad and Forrest Cole, entered a curtained-off shuttle compartment and, moments later, passed out. They were followed by fellow technician, William Wolford, who called for help, but then also passed out. Finally technicians Jimmy Harper and Nick Mullon arrived and helped remove the other men from the deadly area, but as they exited they too lost consciousness. Two of these men died that day, while the three that lived were severely brain damaged from anoxia due to the shuttle compartment being filled with deadly pure nitrogen (Burlison, 2013).

One might ask: Why didn't the men realize that they were breathing poisonous gas, why didn't their perceptual systems warn them that they were encountering a deadly threat? In one sense, this can be considered an extreme illusion, being unable to distinguish between common and deadly types of air. Usually our perceptual system warns us of severe threats – for example, bad food typically looks and smells odd or revolting, and dangerous predators usually look and sound scary and formidable. The evolutionary explanation is that having breathable, oxygenated air has historically been so reliable, and conditions of pure nitrogen gas so rare, that there was no selective pressure to evolve a mechanism to recognize pure nitrogen as toxic; such conditions basically never occurred in past human habitats.

Breathable air can be considered a type of environmental "universal," something so reliable that we and other oxygen-breathing species evolved autonomic biological action-mechanisms to maintain breathing without needing to think about it. Occasionally there are impurities like smoke in the air that we sense and react to avoid, but exceedingly uncommon gas conditions like pure nitrogen are simply assumed to not exist and are not sensed.

The Ubiquity of Light Wave Electromagnetic Frequencies

Another example of a universal is the ubiquity of electromagnetic energy at light wave frequencies (wavelengths of 380–760 nm) compared to other frequencies such as radio waves and microwaves. The physics of our universe results in a lightwave-centered peak of electromagnetic frequencies emitted into our atmosphere, in particular by our sun (Pizlo, Li, Sawada, & Steinman, 2014). Like most other terrestrial life forms, we evolved a hard-wired bias to only perceive or see electromagnetic energy at frequencies that match this environmental universal. In summary, breathable air and the ubiquity of light wave electromagnetic frequencies are universal properties of our terrestrial environment, and we evolved hard-wired perception-action biases to assume their presence, and to not process rare and non-occurring alternatives. These universals are so reliable, and our biases so strongly coupled to them, that these biases are generally not framed as types of distortion or illusion. Universals are the strongest type of natural regularities.

Universal or Invariant Three-Dimensional Geometric Properties

Other universals that directly relate to spatial perception are properties of three-dimensional (3D) geometry of our world, such as an accelerating flow field for approaching

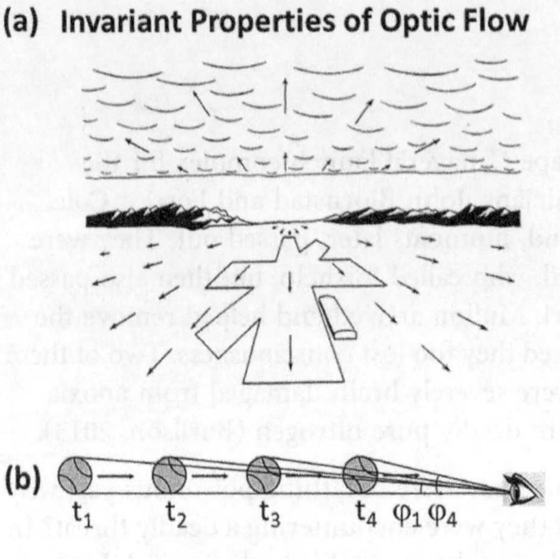

Figure 19.1 Invariant properties in a Gibsonian flow field. (a) When navigating through a terrain, the rate and direction of change in elements provides invariant information depicting characteristics of object shape and the direction of movement of the observer. Examples of invariant properties include that the horizon is at eye height, the point of zero flow specifies direction of heading, and the faster rates of flow are nearer to the observer. (b) Another example of an invariant property, when an object such as a ball approaches an observer (over times t_1 to t_4), the optical angle systematically increases (from φ_1 to φ_4). The fielder may not be aware that the image of the ball keeps getting bigger, but then it hits him.

scenery (Figure 19.1a). These properties also include such aspects as the systematic increase in optical angle (observed size) of an approaching rigid object (e.g., a ball), from the perspective of an active observer (e.g., a baseball fielder) (Figure 19.1b). Gibson (1979) referred to these universal properties as "invariants," because they do not change or vary under different viewing conditions. Approaching rigid objects, such as balls, always increase in optical angle anywhere they are encountered, and by any valid type of measuring device (Sugar & McBeath, 2001; Sugar, McBeath, Suluh, & Mundhra, 2006; Wang, McBeath, & Sugar, 2015b). Gibson emphasized the centrality of geometric principles like the regularity of flow fields when moving through an environment, and he noted how angles between elements in a flow field systematically change in ways that allow observers to glean object and terrain shapes and locations during movement.

Spatial Illusions due to Perceptual Biases Coupled to Universals

The Illusions of the Rising Fastball and the Stop-and-Drop Sinkerball

Many geometric invariants are based on angular properties defined from an egocentric perspective in a changing environment. For example, when a rigid object, such as a ball, approaches an observer at a constant velocity (as shown in Figure 19.1b), the rate at which the visual angle increases can inform the observer of the predicted time to arrival (Lee, 1998). This type of timing estimation is hard-wired at an autonomic level, and humans are remarkably good at sensing when collisions are destined to happen in tasks such as grasping objects and catching balls. Yet, timing errors start to occur when objects move as fast as or faster than humans are capable of reacting, as occurs in sporting activities like baseball. In addition, timing estimation errors also can lead to spatial distortions, such as in the cases of the illusions of the rising fastball and the stop-and-drop sinkerball (McBeath, 1990; Watts & Bahill, 2000). As shown in Figure 19.2a, when a ball player underestimates the speed of an approaching fastball (dark trajectory), he or she can experience an alternate illusory approaching trajectory that is somewhat lower and slower (light trajectory). Then when the ball eventually crosses the plate and the motion blur of its position is experienced veridically, it will appear to accelerate and hop up. This geometric explanation of the

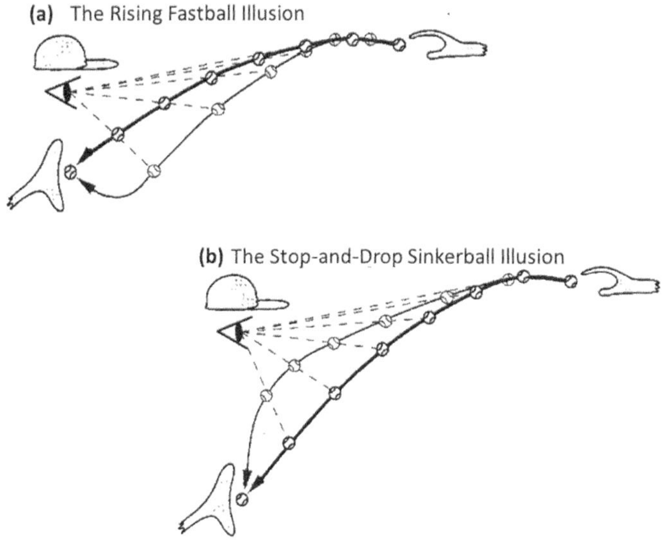

Figure 19.2 Illusion of the rising fastball and the stop-and-drop sinkerball. (a) When a player maintains line-of-sight on an approaching ball (dark trajectory), but underestimates the approach speed, the player can experience an alternate approach path (light trajectory) that is lower and initially slower. At the end, when the player veridically experiences the ball's time and location, it appears to accelerate and hop up (McBeath, 1990). (b) The reverse can also occur when a player overestimates the approach speed, resulting in an experienced trajectory that appears to approach more rapidly than it actually does and then dive downward at the end. Players describe this illusion as looking like the ball rolled off a table.

illusion produces about 1 inch of illusory rise for each mph of speed underestimation, typically resulting in a perceived upward hop of several inches, but sometimes ones as large as a foot (McBeath, 1990; Watts & Bahill, 2000). This explanation has been empirically verified with players as young as Little Leaguers (Bahill & Karnavas, 1993). Similarly, if the speed of the approaching ball is overestimated, as shown in Figure 19.2b, this can cause the ball to appear to slow down dramatically near the end of the trajectory and appear to "roll off the table." Physics research verifies that balls can notably curve in the direction of spin, but not nearly as much or as dramatically as is perceived due to these illusions (Watts & Bahill, 2000). In summary, misestimating the speed of a rapidly approaching object such as a baseball can lead to large, dramatic spatial illusions such as the hop of the rising fastball and the perceived rolling off the table of the stop-and-drop sinkerball.

Invariant Geometric Control Algorithms Like Maintenance of Angular Constancy during Interception

Another example of geometric invariants is use of the principle of optical angular constancy in order to maintain an interception course, such as in the case of running to catch a ball or other moving targets (McBeath, Shaffer, & Kaiser, 1995a). There are several different optical angles that are used by active observers in different kinds of settings. These include maintaining a constant lateral approach or bearing angle, maintaining constant lateral alignment with the target, maintaining constant rate of change in the tangent of the vertical optical angle (known as "optical acceleration cancelation" [OAC]), and maintaining a "linear optical trajectory" (LOT) with respect to the target

(Wang, McBeath, & Sugar, 2015a). Figure 19.3 illustrates the maintenance of a LOT. Here the fielder selects one of a family of LOT interception solutions in which the egocentric optical trajectory is maintained so that the image of the ball rises monotonically along a linear path at a constant optical projection angle, Ψ (McBeath, Shaffer, & Kaiser, 1995a, 1995b, 1996; Shaffer, McBeath, Roy, & Krauchunas, 2003; Sugar, McBeath, & Wang, 2006). When fielders do not select a path that maintains a constant angle Ψ, as shown in the inset, then they reliably do not converge on and successfully intercept the ball (Shaffer & McBeath; 2002; Wang, McBeath, & Sugar, 2015b). The same control strategy of a pursuer maintaining a linear optical trajectory (LOT) has been confirmed with several other species including dogs catching Frisbees (Shaffer, Krauchunas, Eddy, & McBeath, 2004). In many sporting situations, such as ball games, players use control strategies like maintaining a LOT to catch balls or otherwise produce collisions. But also in sports and many other real-world situations, people often intentionally move to eliminate angular constancy in order to avoid collisions.

Overreliance on Angular Constancy: Misjudging Apex of an Approaching Trajectory and Missing Pop-Ups

Geometric optical control strategies, such as LOT, couple perception-action behaviors to invariant natural regularities, such as angular constancies. One aspect of using such strategies is that their high level of success may delude observer-actors into thinking they have more accurate access to assessing object locations and trajectory destinations than they actually do. This can lead to certain types of misrepresentations that are experienced as spatial illusions. An example is the inability of fielders to correctly ascertain the time and

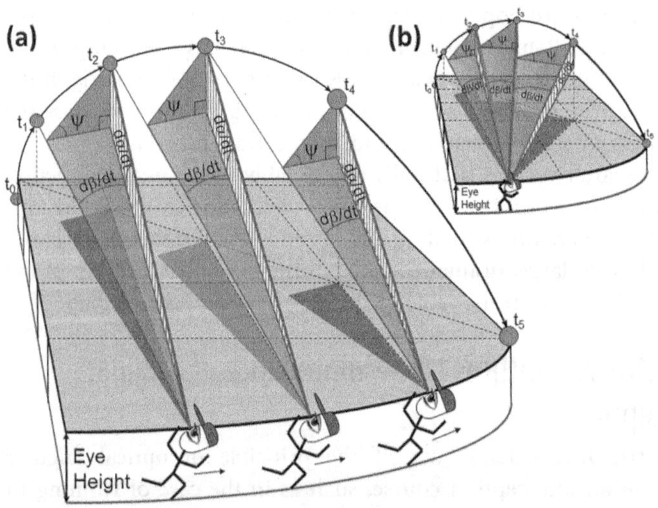

Figure 19.3 The linear optical trajectory (LOT) interception control strategy: an example of the invariant property of maintaining angular constancy. (a) The fielder runs along a path to keep the ball image monotonically rising along a constant projected optical angle, Ψ. (b) If the fielder remains stationary or selects a path that does not converge on the ball (as shown in the inset figure), Ψ will change, and the optical trajectory will curve toward the image of the ground. Continuously cancelling out optical curvature to keep the image of the ball rising along a straight optical line insures interception, because the ball image will never approach that of the ground. A black and white version of this figure will appear in some formats. For the colour version, please refer to the plate section.

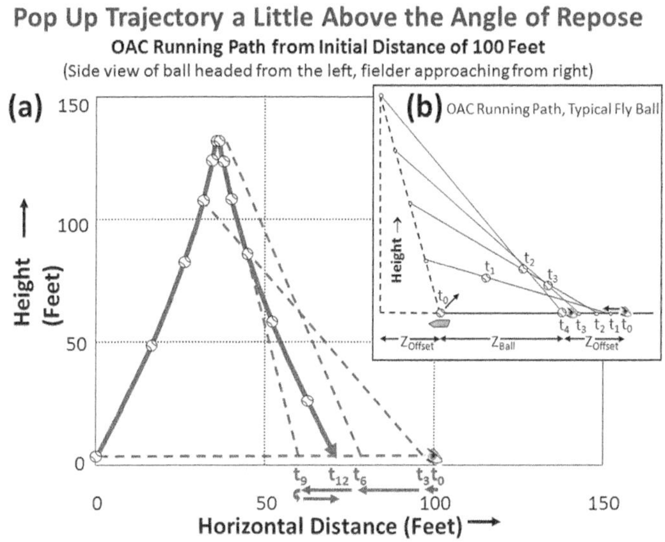

Figure 19.4 Misjudging the destination of a pop-up. The high degree of ball spin on pop-ups causes them to have odd trajectories, such as this one following an inverted-V, or occasionally even a loop-the-loop. As shown in the inset, usage of OAC and LOT angular control algorithms typically produce efficient paths to the desired destination (times t_0 – t_4). But in the current example, they lead fielders to run in too far and then backtrack at time t_9. Fielders can experience this as a spatial illusion in which the ball initially appears to head further in toward the plate than it actually does. (McBeath, Nathan, Bahill, & Baldwin, 2008)

location of the apex of the trajectory of a ball that is headed generally toward them (Shaffer & McBeath, 2005). When a ball trajectory is viewed from the side, observers accurately judge the physical apex, which occurs in sync with the optical apex (highest observed angle to the ball). But from the perspective of a fielder, the optical apex is progressively later as ball destinations get closer. In a study by Shaffer and McBeath (2005), we demonstrated that when a fly ball with a 4-second airtime heads directly to observers, their average time estimates for when the apex occurs are 1.5 seconds late. Here, observers typically reported that the apex occurred only fractions of a second before the ball actually landed just feet away from them. Observers were generally unaware that such a delayed apex requires a tremendously odd trajectory, and that they were experiencing a dramatic illusion. When a fly ball heads toward observers, they see an optical trajectory that continues to rise in the same manner as for fielders trying to position themselves at the interception point (as shown in Figure 19.3). The findings provide strong evidence that the observers are generally unaware that their spatial perception of ball location can be quite distorted.

A second ramification of people being reliant on optical control strategies for interception is that there are occasions when ball trajectories are so odd that these generally reliable strategies mislead them. In baseball, this commonly occurs with pop-ups, which almost universally have a huge amount of backspin due to their glancing upward direction relative to the prior ball and bat movement directions (McBeath, Nathan, Bahill, & Baldwin, 2008). As shown in Figure 19.4, this produces trajectories that can be shaped very much like an inverted V rather than the usual parabolic curve, and sometimes even ones in which the ball travels along a loop-the-loop. As a result, angular control strategies such as OAC and LOT can cause fielders to overrun the ball and then need to backtrack. As shown in Figure 19.4, this can be viewed as a type of spatial illusion or misestimation error by the fielder, as it appears the fielder initially pursues the ball like he or she thinks it will remain closer to home plate than it actually does. Some observing fans might assume that the ball must have been blown off course by the wind, but this type of over-pursuit can be entirely explained as an unusual case in which the angular constancy based control strategies do not work well and often push fielders to overcommit.

Summary of Universals (Invariants), and Illusions due to the Biases Coupled to Universals

In summary, the strongest types of natural regularities are ones that are essentially always true, referred to as universals or invariants. Examples outlined in this section include the ubiquity of breathable air and of light wave electromagnetic frequencies. This chapter reviewed how perceptual biases coupled to these universals are so strong that we are essentially unable to process and react to alternatives. The other types of universals outlined are geometric invariants. These include properties of flow fields like the increase in optical size of approaching objects, and geometric optical control strategies like maintenance of a linear optical trajectory (LOT) to intercept objects such as balls. Estimation errors in one's ability to couple to and utilize such invariant properties can lead to distorted perceptions like the illusion of the rising fastball, the misperception of the location of ball-trajectory apexes, and misjudgments in the destination of pop-flies. All of these errors can be construed as consequences of evolved, hard-wired, short-cut biases that normally help us efficiently and accurately perceive and act with respect to given invariants, but that occasionally fail to help us or even lead us astray in the rare, extreme cases when our assumptions do not hold.

In his later work, Roger Shepard (1994, 2001) used the term "universals" to also describe elegant geometric transformation principles. These include simple translations and rotations of kinematic geometry that can be used to specify paths of travel between differing positions of an object (McBeath & Shepard, 1989). Others cogently argued that perceptual biases for favored paths of transformation might be better understood in terms of real-world dynamics (Wilson & Bingham, 2001; see Hubbard, Chapter 8 in this volume). Real-world dynamics are based not just on geometry and simple Newtonian physics principles, but also include complex forces, particularly in cases of self-guided movements of biological organisms. For this reason, real-world motion paths are generally not fully specified by geometry and are part of a class of natural regularities that are somewhat less reliable than the fully invariant ones described so far. The next section introduces this next level of reliable but not fully invariant regularities.

Natural Regularities of Biological Objects and Scenes

Gravity and the Biological Natural Regularity of Symmetry

Another universal of our terrestrial environment is downward gravity (see Zago, Chapter 10 in this volume). The universality of downward gravity has led to a number of secondary biological natural regularities that are discussed in this section. One example of such a secondary regularity is biological symmetry, the ubiquity of symmetric plants and animals (McBeath, Schiano, & Tversky, 1997; Pizlo, Li, Sawada, & Steinman, 2014). Biological

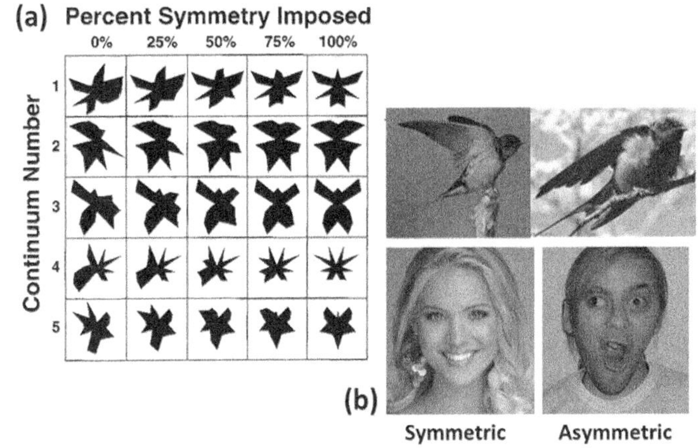

Figure 19.5 The natural regularity of object symmetry. (a) The five 18-sided random polygons in the left column were created and then either the left or right half was folded over to create the five symmetric figures in the right column. The figures in between were made by successively morphing to create a set of random shapes that varied in degree of outline symmetry. Participants were asked to indicate what the figures look like and which direction they faced. Over 95% of selections were symmetric objects such as animals or planes, so degree of outline symmetry could be used to specify their orientation. Virtually all of left-column figures were described as animals or objects at skewed or profile views, and virtually all of right-column figures were described as ones at straight-on views. (b) Symmetry is naturally selected because it is more stable and balanced in actions such as locomotion, and because of sexual selection, with symmetric individuals viewed as more attractive in a number of species including swallows and humans. A black and white version of this figure will appear in some formats. For the colour version, please refer to the plate section.

regularities like symmetry of life-forms are not always true, so they are not universals or invariants. Occasionally we see asymmetric species like crabs that have a much bigger dominant claw on one side, or cacti that have distinctly asymmetric arms, or disfigured or injured organisms that are missing limbs. Yet, to a high degree of reliability, most terrestrial life-forms have been symmetric, with stationary plants being generally circularly symmetric, and moving animals and insects being bilaterally symmetric. It is true that biological organisms typically exhibit some informative minor deviations from perfect symmetry, such as one side of the face revealing more emotion than the other (Sackeim & Gur, 1978), and observers are often sensitive to such information indicating health and history (Leyton, 1989, 1992). Yet, to a first approximation, the general symmetry of pairs of eyes, ears, arms, etc. holds true. Symmetry has been naturally selected both because it produces stronger, more stable shapes and because it is sexually selected, as shown in Figure 19.5 (Møller, 1992). The principle coupled bias regarding symmetry is that observers tend to assume that all shapes are generally symmetric objects, and we possess shape-finding algorithms that represent shapes with their directionality specified by orientation of symmetry (Pizlo et al., 2014). Figure 19.5a shows an array of shapes varying in degree of outline symmetry initially used to test this bias (McBeath, Schiano, & Tversky, 1997). When asked what the shapes look like and what direction they faced, more than 95% of the responses were of symmetric objects such as animals or human-made artifacts such as chairs or airplanes. When the outline shape was asymmetric, observers almost exclusively specified interpretations of symmetric objects viewed at a skewed or profiled angle, while when the outline shape was symmetric, they almost exclusively specified symmetric objects viewed straight on. The study confirmed that observers almost always assume the natural regularity of object symmetry, and are able to use that prior to help identify objects and glean their overall shape and orientation. This appears to be a bias with high payoff when viewing potential life-forms and human-made artifacts that generally are symmetric, while having little cost when viewing inanimate objects that are not necessarily symmetric, such as rocks or clouds. The assumption of convex 3D symmetry has proven to be a useful, fundamental building block for machine vision algorithms that determine 3D shape and recognize objects (Marr, 1982; Pizlo et al., 2014).

Symmetry verticality (the property that an object's natural axis of symmetry is vertical rather than some other non-vertical orientation) is a second, somewhat less reliable natural regularity that is also the result of downward gravity and the ubiquity of vertically symmetric life-forms in our environment. Symmetry verticality implies a natural vertical orientation such that horizontally symmetric objects are frequently interpreted as objects lying on their sides (Schiano, McBeath, & Chambers, 2008). Vertical symmetry appears to occur more frequently in nature because a vertical symmetric axis provides gravitational balance and stability. The coupled bias associated with the natural regularity of symmetry verticality is the tendency for observers to perceptually favor vertical symmetry over other orientations. This favoring is reflected in faster recognition of vertical symmetry compared to horizontal (the Goldmeier effect) (Corballis & Roldan, 1975). In general, observers tend to exhibit a bias to interpret vertically symmetric objects as being in their natural orientation, while horizontally symmetric ones are much more often interpreted as vertically symmetric objects lying down (Schiano, McBeath, & Chambers, 2008).

A number of spatial illusions are associated with our bias to interpret objects as symmetric, with most typically following the pattern of observers accentuating or adding more symmetry to their interpretations and memories of object shapes (Tversky & Schiano, 1989; see Tversky, Chapter 16 in this volume). This can be observed when flying by a cloud that initially looks like an animal, but that later is seen to not possess the symmetry that was assumed from the original vantage, clarifying that it is not at all shaped like the original interpretation. Another example is the tendency for observers to remember road intersection and geographical orientations as more aligned than they actually are. Examples of this include the tendency to assume that South America is more directly lined up south of North America than it actually is, or that Miami is more directly south of New York City than it actually is (Tversky & Schiano, 1989). In short, observers tend to have a spatial illusion to represent and remember things as being more symmetrical than actually observed.

The Natural Regularity of Top Salience

A third biologically selected natural regularity related to gravity is top salience. Top salience is the tendency for life-forms and objects with which we interact to have top portions that are more informative or likely to be interacted with than their bottom portions. This can be observed with most higher life-forms that possess their primary sensory apparatus on their heads, which are typically located at or near their top. Similarly, objects that we manipulate and interact with in our environment tend to be below our heads, where we possess a better vantage and ability to manipulate their tops. The bias that is coupled to this natural regularity is the tendency to attend more to tops of objects than to their bottoms. This bias is regularly observed in eye-tracking studies that confirm observers favor viewing heads and faces over other lower body parts, which can be attributed to the higher salience of heads and faces. To fully test for a pure top salience bias, Chambers, McBeath, Schiano, and Metz (1999) created random asymmetric and symmetric shapes, as shown in Figure 19.6,

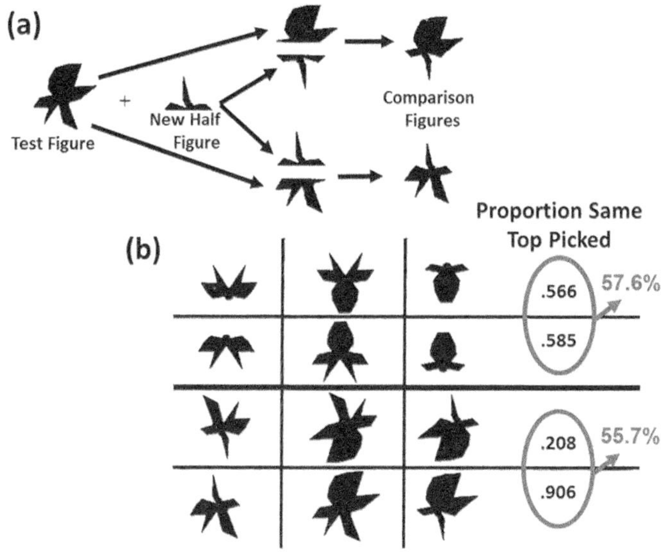

Figure 19.6 The natural regularity of top salience. (a) Shapes were created by taking random test polygons, chopping them into top and bottom halves, and adding a third random half to each original half. (b) Observers were asked to indicate which shape (in the left or right column) was more similar to the initial one (in the center) (i.e., the shape with the same top-half or one with the same bottom-half). On average, observers selected shapes with the same top-half 57% of the time.

and had observers compare them to sibling shapes that contained either the same top or same bottom. The findings confirmed that even for these random shapes, with no heads or any other type of systematic distributional information, in overall judgments of shape similarity, observers still exhibited a reliable tendency to attend to and more heavily weight similarities in tops than in bottoms. The top salience bias may also be described as an illusion, given that observers interpret these object similarities differently when the objects are simply spun over to switch top and bottom. A systematic top-processing advantage also manifests in reaction times when linguistically judging if an object is above versus below another object (Clark & Chase, 1972), further confirming a generic attentional bias to favor tops.

The Natural Regularity of Lighting from Above

Another well-known example of a natural regularity related to gravitational orientation is the tendency for light to shine from above. Historically, this can be ascribed to the principal daylight source being our sun, and principal nightlight source being from stars and sunlight reflecting off the moon, all of which arrive from above. The coupled bias associated with this natural regularity is the assumption of lighting from above. This bias guides determination of shape-from-shading when observing ambiguous convex versus concave surfaces like bumps and divots (Ramachandran, 1988). The bias is manifest in the famous concave or hollow face illusion, which is optimized by shining a light from below into a concave face or the inside of a mask (which makes the shading identical to a convex face lit from above) (Gregory, 1970). The combination of several consistent natural regularities, first the favoring convex interpretations of objects, especially familiar ones like faces, and second the assumption that lighting normally comes from above, together create a very compelling illusion that the concave face is actually convex and lit from above. This illusion is particularly compelling because there are several natural regularities used congruently to strengthen the overall likelihood of observing the expected convex face, lit from above. Though biologically selected natural regularities like symmetry, symmetry verticality, top salience, and lighting from above by themselves are not as reliable as the previously described universal invariants, when they are combined in a consistent manner, they increase the joint probability of a likely 3D interpretation to a very compelling level.

Moderately Reliable Natural Regularities of Biological Motion and Complex Biological Features

Axis-Aligned vs. Off-Axis Motion and Judgments of Motion Direction of American Footballs

The natural regularity of biological symmetry appears to be a consequence of gravity, but there are also other, somewhat less reliable regularities that appear to be a consequence of biological symmetry. Examples include patterns of biological motion like the tendency for mobile bilaterally symmetric organisms to favor moving in alignment with their axes. The natural regularity of axis-aligned motion leads observers to exhibit a corresponding perceptual bias to assume that objects move in directions aligned with their axes. Several examples of axis-aligned motion bias are shown in Figure 19.7. First, research confirms an axis-aligned motion bias that produces greater observer accuracy when estimating direction of motion for symmetric shapes moving in alignment with their axes compared to moving off axis. Second, the axis-aligned motion bias produces less variance for symmetric shapes moving in alignment with their axes compared to asymmetric shapes (Dolgov, Birchfield, McBeath, Thornburg, & Todd, 2009a, 2009b). This bias has been confirmed both with motion direction judgments on a computer screen (Dolgov, Birchfield, McBeath, Thornburg, & Todd, 2009a; Morikawa, 1999) and with judgments of floor-projected shapes in a large-scale virtual environment (Dolgov, Birchfield, McBeath, Thornburg, & Todd, 2009b). The directionality of representational momentum has also been found to favor alignment with object axes (Munger, Solberg, Horrocks, & Preston, 1999; and see Hubbard, Chapter 8 in this volume).

Third, the axis-aligned motion bias has been confirmed with 3D motion in a real-world setting for judgments of the destination of passed American footballs (Dolgov, McBeath, & Sugar, 2009). American football passes are an unusual example of a systematic motion pattern that violates the natural regularity of axis-aligned motion. Because of the biomechanics for throwing football passes, the wrist snap that produces ball spin needs to be unaligned with the arm-movement direction of the overall throw. This results in the front of the ball being tilted to the side of the handedness of the thrower. Thus for football passes, the axis-aligned motion bias systematically produces a robust spatial illusion, with observers incorrectly experiencing that thrown footballs travel along their elongated axis. This manifests by observers having an illusory experience of alignment that leads to a systematic misestimation of motion directionality toward the lateral tilt of the football. In this study, shown in Figure 19.7, 100% of football passes were found to laterally tilt toward the handedness side of the thrower, and observers watching thrown

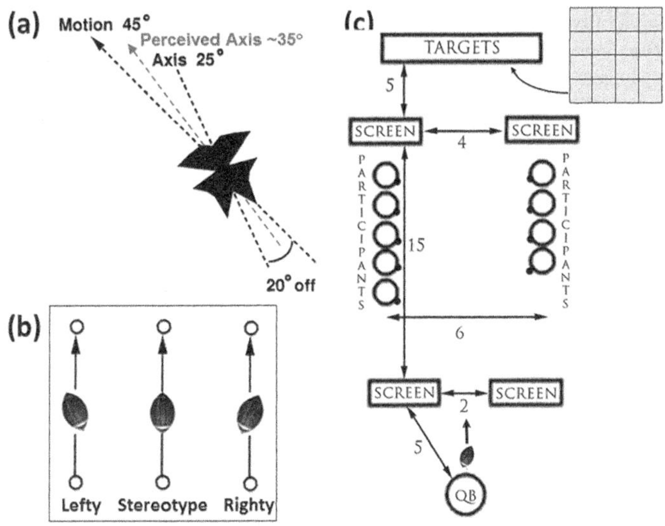

Figure 19.7 The natural regularity of axis-aligned motion. (a) Most mobile life-forms are bilaterally symmetric and favor moving in alignment with their axis of symmetry, and observers have a corresponding bias to assume motion occurs in that direction. When a bilaterally symmetric object moves off-axis, observers typically experience it to move in a direction intermediate to its actual direction of motion and its symmetry orientation. (b) American football passes are an unusual exception to the regularity of axis-aligned motion. Because of the biomechanics of throwing, passes by left-handers generally tilt to the left, and ones thrown by right-handers tilt to the right, while most observers tend to experience stereotypical passes as moving in alignment with their axis of symmetry. (c) Top view of setup in a study to estimate football pass destinations with left- and right-handed varsity quarterback throwers (Dolgov, McBeath, & Sugar, 2009). Observers see a series of football passes from the side and estimate where the ball hits the far wall that contains a large projected array of 4-by-4 boxes (shown in gray). Results revealed a destination bias of several meters in the direction of quarterback handedness (confirming observers experience axis-aligned motion bias for thrown footballs in their direction of tilt).

footballs fly by them systematically estimated that the destinations were a meter or more toward the side in which the footballs were tilted.

Forward-Facing Motion

Most mobile organisms possess orientation-defining features in addition to symmetry, which specify a forward-face to the object. In the case of most animals, the side of the head containing the eyes and mouth is typically interpreted to be the front or forward-face, and there is a natural regularity that animals favor moving in the direction they face. The general salience of faces and the tendency for forward-facing motion lead observers to have a generic forward-facing motion bias. Findings confirm that observers favor fixating faces on the front of heads, and have a bias to experience motion in that direction (McBeath, Morikawa, & Kaiser, 1992; McBeath & Morikawa, 1997; see Hubbard, Chapters 8 and 15 in this volume). Motion directionality biases were verified using bi-stable, two-frame apparent motion. A row of identical shapes was replaced by a row of the same shapes, but shifted so that the replacement ones were located in the spaces between the original shapes, which produces experienced motion in either leftward or rightward direction. Forward-facing motion bias was demonstrated to occur with forward-facing directionality that is defined geometrically (triangles and arrows) and biologically (faces and stickmen features), but not calligraphically (direction of letters like R). The strength of the motion bias increases when shapes have stronger-rated directionality, such as triangles that are successively more pointed.

The forward-facing motion bias can occur entirely as a function of top-down control, as was confirmed with a row of ambiguous duck-rabbits (Jastrow, 1900). When ambiguous duck-rabbits are interpreted to be ducks, observers favor motion in the duck-facing direction, but when the identical figures are interpreted to be rabbits, observers favor the rabbit-facing direction. Since the identical row of shapes produces both opposing results, the bias for forward-facing motion appears to be based entirely on top-down factors such as priming which side to interpret as the face, and the tendency to fixate on that side. Forward-facing motion bias can produce an illusion in real-world cases in which an animal or object quickly moves in a direction other than forward. Like axis-aligned motion bias, it tends to distort perceived direction of motion toward the direction the object faces.

Curved Biological Inertia

One of the best-known examples of naïve physics is judging the direction that a ball will roll after it emerges from a curved or spiral tube (McCloskey, Caramazza, & Green, 1980). Newtonian principles of physics specify that physical inertia is straight, so once the ball leaves the tube and there are no more forces causing curvature, it should maintain a straight trajectory. In this case, it might seem like a natural regularity of straight-inertia should help prevent judgment errors, yet typically more than half the population estimates that the ball will keep curving after it emerges from the curved tube. McBeath, Brimhall, Miller, and Holloway (2010) suggested that this is because there is a natural regularity of biological inertia in which people or other animals running out of curved mazes continue to curve for some time period after they emerge. Standard physical inertia stipulates that, absent other forces, objects maintain their prior state, continuing in a linear direction. In this work, we defined biological inertia as maintenance of the prior biologically motivated state that produces continuation of curved motion. This study measured the size of the curved inertia bias for estimates of balls emerging from a spiral maze, and compared this to the average amount of curvature exhibited by preschoolers and college students running and emerging from the same shaped spiral maze. The findings confirmed that the average estimated curvature for balls and actual running curvature for preschoolers and college students were all essentially identical, at approximately 11° offset from a straight trajectory. The findings support that the curved inertia bias is not necessarily a costly physics error, but rather a beneficial assumption regarding biological behavior consistent with the more impactful natural regularity of curved biological inertia. A standard argument regarding naïve physics has been that educators can teach people to overcome biases like the one for curved inertia. In a survey of physics teachers and educators, we found that they overwhelmingly correctly indicated that balls emerging from a spiral tube would go straight, but also incorrectly indicated that people emerging from a spiral maze would go straight (McBeath & Dye, 2015). So teaching people to overcome a bias due to a natural regularity such as biological inertia may have the benefit of helping them better understand some domains like Newtonian physics. But it appears to also have the cost of leading them to err more regarding generally useful intuitions for biological inertia and the tendency for biological agents to maintain curvature. Ironically, while physics experts learn to overcome the curved inertia bias in the context of physics problems, in so doing they acquire their own bias to no longer correctly anticipate curved biological inertia for moving organisms. In any case, it appears that the expectation of curved inertia may be best framed not as a cognitive illusion of naïve physics, but rather as a generally advantageous bias to expect normal biological behavior (McBeath & Shepard, 1989).

A Greater Likelihood of Looming vs. Receding Motion

In an active environment, observers can frequently observe objects moving around them. Of the infinite possible movement directions, there is a subset that correspond to an object directly approaching (looming toward) an observer, and a much larger set of movement directions in which the object heads off in other, avoidant directions. This results in a natural regularity that if the direction of movement of an object is lined up in depth with an observer, it is more likely that it is approaching on a collision course than that it is receding with the arbitrary constraint that its direction of motion happens to remain aligned with the observer. The matching coupled bias to this regularity is a looming bias, observers are more likely to experience depth-aligned motion to be approaching than receding (Glenberg et al., 2010; Lewis & McBeath, 2004). This was verified using a bilateral apparent motion test similar to the one described earlier for determining forward-facing motion bias, except using stimuli that are aligned in the depth dimension. Here, observers were found to exhibit a bias to experience looming motion more robustly than receding motion, consistent with the natural regularity. The looming bias can also be increased with additional looming-consistent cues such as having elements point toward the viewer and/or having sounds that get louder or quieter at the point in time when the motion occurs (see Neuhoff, Chapter 12 in this volume).

The Natural Regularity of Neoteny, and Use of Head Shape to Judge Age

Other biological natural regularities of moderate strength include ones in which a number of complex features reliably occur together. One example of this is the natural correlation of head shape and animal age that holds for many mammal species, including humans. As noted by Pittenger and Shaw (1975), there is a general pattern of neoteny among people and other mammals in which youngsters tend to have more extended foreheads and more recessed noses and ears, what they referred to as a difference in cardioidal stress. This pattern is shown in Figure 19.8, where flattening the face and extending the forehead makes a variety of animal species appear cuter and more like babies. Further research confirmed that this same cardioidal stress transformation can be applied to inanimate objects such as cars and cartoons to produce an impression of maturity (Pittenger, Shaw, & Mark, 1979). Volkswagen Beetles with their small rounded hoods have a baby car appearance, while Cadillacs with big long hoods tend to appear more grown up. This pattern was noted by the famous evolutionary biologist Steven Jay Gould (1979) in a classic article about the evolution of Mickey Mouse. When Walt Disney first created him, Mickey was an impish troublemaker who looked like a young adult. As decades passed, Disney gradually changed Mickey, widening him, increasing his relative eye size, and extending out his forehead to make him appear more childlike and likable. Gould noted that humans possess more of these childlike facial characteristics than do the other great ape primates, such as gorillas, and argued that our appearance and longer development times are evidence of neoteny, an extension of youthfulness and plasticity that allows our species to adapt more after birth.

An illusion occurs when we observe animals, such as koala bears, that possess an abundance of these childlike appearance qualities that make them look cute, when they actually are often rather nasty in their behavior. Another example of multi-feature regularities is the characteristic facial expressions that indicate basic emotions (Eckman et al., 1987). These can produce a kind of spatial distortion in that observer sensitivity to facial expressions can lead to exaggerations and caricatures of how different people's facial features are remembered.

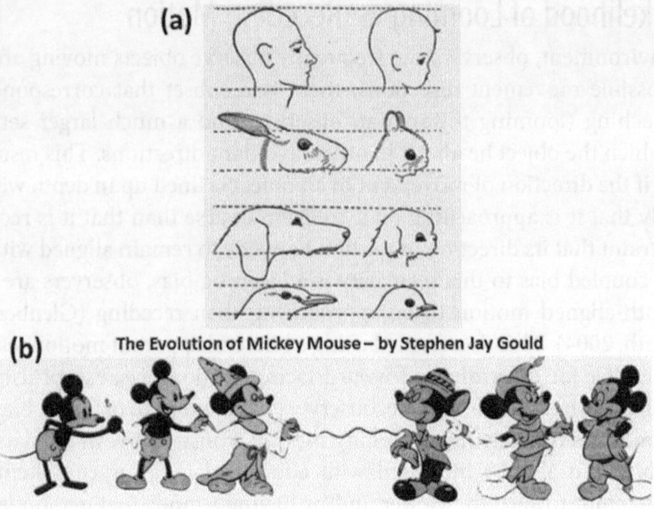

Figure 19.8 The natural regularity of neoteny in head shape as a cue of age. (a) Humans and other animals typically exhibit a characteristic change in head shape as they age. Babies have more extended foreheads and flattened facial features. Observers process these features in judgments of age, and are occasionally fooled by individuals or species in which the regularity does not hold. (b) The evolution of Mickey Mouse, as noted by Stephen Jay Gould (1979), is an example where Walt Disney changed Mickey's appearance and personality to make him look more youthful and likeable. As can be seen in the successive drawings of Mickey over the years, his snout got pushed in, forehead extended, eyes became bigger, and body widened to better fit characteristics associated with more youth and innocence.

In summary, there are a number of natural regularities that occur with moderate reliability for prediction. Examples described here include biological motion patterns such as axis-aligned motion, forward-facing motion, looming motion, curved biological inertia, and complex environmental correlations such as the variety of head-shape features that somewhat reliably change in a systematic manner as individuals age. In turn, observers possess matched, moderate biases to assume that these regularities are likely to hold, which occasionally result in systematic distortions. Thus, we have illusions in which experienced or expected direction of motion is systematically pushed toward a moving object's axis, toward its forward-face, in continuation of path curvature, and in the looming direction. Presumably, the frequency of exceptions when these moderate regularities do not hold and the cost-benefits of different types of errors lead observers to reach some balance point regarding the strength of each corresponding bias.

Weak, Culturally or Behaviorally Learned Natural Regularities

Learned Leftward Attention and Motion Direction Biases in Left-to-Right Readers

The final category of natural regularities includes patterns that are reliable within a culture and that can be deeply socially ingrained, but because they differ across cultures, any bias to expect these regularities is clearly learned. One example of this type of weak regularity is the tendency for new information to be located to the left visual side of an observer in left-to-right reading cultures. In English, the beginning of each title and new paragraph is located on the left, and readers read left-to-right, so their attention is trained to initially jump to the left when exploring anything that may include written language. The accompanying coupled bias is that native English speakers generally exhibit an attentional bias toward the left, while native speakers of languages that read right-to-left tend to exhibit an opposite attentional bias toward the right (see Hubbard, Chapter 15 in this volume). These lateral attentional biases are supported by research that shows native English speakers are more apt to see and remember items flashed on the left side of the monitor. This lateral attentional bias is often utilized in web page designs which have layout standards that are more likely to place important menu guidance links to the left than to the right, and can result in spatial illusions in assessments of lateral area.

Another weak, reading-related natural regularity leads to a bias to favor rapid leftward motion over rightward motion. This was verified by Morikawa and McBeath (1992) in a study that examined directional preferences for bi-stable lateral apparent motion using a methodology similar to that described earlier for the forward-facing motion bias with duck-rabbit figures. In this study, a significant leftward motion bias was found for both right- and left-handed Americans and for both American right-side and Japanese left-side drivers. Finally, the leftward motion bias was found for native English-speaking, left-to-right readers, but not for native Arabic-speaking, right-to-left readers. Furthermore, among the Arabic-speaking sample, there was a significant inverse correlation between amount of exposure individuals had to left-to-right reading languages such as English and strength of their rightward motion bias. The logic of this bias is that during left-to-right reading, most of the time gaze moves rightward, resulting in a preponderance of visual flow to the left. This exposure imbalance appears to train the perceptual system to favor leftward motion. A similar exposure imbalance also holds for looming motion, with forward self-movement being much more common than backward. The lateral motion bias can also lead observers to experience a spatial illusion of lateral lightning traveling from right to left, even though lightening actually flashes on too fast for humans to accurately ascertain the true direction of movement.

Cultural Turning and Walking-Side Tendencies and Learned Biases to Follow These Navigational Norms

Another clearly learned, weak natural regularity is the tendency for Americans to favor turning and locomoting to the right. Scharine and McBeath (2002) demonstrated that American pedestrians and right-handers favor turning to the right, while British and left-handers are relatively more likely to favor turning to the left. This study also verified that other potential controlling variables like eye-dominance and reading-direction experience do not significantly influence turning and walking-side preferences. A further study by Dye, Crawford, and McBeath (2014) confirmed that the American turning bias extends to a preference for walking on the right side of objects or approaching pedestrians, and that this bias is not present in preschool- and primary-school-age children, and so needs to be learned. These findings, along with a pilot study done with American high school students, confirm that only by high school, the age at which most learn to drive, do American students begin to exhibit a significant right-side walking bias that extends into adulthood. This type of bias can lead to spatial illusions regarding perception of spatial layout and choice of the shortest navigational path (Stone & McBeath, 2010).

The commonality for environmental planners to build roads heading in cardinal directions, north–south or east–west, is another weak regularity that also likely contributes to navigational biases and related spatial distortions. People acquire a bias to expect such cardinality in road orientation, and this could contribute to the spatial illusion in which intersections are typically experienced and recalled to be close to 90° right angles, over a wide range of actual intersection angles (Tversky & Schiano, 1989). This bias may assist in reducing cognitive load by not requiring a precise analog representation of intersection angles when a generic 90° representation is perfectly functional most of the time.

These types of cultural norms, such as walking-side preferences and cardinality of direction, can be passed on for many generations in a meme-like fashion, but may not emerge in individuals without years of training. Yet, some types of spatial learning, such as priming, can occur after mere minutes of exposure. Primed expectations can mislead individuals in judgments of shortest path by temporarily redirecting attention toward suboptimal solutions (Stone & McBeath, 2010). Thus both long-term and short-term habits impact everyday activities like choice of pedestrian or driving route, and both can lead to spatial distortions.

In summary, there is a class of weak natural regularities that are culturally based, such as reading direction, lateral navigational customs for driving and walking, and a tendency for roughly orthogonal cardinal road orientations. These are clearly learned and may lead to route choice biases and spatial illusions such as mis-assessments of navigational path lengths or intersection angles (see Tversky, Chapter 16 in this volume). The leftward-oriented attention and motion biases of American left-to-right readers is also consistent with left sides being the preferred real estate on displays, and the superior locations to place more important information such as computer application menus and links on web pages. The same holds for the top salience bias, with the tops of displays being superior at drawing attention, along with the accompanying potential spatial distortions related to vertical position and uneven attention. Attention also appears to be one of many factors that likely contributes to the moon illusion, as well as other related illusions in which elevated street signs, traffic lights, basketball hoops, and street lamps appear smaller than when viewed close up and lower down, nearer to the horizon in elevation. Observers normally focus attention on things near the ground plane, so things outside of our focus of attention may be minimized, especially if we almost never approach them closely enough for them to loom large. The main points are that weak natural regularities, such as driving- and walking-side patterns, and orthogonal cardinal intersection angles all end up having correspondingly weak coupled biases, like the tendency for American pedestrians to head to the right side of stationary obstacles and other moving pedestrians.

There are also natural regularities in such areas as acoustics, odor patterns, and balance physics that produce coupled biases in the nonvisual senses of audition, gustation (smell), and vestibular sensing. Many of these appear to focus on specialties of each sense, such as identifying acoustic or smell sources, and maintaining body balance. Some also contribute to an observer's multisensory representation of space, and can produce or contribute to spatial illusions, but these will not be covered in this chapter. While other senses also have biases coupled to natural regularities that produce perceptual and cognitive spatial illusions, vision typically is our dominant spatial sense. Thus, the major concept of coupled biases has been described here in detail with visual examples.

General Discussion and Conclusions

This chapter takes Gibson's (1979) ideas of perception-action properties that are directly coupled to universal invariants and expands them to include a larger set of perceptual and cognitive biases that are coupled to a wider array of increasingly weaker natural regularities. These expanded regularities were divided into categories of reliability, varying from the universals and invariants of Gibson, Shepard, and others (which essentially always hold), down to more moderate regularities (which are quite likely to hold), and further down to weak, only somewhat likely environmental patterns (which merely tend to hold). At each level of regularity there are coupled biases, which help observers more accurately anticipate and predict environmental interactions, but as the regularities get weaker and less reliable, the respective coupled biases also get weaker and more plastic. The multiple levels of matching reliability between regularities and biases in effect produce Bayesian priors that allow observer-actors to more optimally weight different sources of information to guide their predictions and behavior.

Because individual biases evolved to match the regularities in our environment, they typically help observer-actors to maximize predictive success over a wide range of normal situations. Nevertheless, there are some abnormal or extreme situations in which the biases can systematically mislead observers in ways that distort how they experience and interact with their environment. Some examples of such distortions are experienced as spatial illusions. Thus, many spatial illusions can be characterized as minor misfit errors in a complex multidimensional curve-fitting algorithm, with successively less weight given to less reliable predictor dimensions.

A final facet of having biases selected to match regularities of differing levels of reliability is that the mechanisms of natural selection also are likely to operate at different levels, each with appropriate amounts of plasticity. Universals and invariants are so reliable that it makes sense for us to have very strong corresponding biases to assume such things as breathable air, light frequency energy, and geometric flow fields. These extremely robust biases appear to be hard-wired in at a deep biological level that is guided by traditional Darwinian evolutionary mechanisms, which occur across many species. More moderate biases that correspond to moderate regularities like biological symmetry and top salience, require somewhat more plasticity so that individuals can overcome these assumptions when they do not hold. These biases seem to be predispositions that also eventually become hard-wired, but with a more developmental trajectory, one that likely requires appropriate sensory input during infant development. Finally, the weakest regularities described include ones that are merely cultural conventions such as reading direction and driving directionality, which can and do differ between different cultures that a single individual may encounter. The coupled biases that direct lateral attention (due to reading habits) or the pedestrian walking-side and turning biases (due to cultural norms) are clearly learned and to varying degrees can be unlearned. These regularities are transmitted culturally, so the natural selection mechanisms to produce the coupled biases must be learned and plastic (McBeath & Sugar, 2005). In short, our environment is composed of a continuum of natural regularities of varying reliability, and we evolved coupled perception-action biases guided by different levels of natural selection mechanisms with varying degrees of plasticity that match the reliability of the regularities.

In conclusion, this chapter described how our perception and cognition is guided by biases that match natural regularities spanning a wide range of reliability. These biases usually help observer-actors rapidly and accurately perceive and interact with their environments with only occasional errors, including spatial illusions. Rather than simply considering spatial illusions as mere errors, a glass-half-empty phenomenon, they can be considered occasional under- or overestimations in a highly tuned perception-action system, a glass half full. Such illusions might be more accurately characterized as optimized compromises rather than misjudgments, clever shortcuts that efficiently reduce cognitive load and time. These errors can be construed as consequences of naturally selected short-cut biases that normally help us efficiently and accurately perceive and act with respect to natural regularities in our environment, but that occasionally fail to help us or even lead us astray in those rare cases when our assumptions do not hold.

Chapter 20

Two "Inhibitions of Return" Bias Orienting Differently

Raymond M. Klein and Ralph S. Redden

Discovery of an Inhibitory Aftereffect Following an Uninformative Peripheral Cue

An inhibitory aftereffect of orienting was described thoughtfully and explored thoroughly by Posner and Cohen (1984). Cited more than 3,000 times,[1] the "components of visual orienting" spawned a large, exciting, and multidisciplinary effort to measure the aftereffect's spatial and temporal properties, determine its causes and effects, understand its functional significance, uncover its neural underpinnings, and explore the effects of psychoactive agents, neurological conditions, and various experimental manipulations on it.

In a prototypical experiment the participant fixates a central stimulus amid two or more equi-eccentric locations, marked by unfilled boxes. Then, the following sequence of events unfolds: An uninformative stimulus, often called a cue, is presented by brightening one of the peripheral boxes; at various times after the cue's appearance, a target is presented randomly in one of the peripheral boxes. The participant is instructed to make a simple manual response as soon at the target is detected while remaining fixated on the central stimulus. Eye position is monitored by some means, and trials on which fixation is not maintained are excluded from analysis. In the original experiments one of two means was used to ensure that attention – assumed to have been captured by the spatially uninformative cue – would be returned to fixation: either targets were more likely at fixation than at peripheral locations or the central location was cued (i.e., there was a "cue-back") prior to the appearance of the target. In the terminology first put forward by Posner (1980), these methods encourage withdrawal of attention from the cued location. When targets are probable at fixation, we assume that participants voluntarily disengage attention from the cued location and return it to fixation, a mode of control we refer to as endogenous. When the cue-back to fixation is used, we assume that attention is returned to fixation reflexively, a mode of control we refer to as exogenous.

The pattern of results that is typically obtained is biphasic.[2] When the interval between the cue's and target's onsets (the cue-target onset asynchrony, or CTOA) is brief, reaction time (RT) to targets at the cued location is faster than to targets at the uncued location. This facilitation of cued targets is presumed to be due to the capture of attention by the cue. When the CTOA is longer, assuming attention is removed from the cued location, RT to

[1] Google Scholar, April 2016.

[2] But not always. For example, failure to disengage attention from the cue may allow attention-based facilitation to overshadow the inhibitory aftereffect. Or, in the presence of an especially rapid disengagement, the early facilitation might not be seen. And, in some cases, early sensory adaptation (what some call "onset detection cost") may overshadow attention-based facilitation.

targets at the cued location is slower than to targets at the uncued location, which reflects a presumed inhibitory aftereffect generated by the cue.

Posner and Cohen's seminal paper described many experiments that allowed quite a few inferences to be made about this inhibitory aftereffect, most of which have been substantiated in the 30 years of follow-up research: it could be generated by increasing or decreasing the luminance of a peripheral box; it is not due to guessing, attentional momentum, masking, or inhibiting a response to the cue; it is coded, not in retinotopic but in environmental coordinates; it is independent of and can coexist with facilitation; it is relatively long lasting; and it is not generated following a spatially informative, centrally presented arrow (endogenous orienting).

Posner and Cohen (1984) identified the cause of inhibition as any peripheral stimulation rather than the exogenous orienting normally generated by peripheral cues. We believe that this may be the only incorrect inference that they drew.[3] They identified the effect of inhibition as one that reduces "the efficiency of target detection." And they proposed that the function of the inhibition was to "maximize sampling of the visual environment."

Inhibition of Return

In the next year, Posner, Rafal, Choate, and Vaughan (1985) explored an inhibitory aftereffect, which they called "inhibition of return" (IOR). In Experiment 1, they tested 4 patient groups using the model task pioneered by Posner and Cohen. Patients with progressive supranuclear palsy, which includes damage to the superior colliculus, a brain region that plays a critical role in exogenously generated saccades, failed to show an inhibitory aftereffect in the direction in which their saccadic behavior was compromised (vertically). In two subsequent experiments, Posner and colleagues made important discoveries about the causes and effects of IOR, discoveries that in retrospect diverged strikingly from what was described by Posner and Cohen.

In Experiment 2, IOR was generated using a version of the model task described earlier, but its effect was measured using a temporal order judgment task with two different response methods. On every trial, a target appeared at each of two locations with a range of short temporal intervals between their onsets. In one condition, participants made a temporal order judgment with a button press signaling on which side the first target appeared – the predicted inhibition did not affect temporal order judgment (a finding later replicated by Gibson & Egeth, 1994). In the other condition, participants were asked instead to make an eye movement in the most comfortable direction after the targets had appeared; here, there was a small bias against making eye movements in the cued direction: "The bias in eye movements is less than 10% at critical intervals, suggesting that it is not the major

[3] Two findings led to this inference: (a) Following a double cue (one on each side of fixation) that would not generate orienting, they found a delay (compared to uncued target targets from single-cue trails), and (b) there was no inhibition following a shift of attention generated endogenously. The first finding seems to show that you can get the inhibition without a shift of attention, while the second shows that the inhibition does not necessarily follow a shift of attention. We now know that inhibition is not generated by stimulation, because a balanced array of cues around fixation does not generate inhibition at cued relative to uncued locations (Klein, Christie, & Morris, 2005). We also know, or at least we believe, that exogenous and endogenous attention shifts are different in many ways (e.g., see Klein, 2009 for a review).

determiner of where the eyes will go, but rather a bias built on top of other factors which affect the eye movement systems" (Posner et al., 1985, p. 223).

In Experiment 3, Posner and colleagues assiduously avoided the asymmetric stimulation of the two peripheral locations, which can generate orienting and oculomotor activity reflexively. Two small digits (too small to resolve in peripheral vision) were presented simultaneously, one in each peripheral box. Thus, unlike the more traditional single-cue experiments, the luminance energy at both locations was equivalent. Shortly afterward, an arrow at fixation called for the participants to move their eyes to the indicated box and report the digit. A cue back to fixation called for gaze and attention to return to the central fixation stimulus. One second later, a peripheral target calling for a manual detection response was presented at the previously fixated or un-fixated peripheral location. Manual detection response times were slower for targets presented at the previously fixated location. This inhibitory aftereffect was inferred to have been caused by the eye movement, because there had been no asymmetric stimulation of the periphery. Reinforcing the novelty-seeking attribution proposed by Posner and Cohen, Posner and colleagues hypothesized that IOR biases attention during the search of complex scenes, a proposal that has since been robustly confirmed (for a review, see Wang & Klein, 2010).

Two Different Inhibitory Biases

Over the past 15 years, findings have emerged to support the implication, which goes back to these two seminal studies, that there are two different forms of IOR. We believe that both forms are caused by activity in or reaching the oculomotor system; both forms bias orienting, and hence, both forms can perform the novelty-seeking function attributed to the inhibition(s) in Posner's seminal papers. Despite these shared properties, the two forms bias orienting in different ways, by operating at different stages of processing between stimulation and response. As we will show, one form does so by biasing perception, whereas the other form does so by biasing action. We will refer to the inhibition that Posner and Cohen described as an "input" form and the form of inhibition that Posner and colleagues described as an "output" form. The possible loci of operation of these two forms of IOR are illustrated by explicitly placing them into a theoretical framework that clearly distinguishes input from output stages of processing (for example, see Figure 20.1). The input form of IOR operates relatively early in the processing of stimuli, perhaps by reducing the salience of stimuli near the inhibited object or location. The output form operates relatively late in processing as a bias against spatial responses (either manual or oculomotor) that may include covert orienting.

Diagnostics for Distinguishing between the Two Forms of Inhibition

In the remainder of this chapter we will use two diagnostics to show that the input effect is generated when the oculomotor system responsible for reflexive saccades is suppressed, whereas the output effect is generated when this reflexive oculomotor system is *not* suppressed. By employing this two-form conceptualization of IOR we are also able to reconcile some divergent findings from the literature, including some from within our own work.

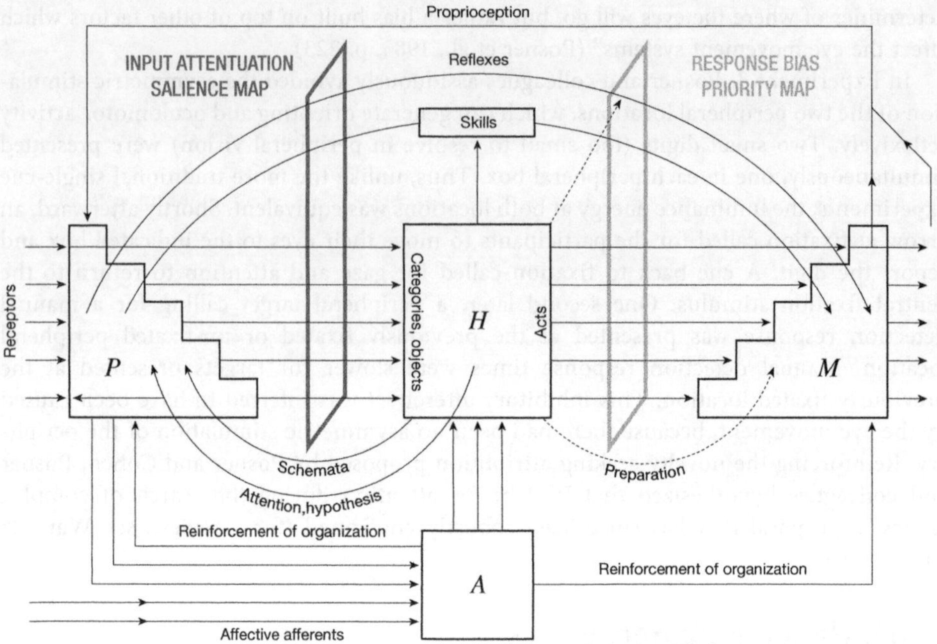

Figure 20.1 A figure illustrating the flow of information between cognitive modules from perceptual (P), input-related processing to motoric (M), output-related processing. A central stage of processing (H), which can be bypassed, is the site of awareness of perceptually derived categories and objects and of the initiation of some actions. The two dashed arrows have been added by RMK and are not in the original figure, which has been redrawn from Attneave (1961). We have inserted into this framework (red and green surfaces) where the two forms of inhibition described in this chapter might be operating. A black and white version of this figure will appear in some formats. For the colour version, please refer to the plate section.

Central vs. Peripheral Arrow Targets

One diagnostic for these two forms of inhibition depends on comparing performance with peripheral onset vs. central arrow targets, with both types of target calling for the same left–right localization response either manually or with a saccadic response, usually in the direction of the peripheral stimulus or the direction indicated by the central arrow. First employed independently by Abrams and Dobkin (1994) and Rafal, Egly, and Rhodes (1994), the logic behind this diagnostic is as follows: If the inhibition exclusively delays or degrades the flow of information from the region of space in which it has been generated, then, while responses to cued peripheral targets should show a RT delay relative to uncued peripheral targets, responses to centrally presented arrows should not be affected by the cue (Figure 20.2A, left panel). On the other hand, if the inhibition exclusively biases responding, then the two types of target should show equal amounts of RT delay in the cued direction (Figure 20.2A, right panel).

When this diagnostic was used by Tracy Taylor in her 1997 dissertation (later described in Taylor & Klein, 2000), she discovered a dramatic dissociation. Six "experiments" were generated from the orthogonal combinations of responses to the first (S1: ignore, manual, saccade) and second (S2: manual, saccade) stimuli, which were separated by a 1 second CTOA and a cue back. In a block of trials, both S1 and S2 were, randomly, either peripheral events or central arrows. S1s were uninformative about

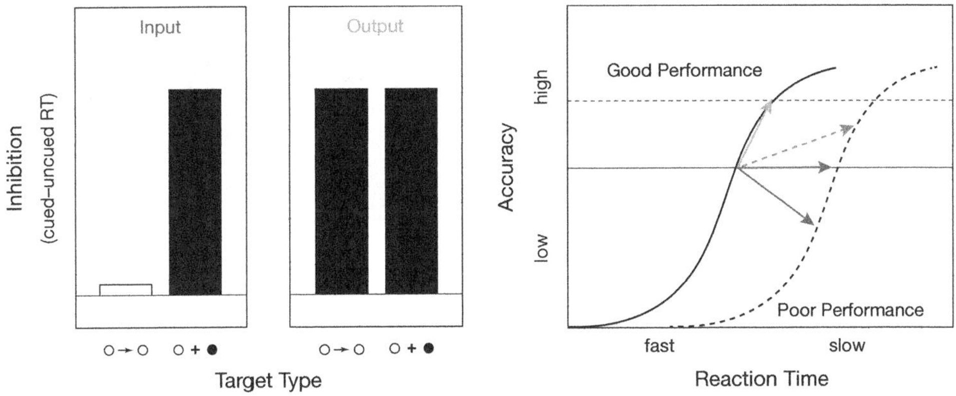

Figure 20.2 Diagnostics for determining whether an effect is early (input) or late (decision/output) in processing. (A) Target-type diagnostic. Two hypothetical patterns of results when the different target types (central arrow or peripheral target, shown on the x-axis) are calling for a localization response are presented about 1 second after an uninformative peripheral cue. (B) Speed–accuracy diagnostic. A nonspatial, two-alternative, forced-choice response is required to a continuous target, and performance is examined in speed–accuracy space. Two hypothetical information accrual functions (solid and dashed curves) and two hypothetical criteria for responding (horizontal lines) are illustrated. See text for explanation. A black and white version of this figure will appear in some formats. For the colour version, please refer to the plate section.

the upcoming S2s. Although the six combinations of what participants were instructed to do in response to S1 and S2 – when orthogonally combined with the two stimulus types of each S1 and S2 – yielded 24 unique conditions for which IOR could be measured, here we will focus only on the conditions with peripheral S1s. The results from these 12 conditions are illustrated in Figure 20.3A, where a significant inhibitory effect is represented by filled (black) bars and the absence of inhibition is represented by unfilled (white) bars. Importantly, two dramatically different forms of inhibition were apparent. The *input* form, observed when there was no eye movement to either S1 or S2, is characterized by inhibition in response to peripheral onset targets but not in response to central arrow targets (as in the left panel of Figure 20.2A). The *output* form, observed when a saccade was made to either S1 or S2 (or both), is characterized by significant, and roughly equivalent amounts of inhibition in response to both peripheral onset and to central arrow targets (as in the right panel of Figure 20.2A).

Examining Results in Speed–Accuracy Space

A second diagnostic was first implied by Ivanoff, Klein, and Lupianez (2002) in their analysis of the interaction between the inhibitory aftereffect of a peripheral cue and the Simon effect. Representing the temporal dynamics of information processing in speed–accuracy space (see Figure 20.2B), Ivanoff and colleagues noted two different mechanisms by which a hypothetical inhibition might cause a delay in RT. Without affecting the accumulation of information about a continuously presented target, one inhibitory effect might raise the threshold or criterion for responding. When such a "response bias" form of inhibition was operating, the change in performance would be characterized as a trade-off between speed and accuracy: that is, performance for cued targets would be slower *but also* more accurate than for uncued targets (a shift along a single SAT function as illustrated by the green arrow in Figure 20.2B). In contrast, the other form of inhibition

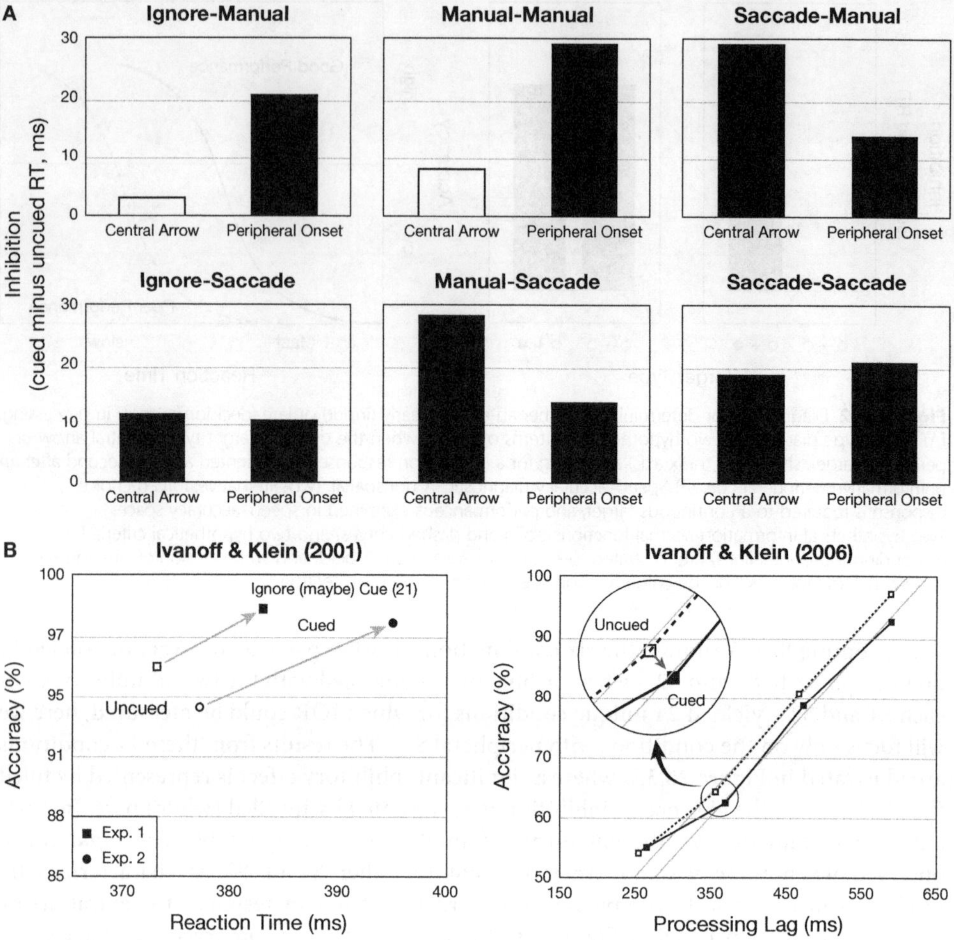

Figure 20.3 (A) Results from the target-type diagnostic as employed in Taylor's dissertation (Taylor & Klein, 2000). Significant IOR is represented by black bars; non-significant IOR is represented by white bars. Each panel represents a different combination of tasks the participant performed in response to S1 and S2 (e.g., Ignore-Manual denotes ignoring S1 and making a manual response to S2). (B) Results from the speed–accuracy diagnostic as employed in Ivanoff's dissertation (Ivanoff & Klein, 2001, 2006). RT and accuracy are plotted on the x- and y-axes, respectively, for targets presented at the cued (filled symbols) and uncued (unfilled symbols) conditions. In the left panel participants performed a go/no-go task with typical reaction time methodology. In the right panel a response-window technique was used to generate full SAT functions. The green arrows signify that IOR is characterized by a speed–accuracy trade-off and the possibility of an output effect. The red arrow signifies a genuine decrease in information processing efficiency. A black and white version of this figure will appear in some formats. For the colour version, please refer to the plate section.

delays or degrades information processing about either the target or the connection between the target and the correct response.

When an input form of inhibition is operating, as represented by the red arrows in Figure 20.2B, performance in response to cued targets should fall on a different and less efficient speed–accuracy trade-off (SAT) function. The inhibited function could be shifted rightward from the performance with uncued targets, as illustrated in Figure 20.2B, or it might begin to rise above chance at the same time while increasing to asymptote more slowly than for uncued targets (not illustrated). RTs to targets suffering from this form of inhibition will be slower than to uncued targets and accompanied by either no change in accuracy or worse accuracy.

This speed–accuracy space diagnostic requires continuously presented peripheral targets calling for a nonspatial discrimination (such as a go/no-go task or two alternative forced choice [2-AFC] task about a property that is orthogonal to the cuing dimension) about the form, identity, or color of a target. The resulting pattern of performance is considered in speed–accuracy space. Using such an RT task, these two different effects on target processing can be distinguished by the direction (or slope) of the line connecting performance (speed and accuracy) from uncued to cued (inhibited) targets. When this line is horizontal or negatively sloped (red lines in Figure 20.2B), there is a genuine effect on the quality of target processing (an RT cost without an increase in accuracy or a cost in both RT and accuracy). When the effect is due to a bias (a change in the criterion for responding), this line will have a positive slope (green line in Figure 20.2B). Whereas the flat or negatively sloping relationship between cued and uncued performance in SAT space is relatively unambiguous in indicating that a shift in the SAT function has been obtained, unfortunately a positively sloping relationship is ambiguous: It can be due to a response bias (green line) or it could be due to a combination of bias and degraded information processing (blue dashed line).

An even more powerful method (but a much more costly one) entails generating full SAT functions (a methodology advocated by Wicklegren, 1977) for cued and uncued targets. The SAT method allows visualization of the temporal dynamics of information processing for cued and uncued targets. And, when the input form of inhibition is being observed, this method allows the scientist to determine which parameter (e.g., start time, rate, asymptote) is affected by the inhibition (cuing). Another benefit of this more powerful method is that it can be used to disambiguate the situation described earlier; is a positive slope coming from a single function or from different functions? Both of these methods were pioneered by Jason Ivanoff in a series of experiments beginning with his 2003 dissertation (Hilchey, Ivanoff, Taylor, & Klein, 2011; Ivanoff & Klein, 2001, 2004, 2006; Ivanoff, Klein, & Lupianez, 2002).

Ivanoff and Klein (2001) explored inhibitory aftereffects of peripheral cues using a go/no-go task. In prior studies using the simple detection task with catch trials, false alarms could be observed, but because nothing was presented on a catch trial, they could not be attributed to the cued versus uncued location. In a go/no-go task, stimuli are presented on every trial at the cued or uncued locations. One stimulus (go) calls for a response; the other one (no-go) does not. Using such a go/no-go task allowed Ivanoff to be able to attribute false alarm rates to the cued vs. uncued locations. In this study (see Figure 20.3B, left panel), he discovered a pattern that was characteristically a response bias: slower responding to cued targets was accompanied by higher accuracy (there were fewer false alarms for cued targets). In two experiments from his dissertation, Ivanoff (presented in Ivanoff & Klein, 2006) used the SAT method (cf. Wicklegren, 1977) to generate full SAT functions for targets presented at cued and uncued locations. Here (see Figures 20.3B, right panel), he demonstrated the input form of inhibitory aftereffect: performance at the cued location was shifted rightward in SAT space.

A few years later it occurred to Klein that if the effect of IOR generated by an eye movement was, as proposed by Posner, Rafal, Choate, and Vaughan (1985) and confirmed by Taylor and Klein (2000), just a bias against responding in the cued direction,

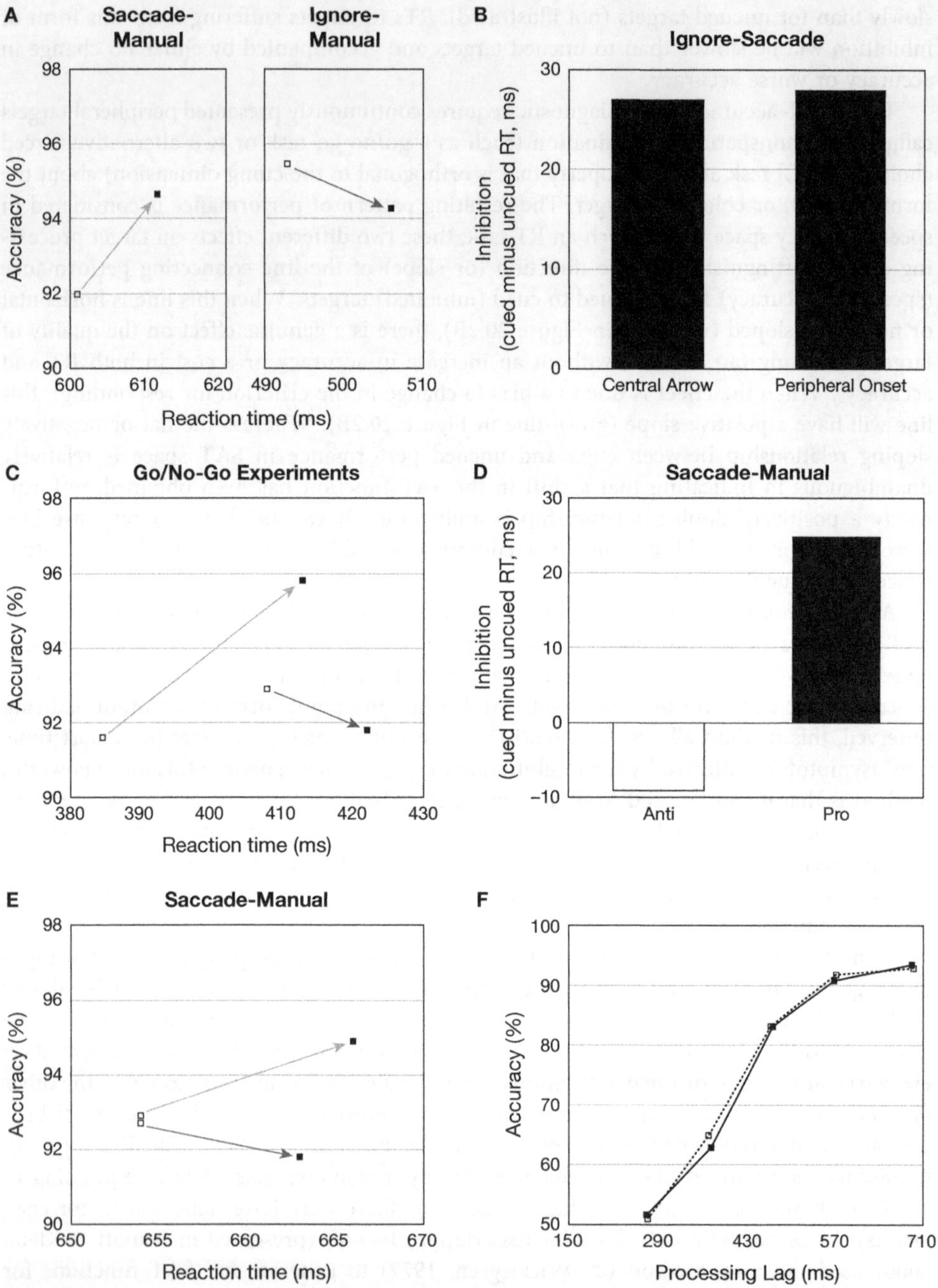

Figure 20.4 Patterns of findings when the two diagnostics illustrated in Figure 20.2 are applied to our studies and to the literature. In A, C, E, and F, data from cued targets are plotted with filled symbols; data from uncued targets with unfilled symbols. (A) Results from Chica et al. (2010). Data connected by the green arrow are from Experiments 1B and 3B; data connected by the red arrow are from 3A. (B) Data from Hilchey, Klein, & Ivanoff (2012). (C) Data are from a go/no-go tasks (errors are false alarms). Data connected by the red arrow are from Hilchey et al. (2014) in which there was careful monitoring of eye position and participants were given feedback whenever an untoward eye movement occurred. Data connected by the green arrow are collapsed from six studies in

then IOR might not be observed when caused by an eye movement and measured with a nonspatial discrimination task. It might seem counterintuitive to propose that something would be caused with no apparent effect. In Klein's thinking, at that time, the hypothetical response bias would have been generated by the eye movement to the signal in the periphery, but it would not have been measurable by the 2-AFC task, because targets at the cued location did not require spatial (localization or detection) responses. Klein suggested this to Ana Chica who was visiting his laboratory from the University of Granada, and Chica, Taylor, Lupianez and Klein (2010) conducted just such an experiment. The results provided a dramatic link between the ideas of Klein's former students, Taylor and Ivanoff.

Chica, Taylor, Lupianez, and Klein (2010) explored two combinations from the matrix illustrated in Figure 20.3A: S1 was either ignored or a saccade target, and S2, presented in the periphery (at the cued or opposite location), required a nonspatial, 2-AFC discrimination based on color. In both cases, eye position was monitored. We expected degraded processing along the input pathway following ignored cues and no such degradation – but possibly a response bias – when saccades were made to the cue. Confirming our predictions, the results (see Figure 20.4A) showed that when participants were instructed to ignore the cue, maintain fixation, and were "beeped" if they made an eye movement, the inhibition was of the input form (right panel). In contrast, when subjects successfully made a saccade to the cue and back to the original fixation before the final target (left panel), the output form (a speed–accuracy trade-off) was observed.

Using the Diagnostics to Resolve Puzzles in the Literature

We have since explored three puzzles concerning the pattern of results from some of conditions illustrated in Figure 20.3A.

Ignore–Saccade: Input + Output or Pure Output?

One puzzle is why Taylor and Klein found equivalent inhibitory effects when saccades were made to peripheral and central arrow targets after an ignored peripheral cue (a pure output form as illustrated in Figure 20.2A, right panel), whereas Abrams and Dobkin (1994) found greater IOR following peripheral than central targets (both input and output forms of IOR). Abrams and Dobkin inferred from this pattern that two components were jointly contributing to the inhibition in this condition: an output component that delayed responses in the cued direction (and that would be operating with both types of target) and an input component that affected only the peripheral target. Klein and Hilchey (2011, p. 480) proposed the following resolution to this conflicting pattern of results:

Figure 20.4 (*cont.*) the literature (Ivanoff & Klein, 2001, 2003, 2004, Exp. 1; Taylor & Ivanoff, 2003; Prime & Ward, 2008, Exp. 3; Prime & Jolicoeur, 2009, Exp. 1). (D) Data are from Hilchey, Dohmen, Crowder, & Klein (2016). In separate blocks participants made pro- or anti-saccades in response to a peripheral luminance increment, and ~1 second after the onset of this stimulus they made a manual response in the direction of a centrally presented arrow. (E) Data are from Redden, Hilchey, & Klein (2016). Different participants made a pro- or anti-saccade in response to a peripheral luminance increment, and ~1 second after the onset of this stimulus they made a nonspatial 2-AFC. (F) From Redden, Hilchey, and Klein (2015a). Participants made a pro-saccade in response to a peripheral luminance increment, and ~1 second after the onset of this stimulus a target was presented calling for a nonspatial 2-AFC response. Importantly, speed–accuracy trade-off functions were obtained for cued and uncued targets using a window method. A black and white version of this figure will appear in some formats. For the colour version, please refer to the plate section.

> Abrams and Dobkin's finding was generated by untoward differences in mental set made possible by their collecting data from arrow and peripheral targets in separate blocks. When the target was never in the periphery (always an arrow at fixation), the participant could adopt an attentional control setting that encouraged effective filtering of the irrelevant peripheral cues. This would not be possible in a block when all (Abrams and Dobkin's peripheral target blocks) or some (Taylor and Klein, 2000) of the targets were peripheral.

Hilchey, Klein, and Ivanoff (2012) confirmed that the Abrams and Dobkin pattern was compromised by an untoward design feature: The two target types were presented in separate blocks of trials. When the two target types are appropriately randomly intermixed, the Taylor and Klein finding of roughly equivalent IOR for the two target types is confirmed (Figure 20.4B).

No Response–Manual: Input or Output?

Another puzzle concerns the fact that in the Ignore-Manual combination with peripheral cues and targets, a speed–accuracy trade-off has often been reported in a go/no-go task even when participants are instructed not to move their eyes. This pattern, first reported by Ivanoff and Klein (2001; see Figure 20.3B, left panel), has been observed in five subsequent studies (see green arrow in Figure 20.4C). We surmised that despite instructions, perhaps the reflexive oculomotor system was not sufficiently inhibited in these studies for the input form to have been generated, because in none of these studies were participants given trial-by-trial feedback about any unwanted eye movements.

We (Hilchey et al., 2014) therefore closely replicated the methods of Ivanoff and Klein (2001) while monitoring eye movements and, importantly, providing trial-by-trial feedback whenever an unwanted eye movement occurred. Under these conditions (see the red arrow in Figure 20.4C), cued targets showed the typical, significant, RT delay, but there was no evidence for a speed–accuracy trade-off; that is, responses to cued targets were not more accurate than those to uncued targets. It is worth mentioning that we found one experiment (Cheal, Chastain, & Lyon, 1998) in which a go/no-go task was combined with assiduous monitoring of eye position and feedback when untoward eye movements were made. In this experiment, there was an RT delay for cued targets and no evidence of a speed–accuracy trade-off. Our pattern of results when considered together with the literature shows the importance of not only monitoring eye position but also giving participants trial-by-trial feedback when unwanted eye movements are executed.

Saccade–Manual: Output or Input?

A third puzzle arises from studies that have asserted that an input form of IOR is generated by saccadic eye movements, whereas, in contrast, Taylor's findings imply an output form. The challenge is rooted in the finding that following an anti-saccade (a saccade executed away from a peripheral "target"), the inhibitory aftereffect is at the location of the stimulus and not the response (e.g., Fecteau, Au, Armstrong, & Munoz, 2004). To resolve the conflict, Klein and Hilchey (2011) proposed that the input form of IOR ought to be generated by an anti-saccade, because in order to successfully look away from a peripheral target, the reflexive oculomotor machinery must be tonically inhibited (Forbes & Klein, 1996; Ignaschenkova et al., 2004). This view, which is critical for the present chapter, represents an important qualification upon Taylor's inference that the output form of IOR is generated whenever eye movements were made.

We (Hilchey, Dohmen, Crowder, & Klein, 2016) first employed the central-target diagnostic in two conditions in which an eye movement was made in response to a peripheral stimulus and then, upon returning gaze to fixation, a manual response was required to respond to a central arrow. Replicating the methods of Taylor and Klein (2000), in one condition, the eye movements were pro-saccades. Differing from Taylor and Klein, however, there were no arrow S1s or peripheral target S2s. If Taylor and Klein's finding of IOR in this condition were due to the fact that their participants experienced both peripheral and central stimuli as both S1s and S2s and if the form of IOR generated by saccades to peripheral targets really is input-based, then no IOR should be observed in this condition. In contrast, if the output form of IOR is generated when saccades are executed to peripheral targets, then we should replicate Taylor and Klein's finding of IOR when it is measured using manual responses to central arrow targets. In a second condition we required participants to make an anti-saccade to the peripheral S1. If Klein and Hilchey's (2011) proposal, that an input form of IOR would be generated in this condition because the reflexive oculomotor machinery must be tonically inhibited for correct anti-saccade performance, were correct, then there would be no IOR observed in manual responses to central arrows. As seen in the "pro-saccade" bar of Figure 20.4D, 22 msec of significant IOR was observed. Replicating Taylor and Klein's finding when this condition was mixed with others, this finding confirms that pro-saccades generate the output form of IOR. And, consistent with Klein and Hilchey's proposal that the input form would be generated when an anti-saccade was used to cause the inhibitory effect, manual responses to centrally presented arrows were not delayed in the cued direction (Figure 20.4D, anti-saccade bar).

Converging evidence for this interpretation is provided by Redden, Hilchey, and Klein (2016) who applied the SAT diagnostic to this cell. As in Hilchey et al. (2016), IOR was generated by a pro- or anti- saccade to a peripheral target. But following the SAT diagnostic, we measured the effect of these two inhibitions in a nonspatial discrimination task. Replicating Chica and colleagues, we found a speed–accuracy trade-off (Figure 20.4E, green arrow) when IOR was generated by a pro-saccade. Reinforcing our interpretation of the results in Figure 20.4D, when generated by an anti-saccade, an input form of inhibition was observed (Figure 20.4E, red arrow). Next, using the SAT method to measure accuracy while controlling RT, Redden, Hilchey and Klein (2015a) generated SAT functions for cued and uncued targets in conditions otherwise the same as those used here with pro-saccades. As would be predicted by an output-based account, when IOR was caused by a pro-saccade, cuing had no effect on the SAT functions (Figure 20.4F). Compare this pattern with SAT functions from the ignore-manual cell of this matrix (Figure 20.3B, right panel), in which Ivanoff and Klein (2006) found that IOR following an ignored cue caused a significant rightward shift of the SAT function, which implies a degradation or delay in the processing of cued targets in the ignore-manual condition of this matrix.

Summary

Using the diagnostic patterns obtained when peripheral onset or central arrow targets are used, we have seen that there are two forms of inhibitory aftereffect. These are distinguished clearly by whether the effect can only be measured by peripheral targets (the input form when the reflexive oculomotor system is suppressed) or by whether there are roughly equivalent delays in response whether the targets are central or peripheral (the output form when the reflexive oculomotor system is not suppressed). When performance on cued and uncued targets is represented in SAT space, the input form is manifest as a shift in the SAT function representing degraded or delayed processing of cued targets, whereas the output form entails NO shift in the function, but instead a movement along it (a response bias).

It has also been shown that the input form of IOR is represented in object-based coordinates (Hilchey, Pratt, & Christie, 2016), whereas the output form of IOR is represented in spatiotopic coordinates (Christie, Hilchey, & Klein, 2013; Redden, Hilchey, & Klein, 2015b), suggesting yet another dissociation at the neural level. This "object-" vs. "space-" representation suggests the input form is likely modulated cortically in the ventral visual stream, whereas the output form is likely modulated by the interaction between subcortical structures (vis superior colliculus) and cortical areas in the dorsal visual stream.

In conclusion, and speaking cognitively, either form of inhibition can perform the function attributed to IOR by Posner – "to maximize sampling of the visual environment." Using Attneave's thoughtful representation of a cognitive architecture (Figure 20.1), input attenuation operates on a salience map that influences what we will attend to and become aware of while a response bias operates on a priority map that influences what behaviors (including orienting) we are likely to engage in.

Chapter 21

Spatial Biases from Action

Jessica K. Witt

Action biases spatial perception. A person's ability to act – such as their skill at hitting a target or the amount of energy required to climb a hill – influences perceived target size and perceived hill slant. For example, softball players who are hitting better than others judge the ball as bigger (Gray, 2013; Witt & Proffitt, 2005). People who are fatigued, burdened by a heavy load, or weigh more than others judge hills as steeper (Bhalla & Proffitt, 1999; Taylor-Covill & Eves, 2013). The optical information specifying the hill is exactly the same, yet it appears different depending on the perceiver's ability to climb the hill.

These effects of a person's ability to act on spatial perception are collectively known as action-specific effects (Witt, 2011a). The action-specific account of perception claims that people perceive the spatial layout of the environment in relation to their ability to perform the intended action. This account can be contrasted to a general-purpose account of perception for which spatial information is processed independently of action. Given the lack of inclusion of action-related factors in most experiments on spatial perception, it is reasonable to conclude that nearly all accounts of spatial perception fall in the general-purpose camp. Yet a wealth of new empirical data challenges the general-purpose account and instead support an action-specific approach.

Historic Roots of the Action-Specific Approach

The action-specific account of perception may seem reminiscent of other approaches to perception. The emphasis on action triggers thoughts of Gibson's ecological approach to perception, and the idea that nonoptical information penetrates visual processes conjures up notions of Bruner's New Look approach. Each will be addressed in turn.

Gibson's Ecological Approach. The action-specific account's emphasis on action for perception builds on Gibson's (1979) theory of affordances. According to Gibson, affordances are the possibilities for action within the environment. For example, an Adirondack chair affords sitting to an adult, climbing to a young child, and shade for a dog. If sturdy enough, the chair also affords stepping for the adult to reach to taller branches on an apple tree. Although affordances exist and are a property of the environment–organism relationship, the affordances that are perceived at any given time depend on the intentions and needs of the perceiver. Gibson argued that affordances are the primary object of perception. Whereas people can judge color or distance when asked, the natural tendency for perception is not to look around and see objects as green or 3 feet away, but rather to see places to sit, places to walk, and objects to grasp.

The action-specific account extends Gibson's theory of affordances by showing that affordances are perceived even when tasked with making judgments about spatial layout. Distances are perceived in terms of their affordances. If the distance affords reaching, it appears shorter than if it does not afford reaching (e.g., Witt, Proffitt, & Epstein, 2005).

If the distance affords effortful jumping, it appears farther than if it affords effortless jumping (Lessard, Linkenauger, & Proffitt, 2009). Similarly, object size is perceived in terms of its affordances. A softball that affords more successful hitting appears larger than when it affords less successful hitting (Gray, 2013; Witt & Proffitt, 2005). Some researchers claim that affordances are such a primary aspect of perception that the reason affordances penetrate spatial judgments is because participants are tacitly reporting on the affordances themselves (Lee, Lee, Carello, & Turvey, 2012).

Despite the shared emphasis on affordances, the action-specific effect is not wholly consistent with Gibson's ecological approach to perception. Gibson also argued that perception is direct, meaning that perception is based entirely on incoming stimulation such as optical information or auditory information (Gibson, 1966, 1979; see also Michaels & Carello, 1981). Gibson rejected the notion that any internal processes such as unconscious inference (Helmholtz, 2000 [1925]) or logical processes (Rock, 1983) shape perception. Gibson's staunch dismissal of such internal processes was driven by his emphasis that the optical information available for perceptual processes is sufficiently rich to fully specify spatial layout. Gibson argued that perception evolved to operate in a full-cue environment for which the perceiver is free to move and visually explore the environment, and in this context, the visual information is sufficient. Internal processes are not necessary and therefore should not be included in a theory of perception. That action-related information influences perception challenges the idea that spatial vision is based purely on optical variables. For this reason, even though the action-specific account emphasizes affordances and ecological environments, many ecological psychologists reject or are skeptical of action-specific claims.

A possible reconciliation between the action-specific account and Gibson's theory of direct perception involves expanding the information for vision from a purely optic array to a global array. The global array contains information that spans across specific energy arrays (e.g., optical, acoustic; Stoffregen & Bardy, 2001). Action-specific effects can be explained by considering action-relevant energy arrays as part of the global array (Witt & Riley, 2014). Information related to action can be directly sensed from proprioceptive receptors, which detect the size and position of the body, and interoceptive receptors, which detect bodily states such as fatigue and hunger (Craig, 2003). Given that information about the body and its readiness for action can be directly sensed, action-specific perception is consistent with an account of direct perception that considers the information as that spanning across multiple energy arrays. Recent evidence supports this proposal (White, Shockley, & Riley, 2013).

In summary, both the ecological and the action-specific accounts emphasize action in perception. Action-specific effects are wholly consistent with the idea that affordances are an important, if not primary, object for perception. Both approaches also consider perception to express the relationship between the perceiver and the environment. And the one primary area of controversy – whether perception is direct – can be resolved by including directly sensed information about action as part of the information for perception.

The New Look. Bruner's New Look approach to perception suggested that top–down effects related to value or need could influence perception (Bruner, 1992; Bruner & Goodman, 1947). For example, children from poorer families judged coins as larger, compared with children from wealthier families. The New Look approach failed for a number of reasons (Carter & Schooler, 1949). Most importantly, many effects failed to replicate when experiments were conducted with more sophisticated methodologies such as the use of counterbalancing and proper controls. Although experiments demonstrating action-specific effects typically involve carefully designed methodologies, the action-specific approach may seem like a reincarnation of the New Look that will eventually collapse under methodological scrutiny. It is important to note several distinctions between action-specific effects and the New Look. The first is methodological. As mentioned in the preceding section, experiments on action-specific effects have taken great care with methodological design. The second relates to the critical factor of interest. Action-specific effects focus entirely on action rather than on influences of motivation such as need and desire.

Action-Specific Perception

Action-specific effects have been demonstrated in a wide range of contexts, for many different aspects of spatial perception, and across several populations of perceivers including athletes, chronic-pain patients, older adults, college students, and community members. Action-specific effects can be divided into several categories of effects including locomotion, reaching and grasping, and hitting.

Locomotion and Far Space. The first demonstration of action-specific effects involved the effect of walking up a hill on perceived hill slant (Bhalla & Proffitt, 1999; Proffitt, Bhalla, Gossweiler, & Midgett, 1995). Although participants were never asked to physically ascend the hill, the ground plane affords locomotion, and is most likely perceived in relation to locomotion. Participants estimated hill slant by verbally estimating the slant in degrees or manipulating the cross-section on a handheld disk so that the visible portion of the cross-section matched the angle of the slant (see Figure 21.1). Both of these measures of perceived hill slant revealed action-specific effects. People who were burdened by carrying a heavy load judged the hill to be steeper compared with people who did not wear a backpack. Runners judged hills to be steeper when they were fatigued from having completed a difficult run than when they judged hill slant prior to the run. Hills were also judged to be steeper to observers who were less physically fit than others and to older adults compared with younger adults. The idea behind all of these effects is that for the people who would

Figure 21.1 A participant performing the visual matching task to estimate hill slant. The participant adjusts the slant of the black line on the disk to match the slant of the hill.

have to exert more energy to ascend the hill, the hill appeared steeper (see also Schnall, Zadra, & Proffitt, 2010; Taylor-Covill & Eves, 2013, 2014, 2016).

In studies on hill slant, a third measure called the haptic measure is sometimes employed. The haptic measure involves rotating one's hand or forearm to be parallel to the hill and to do this without looking at the hand or arm. The haptic measure does not reveal effects of action on estimated hill slant. One possible explanation is that the haptic measure might be driven by unconscious visual information such as that processed by the dorsal stream (cf. Milner & Goodale, 1995). This information does not appear to be biased by visual illusions (Aglioti, DeSouza, & Goodale, 1995; Ganel, Tanzer, & Goodale, 2008), and might also be unbiased by the potential for action. Another explanation is that responses in the haptic measure are limited by biomechanical constraints and poor proprioceptive feedback of wrist flexion (Durgin, Hajnal, Li, Tonge, & Stigliani, 2010). Such limitations could prevent any genuine effects of action potential on perceived slant to emerge. Issues regarding action's influence on dorsal stream processing in general, and the haptic measure in particular, have yet to be resolved.

Effort required to walk also influences perceptual judgments of distance. Targets placed on the ground appear farther away when the perceiver anticipates having to exert more effort. In several experiments, anticipated effort to walk was manipulated using a treadmill to dissociate forward walking effort with resulting optic flow (Rieser, Pick, Ashmead, & Garing, 1995). On a treadmill, a person has to exert forward walking effort, but the resulting optic flow indicates that the person is not moving. Thus, the system learns that it takes a lot of effort to go nowhere. After the perceiver recalibrates to this relationship, targets on the ground appear farther away (Proffitt, Stefanucci, Banton, & Epstein, 2003; White, Shockley, & Riley, 2013; Witt, Proffitt, & Epstein, 2004, 2010). Other studies that increased physical effort for walking also showed effects on perceived distance. Carrying a heavy physical burden such as a backpack leads to increased estimates of distance (Proffitt et al., 2003). Distances appear farther when presented up hills, which require more effort to traverse, than on flat ground (see Figure 21.2; Stefanucci, Proffitt, Banton, & Epstein, 2005;

Figure 21.2 Targets presented up a hill appear farther away than when presented on flat ground. (Stefanucci, Proffitt, Banton, & Epstein, 2005; Tenhundfeld & Witt, 2017)

Tenhundfeld & Witt, 2017). Natural variation in physical body weight and physical fitness also influences estimated distance and slant (Bhalla & Proffitt, 1999; Sugovic, Turk, & Witt, 2016; Taylor-Covill & Eves, 2013, 2016). Combined, these studies also show that short-term changes (such as donning a backpack) and long-term variations in the body both impact spatial perception.

Not all locomotion involves walking, and other types of locomotion also influence perceptual reports. In one study, swimmers ducked their heads underwater and judged the distance to targets floating just below the surface. The more skilled swimmers estimated the targets as closer compared with the less skilled swimmers (Witt, Schuck, & Taylor, 2011). In addition, some swimmers wore fins, which allow the swimmer to move faster while exerting less effort. The swimmers who wore the fins estimated the targets to be closer compared with swimmers not wearing fins. It is interesting to note that swimming skill and wearing fins had separate, additive effects, with skill having twice the effect of the fins. Little research has examined how various manipulations of action interact with each other, and this study was among the first to do so. The finding of additive effects suggests multiple paths by which action influences spatial perception.

Jumping ability also influences perceived distance across gaps and perceived height of walls. Observers judged gaps as farther when wearing ankle weights compared to when not wearing the weights (Lessard, Linkenauger, & Proffitt, 2009). Notably, for the gap study, ankle weights only influenced perceived gap width for gaps that could be jumped across and not gaps that were too wide to be jumpable. Thus, the effects of action on perception are limited to perception within spaces that afford the action. In another study, perceivers who were trained in parkour and stated that they could successfully execute a wall-jump maneuver estimated the walls to be shorter compared with novices not trained in parkour (Taylor, Witt, & Sugovic, 2011). When the space afforded locomotion (in this case, a wall affords jumping to those trained in parkour), the space appeared compressed compared to when the action was not afforded.

In summary, when locomoting from one location to another is easier, distances look closer, hills look shallower, and walls look shorter. In all of these experiments, the optical information specifying hill slant, target distance, and wall height was the same across conditions, yet the perception of these spaces varied as a function of the perceiver's ability to traverse the space.

Reaching and Grasping in Near Space. Actions involving objects in near, or peripersonal (Cutting & Vishton, 1995), space also reveal an influence on spatial perception. Objects that are easier to reach and grasp appear closer and smaller than objects that are more difficult to reach and grasp. Obviously objects that are reachable tend to be physically closer than objects that are not reachable. In order to maintain the same physical distance and only manipulate the perceiver's ability to reach, a reach-extending tool is frequently used (e.g., Witt, Proffitt, & Epstein, 2005). Tool use allows variations in reachability while at the same time keeping physical distance to the target constant.

To assess perception of distance, many studies used a visual matching task (see Figure 21.3). A central circle was presented and was the target for reaching both with and without the tool. Two comparison circles were placed on either side of the target circle, and by pressing buttons on a keyboard, participants manipulated the distance between the comparison circles to match the distance to the target. The idea is that if the target appears closer, the participant would move the comparison circles to be closer together. Participants estimated the distance to a given target, then reached to it. During one block of trials, participants reached with a baton, which extended their reach by nearly 40 cm. In another

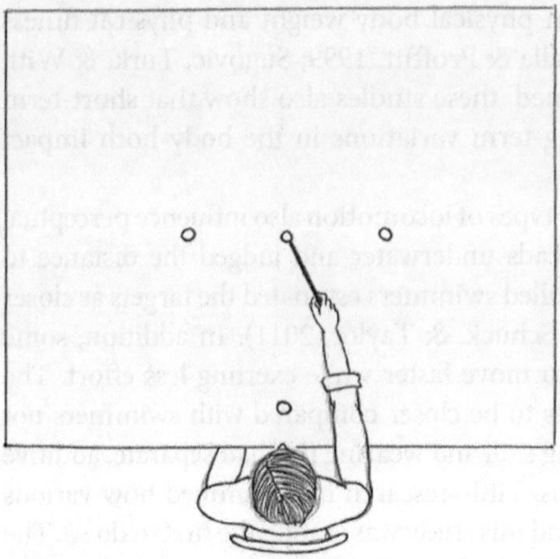

Figure 21.3 Bird's-eye view of visual matching task used to assess perceived distance to objects beyond arm's reach but within reach when using a tool. Participants adjusted the outer circles to be the same distance apart from each other as the distance between the perceiver and the target circle to which the perceiver is reaching and then reached to the target with and without the tool. (Witt et al., 2005)

block of trials, participants reached without the tool. The targets were presented to be just beyond arm's reach but easily within reach with the tool. The targets were judged to be closer when participants used the tool than when they reached with just their hands (see also Bloesch, Davoli, Roth, Brockmole, & Abrams, 2012; Costello et al., 2015; Linkenauger, Bulthoff, & Mohler, 2015). Although the tool was 40 cm, the compression in estimated distance was only 7 cm. That is because information about reaching only serves as one source of information for distance perception, and optic cues still play a large role in how distance is perceived. Although tool length has not been systematically studied, I suspect that the relevant factor is whether or not a target can be reached, and not the length of the tool per se. According to this prediction, a longer tool will extend the distances for which compression will be experienced, but will not lead to *additional* compression for the space just beyond arm's reach.

Several interesting variations on this basic effect have revealed some notable constraints on action-specific effects. First, the use of a tool to reach to targets does not make the targets appear as closer when the tool is too short to be effective for reaching (Osiurak, Morgado, & Palluel-Germain, 2012). Thus, only variations in action that impact the outcome of the action seem to influence perception. Second, the mere holding of the tool but without the intention to use it to reach to the targets also does not make the targets appear closer (Witt et al., 2005). The tool is only influential for perception when the intent is to use it. Similarly, if instead of a tool, virtual reality is used to make the arm appear longer, experience with the virtual arm is necessary for the change to influence distance perception (Linkenauger, Bulthoff, & Mohler, 2015). Third, anticipation of using the tool makes the targets appear closer even if the tool is not actually held during the perceptual judgments (Witt & Proffitt, 2008). Thus, these action-specific effects seem to be primarily about anticipated outcomes rather than current states per se.

Just as reaching affects apparent distance, grasping affects apparent size. Objects appear smaller when the intent is to grasp them with the dominant and skilled right hand compared with the nondominant left hand (Linkenauger, Witt, & Proffitt, 2011). Objects also look smaller when one's hand is rendered bigger than when it is rendered smaller. Several techniques have been used to manipulate apparent hand size. One technique was to use a magnifying lens (Linkenauger, Ramenzoni, & Proffitt, 2010). Another technique was to use virtual reality to manipulate apparent hand size (Linkenauger, Leyrer, Buelthoff, & Mohler, 2013). In both of these cases, when the hand appeared bigger, the target appeared smaller. Importantly, the same effects of hand size on object size are not apparent when the hand is not one's own, suggesting that these perceptual biases are not simply contrast effects or effects based on familiar size. Instead, they are body-specific effects on apparent size. Similar findings on perceived size have also been found after rendering the entire body as bigger or smaller (Van der Hoort & Ehrsson, 2014; van der Hoort, Guterstam, & Ehrsson, 2011). These studies show that perceived distance to and size of objects within near space are influenced by the perceiver's ability to reach and grasp them. Although these effects have been described as action-specific, it is true that in many of these virtual reality experiments, no action was taken toward the object. This raises the possibility that these categories of effects are due to entirely different mechanisms. However, it is also possible that although no explicit action was taken, perceivers viewed the targets with an implicit intention to act. Resolving this discrepancy is an important avenue for future research.

Hitting. A person's ability to hit a target – such as hitting a softball, an archery target, a field goal, or a golf hole – influences apparent size of the target. In one study, softball players estimated softball size by selecting from a range of differently sized circles presented on a posterboard (see Figure 21.4). The size of the selected circle positively correlated with their batting averages for the games played that night (Witt & Proffitt, 2005). Using a similar design, we found that apparent hole size correlated with golf performance: those with better course scores judged the hole as bigger (Witt, Linkenauger, Bakdash, & Proffitt, 2008).

Correlations cannot specify causality, leaving open the possibility that softball players who saw the ball as bigger subsequently hit better. However, a study on American football field goal kicking suggests that performance can influence perception (Witt & Dorsch, 2009). Athletes judged the size of a field goal before and after attempting 10 kicks. Those who kicked the ball more successfully than others judged the goal as bigger *after* kicking but not before. Estimated size prior to kicking attempts did not relate to subsequent performance.

Similar findings have been observed across a range of athletic endeavors. Archers who hit the target more successfully than others see the target as bigger (Lee, Lee, Carello, & Turvey, 2012). Dart throwers who hit the target quicker or more successfully than others see the target as bigger (Canal-Bruland, Pijpers, & Oudejans, 2010; Wesp, Cichello, Gracia, & Davis, 2004). Of interest, better performance does not lead to general effects of seeing

Figure 21.4 Softball players selected the circle that matched the size of the softball prior to filling out a survey on recent batting performance. (Witt & Proffitt, 2005)

everything as bigger. In the case of an obstacle, better performance leads to small perceptual judgments. Tennis players who returned the ball more successfully than others judged the net to be shorter (Witt & Sugovic, 2010).

The effects of hitting on spatial perception are not limited to athletes. In a modified version of the classic computer game Pong, participants attempted to block balls moving at various speeds across a screen. They used a paddle that they controlled with a joystick, and the task was made easier by increasing the size of the paddle or was made more difficult by decreasing the size of the paddle. Participants estimated the ball as moving faster when they played with the smaller paddle compared with when they played with the bigger paddle (Witt & Sugovic, 2010).

In summary, action-specific effects have been found for a variety of actions including various forms of locomotion, near space actions of reaching and grasping, and performance-based actions such as hitting and blocking.

Evidence for Genuine Effects in Perception

An important theoretical issue is whether these action-specific effects reflect genuine differences in perception, as opposed to changes in the judgments themselves. Participants in psychology experiments tend to be "good subjects" (Orne, 1682). This means that they tend to want to appease the experimenter and perform as expected. The consequence is that when making judgments, participants can be susceptible to demand characteristics. Demand characteristics refer to pressures created by the experimental context to respond in a particular way. For instance, when asked to don a backpack and estimate the slant of a hill, participants might infer that the backpack is supposed to alter their judgments (Durgin et al., 2009). Given that the task of assigning a numeric value to the hill's angle is challenging, participants might be slightly biased to report a larger number when wearing a backpack. This would give rise to a purported action-specific effect, but would be due to a change in the judgments, not a difference in perception.

Differentiating genuine effects on perception from judgment-based effects is quite difficult because perception cannot be measured directly. Researchers can only measure judgments or behaviors and from these infer the underlying perceptual process. Judgments and behaviors are influenced by a host of processes other than perception, and any one of these processes could be responsible for a given effect. However, there are a number of techniques that can be used to help build a case that a given effect is (or is not) perceptual.

Signal Detection Theory. A well-known technique for separating out response bias is signal detection theory (Green & Swets, 1966; Macmillan & Creelman, 2008; Tanner & Swets, 1954), and it might seem that the just-mentioned study is well situated to be analyzed using signal detection theory. However, signal detection theory cannot differentiate perceptual bias from response bias. Instead, it differentiates discriminability from bias. Just as discriminability can be perceptual discriminability or memory discriminability, bias can be response bias but it can also be perceptual bias or memory bias. In the case of a perceptual effect that is a shift in one's perception, the effect will manifest itself in signal detection theory's measures of bias, such as *c*, and not in measures of discriminability, such as *d'*. This was recently demonstrated using simulated data from the Müller–Lyer illusion. The Müller–Lyer illusion is a perceptual bias: the orientation of the tails biases perceived line length. Although the Müller–Lyer illusion is perceptual (and was simulated as a perceptual effect without any response bias), the effect produced large changes in *c* and no changes in *d'*

(Witt, Taylor, Sugovic, & Wixted, 2015, 2016). The study highlights that an effect in *c* cannot be interpreted as a response bias (unless the study involves a true signal-present versus signal-absent comparison). Action-specific effects are like the Müller–Lyer illusion in that the theorized change is a perceptual shift or a perceptual bias. Thus, it is expected that action-specific effects would manifest themselves in measures of bias, such as *c*, and that changes in *c* cannot be interpreted as a response bias, just as with the Müller–Lyer illusion.

Reduced Demand Characteristics. Given concerns regarding the role of demand characteristics, effective strategies are those that reduce demand characteristics. One such strategy is the use of cover stories. For instance, in the original backpack study (Bhalla & Proffitt, 1999), participants were given a series of questions designed to make the participants think the study was about estimating weights, distance, and angles in general, and not specifically about backpack weight and hill slant. Participants were asked several questions before being asked about hill slant. In this study, it was found that those who wore the backpack estimated the hill to be steeper, compared with participants who did not wear the backpack.

However, another study that used a more elaborate cover story found no effect of the backpack on apparent slant (Durgin et al., 2009). Participants were divided into three groups: no backpack, backpack, and backpack with cover story. In the backpack-with-cover-story condition, participants were told that the backpack contained special equipment for recording ankle rotation, which made the backpack seem incidental to the task. Participants in this condition estimated the slant of a short, 2m-long ramp as being similar in steepness compared with the no backpack condition, whereas those in the backpack condition without a cover story estimated the ramp as steeper.

Had this cover story failed to eliminate the backpack effect, the data would have offered a compelling case for a perceptual effect. Yet the lack of effect with the cover story should not be accepted as evidence for a non-perceptual effect without undergoing further scrutiny (Proffitt, 2009; Witt, 2011a). First, showing a potential judgment-based effect in the case of a short ramp does not necessarily generalize to what may be a genuine effect of carrying a heavy load on perceived slant of a large hill. Indeed, many studies have shown that action-specific effects are only apparent when there is a notable difference in one's ability to perform an action in one case compared with another. A short ramp might not be sufficient to lead to these changes, thereby raising questions about the generalizability of these findings. Second, the cover story was so elaborate that it may have conjured up its own demand characteristics. Participants in this condition could have seen the ramp as steeper but felt pressure to respond that it was shallower. Third, the cover story may have been effective at making the backpack seem so incidental that participants anticipated walking up the ramp without the backpack. Indeed, when participants were asked to estimate the weight of the backpack, those in the cover story condition estimated the backpack to be substantially lighter compared with participants who were not given a cover story.

Given that cover stories can be ineffective or even create their own demand characteristics, alternative strategies might offer better solutions for reducing demand characteristics. One strategy is to use opaque manipulations so that the participant is unaware of the manipulation. For instance, the use of a treadmill is not an obvious manipulation of increased effort for walking. Most people experience the sensation that they are moving very quickly after having walked on a treadmill, so if anything, the treadmill manipulation is likely to lead to the impression that walking should feel easier, not harder. Indeed, after describing the treadmill experiment to a group of participants and asking them to predict the outcomes on perceived distance, 47% of participants anticipated that targets should appear closer, which is the opposite prediction of what we find. Only 28% guessed the direction of the effect correctly, and 26% predicted no difference in estimated distance (Tenhundfeld & Witt, unpublished data). Thus, treadmill manipulations are likely to be an effective way to manipulate locomotion without creating a demand characteristic (or at least to create a demand characteristic that works opposite to the perceptual predictions).

Several clever ideas have also been used. In one, participants ingested juice with sugar or with fake sweetener (Schnall, Zadra, & Proffitt, 2010). The participants could not differentiate the drinks based on task, but only the juice with sugar contained energy in the form of calories. These participants estimated the hill as less steep compared with participants who drank the juice with fake sweetener. In a somewhat convoluted challenge, some researchers claimed that the sugary calories gave participants the needed energy to be able to resist demand characteristics and that is why this group judged the hill to be less steep (Shaffer, McManama, Swank, & Durgin, 2013). This challenge highlights the nuanced ways in which demand characteristics can operate. Thus, perhaps the best design to date is one for which participants unknowingly self-selected into groups of high and low energy. Participants in this study estimated the steepness of a long staircase and, as a reward for participating, were given a choice of fruits and drinks (Taylor-Covill & Eves, 2014). One fruit and drink had fast-acting sugar release whereas the other fruit and drink had slow-acting sugar release. The idea was that those low in energy would be more likely to select the fast-acting items. Participants who selected these items, and were thus presumed to have less energy, estimated the staircase to be steeper compared with participants who selected the slow-release items and were thus presumed to have more energy.

Another effective way to reduce or eliminate demand characteristics is to provide feedback on the perceptual judgments. Feedback creates pressure on the participant to respond accurately rather than to be biased by other aspects of the experiment. Feedback on perceptual judgments has rarely been used in action-specific research. In one series of experiments, participants performed the Pong task for which they attempted to block moving balls with small and big paddles and then estimated the speed of the ball. The speed estimation task consisted of classifying each ball as moving more like the slow anchor speed or more like the fast anchor speed. They had been trained on these anchor speeds at the start of the experiment. When participants misclassified the ball (i.e., classified a slow ball as fast or vice versa), they received feedback that their judgment was incorrect. Despite this feedback, paddle size continued to influence estimated ball speed (King, Tenhundfeld, & Witt, 2018). Thus, eliminating demand characteristics via implementation of feedback did not eliminate or even reduce this action-specific effect. This suggests a genuine perceptual effect rather than one based on demand characteristics.

Accounting for Demand Characteristics. Instead of eliminating demand characteristics, another strategy is to explicitly measure and account for their effects. For demand characteristics to be responsible for action-specific effects, the participant must be sensitive to them and willing to alter her response accordingly. Not all participants will fall prey to demand characteristics, and some may even go so far to resist them as to demonstrate effects in the opposite direction.

To explicitly measure participants' willingness to alter their responses, participants in a Pong experiment were given a special set of instructions (Witt & Sugovic, 2013b). One group of participants was told to make sure to classify all of the balls moving at the slower speeds as slow. They were told to not classify any slow balls as fast. Another group of participants were given the opposite directions with an emphasis on correctly classifying fast balls. As hypothesized, these instructions biased responses so that those in the slow group estimated the balls as moving slower compared with those in the fast group. Despite individual differences in perceived ball speed, the instructions produced its own effect on how people estimated ball speed. In a follow-up analysis, participants were classified based on whether their responses complied with the instructions. Those in the slow group who estimated the balls as being slower were classified as compliant, whereas those in the slow group who estimated the balls as being faster were classified as noncompliant (and vice versa for the fast group). Not surprisingly, participants who were classified as compliant revealed the effect of paddle size on apparent ball speed: they judged the ball as moving faster when the paddle was small. The interesting test was for the noncompliant participants. If this action-specific effect is driven largely by demand characteristics, participants who are unwilling to alter their responses accordingly should show no effect of paddle size. In contrast, the noncompliant participants showed the same effect of paddle size on apparent speed. This result suggests that this particular action-specific effect is independent of demand characteristics.

Another technique to assess the potential role of demand characteristics was to use post-experiment surveys to assess participants' awareness of the experimental hypothesis (Witt, Tenhundfeld, & Tymoski, 2018). Participants completed the Pong task in which the background color was manipulated to try to distract participants from guessing the correct hypothesis. Afterwards, they were given a survey that started with an open-ended question about the study's purpose. Participants who accurately guessed the hypothesis at this stage were labeled *discerning*. A follow-up question asked what factors might influence perceived ball speed, and those who correctly guessed paddle size (but had not made this guess for the open-ended question) were labeled *suspicious*. All other participants were labeled *naïve*. First, it is interesting to note that only 25% of the participants correctly guessed paddle size, which suggests that even manipulations that seem obvious to researchers might not be so obvious to first-year undergraduate students. In the study, 50% of participants did not even guess the critical experimental factor when prompted with the follow-up question. All three groups (discerning, suspicious, and naïve) revealed a significant effect of paddle size on estimated speed, and there was no difference in the magnitude of the effect across groups. This data provides additional evidence that this Pong effect is not due to demand characteristics. Furthermore, in a follow-up experiment, participants were explicitly told that people tend to report the ball as faster when the paddle is small and that we wanted them to resist this tendency and to report on how the ball appears without allowing the responses to be biased. Even when given these instructions, the Pong effect was significant and just as strong as in the other experiment. Not only were participants unaware of the hypothesis (and thus makes the effect unlikely to be due to a demand characteristic); they were unable to resist the Pong effect even when explicitly instructed to do so.

The Use of Unbiased Measures. Demand characteristics are more likely to influence some responses than others. Verbal reports may be particularly susceptible, and many action-specific effects have been documented using verbal reports. Thus, it is important to use less susceptible measures such as visual matching tasks, action-based measures, and indirect measures.

In visual matching tasks, participants do not have to assign a numeric value to their perception. Instead, they simply compare two objects and judge which is bigger, farther, steeper, or faster. Or they adjust a comparison object until it is the same as the target object. Visual matching tasks have been effectively used to demonstrate action-specific effects in a variety of contexts such as with slant perception and perceived distance to reachable objects (see Figures 21.1 and 21.3).

Potentially more compelling is evidence using action-based measures and indirect measures. Action-based measures require careful consideration because the action itself is frequently manipulated in these studies. For example, in one study participants threw either a heavy ball or a light ball to targets placed in a grassy field and estimated the distance to the targets (Witt, Proffitt, & Epstein, 2004). Had the researchers measured and found differences in thrown distance across the two ball conditions, few would have taken this as evidence for a perceptual effect as opposed to differences in ability to throw the two types of balls. That said, action-based measures can be used effectively at times.

One strategy to use an action-based measure required a complicated experimental design (Witt, Proffitt, & Epstein, 2010). Participants were divided into two groups. One group of participants was told that they would walk on a treadmill, view a target in a hallway, and close their eyes and walk to it. Another group was told that they would walk on a treadmill, view a target in a hallway, and close their eyes and throw a beanbag to it. Both groups walked on the treadmill, viewed a target 6 m away, then donned a blindfold. At this point, both groups were told to walk to the target. They were not given another opportunity to view the target. As predicted, those who intended to walk blindwalked farther than did those who intended to throw. The idea is that those who viewed the target with the intention to walk anticipated having to exert more effort to walk (having just recalibrated to the treadmill) and therefore perceived the target to be farther away. In contrast, those who viewed the target with the intention to throw did not anticipate having to exert more effort because the treadmill had no impact on the effort required to throw a beanbag (cf. Rieser, Pick, Ashmead, & Garing, 1995). So for this group, the target did not look as far as it did for the group that intended to walk, and this difference in initial perceptions was revealed by their blindwalked distance. Thus, the effect of anticipated effort for the intended action was found with both verbal estimates (Witt et al., 2004) and blindwalking measures (Witt et al., 2010), lending credence to the claim that the effect is perceptual.

Not all action-based measures require such complicated designs. In one series of studies using a treadmill, participants intended to walk in all cases, and effort for walking was manipulated using several variations such as recalibration to optic flow, increased effort due to walking up a hill, and increased effort due to walking faster. Regardless of how effort for walking was increased, distances appeared farther away as measured by having participants reproduce the walked distance (White, Shockley, & Riley, 2013).

In another series of studies, participants estimated the distance to virtual objects placed on a table in virtual reality (Linkenauger et al., 2015). Participants' arm lengths were rendered as 25% larger or 85% of the length of actual arm. When the arm appeared longer, the object appeared closer. In one study, participants estimated the distance to the target by pushing a beanbag across the table. The idea is that if the target looks closer, the participant will not push the beanbag as far. Using this as a measure of perceived distance, the researchers replicated the effect of arm length on apparent target distance. Recall that although it might seem that these body-related manipulations in virtual reality could produce similar effects through non-action-related mechanisms (such as contrast effects or effect based on familiar size), these other mechanisms have been ruled out by showing that the same effects on perceived size are not observed when participants have the same visual cues about arm length but are not able to physically control the arm. That the arm needs to be controllable by the perceiver suggests a role for action in these effects.

Another example of the use of action-based measures come from a modification of the Pong experiments called the fishing task (Witt & Sugovic, 2013a; Witt, Sugovic, & Dodd, 2016). A fish moved across a computer screen at one of six speeds, and participants had to release a net to try to catch the fish. The net always moved at the same speed, but it varied in size, making the task of catching the fish easier or harder. When verbally estimating the speed of the fish, participants judged the fish as faster when the net was smaller than when it was bigger. Unknown to the participants, we were also able to use the timing of the net release as a measure of perceived speed. The idea is that if the fish really looks faster, participants should release the net sooner than if the fish appears slower. We found that participants released the small net earlier than they released the big net. Follow-up studies suggested that the difference in net release time was not due to strategies such as the strategy to catch the fish with the top part of the big net and the bottom part of the small net (which is the strategy that would have led to the same pattern of results). This measure of perceived speed is particularly compelling because it is not obvious that net release time is a measure of perceived speed. Truth be told, it did not even occur to us (the researchers) to analyze this measure until long after the data was collected.

Indirect measures can be equally as compelling for demonstrating genuine perceptual effects as action-based measures. One example of an indirect measure is to have participants estimate the shape of an object as a way to measure perceived distance (Witt, 2011b). In one study, a target circle was placed beyond reach of the arm but within reach when a baton was used. Two circles were placed within reach of both (and therefore are not theorized to be influenced by the use of the tool). The three circles comprised a triangle, and the participants' task was to position the base circles so that the three formed an equilateral triangle (see Figure 21.5). If the target circle genuinely looked closer when reaching with the baton, the participants should move the base circles to be closer together. This is the result that was found. The finding was replicated using slight variations of the design. To date, few studies have used indirect measures, and yet they can be relatively easy to implement and can provide quite convincing data for a genuine perceptual effect.

Memory-Based Effects. In addition to differentiating perception from judgments that are biased by demand characteristics, it is also necessary to account for potential effects of memory. As an example, in one study, participants attempted to throw marbles into an opening, then estimated the size of the opening. Judgments of size were biased but only

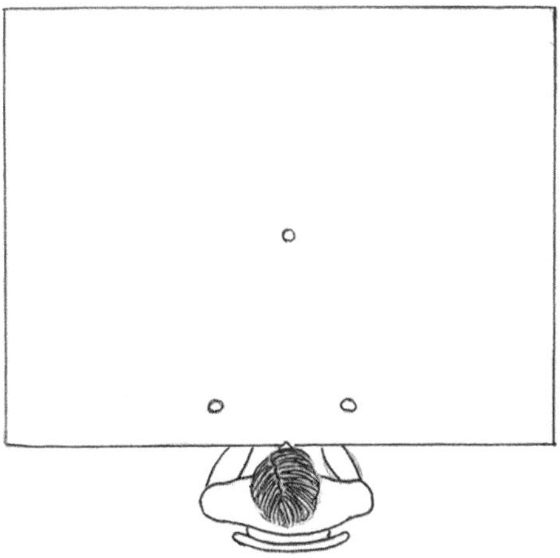

Figure 21.5 Triangle adjustment task used as an indirect measure of perceived distance (from Witt, 2011b). Participants adjusted the width between the base circles until the triangle is equilateral, then reached to the far circle either with or without the tool.

when they were made after the opening was occluded, and so the judgment was based on memory and not perception (Cooper, Sterling, Bacon, & Bridgeman, 2012). However, the majority of action-specific research involves judgments while the object is within view. This includes all of the hill and distance studies and all of the reaching and grasping studies. The only studies for which this criticism is relevant is the hitting studies, and even many of those were conducted with the target within view.

Summary of Evidence for Genuine Effects in Perception. Despite the challenge of differentiating effects based on perception versus judgments, a plethora of research now provides convincing evidence that a person's potential for action can influence spatial perception. This is not to say that all purported action-specific effects are perceptual. Some may reflect perceptual effects, while others reflect judgment-based effects. For many action-specific effects, there is not yet sufficient research to make this determination. Where there is sufficient research, in many cases the evidence favors a perceptual account (Witt, 2017; Witt, Sugovic, Tenhundfeld, & King, 2016). These reviews highlight both strategies that can be used and results consistent with a perceptual account. There are likely other strategies that can be used or developed to help address this issue (Philbeck & Witt, 2015).

To date, there exists no gold standard by which one can access whether a given effect is perceptual. To some extent, this lack of a standard reflects the nuanced ways in which judgments can be altered. For better or for worse, the lack of standard also reflects prior theoretical commitments to accept versus reject top–down effects in spatial perception. Given that bottom–up processes refer to optical (and perhaps oculomotor) information, top–down effects refer to any influence on vision that is not based on the optical information. By this definition, action-specific effects are top–down effects because the optical information is held constant but the perception varies depending on action.

Theoretical approaches such as Gibson's (1979) ecological approach deny any top–down effect. Other approaches such as Fodor's (1983) modularity of mind or Pylyshyn's (1999) distinction between seeing and believing reject all but a few top–down effects. For proponents of these (albeit disparate) accounts, in attempt to preserve the main claims of one's own theoretical approach, it is beneficial to seek alternative, non-perceptual explanations for action-specific effects. As an example, proponents of the ecological approach have argued that the reason action penetrates perceptual judgments of spatial layout is that perception of affordances is so primary that people cannot help but report on the affordances rather than on the geometric properties of the space (Fajen & Phillips, 2012; Lee, Lee, Carello, & Turvey, 2012).

As another example, proponents of the Fodor/Pylyshyn camp have campaigned that all action-specific effects are due to task demands or one of several other non-perceptual explanations (Firestone & Scholl, 2016). To the credit of these authors, they have laid out their own set of gold standards that can now be used to evaluate the perceptual nature of a given effect. Firestone and Scholl (2016) propose six pitfalls that, they argue, can account for all top–down effects on perception including action-specific effects. These pitfalls are (1) having an overly confirmatory line of research, (2) effects of judgment rather than perception, (3) effects due to demand characteristics, (4) effects due to low-level (i.e., optical) differences, (5) effects due to attention, and (6) effects due to memory. Most of these concerns are not new; hence the plethora of research already in place that addresses pitfalls 2, 3, 4, and 6, as reviewed earlier.

The pitfall of an overly confirmatory line of research is quite timely given increased understanding of issues using null-hypothesis statistical testing (NHST) such as p-hacking, low power, and lack of file-drawer disclosures (Neuroskeptic, 2012). Action-specific researchers, like most researchers across the field, have been overly focused on finding significant effects for a variety of reasons including continued attempts to demonstrate the existence of these effects and because significant effects are far easier to publish than are non-significant effects. However, in many of the paradigms, non-significant effects have been reported (or inferred) due to addressing the other pitfalls. For example, to determine that low-level differences did not account for the effect of paddle size on perceived ball speed, we ran several tests for which the visual differences between paddle sizes were maintained but the effective differences in ball-blocking ability across paddle sizes were removed. In this case, we found the predicted null effect, suggesting that effectiveness, and not visual differences, was responsible for the difference in perceived speed (Witt & Sugovic, 2012).

Of the six proposed pitfalls, only the one on attention will be a contentious issue as to whether effects based on attention should or should not count as genuine perceptual effects. All will agree that trivial attention-based effects should not count. An example of a trivial effect is how perception changes when one closes his eyes. What was less clear in the description of this pitfall is how to differentiate trivial from nontrivial attention-based effects. If performing a task well leads to (or was helped by) fixating the target (as suggested by research on the Quiet Eye; Vickers, 1992), and fixating a target makes it appear bigger (Newsome, 1972), does this mean that performance does not influence perception, even if indirectly via attention? For some action-specific effects, the distinction is a moot point, because evidence demonstrates the effects are not driven by attention. To date, these effects include the effect of virtual body size on perceived object size (Van der Hoort & Ehrsson, 2014) and of paddle size on perceived ball speed (Witt, Sugovic, & Dodd, 2016). For other action-specific effects, attention has been shown to play some role in the action-specific effects. These include the effect of golf putting success on perceived hole size (Canal-Bruland, Zhu, van der Kamp, & Masters, 2011) and airplane landing success on perceived runway size in a flight simulator (Gray, Navia, & Allsop, 2014). Thanks to the pitfalls offered by Firestone and Scholl (2016), the issue of attention is likely to dominate the field in the upcoming years. This line of research will help the field make great strides in understanding the underlying mechanism driving action-specific effects, which is an area that has been largely neglected (but see Proffitt & Linkenauger, 2013) due to the tunnel-vision focus on demonstrating the existence of action-specific effects, uncovering their boundary conditions, and differentiating perceptual from judgment-based effects.

Outstanding Issues

Because of the intense scrutiny on the perceptual nature of action-specific effects, researchers such as myself have focused primarily on investigating alternative explanations. As such, this has left less time to address many of the outstanding issues within the field. I will highlight a few of these here. Examining these issues will be important as the field moves forward in uncovering the underlying mechanism(s) that drives action-specific effects.

The action-specific account of perception emphasizes the future over the past. The account claims that a person's ability to perform future actions will be more relevant than past actions. As an example, in one study, two groups of participants threw a heavy ball to targets placed 4–10 m away in a grassy field and then verbally estimated the distance to the targets. After each estimate, one group threw the heavy ball again whereas the other group closed their eyes and blindwalked to the target. Even though both groups had just thrown the heavy ball three times, the group that made their estimate while intending to throw again perceived the targets to be farther away than did the group that intended to walk (Witt et al., 2004). As a counterexample, in the field goal study, performance correlated with perceptual judgments made after kicking but not perceptual judgments made before kicking (Witt & Dorsch, 2009). The issue to be resolved is to understand when prior performance versus future performance will impact perception. Something to consider in addressing this question will be the extent to which future performance can be anticipated by the motor system. In the field goal kicking study, only one participant had had experience kicking field goals. Consequently, it is unclear what information participants had that would have allowed them to anticipate the extent of their own upcoming success. One prediction is that experts (or at least people who do have experience) will be better able to anticipate performance, and for them perceived size prior to kicking will relate to upcoming performance.

A second issue that will need to be resolved concerns the distinction between moment-to-moment performance (trial-by-trial performance in an experiment), mean performance, and skill level. For example, in the golf study we examined putting performance on the last hole (moment-to-moment), course score (mean performance), and handicap (skill). We found effects on perceived hole size for moment-to-moment and mean performance but not skill (Witt et al., 2008). However, we have since been informed that self-reported handicap scores should not necessarily be trusted, and other studies have found effects of skill, but not moment-to-moment performance (Cooper, Sterling, Bacon, & Bridgeman, 2012; Kwon & Kim, 2012; Witt & Sugovic, 2010, 2013a). These various effects should be systematically examined.

A third issue concerns effects, or lack thereof, in two-dimensional displays. Action-specific research has largely taken place in ecological contexts such as big grassy hills and fields or athletic pitches and arenas. But addressing many of the concerns raised earlier requires systematic control that is best suited for lab-based research. The pong task (Witt & Sugovic, 2010) has the advantage of being robust and easy to do in the lab, and thus much research has been done with this paradigm to resolve concerns over the perceptual nature of its effect (Witt, 2017; Witt, Sugovic, Tenhundfeld, & King, 2016). Although unreported, I have been unable to find robust effects of action on perceived size on a computer screen. The Pong experiments demonstrate effects on perceived speed on a computer monitor, and much research has found effects on perceived size in virtual environments (Linkenauger et al., 2013), which is technically a 2D display rendered to appear 3D. To be resolved is why action does not influence perceived size of 2D objects presented on a monitor.

A fourth unresolved issue concerns the role of intention. The intention to perform a specific action can be necessary (Witt et al., 2005) and a critical factor for selecting which aspect of effort is relevant for perception (Witt et al., 2004, 2010). Yet many experiments do not involve explicit intentions, such as experiments on perceived hill slant. Does the ground plane generate a default intent to walk? Do multiple intentions both influence perception, or is only one selected at a time? If only one, how much delay is necessary to switch from one to the other? As made clear by these questions, there is much to learn about the role of intention.

A final unresolved issue concerns the paradox on whether easier action makes objects look bigger versus smaller. In sports-related experiments, athletes performing better than others see the target as bigger. In grasp-related experiments, easier grasping (such as that performed with the dominant hand) makes the objects appear smaller. Action-specific researchers need to be explicit in predicting the direction of the effects. If the approach is so flexible that it can accommodate effects in any direction, the account will fail to have explanatory power (Goldinger, Papesh, Barnhart, Hansen, & Hout, 2016). However, to some extent, this paradox makes sense. Bigger objects are easier to hit (Chauvel, Wulf, & Maquestiaux, 2015; Witt, Linkenauger, & Proffitt, 2012; Wood, Vine, & Wilson, 2013) and smaller objects are easier to grasp. Consequently, it is sensible that objects appear bigger when easier to hit and small when easier to grasp.

Conclusion

In conclusion, a person's ability to perform the intended action influences perception of the spatial layout of the environment, including perceived size, distance, slant, height, and speed. The action-specific account has gone to great lengths to examine the perceptual nature of the effects. To date, standards have been set as to how to evaluate the perceptual nature of these effects, and several types of action-specific effects meet these standards. The future of the field is to further examine other purported effects using these standards, determine the boundary conditions for which action-specific effects exist, and resolve the issues highlighted earlier. These steps will help achieve the overarching goal of uncovering the mechanism or mechanisms driving these effects. Indeed, it will be difficult to fully argue for a perceptual effect until a perceptual mechanism can be shown (Firestone, 2013; Philbeck & Witt, 2015; Witt, 2015).

In addition, research has and should continue to understand the practical applications of action-specific effects (Gray, 2014; Witt, Linkenauger, & Wickens, 2016). For example, action-based biases in spatial perception are likely to contribute to unhealthy lifestyles such as choosing to use escalators instead of stairs (Eves, Thorpe, Lewis, & Taylor-Covill, 2014), especially in people who have greater percent of body mass from fat, who see stairs as steeper (Taylor-Covill & Eves, 2016). In some cases, action-specific effects are beneficial because they could help keep people a safe distances away from the edge of a cliff (Stefanucci & Proffitt, 2009). But when people must overcome their perceptions of impossible terrain and force themselves to be more active, action-specific perception may be working against their overall goals.

Author Note

The author would like to thank Nicole Versace for her artistic renderings of the experimental paradigms. The writing of this chapter was supported by the National Science Foundation (BCS-1348916 and BCS-1632222).

Chapter 22 Spatial Biases in Navigation and Wayfinding

Jan M. Wiener and Tobias Meilinger

Imagine planning a route between two locations in your hometown. Even though you may have never navigated between the specific start and target location, you will be able to plan a route and navigate between these locations. Chances are that several alternative routes exist that are similar in terms of their length and the time needed to navigate them. So why did you decide on that particular route? While such choices (or solutions to the planning problem) may be random, the literature describes several situations in which participants exhibit systematic biases for one route over alternatives, even if alternatives are shorter or faster to navigate. In this chapter we provide examples for systematic biases in navigation and discuss the psychological mechanisms that underlie these biases.

The example that opened this chapter describes a navigation behavior in a large-scale environmental space (Montello, 2003). Such spaces are too large to be apprehended from a single vantage point, and navigating environmental scale spaces typically requires movement between different "nested" subspaces. Navigating nested subspaces poses a number of challenges that are not present when navigating vista scale spaces – i.e., spaces that can be apprehended from a single vantage point (Wolbers & Wiener, 2014) – and that may explain some of the spatial biases observed in navigation. For example, as navigators cannot oversee the entire environment without movement, they need to integrate information across space and time while moving through the environment in order to learn it. This integration can result in systematic distortions in spatial memory, such as overestimation of distances between different parts of the environment (McNamara, 1986), which in turn can systematically affect navigation behavior.

Spatial biases in navigation and wayfinding, however, are not restricted to the generation of movement patterns, but may also manifest themselves in the selection of information that is encoded in spatial memory or the reference frames that is used for encoding information. For example, when navigating and learning large environmental scale spaces, navigators will typically memorize only a small subset of the abundant information they perceive: they will not memorize all environmental cues that may function as landmarks, but will concentrate on cues at navigationally relevant locations such as intersections (Aginsky, Harris, Rensink, & Beusmans, 1997). In order to explain biases in navigation and wayfinding behavior, it is therefore important to consider which information navigators take into account.

Generally speaking, biases in navigation behavior, i.e., the systematic preference for particular paths, routes, or specific movement patterns, result from the strategies that navigators apply. These strategies, in turn, depend on the knowledge that the navigator has about the environment and on the specific task that a navigator tries to solve (Wiener, Büchner, & Hölscher, 2009). In unfamiliar environments, for example, in which navigators lack detailed knowledge about the structure of the environment and the location of the goal, they have to resort to search behavior. Search strategies not only aim to find the target location as quickly and efficiently as possible; navigators are often also concerned with ensuring that they remain oriented and do not get lost. In familiar environments navigators may face a number of different tasks. They may search for a target whose current location is unknown. In this case, navigators will need to search through a familiar space, which means the risk of getting lost is minimal. Alternatively, and as in the scenario introduced earlier, they may know the environment and the location of the target but have never navigated that particular route, in which case they are faced with the task of planning a novel route from their current location to the destination. Finally, human navigators often use external representations of space such as maps when planning routes. Maps provide navigators with the spatial information about an unfamiliar environment required to plan routes. Spatial biases in route choices then depend on how information from maps is interpreted.

Given the importance of the navigator's knowledge about the environment for navigation behavior, we have structured our discussion of spatial biases in navigation and wayfinding as follows.

We will first consider navigation behavior in unfamiliar environments in which the location of targets is unknown. In such situations, navigation behavior in both animals and humans typically aims at exploring the environment or at foraging or finding specific items within the environment. We will highlight that search, foraging, and exploratory behavior is not random, but guided by strategies that result in systematic biases or movement patterns. In doing so, we will discuss behavior at the level of the entire environment, as well as on the level of single movement decision informed by the information that is available locally. Specifically, we will highlight how single movement decisions are informed and influenced by local cues that are available at the decision point.

We will then discuss systematic biases in route planning where environmental knowledge is available, i.e., in choosing between multiple alternative routes through an environment. In everyday life human navigators plan routes from memory (unaided route planning) as well as with the help of external maps (aided route planning). We will consider the psychological mechanisms and memory structures affecting aided and unaided route planning, and we will argue that the interaction of working memory and long-term memory plays a crucial role in explaining spatial biases in route choice.

A summary of the navigation strategies discussed in this chapter can be found in Table 22.1.

Search in Open, Unstructured Environments

Most animal species spend a considerable proportion of their day foraging for food or searching to encounter other organisms. This foraging behavior is typically carried out without prior knowledge of the environment and often results in very specific movement patterns (see Figure 22.1). Similar trajectories have been reported in different animal species and they have particular regularities: there are times when the foragers intensify their search, changing movement directions frequently and therefore remaining in the same area of the environment; at other times they cover relatively large distances before intensifying search in another area of the environment. Such movement patterns have been described in many animal species and are thought to be optimal in situations in which targets are randomly distributed and scarce (Viswanathan et al., 2001; Viswanathan et al., 1999).

Table 22.1 Overview of route choice strategies described in this chapter

Strategy	Description
Search in open, unstructured environments with many goals	
Lévy walks	Randomly choose direction of next segment. Segment lengths are likely short and rarely long
Area restricted search	Increase turning frequency when finding much food (goals) and decrease when finding little food (goals)
Search in built environments	
Perimeter strategy	Move along the outer limits of an enclosed environment (e.g., building)
Central point	Stick to central locations in a building and explore from there
Information gain	Choose street that provides most future information
Route selection with limited environmental knowledge	
Edge following	If the goal is on an edge (street, river, coast), follow this edge
Central point	Stick to familiar (central) areas
Least angle	Always approach the target as directly as possible
Route selection with extensive knowledge – terrain and direction on map	
Flat trails	Prefer flat over hilly routes
South preference	Choose routes yielding initially or generally south (downward) on a map
Route selection with extensive knowledge – reduce planning and memory costs	
Nearest neighbor	Move to the closest of multiple goals next
Longest initial segment	Among the routes leading toward the goal choose the one with the longest straight beginning
Focus on decision points	Only consider intersections where an alternative route might be selected
When in doubt, follow your nose	Only consider turns and move straight by default
Defer path choice	Turn as late as possible
Least decision load	Choose the route with the fewest number of turns
Hierarchical planning	Plan a route the target region first and then the within the target region to the goal

How can we explain such movement patterns? Foraging is often described as being based on a Lévy process (for an overview, see Viswanathan, 2011). A Lévy process is a stochastic motion in which successive movements are random and independent from each other. In other words, the lengths of single steps – here defined as straight-line segments along the trajectory – are randomly sampled from a probability distribution. The shape of the foraging trajectory is defined by the exact nature of this probability distribution. Let us look at the trajectories in Figure 22.1 again: if they are divided into straight-line segments, we find many short segments, few segments of medium length, and even fewer long segments. Analyses of trajectories from different animal species suggest that foraging movement patterns are power law distributed. Specifically, the frequency of step lengths closely fits the probability distribution function $P(l) \sim l^{-\mu}$ with $1 < \mu < 3$. Foraging

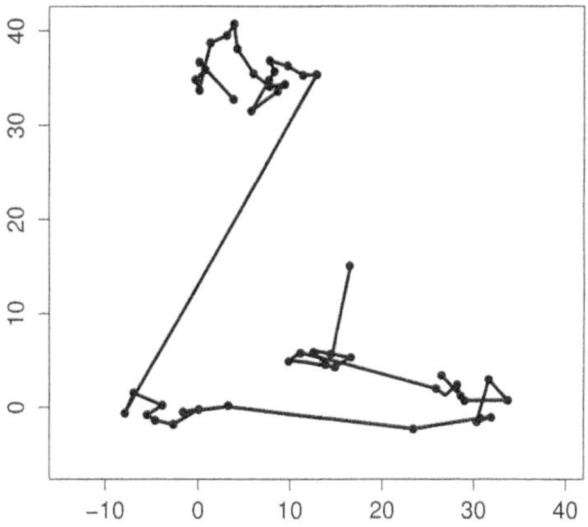

Figure 22.1 Example of a movement pattern generated by a Lévy process.

trajectories that are generated by Lévy processes, so-called Lévy walks and Lévy flights, randomly sample the next step length from such a power law probability distribution.

What is the advantage of Lévy walks or Lévy flights over other types of random walk/flight when foraging in an environment? As compared to other forms of random walk, such as Brownian walks, the few very long steps bring the forager to a different part of the environment and therefore reduce the risk that the same part of the environment is repeatedly searched.

The question whether human movements can be described by Lévy walks has been addressed using a variety of procedures. Brown, Liebovitch, and Glendon (2007) analyzed foraging patterns of Ju/'hoansi hunter-gatherer in the Kalahari Desert in Botswana and Namibia. Gonzalez, Hidalgo, and Barabasi (2008) studied the trajectories of 1 million mobile phone users. Rhee, Shin, Hong, Lee, and Chong (2008) analyzed 1,000 hours of GPS traces of 44 individuals, and Brockmann, Hufnagel, and Geisel (2006) traced bank note circulation as a proxy for people's movement. Data from all these studies suggest that humans move in a manner similar to Lévy walks.

It is important to note that foraging behavior that is based on Lévy processes does not take into account when and where resources or targets are encountered. However, is it reasonable to assume that (human) foragers ignore such information? Imagine searching for mushrooms in a forest. As an experienced forager you know that many mushroom species tend to grow in patches. In other words, the distribution of mushrooms is spatially correlated; finding one mushroom therefore increases the chances of finding another mushroom of the same species nearby. In such situations it makes sense to respond to the encounter with a resource and to intensify the search in the area nearby.

Search behavior in which resource encounters result in increased turning frequency that slowly decreases without further encounters is referred to as *Area Restricted Search* (Hills, Brockie, & Maricq, 2004; Tinbergen, Impekoven, & Franck, 1967). Similar to foraging behavior that is based on Lévy processes, area restricted search can lead to patterns in which agents seemingly switch between two different movement patterns: an intensive search with high frequency of turning, and extensive search with low turning (e.g., Boyd, 1996; Hills, Kalff, & Wiener, 2013). During the intensive search phase, a relatively small area is covered in great detail; during the extensive search, agents cover a larger distance without sampling the environment in great detail.

Area restricted search is particularly beneficial in environments in which food resources are distributed in patches or clumps (Walsh, 1996). Increased turning rate and slowing down after a food encounter means that animals are intensifying their search around the area where food was last encountered. If no more food items are encountered – i.e., if the food patch is depleted – the turning rate decreases and animals move away and expand their search to new areas.

Search in Built Environments

The models describing foraging and search behavior that we discussed above typically do not consider environmental factors other than the distribution of resources or target. However, much human navigation takes place in built environments such as cities and complex indoor architectures, and it is obvious that the form and structure of the built environment affect our navigation behavior: first, simply by restricting movements to the available path options; second, and more importantly in the context of this chapter, through systematic biases in search behavior that result from physical or architectural properties of the environment.

So far, only few case studies investigated search strategies in unfamiliar complex built spaces. Büchner, Wiener, and Hölscher (2009), for example, studied the influence of target location on search behavior and search performance in large unfamiliar and complex architectural spaces. Figure 22.2 shows two example trajectories, highlighting a systematic bias in search behavior. Specifically, participants did not perform random searches, but tended to first explore the perimeter of unfamiliar environments and only eventually explore more central parts of the environment. This so-called perimeter strategy results in superior search performance when targets were located along the perimeter of the environment as compared to more central target locations, even if the central targets were closer to the start location of the search.

Another search strategy in unfamiliar complex architectural spaces that yields systematic biases in navigation behavior is the reference-point strategy (Hölscher, Meilinger, Vrachliotis, Brösamle, & Knauff, 2006). If asked to search for and navigate to several locations in an unfamiliar complex building, a strategy often used is to search and frequently return to the initial starting point (central point). In contrast to the perimeter strategy, such a central point strategy results in more star-like exploration or search patterns. The perimeter strategy and the central point strategy are likely to result from limited knowledge about the environment, as both strategies reduce the risk of getting lost in unfamiliar environments.

The Impact of Spatial Structure on Search Behavior

While the case studies introduced in the preceding sections reveal systematic biases in navigation behavior, they are difficult to compare and to integrate, as the behavior strongly depends on the specific architectural form and structure of the specific environment. To explore how the environment affects navigation behavior more systematically and to compare navigation behavior between arbitrarily shaped environments, it is important to have objective and quantitative descriptions of environmental properties. Space syntax (Hillier & Hanson, 1984) and isovist analysis (Benedikt, 1979) are techniques from architecture that allow the derivation of objective quantitative measures of global and local environmental properties. Both space syntax and isovist analysis have always been seen as closely related to human behavior.

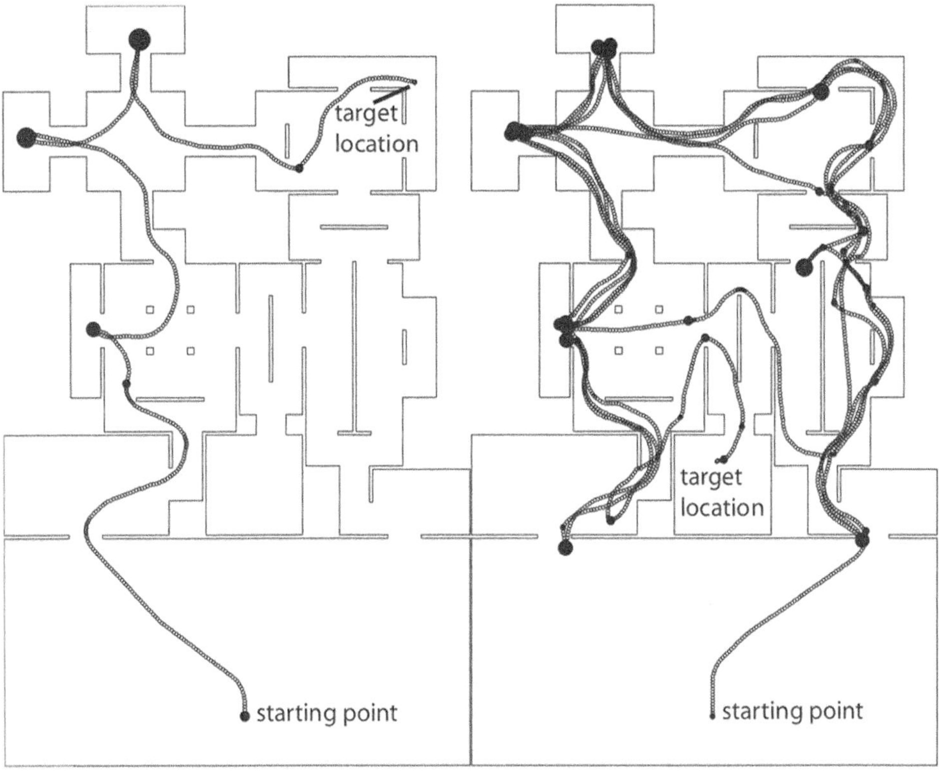

Figure 22.2 Two sample search trajectories. Left: with a target located along the perimeter; right: with a target located more centrally. (from Büchner et al., 2009)

Case studies have demonstrated that space syntax, which provides information about connectivity, centrality, and control level of different parts of the environment, is closely related to pedestrian dispersion in the built environments (Hillier, 1996). Specifically, pedestrians display a systematic bias toward navigating along well-integrated paths. Integration is a measure that describes how many turns a navigator has to make from one part of the environment (e.g., a street segment) in order to reach all other street segments in the environment.

While space syntax provides information about the structure of the entire environment, isovist analysis provides quantitative measures, such as the line of sight, about the visible space at specific locations within the environments. It is therefore better suited to explain behavior at single decision points that may lead to systematic biases in navigation behavior. A number of isovist measures have been shown to be associated with movement decisions (Emo, 2012, 2014; Wiener, Hölscher, Büchner, & Konieczny, 2012). When presented with images of decisions points and asked which of two path alternatives they would explore when searching through complex and unfamiliar built spaces, participants preferred to "navigate" along paths that offered a longer line of sight and that had a higher local spatial complexity, i.e., that suggested that more decision points would be encountered along these path options. A possible explanation for these systematic biases is that these preferred path options promise to reveal more about the unknown space than do the alternatives. In other words, by choosing path options with long lines of sight, searchers maximize the information gain when navigating and exploring unfamiliar environments.

Route Selection with Limited Environmental Knowledge

In case navigators have limited knowledge about the environment, the chance of getting lost during navigation is rather high. Getting lost can have a severe impact on the overall distance traveled and/or time taken to navigate to the destination – more so than differences in route lengths between alternatives (e.g., Meilinger, Hölscher, Büchner, & Brösamle, 2007). Therefore, it makes sense to select or plan routes that minimize the risk of getting lost.

The minimal amount of spatial information that allows navigators to use route selection or planning strategies is information about the target location without information about the environment (uninformed search; see Wiener et al., 2009). For example, navigators may know that the destination is located along a river, coastline, or street but they are otherwise unfamiliar with the environment. In such situations navigators may use a strategy that is referred to as *edge following* (Allen, 1999; Hutchins, 1995). Essentially, they only need to decide on the direction in which they head off and then follow the edge until reaching the destination. While this may result in a route that is substantially longer than the shortest possible route – for example, if the river bends – it is a relatively safe route and the risk of getting lost is minimal.

Another simple navigation strategy that minimizes the risk of getting lost when navigators have limited information about the environment is to heavily rely on parts of the environment encountered before. For example, if some routes within an environment are known, this route network can be used to plan "safe" routes, even if these routes require substantial detours. The central point strategy mentioned before is an example of such a strategy (Hölscher et al., 2006). Navigators tend to frequently return to already known parts of the environment in order to remain oriented rather than trying out different alternatives that might yield shorter routes. Here areas explored first – in buildings, mainly the central points – become known, and navigators stick to them.

In some cases, navigators may have knowledge about the direction and distance to the destination. Imagine, for example, navigating toward a tower that was visible from a distance, or pointed out by a pedestrian, without knowledge of the structure or street network of the environment. In this situation, navigators can use the *least angle* strategy (Hochmair & Frank, 2002). At each decision point, i.e., street intersection, navigators select the path option that is most aligned with the direction of the destination. Note, however, that without knowledge of the street network, this form of local optimization or hill-climbing strategy can lead to non-optimal paths, for example, when selecting a street that approaches the goal directly in the beginning but then continues in a different direction or passes the goal requiring a detour.

Route Selection with Knowledge about the Environment

When navigators possess good spatial knowledge about an environment, either in form of internal representations of space (cognitive maps) or by using externalized representations of space such as maps, they do not have to perform uninformed searches or focus on limited information, but can plan a complete route towards a target location. Systematic biases in route selection can result from different factors and may vary between navigators and environments depending on the exact planning criteria. For example, navigators may choose the route that is easiest to remember, routes that minimize the effort of navigation by choosing short routes or routes that go over terrain that is easy to traverse, or they may prefer to walk straight most of the time or to approach their target as directly as possible.

Influence of Terrain and Direction on the Map

A series of studies investigating route selection from maps have suggested that terrain elevation and route orientation on a map systematically affect route choices.

Terrain elevation can have a strong impact on the effort required to navigate along a route and has been shown to affect route planning. Traversing hilly trails will typically be more effortful than paths of similar length trails along a flat terrain. Indeed, when selecting between alternative routes from aerial photographs, participants tend to prefer *flat trails* (Brunyé et al., 2015).

When selecting routes from maps, participants show a bias for routes traveling initially or generally south over routes traveling north (Brunye, Mahoney, Gardony, & Taylor, 2010; Brunyé et al., 2012, 2015). This south preference bias was explained with a mental bias of estimating north locations as being of higher elevation as compared to south locations, in which case more southerly routes appear to be going downhill and therefore easier to traverse. Note, however, that even though south routes are estimated as the "better routes" over north routes, navigators, when required to actually walk a route with help of a map, show bias to select north routes and exhibit better navigation performance than on south routes (Wen & Kawabata, 2013). The most likely explanation for these preferences are alignment effects (Hölscher, Büchner, Meilinger, & Strube, 2009; Levine, Marchon, & Hanley, 1984). Specifically, north routes are leading upward in the presented map and are therefore easier to translate to the egocentric perspective required for actual navigation. South routes, in contrast, have to be mentally rotated in order to align with the egocentric travel direction, which affects performance. Interestingly, this example highlights a trade-off between the route presumably least effortful to walk and the route easiest to remember and use during navigation.

Reducing Planning and Memory Costs

In route planning, navigators will typically aim for solutions, i.e., routes between their current location and the target(s), that are short, even though this is not necessarily the only criterion (see discussion earlier and later in this chapter). Planning short routes through a network of streets or locations is considered an optimization problem. This is best demonstrated when assuming a typical street network that features an infinite number of paths between any two locations. Planning the shortest possible path in such a network requires precise knowledge of the environment either from memory or from external representations of the environment (such as maps). Planning the optimal path also requires the consideration and comparison of a large number of route alternatives, which is cognitively demanding. Navigators, therefore, use a number of planning heuristics and strategies to reduce the search space. While these planning heuristics do not always result in the optimal solution to the planning problem, they typically allow generating a reasonable solution quickly. In the following, we discuss planning heuristics and strategies that result in biases for particular solutions.

Heuristics for Short Routes

Rather than searching for the optimal global solution, navigators often use more local strategies or heuristics when planning short routes. One such local strategy that is particularly suited when planning and navigating multi-stop routes is the *nearest neighbor* strategy: Instead of planning complex routes in taking all destinations into account, human navigators often simplify the planning problem by simply approaching the nearest target locations repeatedly. Specifically, Gärling and Gärling (1988) demonstrated that pedestrian shoppers tend to first visit the most distant shop, most likely to minimize the effort of carrying bought goods, and then repeatedly select and navigate to the nearest target shop until all shops have been visited. It is important to note that this nearest neighbor heuristic does not guarantee to result in the optimal, i.e., shortest possible, solution, but will typically result in solutions that are not far off the optimum while reducing planning effort considerably.

Another approximation to the shortest route to a single target location is the longest *initial segment* strategy. When confronted with path alternatives presented on maps, participants preferred the paths with the longest initial straight segment in the direction of the goal (Bailenson, Shum, & Uttal, 1998, 2000; Brunyé et al., 2015). Although later route parts of the preferred paths may render two alternatives as equally long and complex, navigators seem to value the initial part of a path as more relevant for route choices.

The initial segment strategy and the least angle strategy (see earlier discussion) were also compared with each other (Hochmair & Karlsson, 2005). In a desktop virtual environment participants chose one of two alternative route segments both approaching a visible target in the distance more or less directly for a shorter or longer distance. Participants switched between least angle and longest initial segment strategies in a way that they minimized the overall distance to the target that also included moving from the end of the segment straight to the target. This suggests that navigators may balance both strategies and in support of the shortest overall route at least for simple routes with all relevant information visible.

Reducing Memory Costs for/during Navigation

When planning and navigating environments without the use of maps, information about the environment has to be retrieved from long-term spatial memory and made available in working memory. Working memory is a capacity-limited system (Anderson, 2000) and may not be able to simultaneously keep all the spatial information required to plan the optimal route. This problem has been illustrated by Hölscher, Tenbrink, and Wiener (2011), who asked participants to first plan and then navigate the shortest possible routes between several start and target places. Routes planned from memory were longer and simpler, mostly relying on main streets, than the routes that participants actually navigated shortly after reporting their route plan. When planning from memory, participants were not able to activate all the relevant information and were often forgetting smaller streets that would have resulted in more efficient routes. The ways navigators deal with limited working memory capacity have been addressed in several studies as well as models of route planning that we will discuss below and that are illustrated in Figure 22.3.

One important way to reduce the memory costs associated with route planning is to focus on information that is essentially required for wayfinding and to discard all other information. As discussed earlier, navigation in built environments and even sometimes in the countryside is restricted to visible paths, streets, or corridors (Allen, 1999). This allows navigators to focus on decision points along their route and largely ignore information in

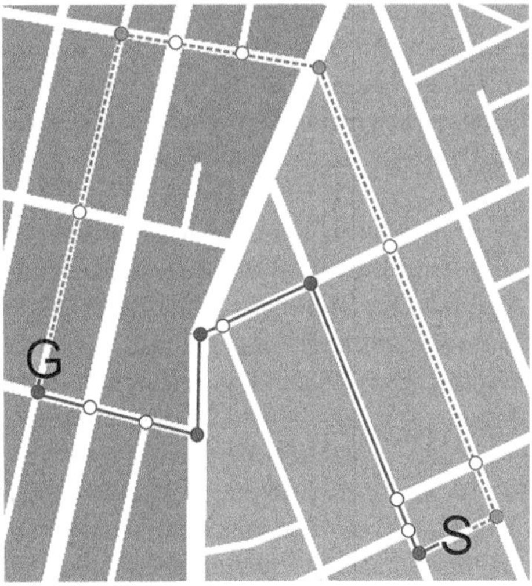

Figure 22.3 Schematic drawing of a navigation task and two alternative solutions. The navigators' task is to select a route from the start S to the goal G. Instead of planning and memorizing the whole route, navigators can focus on the 10 decision points (circles) along the solid blue route. When applying a "when in doubt, follow your nose" strategy, navigators can reduce the required memory by concentrating on the five turns (filled circles). The least decision load strategy would suggest a different route (red dotted line) with even fewer turns, namely three. Note, however, that this alternative is longer. These strategies can be combined with hierarchical planning, which states that navigators plan the route that allows for fastest access to the target region (dark grey houses). This would require three turns along the solid blue route and then an additional two turns to plan the rest of the route to the goal within the target region.

between. Consistent with such a strategy, participants, when asked to describe environments or routes, mention objects (landmarks) located at decision points more frequently and recognized them more reliably than objects located elsewhere in the environment (Aginsky, Harris, Rensink, & Beusmans, 1997; Appleyard, 1969; Cohen & Schuepfer, 1980; Denis, Pazzaglia, Cornoldi, & Bertolo, 1999; Janzen, 2006). These results suggest that navigators primarily encode information at locations that require them to make decisions, i.e., at navigationally relevant locations.

Encoding environmental information primarily around decision points will considerably reduce memory load. Memory load can be even further reduced by focusing on decision points with turns rather than considering all decision points. The "when in doubt, follow your nose" strategy, i.e., to walk straight at decision points by default (Dalton, 2003; Meilinger, Frankenstein, & Bülthoff, 2014; Meilinger, Franz, & Bülthoff, 2012), is an example of focusing on decision points with turns. The "when in doubt, follow your nose" strategy has been described when exploring novel virtual environments (Dalton, 2003), when memorizing and retracing novel routes (Meilinger, Franz, & Bülthoff, 2012), and when selecting and indicating routes within highly familiar environments (Meilinger, Frankenstein, & Bülthoff, 2014). In these experiments, navigators recalled turns more often than straight intersections. Moreover, when asked to indicate how a route continued at single intersections, navigators performed better at intersections that required a straight continuation as compared to a turn. In other words, while navigators focus on turns when recalling environments and when planning routes, they have a bias to walk straight at intersections and have to override this default response when executing routes.

Another navigation strategy that can be interpreted as an implementation of the "when in doubt follow your nose" strategy is the *defer path choice* strategy (Christenfeld, 1995). This strategy states that navigators have a tendency to turn at the last possible turning point. For example, if navigators walk toward a goal that is located in a parallel street, they will move straight as long as possible before making the required turn in order to get to the street with the target. Such a bias to delay turns as long as possible was shown in route choices on maps, in vista, and in environmental spaces when route alternatives were identical in terms of their length and number of turns (Christenfeld, 1995).

In addition to concentration on turns, participants often also choose the alternative with the fewest number of turns (Golledge, 1995). Such a route can also be considered the simplest route. One possible explanation for this strategy, which has also been called *least decision load* strategy (Wiener, Schnee, & Mallot, 2004), is that simple routes with few turns (i.e., route segments) minimize cognitive or memory load and reduce the chance of making errors during navigation.

Another way to reduce cognitive effort during route planning and navigation is to not plan and represent the entire route in detail, but only the parts that are immediately relevant. Wiener and Mallot (2003) suggested a *hierarchical planning* heuristic that exploits the fact that people represent spatial information at different levels of detail (i.e., hierarchical spatial memory; Hirtle & Jonides, 1985). Specifically, they suggest that during planning, navigators form a working memory representation of space in which the immediate surrounding is represented at great detail, while distant places are represented at higher levels of abstraction (regions). Planning a route to a location in a different region using such a *focal representation* will result in a detailed route plan for the immediate movement decisions only. The remainder of the route plan is represented at a higher level of abstraction. Route plans then need to be updated during navigation (see also Wiener, Schnee, & Mallot, 2004). Similar strategies have also been suggested for building or building complexes in which navigators approach the target area (i.e., a building or a floor in a building) as directly as possible (Hölscher, Büchner, Meilinger, & Strube, 2009; Hölscher et al., 2006). Such hierarchical planning strategies reduce working memory load by only making available detailed information for the current surrounding. At the same time, planning costs are reduced, as the search space is reduced.

Influence of the Structure of Long-Term Memory on Route Choices

Biases in route choice can result from planning strategies or heuristics. Biases in route choice can, however, also result from the structure of long-term memory itself. We have already introduced the hierarchical planning strategy that depends on the hierarchical structure of spatial long-term memory (Tversky, Chapter 16 in this volume). It should be noted here that different navigators, all relying on hierarchical route planning, may still come up with different routes, depending on how they cluster locations into regions. Specifically, dividing environments differently into multiple regions will also yield different route choices when planning the same route.

Another source for biases in route choices are *biases in long-term memory* itself. Such biases are common as laid out in detail in Tversky (Chapter 16 in this volume). For example, navigators recall edges such as rivers or streets as straighter (Byrne, 1979) and more aligned with the vertical or horizontal map orientations than they actually are (Tversky, 1981). Also, intersections are often recalled having orthogonal street layout even when there is a huge deviation from that in reality (Gillner & Mallot, 1998). When trying to directly approach a goal or planning the shortest route, such memory biases will yield route choice biases.

Final Considerations

In this chapter we highlighted the main sources for biases in human wayfinding and navigation. Our primary organizing principle for introducing biases in route choice was knowledge about the environment. Without knowledge about the environment, navigators have to search for goals. They might do so systematically: for example, following the perimeter of a building, or choosing streets that promise larger information gain about an environment. Limited knowledge about the environment or the destination can be exploited when walking along edges or familiar areas or when directly approaching a goal location. Within familiar environments or when using external representations such as maps, navigators have all the required information to plan and choose between alternative routes. While most planning strategies aim to produce short routes or routes that minimize locomotion effort, we have discussed how other constraints such as limited working memory capacity or considerations such as minimizing the risk of getting lost affect route planning and can result in systematic biases.

While we have discussed a number of factors that may result in systematic navigation biases, the list is not comprehensive, and we will briefly mention further factors that we cannot discuss in detail here. First, gender has been suggested to impact environmental learning and therefore may result in systematic biases. Specifically, men typically report focusing more strongly on configurations as used in a least angle strategy, whereas women often report relying more on properties of the route (Lawton, 1994, 1996; Lawton & Kallai, 2002). Second, environmental properties have been suggested to influence navigation strategies. For example, in planning routes, navigators aim to minimize the number of turns more in environments with a grid layout than in environments that consist of diagonal or curvilinear streets (Golledge, 1995). Third, even route direction of the same start-target pairs matters: Participants may choose different routes on the way out and back (Golledge, 1995; Stern & Leiser, 1988). Fourth, route selection criteria also differ when planning a route between start and a single goal as compared to planning a route via a third location (Golledge, 1995) and change across the life span (Wiener, de Condappa, Harris, & Wolbers, 2013). Finally, the way we acquire spatial knowledge may affect navigation: Experiments suggest that configurational knowledge acquired through studying maps and route knowledge acquired through active navigation are memorized independently, even for highly familiar environments such as one's home town (Meilinger, Frankenstein, & Bülthoff, 2013).

It should also be noted at this point that route choices and/or biases in everyday navigation can often be explained by various strategies or a combination of strategies. For example, choosing to walk along a particular long and straight street rather than choosing a more complex and winding path could be explained by the initial segment strategy, the defer path choice strategy, and the least angle strategy. Moreover, route choices may also be informed by criteria that are not related to minimizing locomotion or planning effort. Navigators, for example, may choose the *most scenic route* or the *safest route* (e.g., Brunyé et al., 2015) even if that results in substantially longer routes. These considerations highlight some of the difficulties in addressing route choice criteria and strategies using experimental approaches.

Chapter 23

Grounding Social Cognition in Space

Caterina Suitner and Thomas W. Schubert

Our social interactions necessarily take place in actual space; even when we interact in social networks, our real bodies are of a certain size, have a real spatial distance to each other, and are oriented toward each other in some way. In addition, metaphoric spatial cues overlay our communication. As a result, there is a ubiquitous involvement of spatial features in interpersonal and intergroup relations (Maass & Suitner, 2011). Often we actively manage space in order to regulate our social interactions. For example, we may reduce the interpersonal distance in order to approach others and to engage a relationship. We may stand up while arguing, as a vertical advantage may imply a power advantage. On the other hand, we use spatial cues to interpret our social reality. For example, we can guess social bonds by assessing spatial distance, inferring that two persons leaning toward each other on the bus are probably affiliated. Finally, we also use position and movement in space to coordinate interactions. We have cultural conventions regulating our communication (e.g., culture establishes the script direction of our written messages), our movements (e.g., about the side of the road we need to keep while driving), and even our personal hygiene (e.g., prescribing whether the right or left side of the tap regulates hot water). In all these examples, social norms grounded in space are necessary and useful for the task we are accomplishing.

Interestingly, we are so accustomed to using spatial information that we likely do so even when it is not important for our goals or even when it is actually detrimental to our performance. In fact, the spatial embedding of social interactions often emerges even when the spatial features are not critical for the task or goal we are dedicating our efforts to. For example, while deciding which presidential candidate to vote for, the candidates' height may unintentionally enter in the evaluating process, biasing it (Stulp, Bunk, Verhulst, & Pollet, 2013).

The social cognitive functions of space are the core topic of the present chapter. We will address the role of space in basic social-cognitive processes, such as group membership definition (differentiation of in-group vs. out-group), stereotyping (the attribution of characteristics on the base of social membership), emergence of prejudice (whether we prefer or dislike people because of their group membership), and creation of social status (see Box 23.1 for definitions of these concepts from a social psychological perspective). We will address the role spatial features play in these social phenomena, specifically focusing on three major spatial aspects, namely *distance*, *verticality,* and asymmetries in the *horizontal axis*. Each of these is linked to primarily one social aspect, namely *social closeness*, *hierarchy*, and *agency*, respectively. These links between aspects of the spatial environment and the social environment direct our cognition, behavior, feeling, and relations. One could say they *bias* them, and thus use the term *biases* instead;

however, we will mainly stick to the term *link* because we believe that they are mostly functional, both for the individual and for cultures. In the following, we will discuss these three links in turn to describe how they play out and why they are important. In the final part of the chapter, we will turn to the theoretical approaches available to explain them, and discuss what they reveal about function or design principles of spatial and social perception and cognition.

The identification of these links or spatial biases in the social field is particularly relevant, because they represent a pervasive but subtle way in which social relations are created and regulated. This includes both positive and negative processes, such as creating social bonds or discriminating others and maintaining unjustified social hierarchies. Spatial features are part of the nonverbal repertoire of communication, which is on the one hand extremely powerful and on the other hand difficult to be always aware of and to keep under

Box 23.1 Definition of the key social process addressed in the chapter in relation to space

Definition of main social processes and key social concepts

Membership definition. We make sense of the complexity of the social world by grouping people in meaningful categories of people, namely social groups (Crisp, & Hewstone, 2007). The definition of social membership can be driven by an infinite number of criteria, among them physical similarity (see gender identity and age), shared beliefs (see political affiliation and religion), or shared passions (see football supporters). Spatial cues contribute to the classification of people into groups.

In-Group vs. Out-Group differentiation. According to the Social Identity Theory (Tajfel, 1974), persons are automatically and effortlessly categorized into members of the own group (defined as in-group) or members of groups to whom the cognizer does not belong (defined as out-groups). The distinction between in-group and out-group partly depends on how we manage space, and vice versa, social membership may modify our perception of spatial information.

Intergroup bias. According to the Social Identity Theory (Tajfel, 1974), simple classification of in-group and out-group members leads to further intergroup processes such as the favoritism of the in-group (or in-group bias) and the derogation of the out-groups (Tajfel, Billig, Bundy, & Flament, 1971). These basic social processes are also evident when only spatial cues are taken into account.

Stereotyping. Stereotypes are mental representations that describe (and prescribe) traits and behavior of people according to their membership in social categories (Allport, 1954). These attributions create expectancies that are reinforced by confirmatory biases and need for coherence (Bodenhausen & Wyer, 1985). Spatial positioning may be affected by the stereotype content describing a social target, and vice versa, spatial cues may influence the attribution of stereotypic traits.

Social hierarchies. Social group and societies are typically organized in a hierarchical structure, with some members being of higher status, authority, and having more power than others. The bases of the hierarchical organization can vary; typically, dominance hierarchies based on force are distinguished from prestige hierarchies based on freely conferred deference (Henrich & Gil-White, 2001). The status quo is socially defined and maintained both by majority and minority members' need for coherence, stability, and significance (Jost, Banaji, & Nosek, 2004). Spatial cues concur to the maintenance of the status quo by offering consistent visual and mental representations of the hierarchical structure of intergroup and interpersonal relations.

control. While we can often avoid derogatory labels or overtly discriminating behavior when we are motivated to do that, it is more difficult to control nonverbal behavior because it has a large automatic component. Spatial cues are typically extremely subtle, for both the sender and the receiver of a message. It is therefore critical to enhance the awareness of this component of social communication in order to manage interpersonal and intergroup dynamics in a wiser and informed way.

Distance

The first spatial dimension we are going to explore in relation to social processes is distance. Distance is probably the most obvious spatial cue that is available to process social reality. We understand the structure and dynamics of groups by observing the spatial arrangements that members spontaneously create. In a room, people sitting closer are likely to be somewhat affiliated, whereas people kept at distance are typically disliked. For example, Markovits and Benenson (2010) showed participants drawings representing several targets in a courtyard. The targets belonged to a group of friends. Participants were asked to recall the spatial positioning of the targets in the courtyard. The accuracy in recalling the exact position of the targets was higher when friends were placed close to rather than far from each other. This suggests that spatial cues can inform understanding of social relations. Linguistic metaphors are extremely rich in this sense, as we feel *close*, we *approach* and *avoid* others, we *reduce distance*, we *leave* a partner or *join* a group.

Distance has become complicated in modern life, given that the advent of new technologies facilitates social communication between individuals that are physically far away. In the past, spatial proximity was a condition that was necessarily fulfilled in frequent social exchange; today the social exchange between persons that are far apart is easier and more frequent, but of course the actual spatial distance is not reduced. On the one hand, current communication devices facilitate communication, making social exchange available between people that would otherwise be geographically unreachable. Nowadays mediated teaching, coaching, and even health services gain from the power of this enriched communication endeavor. On the other hand, the large spatial distances between communicators may have important consequences for the nature and quality of the social and interpersonal interactions. In fact, spatial distance is critical in both interpersonal relations and group processes because it is highly associated with psychological and social distance (see the Construal Level Theory by Trope & Liberman, 2010, to which we return in the final part on theoretical approaches).

At the interpersonal level, persons that are mentally construed as physically distant are also perceived as more affectively and socially distant: *out of sight, out of mind* (or in the Italian, warmer version, *distant from eyes, distant from heart*). For example, participants primed with spatial distance with a task consisting in plotting two distant (vs. close) points in a chart reported a subsequent weaker emotional attachments to family members and hometowns (Williams & Bargh, 2008, Study 4; but see Pashler, Coburn, & Harris, 2012). It is important to note that distance may also be used to the advantage of the interpersonal relationship, as long-distance relationship favor an adaptive self-disclosure and couple idealization (Crystal Jiang & Hancock, 2013). Indeed, larger physical distance may lead to higher construal in the communicative exchange, with a congruent focus on higher interpersonal goals (Carstensen, Isaacowitz, & Charles, 1999; Stafford, 2010).

Distance is also used to regulate hierarchies among persons. Classic studies of nonverbal behavior assessed the use of proxemic cues (i.e., the distance between persons) in relation to social status and showed that interpersonal distance varies as a function of status differences (Dean, Willis, & Hewitt, 1975). In this case, status creates social distance, which in turn begets spatial distance.

At the group level, spatial closeness is a critical element to identify a group (and differentiate from other groups) and to assess its cohesiveness (Forsyth, 1983). Many social memberships are defined by a shared nearness, such as geographical identities (e.g., national or regional identities), social status identity (a neighborhood is typically assembling people sharing similar socioeconomic status), linguistic identities (people living close typically share the same language), and so on. Interestingly, a similar use of space is observed in virtual reality, where spatial cues signal affiliation, and spatial positioning in the virtual environment is regulated by social norms with patterns that are coherent with those observed in the physical world (Becker & Mark, 1998).

In addition to influences of distance on social cognition, and the active use of distance to regulate relations, the estimation of physical distance may also be affected by social processes such as the activation of stereotypes or the social categorization of a location. A prime of a stereotype may indeed affect the judgment of an observed distance. For instance, in a study by Chambon (2009), participants were primed with the stereotype of elderly in a scrambled-sentence task, including words such as *retired, solitude, wisdom* (vs. words unrelated with elderly). Participants primed with the elderly stereotype judged the distances to be longer compared to the control participants. The activation of the elderly stereotype is assumed to activate the expectancy that a distance is covered with more effort and time, influencing the distance estimation.

Following the equation of social distance with spatial distance, one is translated frequently into the other. Burris and Branscombe (2005) found that the distance between a domestic (in-group) and a foreign (out-group) location was perceived as bigger than the distance between two domestic locations (Studies 1–3). This may be interpreted as a mere effect of categorization, in line with the finding that two points are perceived as more distant when they belong to different (vs. the same) perceptual units (Coren & Girgus, 1980). However, the overestimation of cross-national distances was observed only when comparing domestic location vs. foreign locations and not two locations belonging to two different foreign countries (Burris & Branscombe, 2005, Study 6). Thus, intergroup processes involving the distinction of in-group and out-group seem at play here.

Moving away from or toward others depending on their group membership is a spontaneous behavior that we engage in an automatic fashion. For instance, when performing speeded classification tasks that involve hand movements toward or away from a screen, participants' performance is modulated by the social membership of the targets represented on the screen. Participants are generally faster in performing movements that resemble the approach of in-group members (i.e., moving toward) or the avoidance of out-group members (i.e., distancing from) than approaching out-groups or avoiding in-groups, even when group membership is actually irrelevant for the task (Paladino & Castelli, 2008).

This process may have critical consequences for social and intergroup relations, since distancing others may feed discrimination. Indeed, persons characterized by abnormalities are more likely kept at greater distance (Heinemann, Pellander, Vogelbusch, & Wojtek, 1981; Park, Faulkner, & Schaller, 2003), which of course does not help social inclusion, if that is the goal. Therefore, spatial avoidance is highly related to psychological avoidance, which may further translate into stigma. In an illustrative study by Fasoli et al. (2015, Study 2), participants were subliminally primed either with a homophobic epithet ("faggot"), category labels ("gay"), or a generic insult ("asshole"). After the priming task, participants were told that they were about to discuss issues relevant to gay people with an ostensibly gay person. Participants were asked to position two chairs wherever they wanted in the room, one for themselves and one for the person they were going to talk with. The results confirmed the link between stigma and spatial distancing, as participants positioned their chair more distant from the chair of the gay interlocutor after being exposed to a homophobic word.

It is important to note that proximity not always means interpersonal sharing or social closeness or bounds, as being spatially close enhances the probability not only to become friends, but also to become enemies (Ebbesen, Kjos, & Konečni, 1976). We are more likely to identify enemies among people that we know very well or live close to us rather than among people that are physically (and psychologically) distant. In addition, people may move closer to a disliked person in specific circumstances. This is the case when the movement toward the target has a negative connotation, such as in anger or aggressive behaviors. Or we may perceive a threatening target as closer when we are motivated to monitor risks. For example, Mexico City was perceived as closer by Americans who felt threatened by Mexican immigrants (Xiao & Van Bavel, 2012, Study 3). Therefore, an out-group may be perceived as closer when believed to be threatening. This process has been specifically tested by manipulating the salience of social identity threat (Xiao & Van Bavel, 2012, Study 2). Participants either from New York University (NYU) or from other universities read an article in which Columbia University (main academic rival of NYU) was defined as better than NYU, or read an article claiming that the two universities were comparable. Participants were then asked to indicate the distance between NYU and Columbia. The results confirmed that space is used to distinguish social groups. Participants not belonging to NYU estimated the distance between NYU and Columbia to be shorter after reading that the two universities were comparable. The opposite pattern was observed when participants experienced social threat: For NYU students, the distance estimations were smaller when the social threat of the opponent group was made salient. Therefore, the meaning of distance in the social domain may assume different semantic properties according to the relevance of the social categorization for the self. This is in line with studies investigating the effects of personal insults on approaching behaviors (Harmon-Jones & Sigelman, 2001; Roidl, Siebert, Oehl, & Hoeger, 2013).

The motivational component therefore plays a key role in defining the direction of the effect of spatial distance in social cognitive processes. According to Balcetis (2015, p. 2), "perceiving objects as closer than others (perceived proximity) could encourage action, while perceiving objects as farther than others (perceived extremity) could inhibit action or promote reevaluation." Consistent with this line of reasoning are the results by Cesario et al. (2010), which confirmed that a distancing behavior elicited by a threatening out-group (i.e., a subliminal prime of black faces) is more likely when participants are tested in an enclosed booth than in an open field. In this case, the distance was not defined in terms of interpersonal space, but rather in terms of experimental setting. The enclosed setting enhanced the association between danger and out-group and inhibited escaping responses, in favor of fight responses, even if the stimulus was always presented on a computer screen, at the same distance from the participants.

Taken together, the results show that distance is a dimension that is inherently socially meaningful. It is defined as distance from one's own body to others, or between others that are perceived or imagined in the environment. Empirical evidence confirms that distance is strategically used by people to manage interpersonal and social interaction according to their motives, and that social categorization in turn affects how we judge distance. This bidirectional link of spatial distance and social closeness is not necessarily useful, nor is it always pertinent to the task at hand; on the contrary, it may be a subtle process that reduces social inclusion, intergroup dialogue, and equality.

Asymmetry in Vertical Dimension

Let us now turn to the next dimension of space, verticality. Two features shall be of special interest here: size (especially tallness) and vertical difference (as in being higher or lower). Both are related to each other, and both are again inherently socially meaningful.

Size in general is a stable cue for dominance for both children and adults (Lourenco, Bonny, & Schwartz, 2016). In everyday life, this is most obvious for tallness. In dyadic interactions, the taller individual is more likely to be dominant (Stulp, Buunk, Verhulst, & Pollet, 2015). Bodily height in particular contributes to social status in various domains: workplace success and income (Judge & Cable, 2004), referential authority (Stulp, Buunk, Verhulst, & Pollet, 2012a), being preferred as a leader (Murray & Schmitz, 2011; Re et al., 2012), and even election outcomes (Stulp, Buunk, Verhulst, & Pollet, 2012b). It has little effect on military rank, where facial dominance is more important (Mazur, Mazur, & Keating, 1984). These effects are also robust if gender differences are controlled for. On average, bodily height is a robust predictor of achieved status and income at least for men (Gawley, Perks, & Curtis, 2009; Hensley & Cooper, 1987; Melamed & Bozionelos, 1992;), and it may confer an evolutionary advantage (Mueller & Mazur, 2001). Because it suggests dominance, tallness can also have negative consequences; for instance, taller soccer players are more readily suspected of committing fouls (van Quaquebeke & Giessner, 2010).

Such differences arise because taller individuals may behave differently, but also because others treat them differently. For instance, taller individuals are more likely to believe they could be good leaders, and be interested in becoming one (Murray & Schmitz, 2011). Larger toddlers are more likely to be aggressive (Raine, Reynolds, Venables, Mednick, & Farrington, 1998), and thus become more dominant and attain higher ranks in dominance-based hierarchies. At the same time, preverbal infants understand that larger agents typically get their will, and are surprised if they do not (Thomsen & Carey, 2013; Thomsen, Frankenhuis, Ingold-Smith, & Carey, 2011). Consequently, behavior in hierarchies will be adjusted due to others' height. Differential treatment by others and different behavior due to being taller are likely to mutually reinforce each other.

Similar to the link between spatial distance and social closeness, the psychological link between bodily height and status is bidirectional: Individuals who are dominant or have authority are also judged to be taller (Dannemaier & Thumin, 1964; Higham & Carment, 1992; Wilson, 1968). Similarly, exhibiting nonverbal cues of power increases apparent bodily size (Marsh, Yu, Schechter, & Blair, 2009), as does holding a weapon (Fessler & Holbrook, 2013). This bidirectionality even seems to apply to cognition about the self: Assuming poses that extend the body in space makes people feel powerful compared to assuming constricted poses (Carney, Cuddy, & Yap, 2010), while being powerful leads people to underestimate others' size (Yap, Mason, & Ames, 2013) and overestimate their own height (Duguid & Goncalo, 2012).

In addition to the bidirectionality of the psychological link, there is a larger bidirectional influence between the ecology created by humans and the psychology of tallness. In a series of content analyses, Giessner and colleagues have shown that visual media use photos taken from a lower angle when portraying powerful individuals than when portraying powerless individuals. When participants are asked to select such pictures in the lab, they do the same, especially when the context renders power salient (Giessner, Ryan, Schubert, & van Quaquebeke, 2011; see also Frimer & Sinclair, 2016).

The literature in this field often does not distinguish between dominance-based and prestige-based hierarchies (Cheng, Tracy, Foulsham, Kingstone, & Henrich, 2013; Henrich & Gil-White, 2001). One reason to attend to this distinction is the possibility that tallness covaries positively with higher muscle mass and thus higher formidability, which in turn is related to dominance, especially if the dominance hierarchy is built through physical contests. Thus, the influence of tallness is potentially mediated by physical formidability. After reviewing data on this issue, Blaker and van Vugt (2014) concluded that height is related to attaining higher positions in both dominance and prestige-based hierarchies, while muscularity has an additional effect in dominance hierarchies.

The evidence discussed so far concerned a very particular vertical spatial cue, that of the human body itself. However, the link between hierarchies and vertical space is broader than that: Verticality is connected to hierarchies also in a schematic, abstract form. Several lines of studies show evidence for this idea. For instance, implicit associations exist between power and high-level topography and the cardinal direction North rather than low levels and South (Gagnon, Brunyé, Robin, Mahoney, & Taylor, 2011). Schematized vertical spatial cues are also used in org charts. In one experimental study, Giessner and Schubert (2007) manipulated such an org chart that was shown to participants; the line connecting the only leader to five subordinates was either short (equivalent the height of the boxes in the chart) or long (about five boxes high). Leaders depicted with the larger vertical difference were judged as more powerful.

The schematization can go still further (Schubert, Waldzus, & Seibt, 2010). Repeatedly comparing vertical lines that differ substantially primes larger power differences than control conditions of comparing vertical lines that differ very little, or comparing horizontal lines of both kinds (Giessner & Schubert, 2007). When a portrait is shown at the top of the screen rather than at the bottom of the screen, more dominance is attributed to the person (Meier, Hauser, Robinson, Friesen, & Schjeldahl, 2007; Schubert, Schubert, & Topolinski, 2013; Schwartz, Tesser, & Powell, 1982). Being associated with such cues can have downstream consequences beyond the hierarchy: When men's pictures are shown at the top rather than the bottom of the screen, they are rated as more attractive; that is not true for women (Meier & Dionne, 2009).

Similar evidence exists from Stroop-like interference paradigms that show the automaticity of this process. Schubert and colleagues showed that judgments of verbal labels of roles and persons (e.g., "master" or "slave") are facilitated, and thus faster and more accurate, when spatial cues are in line with the suggested hierarchy position. For instance, judgments are facilitated when powerful labels are presented above another label, alone at the top of the screen, in large font, or when their judgment requires pressing the "up" cursor button (Schubert, 2005; Schubert, Waldzus, & Giessner, 2009; Schubert, Waldzus, & Seibt, 2008). Finally, judging a group as powerful drives attention to upper regions of space, compared to judging a group as powerless (Zanolie et al., 2012).

Equivalent findings from reaction time paradigms have been reported for studies run in Mandarin, and extended to judgments that did not explicitly involve power but required participants to indicate whether the target word labeled a person or an animal (Jiang, Sun, & Zhu, 2015; Lu, Zhang, He, Zheng, & Hodges, 2014; see also Jiang & Zhu, 2015); they thus do not rely on Western cultural norms. Interestingly, a similar effect has been reported in chimpanzees: When they needed to discriminate face identities, they performed better when high-ranked individuals were presented higher than low-ranked individuals were. This suggests that the effect does not depend on language and metaphors (Dahl & Adachi, 2013).

However, it should be noted that finding such effects tests specific cognitive processes under very specific conditions. It should not be assumed that individuals who are larger or elevated in space are automatically and indiscriminately attributed more power and dominance. If the context does not make hierarchies salient, the cue might have no influence (Giessner, Ryan, Schubert, & van Quaquebeke, 2011). If other information contradicts the spatial cue and its social implications, contrast effects are possible. For instance, if an underachieving target is encountered elevated on a pedestal, respect for the target is reduced rather than increased compared to no elevation (Schubert et al., 2013; this study used a cardboard figure of the target).

In summary, there is comprehensive evidence that height and elevation are tied to social status in hierarchies and associated concepts such as dominance and power, sometimes collectively referred to as "V" (Hall, Coats, & LeBeau, 2005). This link is bidirectional in the sense that spatial perception influences social judgment and social judgment influences judgment of spatial dimensions (although perhaps not spatial perception itself). It is also bidirectional in the sense that naturally occurring features such as tallness of bodies are mimicked by making bodies larger using postures, clothing, or taking pictures from certain angles, elevating bodies in space, or even building large structures (Joye & Verpooten, 2013).

Asymmetry in the Horizontal Dimension: Direction

Within the horizontal dimension, a comprehensive bias known as the Spatial Agency Bias (SAB; for an overview, see Suitner & Maass, 2016) links the direction of the written language with the horizontal representation of *agency*. Agency is defined in general as the actual or potential capability of acting, and within the social domain as the *stereotypic* attribution of agency. Agency (often labeled as competence or as masculinity) is one of the core dimensions (together with warmth; Bakan, 1966) of social judgments, and it is widely used to differentiate the stereotypic description of social groups (Cuddy, Fiske, & Glick, 2007), as well the self, from others in general (Abele & Wojciszke, 2007). For example, agency is stereotypical of social groups such as males (vs. females; Spence & Helmerich, 1974) or young (vs. older) persons (Cuddy, Norton, & Fiske, 2005).

According to the model behind the SAB, the direction of written texts is so frequently experienced that it ends up shaping the mental representation of action, or agency. Indeed, rightward readers/writers such as Italian or English speakers usually represent action with a rightward trajectory, placing the agent to the left of the recipient (Maass & Russo, 2003; Chatterjee, 2002). To the contrary, Arabic speakers (whose language is written from right to left) typically show a reversed or reduced bias, according to their exposure to rightward written languages (Maass & Russo, 2003) and to the standard word order of their language (Maass, Suitner, & Nadhmi, 2014). The SAB emerges in a variety of tasks. For example, when drawing interpersonal interactions, the agent is positioned at the spatial starting point of the written language (Chatterjee, 2002), and the direction of stroke movements in drawing follow writing direction, but only for literate participants (Vaid, Singh, Sakhuja, & Gupta, 2002; for a similar comparison of literacy and writing direction in the SNARC effect, see Zebian, 2005); in memory tasks, the performance is facilitated by placing the dominant element on the left for English speakers and on the right for Farsi speakers (von Hecker, Klauer, Wolf, & Fazilat-Pour, 2016); when evaluating the power or beauty of a stimulus such as a soccer goal, it is perceived as more beautiful when rightward oriented by Italian participants and when leftward oriented by Arabic participants (Maass, Pagani, & Berta, 2007); in archival studies comparing productions with opposite writing direction systems as cultural backgrounds (i.e., from Spanish vs. Iranian artists), artworks' spatial layouts are in line with writing text (Pèrez Gonzàlez, 2012). Despite the potential differences between the methodologies of these studies, they share a critical common ground linking spatial bias to writing direction, since in each of these tasks the grounding role of writing direction has been tested and confirmed. Acquisition of writing direction is therefore critical for the emergence of the SAB, with more intense motor acquisition of direction resulting in stronger subsequent bias (Suitner, Maass, Bettinsoli, Carraro, & Kumar, 2017).

The relation between the spatial horizontal features of a stimulus and social processes is not straightforward, since it requires two conventionalized prerequisites: namely, that an asymmetric motion is familiar because of the experience with an asymmetric writing direction, and that this familiarity is used to make sense of social reality. The process behind the SAB is that the asymmetric experience with texts is generalized to the mental representation of the social world, with stereotypically agentic targets mentally moving in line with the direction of the written text, or placed at the starting point of the text vector.

This link has been investigated in left-to-right writing system countries in the context of gender stereotypes; according to such stereotypes, women are less agentic than men are (Spence & Helmerich, 1974). When asked to position male and female players in a sport court (e.g., a tennis field), Italian participants position males to left and female to the right (Maass, Suitner, Favaretto, & Cignacchi, 2009). When asked to express a preference for name orders of gender couples, participants typically prefer the order envisaging the male name to the left of the female name, but this was evident only when the male was described according to gender role expectations, namely having a high-status job (Carnaghi, Piccoli, Brambilla, & Bianchi, 2014). Even in the academic context, and in particular in the representation of gender differences within graphs published in American Psychological Association journals, the spatial layout confirms gender stereotypes, with male bars being systematically (namely in the 72% of the coded graphs) positioned to the left of female bars (Hegarty & Buechel, 2006; for converging evidence in experimental studies, see Hegarty, Lemieux, & McQueen, 2010).

The literature on the spatial layout of artistic production provides further evidence for the SAB in relation to gender stereotypes. Archival studies on Western artworks investigating the direction faced by portrayed individuals has shown that there is a general bias for depicting the targets facing toward the left of the observer (Gordon, 1974; McManus & Humphrey, 1973; however, a rightward bias has been recently reported for Western advertisement by Burkitt, Saucier, Thomas, & Ehresman, 2006), and this has been interpreted in relation to the passive role of the sitter that is posing for the artist (see Chatterjee, 2002; Suitner & McManus, 2011). Interestingly, this leftward bias is generally stronger for female targets (whereas in Burkitt, Saucier, Thomas, & Ehresman, 2006, the rightward bias is stronger for products targeting male consumers, which is conceptually coherent with the content of gender stereotypes, as we will discuss in a moment). Interestingly, if the gender of the sitter is taken into account, the leftward bias for female sitters disappears, suggesting that space is used to differentiate in-group members from out-group members.

Given that in gender roles the intergroup process is inherently linked with the content of stereotypes, it is worthwhile to discuss an unpublished study that directly tested the strategic use of space in the intergroup context. The intergroup context was football, which is in Italy a social membership that is extremely important among football supporters. Italian football fans were involved in the study and were asked to position two very famous football teams in a soccer field. One of the two teams was the team they supported; the other was its main opponent. The two teams were to be placed in the soccer field so that the team placed in the left hemi-field would run rightward toward the goal (and vice versa). Participants overwhelmingly positioned the team they supported to left side, moving rightward, and the antagonist team to the left side moving leftward. This bias was replicated by asking participants to position the football team after a victory or a defeat. Again, the strategy was to position their own team to the left, independently of the outcome of the game. Therefore, it is the intergroup process and the differentiation between in-group and out-group, rather than the strength attributed to the team, that is driving participants' decision. Moreover, this bias was stronger among those participants who highly identified with the supported soccer team, confirming the strategic use of horizontal spatial features of the visual field to differentiate the in-group from the out-group.

The SAB may also emerge in self-presentation contexts, where either the stereotypes or the communication goals may drive the spatial layout chosen to represent the self. For example, in self-portraits, male Western artists tend to portray themselves facing rightward (from the observer perspective), whereas Western female artists tend to portray themselves facing leftward. Interestingly, the bias among female painters has diminished over time, suggesting a process of female emancipation (Suitner & Maass, 2007). The most striking evidence of the strategic use of spatial cues in self-presentation has been provided by Nicholls, Clode, Wood, and Wood (1999), who asked participants to pose either for a family picture or an academic picture. Participants exposed their right cheek (therefore facing rightward in the picture) to show their academic value, and put forward their left cheek to express warmth for the family context (Nicholls, Clode, Lindell, & Wood, 1999; see Lindell, 2013, for a recent review of the relation between profile direction and emotion). This result has been confirmed by evidence showing that participants scoring higher in emotion expressiveness – which were mainly females – were more likely to pose for a photograph putting their left cheek forward (Nicholls, Clode, Lindell, & Wood, 2002).

The evidence reported so far suggests that the horizontal space is used to confirm gender stereotypes, and in turn that gender stereotypes make a biased use of spatial cues more likely. The final issue we would like to discuss is whether space may be used to change attitudes. This has been explored by Suitner, Maass, and Ronconi (2017) in a paradigm of spatial conditioning. Participants were asked to perform a gender categorization task, in which they were exposed to pictures of male and female actors with neutral emotionality. The targets were portrayed in profile, and the proportion of male and female targets facing leftward vs. rightward was experimentally manipulated. The results confirmed the general SAB, with rightward male profile and leftward female profile being processed more easily even when appearing in equal numbers. More importantly, the exposure to an increased proportion of rightward-facing female targets reduced the subsequent beliefs in benevolent sexism, suggesting a potential role of spatial association in modifying our worldview.

Of course, the SAB is related not only to gender stereotypes. Stereotypes about age (Maass, Suitner, Favaretto, & Cignacchi, 2009), or indeed any stereotype addressing agency, may be linked to horizontal space. The use of the horizontal mental schema for representing social reality is subtle but pervasive, since it emerges in various tasks and contexts, with a coherent pattern. Any target that is eventually defined in terms of agency may benefit or suffer from this link between the horizontal spatial dimension, writing direction, and agency.

Theoretical Accounts

In the preceding sections, we have presented three areas of evidence on the links between space and social meaning: (1) distance is connected to social closeness; (2) size and elevation are connected to status in hierarchies; and (3) horizontal orientation or movements are connected to agency. The three areas of evidence share several features. There is typically evidence for bidirectionality in the sense that spatial cues impact social judgment, and that social representations impact spatial judgment. Moreover, the evidence shows that people actively reify and embody social representations in actual space and through artifacts and their own behavior, both spontaneously in the lab and in large-scale coordinated actions (e.g., building monumental architecture). The three links can be analytically separated, but in (spatial) reality they can all operate simultaneously (e.g., Frimer & Sinclair, 2016).

So far, we have for the most part focused on presenting the evidence. We will use the remainder of the chapter to explore possible theoretical explanations for these effects. A host of social-cognitive theories have been used to guide the cited studies or to interpret their findings. They are for the most part not mutually exclusive, and no comprehensive accounts of this large array of findings exists, or even for one of the areas alone. Some accounts focus on the mere existence of the link between spatial dimensions and social meaning, explaining little else. Some theories focus on explaining where the link originates, taking evolutionary or ontogenetic perspectives. Other theories focus on explaining how the link influences social perception and judgment.

Spatial Distance and Psychological Distance: A Construal Level Approach

Construal Level Theory (Trope & Liberman, 2003) can be used to explain many of the findings on distance in the first part of our chapter. It introduces the concept of psychological distance and proposes that it is linked to a specific construal of events and actions. More distance is linked to more abstract construal, and less distance is linked to more concrete construal. The link is assumed to be bidirectional. Psychological distance is assumed to come in four varieties: spatial distance, social distance (linking the concepts we introduced in our first part), temporal distance, and hypotheticality. All dimensions are egocentric: They depart from the self in the here and now, which is zero on all four dimensions. Psychological distance is thought to be the common meaning of all four dimensions, and to be automatically assessed when one of the dimensions is processed cognitively. Through this common meaning, cues of distance on one dimension can affect perceived distance on other dimensions (for a review, see Trope & Liberman, 2003).

Construal Level Theory can explain many of the interactions between social closeness and spatial closeness that we have cited, although it was not the theoretical frame of reference for many of these studies. It also has the advantage of widening the focus to temporal distance and hypotheticality, and to bring in construal level as an explanatory process. However, its focus is actually on construal level, through which it explains effects on judgments, and less on the associations among the dimensions themselves. The theory suggests that associative learning underlies the links among the four distance dimensions (Maglio, Trope, & Liberman, 2013), as well as the link between distance and construal level (Henderson, Wakslak, Fujita, & Rohrbach, 2011). As a primarily cognitive theory assuming overgeneralization and association rather than functionality behind the interaction of the dimensions, it may be less well equipped to theorize influences of motivation (e.g., Xiao & Van Bavel, 2012). Because of its focus on distance (and thus difference), it also is less well equipped to reason about magnitudes (Kanten & Teigen, 2015). This also restricts its application to our other two biases (but see Smith & Trope, 2006 for a conceptualization of power as a cause of distance; for a discussion of the relation between horizontal bias and Construal Level see Suitner & Giacomantonio, 2012).

Size and Verticality and the Hierarchical Status

Theoretical approaches to the link between hierarchy and verticality differ mainly with respect to the role learning is thought to play. Very often, the associations between spatial and other embodied cues and social concepts have been called *metaphors*, which suggests that they are based first and foremost on language. However, not all associations are reflected in language. For instance, no metaphor exists that links agency to rightward movement (i.e., we are not saying "You are very rightward moving today" to mean "You are very active today"). Instead, it is useful to distinguish between linguistic metaphors and the associations themselves.

Early accounts emphasized learning, especially through parent–child interactions that provide plenty of experiences where smaller individuals are dominated by larger individuals (Lakoff & Johnson, 1980; Schwartz, Tesser, & Powell, 1982; for an overview, see Casasanto, 2014). Learning is also the basis for theories of embodied cognition (Barsalou, 1999, 2008) that have been used to explain links between space and hierarchy (Schubert, 2005).

However, it seems unlikely that experiential learning is the exclusive source of the link. Language and linguistic experience are likely to play a role in the acquisition of many embodied concepts. Plenty of conceptual metaphors link social hierarchies and power to spatial relations and distance to social closeness (Lakoff & Johnson, 1980). Language can evoke spatial imagery that then gets associated with other concepts. For instance, the association of conservative politics with spatially right might be mediated by consistent use of the linguistic label "right" (Casasanto, 2014; van Elk, van Schie, & Bekkering, 2010). It is quite possible that additional meaning is introduced by reasoning using linguistic metaphors, for instance through adding knowledge about gravity to spatial understanding of power (e.g., as in "pride always goes before a fall").

Nevertheless, in the particular case of the power-size metaphor, the evidence suggests that language is not *necessary*: Both preverbal infants and chimpanzees show versions of linking size or elevation and hierarchy (Dahl & Adachi, 2013; Thomsen & Carey, 2013; Thomsen, Frankenhuis, Ingold-Smith, & Carey, 2011). Evolutionary accounts have emphasized that bodily size in particular should be a ubiquitous cue to dominance in species in which dominance is based on physical contests (Blaker & van Vugt, 2014). Others have proposed that not just the use of bodily size for dominance judgments but also the use of size in general is an inherited, adaptive process; for instance, Joye and Verpooten (2013) proposed that a "sensitivity for bigness" was exploited by the builders of prehistoric monumental architecture in order to foster social coordination through cueing awe and building hierarchies.

An integrated account of inherited tendencies and learning is offered by Conformation Theory (Fiske, 2004; see also Fiske, 2000). Its rationale is that both children and adults are equipped with a sensitivity to attend to and to interpret magnitude cues of various kinds as cues for hierarchies. The magnitude cues listed by Fiske include size and elevation but extend to mass, force, number, time, and loudness; brightness can also be added. Together, these make up an iconic physics of hierarchies. Crucially, the theory argues that these cues are implemented and elaborated upon by culture, which specifies which magnitude cues are meaningful for which hierarchical relations, and directs appropriate behavior and motivation in the service of coordination in authority relations. These cultural complements have to be *learned*, but this learning is scaffolded by *innate* sensitivity to specific media for specific relations. The process should be partly mediated by language. Fiske proposed that this conformation system for authority ranking in hierarchies is one of four such systems.

In line with Fiske's generalization of spatial cues to magnitudes, recent work has tested whether effects that are typical for magnitude processing in general occur for hierarchy processing as well. This would confirm that hierarchy in the form of status, power, and dominance is processed as a magnitude in the human mind (Immordino-Yang, Chiao, & Fiske, 2010). Similar to other magnitude dimensions, hierarchy processing shows the distance effect (Chiao, Bordeaux, & Ambady, 2004; Chiao et al., 2009; von Hecker, Klauer, & Sankaran, 2013) and may recruit the inferior parietal cortex (Chiao, 2010; Chiao et al., 2009; Quadflieg et al., 2011). This suggests that the magnitude system, which underlies processing of all kinds of magnitudes and is one basis of numerical cognition, could be crucial for forming the link between verticality and hierarchy.

Any promising theoretical account for the link between hierarchies and verticality requires an integration of inherited preparedness, an explanation of how representational systems are shared or linked, and an understanding of how culture-specific implementations of relevant cues are learned. It seems to us that similar integrative accounts would be fruitful for other links between space and social meaning as well. Regarding distance, an evolutionary perspective suggests that physical distance is a fundamental defensive device to keep away persons that are potentially dangerous, in order to avoid threats such as conflicts or even contagion. At the same time, if physical interaction is the goal, either because of positive attitudes or because of one's own aggressive intent, physical closeness is necessary. Closeness in general facilitates cooperation within one's own group. It would be surprising if the understanding of closeness and its regulation through approach and avoidance were not evolutionarily prepared. At the same time, it is clear that learning is required to understand how distance is regulated, and that a certain cultural variability exists regarding what specific distances are kept with whom.

Horizontal Trajectory and Agency

Does a similar evolutionarily prepared foundation underlie the spatial agency bias? At first glance, it might seem as if the spatial agency bias differs because it relies on an acquired cultural technique (reading and writing). This is true for the particular difference of comparing left and right in a specific culture. However, behind the directional effect is the larger bias of associating agency with actual movement or implied movement (through facing direction) in the first place: Moving objects, especially if their movement is self-propelled, get attributed agency, and this attribution is comparable to the association of spatial and social closeness and verticality and hierarchy (Carey, 2009; Leslie, 1994). It seems to be one of the few evolutionarily prepared core concepts, is present in other mammals, and is active in preverbal infants. This system is likely to underlie the learning of the spatial agency bias, which in effects associates one overlearned *direction* of movement in particular with agency.

Conclusions

In summary, we suggest that the meaningfulness of links between space and social meaning has to be understood from a variety of perspectives. Phylogenetic and ontogenetic processes most likely combine to form particular manifestations. Various cognitive processes from language to embodied associations can be involved in a particular observable manifestation of the links. A possible explanation for how phylogenetic and ontogenetic processes combine may be the recruitment of particular modules of the mind, such as the magnitude system or the agent recognition. The three links we have discussed may vary in how evolutionary preparedness and cultural influence are balanced. The link of spatial distance to social closeness is the most obvious, and although there is some cultural variation, it is likely to be mostly predetermined. The hierarchy–verticality link is also immediately intuitive, but its cultural implementation can vary widely, stretch to new applications, and sometimes even reverse (e.g., in the high status of small objects). Finally, the spatial agency bias is completely driven by cultural practices and thus varies across cultures (Leung, Qiu, Ong, & Tam, 2011), although it probably rests on an evolutionary foundation.

Acknowledgments

We thank Alf Børre Kanten for helpful input and Timothy Hubbard for helpful comments on an earlier version of this chapter.

Chapter 24 Forms of Bias in Cognitive Science

Moving beyond Perception, Action, and Cognition

J. Scott Jordan, Vincent Cialdella, Daniel S. Schloesser, and Jiuyang Bai

It has been known for some time that the perceived vanishing point of a moving stimulus is displaced in the direction of motion, beyond the actual vanishing point (for review, see Hubbard, 2005b, 2014a, Chapter 8 in this volume). Jordan and Hunsinger (2008) discovered that the magnitude of the displacement was larger for participants who had previous experience controlling the stimulus. This finding, that participants with previous control experience perceive events *further ahead* than do novices, has since been replicated in a host of practical settings. Expert basketball players perceive other players as being further ahead in their trajectories than do novice basketball players (Gorman, Abernethy, & Farrow, 2011, 2012; Nakamoto, Mori, Ikudome, Unenaka, & Imanaka, 2015). Expert pianists perceive the second of two ambiguous piano tones as being higher than the first if the two are produced by piano key presses that are consistent with increasing pitch (i.e., the second key is to the right of the first; Repp & Knoblich, 2009). Experienced drivers experience road scenes as being further ahead than do novice controllers (Blättler, Ferrari, Didierjean, van Elslande, & Marmèche, 2010). And finally, expert pilots percieve landing strip scenarios as being further ahead than do novices (Blättler, Ferrari, Didierjean, & Marmèche, 2011).

While it is common practice in the field to account for spatial displacements in perception in terms of spatial bias, and to assert that experts have more of it, the purpose of the present chapter is to examine two forms of theoretical bias we feel to be at work in theories of perception, action, and cognition in general and in the field of spatial bias specifically. These two forms of bias are the *efficient-cause* bias and the *realism* bias. In what follows, we describe these types of bias and potential alternatives to each. Our objective is not to make the case that cognitive scientists are wrong when it comes to the notion of bias. Rather, our point is to show that the assumptions underlying our current thinking about action, perception, and cognition, and the presence of bias within each, are driven by another set of strongly held assumptions (i.e., biases) that one is not logically obliged to hold.

Forms of Bias in Cognitive Science

In everyday language, the assumptions underlying the use of the concept *bias* are pretty straightforward; someone perceives an event, holds a belief, or makes a decision that seems to be motivated by something other than the true state of affairs. In cognitive science, the notion of bias is used in much the same way; someone generates an action, perception, or cognition that seems to correspond to something other than the way things *truly are*, and we assert that the action, perception, or cognition entails bias. It is our contention that this approach to bias makes its way into cognitive science because of our commitment to the assumptions of *efficient cause* and *realism.*

The *Efficient-Cause* Bias

Experimental psychology has a history of designing experiments that start the clock, as it were, with the presentation of a stimulus. Of course, such a methodology historically emerged for sound scientific reasons. When behaviorists such as Watson, Weiss, and Holt gave up on first-person phenomenology (i.e., consciousness) as a dependent variable, and decided psychology should be a science described solely in terms of third-person observables (Boring, 1950), researchers focused on discovering lawful relationships between environmental events (i.e., stimuli) and observable body events (i.e., behavior). At the root of S-R experimental psychology was an *efficient-cause* approach to causality, in which *stimuli* were modeled as causing *behavior* in much the same way a particle imparts force on another particle when they collide. To be sure, Skinner later changed the focus of efficient-cause theorizing by referring to *behavior* as the cause of *stimuli* (i.e., reinforcers or punishers). Regardless of their differences, however, common to both classical and operant conditioning was a commitment to an efficient-cause approach to describing organism–environment coordinations.

To be sure, there were scientific psychologists who criticized behaviorism's stimulus–response, efficient-cause theorizing. Dewey (1986), for example, argued very early on for a different approach to defining the concepts *stimulus* and *response*:

> The stimulus is that phase of the forming coordination which represents the conditions which have to be met in bringing it to a successful issue; the response is that phase of one and the same forming coordination which gives the key to meeting these conditions, which serves as instrument in effecting the successful coordination. (p. 379)

According to Dewey, reflex arcs actually constituted a unity that could not be coherently deconstructed into independent sensory and motor components. Dewey therefore avoided efficient-cause theorizing and, instead, proposed a *transactional* approach to organism–environment coordinations that is rather consistent with contemporary, dynamical systems approaches (Jordan, 1998; Jordan & Ghin, 2007).

Despite the theoretical concerns expressed by Dewey (1986), behaviorism flourished in the first half of the 20th century, and it did so because its efficient-cause-driven methodology was (1) consistent with the neuroscience of the time and (2) empirically fruitful. In regards to neuroscience, behaviorism's commitment to efficient-cause theorizing was consistent with the input–output theorizing inspired by the Bell-Magendie law: the discovery that "the posterior (dorsal) roots of the spinal cord contain only sensory fibers and the anterior (ventral) roots only motor fibers" (Boring, 1950, p. 32). Given the sensory fibers ascend through the spinal cord to the brain, while the motor fibers descend from the brain to the spinal cord, the sensorimotor conceptualization of the nervous system as a reflex-arc circuit was completely consistent with the efficient-cause theorizing of the behaviorists.

As regards behaviorism's empirical fruitfulness, researchers could agree on the meaning of concepts such as *number of bar presses, percentage correct trials,* and *trials-to-criterion,* and coherent studies could be developed to follow up on previous findings. In short, behaviorism's efficient-cause approach to experimental psychology was successful because it worked as a method of doing science.

As behaviorism's influence grew, so did experimental psychology's commitment to the efficient-cause approach to the nervous system (i.e., sensorimotor circuits) and organism–environment coordinations (i.e., S-R and R-S relationships). As a matter of fact, the commitment to efficient-cause theorizing was carried over into cognitive psychology. Specifically, advocates of the cognitive revolution were inspired by findings such as Tolman's (1948) discovery that rats could make behavioral choices that strongly indicated the use of internal cognitive maps, and Cherry's (1953) discovery that people could selectively attend to one of two simultaneously presented auditory messages, as long as the two were sent to different ears. In order to create a theoretical framework that allowed internal states to play a causal role in scientific explanations of behavior, cognitive scientists utilized Shannon's (1948) information processing theory (IPT), which conceptualizes information transfer in terms of discrete bits. Within IPT, bits were and are conceptualized as discrete vehicles of content (i.e., carriers of information) that transfer information from one processing module to the next (Clark, 2001). Implicit in this approach to information is the assumption that such bits are *efficacious*. That is, bits causally influence the content values within the modules they influence, and they do so in an efficient-cause manner (e.g., a stimulus strikes a receptor neuron, which then sends *information* to another neuron, and so on and so on, until the chain of causal neural activity comes together in a location whose content entails a perception, a cognition, or an action plan). Echoing Dennett (1971), Jordan (2008) referred to such a conceptualization of information processing as *causal content* in order to draw attention to the assumptions that such vehicles (1) contain content and (2) are efficacious.

Just as was the case with behaviorism, the IPT approach to experimental psychology has sustained itself because it has been both consistent with the neuroscience of its time and empirically fruitful. ERP and fMRI data, and their ability to detect neural activity occurring at specific times in specific locations, are consistent with the idea that the brain is an information-processing system that processes information of different types in different locations (Brodersen, 2009; Haynes, 2015; Heeger & Rees, 2002). And experiment upon experiment has been motivated by IPT theorizing and its ability to spawn new experiments. In short, IPT, just like behaviorism, has sustained itself as a scientific account of organism–environment coordinations because it works as a method of doing science.

We have highlighted the scientific success of IPT because we would like to critique its commitment to efficient-cause theorizing, while simultaneously acknowledging its scientific utility. As previously stated, our intent is not to claim that the efficient-cause theorizing inherent in behaviorism and IPT is wrong. Rather, our plan is to describe (1) how the bias toward efficient-cause theorizing has made its way into cognitive science, (2) what the implications of holding such a bias might be for cognitive science generally and the issue of spatial bias specifically, and (3) what alternatives are available.

How the Efficient-Cause Bias Influences Theorizing. When we develop causal explanations of the results of our experiments, particularly those involving spatial bias, our efficient-cause bias leads us to propose theories that account for spatial bias in terms of internal, causal factors such as *top-down*, and/or *bottom-up processing*. *Implicit* in such theorizing is the assumption that there exists a *pure, unbiased* signal, either inside or outside of the organism, that is transformed by the organism's internal processes, resulting in an action, perception, or cognition that is different from the true state of affairs. This assumption of a *true* signal that is biased by the organism proves problematic because it is becoming increasingly difficult to find a "pure" signal either inside or outside the organism. In what follows, we describe recent developments in neuroscience that challenge the notion of there being a pure, non-biased signal anywhere in the nervous system.

In its simplest form, the idea of a pure internal signal is reflected in the idea that there exist sensory events in the organism's nervous system that clearly correspond to the stimulus events that *caused* them (e.g., changes in rod and cone activity in the retina following a brief flash of light, changes in neurotransmitter release in pressure sensors in the fingers after grasping an object, or changes in activity in hair cells following the arrival of an acoustic signal at the cochlea). Despite the seemingly straightforward notion of clear correspondence between external stimuli and internal sensory events, more and more data indicate that even sensory neurons find themselves modulated by feedback influences from multiple neural centers. Auerbach, Beiler, Henkes, and Goldhaber (1961), for example, reported that monocular illumination of a cat's eye resulted in ERG potentials in the other, non-illuminated eye. van Hasselt (1972) published a book entitled *The Centrifugal Control of Retinal Function*, in which he described many such findings. More recently, Kolb (2007) published a review regarding feedback loops in the retina, in which she said:

> At every level of the retina, there are reciprocal or feedback loops in the circuitry so that certain neurons can interact laterally within the same layer, vertically from one layer to the other, and indeed from the brain to the retina. (p. 1)

Given that neural activity in the retina is influenced by recursive feedback from retinal and non-retinal neural sources, continuously, it seems difficult to sustain the idea of a true, unbiased neural signal. The situation is even worse for neurons located more centrally. For example, in a discussion of the developmental changes that take place in the connections between different brain areas, Fair et al. (2007) reported that during certain parts of neural development, short-range connections between neurons that are located in adjacent brain regions *grow down* (i.e., decrease), while long-range connections between neurons located in different systems *grow up* (i.e., increase). This developmental increase in the number of long-range neural connections between brain areas sheds light on the fact that the activity in a neural area is continuously modulated by its long- and short-range connections. This reveals that the activity in a cortical area is always

> an emergent, dynamic compromise among all the forces impinging upon the neurons that constitute that cortical area. In short, the extreme level of recursion in brain organization makes if difficult, if not impossible, to make coherent "efficient-cause" assertions regarding brain dynamics. (Jordan, 2013, p. 7)

This critique of the idea of unbiased neural activity is consistent with Van Orden and Holden's (2002) assertion that the brain does not contain a causally isolated level of activity that is able to manage efficient cause relationships between independent content vehicles (i.e., a *cognitive* band; Clark, 2001). According to Van Orden and Holden, brain dynamics are *interaction dominant*, as opposed to *component dominant*, because the activity of all neurons and neural areas is continuously, recursively modulated by activity occurring in other neurons and other neural areas.

The notion of interaction-dominant neural dynamics has implications for some of the explanatory concepts we utilize in accounts of spatial bias: *top-down*, *bottom-up*, *planning*, *anticipation*, and *expectation*. The concepts *top-down* and *bottom-up* seem to apply to the spatial configuration of neurodynamics in terms of how far they are, in number of synapses, from sensory systems. A problem with using these concepts as explanations of spatial bias is that if every level of *bottom-up* neural activity is continuously, recursively modulated by higher-level neural activity, and visa-versa, it seems there is no true *bottom* level in the sense there is no level of neural activity that is only *about* something outside of the system. There is bottom-up activity in the sense that activity is occurring in neural systems that are 'closer' to sensory systems in terms of synapse distance, but not in the sense there exists an unbiased neural signal at or near the sensory systems that is then biased by neural systems further removed from the sensory systems. This latter distinction is important, for as accounts of spatial bias become increasingly steeped in neuroscience, it will become increasingly important to replace concepts such as *bottom-up* and *top-down* with clearly specified neurodynamics. If it is the case there are no unbiased (i.e., pure) neural dynamics, however, the notion of "bias" may very well come to be replaced by the language of neurodynamics. Of course, this is exactly the sort of neuro-reductive eliminativism Patricia and Paul Churchland (1981, 1986) have been proposing for years.

While concepts such as *bottom-up* and *top-down* refer to a system's synaptic distance from sensory systems, concepts such as *planning*, *anticipation*, and *expectation* seem to have more to do with the temporal distance between an event (e.g., grabbing a pen) and the moment in time represented by the content in the biasing neurodynamics. That is, content in one neural area is assumed to bias content in another area because the content in the former is about a point in time (e.g., the future or the past) that is different from that represented in the latter (i.e., the present). Within the interaction-dominant approach to neurodynamics, however, in which all neural systems find themselves continuously, recursively modulated by activity in multiple neural areas, it becomes very difficult, if not logically impossible, to clearly state which levels of neural activity constitute planning, anticipation, and/or expectation and which ones do not. As was the case with the concepts of bottom-up and top-down, there does not seem to be a level of neurodynamics that is unbiased, in the sense that it is "only" about the present. Thus, just as was the case with the concepts *bottom-up* and *top-down*, time-related concepts such as *planning*, *anticipation*, and *expectation* may come to be increasingly replaced by the language of neurodynamics. In short, "bias" will no longer be considered a sufficient scientific explanation of "bias."

To be sure, all of us have had the experience of planning, anticipating, and/or expecting an event (e.g., planning to reach for a pen, anticipating the consequences of grabbing the pen, and/or expecting the pen to give rise to marks on paper as we write). In research on spatial bias, theorists use the concepts *anticipation*, *expectation*, and *planning* in the same way these concepts are used in daily life; i.e., the plans, anticipations, and expectations we have of an event bias our perceptions of that event. And while it may be difficult, if not logically impossible, to find neural activity that is temporally unbiased (i.e., it is solely about the present) that can then be biased by neurodynamics containing content about a different moment in time (e.g., the future), there are certain patterns in brain connectivity that are consistent with how we use these temporally indexed concepts. Specifically, there are neurons in motor cortex that synapse with neurons in the spinal cord, while also projecting collaterals that ultimately synapse with neurons in the cerebellum. These cerebellar neurons are also innervated by sensory neurons located in the body parts that are made to move by the activity in the motor neurons (Kawato, Furukawa, & Suzuki, 1987). These very same cerebellar neurons ultimately project back to cortex. This implies that as one learns a particular movement (e.g., an infant learning to shake a rattle), and repeats successful movements, patterns of successful *command–feedback patterns* become embodied in the these cortical-cerebellar networks. As a result, as one later initiates such a movement, cerebellar neurons, when activated by activity in motor neurons, recursively prime the motor neurons in terms of remembered (i.e., embodied) command–feedback patterns. Such recursive, memory-based priming occurs at a time-scale faster than the time-scale of the feedback returning from the moving limb, because cortical-cerebellar loops have a time-cycle of 10–20 msec, while sensory feedback loops have a time-cycle of 120 msec.

Many researchers refer to such cortico-cerebellar loops as *forward models* (Golfinopoulos, Tourville, & Guenther, 2009; Ito, 1993, 2008; Koziol & Lutz, 2013; Shadmehr & Krakaver, 2008; Wolpert, Doya, & Kawato, 2003), and/or *cerebellar control models* (Koziol, Budding, & Chidekel, 2011). Common to both concepts is the idea that (1) cerebellar-cortical networks are anticipatory, and (2) the anticipation derives from previous experience.

While the use of the concept *anticipatory* in relation to the functions addressed by motor-cortical forward models (e.g., reaching for an object) seems consistent with how we use the notion of anticipation in everyday life, it has recently been discovered the cerebellum shares recursive innervation with almost all areas of cortex (Schmahmann, 2001), except for maybe the inferior temporal cortex (Koziol et al., 2013), and/or very early vision and audition centers (Buckner, Krienen, Castellanos, Diaz, & Yeo, 2011). Such ubiquitous cortico-cerebellar recursion implies that all cortical activity is to some extent continuously primed by past patterns of cortico-cerebellar activity. Given such recursions build up over the life course as a function of experience, it seems reasonable to conclude that cortico-cerebellar recursion renders the involved cortical areas inherently prospective, in the sense that past dynamics are being regenerated in the present context.

Ito (1993) recognized the ubiquity of cortico-cerebellar recursion decades ago, and proposed that the neurodynamics involved in language were homologous to those involved in movement control. Many researchers now espouse the notion of a neurodynamic homology between cognition and action (Ito, 2008; Kinsbourne & Jordan, 2009; Koziol, Budding, & Chidekel, 2011; Koziol & Lutz, 2013; Schmahmann, 2001).

Given the ubiquity of prospectivity in neural dynamics, it seems the use of time-indexed concepts as explanations of bias might no longer be consistent with contemporary neuroscience. The phenomena we refer to as action, perception, and cognition all involve neuro-architecture that is inherently prospective, and any given moment of action, perception, and cognition finds itself nested within a multi-scale field of events the individual is working to produce and maintain (e.g., maintain balance against gravity, avoid obstacles in the environment, and carry on a conversation while simultaneously completing the previous two levels of control). While we can pick one scale (e.g., perception) and assert that it contained a bias that explains our experimental results, the ubiquity of prospectivity in brain dynamics and organism–environment coordinations indicates there may be no such thing as *unbiased* action, perception, or cognition. Every level of neurodynamics and, therefore, organism–environment dynamics is inherently *biased* in the sense that no level is solely about one time-scale of event. On the one hand, one can simply accept this position and assert that brains are inherently biased. But if one does, then it is no longer clear what is being discriminated by the concept *bias*.

Alternatives to efficient-cause theorizing. As an alternative to efficient cause models of perception, action, and cognition, many researchers now propose interaction-dominant approaches. Such approaches tend to rely on the notion of *self-organizing* dynamics. Keslo (1995) describes the self-organizing dynamics of convection rolls:

> The resulting convection rolls are what physicists call a collective or cooperative effect, which arises without any external instructions... Such spontaneous pattern formation is exactly what we mean by self-organization: the system organized itself, but there is no 'self', no agent inside the system doing the organizing. (pp. 7–8)

One of the most radical alternatives to efficient-cause theorizing based on the notion of self-organizing dynamics was proposed by van Gelder (1998), who strongly rejected the idea that a scientific account of organism–environment coordinations necessitates the use of IPT concepts such as *codes, plans,* and *information processing.* He presented the Watt Governor as an example.

> A Watt Governor is a mechanism that keeps the velocity of a fan constant. It does so because the velocity of the fan's flywheel influences itself recursively. That is, the faster the flywheel moves, the more it slows itself down. This is because the movements of the flywheel cause two weighted arms, attached to the flywheel via a spindle, to rise as the flywheel moves faster. These movements cause a valve, connected to a steam-engine driving the flywheel movement, to close, thereby reducing the amount of steam released which, in turn, slows the flywheel and allows the weighted arms to lower. Because of the constant, mutual influence all the components have on one another, fan velocity can be held fairly constant. Van Gelder proposes the Watt Governor as a prime example of regulated behavior emerging out of purely self-organizing processes. At no point in the Watt Governor systems is there a 'code', a 'representation', or information being processed. As a result, he argues it should be possible to model cognitive systems solely in terms of self-organizing dynamics. (Jordan & Ghin, 2007, pp. 183–184)

Most interaction-dominant theories do not represent as radical a departure from the assumptions of IPT as van Gelder's. Rather, most tend to be framed in dynamical systems language, and come in two varieties: what we will refer to as the *internal content* camp and the *external content* camp. Theorists in the *internal content* camp (i.e., content is located in the brain) conceptualize organism–environment coordinations in terms of interaction-dominant dynamics, while simultaneously continuing the IPT tradition of theoretically locating *content* in the brain. Clark and Toribio (1994), for example, while sympathetic to the interaction-dominant approach to organism–environment coordinations, nonetheless argue that dynamical systems approaches to cognition that reject the use of concepts such as *code, representation,* and *information processing* do not make good models of cognition because they cannot account for the ability of cognitive agents to engage in *representation-hungry* tasks (i.e., tasks that seem to require information not available in the current context) such as predicting outcomes and considering counterfactual events when planning actions. Clark (2001) states:

> The image of brain, body, and world as a single, densely coupled system threatens to eliminate the idea of purposive agency unless it is combined with some recognition of the special way goals and knowledge figure in the origination of some of our bodily motions. The computational/information-processing approach provides such recognition by embracing a kind of dual aspect account in which certain inner states and processes act as the vehicles of knowledge and information. (p. 135)

Kelso (1995, 2003) also proposes an internal content alternative, and focuses on how internal *intentions* influence the system's dynamics:

> Formally, an intention is conceived as specific information acting on the dynamics, attracting the system toward the intended pattern. This means that intentions are an intrinsic aspect of the pattern dynamics, stabilizing or destabilizing the organization that is already there. (Kelso, 2003, p. 140)

In contrast to internal content theories, the external content camp, comprised mainly of ecological psychologists (Gibson, 1979; Shaw, Turvey & Mace, 1982; Turvey & Shaw, 1999), conceptualizes organisms as being embedded in *relations* with their environments, such that the two mutually *constrain* each other instead of *causing* things to happen to each other. That is, instead of conceptualizing organisms and environments as systems that seem to *collide* with each other in an efficient-cause fashion, ecological psychologists see the two as nested and synergistically yoked.

We refer to the ecological position as an *external content* position because it places content (i.e., information) in the environment, not the organism. Ecological theorists argue organisms detect information in the ambient array that precisely specifies the behavioral possibilities the environment affords the organism at that moment. In short, organisms detect *affordances*, and these affordances constitute the ecological equivalent of the content vehicles IPT theorists conceptualized as existing in the brain.

Issues with Internal- and External-Content Options. Although internal- and external-content theories address the interaction-dominant nature of organism–environment coordinations, they both still tend to conceptualize perception, action, and cognition in ways that do not acknowledge the multi-scale prospectivity inherent in organism–environment coordinations and neural dynamics. For internal content theorists, perception, cognition, and action are conceptualized as rather different functions (i.e., one is an input, one is a mediating processor, and the other is an output), and such disparate functionality does not take into account their shared, cortico-cerebellar-driven prospectivity.

As regards external content theories, they too conceptualize perception, action, and cognition in ways that do not acknowledge their homological prospectivity. This is because ecological theories make little, if any, reference to brain dynamics (cf. Coey, Varlet, & Richardson, 2012). To be sure, they do acknowledge the prospective nature of organism–environment coordinations by asserting that affordances are future-looking because they entail possible behaviors for the organism. As a result, ecological theorists might be inclined to agree that perception is inherently *biased* toward the future. But the ecological decision not to include brain dynamics into their scientific explanation of organism–environment coordinations renders their theory currently incapable of thoroughly addressing the organism's contributions to its own coordinations. While the organism is nested within its environment, and the two are causally yoked, the type of influence the two have on each other is not the same. It is the organism, not the environment, that actively reorganizes itself in order to maintain certain coordinations. And the multi-scale prospective neural dynamics mentioned previously constitute a major aspect of the explanation of how organisms prespecify and maintain such coordinations.

Another Alternative: Multi-Scale Effect Control (MSEC). Given that contemporary internal- and external-content theories do not address the multi-scale prospectivity inherent in perception, action, and cognition, Jordan (2003, 2013) models organism–environment coordinations in terms of multi-scale effect control (MSEC). According to MSEC, organisms are able to control multiple events at multiple time-scales simultaneously because the prospectivity inherent in their cortico-cerebellar systems allows them to (1) prespecify outcomes at multiple time-scales simultaneously (e.g., maintain balance, keep head in certain position, complete the trial correctly, collect extra credit, and achieve a good grade in a class) and (2) control multi-scale outcomes simultaneously. In relation to perception, action, and cognition, this means that what we traditionally refer to as *action* is referred to as *proximal effect control*, where proximal refers to the somatosensory, kinesthetic effects of body movements. Perception, then, is referred to as *distal effect control*, where distal refers to the effects of movement that result from moving the body in relation to events beyond the body, within one's immediate context (e.g., seeing and feelings oneself make contact with a soccer ball).

While it seems odd to refer to perception as distal effect control, it has been known for some time that areas of the brain involved in what we traditionally refer to as *motor planning* (i.e., area F5, which is also part of a cortico-cerebellar network) are also involved in what we traditionally refer to as *perception* (Hommel, Müsseler, Aschersleben, & Prinz, 2001; Rizzolatti, Fadiga, Gallese, & Fogassi, 1996). Given the same neurodynamics are involved in two supposedly different functions, it may be the case that we are somewhat biased against the notion of distal effect control because of our commitment to the IPT, sensorimotor interpretation of neural functionality, and have been led by this bias to believe that perception and action constitute orthogonal functions (i.e., one is an input, the other is an output). In the attempt to propose distal event control as an alternative approach to perception, Jordan (2013, p. 6) states:

> [I]t seems evolution has left us with a rather elegant solution to controlling distal effects. Instead of developing one system for *doing* and another for *seeing*, as is assumed in traditional approaches to psychological functionality, evolution has endowed us with a host of systems that are able to pre-specify and detect distal events simultaneously.

MSEC then refers to cognition as *virtual* or *abstract* effect control. Abstract refers to our ability to prespecify and control effects that are beyond our current, immediate context. Examples include walking to a café to meet a friend, saving money to buy a new car, or attending classes to become a college graduate. In all three cases, current behaviors (i.e., walking, saving, and attending) are being generated in order to produce a later, currently non-present effect (i.e., meet a friend, buy a car, and graduate college).

In addition to acknowledging multi-scale prospectivity, MSEC also takes an interaction-dominant approach to accounting for the interactions between the organism, the environment, and different levels of effect control. Thus, no single environmental, muscular, or neural event is theorized as *causing* things to happen in an efficient-cause fashion. Every aspect of the multi-scale system is *constraining*, either directly or indirectly, all other aspects of the system

Given MSEC's commitment to an interaction-dominant approach to causality, as well as its to ability to address the multi-scale prospectivity inherent in cortico-cerebellar networks – hence Clark's (2001) and Kelso's (2003) concerns about interaction dominant dynamics and agency and intentionality – it seems MSEC may provide a theoretical framework that moves us beyond the biases we hold in the way we discuss bias. The phenomena we traditionally refer to as action, perception, and cognition are not functionally different. They share a common functionality of effect control, and what differs between them is the scale of effects they actually control.

In addition to being consistent with contemporary neuroscience, MSEC, just as behaviorism and IPT, has to prove its scientific utility. That is, MSEC has to give rise to testable research questions, and it has to inspire experiments other theories do not. To some extent, this has already happened. For example, the Jordan and Hunsinger (2008) experiments, as well as others (Gill, Jordan, & Cutting, 2014; Jordan, Gill, & DuPuis, 2014; Jordan, Kenning, Clinton, & Durtschi, 2010; Jordan & Knoblich, 2004; Jordan, Stork, Knuf, Kerzel, & Müsseler, 2002; Kerzel, Jordan, & Müsseler, 2001; Vinson, Jordan, & Hund, 2017), were inspired by MSEC and the idea that perception constitutes distal effect control (Jordan, 2003). As a result, an MSEC account of the difference in spatial displacement between an observer and a controller (Jordan & Hunsinger, 2008) asserted that because perception and action planning (i.e., distal effect detection and distal effect planning, respectively) share overlapping neural systems and, therefore, constitute a single distal effect control system, experience controlling the stimulus movements would result in the development of control memories that would be later activated while watching stimulus movements being controlled by another.

As another example of MSEC's scientific utility, Kumar and Srinivasan (2013) developed an experiment, inspired by MSEC, in which participants used a joystick to aim a cursor at targets on a computer monitor. Target trajectory was influenced by one of three levels of random noise. When the participant pulled the joystick trigger, a stimulus appeared at the targeted location, and the participant indicated (1) the amount of time that passed between the pull of the trigger and the appearance of the stimulus and (2) how robustly the participant had experienced herself or himself as the cause of the action. If participants did not achieve the distal effect (i.e., they missed the target), their trigger-stimulus temporal estimates were correlated with the degree of noise in the target movements. That is, as trajectory noise decreased, so did trigger-stimulus estimates. This attraction of the timing of a distal effect (i.e., post-trigger stimulus appearance) toward the moment of the proximal effect (i.e., the trigger pull) has been referred to as *intentional binding* (Haggard, Clark, & Kalogeras, 2002; Jordan, 2000), and it is believed to be an *implicit* measure of sensed agency, in that one experiences the effect of the action as being bound with (i.e., caused by) the action. As regards participants' ratings of how robustly they felt they had caused the post-trigger stimulus to appear (i.e., an explicit measure of sensed agency), these values also increased significantly with the participant's actual level of control over the stimulus.

If the participant hit the target, the pattern was different. Specifically, the explicit sense of agency once again increased with their degree of actual stimulus control. However, the trigger-stimulus estimates did not vary with the trajectory noise (i.e., intentional binding did not vary with the amount of actual control the participant had over the target).

Collectively, these data imply that the participants' conscious, explicit sense of agency varied with their actual level of stimulus control, regardless of whether or not they hit the target. However, their implicit sense of agency (i.e., intentional binding) depended on whether or not they hit the target. If they achieved the distal effect (i.e., hit the target), intentional binding was not influenced by the amount of actual control. If they did not achieve the distal effect (i.e., they did not hit the target), intentional binding *was* influenced by the actual level of control. These results are consistent with the MSEC assertion that one's sense of agency moves back and forth across different levels of effect control as issues in one level become important at any given moment (Jordan, 2003). For example, if one is talking to a friend while walking down a flight of stairs and suddenly trips, one's conscious awareness will switch rapidly from the content of the conversation to the content regarding one's foot location. In addition, the moment of surprise regarding foot location reveals that while the person was walking and talking, unconscious predictions regarding anticipated foot pressure were continuously being generated, and the individual did not become aware of these unconscious expectations until they were violated. This idea has much in common with IPT theories such as Vallacher, Wegner, and Somoza's (1989) action-identity theory (AIT) and Baars's (1988) global workspace hypothesis. While AIT agrees with the MSEC assertion that people engage in multiple levels of behavior simultaneously, it further asserts that only one level of behavior (i.e., level of effect control) tends to be prepotent for a person at any one time:

> [A]lthough talking, for example, could be identified as sharing information, expressing an opinion, influencing someone, passing time, or choosing words, the actor is likely to have in mind only one of these identifications. (Vallacher et al., 1989, p. 199)

The AIT notion that only one behavior is prepotent at a time is consistent with Baars's (1988) global workspace hypothesis (GWH), which claims that the purpose of consciousness is to make specific information massively available to a host of unconscious brain processes so that these latter processes can be utilized to solve an immediate problem (i.e., help manage disturbances at a particular level of event control). From this perspective, which is consistent with MSEC, the contents of consciousness vary from moment to moment as different problems emerge in different levels of effect control in real time. This approach to consciousness is also consistent with Morsella's (2005) proposal that the function of conscious states is to determine which of many possible action plans actually expresses itself within the skeletal muscular system at any given moment (i.e., PRISM – parallel responses into skeletal muscle).

Although MSEC, PRISM, GWH, and AIT share a common approach to consciousness and multi-scale problem solving, they differ in their theorizing. PRISM, GWH, and AIT emerged within the framework of IPT, and would probably assert that certain conscious states *cause* a particular action to happen, in an efficient cause manner. MSEC, on the other hand, would argue that at any given moment,

> the pattern of multi-scale effects one works to control emerges spontaneously and continuously out of both exogenous influences that activate pre-specifications of past effect-control episodes via cortico-cerebellar systems, and the endogenous constellation of constraint that builds up over the life course across different levels of effect control. (Jordan, 2013, p. 10)

While both the IPT and MSEC accounts are consistent with the data, only the latter (1) acknowledges the interaction-dominant nature of organism–environment coordinations and nervous systems and (2) is maximally consistent with contemporary neuroscience. Only time will tell if MSEC continues to inspire new experiments and increase its scientific utility. Regardless if it does, however, it will have, at a minimum, provided a coherent alternative theoretical framework in which we, as a field, can assess the theoretical biases inherent in our conversation regarding bias.

The *Realism* Bias

Part and parcel to cognitive science's *efficient-cause* bias is its *realism* bias. Although there are many varieties of realism (Anderson, 2007; Atmanspacher & Kronz, 1998), it is most often expressed in cognitive science as the belief in a clear division between *objective* and *subjective* phenomena. The former refers to phenomena that exist, as they are, independently of the presence of an observer, while the latter refers to phenomena whose existence depends on the presence of an observer.

A commitment to realism influences how we describe the functions of action, perception, and cognition, and, as a result, the manner in which we describe *bias*. There are two types of realism in contemporary cognitive science. *Indirect* realists (Jordan & Day, 2015) tend to hold an IPT-driven, representational approach to perception, and assert that we are able to learn about, and know, external, objective reality because we create representations of it in our nervous system. It is referred to as *indirect* realism because its reliance on internal representations implies that observers perceive external reality indirectly, via internal, *mediating* representations. As a result of the mediating role of representations, indirect realism is also referred to as a *mediation* theory of perception. An implication of such mediation is that organisms are not in *direct* contact with external reality, as it is. Rather, organisms are in *direct* contact with their internal representations of external reality.

What is known as *direct* realism developed as a counterargument to *indirect* realism. Specifically, direct realists as far back as the Six Realists (Holt et al., 1910) have reasserted Hume's (Hume & Selby-Bigge, 1967) critique of Locke's mediated theory of perception because the notion of *mediated perception* logically denies one access to external reality. That is, if our contact with external reality is mediated by representations, then all we can ever really know, and all we can ever be in contact with, are our internal, mediating representations. Put another way, any test we could ever run to determine whether or not our internal representations correspond to external reality would rely on the use of our internal representation. In short, we cannot step out of our representations and see reality as it is, for all we actually have access to is our representations. As a result, because of such logic, we can never truly know or come in contact with external reality. In short, the indirect-realist commitment to a mediated theory of perception logically denies perceivers access to external reality and, as a result, renders indirect realism self-contradictory (i.e., logically incoherent).

Given their dissatisfaction with indirect realism, contemporary *direct*-realists assert that organisms are actually in direct contact with external reality, as it is, because the ambient array is comprised of invariant energy structure – *information* – that directly specifies its source (Gibson, 1979; Turvey & Shaw, 1999). Within this framework, which is most often espoused by ecological psychologists, perception is the act of *detecting* such information.

While indirect and direct realism seem quite different, their shared commitment to realism leads both to the realist belief in an external, *objective,* reality that (1) is perceived by observers via either *information detection* or *representation construction* and (2) exists, as it does, independently of observers. In what follows, we discuss these assumptions and assess their impact on the notion of bias.

Perception as Detection and/or Representation Construction. Given realism's commitment to a clear distinction between objective and subjective phenomena, indirect- and direct-realist accounts of perception are biased to treat perception as if its function is to *detect* or *construct* information structures that serve to connect the organism to the environment. It is our contention that describing perception in terms of detection and construction *biases* us to think of perception as a *reaction* to the environment. In the indirect case, perception is conceptualized as the post-stimulus construction of representations. In the direct case, it is seen as the post-stimulus detection of information.

As was mentioned in the *efficient-cause bias* section of the present chapter, conceptualizing perception, action, and/or cognition in a post-stimulus manner is inconsistent with the multi-scale prospectivity that exists in nervous systems. While one can always select a particular environmental event (e.g., the presentation of a stimulus in an experiment), and describe the ensuing activities of the organism as constituting post-stimulus *information processing* or *information detection,* MSEC and the notion of multi-scale prospectivity imply that any stimulus events or information structures that are processed or detected always encounter a nervous systems whose dynamics have already been heavily, if not entirely, pre-contextualized by multi-scale prospectivity (i.e., the brain's ubiquitous ability to continuously prime itself via cortico-cerebellar recursion). And given these networks' share overlap with those involved in planning (i.e., distal effect control), it seems more coherent to assert that external stimuli and information structures *modulate* the organism's ongoing, multi-scale control of its relationship with its environment (Jordan & Day, 2015; Jordan & Vinson, 2012). In short, as previously mentioned, MSEC defines perception as the control of distal effects, not the construction of representations or the detection of information.

To be sure, activity in the nervous system *is* modulated by activity in the external environment (i.e., what indirect realists refer to as representation construction), and patterns in the ambient array *do* modulate neural activity in reliable, predictable ways (i.e., what ecological psychologists might refer to as information detection). But since neither of these descriptions acknowledges the widespread prospectivity such modulations encounter when arriving at the brain, it seems the idea of perception as information processing or information detection is only half of the story.

Objectivity as Being Independent of Subjectivity. Another consequence of cognitive science's commitment to realism is its assumption that the only way to talk about reality is in terms of objective and subjective phenomena. Jordan and Day (2015) propose that (1) the objective–subjective assumption gives rise to an *epistemic gap* between the organism and the environment, and (2) once such a gap is assumed, phenomena such as *information processing* and *information detection* need to be invoked in order to conceptually bridge the gap.

As an alternative to realism's objective–subjective divide, and its resultant epistemic gap, Jordan and colleagues (Jordan, 2013; Jordan & Day, 2015; Jordan & Ghin, 2006; Jordan & Heidenreich, 2010) propose Wild Systems Theory (WST), a recently developed framework for cognitive science that conceptualizes organisms as *self-sustaining, multi-scale embodiments of context. Self-sustaining* refers to the idea that the actual energy exchanges that constitute an organism produce products that feed back into and sustain the work that produced the product in the first place. As an example, when a person reaches to pick an apple from a tree, and then eats the apple, all of the work (i.e., energy transformations) that went into picking and chewing the apple produces products (e.g., chemical nutrients) that feed back into and sustain, via energy transformation, the systems that made the picking and chewing possible in the first place.

Multi-scale refers to the idea that an organism is constituted of multiple, nested scales of self-sustaining systems. Individual cells, for example, constitute micro–macro synergies (Jordan & Ghin, 2006) within which the internal chemical transformations (i.e., micro level) give rise to and sustain the integrity of the cell as a whole (i.e., the macro level), while the cell as a whole recursively influences the internal dynamics in ways that give rise to sustainment.

> The micro-level cell-wall proteins that allow food particles into the cell, for example, are kept in a state of transformation we refer to as food-intake via the macro-level transformations of the whole cell – what we refer to as swimming or tumbling, depending upon the species. These macro-level phase transitions emerge out of the micro-level transformation states that comprise the cell as a whole. As a result of this micro-macro synergy, the cell is capable of regulating the inputs to its cell-wall proteins. (Jordan & Ghin, 2006, p. 62)

Different researchers have discovered various levels of self-sustaining work at different points in time, in different disciplines. Kauffman (1995) proposed the notion of self-sustaining systems at the chemical level, and referred to such systems as *autocatalytic* and *self-metabolizing* in order to capture the fact that the work of such systems *produces its own catalysts*, which, in turn, affords these systems the ability to *metabolize themselves.* Maturana and Varela (1980) referred to the self-sustaining nature of single-cell organisms as *autopoiesis* (i.e., a system's ability to reproduce and sustain itself). Hebb (1949) referred to the self-sustaining nature of neurons (i.e., *"neurons that fire together, wire together"*) as the *cell-assembly.* Skinner (1976) referred to the self-sustaining nature of behavioral work (i.e., the generation of behaviors) in terms of reinforcement theory, in that types of behavioral work are either sustained within or deselected from one's behavioral repertoire as a function of the consequences (i.e., products) such work produces. Streeck and Jordan (2009) conceptualize communication as being constituted of multi-scale work (e.g., postural alignment, gaze, gesture, and speech). And Odum (1988) and Vandervert (1995) conceptualize ecologies and societies in terms of self-sustaining work.

Finally, *embodiment of context* refers to the WST assertion that organisms constitute multi-scale embodiments of multi-scale contexts from which they emerged, and within which they sustain themselves. Jordan and Vinson (2012, p. 236) state:

> Plants constitute embodiments of the wide-scale availability of electromagnetic radiation (i.e., sunlight) and the rich chemical context of the soil from which they take in the chemicals necessary to sustainment. Herbivores constitute embodiments of the constraints a system needs to address in order to sustain itself on plant life. And carnivores embody the constraints necessary to sustain oneself on the energy encapsulated in herbivores (i.e., nervous systems, muscles and bones). At every level in this multi-scale hierarchy of embodied context, the contextually emergent systems (Atmanspacher, 2007; Bishop & Atmanspacher, 2006) are necessarily 'about' the multi-scale context they embody and in which they sustain themselves.

WST's conceptualization of organisms as self-sustaining, multi-scale embodiments of context provides a means of describing organism–environment coordinations that do not rely on the distinction between objective and subjective phenomena and, as a result, do not give rise to an epistemic gap that needs to be crossed via concepts such as *information processing* or *information detection.* Self-sustaining embodiments of context are naturally and necessarily *about* the contexts they embody. As a result, they are inherently *meaning-full* in the sense that they constitute *embodied aboutness.* They do not need to process or detect information in order to become *about* their context because they naturally and necessarily *are* so. Indirect- and direct-realists, and the internal- and external-content theories they propose, respectively, are biased to assume the need for epistemic bridges because of their commitment to realism's objective–subjective divide.

While WST's notions of embodied context and mutual modulation negate the need for epistemic bridges between objective and subjective phenomena, they simultaneously provide direction for ways to discuss the nature of reality that do not require the use of dialectic counterparts such as objective–subjective, or physical–mental. Specifically, by conceptualizing organisms as embodiments of context, WST allows us to consider the possibility that the important question about reality may not be which aspects of it exist independently of observers, but rather which aspects of it exist independently of all *context.* WST asserts that the question of context dependence versus context independence is important for cognitive science specifically, and scholarship in general, because it gets us to address the deeper, more entrenched distinction between *intrinsic* and *relational* properties (Jordan, Bai, Cialdella, & Schloesser, 2015; Jordan & Day, 2015; Jordan & Vinson, 2012). Specifically, for decades, scholars have conceptualized weight as a *relative* property because its value depends on gravity, while mass is described as an *intrinsic* property because its value does not. Implicit in the relative–intrinsic distinction is the assumption that intrinsic properties exist, as they are, independently of all context, and it is their lack of dependence on context that renders them intrinsic, and not relative.

Despite the long-standing nature of the extrinsic–intrinsic distinction, the notion of intrinsic properties has come under increasing scrutiny by philosophers of science. Jammer (2000), for example, argues that inertial mass emerges out of the manner in which particles interact with the Higgs field, a scalar field that "permeates all of space" and "endows particles with mass" (p. 162). Bauer (2011) argues such interactive dependence renders the property of mass *externally grounded*, which means the particle's mass is dependent on its context. Other such critiques utilize concepts such as *ultra-grounding* (Harré, 1986) and *global groundedness* (Prior, Pargetter, & Jackson, 1982) to make much the same point, while Schaffer (2003) and Dehmelt (1989) assert there may not be a fundamental level to reality at all. Rather, they propose reality might ultimately be constituted of infinite levels of microstructure.

Deconstructing the notion of intrinsic properties provides us with a new means of describing reality that does not lead to the creation of objective–subjective divides. Specifically, if all properties are context-dependent and, therefore, inherently relational, it seems we have to give up the intrinsic–extrinsic distinction and, instead, conceptualize all of reality as being inherently relational. In short, all things are *about* all things, and nothing is *about* itself.

To be honest, such a view is in no way new. In fact, it is rather similar to approaches to reality initiated by the likes of Kant, Hegel, and Oakeshott (Gardner, 2007; Priest, 1991; Tseng, 2003) as they created ontologies (i.e., philosophies of reality) that took Hume's critique of mediated theories of perception seriously and then attempted to describe reality in ways that avoided objective–subjective, physical–mental divides while simultaneously retaining the reality of phenomena such as phenomenology, meaning, and value. Jordan and Day (2015) present WST as a 21st-century, scientifically informed reformulation of these centuries-old philosophies in the hope of clarifying the biases to which we are led by our commitment to realism and its habit of conceptualizing reality, as well as perception, action, and cognition, in terms of intrinsic–extrinsic, objective–subjective divides. According to WST, there are no epistemic divides. Rather, such divides are the product of realist-induced bias.

Conclusions

In addition to providing a means of overcoming the notion of epistemic gaps, WST also provides us with a unique way of describing bias in perception, action, and cognition. While our commitments to *efficient cause* and *realism* leads us to believe that phenomena such as *representational momentum* and *forward displacement* constitute *errors*, or *biases*, because the observer's subjective perceptions do not correspond to objective reality, as it truly is, the arguments of the present chapter indicate the notion of non-biased information, either inside or outside of the organism, may be a logical impossibility. As regards internal "information," the notion of interaction-dominant dynamics indicates there may be no such thing as a "pure" neural signal that clearly corresponds to any one event. As regards external "information," the notion of context dependence leads to the assertion that there exists nothing outside of an organism that is purely or truly only about itself. In short, if we are to use the concept bias as an explanation, we may have to eventually assert that all of reality is inherently biased. Of course, such a statement entails a logical contradiction, for bias can only exist if it exists in contrast to something that is non-biased.

As a means of overcoming our biases regarding bias, WST proposes that the phenomena we measure in our laboratories emerge naturally, spontaneously, and contingently out of the ongoing, interaction-dominant relationships between embodiments of contexts and the contexts in which they are embedded. Within such interaction-dominant relations, in which all properties are inherently relational, the truth status of a perception is not, and cannot be, determined via a comparison of observer-independent objectivity to observer-dependent subjectivity, for such a test is simply, as stated previously, logically incoherent (i.e., self-contradictory). Rather, what we can do is compare one phenomenon (e.g., what we refer to as the *actual vanishing point*) with another phenomenon (e.g., what we refer to as the *perceived vanishing point*) and attempt to quantify the relationship between them. In doing so we are not culling out the objective, intrinsic properties of observer-independent reality, as realism tells us we are doing. Rather, what we are doing is what all self-sustaining systems do: specifically, modulate the context in which we are embedded, as it recursively modulates us, in the hope of revealing yet-to-be-discovered degrees of freedom in the manner in which we are able to sustain ourselves in context. In short, we are becoming more and more *about* where and what we are.

References

Abele, A. E., & Wojciszke, B. (2007). Agency and communion from the perspective of self versus others. *Journal of Personality and Social Psychology, 93*, 751–763.

Abrahamse, E., van Dijck, J. P., & Fias, W. (2016). How does working memory enable number-induced spatial biases? *Frontiers in Psychology, 7*, 977.

Abrams, J., Nizam, A., & Carrasco, M. (2012). Isoeccentric locations are not equivalent: The extent of the vertical meridian asymmetry. *Vision Research, 52*(1), 70–78.

Abrams, R. A., & Dobkin, R. S. (1994). Inhibition of return: Effects of attentional cuing on eye movement latencies. *Journal of Experimental Psychology: Human Perception and Performance, 20*, 467–477.

Actis-Grosso, R., Bastianelli, A., & Stucchi, N. (2008). Direction of perceptual displacement of a moving target's starting and vanishing points: The key role of velocity. *Japanese Psychological Research, 50*, 253–263.

Actis-Grosso, R., & Stucchi, N. (2003). Shifting the start: Backward mislocation of the initial position of a motion. *Journal of Experimental Psychology: Human Perception and Performance, 29*(3), 675–691.

Adamian, N., & Cavanagh, P. (2017). Fröhlich effect and delays of visual attention. *Journal of Vision, 17*, 1–14.

Aginsky, V., Harris, C., Rensink, R., & Beusmans, J. (1997). Two strategies for learning a route in a driving simulator. *Journal of Environmental Psychology, 17*, 317–331.

Aglioti, S. M., DeSouza, J. F. X., & Goodale, M. A. (1995). Size-contrast illusions deceive the eye but not the hand. *Current Biology, 5*, 679–685.

Aiello, M., Jacquin-Courtois, S., Merola, S., Ottaviani, T., Tomaiuolo, F., Bueti, D., et al. (2012). No inherent left and right side in human mental number line: Evidence from right brain damage. *Brain, 135*(8), 2492–2505.

Akutsu, H., McGraw, P. V., & Levi, D. M. (1992). Alignment of separated patches: Multiple location tags. *Vision Research, 39*, 789–801.

Alais, D., & Burr, D. (2003). The "flash-lag" effect occurs in audition and cross-modally. *Current Biology, 13*, 59–63.

Alais, D., & Burr, D. (2004a). No direction-specific bimodal facilitation for audiovisual motion detection. *Cognitive Brain Research, 19*, 185–194.

Alais, D., & Burr, D. (2004b). The ventriloquist effect results from near-optimal bimodal integration. *Current Biology, 14*, 257–262.

Alais, D., Newell, F., & Mamassian, P. (2010). Multisensory processing in review: From physiology to behavior. *Seeing and Perceiving, 23*, 3–38.

Albouy, P., Lévêque, Y., Hyde, K. L., Bouchet, P., Tillmann, B., & Caclin, A. (2015). Boosting pitch encoding with audiovisual interactions in congenital amusia. *Neuropsychologia, 67*, 111–120.

Alink, A., Singer, W., & Muckli, L. (2008). Capture of auditory motion by vision is represented by an activation shift from auditory to visual motion cortex. *Journal of Neuroscience, 28*, 2690–2697.

Allen, G. L. (1999). Spatial abilities, cognitive maps, and wayfinding. In R. G. Golledge (Ed.), *Wayfinding behavior: Cognitive mapping and other spatial processes* (pp. 46–80). Baltimore: Johns Hopkins University Press.

Allen, G. L., & Kirasic, K. C. (1985). Effects of the cognitive organization of route knowledge on judgements of macrospatial distance. *Memory & Cognition, 13*, 218–227.

Allen, P. G., & Kolers, P. A. (1981). Sensory specificity of apparent motion. *Journal of Experimental Psychology: Human Perception and Performance*, 7, 1318–1326.

Allport, G. W., 1954. *The nature of prejudice.* Reading, MA: Addison-Wesley.

Allred, S. R., Crawford, L. E., Duffy, S., & Smith, J. (2016). Working memory and spatial judgments: Cognitive load increases the central tendency bias. *Psychonomic Bulletin & Review.* Advance online publication. http://dx.doi.org/10.3758/s13423-016-1039-0

Alvarez, G. A. (2011). Representing multiple objects as an ensemble enhances visual cognition. *Trends in Cognitive Sciences*, 15(3), 122–131.

Alvarez, G. A., & Oliva, A. (2009). Spatial ensemble statistics are efficient codes that can be represented with reduced attention. *Proceedings of the National Academy of Sciences, 106*(18), 7345–7350.

Amedi, A., von Kriegstein, K., van Atteveldt, N. M., Beauchamp, M. S., & Naumer, M. J. (2004). Functional imaging of human crossmodal identification and object recognition. *Experimental Brain Research, 166*, 559–571.

Amirshahi, S. A., Hayn-Leichsenring, G. U., Denzler, J., & Redies, C. (2014). Evaluating the rule of thirds in photographs and paintings. *Art & Perception, 2*, 163–182.

Amorapanth, P., Kranjec, A., Bromberger, B., Lehet, M., Widick, P., Woods, A. J., et al. (2012). Language, perception, and the schematic representation of spatial relations. *Brain and Language, 120*, 226–236.

Amorapanth, P. X., Widick, P., & Chatterjee, A. (2010). The neural basis for spatial relations. *Journal of Cognitive Neuroscience, 22*(8), 1739–1753.

Amorim, M. A., Lang, W., Lindinger, G., Mayer, D., Deecke, L., & Berthoz, A. (2000). Modulation of spatial orientation processing by mental imagery instructions: A MEG study of representational momentum. *Journal of Cognitive Neuroscience, 12*, 569–582.

Anderson, D. (2007). Consciousness and realism. *Journal of Consciousness Studies, 14*, 1–17.

Anderson, J. R. (2000). *Learning and memory.* New York: John Wiley.

Angelaki, D., McHenry, M., Dickman, J. D., Newlands, S., & Hess, B. (1999). Computation of inertial motion: Neural strategies to resolve ambiguous otolith information. *Journal of Neuroscience, 19*, 316–327.

Angelaki, D. E., Shaikh, A. G., Green, A. M., & Dickman, J. D. (2004). Neurons compute internal models of the physical laws of motion. *Nature, 430*(6999), 560–564.

Ansorge, U., Carbone, E., Becker, S. I., & Turatto, M. (2010). Attentional capture by motion onset is spatially imprecise. *European Journal of Cognitive Psychology, 22*, 62–105.

Anstis, S. (2007). The flash-lag effect during illusory chopstick motion. *Perception, 36*, 1043–1048.

Anstis, S. (2010). Illusions of time, space, and motion: Flash-lag meets chopsticks and reversed phi. In R. Nijhawan & B. Khurana (Eds.), *Space and time in perception and action* (pp. 408–421). New York: Cambridge University Press.

Antović, M., Bennett, A., & Turner, M. (2013). Running in circles or moving along lines: Conceptualization of musical elements in sighted and blind children. *Musicae Scientiae, 17*, 229–245.

Appelle, S. (1972). Perception and discrimination as a function of orientation: The "oblique effect" in man and animals. *Psychological Bulletin, 78*, 266–278.

Appleyard, D. (1969). Why buildings are known. *Environment and Behavior, 1*, 131–156.

Arduino, L. S., Previtali, P., & Girelli, L. (2010). The centre is not in the middle: Evidence from line and word bisection. *Neuropsychologia, 48*, 2140–2446.

Ariely, D. (2001). Seeing sets: Representation by statistical properties. *Psychological Science, 12*(2), 157–162.

Arnheim, R. (1974). *Art and visual perception: A psychology of the creative eye.* Berkeley: University of California Press.

Arnheim, R. (1982). *The power of the center.* Berkeley: University of California Press.

Arnheim, R. (1986). *New essays on the psychology of art.* Berkeley: University of California Press.

Arnheim, R. (1988). Visual dynamics. *Scientific American, 76*, 585–591.

Arnheim, R. (2006). *Film as art (50th anniversary edition)*. Berkeley: University of California Press.

Arnold, D. H., Clifford, C. W. G., & Wenderoth, P. (2001). Asynchronous processing in vision: Color leads motion. *Current Biology, 11*, 596–600.

Arnold, D. H., Durant, S., & Johnston, A. (2003). Latency differences and the flash-lag effect. *Vision Research, 43*, 1829–1835.

Arnold, D. H., Ong, Y., & Roseboom, W. (2009). Simple differential latencies modulate, but do not cause the flash-lag effect. *Journal of Vision, 9*(5), 4.

Arrighi, R., Alais, D., & Burr, D. (2005). Neural latencies do not explain the auditory and audio-visual flash-lag effect. *Vision Research, 45*, 2917–2925.

Arrighi, R., Marini, F., & Burr, D. (2009). Meaningful auditory information enhances perception of visual biological motion. *Journal of Vision, 9*, 25.

Arrington, C. M., Carr, T. H., Mayer, A. R., & Rao, S. M. (2000). Neural mechanisms of visual attention: Object-based selection of a region in space. *Journal of Cognitive Neuroscience*, 12(supplement 2), 106.

Arzy, S., Adi-Japha, E., & Blanke, O. (2009). The mental time line: An analogue of the mental number line in the mapping of life events. *Consciousness and Cognition, 18*, 781–785.

Aschersleben, G., & Müsseler, J. (1999). Dissociations in the timing of stationary and moving stimuli. *Journal of Experimental Psychology: Human Perception and Performance, 25*, 1709–1720.

Asher, M. F., Tolhurst, D. J., Troscianko, T., & Gilchrist, I. (2013). Regional effects of clutter on human target detection performance. *Journal of Vision, 13*(5): 1–15.

Ashida, H. (2004). Action-specific extrapolation of target motion in human visual system. *Neuropsychologia, 42*, 1515–1524.

Atmanspacher, H. (2007) Contextual emergence from physics to cognitive neuroscience. *Journal of Consciousness Studies, 14*, 18–36.

Atmanspacher, H., & Kronz, F. (1998). Many realisms. *Acta Polytechnica Scandinavica, 91*, 31–43.

Attneave, F. (1959). Comments: In defense of homunculi. In W. A. Rosenblith (Ed.), *Sensory communication: Contributions to the symposium on principels of sensoty comunication, July 19–August 1* (pp. 777–781). Cambrdidge, MA: MIT Press.

Au, R. K. C., & Watanabe, K. (2013). Object motion continuity and the flash-lag effect. *Vision Research, 92*, 19–25.

Aubert, H. (1861). Eine scheinbare bedeutende Drehung von Objekten bei Neigung des Kopfes nach rechts oder links. *Virchows Archiv fur pathologische Anatomie und Physiologic und fur klinische Medizin, 20*, 381–393.

Auerbach, E., Beller, A. J., Henkes, H. E., & Goldhaber, G. (1961). Electric potentials of retina and cortex of cats evoked by monocular and binocular photic stimulation. *Vision Research, 1*(1–2), 166IN5–182IN8.

Avant, L. L. (1965). Vision in the Ganzfeld. *Psychological Bulletin, 64*, 246–258.

Avery, G., & Day, R. (1969). Basis of the horizontal vertical illusion. *Journal of Experimental Psychology, 81*, 376–380.

Avrahami, J., Argaman, T., & Weiss-Chasum, D. (2004). The mysteries of the diagonal: Gender-related perceptual asymmetries. *Perception & Psychophysics, 66*, 1405–1417.

Awh, E., Armstrong, K. M., & Moore, T. (2006). Visual and oculomotor selection: Links, causes and implications for spatial attention. *Trends in Cognitive Science, 10*, 124–130.

Awh, E., & Pashler, H. (2000). Evidence for split attentional foci. *Journal of Experimental Psychology: Human Perception and Performance, 26*, 834–846.

Aznar-Casanova, J. A., Torrents, A., & Torro-Alves, N. T. (2008). The role of the vertical disparities in the Oblique Effect. *Psychology & Neuroscience, 1*(2), 85–98.

Baars, B. (1988). *A cognitive theory of consciousness*. Cambridge: Cambridge University Press.

Bach, D. R., Buxtorf, K., Strik, W. K., Neuhoff, J. G., & Seifritz, E. (2011). Evidence for impaired sound intensity processing in schizophrenia. *Schizophrenia Bulletin*, *37*(2), 426–431.

Bach, D. R., Furl, N., Barnes, G., & Dolan, R. J. (2015). Sustained magnetic responses in temporal cortex reflect instantaneous significance of approaching and receding sounds. *PLoS One*, *10*(7). doi:10.1371/journal.pone.0134060

Bach, D. R., Neuhoff, J. G., Perrig, W., & Seifritz, E. (2009). Looming sounds as warning signals: The function of motion cues. *International Journal of Psychophysiology*, *74*(1), 28–33.

Bach, D. R., Schachinger, H., Neuhoff, J. G., Esposito, F., Di Salle, F., Lehmann, C., . . . Seifritz, E. (2008). Rising sound intensity: An intrinsic warning cue activating the amygdala. *Cerebral Cortex*, *18*(1), 145–150.

Bachmann, T. (1984). The process of perceptual retouch: Nonspecific afferent activation dynamics in explaining visual masking. *Perception & Psychophysics*, *35*, 69–84.

Bachmann, T. (1999). Twelve spatiotemporal phenomena and one explanation. *Advances in Psychology*, *129*, 173–206.

Bachmann, T. (2007). Binding binding: Departure points for a different version of the perceptual retouch theory. *Advances in Cognitive Psychology*, *3*, 41–55.

Bachmann, T. (2010). Priming and retouch in flash-lag and other phenomena of the streaming perceptual input. In R. Nijhawan & B. Khurana (Eds.), *Space and time in perception and action* (pp. 536–557). New York: Cambridge University Press.

Bachmann, T., & Kalev, K. (1997). Adjustment of collinearity of laterally moving, vertically separated lines reveals compression of subjective distance as a function of aperture size and speed of motion. *Perception*, *26*, 119–120.

Bachmann, T., Luiga, I., Põder, E., & Kalev, K. (2003). Perceptual acceleration of objects in stream: Evidence from flash-lag displays. *Consciousness and Cognition*, *12*, 279–297.

Bachmann, T., Murd, C., & Põder, E. (2012). Flash-lag effect: Complicating motion extrapolation of the moving reference-stimulus paradoxically augments the effect. *Psychological Research*, *76*, 654–666.

Bachmann, T., & Põder, E. (2001). Change in feature space is not necessary for the flash-lag effect. *Vision Research*, *41*, 1103–1106.

Bächtold, D., Baumüller, M., & Brugger, P. (1998). Stimulus-response compatibility in representational space. *Neuropsychologia*, *36*, 731–735.

Bahill, A. T., & Karnavas, W. J. (1993). The perceptual illusion of baseball's rising fastball and breaking curveball. *Journal of Experimental Psychology: Human Perception and Performance*, *19*(1), 3–14.

Baier, B., Zu Eulenburg, P., Best, C., Geber, C., Müller-Forell, W., Birklein, F., & Dieterich, M. (2013). Posterior insular cortex – a site of vestibular-somatosensory interaction? *Brain and Behavior*, *3*(5), 519–524.

Bailenson, J. N., Shum, M. S., & Uttal, D. H. (1998). Road climbing: Principles governing asymmetric route choices on maps. *Journal of Environmental Psychology*, *18*(3), 251–264.

Bailenson, J. N., Shum, M. S., & Uttal, D. H. (2000). The initial segment strategy: A heuristic for route selection. *Memory & Cognition*, *28*(2), 306–318.

Baird, J., Merrill, A., & Tannenbaum, J. (1979). Studies of the cognitive representation of spatial relations. II. A familiar environment. *Journal of Experimental Psychology: General*, *108*, 92–98.

Baird, J. C., Wagner, M., & Noma, E. (1982). Impossible cognitive spaces. *Geographical Analysis*, *14*(3), 204–216.

Bakan, D. (1966). *The duality of human existence: An essay on psychology and religion*. Chicago: Rand McNally.

Balcetis, E. (2016). Approach and avoidance as organizing structures for motivated distance perception. *Emotion Review*, *8*(2), 115–128.

Baldo, M. V. C., & Caticha, N. (2005). Computational neurobiology of the flash-lag effect. *Vision Research*, *45*, 2620–2630.

Baldo, M. V. C., Kihara, A. H., Namba, J., & Klein, S. A. (2002). Evidence for an attentional component of the perceptual misalignment between moving and flashing stimuli. *Perception, 31*, 17–30.

Baldo, M. V. C., & Klein, S. A. (1995). Extrapolation or attention shift? *Nature, 378*, 565–566.

Baldo, M. V. C., & Klein, S. A. (2008) Shifting attention to the flash-lag effect. *Behavioral and Brain Sciences, 31*(2), 198–199.

Baldo, M. V. C., & Klein, S. A. (2010). Paying attention to the flash-lag effect. In R. Nijhawan & B. Khurana (Eds.), *Space and time in perception and action* (pp. 396–407). New York: Cambridge University Press.

Baldo, M. V. C., & Namba, J. (2002). The attentional modulation of the flash-lag effect. *Brazilian Journal of Medical and Biological Research, 35*, 969–972.

Baldo, M. V. C., Ranvaud, R. D., & Morya, E. (2002). Flag errors in soccer games: The flash-lag effect brought to real life. *Perception, 31*, 1205–1210.

Ball, K., & Sekuler, R. (1980). Human vision favors centrifugal motion. *Perception, 9*(3), 317–325.

Ball, K., & Sekuler, R. (1987). Direction-specific improvement in motion discrimination, *Vision Research* 27, 953–965.

Ball, W., & Tronick, E. (1971). Infant responses to impending collision – optical and real. *Science, 171*(3973), 818–820.

Banich, M. T., Heller, W., & Levy, J. (1989). Aesthetic preference and picture asymmetries. *Cortex, 25*, 187–195.

Bar, M. (2004). Visual objects in context. *Nature Reviews Neuroscience, 5*, 617–629.

Bar, M., & Neta, M. (2006). Humans prefer curved visual objects. *Psychological Science*, 17, 645–648.

Bar, M., & Neta, M. (2007). Visual elements of subjective preference modulate amygdala activation. *Neuropsychologia*, 45, 2191–2200.

Barlow, H. B. (1980). Critical limiting factors in the design of the eye and visual cortex. *Proceedings of the Royal Society B*. doi: 10.1098/rspb.1981.0022.

Barnett-Cowan, M., Fleming, R. W., Singh, M., & Bülthoff, H. H. (2011). Perceived object stability depends on multisensory estimates of gravity. *PLoS ONE 6*(4), e19289.

Barra, J., Marquer, A., Joassin, R., Reymond, C., Metge, L., Chauvineau, V., & Pérennou, D. (2010). Humans use internal models to construct and update a sense of verticality. *Brain*, 133(Pt 12), 3552–3563.

Barsalou, L. W. (1999). Perceptual symbol systems. *The Behavioral and Brain Sciences, 22*(4), 577–609.

Barsalou, L. W. (2008). Grounded cognition. *Annual Review of Psychology, 59*, 617–645.

Barth, H., Lesser, E., Taggart, J., & Slusser, E. (2015). Spatial estimation: A non-Bayesian alternative. *Developmental Science, 18*(5), 853–862.

Bartolomeo, P. (1987). Aspetti dell'emi-inattenzione spaziale nelle lesioni emisferiche: fattori che influenzano la bisezione di linee [Aspects of spatial hemi-inattention: factors influencing line bisection performance]. Unpublished MD thesis, Università Cattolica, Roma.

Bartolomeo, P. (1997). The novelty effect in recovered hemineglect. *Cortex, 33*(2), 323–332.

Bartolomeo, P. (2000). Inhibitory processes and compensation for spatial bias after right hemisphere damage. *Neuropsychological Rehabilitation, 10*, 511–526.

Bartolomeo, P. (2002). The relationship between visual perception and visual mental imagery: A reappraisal of the neuropsychological evidence. *Cortex, 38*(3), 357–378.

Bartolomeo, P. (2006). A parieto-frontal network for spatial awareness in the right hemisphere of the human brain. *Archives of Neurology, 63*, 1238–1241.

Bartolomeo, P. (2007). Visual neglect. *Current Opinion in Neurology, 20*(4), 381–386.

Bartolomeo P. (2014). *Attention disorders after right brain damage: Living in halved worlds.* London: Springer-Verlag.

Bartolomeo, P., Bachoud-Lévi, A.-C., Azouvi, P., & Chokron, S. (2005). Time to imagine space: A chronometric exploration of representational neglect. *Neuropsychologia, 43*(9), 1249–1257.

Bartolomeo, P., & Chokron, S. (1999). Left unilateral neglect or right hyperattention. *Neurology, 53*(9), 2023–2027.

Bartolomeo, P., Chokron, S., & Gainotti, G. (2001). Laterally directed arm movements and right unilateral neglect after left hemisphere damage. *Neuropsychologia, 39*, 1013–1021.

Bartolomeo, P., Chokron, S., & Siéroff, E. (1999). Facilitation instead of inhibition for repeated right-sided events in left neglect. *NeuroReport, 10*, 3353–3357.

Bartolomeo, P., D'Erme, P., & Gainotti, G. (1994). The relationship between visuospatial and representational neglect. *Neurology, 44*, 1710–1714.

Bartolomeo, P., Thiebaut de Schotten, M., & Chica, A. B. (2012). Brain networks of visuospatial attention and their disruption in visual neglect. *Frontiers in Human Neuroscience, 6*, 110.

Bartolomeo, P., Urbanski, M., Chokron, S., Chainay, H., Moroni, C., Siéroff, E., et al. (2004). Neglected attention in apparent spatial compression. *Neuropsychologia, 42*(1), 49–61.

Bates, E., Wilson, S. M., Saygin, A. P., Dick, F., Sereno, M. I., Knight, R. T., & Dronkers, N. F. (2003). Voxel-based lesion–symptom mapping. *Nature Neuroscience, 6*(5), 448–450.

Battaglia, P. W., Hamrick, J. B., & Tenenbaum, J. B. (2013). Simulation as an engine of physical scene understanding. *Proceedings of the National Academy of Science of the United States of America, 110*, 18327–18332.

Baumann, O., & Greenlee, M. W. (2007). Neural correlates of coherent audiovisual motion perception. *Cerebral Cortex, 17*, 1433–1443.

Bauer, W. (2011). An argument for the extrinsic grounding of mass, *Erkenntnis, 74*(1), 81–99.

Bays, P. M., & Husain, M. (2008). Dynamic shifts of limited working memory resources in human vision. *Science, 321*(5890), 851–854.

Beaumont, J. G. (1985). Lateral organization and aesthetic preference. *Neuropychologia, 23*, 103–113.

Becker, B., & Mark, G. (1998, June). Social conventions in collaborative virtual environments. In *Proceedings of Collaborative Virtual Environments* (pp. 17–19).

Becker, S. I., Ansorge, U., & Turatto, M. (2009). Saccades reveal that allocentric coding of the moving object causes mislocalization in the flash-lag effect. *Attention, Perception, & Psychophysics, 71*, 1313–1324.

Beecham, R., Reeve, R. A., & Wilson, S. J. (2009). Spatial representations are specific to different domains of knowledge. *PlosOne, 4*, e5543.

Beis, J. M., Keller, C., Morin, N., Bartolomeo, P., Bernati, T., Chokron, S., et al. (2004). Right spatial neglect after left hemisphere stroke: Qualitative and quantitative study. *Neurology, 63*, 1600–1605.

Ben-Artzi, E., & Marks, L. E. (1995). Visual-auditory interaction in speeded classification: Role of stimulus difference. *Perception & Psychophysics, 57*, 1151–1162.

Benedikt, M. L. (1979). To take hold of space: Isovists and isovist fields. *Environment and Planning B, 6*, 47–65.

Benjafield, J. G. (2010). The golden section and American psychology 1892–1938. *Journal of the History of the Behavioral Sciences, 46*, 52–71.

Benson, A. J. (1988). Spatial disorientation – general aspects. In J. Ernsting & P. King (Eds.), *Aviation medicine* (pp. 277–296). London: Butterworth.

Berlucchi, G. (2006). Inhibition of return: A phenomenon in search of a mechanism and a better name. *Cognitive Neuropsychology, 23*, 1065–1074.

Berlyne, D. E. (1971). *Aesthetics and psychobiology*. New York: Appleton-Century-Crofts.

Bernstein, I. H., & Edelstein, B. A. (1971). Effects of some variations in auditory input upon visual choice reaction time. *Journal of Experimental Psychology, 87*, 241–247.

Berry, II, M. J., Brivanlou, I. H., Jordan, T. A., & Meister, M. (1999). Anticipation of moving stimuli by the retina. *Nature, 398*, 334–338.

Berryhill, M. E., Fendrich, R., & Olson, I. R. (2009). Impaired distance perception and size constancy following bilateral occipitoparietal damage. *Experimental Brain Research, 194*(3), 381–393.

Bertamini, M. (1993). Memory for position and dynamic representations. *Memory & Cognition, 21*, 449–457.

Bertamini, M., Bennett, K. M., & Bode, C. (2011). The anterior bias in visual art: The case of images of animals. *Laterality, 16*, 673–689.

Bertamini, M., Bode, C., & Bruno, N. (2011). The effect of left-right reversal on film: Watching Kurosawa reversed. *i-Perception, 2*, 528–540.

Bertamini, M., Jones, L. A., Spooner, A., & Hecht, H. (2005). Boundary extension: The role of magnification, object size, context, and binocular information. *Journal of Experimental Psychology: Human Perception and Performance, 31*, 1288–1307.

Bertamini, M., Latto, R., & Spooner, A. (2003). The Venus effect: People's understanding of mirror reflections in paintings. *Perception, 32*, 593–599.

Bertamini, M., Lawson, R., Jones, L., & Winters, M. (2010). The Venus effect in real life and in photographs. *Attention, Perception, & Psychophysics, 72*, 1948–1964.

Bertamini, M., Lawson, R., & Liu, D. (2008). Understanding 2D projections on mirrors and on windows. *Spatial Vision, 21* (3–5), 273–289.

Bertamini, M., & Parks, T. E. (2005). On what people know about images on mirrors. *Cognition, 98*, 85–104.

Bertamini, M., Spooner, A., & Hecht, H. (2003). Naïve optics: Predicting and perceiving reflections in mirrors. *Journal of Experimental Psychology: Human Perception and Performance, 29*(5), 982–1002.

Bertamini, M., & Wynne, L. A. (2009). The tendency to overestimate what is visible in a planar mirror amongst adults and children. *European Journal of Cognitive Psychology, 22*, 516–528.

Bertamini, M., Yang, T. L., & Proffitt, D. R. (1998). Relative size perception at a distance is best at eye level. *Attention, Perception, & Psychophysics, 60*, 673–682.

Berthoz, A. (2000). *The brain's sense of movement*. Cambridge, MA: Harvard University Press.

Berti, A., & Frassinetti, F. (2000). When far becomes near: Remapping of space by tool use. *Journal of Cognitive Neuroscience, 12*(3), 415–420.

Beschin, N., Cubelli, R., Della, S., & Spinazzola, L. (1997). Left of what? The role of egocentric coordinates in neglect. *Journal of Neurology, Neurosurgery, and Psychiatry, 63*(4), 483–489.

Bestmann, S., Ruff, C. C., Blakemore, C., Driver, J., & Thilo, K. V. (2007). Spatial attention changes excitability of human visual cortex to direct stimulation. *Current Biology, 17*, 134–139.

Bex, P. J., & Makous, W. (1997). Radial motion looks faster. *Vision Research, 37*(23), 3399–3405.

Bhalla, M., & Proffitt, D. R. (1999). Visual-motor recalibration in geographical slant. *Journal of Experimental Psychology: Human Perception and Performance, 25*(4), 1076–1096.

Bian, Z., & Andersen, G. J. (2010). The advantage of a ground surface in the representation of visual scenes. *Journal of Vision, 10*(8), 1–19.

Bian, Z., Braunstein, M. L., & Andersen, G. J. (2005). The ground dominance effect in the perception of 3-D layout. *Attention, Perception, & Psychophysics, 67*, 815–828.

Bian, Z., Braunstein, M. L., & Andersen, G. J. (2006). The ground dominance effect in the perception of relative distance in 3-D scenes is mainly due to characteristics of the ground surface. *Attention, Perception, & Psychophysics, 68*, 1297–1309.

Bianchi, I., Bertamini, M., & Savardi, U. (2015). Differences between predictions of how a reflection behaves based on the behaviour of an object. *Acta Psychologica, 161*, 54–63.

Bianchi, I., Burro, R., Torquati, S., & Savardi, U. (2013). The middle of the road: Perceiving intermediates. *Acta Psychologica, 144*(1), 121–135.

Bianchi, I., Paradis, C, Burro, R., van de Weijer, J., Nyström, M., & Savardi, U. (2017). Identification of poles and intermediates by eye and by hand. *Acta Psychologica*, 180, 175–189.

Bianchi, I., & Savardi, U. (2008a) *The perception of contraries*. Rome, IT: Aracne.

Bianchi, I., & Savardi, U. (2008b). The relationship perceived between the real body and the mirror image. *Perception, 5*, 666–687.

Bianchi, I., & Savardi U. (2009). Contrariety in plane mirror reflections. In U. Savardi (Ed.), *The perception and cognition of contraries* (pp. 113–128). Milan: McGraw Hill.

Bianchi, I., & Savardi, U. (2012a). What fits in into a mirror: Naïve beliefs on the field of view of mirrors. *Journal of Experimental Psychology: Human Perception and Performance, 38*(5), 1144–1158.

Bianchi I., & Savardi, U. (2012b). The cognitive dimensions of contrariety. In J.-Y. Bezieau & G. Payette (Eds.), *The square of opposition: A general framework for cognition* (pp. 443–470). New York: Peter Lang Publishing Group.

Bianchi, I., & Savardi, U. (2014). Grounding naive physics and optics in perception. *The Baltic International Yearbook for Cognition Logic and Communication, 6*, 1–15.

Bianchi, I., Savardi, U., & Burro, R. (2011). Perceptual ratings of opposite spatial properties: Do they lie on the same dimension? *Acta Psychologica, 138*(3), 405–418.

Bianchi, I., Savardi, U., Burro, R., & Martelli, M.F. (2014). Doing the opposite to what another person is doing. *Acta Psychologica, 151*, 117–133.

Bianchi, I., Savardi, U., & Kubovy, M. (2011). Dimensions and their poles: A metric and topological theory of opposites. *Language and Cognitive Processes, 26*(8), 1232–1265.

Biederman, I., Mezzanotte, R. J., & Rabinowitz, J. C. (1982). Scene perception: Detecting and judging objects undergoing relational violations. *Cognitive Psychology, 14*, 143–177.

Billig, M., & Tajfel, H. (1973). Social categorization and similarity in intergroup behavior. *European Journal of Social Psychology, 3*, 27–52.

Binetti, N., Hagura, N., Fadipe, C., Tomassini, A., Walsh, V., & Bestmann, S. (2015). Binding space and time through action. *Proceedings of the Royal Society of London B: Biological Sciences, 282*(1805), 20150381.

Bingham, G. P., Schmidt, R. C., & Rosenblum, L. D. (1995). Dynamics and the orientation of kinematic forms in visual event recognition. *Journal of Experimental Psychology: Human Perception and Performance, 21*, 1473–1493.

Bishop, R., & Atmanspacher, H. (2006) Contextual emergence in the description of properties. *Foundations of Physics, 36*, 1753–1777.

Bisiach, E., & Luzzatti, C. (1978). Unilateral neglect of representational space. *Cortex, 14*(1), 129–133.

Bisiach, E., Perani, D., Vallar, G., & Berti, A. (1986). Unilateral neglect: Personal and extra-personal. *Neuropsychologia, 24*, 759–767.

Bisiach, E., Pizzamiglio, L., Nico, D., & Antonucci, G. (1996). Beyond unilateral neglect. *Brain, 119*, 851–857.

Bisiach, E., Rusconi, M. L., Peretti, V. A., & Vallar, G. (1994). Challenging current accounts of unilateral neglect. *Neuropsychologia, 32*, 1431–1434.

Blakemore, C., Carpenter, R. H. S., & Georgeson, M. A. (1970). Lateral inhibition between orientation detectors in the human visual system. *Nature, 228*, 37–39.

Blaker, N. M., & van Vugt, M. (2014). The status-size hypothesis: How cues of physical size and social status influence each other. In J. T. Cheng, J. L. Tracy, & C. Anderson (Eds.), *The psychology of social status* (pp. 119–137). New York: Springer Science +Business Media.

Blättler, C., Ferrari, V., Didierjean, A., & Marmèche, E. (2011). Representational momentum in aviation. *Journal of Experimental Psychology: Human Perception and Performance, 37*, 1569–1577.

Blättler, C., Ferrari, V., Didierjean, A., van Elslande, P., & Marmèche, E. (2010). Can expertise modulate representational momentum? *Visual Cognition, 18*, 1253–1273.

Bloesch, E. K., Davoli, C. C., Roth, N., Brockmole, J. R., & Abrams, R. A. (2012). Watch this! Observed tool use affects perceived distance. *Psychonomic Bulletin & Review, 19*(2), 177–183.

Bloom, P., Peterson, M., Nadel, L., & Garrett, M. (1996). *Language and space*. Cambridge, MA: The MIT Press.

Blumenfeld, W. (1913). Untersuchungen über die scheinbare Grösse im Sehraume. *Zeitschrift für Psychologie, 65*, 241–404.

Bocianski, D., Müsseler, J., & Erlhagen, W. (2008). Relative mislocalization of successively presented stimuli. *Vision Research, 48*, 2204–2212.

Bocianski, D., Müsseler, J., & Erlhagen,W. (2010). Effects of attention on a relative mislocalization with successively presented stimuli. *Vision Research, 50*, 1793–1802.

Bock, G. R., & Good, J. (1994) *Higher-order processing in the visual system*. Oxford: Wiley.

Bodenhausen, G. V., & Wyer, R. S. (1985). Effects of stereotypes in decision making and information-processing strategies. *Journal of Personality and Social Psychology, 48*(2), 267.

Boeschoten, M. A., Kemner, C., Kenemans, J. L., & Engeland, H. (2005). The relationship between local and global processing and the processing of high and low spatial frequencies studied by event-related potentials and source modeling. *Cognitive Brain Research, 24*, 228–236.

Bonato, M. (2012). Neglect and extinction depend greatly on task demands: A review. *Frontiers in Human Neuroscience, 6*, 195.

Bonato, M., & Cutini, S. (2015). Increased attentional load moves the left to the right. *Journal of Clinical and Experimental Neuropsychology, 38*, 158–170.

Bonato, M., Saj, A., & Vuilleumier, P. (2016). *Neural Plasticity*, Volume *2016*, Article ID 2716036.

Bonato, M., & Umiltà, C. (2014). Heterogeneous timescales are spatially represented. *Frontiers in Psychology* 5: 542.

Bonato, M., Zorzi, M., & Umiltà, C. (2012). When time is space: Evidence for a mental time line. *Neuroscience and Biobehavioural Reviews 36*, 2257–2273.

Boring, E. G. (1950). *A history of experimental psychology*. New York: Appelton-Century-Crofts.

Boroditsky, L. (2000). Metaphoric structuring: Understanding time through spatial metaphors. *Cognition, 75*, 1–28.

Boroditsky, L. (2011). How languages construct time (pp. 333-342). In S. Dehaene & E. Brannon (Eds.), *Space, time and number in the brain: Searching for the foundations of mathematical thought*. London: Elsevier.

Bortolami, S. B., Pierobon, A., DiZio, P., & Lackner, J. R. (2006). Localization of the subjective vertical during roll, pitch, and recumbent yaw body tilt. *Experimental Brain Research, 173*, 364–373.

Bosco, G., Delle Monache, S., Gravano, S., Indovina, I., La Scaleia, B., et al. (2015). Filling gaps in visual motion for target capture. *Frontiers Integrative Neuroscience*, 9:13.

Bosco, G., Delle Monache, S., & Lacquaniti, F. (2012). Catching What We Can't See: Manual Interception of Occluded Fly-Ball Trajectories. *PLoS ONE, 7*(11), e49381.

Boselie, F. (1984). The aesthetic attractivity of the golden section. *Psychological Research, 45*, 367–375.

Boselie, F. (1997). The golden section and shape of objects. *Empirical Studies of the Arts, 15*, 131–141.

Bouleau, C. (2014). *The painter's secret geometry*. Allegro Editions.

Bourgeois, A., Chica, A. B., Migliaccio, R., Bayle, D. J., Duret, C., Pradat-Diehl, P., et al. (2015). Inappropriate rightward saccades after right hemisphere damage: Oculomotor analysis and anatomical correlates. *Neuropsychologia, 73*, 1–11.

Bourgeois, A., Chica, A. B., Migliaccio, R., Thiebaut de Schotten, M., & Bartolomeo, P. (2012). Cortical control of inhibition of return: Evidence from patients with inferior parietal damage and visual neglect. *Neuropsychologia, 50*(5), 800–809.

Bourlon, C., Duret, C., Pradat-Diehl, P., Azouvi, P., Loeper-Jeny, C., Merat-Blanchard, M., et al. (2011). Vocal response times to real and imagined stimuli in spatial neglect: A group study and single-case report. *Cortex*, *47*(5), 536–546.

Bourlon, C., Oliviero, B., Wattiez, N., Pouget, P., & Bartolomeo, P. (2011). Visual mental imagery: What the head's eye tells the mind's eye. *Brain Research*, *1367*, 287–297.

Bourlon, C., Pradat-Diehl, P., Duret, C., Azouvi, P., & Bartolomeo, P. (2008). Seeing and imagining the "same" objects in unilateral neglect. *Neuropsychologia*, *46*(10), 2602–2606.

Bourne, V. J. (2006). The divided visual field paradigm: Methodological considerations. *Laterality*, *11*(4), 373–393.

Bower, T. G. R., Broughto, J. M., & Moore, M. K. (1971). Infant responses to approaching objects – indicator of response to distal variables. *Perception & Psychophysics*, 9(2B), 193–199.

Bowerman, M., & Choi, S. (2001). Shaping meanings for language: Universal and language-specific in the acquisition of spatial semantic categories. In M. Bowerman & S. C. Levinson (Eds.), *Language acquisition and conceptual development* (pp. 475–511). Cambridge: Cambridge University Press.

Bowers, D., & Heilman, K. M. (1980). Pseudoneglect: Effects of hemispace on a tactile line bisection task. *Neuropsychologia*, *18*, 491–498.

Boyd, I. L. (1996). Temporal scales of foraging in a marine predator. *Ecology*, *77*, 426–434.

Braaten, R. (1993, November). *Synesthetic correspondence between visual location and auditory pitch in infants.* Paper presented at the Annual Meeting of the Psychonomic Society, Washington, DC.

Bradburn, N. M., Rips, L. J., & Shevell, S. K. (1987). Answering autobiographical questions: The impact of memory and inference on surveys. *Science*, *236*, 158–161.

Brainard, D. H. (1997). The psychophysics toolbox. *Spatial Vision*, *10*, 433–436.

Branchini, E., Bianchi, I., Burro, R., Capitani, E., & Savardi, U. (2016). Can contraries prompt intuition in insight problem solving? *Frontiers in Psychology*, 7, 1962.

Branchini, E., Burro, R., Bianchi, I., & Savardi, U. (2015). Contraries as an effective strategy in geometrical problem solving. *Thinking & Reasoning*, *21*(4), 397–430.

Brandt, T., Dieterich, M., & Danek, A. (1994). Vestibular cortex lesions affect the perception of verticality. *Annals of Neurology*, *35*(4), 403–412.

Brehaut, J. C., & Tipper, S. P. (1996). Representational momentum and memory for luminance. *Journal of Experimental Psychology: Human Perception and Performance*, *22*, 480–501.

Brendel, E., DeLucia, P. R., Hecht, H., Stacy, R. L., & Larsen, J. T. (2012). Threatening pictures induce shortened time-to-contact estimates. *Attention Perception & Psychophysics*, *74*(5), 979–987.

Brendel, E., Hecht, H., DeLucia, P. R., & Gamer, M. (2014). Emotional effects on time-to-contact judgments: Arousal, threat, and fear of spiders modulate the effect of pictorial content. *Experimental Brain Research*, *232*(7), 2337–2347.

Brenner, E., van Beers, R. J., Rotman, G., & Smeets, J. B. J. (2006). The role of uncertainty in the systematic spatial mislocalization of moving objects. *Journal of Experimental Psychology: Human Perception and Performance*, *32*, 811–825.

Brenner, E., & Smeets, J. B. J. (2000). Motion extrapolation is not responsible for the flash-lag effect. *Vision Research*, *40*, 1645–1648.

Brenner, E., Smeets, J. B. J. & Landy, M. S. (2001). How vertical disparities assist judgements of distance. *Vision Research*, *41*, 3455–3465.

Brenner, E., van Beers, R. J., Rotman, G., & Smeets, J. B. J. (2006). The role of uncertainty in the systematic spatial mislocalization of moving objects. *Journal of Experimental Psychology: Human Perception and Performance*, *32*, 811–825.

Bresciani, J-P., Dammeier, F., & Ernst, M. O. (2006). Vision and touch are automatically integrated for the perception of sequences of events. *Journal of Vision*, *6*, 554–564.

Bridgeman, B. (2006). Contributions of lateral inhibition to object substitution masking and attention. *Vision Research*, *46*(24), 4075–4082.

Briggs, R. (1973) Urban cognitive distance. In R. Downs & D. Stea (Eds.), *Image and environment* (pp. 361–388). Chicago: Aldine.

Brockmann, D., Hufnagel, L., & Geisel, T. (2006). The scaling laws of human travel. *Nature*, *439*(7075), 462–465.

Brodersen, K. H. (2009). Decoding mental activity from neuroimaging data – the science behind mind-reading. *New Collection Oxford*, *4*, 50–61.

Brodie, E. E. (2010). Reflecting upon line bisection: Mirror reversal increases the magnitude of pseudoneglect. *Neuropsychologia*, *48*, 1517–1520.

Bromberger, B., Sternschein, R. Widick, P, Smith, II, S., & Chatterjee, A. (2011). The right hemisphere in esthetic perception. *Frontiers in Human Neuroscience*, *5*, 109; 1–8.

Brooks, A., van der Zwan, R., Billard, A., Petreska, B., Clarke, S., & Blanke, O. (2007). Auditory motion affects visual biological motion processing. *Neuropsychologia*, *45*, 523–530.

Brouwer, A. M., Franz, V. H., & Thornton, I. M. (2004). Representational momentum in perception and grasping: Translating versus transforming objects. *Journal of Vision*, *4*, 575–584.

Brouwer, A. M., Thornton, I. M., & Franz, V. H. (2005). Forward displacement in grasping and judging pliers. *Visual Cognition*, *12*, 800–816.

Brown, C., Liebovitch, L., & Glendon, R. (2007). L'evy Flights in Dobe Ju/'hoansi Foraging Patterns. *Human Ecology*, *35*(1), 129–138.

Brown, L. (1979). *The story of maps*. New York: Dover.

Brown, N. R., Rips, L. J., & Shevell, S. K. (1985). The subjective dates of natural events in very-long-term memory. *Cognitive Psychology*, *17*, 139–177.

Brown, T. A., & Munger, M. P. (2010). Representational momentum, spatial layout, and viewpoint dependency. *Visual Cognition*, *18*, 780–800.

Brunetti, R., Indraccolo, A., Del Gatto, C., Spence, C., & Santangelo, V. (2018). Are crossmodal correspondences absolute or relative? Context effects on speeded classification. *Attention, Perception, & Psychophysics*, 80(2), 527–534.

Bruner, J. S. (1992). Another look at New Look 1. *American Psychologist*, *47*(6), 780–783.

Bruner, J. S., & Goodman, C. C. (1947). Value and need as organizing factors in perception. *Journal of Abnormal and Social Psychology*, *42*, 33–44.

Brunyé, T. T., Andonova, E., Meneghetti, C., Noordzij, M. L., Francesca Pazzaglia, P., Wienemann, R., Mahoney, C. R., & Taylor, H. A (2012). Planning routes around the world: International evidence for southern route preferences. *Journal of Environmental Psychology*, *32*, 297–304.

Brunyé, T. T., Collier, Z. A., Cantelon, J., Holmes, A., Wood, M. D., Linkov, I., & Taylor, H. A. (2015). Strategies for selecting routes through real-world environments: relative topography, initial route straightness, and cardinal direction. *PLoS ONE*, *10*(5), 1–16.

Brunyé, T. T., Mahoney, C. R., Gardony, A. L., & Taylor, H. A. (2010). North is up (hill): Route planning heuristics in real-world environments. *Memory & Cognition*, *38*, 700–712.

Bryant, D. J., & Subbiah, I. (1994). Subjective landmarks in perception and memory for spatial location. *Canadian Journal of Experimental Psychology*, *48*, 119–139.

Buchanan-Smith, H. M., & Heeley, D. W. (1993). Anisotropic axes in orientation perception are not retinotopically mapped. *Perception*, *22*, 1389–1402.

Buckner, R. L., Krienen, F. M., Castellanos, A., Diaz, J. C., & Yeo, B. T. T. (2011). The organization of the human cerebellum estimated by intrinsic functional connectivity. *Journal of Neurophysiology*, *106*, 2322–2345.

Büechner S. J., Hölscher, C., & Wiener, J. (2009). Search strategies and their success in a virtual maze. In N. A. Taatgen & H. van Rijn (Eds.), *Proceedings of the 31th Annual Conference of the Cognitive Science Society* (pp. 1066–1071). Austin, TX: Cognitive Science Society.

Bueti, D., & Walsh, V. (2009). The parietal cortex and the representation of time, space, number and other magnitudes. *Philosophical Transactions of the Royal Society B: Biological Sciences, 364*, 1831–1840.

Buetti, S., Lleras, A., & Moore, C. M. (2014). The flanker effect does not reflect the processing of "task-irrelevant" stimuli: Evidence from inattentional blindness. *Psychological Bulletin & Review, 21*, 1231–1237.

Buhlman, I., Umiltà, C., & Wascher, E. (2007). Response coding and visuomotor transformation in the Simon task: The role of action goals. *Journal of Experimental Psychology: Human Perception and Performance, 33*, 1269–1282.

Burbeck, C. A. (1987). On the locus of spatial frequency discrimination. *Journal of the Ophthalmology Society of America A, 4*, 1807–1813.

Burkitt, J. A., Saucier, D. M., Thomas, N. A., & Ehresman, C. (2006). When advertising turns "cheeky"! *Laterality, 11*(3), 277–286.

Burlison, T. (2013). *Columbia's first victims.* New York: Baen Publishing.

Burr, D. C., & Wijesundra, S. A. (1991) Orientation discrimination depends on spatial frequency. *Vision Research, 31*(7–8), 1449–1952.

Burris, C. T., & Branscombe, N. R. (2005). Distorted distance estimation induced by a self-relevant national boundary. *Journal of Experimental Social Psychology, 41*(3), 305–312.

Buss, D. M. (1995). Psychological sex-differences – origins through sexual selection. *American Psychologist, 50*(3), 164–168.

Butterworth, B. (2005). The development of arithmetical abilities. *Journal of Child Psychology and Psychiatry, 46*, 3–18.

Byrd-Craven, J., & Geary, D. C. (2007). Biological and evolutionary contributions to developmental sex differences. *Reproductive Biomedicine Online, 15*, 12–22.

Byrne, R. W. (1979) Memory for urban geography. *Quarterly Journal of Experimental Psychology, 31*, 147–154.

Byrne, R. M. J. (2005). *The rational imagination: How people create alternatives to reality.* Cambridge, MA: MIT Press.

Cai, R. H., Jacobson, K., Baloh, R., Schlag-Rey, M., & Schlag, J. (2000). Vestibular signals can distort the perceived spatial relationship of retinal stimuli. *Experimental Brain Research, 135*, 275–278.

Cai, R. H., & Schlag, J. (2001). Asynchronous feature binding and the flash-lag illusion. *Investigative Ophthalmology & Visual Science, 42*(4), S711

Calabria, M., Jacquin-Courtois, S., Miozzo, A., Rossetti, Y., Padovani, A., Cotelli, M., & Miniussi, C. (2011). Time perception in spatial neglect: A distorted representation? *Neuropsychology, 25*, 193–200.

Calvert, G. A., Spence, C., & Stein, B. E. (Eds.). (2004). *The handbook of multisensory processing.* Cambridge, MA: MIT Press.

Campbell, F. W., & Kulikowski, J. J. (1966). Orientational selectivity of the human visual system. *Journal of Physiology, 187*, 437–445.

Canal-Bruland, R., Pijpers, J. R., & Oudejans, R. R. (2010). The influence of anxiety on action-specific perception. *Anxiety, Stress, & Coping, 23*, 353–361.

Canal-Bruland, R., Zhu, F. F., van der Kamp, J., & Masters, R. S. W. (2011). Target-directed visual attention is a prerequisite for action-specific perception. *Acta Psychologica, 136*, 285–289.

Candel, I., Merckelbach, H., Houben, K., & Vandyck, I. (2004). How children remember neutral and emotional pictures: Boundary extension in children's scene memories. *American Journal of Psychology, 117*, 249–257.

Canevet, G., Scharf, B., Schlauch, R. S., Teghtsoonian, M., & Teghtsoonian, R. (1999). Perception of changes in loudness. *Nature, 398*(6729), 673.

Cantlon, J. F., Platt, M. L., & Brannon, E. M. (2009). Beyond the number domain. *Trends in Cognitive Sciences, 13*(2), 83–91.

Cantor, C. R. L., & Schor, C. M. (2007). Stimulus dependence of the flash-lag effect. *Vision Research*, *47*, 2841–2854.

Canzoneri, E., Magosso, E., & Serino, A. (2012). Dynamic sounds capture the boundaries of peripersonal space representation in humans. *PLoS ONE*, *7*(9). doi:10.1371/journal.pone.0044306

Cappe, C., Thelen, A., Romei, V., Thut, G., & Murray, M. M. (2012). Looming signals reveal synergistic principles of multisensory integration. *Journal of Neuroscience*, *32*(4), 1171–1182.

Cappe, C., Thut, G., Romei, V., & Murraya, M. M. (2009). Selective integration of auditory-visual looming cues by humans. *Neuropsychologia*, *47*(4), 1045–1052.

Carbon, C. C. (2010). The cycle of preference: Long-term dynamics of aesthetic appreciation. *Acta Psychologica*, *134*, 233–244.

Carbon, C. C. (2011). Cognitive mechanism for explaining dynamics of aesthetic appreciation. *i-Perception*, *2*, 708–719.

Carbone, E., & Ansorge, U. (2008). Investigating the contribution of metacontrast to the Fröhlich effect for size. *Acta Psychologica*, *128*(2), 361–367.

Carbone, E., & Pomplun, M. (2007). Motion misperception caused by feedback connections: A neural model simulating the Fröhlich effect. *Psychological Research*, *71*(6), 709–715.

Carey, S. (2009). *The origin of concepts*. Oxford: Oxford University Press.

Carlson, T. A., Hogendoorn, H., Fonteijn, H., & Verstraten, F. A. J. (2011). Spatial coding and invariance in object-selective cortex. *Cortex*, *47*(1), 14–22.

Carnaghi, A., Piccoli, V., Brambilla, M., & Bianchi, M. (2014). Gender hierarchy in the space: the role of gender status in shaping the spatial agency bias. *The Journal of Social Psychology*, *154*(2), 105–114.

Carnevale, M. J., & Harris, L. R. (2016). Which direction is up for a high pitch? *Multisensory Research*, *29*, 113–132.

Carney, D. R., Cuddy, A. J. C., & Yap, A. J. (2010). Power posing: Brief nonverbal displays affect neuroendocrine levels and risk tolerance. *Psychological Science*, *21*, 1363–1368.

Carpenter, R. H. S., & Blakemore, C. (1973). Interaction between orientations in human vision. *Experimental Brain Research*, *18*, 287–303.

Carrasco, M., Talgar, C. P., & Cameron, E. L. (2001). Characterizing visual performance fields: Effects of transient covert attention, spatial frequency, eccentricity, task and set size. *Spatial Vision*, *15*, 61–75.

Carstensen, L. L., Isaacowitz, D. M., & Charles, S. T. (1999). Taking time seriously: A theory of socioemotional selectivity. *American Psychologist*, *54*, 165–181.

Carter, L. F., & Schooler, K. (1949). Value, need, and other factors in perception. *Psychological Review*, *56*(4), 200–207.

Casasanto, D. (2014). Experiential origins of mental metaphors: Language, culture, and the body. In M. Landau, M. D. Robinson, & B. P. Meier (Eds.), *The power of metaphor: Examining its influence on social life* (pp. 249–268). Washington, DC: American Psychological Association.

Casasanto, D., & Boroditsky, L. (2008). Time in the mind: Using space to think about time. *Cognition*, *106*, 579–593.

Casasanto, D., & Bottini, R. (2014). Mirror-reading can reverse the flow of time. *Journal of Experimental Psychology: General*, *143*, 473–479.

Casasola, M. (2008). The development of infants' spatial categories. *Current Directions in Psychological Science*, *17*, 21–25.

Casasola, M., Cohen, L. B., & Chiarello, E. (2003). Six-month-old infants' categorization of containment spatial relations. *Child Development*, *74*, 679–693.

Catteeuw, P., Gilis, B., Jaspers, A., Wagemans, J., & Helsen, W. (2010). Training of perceptual-cognitive skills in offside decision making. *Journal of Sport & Exercise Psychology*, *32*, 845–861.

Catteeuw, P., Gilis, B., Wagemans, J., & Helsen, W. (2010). Perceptual cognitive skills in offside decision making: Expertise and training effects. *Journal of Sport & Exercise Psychology*, *32*, 828–844.

Catteeuw, P., Helsen, W., Gilis, B., van Roie, E., & Wagemans, J. (2009). Visual scan patterns and decision-making skills of expert assistant referees in offside situations. *Journal of Sport & Exercise Psychology, 31*, 786–797.

Cavanagh, P., & Anstis, S. (2013). The flash grab effect. *Vision Research, 91*, 8–20.

Cave, K. R., Bush, W. S., & Taylor, T. G. G. (2010). Split attention as part of a flexible attentional system for complex scenes: Comment on Jans, Peters, and De Weerd (2010). *Psychological Review, 117*, 685–696.

Cellini, C., Scocchia, L., & Drewing, K. (2016). The buzz-lag effect. *Experimental Brain Research, 234*, 2849–2857.

Cesario, J., Plaks, J. E., Hagiwara, N., Navarrete, C. D., & Higgins, E. T. (2010). The ecology of automaticity how situational contingencies shape action semantics and social behavior. *Psychological Science, 21*, 1311–1319.

Cesqui, B., d'Avella, A., Portone, A., & Lacquaniti, F. (2012). Catching a ball at the right time and place: Individual factors matter. *PLoS ONE*, 7(2), e31770.

Chadwick, M. J., Mullally, S. L., & Maguire, E. A. (2013). The hippocampus extrapolates beyond the view in scenes: An fMRI study of boundary extension. *Cortex, 49*, 2067–2079.

Chambers, K. W., McBeath, M. K., Schiano, D. J., & Metz, E. (1999). Tops are more salient than bottoms. *Perception & Psychophysics, 61*(4), 625–635.

Chambon, M. (2009). Embodied perception with others' bodies in mind: Stereotype priming influence on the perception of spatial environment. *Journal of Experimental Social Psychology, 45*(1), 283–287.

Champion, R. A., & Warren, P. A. (2008). Rapid size scaling in visual search. *Visual Research, 48*, 1820–1830.

Chang, D. H., Harris, L. R., & Troje, N. F. (2010). Frames of reference for biological motion and face perception. *Journal of Vision, 10*(6), 1–11.

Chang, D. H. F., & Troje, N. F. (2009). Acceleration carries the local inversion effect in biological motion perception. *Journal of Vision, 9*(1), 1–17.

Chang, S., & Cho, Y. S. (2015). Polarity correspondence effect between loudness and lateralized response set. *Frontiers in Psychology, 6*, 683.

Chapanis, A., & Mankin, D. A. (1967). The vertical-horizontal illusion in a visually-rich environment. *Attention, Perception & Psychophysics, 2*(6), 249–255.

Chapman, B., & Bonhoeffer, T. (1998). Overrepresentation of horizontal and vertical orientations preferences in developing ferret area 17. *Proceedings of the National Academy of Sciences, 95*, 2609–2614.

Chapman, P., Ropar, D., Mitchell, P., & Ackroyd, K. (2005). Understanding boundary extension: Normalization and extension errors in picture memory among adults and boys with and without Asperger's syndrome. *Visual Cognition, 12*, 1265–1290.

Chappell, M., & Hinchy, J. (2014). Turning the corner with the flash–lag illusion. *Vision Research, 96*, 39–44.

Chappell, M., & Hine, T. J. (2004). Events before the flash do influence the flash-lag magnitude. *Vision Research, 44*, 235–239.

Chappell, M., Hine, T. J., Acworth, C., & Hardwick, D. R. (2006). Attention "capture" by the flash-lag flash. *Vision Research, 46*, 3205–3213.

Chappell, M., & Mullen, K. T. (2010). The magnocellular visual pathway and the flash-lag illusion. *Journal of Vision, 10*(11), 24.

Charles, J., Sahraie, A., & McGeorge, P. (2007). Hemispatial asymmetries in judgement of stimulus size. *Attention, Perception & Psychophysics, 69*(5), 687–698.

Charras, P., Brod, G., & Lupianez, J. (2012). Is 26 + 26 smaller than 24 + 28? Estimating the approximate magnitude of repeated versus different numbers. *Attention, Perception, & Psychophysics, 74*(1), 163–173.

Charras, P., & Lupianez, J. (2009). The relevance of symmetry in line length perception. *Perception, 38*(10), 1428–1438.

Charras, P., & Lupiáñez, J. (2010). Length perception of horizontal and vertical bisected lines. *Psychological Research, 74*(2), 196–206.

Charras, P., Lupianez, J., & Bartolomeo, P. (2010). Assessing the weights of visual neglect: A new approach to dissociate defective symptoms from productive phenomena in length estimation. *Neuropsychologia, 48*(11), 3371–3375.

Charras, P., Lupianez, J., Migliaccio, R., Toba, M., Pradat-Diehl, P., Duret, C., & Bartolomeo, P. (2012). Dissecting the component deficits of perceptual imbalance in visual neglect: Evidence from horizontal-vertical length comparisons. *Cortex, 48*(5), 540–552.

Charras, P., Molina, E., & Lupianez, J. (2014). Additions are biased by operands: Evidence from repeated versus different operands. *Psychological Research, 78*(2), 248–265.

Chase, W. G. (1983). Spatial representations of taxi drivers. In R. Rogers & J. A. Sloboda (Eds.), *Acquisition of symbolic skills* (pp. 111–136). New York: Plenum Press.

Chatterjee, A. (2001). Language and space: Some interactions. *Trends in Cognitive Sciences, 5*, 55–61.

Chatterjee, A. (2002). Portrait profiles and the notion of agency. *Empirical Studies of the Arts, 20*, 33–41.

Chatterjee, A. (2008). The neural organization of spatial thought and language. *Seminars in Speech and Language, 29*, 226–238.

Chatterjee, A. (2012). Neuroaesthetics: Growing pains of a new disciple. In A. P. Shimamura & S. E. Palmer (Eds.), *Aesthetic science: Connecting minds, brains, and experience* (pp. 299–317). Oxford: Oxford University Press.

Chatterjee, A. (2014). *The aesthetic brain*. New York: Oxford University Press.

Chauvel, G., Wulf, G., & Maquestiaux, F. (2015). Visual illusions can facilitate sport skill learning. *Psychonomic Bulletin & Review, 22*, 717–721.

Cheal, M., Chastain, G., & Lyon, D. R. (1998). Inhibition of return in visual identification tasks. *Visual Cognition, 5*, 365–388.

Chen, L., & Vroomen, J. (2013). Intersensory binding across space and time: A tutorial review. *Attention, Perception, & Psychophysics, 75*, 790–811.

Chen, Y. C., & Scholl, B. J. (2014). Seeing and liking: Biased perception of ambiguous figures consistent with the "inward bias" in aesthetic preferences. *Psychonomic Bulletin & Review, 21*, 1444–1451.

Cheng, J. T., Tracy, J. L., Foulsham, T., Kingstone, A., & Henrich, J. (2013). Two ways to the top: Evidence that dominance and prestige are distinct yet viable avenues to social rank and influence. *Journal of Personality and Social Psychology, 104*(1), 103–125.

Cheng, K., Shettleworth, S. J., Huttenlocher, J., & Rieser, J. J. (2007). Bayesian integration of spatial information. *Psychological Bulletin, 133*(4), 625–637.

Cherry, E. C. (1953). Some experiments on the recognition of speech, with one and with two ears. *The Journal of the Acoustical Society of America, 25*, 975–979.

Cheung, B., Hofer, K., Brooks, C. J., & Gibbs, P. (2000). Underwater disorientation as induced by two helicopter ditching devices. *Aviation, Space, and Environmental Medicine, 71*(9), 879–888.

Chiao, J. Y. (2010). Neural basis of social status hierarchy across species. *Current Opinion in Neurobiology, 20*(6), 803–809.

Chiao, J. Y., Bordeaux, A. R., & Ambady, N. (2004). Mental representations of social status. *Cognition, 93*(2), B49–B57.

Chiao, J. Y., Harada, T., Oby, E. R., Li, Z., Parrish, T., & Bridge, D. J. (2009). Neural representations of social status hierarchy in human inferior parietal cortex. *Neuropsychologia, 47*(2), 354–363.

Chica, A. B., Bartolomeo, P., & Valero-Cabré, A. (2011). Dorsal and ventral parietal contributions to spatial orienting in the human brain. *Journal of Neuroscience, 31*(22), 8143–8149.

Chica, A. B., Taylor, T. L., Lupianez, Y., & Klein, R. M. (2010). Two mechanisms underlying inhibition of return. *Experimental Brain Research, 201*, 25–35.

Chieffi, S. (1996). Effects of stimulus asymmetry on line bisection. *Neurology, 47*(4), 1004–1008.

Chien, S. E., Ono, F., & Watanabe, K. (2013). A transient auditory signal shifts the perceived offset position of a moving visual object. *Frontiers in Psychology, 4*, 70.

Chiou, R., & Rich, A. N. (2012). Cross-modality correspondence between pitch and spatial location modulates attentional orienting. *Perception, 41*, 339–353.

Chiou, R., & Rich, A. N. (2015). Volitional mechanisms mediate the cuing effect of pitch on attention orienting: The influences of perceptual difficulty and response pressure. *Perception, 44*, 169–182.

Cho, Y. S., Bae, G. Y., & Proctor, R. W. (2012). Referential coding contributes to the horizontal SMARC effect. *Journal of Experimental Psychology: Human Perception and Performance, 38*, 726–734.

Cho, Y. S., & Proctor, R. W. (2003). Stimulus and response representations underlying orthogonal stimulus-response compatibility effects. *Psychonomic Bulletin and Review 10*, 45–73.

Choi, H., & Scholl, B. J. (2006). Measuring causal perception: Connections to representational momentum? *Acta Psychologica, 123*, 91–111.

Chokron, S., Bartolomeo, P., Perenin, M. T., Helft, G., & Imbert, M. (1998). Scanning direction and line bisection: A study of normal subjects and unilateral neglect patients with opposite reading habits. *Cognitive Brain Research, 7*(2), 173–178.

Chokron, S., & de Agnostini, M. (2000). Reading habits influence aesthetic preference. *Cognitive Brain Research, 10*, 45–49.

Chong, S. C., & Treisman, A. (2003). Representation of statistical properties. *Vision Research, 43*, 393–404.

Chowdhury, S. A., Christiansen, D. L., Morgan, M. L., & DeAngelis, G. C. (2008). Effect of vertical disparities on depth representation in macaque monkeys: MT physiology and behavior. *Journal of Neurophysiology*, 99, 876–887.

Chrastil, E. R., & Warren, W. H. (2013). Active and passive spatial learning in human navigation: Acquisition of survey knowledge. *Journal of Experimental Psychology: Learning, Memory, and Cognition, 39*(5), 1520–1537.

Chrastil, E. R., & Warren, W. H. (2014). From cognitive maps to cognitive graphs. *PLoS ONE, 9*(11), e112544.

Christenfeld, N. (1995). Choices from identical options. *Psychological Science, 6*(1), 50–55.

Christie, J., Hilchey, M. D., & Klein, R. M. (2013). Inhibition of return is at the midpoint of simultaneous cues. *Attention, Perception, & Psychophysics, 75*(8), 1610–1618.

Christman, S. D., & Niebauer, C. L. (1997). The relation between left–right and upper–lower visual field differences. In S. Christman (Ed.), *Cerebral asymmetries in sensory and perceptual processing* (pp. 263–298). Amsterdam: Elsevier.

Churchland, P. (1981). Eliminative materialism and the propositional attitudes. *The Journal of Philosophy, 78*, 67–90.

Churchland, P. (1986). *Neurophilosophy*. Cambridge, MA: MIT Press.

Cipora, K., Hohol, M., Nuerk, H.-C., Willmes, K., Brożek, B., Kucharzyk, B., & Nęcka, E. (2016). Professional mathematicians differ from controls in their spatial-numerical associations. *Psychological Research, 80*, 710–726.

Cipora, K., & Nuerk, H.-C. (2013). Is the SNARC effect related to the level of mathematics? No systematic relationship observed despite more power, more repetitions, and more direct assessment of arithmetic skill. *The Quarterly Journal of Experimental Psychology, 66*(10), 1974–1991.

Cipora, K., Patro, K., & Nuerk, H.-C. (2015). Are spatial-numerical associations a cornerstone for arithmetic learning? The lack of genuine correlations suggests no. *Mind, Brain, and Education, 9*(4), 190–206.

Clark, A. (2001). *Mindware: An introduction to the philosophy of cognitive science*. New York: Oxford University Press.

Clark, A., & Toribio, J. (1994). Doing without representing? *Synthese, 101*, 401–431.

Clark, H. H. (1973). Space, time, semantics, and the child. In T. E. Moore (Ed.), *Cognitive development and the acquisition of language* (pp. 27–64). New York: Academic Press.

Clark, H. H., & Chase, W.G. (1972). On the process of comparing sentences against pictures. *Cognitive Psychology, 3*, 472–517.

Clemens, I. A., De Vrijer, M., Selen, L. P., Van Gisbergen, J. A., & Medendorp, W. P. (2011). Multisensory processing in spatial orientation: an inverse probabilistic approach. *Journal of Neuroscience, 31*(14), 5365–5377.

Clément, G., & Reschke, M. F. (2008). *Neuroscience in space*. New York: Springer.

Clery, J., Guipponi, O., Odouard, S., Wardak, C., & Ben Hamed, S. (2015). Impact prediction by looming visual stimuli enhances tactile detection. *Journal of Neuroscience, 35*(10), 4179–4189.

Coey, C., Varlet, M., & Richardson, M. (2012). Coordination dynamics in a socially situated nervous system. *Frontiers in Human Neuroscience, 6*, 164.

Cohen, R., & Schuepfer, T. (1980). The representation of landmarks and routes. *Child Development, 51*(4), 1065–1071.

Cohen, W. (1957). Spatial and textural characteristics of the Ganzfeld. *American Journal of Psychology, 70*, 403–410.

Colby, C. L. (1998). Action-oriented spatial reference frames in cortex. *Neuron, 20*(1), 15–24.

Colby, C. L., & Duhamel, J. R. (1996). Spatial representations for action in parietal cortex. *Brain Research Cognitive Brain Research, 5*(1), 105–115.

Collyer, C. E. (1977). Discrimination of spatial and temporal intervals defined by three light flashes: Effect of spacing on temporal judgments and of timing on spatial judgments. *Perception & Psychophysics, 21*, 357–364.

Conners, F. A., Wyatt, B. S., & Dulaney, C. L. (1998). Cognitive representation of motion in individuals with mental retardation. *American Journal on Mental Retardation, 102*, 438–450.

Cooper, A. D., Sterling, C. P., Bacon, M. P., & Bridgeman, B. (2012). Does action affect perception or memory? *Vision Research, 62*, 235–240.

Cooper, L. A., & Munger, M. P. (1993). Extrapolations and remembering positions along cognitive trajectories: Uses and limitations of analogies to physical momentum. In N. Eilan, R. McCarthy, & B. Brewer (Eds.), *Spatial representation: Problems in philosophy and psychology* (pp. 112–131). Cambridge, MA: Blackwell.

Coppola, D. M., White, L. E., Fitzpatrick, D., & Purves, D. (1998). Unequal representation of cardinal and oblique contours in ferret visual cortex. *Proceedings of the National Academy of Sciences, 95*(5), 2621–2623.

Corballis, M. C. (2000). Much about mirrors. *Psychonomic Bullettin and Review, 7*(1), 163–169.

Corballis, M. C., & Roldan, C. E. (1975). Detection of symmetry as a function of angular orientation. *Journal of Experimental Psychology: Human Perception and Performance, 1*(3), 221–230.

Corbetta, M., Kincade, J. M., Ollinger, J. M., McAvoy, M. P., & Shulman, G. L. (2000). Voluntary orienting is dissociated from target detection in human posterior parietal cortex. *Nature Neuroscience, 3*, 292–297.

Corbetta, M., Kincade, M. J., Lewis, C., Snyder, A. Z., & Sapir, A. (2005). Neural basis and recovery of spatial attention deficits in spatial neglect. *Nature Neuroscience, 8*(11), 1603–1610.

Corbetta, M., Patel, G., & Shulman, G. L. (2008). The reorienting system of the human brain: From environment to theory of mind. *Neuron, 58*, 306–324.

Corbetta, M., & Shulman, G. L. (2002). Control of goal-directed and stimulus-driven attention in the brain. *Nature Reviews Neuroscience, 3*(3), 201–215.

Corbetta, M., & Shulman, G. L. (2011). Spatial neglect and attention networks. *Annual Reviews Neuroscience, 34*, 569–599.

Coren, S., & Girgus, J. S. (1978). *Seeing is deceiving: The psychology of visual illusions*. Hillsdale, NJ: Erlbaum.

Coren, S., & Girgus, J. S. (1980). Principles of perceptual organization and spatial distortion: The gestalt illusions. *Journal of Experimental Psychology: Human Perception and Performance, 6*, 404–412.

Coren, S., Ward, L. M., Porac, C., & Fraser, R. (1978). The effect of optical blur on visuo-geometric illusions. *Bulletin of the Psychonomic Society, 11*, 390–392.

Coslett, H. B. (1997). Neglect in vision and visual imagery: A double dissociation. *Brain, 120*, 1163–1171.

Coslett, H. B., Bowers, D., Fitzpatrick, E., Haws, B., & Heilman, K. M. (1990). Directional hypokinesia and hemispatial inattention in neglect. *Brain, 113*(2), 475–486.

Costello, M. C., Bloesch, E. K., Davoli, C. C., Panting, N. D., Abrams, R. A., & Brockmole, J. R. (2015). Spatial representations in older adults are not modified by action: Evidence from tool use. *Psychology and Aging, 30*(3), 656–668.

Couclelis, H., Golledge, R. G., Gale, N., & Tobler, W. (1987) Exploring the anchor-point hypothesis of spatial cognition. *Journal of Environmental Psychology, 7*, 99–122.

Coull, J. T., Frackowiak, R. S., & Frith, C. D. (1998). Monitoring for target objects: activation of right frontal and parietal cortices with increasing time on task. *Neuropsychologia, 36*, 1325–1334.

Courtney, J. R., & Hubbard, T. L. (2008). Spatial memory and explicit knowledge: An effect of instruction on representational momentum. *The Quarterly Journal of Experimental Psychology, 61*, 1778–1784.

Courtney, J. R., Motes, M. A., & Hubbard, T. L. (2007). Multi- and unisensory visual flash illusions, *36*, 516–524.

Cowey, A., Small, M., & Ellis, S. (1994). Left visuo-spatial neglect can be worse in far than in near space. *Neuropsychologia, 32*(9), 1059–1066.

Craig, A. D. (2003). Interoception: The sense of the physiological condition of the body. *Current Opinion in Neurobiology, 13*, 500–505.

Cravo, A. M., & Baldo, M. V. C. (2008). A psychophysical and computational analysis of the spatio-temporal mechanisms underlying the flash-lag effect. *Perception, 37*, 1850–1866.

Crawford, L. E., & Duffy, S. (2010). Sequence effects in estimating spatial location. *Psychonomic Bulletin & Review, 17*(5), 725–730.

Crawford, L. E., Huttenlocher, J., & Engebretson, P. H. (2000). Category effects on estimates of stimuli: Perception or reconstruction? *Psychological Science, 11*(4), 280–284.

Crawford, L. E., & Jones, E. L. (2011). The flexible use of inductive and geometric spatial categories. *Memory & Cognition, 39*(6), 1055–1067.

Crawford, L. E., Landy, D., & Salthouse, T. A. (2016). Spatial working memory capacity predicts bias in estimates of location. *Journal of Experimental Psychology: Learning, Memory, and Cognition*. http://dx.doi.org/10.1037/xlm0000228

Crawford, L. E., Margolies, S. M., Drake, J. T., & Murphy, M. E. (2006). Affect biases memory of location: Evidence for the spatial representation of affect. *Cognition and Emotion, 20*, 1153–1169.

Crawford, L. E., Regier, T., & Huttenlocher, J. (2000). Linguistic and non-linguistic spatial categorization. *Cognition, 75*(3), 209–235.

Crisinel, A.-S., & Spence, C. (2009). Implicit association between basic tastes and pitch. *Neuroscience Letters, 464*, 39–42.

Crisinel, A.-S., & Spence, C. (2012). A fruity note: Crossmodal associations between odors and musical notes. *Chemical Senses, 37*, 151–158.

Crisp, R. J., & Hewstone, M. (2007). Multiple social categorization. *Advances in experimental social psychology, 39*, 163–254.

Croft, W., & Cruse, D. A. (2004). *Cognitive linguistics*. Cambridge: Cambridge University Press.

Crollen, V., Dormal, G., Seron, X., Lepore, F., & Collignon, O. (2013). Embodied numbers: The role of vision in the development of number–space interactions. *Cortex, 49*, 276–283.

Croucher, C. J., Bertamini, M., & Hecht, H. (2002). Naive optics: Understanding the geometry of mirror reflections. *Journal of Experimental Psychology: Human Perception and Performance, 28*(3), 546–562.

Cruse, D. A., & Pagona, T. (1995). Towards a cognitive model of antonymy. *Lexicology, 1*, 113–141.

Crystal Jiang, L., & Hancock, J. T. (2013). Absence makes the communication grow fonder: Geographic separation, interpersonal media, and intimacy in dating relationships. *Journal of Communication, 63*(3), 556–577.

Cuddy, A. J., Norton, M. I., & Fiske, S. T. (2005). This old stereotype: The pervasiveness and persistence of the elderly stereotype. *Journal of Social Issues, 61*(2), 267–285.

Cuddy, A. J. C., Fiske, S. T., & Glick, P. (2007). The BIAS map: Behaviors from intergroup affect and stereotypes. *Journal of Personality and Social Psychology, 92*(4), 631–648.

Cuijpers, R. H., Kappers, A. M. L., & Koenderink, J. J. (2000a). Investigation of visual space using an exocentric pointing task. *Attention, Perception, & Psychophysics, 62*, 1556–1571.

Cuijpers, R. H., Kappers, A. M. L. & Koenderink, J. J. (2000b). Large systematic deviations in visual parallelism. *Perception, 29*(12), 1467–1482

Cuijpers, R. H., Kappers, A. M. L. & Koenderink, J. J. (2002). Visual perception of collinearity. *Perception, 64*(3), 392–404.

Curcio, C. A., & Allen, K. A. (1990). Topography of ganglion cells in human retina. *Journal of Comparative Neurology, 300*, 5–25.

Cutsuridis, V., Hussain, A., & Taylor, J. G. (2011). *Perception-action cycle: Models, architectures, and hardware*. New York: Springer-Verlag.

Cutting J. E. (1986). *Perception with an eye for motion*. Cambridge, MA: MIT Press.

Cutting, J. E., & Vishton, P. M. (1995). Perceiving layout and knowing distances: The integration, relative potency, and contextual use of different information about depth. In W. Epstein & S. Rogers (Eds.), *Perception of space and motion* (Vol. 5 of *Handbook of perception and cognition*, pp. 69–117). San Diego, CA: Academic Press.

Da Silva, J. A. (1985). Scales for perceived egocentric distance in a large open field: Comparison of three psychophysical methods. *American Journal of Psychology, 98*(1), 119–144.

Dackermann, T., Huber, S., Bahnmueller, J., Nuerk, H.-C., & Moeller, K. (2015). An integration of competing accounts on children's number line estimation. *Frontiers in Psychology, 6*, 884.

Dahl, C. D., & Adachi, I. (2013). Conceptual metaphorical mapping in chimpanzees (Pan troglodytes). *eLife, 2013*(2), 1–7.

Daini, R., Angelelli, P., Antonucci, G., Cappa, S. F., & Vallar, G. (2002). Exploring the syndrome of spatial unilateral neglect through an illusion of length. *Experimental Brain Research, 144*, 224–237.

Dakin, S. C. (2001). Information limit on the spatial integration of local orientation signals. *Journal of the Optometry Society of America A, 18*, 1016–1026.

Dakin, S. C., Tibber, M. S., Greenwood, J. A., Kingdom, F. A., & Morgan, M. J. (2011). A common visual metric for approximate number and density. *PNAS, 108*(49), 19552–19557.

Dalton, R. C. (2003). The secret is to follow your nose: Route path selection and angularity. *Environment and Behavior, 35*(1), 107–131.

Damasio, H., Grabowski, T. J., Tranel, D., Ponto, L. L., Hichwa, R. D., & Damasio, A. R. (2001). Neural correlates of naming actions and of naming spatial relations. *Neuroimage, 13*(6 Pt 1), 1053–1064.

Dannemaier, W. D., & Thumin, F. J. (1964). Authority status as a factor in perceptual distortion of size. *The Journal of Social Psychology, 63*, 361–365.

Das, A., & Gilbert, C. D. (1997). Distortions of visuotopic map match orientation singularities in primary visual cortex. *Nature, 387*, 594–598.

Datta, R., Joshi, D., Li, J., & Wang, J. Z. (2006). Studying aesthetics in photographic images using a computational approach. In *Proceedings of Computer Vision-ECCV 2006*, Graz, Austria, pp. 288–301.

Davis, S. T., & Jahnke, J. C. (1991). Unity and the golden section: Rules for aesthetic choice? *American Journal of Psychology, 104*, 257–277.

Day, R. H. (1988). Reduction and elimination of the Poggendorff misalignment effect by minor changes at intersections: Implications for the perceptual-compromise explanation. *Psychological Research, 50*(1), 7–11.

Day, R. H., & Dickenson, R. G. (1976). The components of the Poggendorff illusion. *British Journal of Psychology, 67*, 537–552.

de Hevia, M. D., Girelli, L., Bricolo, E., & Vallar, G. (2008b). The representational space of numerical magnitude: Illusions of length. *The Quarterly Journal of Experimental Psychology, 61*(10), 1496–1514.

de Hevia, M. D., Girelli, L., & Vallar, G. (2006). Numbers and space: A cognitive illusion? *Experimental Brain Research, 168*(1–2), 254–264.

de Hevia, M. D., Izard, V., Coubart, A., Spelke, E. S., & Streri, A. (2014). Representations of space, time, and number in neonates. *Proceedings of the National Academy of Sciences, 111*(13), 4809–4813.

de Hevia, M. D., & Spelke, E. S. (2009). Spontaneous mapping of number and space in adults and young children. *Cognition, 110*(2), 198–207.

de Hevia, M. D., & Spelke, E. S. (2010). Number-space mapping in human infants. *Psychological Science, 21*(5), 653–660.

de Hevia, M. D., Vallar, G., & Girelli, L. (2008a). Visualizing numbers in the mind's eye: The role of visuo-spatial processes in numerical abilities. *Neuroscience & Biobehavioral Reviews, 32*(8), 1361–1372.

De Jong, R., Liang, C.-C., & Lauber, E. (1994). Conditional and unconditional automaticity: A dual-process model of effects of spatial stimulus–response correspondence. *Journal of Experimental Psychology: Human Perception and Performance, 20*, 731–750.

de la Fuente, J., Santiago, J., Román, A., Dumitrache, C., & Casasanto, D. (2014). When you think about it, your past is in front of you: How culture shapes spatial conceptions of time. *Psychological Science, 25*, 1682–1690.

de la Malla, C., & López-Moliner, J. (2015). Predictive plus online visual information optimizes temporal precision in interception. *Journal of Experimental Psychology: Human Perception and Performance, 41*(5), 1271.

De Renzi, E., Colombo, A., Faglioni, P., & Gibertoni, M. (1982). Conjugate gaze paresis in stroke patients with unilateral damage: An unexpected instance of hemispheric asymmetry. *Archives of Neurology, 39*, 482–486.

de sá Teixeira, N. (2014). Fourier decomposition of spatial localization errors reveals an idiotropic dominance of an internal model of gravity. *Vision Research, 105*, 177–188.

de sá Teixeira, N. A. (2016). How fast do objects fall in visual memory? Uncovering the temporal and spatial features of representational gravity. *PLoS ONE, 11*(2), e0148953. doi:10.1371/journal.pone.0148953

de sá Teixeira, N., & Hecht, H. (2014). Can representational trajectory reveal the nature of an internal model of gravity? *Attention, Perception, & Psychophysics, 76*(4), 1106–1120.

de sá Teixeira N. A, & Hecht, H. (2014). The dynamic representation of gravity is suspended when the idiotropic vector is misaligned with gravity. *Journal of Vestibular Research, 24*(4), 267–279.

de sá Teixeira, N., Hecht, H., & Oliveira, A. M. (2013). The representational dynamics of remembered projectile locations. *Journal of Experimental Psychology: Human Perception and Performance, 39*, 1690–1699.

de sá Teixeira, N., & Oliveira, A. M. (2011). Disambiguating the effects of target travelled distance and target vanishing point upon representational momentum. *Journal of Cognitive Psychology, 23*, 650–658.

de sá Teixeira, N., Oliveira, A. M., & Amorim, M. A. (2010). Combined effects of mass and velocity on forward displacement and phenomenological ratings: A functional measurement approach to the momentum metaphor. *Psicologica, 31*, 659–676.

de sá Teixeira, N., Oliveira, A. M., & Viegas, R. (2008). Functional approach to the integration of kinematic and dynamic variables in causal perception: Is there a link between phenomenology and behavioral responses. *Japanese Psychological Research*, 50, 232–241.

de sá Teixeira, N., Pimenta, S., & Raposo, V. (2013). A null effect of target's velocity in the visual representation of motion with schizophrenic patients. *Journal of Abnormal Psychology, 122*, 223–230.

De Valois, R. L., & De Valois, K. K. (1991). Vernier acuity with stationary moving Gabors. *Vision Research, 31*, 1619–1626.

De Valois, R. L., Yund, E. W., & Hepler, N. (1982). The orientation and direction selectivity of cells in macaque visual cortex. *Vision Research, 22*, 531–544.

De Vrijer, M., Medendorp, W. P., & Van Gisbergen, J. A. M. (2008). Shared computational mechanism for tilt compensation accounts for biased verticality percepts in motion and pattern vision. *Journal of Neurophysiology*. 99, 915–930.

de Winkel, K. N., Clément, G., Groen, E. L., & Werkhoven, P. J. (2012). The perception of verticality in lunar and Martian gravity conditions. *Neuroscience Letters, 529*, 7–11.

Deacon, T. (1997). *The symbolic species*. New York: Norton.

Dean, L. M., Willis, F. N., & Hewitt, J. (1975). Initial interaction distance among individuals equal and unequal in military rank. *Journal of Personality and Social Psychology, 32*(2), 294–299.

Dehaene, S. (1997). *The number sense: How the mind creates mathematics*. Oxford: Oxford University Press.

Dehaene, S., Bossini, S., & Giraux, P. (1993). The mental representation of parity and number magnitude. *Journal of Experimental Psychology: General, 122*, 371–396.

Dehaene, S., Piazza, M., Pinel, P., & Cohen, L. (2003). Three parietal circuits for number processing. *Cognitive Neuropsychology, 20*, 487–506.

Dehmelt, H. (1989). Triton, . . . Electron, . . . Cosmon . . .: An infinite regression?, *Proceedings of the National Academy of Sciences, 86*, 8618–8619.

Deininger, R. L., & Fitts, P. M. (1955). Stimulus–response compatibility, information theory, and perceptual-motor performance. In H. Quastler (Ed.), *Information theory in psychology: Problems and methods* (pp. 316–341). Glencoe, IL: Free Press.

Della Sala, S., Foley, J. A., Beschin, N., Allerhand, M., & Logie, R. H. (2010). Assessing dual-task performance using a paper-and-pencil test: Normative data. *Archives of Clinical Neuropsychology, 25*(5), 410–419.

DeLucia, P. R. (1991). Pictorial and motion-based information for depth-perception. *Journal of Experimental Psychology-Human Perception and Performance, 17*(3), 738–748.

DeLucia, P. R. (2004). Multiple sources of information influence time-to-contact judgments: Do heuristics accommodate limits in sensory and cognitive processes? In H. Hecht & G. J. P. Savelsbergh (Eds.), *Time-to-contact* (pp. 243–286). Amsterdam: Elsevier Science Publishers.

DeLucia, P. R., & Maldia, M. M. (2006). Visual memory for moving scenes. *The Quarterly Journal of Experimental Psychology, 59*, 340–360.

DeLucia, P. R., Preddy, D., & Oberfeld, D. (2015). Audiovisual integration of time-to-contact information for approaching objects. *Multisensory Research*. doi:10.1163/22134808-00002520

Denis, M., Pazzaglia, F., Cornoldi, C., & Bertolo, L. (1999). Spatial discourse and navigation: An analysis of route directions in the city of Venice. *Applied Cognitive Psychology, 13*(2), 145–174.

Dennett, D. C. (1971). Intentional systems. *The Journal of Philosophy, 68*(4), 87–106.

Deregowski J. B., & Parker D. M. (1988). On a changing perspective illusion within Vermeer's "The Music Lesson". *Perception*, 17, 13–21.

Deroy, O., Fasiello, I., Hayward, V., & Auvray, M. (2016). Differentiated audio-tactile correspondences in sighted and blind individuals. *Journal of Experimental Psychology: Human Perception and Performance, 42*(8), 1204–1214.

Deroy, O., & Spence, C. (2013). Weakening the case for "weak synaesthesia": Why crossmodal correspondences are not synaesthetic. *Psychonomic Bulletin & Review, 20*, 643–664.

Deroy, O., & Spence, C. (2016). Crossmodal correspondences: Four challenges. *Multisensory Research, 30*, 29–48.

Desimone, R., & Duncan, J. (1995). Neural mechanisms of selective visual attention. *Annual Review of Neuroscience, 18*, 193–222.

Dewey, J. (1986). The reflex arc concept in psychology, *Psychological Review*, *3*, 357.

Di Lollo, V., Enns, J. T., & Rensink, R. A. (2000). Competition for consciousness among visual events: The psychophysics of reentrant visual processes. *Journal of Experimental Psychology: General*, *129*(4), 481–507.

Di Luca, S., & Pesenti, M. (2011). Finger numeral representations: More than just another symbolic code. *Frontiers in Psychology*, *2*, 272.

Dickinson, C. A., & Intraub, H. (2008). Transsaccadic representation of layout: what is the time course of boundary extension? *Journal of Experimental Psychology: Human Perception and Performance*, *34*, 543–555.

Dickinson, C. A., & Intraub, H. (2009). Spatial asymmetries in viewing and remembering scenes: Consequences of an attentional bias? *Attention, Perception, & Psychophysics*, *71*(6), 1251–1262.

DiGiovanni, J. J., & Schlauch, R. S. (2007). Mechanisms responsible for differences in perceived duration for rising-intensity and falling-intensity sounds. *Ecological Psychology*, *19*(3), 239–264.

Dodds, C. M., van Belle, J., Peers, P. V., Dove, A., Cusack, R., Duncan, J., et al. (2008). The effects of time-on-task and concurrent cognitive load on normal visuospatial bias. *Neuropsychology*, *22*(4), 545–552.

Dolgov, I., Birchfield, D. A., McBeath, M. K., Thornburg, H., & Todd, C. G.. (2009a). Amelioration of axis-aligned motion bias for active versus stationary judgments of bilaterally symmetric shapes' final destinations. *Attention, Perception, & Psychophysics*, *71*(3), 523–529.

Dolgov, I., Birchfield, D. A., McBeath, M. K., Thornburg, H., & Todd, C. G. (2009b). Perception of approaching and retreating shapes in a large, immersive, multimedia learning environment. *Perceptual & Motor Skills*, *108*(2), 623–630.

Dolgov, I., McBeath, M. K., & Sugar, T. G. (2009) Evidence for axis-aligned motion bias: Football axis-trajectory misalignment causes systematic error in projected final destinations of thrown American footballs. *Perception*, *38*(3), 399–410.

Dolscheid, S., & Casasanto, D. (2015). Spatial congruity effects reveal metaphorical thinking, not polarity correspondence. *Frontiers in Psychology*, 6: 1836.

Dolscheid, S., Hunnius, S., Casasanto, D., & Majid, A. (2014). Prelinguistic infants are sensitive to space-pitch associations found across cultures. *Psychological Science*, *25*, 1256–1261.

Domhas, F., Zamarian, L., & Delazer, M. (2008). Sound arithmetic: Auditory cues in the rehabilitation of impaired fact retrieval. *Neuropsychological Rehabilitation* 18, 160–181.

Donders, F. C. (1868/1969). On the speed of mental processes. In W. G. Koster (Ed.), *Acta Psychologica, 30, Attention and Performance II* (pp. 412–431). Amsterdam: North-Holland.

Doppler, C. (1842). *Ueber das farbige Licht der Doppelsterne und einiger anderer Gestirne des Himmels: Versuch einer das Bradley'sche Aberrations-Theorem als integrirenden Theil in sich schliessenden allgemeineren Theorie.* Prague: K. Bohnm, Association of Science.

Doricchi, F., & Galati, G. (2000). Implicit semantic evaluation of object symmetry and contralesional visual denial in a case of left unilateral neglect with damage of the dorsal paraventricular white matter. *Cortex*, *36*(3), 337–350.

Doricchi, F., Guariglia, P., Gasparini, M., & Tomaiuolo, F. (2005). Dissociation between physical and mental number line bisection in right hemisphere brain damage. *Nature Neuroscience*, *8*(12), 1663–1665.

Doricchi, F., Merola, S., Aiello, M., Guariglia, P., Bruschini, M., Gevers, W., et al. (2009). Spatial orienting biases in the decimal numeral system. *Current Biology*, *19*(8), 682–687.

Doricchi, F., Thiebaut de Schotten, M., Tomaiuolo, F., & Bartolomeo, P. (2008). White matter (dis)connections and gray matter (dys)functions in visual neglect: Gaining insights into the brain networks of spatial awareness. *Cortex*, *44*, 983–995.

Douglas, K. M., & Bilkey, D. K. (2007). Amusia is associated with deficits in spatial processing. *Nature Neuroscience, 10*, 915–921.

Doumen, M. J., Kappers, A. M., & Koenderink, J. J (2006). Horizontal-vertical anisotropy in visual space. *Acta Psychologica, 123*, 219–239.

Downar, J., Crawley, A. P., Mikulis, D. J., & Davis, K. D. (2000). A multimodal cortical network for the detection of changes in the sensory environment. *Nature Neuroscience, 3*, 277–283.

Driver, J., & Husain, M. (2002). The role of spatial working memory deficits in pathological search by neglect patients. In H.-O. Karnath, A. D. Milner & G. Vallar (Eds.), *The cognitive and neural bases of spatial neglect* (pp. 351–364). Oxford: Oxford University Press.

Driver, J., & Pouget, A. (2000). Object-centered visual neglect, or relative egocentric neglect? *Journal of Cognitive Neuroscience, 12*(3), 542–545.

Driver, J., & Vuilleumier, P. (2001). Perceptual awareness and its loss in unilateral neglect and extinction. *Cognition, 79*, 39–88.

Du, F., & Abrams, R. A. (2010). Visual field asymmetry in attentional capture. *Brain and Cognition, 72*, 310–316.

Duffy, S., & Crawford, L. E. (2008). Primacy or recency effects in forming inductive categories. *Memory & Cognition, 36*(3), 567–577.

Duffy, S., Huttenlocher, J., & Crawford, L. E. (2006). Children use categories to maximize accuracy in estimation. *Developmental Science, 9*(6), 597–603.

Duffy, S., Huttenlocher, J., Hedges, L. V., & Crawford, L. E. (2010). Category effects on stimulus estimation: Shifting and skewed frequency distributions. *Psychonomic Bulletin & Review, 17*(2), 224–230.

Duguid, M. M., & Goncalo, J. A. (2012). Living large the powerful overestimate their own height. *Psychological Science, 23*(1), 36–40.

Dumas, D., & Alexander, P. A. (2016). Calibration of the test of relational reasoning. *Psychological Assessment, 10*, 1303–1318.

Durant, S., & Johnston, A. (2004). Temporal dependence of local motion induced shifts in perceived position. *Vision Research, 44*, 357–366.

Durgin, F. H., Baird, J. A., Greenburg, M., Russell, R., Shaughnessy, K., & Waymouth, S. (2009). Who is being deceived? The experimental demands of wearing a backpack. *Psychonomic Bulletin & Review, 16*(5), 964–969.

Durgin, F. H., Hajnal, A., Li, Z., Tonge, N., & Stigliani, A. (2010). Palm boards are not action measures: An alternative to the two-systems theory of geographical slant perception. *Acta Psychologica, 134*(2), 182–197.

Durgin, F. H., & Li, Z. (2011). The perception of 2D orientation is categorically biased. *Journal of Vision, 11*(8), 1–10.

Durso, F. T., & Johnson, M. K. (1980). The effects of orienting tasks on recognition, recall, and modality confusion of pictures and words. *Journal of Verbal Learning and Verbal Behavior, 19*, 416–429.

Dyde, R. T., Jenkin, M. R., & Harris, L. R. (2006). The subjective visual vertical and the perceptual upright. *Experimental Brain Research, 173*, 612–622.

Dye, R. A., Crawford, T. M., & McBeath, M. K. (2014). Absence of lateral navigational bias in young children. *Perceptual and Motor Skills: Physical Development & Measurement, 119*(1), 292–300.

Eagleman, D. M. (2008). Human time perception and its illusions. *Current Opinion in Neurobiology*, 18(2), 131–136.

Eagleman, D. M., & Pariyadath, V. (2009). Is subjective duration a signature of coding efficiency? *Philosophical Transactions of the Royal Society B: Biological Sciences, 364* (1525), 1841–1851.

Eagleman, D. M., & Sejnowski, T. J. (2000a). Flash-lag effect: Differential latency, not postdiction: Response. *Science, 290*, 1051a.

Eagleman, D. M., & Sejnowski, T. J. (2000b). Motion integration and postdiction in visual awareness. *Science, 287*, 2036–2038.

Eagleman, D. M., & Sejnowski, T. J. (2000c). The position of moving objects: Response. *Science, 289*, 1107a.

Eagleman, D. M., & Sejnowski, T. J. (2002). Untangling spatial from temporal illusions. *Trends in Neurosciences, 25*, 293.

Eagleman, D. M., & Sejnowski, T. J. (2007). Motion signals bias localization judgments: A unified explanation for the flash-lag, flash-drag, flash-jump, and Fröhlich illusions. *Journal of Vision, 7*(4), 3.

Ebbesen, E. B., Kjos, G. L., & Konečni, V. J. (1976). Spatial ecology: Its effects on the choice of friends and enemies. *Journal of Experimental Social Psychology, 12*(6), 505–518.

Eckman, P, Friesen, W.V., O'Sullivan, M., Chan, A., Diacoyanni-Tarlatzis, I., et al. (1987). Universals and cultural differences in the judgments of facial expressions of emotion. *Journal of Personality and Social Psychology, 53*(4), 712–717.

Edwards, M., & Badcock, D. R. (1993). Asymmetries in the sensitivity to motion in-depth – a centripetal bias. *Perception, 22*(9), 1013–1023.

Edwards, M., & Ibbotson, M. R. (2007). Relative sensitivities to large-field optic-flow patterns varying in direction and speed. *Perception, 36*(1), 113–124.

Eerland, A., Guadalupe, T. M., & Zwaan, R. A. (2011). Leaning to the left makes the Eiffel Tower seem smaller. Posture-modulated estimation. *Psychological Science, 22*(12), 1511–1514.

Eimer, M. (1998). The lateralized readiness potential as an on-line measure of central response activation processes. *Behavior Research Methods, Instruments, & Computers, 30*,146–156.

Eitan, Z., & Granot, R. Y. (2006). How music moves: Musical parameters and listeners' images of motion. *Music Perception, 23*, 221–247.

Eitan, Z., & Granot, R. Y. (2007). Intensity changes and perceived similarity: Inter-parametric analogies. *Musicae Scientiae*, Discussion Forum, 4a, 99–133.

Eitan, Z., Ornoy, E., & Granot, R. Y. (2012). Listening in the dark: Congenital and early blindness and cross-domain mapping in music. *Psychomusicology: Music, Mind, & Brain, 22*, 33–45.

Eitan, Z., Schupak, A., Gotler, A., & Marks, L. E. (2014). Lower pitch is larger, yet falling pitches shrink. *Experimental Psychology, 61*, 273–284.

Eitan, Z., & Timmers, R. (2010). Beethoven's last piano sonata and those who follow crocodiles: Cross-domain mappings of auditory pitch in a musical context. *Cognition, 114*, 405–422.

Emmorey, K., Damasio, H., McCullough, S., Grabowski, T., Ponto, L. L., Hichwa, R. D., & Bellugi, U. (2002). Neural systems underlying spatial language in American Sign Language. *Neuroimage, 17*(2), 812–824.

Emo, B. (2012). Wayfinding in real cities: Experiments at street corners. In C. Stachniss, K. Schill, & D. Uttal (Eds.), *Spatial Cognition 2012* (pp. 461–477). *LNAI*7463.

Emo, B. (2014). Seeing the axial line: Evidence from wayfinding experiments. *Behavioral Science 4*(3), 167–180.

Engebretson, P. H., & Huttenlocher, J. (1996). Bias in spatial location due to categorization: Comment on Tversky and Schiano. *Journal of Experimental Psychology: General, 125*(1), 96–108.

Enns, J. T., & Girgus, J. S. (1985). Perceptual grouping and spatial distortion: A developmental study. *Developmental Psychology, 21*, 241–246.

Epstein, R. A. (2008). Parahippocampal and retrosplenial contributions to human spatial navigation. *Trends in Cognitive Sciences, 12*, 388–396.

Epstein, R. A. (2011). Cognitive neuroscience: Scene layout from vision and touch. *Current Biology, 21*, R437–R438.

Epstein, W., Park, J., & Casey, A. (1961). The current status of the size-distance hypotheses. *Psychological Bulletin, 58*(6), 491–514.

Epstein, W., & Rogers, S. (1995). *Perception of space and motion*. New York: Academic Press.

Eriksen, B. A., & Eriksen, C. W. (1974). Effects of noise letters upon the identification of a target letter in a nonsearch task. *Perception & Psychophysics, 16*, 143–149.

Erlhagen, W. (2003). Internal models for visual perception. *Biological Cybernetics, 88*, 409–417.

Ernst, M. O. (2005). A Bayesian view on multimodal cue integration. In G. Knoblich, I. M. Thornton, M. Grosjean, & M. Shiffrar (Eds.), *Human body perception from the inside out* (pp. 105–131). Oxford: Oxford University Press.

Ernst, M. O. (2007). Learning to integrate arbitrary signals from vision and touch. *Journal of Vision*, 7, 7.

Ernst, M. O. (2012). Optimal multisensory integration: Assumptions and limits. In B. E. Stein (Ed.), *The new handbook of multisensory processes* (pp. 1084–1124). Cambridge, MA: MIT Press.

Ernst, M. O., & Banks, M. S. (2002). Humans integrate visual and haptic information in a statistically optimal fashion. *Nature, 415*, 429–433.

Ernst, M. O., & Bülthoff, H. H. (2004). Merging the senses into a robust percept. *Trends in Cognitive Sciences, 8*, 162–169.

Essock, E. A., DeFord, J. K., Hansen, B. C., & Sinai, M. J. (2003). Oblique stimuli are seen best (not worst!) in naturalistic broad-band stimuli: A horizontal effect. *Vision Research, 43*, 1329–1335.

Esterman, M., McGlinchey-Berroth, R., & Milberg, W. P. (2000). Parallel and serial search in hemispatial neglect: Evidence for preserved preattentive but impaired attentive processing. *Neuropsychology, 14*, 599–611.

Estner, B. (2014). *Stimulus-response compatibility of auditory stimulus features: Timbre, pitch, and number*. Technische Universität Kaiserslautern. Retrieved from urn:nbn:de:hbz:386-kluedo-37812

Euclid (ed. 1996) *Ottica. Immagini di una teoria della visione* [Optics. Images of a theory of vision] (Ed. F. Incardona). Roma: Di Renzo Editore.

Evans, K. K., & Treisman, A. (2010). Natural cross-modal mappings between visual and auditory features. *Journal of Vision, 10*(1), 6.

Evans, V., & Chilton, P. (2007). *Language, cognition and space*: Sheffiled, UK: Equinox.

Eves, F. F., Thorpe, S. K. S., Lewis, A., & Taylor-Covill, G. A. H. (2014). Does perceived steepness deter stair climbing when an alternative is available? *Psychonomic Bulletin & Review, 21*(3), 637–644.

Fair, D., Dosenbach, N., Church, J., Cohen, Al., Brahmbhatt, S., Miezin, F., Barch, D., Raichle, M., Petersen, S., & Schlaggar, B. (2007). Development of distinct control networks through segregation and integration. *Proceedings of the National Academy of Sciences of the United States of America, 104*, 13507–13512.

Fairhurst, M., & Deroy, O. (2017). Magnitude-space mapping for auditory and audio-visual intensity: A test of the shared spatial representation of magnitude. *Journal of Experimental Psychology Human Perception and Performance, 43*(3), 629–637.

Fajen, B. R., & Phillips, F. (2012). Spatial perception and action. In D. A. Waller & L. Nadel (Eds.), *Handbook of spatial cognition* (pp. 67–80). Washington, DC: American Psychological Association.

Fantoni, C., Hilger, J. D., Gerbino, W., & Kellman, P. J. (2008). Surface interpolation and 3D relatability. *Journal of Vision, 8*, 29 21–19.

Farran, E. K., Whitaker, A., & Patel, N. (2009). The effect of pictorial depth information on projected size judgments. *Attention, Perception, & Psychophysics, 71*(1), 207–214.

Fasoli, F., Paladino, M. P., Carnaghi, A., Jetten, J., Bastian, B., & Bain, P. G. (2016). Not "just words": Exposure to homophobic epithets leads to dehumanizing and physical distancing from gay men. *European Journal of Social Psychology, 46*(2), 237–248.

Fattorini, E., Pinto, M., Rotondaro, F., & Doricchi, F. (2015). Perceiving numbers does not cause automatic shifts of spatial attention. *Cortex, 73*, 298–316.

Faust, M. (1990). *Representational momentum: A dual process perspective* (Unpublished doctoral dissertation). University of Oregon, Eugene.

Favretto, A. (2002). *Displaced representations of targets undergoing luminance transformations* (Unpublished doctoral dissertation). University of Trieste, Italy.

Fechner, G. T. (1876). *Vorschule de Aesthetik*. Leipzig: Brietkopt and Hatrtel.

Fecteau, J. H., Au, C., Armstrong, I. T., & Munoz, D. P. (2004). Sensory biases produce alternation advantage found in sequential saccadic eye movement tasks. *Experimental Brain Research*, 159, 84–91.

Fendrich, R., & Corballis, P. M. (2001). The temporal cross-capture of audition and vision. *Perception & Psychophysics*, 63, 719–725.

Fernandez, C., & Goldberg, J. M. (1976). Physiology of peripheral neurons innervating otolith organs of the squirrel monkey. I. Response to static tilts and to long-duration centrifugal force. *Journal of Neurophysiology*, 39, 970–984.

Fernández-Prieto, I., & Navarra, J. (2017). Spatial encoding of pitch in frequency sweeps. *Psychology of Music*, 0305735616684205

Fernández-Prieto, I., Navarra, J., & Pons, F. (2015). How big is this sound? Crossmodal association between pitch and size in infants. *Infant Behavior and Development*, *38*, 77–81.

Fernández-Prieto, I., Navarra, J., & Pons, F. (2016). *Crossmodal correspondence between pitch and height in childhood.* Poster presented at the International Meeting of the Psychonomic Society, Granada, Spain.

Ferri, F., Tajadura-Jimenez, A., Valjamae, A., Vastano, R., & Costantini, M. (2015). Emotion-inducing approaching sounds shape the boundaries of multisensory peripersonal space. *Neuropsychologia*, *70*, 468–475.

Fessler, D. M. T., & Holbrook, C. (2013). Friends shrink foes: The presence of comrades decreases the envisioned physical formidability of an opponent. *Psychological Science*, *24*, 797–802.

Fias, W., Bryspaert, M, Geypens, F., & d'Ydewalle, G. (1996). The importance of magnitude information in numerical processing: Evidence from the SNARC effect. *Mathematical Cognition*, *2*(1), 95–110.

Fias, W., & van Dijck, J. P. (2016). The temporary nature of number – space interactions. *Canadian Journal of Experimental Psychology*, *70*(1), 33–40.

Fick, A. (1851). *De errore quodam optico asymetrica bulbi effecto*. Marburg: J. A. Kochin.

Field, D. J., Hayes, A., & Hess, R. F. (1993). Contour integration by the human visual system: evidence for a local "association field." *Vision Research*, *33*, 173–193.

Findlay, J. M. (1973). Feature detectors and vernier acuity. *Nature*, *241*, 135–137.

Findlay, J. M. (1982). Global visual processing for saccadic eye movement. *Vision Research*, *22*, 1033–1045.

Finger, F. W., & Spelt, D. K. (1947). The illustration of the horizontal vertical illusion. *Journal of Experimental Psychology*, *37*, 243–250.

Finisguerra, A., Canzoneri, E., Serino, A., Pozzo, T., & Bassolino, M. (2015). Moving sounds within the peripersonal space modulate the motor system. *Neuropsychologia*, *70*, 421–428.

Fink, G. R., Marshall, J. C., Weiss, P. H., & Zilles, K. (2001). The neural basis of vertical and horizontal line bisection judgments: An fMRI study of normal volunteers. *Neuroimage*, *14*, S59–S67.

Finke, R. A., & Freyd, J. J. (1985). Transformations of visual memory induced by implied motions of pattern elements. *Journal of Experimental Psychology: Learning, Memory, and Cognition*, *11*, 780–794.

Finke, R. A., Freyd, J. J., & Shyi, G. C. W. (1986). Implied velocity and acceleration induce transformations of visual memory. *Journal of Experimental Psychology: General*, *115*, 175–188.

Finke, R. A., Johnson, M. K., & Shyi, G. C. W. (1988). Memory confusions for real and imaged completions of symmetrical visual patterns. *Memory & Cognition*, *16*, 133–137.

Finke, R. A., & Shyi, G. C. W. (1988). Mental extrapolation and representational momentum for complex motions. *Journal of Experimental Psychology: Learning, Memory, and Cognition*, *14*, 112–120.

Firestone, C. (2013). How "paternalistic" is spatial perception? Why wearing a heavy backpack doesn't – and *couldn't* – make hills appear steeper. *Perspectives on Psychological Science*, *8*(4), 455–473.

Firestone, C., & Scholl, B. J. (2016). Cognition does not affect perception: Evaluating the evidence for "top-down" effects. *Behavioral and Brain Sciences*, *39*, e229.

Fischer, M. H. (2001). Number processing induces spatial performance biases. *Neurology*, *57*(5), 822–826.

Fischer, M. H. (2012). A hierarchical view of grounded, embodied, and situated numerical cognition. *Cognitive Processing*, *13*, 161–164.

Fischer, M. H. (2013). The spatial mapping of numbers: Its origin and flexibility. In Y. Coello & A. Bartolo (Eds.), *Language and action in cognitive neuroscience* (p. 225–242). London: Psychology Press.

Fischer, M. H., & Brugger, P. (2011). When digits help digits: Spatial–numerical associations point to finger counting as prime example of embodied cognition. *Frontiers in Psychology*, *2:260*

Fischer, M. H., Castel, A. D., Dodd, M. D., & Pratt, J. (2003). Perceiving numbers causes spatial shifts of attention. *Nature Neuroscience*, *6*(6), 555–556.

Fischer, M. H., & Hill, R. (2004, January). *A SNARC in the dark: Input modality affects number representation*. A poster presented at the 22nd European Workshop of Cognitive Neuropsychology, Bressanone, Italy.

Fischer, M. H., Mills, R. A, & Shaki, S. (2010). How to cook a SNARC: Number placement in text rapidly changes spatial-numerical associations. *Brain and Cognition*, *72*, 333–336.

Fischer, M. H., Riello, M., Giordano, B. L., & Rusconi, E. (2013). Singing numbers . . . in cognitive space – a dual-task study of the link between pitch, space, and numbers. *Topics in Cognitive Sciences*, *5*, 354–366.

Fischer, M. H., & Shaki, S. (2014). Spatial associations in numerical cognition – from single digits to arithmetic. *The Quarterly Journal of Experimental Psychology 67*, 1461–1483.

Fischer, M. H., & Shaki, S. (2015). Two steps to space for numbers. *Frontiers in Psychology*, *6*, 612.

Fischer, M. H., & Shaki, S. (2016). Measuring spatial–numerical associations: Evidence for a purely conceptual link. *Psychological Research*, *80*(1), 109–112.

Fischer, M. H., Shaki, S., & Cruise, A. (2009). It takes just one word to quash a SNARC. *Experimental Psychology*, *56*(5), 361–366.

Fischer, U., Moeller, K., Bientzle, M., Cress, U., & Nuerk, H.-C. (2011). Embodied spatial numerical training of number magnitude representation: An intervention study. *Psychonomic Bulletin & Review*, *18*, 177–183.

Fiske, A. P. (2000). Complementarity theory: Why human social capacities evolved to require cultural complements. *Personality and Social Psychology Review*, *4*(1), 76–94.

Fiske, A. P. (2004). Four modes of constituting relationships: Consubstantial assimilation; space, magnitude, time, and force; concrete procedures; abstract symbolism. In N. Haslam (Ed.) *Relational models theory: A contemporary overview* (pp. 61–146). Mahwah, NJ: Erlbaum.

Fitting, S., Wedell, D. H., & Allen, G. L. (2007). Memory for spatial location: Cue effects as a function of field rotation. *Memory & Cognition*, *35*(7), 1641–1658.

Fitts, P. M., & Deininger, R. L. (1954). S-R compatibility: Correspondence among paired elements within stimulus and response codes. *Journal of Experimental Psychology*, *48*, 483–492.

Fitts, P. M., & Seeger, C. M. (1953). S-R compatibility: Spatial characteristics of stimulus and response codes. *Journal of Experimental Psychology*, *46*, 199–210.

Fletcher, P. D., Nicholas, J. M., Shakespeare, T. J., Downey, L. E., Golden, H. L., et al. (2015). Dementias show differential physiological responses to salient sounds. *Frontiers in Behavioral Neuroscience*, *9*. doi:10.3389/fnbeh.2015.00073

Flöel, A., Jansen, A., Deppe, M., Kanowski, M., Konrad, C., Sommer, J. & Knecht, S. (2005). Atypical hemispheric dominance for attention: Functional MRI topography. *Journal of Cerebral Blood Flow & Metabolism*, *25*, 1197–1208.

Fodor, J. A. (1983). *The modularity of mind*. Cambridge, MA: MIT Press.

Foley, J. M. (1972). The size-distance relation and intrinsic geometry of visual space: Implications for processing. *Vision Research, 12*, 323–332.

Foley, J. M. (1977). Effect of distance information and range on two indices of visually perceived distance. *Perception, 6*, 449–460.

Foley, J. M., Ribeiro-Filho, N. P. & Da Silva, J. A. (2004). Visual perception of extent and the geometry of visual space. *Vision Research, 44*, 147–156.

Foley, M. A., & Johnson, M. K. (1985). Confusions between memories for performed and imagined actions: A developmental comparison. *Child Development, 56*, 1145–1155.

Foo, P., Warren, W. H., Duchon, A., & Tarr, M. J. (2005). Do humans integrate routes into a cognitive map? Map- versus landmark-based navigation of novel shortcuts. *Journal of Experimental Psychology: Learning, Memory, and Cognition, 31*(2), 195–215.

Forbes, K., & Klein, R. M. (1996). The magnitude of the fixation offset effect with endogenously and exogenously controlled saccades. *Journal of Cognitive Neuroscience, 8*(4), 344–352.

Forbus, K. D. (2011). Qualitative modeling. *Cognitive Science, 2*, 374–391.

Forsyth, D. R. (1983). *An introduction to group dynamics*. Pacific Grove, CA: Brooks/Cole.

Foster, D. H., & Gravano, S. (1982). Overshoot of curvature in visual apparent motion. *Perception & Psychophysics, 31*, 411–420.

Foster, N. E. V., & Zatorre. R. J. (2010). Cortical structure predicts success in performing musical transformation judgments. *NeuroImage, 53*, 26–36.

Foulsham, T., & Kingstone, A. (2010). Asymmetries in the direction of saccades during perception of scenes and fractals: Effects of image type and image features. *Vision Research, 50*(8), 779–795.

Foxe, J. J., McCourt, M. E., & Javitt, D. C. (2003). Right hemisphere control of visuospatial attention: Line-bisection judgments evaluated with high-density electrical mapping and source analysis. *Neuroimage, 19*(3), 710–726.

Franconeri, S. L., Alvarez, G. A., & Enns, J. T. (2007). How many locations can you select? *Journal of Experimental Psychology: Human Perception and Performance, 33*, 1003–1012.

Franconeri, S. L., & Simons, D. J. (2003). Moving and looming stimuli capture attention. *Perception & Psychophysics, 65*(7), 999–1010.

Franklin, N., Henkel, L. A., & Zangas, T. (1995). Parsing surrounding space into regions. *Memory & Cognition, 23*, 397–407.

Fraser, J. (1908). A new visual illusion of direction. *British Journal of Psychology, 2*, 307–337.

Frassinetti, F., Magnani, B., & Oliveri, M. (2009). Prismatic lenses shift time perception. *Psychological Science, 20*, 949–954.

Frederici, A. (1981). Production and comprehension of prepositions in aphasia. *Neuropsychologia, 19*, 191–199.

Freeman, E., & Driver, J. (2008). Direction of visual apparent motion driven solely by timing of a static sound. *Current Biology, 18*, 1262–1266.

Freeman, J., Brouwer, G. J., Heeger, D. J., & Merriam, E. P. (2011). Orientation decoding depends on maps, not columns. *Journal of Neuroscience, 31*(13), 4792–4804.

Freiberg, K., Tually, K., & Crassini, B. (2001). Use of an auditory looming task to test infants' sensitivity to sound pressure level as an auditory distance cue. *British Journal of Developmental Psychology, 19*, 1–10.

Freyd, J. J. (1983). The mental representation of movement when static stimuli are viewed. *Perception & Psychophysics, 33*, 575–581.

Freyd, J. J. (1987). Dynamic mental representations. *Psychological Review, 94*, 427–438.

Freyd, J. J. (1993). Five hunches about perceptual processes and dynamic representations. In D. Meyer & S. Kornblum (Eds.), *Attention and Performance XIV: Synergies in experimental psychology, artificial intelligence, and cognitive neuroscience* (pp. 99–119). Cambridge, MA: MIT Press.

Freyd, J. J., & Finke, R. A. (1984). Representational momentum. *Journal of Experimental Psychology: Learning, Memory, and Cognition, 10*, 126–132.

Freyd, J. J., & Finke, R. A. (1985). A velocity effect for representational momentum. *Bulletin of the Psychonomic Society, 23*, 443–446.

Freyd, J. J., & Johnson, J. Q. (1987). Probing the time course of representational momentum. *Journal of Experimental Psychology: Learning, Memory, and Cognition, 13*, 259–268.

Freyd, J. J., & Jones, K. T. (1994). Representational momentum for a spiral path. *Journal of Experimental Psychology: Learning, Memory, and Cognition, 20*, 968–976.

Freyd, J. J., Kelly, M. H., & DeKay, M. L. (1990). Representational momentum in memory for pitch. *Journal of Experimental Psychology: Learning, Memory, and Cognition, 16*, 1107–1117.

Freyd, J. J., & Miller, G. F. (1992, November). Creature motion. Paper presented at the 33rd Annual Meeting of the Psychonomic Society, St. Louis, MO.

Freyd, J. J., & Pantzer, T. M. (1995). Static patterns moving in the mind. In S. M. Smith, T. B. Ward, & R. A. Finke (Eds.), *The creative cognition approach* (pp. 181–204). Cambridge, MA: MIT Press.

Freyd, J. J., Pantzer, T. M., & Cheng, J. L. (1988). Representing statics as forces in equilibrium. *Journal of Experimental Psychology: General, 117*, 395–407.

Freyd, J. J., & Tversky, B. (1984). The force of symmetry in form perception. *American Journal of Psychology, 97*, 109–126.

Friedenberg, J. (2012). Aesthetic judgment of triangular shape: compactness and not the golden ratio determines perceived attractiveness. *i-Perception, 3*, 163–175.

Friedenberg, J., & Bertamini, M. (2015). Aesthetic preference for polygon shape. *Empirical Studies of the Arts, 33*, 144–160.

Friedman, A. (2009). The role of categories and spatial cuing in global-scale location estimates. *Journal of Experimental Psychology: Learning, Memory, and Cognition, 35*(1), 94–112.

Friedman, A., & Brown, N. R. (2000a). Reasoning about geography. *Journal of Experimental Psychology: General, 129*(2), 193–219.

Friedman, A., & Brown, N. R. (2000b). Updating geographical knowledge: Principles of coherence and inertia. *Journal of Experimental Psychology: Learning, Memory, and Cognition, 26*(4), 900–914.

Friedman, A., Brown, N. R., & McGaffey, A. P. (2002). A basis for bias in geographical judgments. *Psychonomic Bulletin & Review, 9*(1), 151–159.

Friedman, A., Kerkman, D. D., Brown, N. R., Stea, D., & Cappello, H. M. (2005). Cross-cultural similarities and differences in North Americans' geographic location judgments. *Psychonomic Bulletin & Review, 12*(6), 1054–1060.

Friedman, A., Ludvig, E. A., Legge, E. L. G., & Vuong, Q. C. (2013). Bayesian combination of two-dimensional location estimates. *Behavior Research Methods, 45*(1), 98–107.

Friedman, A., & Montello, D. R. (2006). Global-scale location and distance estimates: Common representations and strategies in absolute and relative judgments. *Journal of Experimental Psychology: Learning, Memory, and Cognition, 32*(2), 333–346.

Friedman, A., Montello, D. R., & Burte, H. (2012). Location memory for dots in polygons versus cities in regions: Evaluating the category adjustment model. *Journal of Experimental Psychology: Learning, Memory, and Cognition, 38*(5), 1336–1351.

Friemuth, M., & Wapner, S. (1979). The influence of lateral organization on the evaluation of paintings. *British Journal of Psychology, 70*, 211–218.

Frimer, J. A., & Sinclair, L. (2016). Moral heroes look up and to the right. *Personality and Social Psychology Bulletin, 42*, 400–410.

Frisén, L. (2010). Deviations of the visual upright in three dimensions in disorders of the brainstem: A clinical exploration. *Brain, 133*, 3541–3551.

Frith, C. D., & Nias, D. K. B. (1974). What determines aesthetic preference? *Journal of General Psychology*, *91*, 163–173.

Fröhlich, F. W. (1923). Über die Messung der Empfindungszeit [Measuring the time of sensation]. *Zeitschrift für Sinnesphysiologie*, *54*, 58–78.

Fröhlich, F. W. (1925). Über die Methoden der Empfindungszeitmessung im Gebiet des Gesichtsinnes [On the measurement of the sensation time in the visual field]. *Pflüger's Archiv für die gesamte Physiologie des Menschen und der Tiere*, *208*(1), 120–134.

Fröhlich, F. W. (1929). Die Empfindungszeit: Ein Beitrag zur Lehre von der Zeit-, Raum- und Bewegungsempfindung [Time of sensation: A contribution on the model of the perception of time, space and motion]. *Jena: Fischer*.

Fröhlich, F. W. (1930). Über die Messung der Empfindungszeit [Measuring the time of sensation]. *Psychologische Forschung*, *13*(1), 285–288.

Fröhlich, F. W. (1932). Bemerkungen zu G. E. Müllers Kritik der Empfindungszeitmessung [Remarks on the criticism of G. E. Müller of the measurement of sensation time]. *Zeitschrift für Psychologie und Physiologie der Sinnesorgane*, *62*, 246–249.

Fu, Y. X., Shen, Y., & Dan, Y. (2001). Motion-induced perceptual extrapolation of blurred visual targets. *The Journal of Neuroscience*, *21*, RC172.

Fuhrman, O., Boroditsky, L. (2010). Cross-cultural differences in mental representations of time: Evidence from an implicit nonlinguistic task. *Cognitive Science*, *34*, 1430–1451.

Fujita, K., Henderson, M. D., Eng, J., Trope, Y., & Liberman, N. (2006). Spatial distance and mental construal of social events. *Psychological Science*, *17*(4), 278–282.

Fukiage, T., & Murakami, I. (2010). The tilt aftereffect occurs independently of the flash-lag effect. *Vision Research*, *50*, 1949–1956.

Fukiage, T., & Murakami, I. (2013). Adaptation to a spatial offset occurs independently of the flash-drag effect. *Journal of Vision*, *13*(2), 1–14.

Fukiage, T., Whitney, D., & Murakami, I. (2011). A flash-drag effect in random motion reveals involvement of preattentive motion processing. *Journal of Vision*, *11*(13), 12.

Fukusima, S. S., & Faubert, J. (2001). Perceived length in the central visual field: Evidence for visual field asymmetries. *Vision Research*, *41*(16), 2119–2126.

Fukusima, S. S., Loomis, J. M., & Da Silva, J. A. (1997). Visual perception of egocentric distance as assessed by triangulation. *Journal of Experimental Psychology: Human Perception and Performance*, *23*(1), 86–100.

Furmanski, C. S., & Engel, S. A. (2000). An oblique effect in human primary visual cortex, *Nature Neuroscience*, *3*, 535–536.

Futterweit, L. R., & Beilin, H. (1994). Recognition memory for movement in photographs: A developmental study. *Journal of Experimental Child Psychology*, *57*, 163–179.

Gaffron, M. (1950). Right and left in pictures. *Art Quarterly*, *13*, 312–331.

Gagnier, K. M. (2010). *Rethinking Boundary Extension: The Role of Source Monitoring in Scene Memory* (Doctoral dissertation). Retrieved from Proquest, 854499379.

Gagnier, K. M., Dickinson, C. A., & Intraub, H. (2013). Fixating picture boundaries does not eliminate boundary extension: Implications for scene representation. *Quarterly Journal of Experimental Psychology*, *66*, 2161–2186.

Gagnier, K. M., & Intraub, H. (2012). When less is more: Line-drawings lead to greater boundary extension than colour photographs. *Visual Cognition*, *20*, 815–824.

Gagnier, K. M., Intraub, H., Oliva, A., Wolfe, J. M. (2011). Why does vantage point affect boundary extension? *Visual Cognition*, *19*, 234–257.

Gagnon, S. A., Brunyé, T. T., Robin, C., Mahoney, C. R., & Taylor, H. A. (2011). High and mighty: Implicit associations between space and social status. *Frontiers in Psychology*, *2*, 1–10.

Gainotti, G., D'Erme, P., & Bartolomeo, P. (1991). Early orientation of attention toward the half space ipsilateral to the lesion in patients with unilateral brain damage.

Journal of Neurology, Neurosurgery and Psychiatry, 54, 1082–1089.

Gale, M., & Ball, L. J. (2012). Contrast class cues and performance facilitation in a hypothesis testing task: Evidence for an iterative counterfactual model. *Memory & Cognition, 40*, 408–419.

Gallace, A., Martelli, M., & Daini, R. (2012). Inconstancy and inconsistency of visual illusory phenomena? The case of the Poggendorff figure. *Psychology, 3*, 257–264.

Gallace, A., & Spence, C. (2006). Multisensory synesthetic interactions in the speeded classification of visual size. *Perception & Psychophysics, 68*, 1191–1203.

Ganel, T., Tanzer, M., & Goodale, M. A. (2008). A double dissociation between action and perception in the context of visual illusions opposite effects of real and illusory size. *Psychological Science, 19*(3), 221–225.

Gärdenfors, P. (2000). *Conceptual spaces: The geometry of thought*. Cambridge, MA: MIT Press.

Gärdenfors, P. (2014). *The geometry of meaning: Semantics based on conceptual spaces*. Cambridge, MA: The MIT Press

Gardner, J. S., Fowlkes, C., Nothelfer, C., & Palmer, S. E. (2008). Exploring aesthetic principles of spatial composition through stock photography. *Journal of Vision, 8*, 337a.

Gardner, S. (2007). The limits of naturalism and the metaphysics of German idealism. In E. Hammer (Ed.), *German idealism: Contemporary perspectives* (pp. 19–49). Abingdon, UK: Routledge.

Gärling, T., & Gärling, E. (1988). Distance minimization in downtown pedestrian shopping. *Environment and Planning A, 20*(4), 547–554.

Gartus, A., & Leder, H. (2013). The small step toward asymmetry: Aesthetic judgment of broken symmetries. *i-Perception*, 4, 352–355.

Gattis, M. (Ed.). (2001). *Spatial schemas and abstract thought*. Cambridge, MA: MIT Press.

Gauch, A., & Kerzel, D. (2008a). Comparison of flashed and moving probes in the flash-lag effect: Evidence for misbinding of abrupt and continuous changes. *Vision Research, 48*, 1584–1591.

Gauch, A., & Kerzel, D. (2008b). Perceptual asynchronies between color and motion at the onset of motion and along the motion trajectory. *Perception & Psychophysics, 70*, 1092–1103.

Gauch, A., & Kerzel, D. (2009). Contributions of visible persistence and perceptual set to the flash-lag effect: Focusing on flash onset abolishes the illusion. *Vision Research, 49*, 2983–2991.

Gauthier, L., Dehaut, F., & Joanette, Y. (1989). The bells test: A quantitative and qualitative test for visual neglect. *International Journal of Clinical Neuropsychology, 11*, 46–54.

Gawley, T., Perks, T., & Curtis, J. (2009). Height, gender, and authority status at work: Analyses for a national sample of Canadian workers. *Sex Roles, 60*, 208–222.

Gazzaniga, M. S., & Ladavas, E. (1987). Disturbances in spatial attention following lesion of the right parietal lobe. In M. Jeannerod (Ed.), *Neurophysiological and neuropsychological aspects of spatial neglect* (pp. 203–213). Amsterdam: Elsevier Science Publishers.

Ge, Y., Wang, L., & He, S. (2015). Perceived illusory orientation from the flash grab effect induces the Tilt Aftereffect. *Journal of Vision, 15*(12), 992.

Gebhard, J. W., & Mowbray, G. H. (1959). On discriminating the rate of visual flicker and auditory flutter. *American Journal of Psychology, 72*, 521–529.

Geer, M., & Schmidt, W. C. (2006). Perception of initial moving target signals: Support for a cumulative lateral inhibition theory. *Journal of Experimental Psychology: Human Perception and Performance, 32*(5), 1185–1196.

Geesaman, B. J., & Qian, N. (1996). A novel speed illusion involving expansion and rotation patterns. *Vision Research, 36*(20), 3281–3292.

Geminiani, G., Camaschella, E., Mariani, C., Alberoni, M., & Farina, E. (2002). Direction and position factors in performance of line extension task by unilateral neglect subjects. *Neuropsychologia, 40*(11), 1834–1840.

Gentner, D. (2003). Why we're so smart. In D. Gentner & S. Goldin-Meadows (Eds.), *Language in mind* (pp. 195–235). Cambridge, MA: MIT Press.

Gentner, D., & Markman, A. B. (1994). Structure mapping in analogy and similarity. *American Psychologist, 52*(1), 45–56.

Georges, S., Schiltz, C., & Hoffmann, D. (2015). Task instructions determine the visuospatial and verbal-spatial nature of number–space associations, *The Quarterly Journal of Experimental Psychology, 68*(9), 1895–1909.

Gershoni, S., & Hochstein, S. (2011). Measuring pictorial balance perception at first glance using Japanese calligraphy. *I-Perception*, 2, 508–527.

Getzmann, S. (2005). Representational momentum in spatial hearing does not depend on eye movements. *Experimental Brain Research, 165*, 229–238.

Getzmann, S. (2007). The effect of brief auditory stimuli on visual apparent motion. *Perception*, 36, 1089–1103.

Getzmann, S., & Lewald, J. (2007). Localization of moving sound. *Perception & Psychophysics, 69*, 1022–1034.

Getzmann, S., & Lewald, J. (2009). Constancy of target velocity as a critical factor in the emergence of auditory and visual representational momentum. *Experimental Brain Research, 193*, 437–443.

Getzmann, S., Lewald, J., & Guski, R. (2004). Representational momentum in spatial hearing. *Perception, 33*, 591–599.

Gevers, W., Caessens, B., & Fias, W. (2005). Towards a common processing architecture underlying Simon and SNARC effects. *European Journal of Cognitive Psychology, 17*(5), 659–673.

Gevers, W., & Lammertyn, J. (2005). The hunt for SNARC. *Psychology Science, 47*, 10–21.

Gevers, W., Ratinckx, E., De Baene, W., & Fias, W. (2006). Further evidence that the SNARC effect is processed along a dual-route architecture: Evidence from the lateralized readiness potential. *Experimental Psychology, 53*(1), 58–68.

Gevers, W., Reynvoet, B., & Fias, W. (2003). The mental representation of ordinal sequences is spatially organized. *Cognition, 87*, B87–95.

Gevers,W., Reynvoet, B., & Fias, W. (2004). The mental representation of ordinal sequences is spatially organized: Evidence from days of the week. *Cortex, 40*(1):171–172

Gevers, W., Santens, S., Dhooge, E., Chen, Q., Van den Bossche, L., Fias, W., & Verguts, T. (2010). Verbal-spatial and visuospatial coding of number–space interactions. *Journal of Experimental Psychology: General, 139*, 180–190.

Ghazanfar, A. A., Neuhoff, J. G., & Logothetis, N. K. (2002). Auditory looming perception in rhesus monkeys. *Proceedings of the National Academy of Sciences of the United States of America, 99*(24), 15755–15757.

Gibb, R., Ercoline, B., & Scharff, L. (2011). Spatial disorientation: Decades of pilot fatalities. *Aviation, Space, and Environmental Medicine 82*, 717–724.

Gibbs, R. W. (2005). *Embodiment and cognitive science*. New York: Cambridge University Press.

Gibson, B. S., & Egeth, H. (1994). Inhibition and disinhibition of return: Evidence from temporal order judgments. *Perception & Psychophysics, 56*(6), 669–680.

Gibson, J. J. (1950). *The perception of the visual world*. Boston, MA: Houghton Mifflin.

Gibson, J. J. (1966). *The senses considered as perceptual systems*. Boston, MA: Houghton Mifflin.

Gibson, J. J. (1979). *The ecological approach to visual perception*. Boston, MA: Houghton Mifflin.

Giessner, S. R., Ryan, M. K., Schubert, T. W., & van Quaquebeke, N. (2011). The power of pictures: Vertical picture angles in power pictures. *Media Psychology, 14*(4), 442–464.

Giessner, S. R., & Schubert, T. W. (2007). High in the hierarchy: How vertical location and judgments of leaders' power are interrelated. *Organizational Behavior and Human Decision Processes, 104*(1), 30–44.

Gilbert, A. L., Regier, T., Kay, P., & Ivry, R. B. (2006). Whorf hypothesis is supported in the right visual field but not the left. *Proceedings of the National Academy of Sciences of the United States of America, 103*(2), 489–494.

Gilbert, A. L., Regier, T., Kay, P., & Ivry, R. B. (2008). Support for lateralization of the Whorf effect beyond the realm of color discrimination. *Brain and Language, 105*(2), 91–98.

Gilinsky, A. S. (1951). Perceived size and distance in visual space, *Psychological Review, 58*(6), 460–482.

Gilis, B., Helsen, W., Catteeuw, P., & Wagemans, J. (2008). Offside decisions by expert assistant referees in association football: Perception and recall of spatial positions in complex dynamic events. *Journal of Experimental Psychology: Applied, 14*, 21–35.

Gill, D., Jordan, J. S., & Cutting, J. C. (2014). *Examining possible perceptual proxies of flow state*. Poster presented at the 55th Annual Meeting of the Psychonomic Society, Long Beach, CA.

Gillam, B. (1971). A depth processing theory of the Poggendorff illusion. *Perception & Psychophysics, 10*, 211–216.

Gillam, B. (1988). Illusions at century's end. In J. Hochberg (Ed.), *Perception and cognition at century's end: History, philosophy, theory* (pp. 98–137). New York: Academic Press.

Gillner, S., & Mallot, H. A. (1998). Navigation and acquisition of spatial knowledge in a virtual maze. *Journal of Cognitive Neuroscience, 10*(4), 445–463.

Ginsburg, V., & Gevers, W. (2015). Spatial coding of ordinal information in short-and long-term memory. *Frontiers in Human Neuroscience, 9:8*

Ginsburg, V., van Dijck, J. P., Previtali, P., Fias, W., & Gevers, W. (2014). The impact of verbal working memory on number–space associations. *Journal of Experimental Psychology: Learning, Memory, and Cognition, 40*(4), 976–986.

Giocomo, L. M., Zilli, E. A., Fransén, E., & Hasselmo, M. E. (2007). Temporal frequency of subthreshold oscillations scales with entorhinal grid cell field spacing. *Science, 315*(5819), 1719–1722.

Girard, T. A., Martius, D. L. M. A., & Cheyne, J. A. (2007). Mental representation of space: Insights from oblique distribution of hallucinations. *Neuropsychologia, 45*, 1257–1269.

Glasauer, S., & Mittelstaedt, H. (1998) Perception of spatial orientation in microgravity. *Brain Research Review, 28*, 185–193.

Glass, L. (1970). Effects of blurring on perception of a simple geometric pattern. *Nature, 228*, 1341–1342.

Gleitman, L., & Papafragou, A. (2012). New perspectives on language and thought. In K. Holyoak & R. Morrison (Eds.), *The Oxford handbook of thinking and reasoning* (pp. 543–567). New York: Oxford University Press.

Glenberg, A. M., Lopez-Mobilia G., McBeath, M. K., Toma, M., Sato, M., & Cattaneo, L. (2010). Knowing beans: Human mirror mechanisms revealed through motor adaptation. *Frontiers in Human Neuroscience, 4*(206), 1–6.

Glicksohn, J. (1994). Rotation, orientation, and cognitive mapping. *American Journal of Psychology, 107*, 39–51.

Gobara, A., Yamada, Y., & Miura, K. (2016). Crossmodal modulation of spatial localization by mimetic words. *i-Perception, 7*(6). doi: 10.1177/2041669516684244

Göbel, S. M. (2015). Up or down? Reading direction influences vertical counting direction in the horizontal plane – a cross-cultural comparison. *Frontiers in Psychology, 6*, 228.

Gogel, W. C., & Da Silva, J. A. (1987). A two-process theory of the response to size and distance. *Attention, Perception, & Psychophysics, 41*(3), 220–238.

Göksun, T., Lehet, M., Malykhina, K., & Chatterjee, A. (2013). Naming and gesturing spatial relations: Evidence from focal brain-injured individuals. *Neuropsychologia, 51*(8), 1518–1527.

Goldinger, S. D., Papesh, M. H., Barnhart, A. S., Hansen, W. A., & Hout, M. C. (2016). The poverty of embodied cognition. *Psychonomic Bulletin & Review, 23*(4), 959–978.

Goldmeier, E. (1936/1972). Similarity in visually perceived forms. *Psychological Issues, 29*, 1–36.

Goldstone, R. L. (1994). The role of similarity in categorization: Providing a groundwork. *Cognition, 52*(2), 125–157.

Goldstone, R. L., & Hendrickson, A. T. (2010). Categorical perception. *Wiley Interdisciplinary Review of Cognitive Science, 1*(1), 69–78.

Golfinopoulos, E., Tourville, J. A., & Guenther, F. H. (2009). The integration of large-scale neural network modeling and functional brain imaging in speech motor control. *Neuroimage, 52*, 862–874.

Golledge, R. G. (1992). Place recognition and wayfinding: Making sense of space. *Geoforum, 23*, 199–214.

Golledge, R. G. (1995). Path selection and route preference in human navigation: A progress report. In *Spatial information theory: A theoretical basis for GIS* (Vol. 988, pp. 207–222). Berlin: Springer.

Golledge, R. G., & Zannaras, G. (1973) Cognitive approaches to the analysis of human spatial behavior. In W. H. Ittelson (Ed.), *Environment and cognition* (pp. 59–94). New York: Seminar.

Gollisch, T., & Meister, M. (2010). Eye smarter than scientists believed: Neural computations in circuits of the retina. *Neuron, 65*, 150–164.

Gombrich, E. H. (1979). *The sense of order.* London: Phaidon Press.

Gonzalez, M. C., Hidalgo, C. A., & Barabasi, A. L. (2008). Understanding individual human mobility patterns. *Nature, 453*(7196), 779–782.

Gooch, B., Reinhard, E. Moulding, C., & Shirley, P. (2001). Artistic composition for image creation. In *Proceedings of Rendering Techniques 2001 Eurographics* (pp. 83–88). London, UK.

Goodale, M. A., & Milner, A. D. (1992). Separate visual pathways for perception and action. *Trends in Neurosciences, 15*(1), 20–25.

Goodhill, G. J., Finch, S., & Sejnowski, T. J. A. (1995) Unifying measure for neighborhood preservation in topographic mappings. Proceedings of the 2nd Joint Symposium on Neural Computation, University of California, San Diego and California Institute of Technology Institute for Neural Computation, La Jolla, CA, 191–202

Goodwin, G. P., & Johnson-Laird, P. N. (2006). Reasoning about the relations between relations. *Quarterly Journal of Experimental Psychology, 59*(6), 1047–1069.

Gordon, I. E. (1974). Left and right in Goya's portraits. *Nature, 249*, 197–198.

Gordon, M. S., & Rosenblum, L. D. (2005). Effects of intrastimulus modality change on audiovisual time-to-arrival judgments. *Perception & Psychophysics, 67*(4). doi:10.3758/bf03193516

Gordon, M. S., Russo, F. A., & MacDonald, E. (2013). Spectral information for detection of acoustic time to arrival. *Attention, Perception, & Psychophysics, 75*(4), 738–750.

Gorea, A., Belkoura, S., & Solomon, J. A. (2014). Summary statistics for size over space and time. *Journal of Vision, 14*(9), 1–14.

Gorman, A. D., Abernethy, B., & Farrow, D. (2011). Investigating the anticipatory nature of pattern perception in sport. *Memory & Cognition, 39*, 894–901.

Gorman, A. D., Abernethy, B., & Farrow, D. (2012). Classical pattern recall tests and the prospective nature of expert performance. *The Quarterly Journal of Experimental Psychology, 65*, 1151–1160.

Gottesman, C. V. (2011). Mental layout extrapolations prime spatial processing of scenes. *Journal of Experimental Psychology: Human Perception and Performance, 37*, 382–395.

Gottesman, C. V., & Intraub, H. (2002). Surface construal and the mental representation of scenes. *Journal of Experimental Psychology: Human Perception and Performance, 28*, 589–599.

Gottwald, V. M., Lawrence, G. P., Hayes, A. E., & Khan, M. A. (2015). Representational momentum reveals visual anticipation differences in the upper and lower visual fields. *Experimental Brain Research, 233*, 2249–2256.

Gould, S. J. (1979). Mickey-mouse meets Konrad Lorenz. *Natural History, 88*(5), 30.

Gould, S. J., & Lewontin, R. C. (1979). The spandrels of San Marco and the panglossian paradigm: A critique of the adaptationist paradigm. *Proceedings of the Royal Society of London, Series B, 205*, 581–598.

Grassi, M. (2010). Sex difference in subjective duration of looming and receding sounds. *Perception, 39*(10), 1424–1426.

Grassi, M., & Darwin, C. J. (2006). The subjective duration of ramped and damped sounds. *Perception & Psychophysics, 68*(8), 1382–1392.

Grau, J. W., & Nelson, D. K. (1988). The distinction between integral and separable dimensions: Evidence for the integrality of pitch and loudness. *Journal of Experimental Psychology: General, 117*, 347–370.

Gray, R. (2013). Being selective at the plate: Processing dependence between perceptual variables relates to hitting goals and performance. *Journal of Experimental Psychology: Human Perception and Performance, 39*(4), 1124–1142.

Gray, R. (2014). Embodied perception in sport. *International Review of Sport and Exercise Psychology, 7*(1), 72–86.

Gray, R., Navia, J. A., & Allsop, J. (2014). Action-specific effects in aviation: What determines judged runway size? *Perception, 43*, 145–154.

Gray, R., & Regan, D. (1998). Accuracy of estimating time to collision using binocular and monocular information. *Vision Research, 38*(4), 499–512.

Gray, R., & Thornton, I. M. (2001). Exploring the link between time to collision and representational momentum. *Perception, 30*, 1007–1022.

Green, C. D. (1995). All the glitters: A review of psychological research on the aesthetics of the golden section. *Perception, 24*, 937–968.

Green, D. M., & Swets, J. A. (1966). *Signal detection theory and psychophysics*. New York: Wiley.

Greene, M. R., & Oliva, A. (2009). The briefest of glances: the time course of natural scene understanding. *Psychological Science, 20*(4), 464–472.

Greenstein, M., Franklin, N., Martins, M., Sewack, C., & Meier, M. A. (2016). When anticipation beats accuracy: Threat alters memory for dynamic scenes. *Memory & Cognition, 44*(4), 633–649.

Gregory, R. L. (1963). Distortion of visual space as inappropriate constancy scaling. *Nature, 199*, 678–680.

Gregory, R. L. (1970). *The intelligent eye*. New York: McGraw Hill.

Gregory, R. L. (1996). *Mirrors in mind*. New York: Freeman Spektrum.

Grieves, R. M., Jenkins, B. W., Harland, B. C., Wood, E. R., & Dudchenko, P. A. (2016). Place field repetition and spatial learning in a multicompartment environment. *Hippocampus, 26*(1), 118–134.

Guariglia, C., & Antonucci, G. (1992). Personal and extrapersonal space: A case of neglect dissociation. *Neuropsychologia, 30*(11), 1001–1009.

Guariglia, C., Padovani, A., Pantano, P., & Pizzamiglio, L. (1993). Unilateral neglect restricted to visual imagery. *Nature, 364*(6434), 235–237.

Gulick, W. L. (1971). *Hearing: Physiology and psychophysics*. New York: Oxford University Press.

Guski, R. (1992). Acoustic tau: An easy analogue to visual tau? *Ecological Psychology, 4*, 189–197.

Haber, R. N. (1985). Toward a theory of the perceived spatial layout of scenes. *Computer Vision, Graphics, and Image Processing, 31*(3), 282–321.

Haberman, J., & Whitney, D. (2007). Rapid extraction of mean emotion and gender from sets of faces. *Current Biology, 17*(17), 751–753.

Haberman, J., & Whitney, D. (2009). Seeing the mean: Ensemble coding for sets of faces. *Journal of Experimental Psychology: Human Perception and Performance, 35*(3), 718–734.

Haberman, J., & Whitney, D. (2011). Efficient summary statistical representation when change localization fails. *Psychonomic Bulletin & Review, 18*(5), 855–859.

Hafting, T., Fyhn, M., Molden, S., Moser, M. B., & Moser, E. I. (2005). Microstructure of a spatial map in the entorhinal cortex. *Nature, 436*(7052), 801–806.

Hagen M.A. (1980) *The perception of pictures, vol. 1. Alberti's window: The projective model of pictorial information*. New York: Academic Press.

Hagenbeek, R. E., & Van Strien, J. W. (2002). Left–right and upper-lower visual field asymmetries for face matching, letter naming, and lexical decision. *Brain and Cognition, 49*, 34–44.

Haggard, P., Clark, S., & Kalogeras, J. (2002). Voluntary action and conscious awareness. *Nature Neuroscience, 5*, 382–385.

Hagler Jr., D. J. (2014). Visual field asymmetries in visual evoked responses. *Journal of Vision, 14*(14), 1–19.

Haig, N. D. (1993). Reflections on inversion and reversion. *Perception, 22*, 863–868.

Hall, D. A., & Moore, D. R. (2003). Auditory neuroscience: The salience of looming sounds. *Current Biology, 13*(3), R91–R93.

Hall, J. A., Coats, E. J., & LeBeau, L. S. (2005). Nonverbal behavior and the vertical dimension of social relations: A meta-analysis. *Psychological Bulletin, 131*(6), 898.

Halligan, P. W., & Marshall, J. C. (1991). Left neglect for near but not far space in man. *Nature, 350*, 498–500.

Halligan, P. W., & Marshall, J. (1994). Toward a principled explanation of unilateral neglect. *Cognitive Neuropsychology, 11*, 167–206.

Halligan, P. W., & Marshall, J. C. (1995). Grounding figural attention in left neglect. *Neurocase, 1*, 79–82.

Halpern, A. R., & Kelly, M. H. (1993). Memory biases in left versus right implied motion. *Journal of Experimental Psychology: Learning, Memory, and Cognition, 19*, 471–484.

Hamburger, K., & Hansen, T. (2010). Analysis of individual variations in the classical horizontal-vertical illusion. *Attention, Perception, & Psychophysics, 72*(4), 1045–1052.

Hamburger, K., Hansen, T., & Gegenfurtner, K. R. (2007). Geometric-optical illusions at isoluminance. *Vision Research, 47*(26), 3276–3285.

Hancock, P. A., & Manser, M. P. (1997). Time-to-contact: More than Tau alone. *Ecological Psychology, 9*(4), 265–297.

Hansen, B. C., & Essock, E. A. (2004). A horizontal bias in human visual processing of orientation and its correspondence to the structural components of natural scenes. *Journal of Vision, 4*, 1044–1060.

Hansen, B. C., & Essock, E. A. (2006). Anisotropic local contrast normalization: The role of stimulus orientation and spatial frequency bandwidths in the oblique and horizontal effect perceptual anisotropies. *Vision Research, 46*, 4398–4415.

Harmon-Jones, E., & Sigelman, J. (2001). State anger and prefrontal brain activity: Evidence that insult-related relative left-prefrontal activation is associated with experienced anger and aggression. *Journal of Personality and Social Psychology, 80*(5), 797–803.

Harrar, V., Tammam, J., Pérez-Bellido, A., Pitt, A., Stein, J., & Spence, C. (2014). Multisensory integration and attention in developmental dyslexia. *Current Biology, 24*, 531–535.

Harré, R. (1986) *Varieties of realism: A rationale for the natural sciences*. Oxford: Blackwell.

Harris, L. R., Duke, P. A., & Kopinska, A. (2006). Flash lag in depth. *Vision Research, 46*, 2735–2742.

Harris, L. R., Jenkin, M., Dyde, R. T., & Jenkin, H. (2011). Enhancing visual cues to orientation: Suggestions for space travelers and the elderly. *Progress in Brain Research, 191*, 133–142.

Harris, J. M., & Morgan, M. J. (1993). Stereo and motion disparities interfere with positional averaging. *Vision Research, 33*(3), 309–312.

Harrison, N. (2012). Auditory motion in depth is preferentially "captured" by visual looming signals. *Seeing and Perceiving, 25*(1), 71–85.

Haselton, M. G., Bryant, G. A., Wilke, A., Frederick, D. A., Galperin, A., Frankenhuis, W. E., & Moore, T. (2009). Adaptive rationality: An evolutionary perspective on cognitive bias. *Social Cognition, 27*(5), 733–763.

Haselton, M. G., & Buss, D. M. (2000). Error management theory: A new perspective on biases in cross-sex mind reading. *Journal of Personality and Social Psychology, 78*(1), 81–91.

Haselton, M. G., & Nettle, D. (2006). The paranoid optimist: An integrative evolutionary model of cognitive biases. *Personality and Social Psychology Review, 10*(1), 47–66.

Hassabis, D., Kumaran, D., Vann, S. D., & Maguire, E. A. (2007). Patients with hippocampal amnesia cannot imagine new experiences. *Proceedings of the National Academy of Sciences, 104*, 1726–1731.

Hatin, B., Tottenham, L. S., & Oriet, C. (2012). The relationship between collisions and pseudoneglect: Is it right? *Cortex, 48*(8), 997–1008.

Haun, D. B. M., Allen, G. L., & Wedell, D. H. (2005). Bias in spatial memory: A categorical endorsement. *Acta Psychologica, 118*(1–2), 149–170.

Hayes, A. E., & Freyd, J. J. (2002). Representational momentum when attention is divided. *Visual Cognition, 9*, 8–27.

Hayes, A. E., Sacher, G., Thornton, I. M., Sereno, M. E., & Freyd, J. J. (1996). Representational momentum in depth using stereopsis [ARVO Abstract 2120]. *Investigative Ophthalmology & Visual Science*, 37(Suppl. 3), S467.

Haynes, J. D. (2015). An information-based approach to consciousness: Mental state decoding. In *Open MIND*. Frankfurt am Main: MIND Group.

Hayward, W. G., & Tarr, M. J. (1995). Spatial language and spatial representation. *Cognition, 55*(1), 39–84.

He, B. J., Snyder, A. Z., Vincent, J. L., Epstein, A., Shulman, G. L., & Corbetta, M. (2007). Breakdown of functional connectivity in frontoparietal networks underlies behavioral deficits in spatial neglect. *Neuron, 53*(6), 905–918.

He, S., Cavanagh, P. & Intriligator, J. (1996). Attentional resolution and the locus of awareness. *Nature, 383*, 334–338.

Hebb, D. O. (1949). *The organization of behavior: A neuropsychological theory.* New York: Wiley.

Hecht, H. (2001). Regularities of the physical world and the absence of their internalization. *Behavioral and Brain Sciences, 24*, 608–617.

Hecht, H., Bertamini, M., & Gamer, M. (2005). Naive optics: Acting on mirror reflections. *Journal of Experimental Psychology: Human Perception and Performance, 31*(5), 1023–1038.

Hedge, A., & Marsh, N. W. A. (1975). The effect of irrelevant spatial correspondences on two-choice response time. *Acta Psychologica, 39*, 427–439.

Heeger, D. J., & Ress, D. (2002). What does fMRI tell us about neuronal activity? *Nature Reviews Neuroscience, 3*(2), 142–151.

Heeley, D. W., Buchanan-Smith, H. M., Cromwell, J. A., & Wright, J. S. (1997). The oblique effect in orientation acuity. *Vision Research, 31*(2), 235–242.

Hegarty, P., & Buechel, C. (2006). Androcentric reporting of gender differences in APA journals: 1965–2004. *Review of General Psychology, 10*(4), 377–389.

Hegarty, P., Lemieux, A. F., & McQueen, G. (2010). Graphing the order of the sexes: Constructing, recalling, interpreting, and putting the self in gender difference graphs. *Journal of Personality and Social Psychology, 98*(3), 375.

Heinemann, W., Pellander, F., Vogelbusch, A., & Wojtek, B. (1981). Meeting a deviant person: Subjective norms and affective reactions. *European Journal of Social Psychology, 11*(1), 1–25.

Held, R., Ostrovsky, Y., de Gelder, B., Gandhi, T., Ganesh, S., Mathur, U., & Sinha, P. (2011). The newly sighted fail to match seen with felt. *Nature Neuroscience, 14*, 551–553.

Helmholtz, H. von (1962). *Physiological optics*, J. P. C. Southall, Trans. New York: Dover. (Original work published 1866).

Helmholtz, H. von (1925/2000). Physiological optics. (Vol. III Concerning the perceptions in general). In S. Yanstis (Ed.), *Visual perception: Essential readings* (pp. 24–44). Philadelphia: Psychology Press.

Helson H. (1964). *Adaptation-level theory.* New York: Harper & Row.

Henderson, J. M., & Hollingworth, A. (1999). High level scene perception. *Annual Review of Psychology, 50*, 243–271.

Henderson, M. D., Fujita, K., Trope, Y., & Liberman, N. (2006). Transcending the "here": The effect of spatial distance on social judgment. *Journal of Personality and Social Psychology, 91*(5), 845.

Henderson, M. D., Wakslak, C. J., Fujita, K., & Rohrbach, J. (2011). Construal level theory and spatial distance. *Social Psychology, 42*(3), 165–173.

Henkel, L. A., Franklin, N., & Johnson, M. K. (2000). Cross-modal source monitoring confusions between perceived and imagined events. *Journal of Experimental Psychology: Learning, Memory, and Cognition, 26*, 321.

Henrich, J., & Gil-White, F. J. (2001). The evolution of prestige: Freely conferred deference as a mechanism for enhancing the benefits of cultural transmission. *Evolution and Human Behavior, 22*, 165–196.

Henrich, J., Heine, S. J., & Norenzayan, A. (2010). The weirdest people in the world? *Behavioral and Brain Sciences, 33*, 61–135.

Hensley, W. E., & Cooper, R. (1987). Height and occupational success: A review and critique. *Psychological Reports, 60*(3, Pt 1), 843–849.

Heron, J., Whitaker, D., & McGraw, P. V. (2004). Sensory uncertainty governs the extent of audio-visual interaction. *Vision Research, 44*, 2875–2884.

Herrera, A., Macizo, P., & Semenza, C. (2008). The role of working memory in the association between number magnitude and space. *Acta Psychologica, 128*(2), 225–237.

Hershenson, M. (1999). *Visual space perception.* Cambridge, MA: MIT Press.

Herz-Fischler, R. (1998). *A mathematical history of the golden number.* Mineola, NY: Dover.

Hespos, S. J., & Spelke, E. S. (2004). Conceptual precursors to language. *Nature, 430*(6998), 453–456.

Hevner, K. (1935). Experimental studies of the affective value of colors and lines. *Journal of Applied Psychology, 19*, 385–398.

Heyes, C. M. (2011). Automatic imitation. *Psychological Bulletin, 137*(3), 463–483.

Hidaka, S., Higuchi, S., Teramoto, W., and Sugita, Y. (2017). Neural mechanisms underlying sound-induced visual motion perception: An fMRI study. *Acta Psychologica, 178*, 66–72.

Hidaka, S., Kawachi, Y., & Gyoba, J. (2009). The representation of moving 3-D objects in apparent motion perception. *Attention, Perception, & Performance, 71*, 1294–1304.

Hidaka, S., Manaka, Y., Teramoto, W., Sugita, Y., Miyauchi, R., Gyoba, J., Suzuki, Y., & Iwaya, Y. (2009). Alternation of sound location induces visual motion perception of a static object. *PLoS ONE, 4*, e8188.

Hidaka, S., Teramoto, W., Keetels, M., & Vroomen, J. (2013). Effect of pitch-space correspondence on sound-induced visual motion perception. *Experimental Brain Research, 231*, 117–126.

Hidaka, S., Teramoto, W., Kobayashi, M., & Sugita, Y. (2011). Sound-contingent visual motion aftereffect. *BMC Neuroscience, 12*, 1–6.

Hidaka, S., Teramoto, W., & Sugita, Y. (2015). Spatiotemporal processing in crossmodal interactions for perception of the external world: A review. *Frontiers in Integrative Neuroscience, 9*, 62.

Hidaka, S., Teramoto, W., Sugita, Y., Manaka, Y., Sakamoto, S., & Suzuki, Y. (2011). Auditory motion information drives visual motion perception. *PLoS ONE*, 6, e17499.

Higashiyama, A. (1996). Horizontal and vertical distance perception: The discorded-orientation theory. *Attention, Perception, & Psychophysics, 58*, 259–270.

Higashiyama, A., & Ueyama, E. (1988). The perception of vertical and horizontal distances in outdoor settings. *Attention, Perception, & Psychophysics, 44*(2), 151–156.

Higham, P. A., & Carment, D. W. (1992). The rise and fall of politicians: The judged heights of Broadbent, Mulroney and Turner before and after the 1988 Canadian federal election. *Canadian Journal of Behavioural Science, 24*(3), 404–409.

Hilchey, M. D., & Christie, J. (2016). The input form of "inhibition of return" requires the presence of objects. revision in preparation.

Hilchey, M. D., Dohmen, D., Crowder, N. A., & Klein, R. M. (2016). When is inhibition of return input- or output-based? It depends on how you look at it. *Canadian Journal of Experimental Psycholology, 70*(4), 325–334.

Hilchey, M. D., Hashish, M., MacLean, G. H., Ivanoff, J., Satel, J., & Klein, R. M. (2014). On the role of eye movement monitoring and discouragement on inhibition of return in a go no-go task. *Vision Research, 96*, 133–139.

Hilchey, M. D., Ivanoff, J., Taylor, T. L., & Klein, R. M. (2011) Visualizing the temporal dynamics of spatial information processing responsible for the Simon effect and its amplification by inhibition of return, *Acta Psychologica, 136*, 235–244.

Hilchey, M. D., Klein, R. M., & Ivanoff, J. (2012). Perceptual and motor IOR: Components or flavors? *Attention, Perception & Psychophysics, 74*, 1416–1429.

Hilchey, M. D., Klein, R. M., & Satel, J. (2014). Returning to "inhibition of return" by dissociating long-term oculomotor IOR from short-term sensory adaptation and other nonoculomotor "inhibitory" cueing effects. *Journal of Experimental Psychology: Human Perception and Performance, 40*(4), 1606–1616.

Hillebrand, F. (1902). Theorie der scheinbaren grösse bei binocularem Sehen (Theory of apparent size in binocular vision). *Denkschriften der Wiener Akademie, Mathematisch Naturwissenschaft Klasse, 72*, 255–307.

Hillier, B. (1996). *Space is the machine.* Cambridge: Cambridge University Press.

Hillier, B., & Hanson, J. (1984). *The social logic of space.* Cambridge: Cambridge University Press.

Hills, T., Brockie, P. J., & Maricq, A. V. (2004). Dopamine and glutamate control area-restricted search behavior in Caenorhabditis elegans. *Journal of Neuroscience, 24*, 1217–1225.

Hills, T. T., Kalff, C., & Wiener, J. M. (2013). Adaptive Lévy processes and area-restricted search in human foraging. *PLoS ONE, 8*(4), e60488.

Hine, T. J., White, A. M. V., & Chappell, M. (2003). Is there an auditory-visual flash-lag effect? *Clinical and Experimental Ophthalmology, 31*, 254–257.

Hinton, G. E., McClelland, J. L., & Rummelhart, D. E. (1986). Distributed representations. In J. L. McClelland & D. E. Rummelhart (Eds.), *Parallel distributed processing.* Cambridge: MIT Press.

Hirtle, S. C., & Jonides, J. (1985). Evidence of hierarchies in cognitive maps. *Memory & Cognition, 13*(3), 208–217.

Hochberg, J. (1986). Representation of motion and space in video and cinematic displays. In K. J. Boff, L. Kaufman, & J. P. Thomas (Eds.), *Handbook of perception and human performance* (Vol. 1, pp. 22.1–22.64). New York: Wiley.

Hochmair, H. H., & Frank, A. U. (2002). Influence of estimation errors on wayfinding-decisions in unknown street networks – analyzing the least-angle strategy. *Spatial Cognition and Computation, 2*, 283–313.

Hochmair, H. H., & Karlsson, V. (2005). Investigation of preference between the least-angle strategy and the initial segment strategy for route selection in unknown environments. In C. Freksa, M. Knauff, B. Krieg-Brückner, B. Nebel, & T. Barkowsky (Eds.), *Spatial cognition IV. Reasoning, action, interaction* (pp. 79–97). Berlin: Springer.

Hoeckner, S. H., Moeller, K., Zauner, H., Wood, G., Haider, C., Gaßner, A., & Nuerk, H.-C. (2008). Impairments of the mental number line for two-digit numbers in neglect. *Cortex, 44*, 429–438.

Hofbauer, M., Wuerger, S. M., Meyer, G. F., Roehrbein, F., Schill, K., & Zetzsche, C. (2004). Catching audiovisual mice: Predicting the arrival time of auditory-visual motion signals. *Cognitive Affective & Behavioral Neuroscience, 4*(2), 241–250.

Höge, H. (1995). Fechner's experimental aesthetics and the golden section hypothesis today. *Empirical Studies of the Arts*, 13, 131–148.

Holden, M. P., Curby, K. M., Newcombe, N. S., & Shipley, T. F. (2010). A category adjustment approach to memory for spatial location in natural scenes. *Journal of*

Experimental Psychology: Learning, Memory, and Cognition, *36*(3), 590–604.

Holden, M. P., Duff-Canning, S. J., & Hampson, E. (2015). Sex differences in the weighting of metric and categorical information in spatial location memory. *Psychological Research*, *79*(1), 1–18.

Holden, M. P., & Hampson, E. (2014). Categorical bias in line angle judgments: Sex differences and the use of multiple categories. *Spatial Cognition and Computation*, *14*(3), 199–219.

Holden, M. P., & Newcombe, N. S. (2013). The development of location coding: An adaptive combination account. In D. Waller & L. Nadel (Eds.), *Handbook of spatial cognition* (pp. 191–209). http://dx.doi.org/10.1037/13936-011

Holden, M. P., Newcombe, N. S., Resnick, I., & Shipley, T. F. (2015). Seeing like a geologist: Bayesian use of expert categories in location memory. *Cognitive Science*. Advance online publication. http://dx.doi.org/10.1111/cogs.12229

Holden, M. P., Newcombe, N. S., & Shipley, T. F. (2013). Location memory in the real world: Category adjustment effects in 3-dimensional space. *Cognition*, *128*(1), 45–55.

Holden, M. P., Newcombe, N. S., & Shipley, T. F. (2015). Categorical biases in spatial memory: The role of certainty. *Journal of Experimental Psychology: Learning, Memory, and Cognition*, *41*(2), 473–481.

Holmes, K. J., & Lourenco, S. F. (2012). Orienting numbers in mental space: Horizontal organization trumps vertical. *Quarterly Journal of Experimental Psychology*, *65*(6), 1044–1051.

Holmes, K. J., & Wolff, P. (2012). Does categorical perception in the left hemisphere depend on language? *Journal of Experimental Psychology: General*, *141*(3), 439–443.

Holmes, K. J., & Wolff, P. (2013). Spatial language and the psychological reality of schematization. *Cognitive Processing*, *14*(2), 205–208.

Holmes, M. C., & Sholl, M. J. (2005). Allocentric coding of object-to-object relations in overlearned and novel environments. *Journal of Experimental Psychology: Learning, Memory, and Cognition*, 31, 1069–1087.

Hölscher, C., Büchner, S. J., Meilinger, T., & Strube, G. (2009). Adaptivity of wayfinding strategies in a multi-building ensemble: The effects of spatial structure, task requirements, and metric information. *Journal of Environmental Psychology*, *29*(2), 208–219.

Hölscher, C., Meilinger, T., Vrachliotis, G., Brösamle, M., & Knauff, M. (2006). Up the down staircase: Wayfinding strategies and multi-level buildings. *Journal of Environmental Psychology*, 26(4), 284–299

Hölscher, C., Tenbrink, T., & Wiener, J. M. (2011). Would you follow your own route description? Cognitive strategies in urban route planning. *Cognition*, *121*(2), 228–247.

Holt, E. B., Marvin, W. T., Montague, W. P., Perry, R. B., Pitkin, W. B., & Spaulding, E. G. (1910). The program and first platform of six realists. *The Journal of Philosophy, Psychology and Scientific Methods*, *7*, 393–401.

Holten, V., Donker, S. F., Stuit, S. M., Verstraten, F. A. J., & van der Smagt, M. J. (2015). Visual directional anisotropy does not mirror the directional anisotropy apparent in postural sway. *Perception*, *44*(5), 477–489.

Holyoak, K. J., & Mah, W. A. (1982). Cognitive reference points in judgments of symbolic magnitude. *Cognitive Psychology*, *14*(3), 328–352.

Hommel, B. (2011a). The Simon effect as a tool and heuristic. *Acta Psychologica*, *136*, 188–201.

Hommel, B. (2011b). Attention and spatial stimulus coding in the Simon task: A rejoinder to Van der Lubbe and Abrahamse. *Acta Psychologica*, *136*, 265–268.

Hommel, B., Müsseler, J., Aschersleben, G., & Prinz, W. (2001). The theory of event coding (TEC): A framework for perception and action planning. *Behavioral and Brain Sciences*, *24*, 849–937.

Hommel, B., Proctor, R. W., & Vu, K.-P. L. (2004). A feature-integration account of sequential effects in the Simon task. *Psychological Research*, *68*, 1–17.

Hommuk, K., Bachmann, T., & Oja, A. (2008). Precuing an isolated stimulus temporarily outweighs in-stream stimulus facilitation. *Journal of General Psychology, 135*, 167–181.

Hopfinger, J. B., Buonocore, M. H., & Mangun, G. R. (2000). The neural mechanisms of top-down attentional control. *Nature Neuroscience, 3*, 284–291.

Hotopf, W. H., & Hibberd, M. C. (1989). The role of angles in inducing misalignment in the Poggendorff figure. *Quarterly Journal of Experimental Psychology, 41A*(2), 355–383.

Hotopf, W. H., & Ollerearnshaw, C. (1972). The regression to right angles tendency and the Poggendorff illusion. I. *British Journal of Psychology 63*(3), 359–367.

Hotopf, W. H., Ollerearnshaw, C., & Brown, S. (1974). The regression to right angles tendency and the Poggendorff illusion. 3. *British Journal of Psychology, 65*(2), 213–231.

Howard, I. P. (1982). *Human visual orientation*. New York: John Wiley & Sons.

Howard, I. P., & Rogers, B. J. (2002). *Seeing in depth* (Vol. 2). Toronto: Toronto University Press.

Howard, I. P., & Templeton, W. B. (1966). *Human spatial orientation*. New York: Wiley.

Howe, C. Q., & Purves, D. (2002). Range image statistics can explain the anomalous perception of length. *Proceedings of the National Academy of Sciences of the United States of America, 99*(20), 13184–13188.

Hubbard, E. M., Piazza, M., Pinel, P., & Dehaene, S. (2005). Interactions between number and space in parietal cortex. *Nature Reviews Neuroscience, 6*(6), 435–448.

Hubbard, T. L. (1990). Cognitive representation of linear motion: Possible direction and gravity effects in judged displacement. *Memory & Cognition, 18*, 299–309.

Hubbard, T. L. (1993). The effect of context on visual representational momentum. *Memory & Cognition, 21*, 103–114.

Hubbard, T. L. (1994). Judged displacement: A modular process? *American Journal of Psychology, 107*, 359–373.

Hubbard, T. L. (1995a). Auditory representational momentum: Surface form, direction, and velocity effects. *American Journal of Psychology, 108*, 255–274.

Hubbard, T. L. (1995b). Cognitive representation of motion: Evidence for friction and gravity analogues. *Journal of Experimental Psychology: Learning, Memory, and Cognition, 21*, 241–254.

Hubbard, T. L. (1995c). Environmental invariants in the representation of motion: Implied dynamics and representational momentum, gravity, friction, and centripetal force. *Psychonomic Bulletin & Review, 2*, 322–338.

Hubbard, T. L. (1996a). Displacement in depth: Representational momentum and boundary extension. *Psychological Research, 59*, 33–47.

Hubbard, T. L. (1996b). Representational momentum, centripetal force, and curvilinear impetus. *Journal of Experimental Psychology: Learning, Memory, and Cognition, 22*, 1049–1060.

Hubbard, T. L. (1997). Target size and displacement along the axis of implied gravitational attraction: Effects of implied weight and evidence of representational gravity. *Journal of Experimental Psychology: Learning, Memory, and Cognition, 23*, 1484–1493.

Hubbard, T. L. (1998). Some effects of representational friction, target size, and memory averaging on memory for vertically moving targets. *Canadian Journal of Experimental Psychology, 52*, 44–49.

Hubbard, T. L. (1999). How consequences of physical principles influence mental representation: The environmental invariants hypothesis. In P. R. Killeen & W. R. Uttal (Eds.), *Fechner Day 99: The end of 20th century psychophysics. Proceedings of the 15th Annual Meeting of the International Society for Psychophysics* (pp. 274–279). Tempe, AZ: The International Society for Psychophysics.

Hubbard, T. L. (2001). The effect of height in the picture plane on the forward displacement of ascending and descending targets. *Canadian Journal of Experimental Psychology, 55*, 325–329.

Hubbard, T. L. (2004). The perception of causality: Insights from Michotte's launching effect, naive impetus theory, and representational momentum. In A. M. Oliveira, M. P. Teixeira, G. F. Borges, & M. J. Ferro (Eds.), *Fechner Day 2004* (pp. 116–121). Coimbra, Portugal: The International Society for Psychophysics.

Hubbard, T. L. (2005a). An effect of target orientation on representational momentum. *Paidéia, 15*, 207–216.

Hubbard, T. L. (2005b). Representational momentum and related displacements in spatial memory: A review of the findings. *Psychonomic Bulletin & Review, 12*, 822–851.

Hubbard, T. L. (2006a). Bridging the gap: Possible roles and contributions of representational momentum. *Psicológica (Valencia), 27*, 1–34.

Hubbard, T. L. (2006b). Computational theory and cognition in representational momentum and related types of displacement: A reply to Kerzel. *Psychonomic Bulletin & Review, 13*, 174–177.

Hubbard, T. (2007). What is mental representation? And how does it relate to consciousness? *Journal of Consciousness Studies, 14*(1–1), 37–61.

Hubbard, T. L. (2008). Representational momentum contributes to motion induced mislocalization of stationary objects. *Visual Cognition, 16*, 44–67.

Hubbard, T. L. (2010). Approaches to representational momentum: Theories and models. In R. Nijhawan & B. Khurana (Eds.), *Space and time in perception and action* (pp. 338–365). Cambridge: Cambridge University Press.

Hubbard, T. L. (2011a). Boundary extension as a new Gestalt principle. In D. Algom, D. Zakay, E. Chajut, S. Shaki, Y. Mama, & V. Shakuf (Eds.), *Fechner Day 2011: Proceedings of the 27th Annual Meeting of the International Society for Psychophysics* (pp. 65–70). Raanana, Israel: International Society for Psychophysics.

Hubbard, T. L. (2011b). Extending pragnanz: Dynamic aspects of mental representation and Gestalt principles. In L. Albertazzi, G. J. van Tonder, & D. Vishwanath (Eds.), *Perception beyond inference: The information content of visual processes* (pp. 75–108). Cambridge, MA: MIT Press.

Hubbard, T. L. (2012). Visual perception of force: Comment on White (2012). *Psychological Bulletin, 138*, 616–623.

Hubbard, T. L. (2013a). Auditory imagery contains more than audition. In S. Lacey & R. Lawson (Eds.), *Multisensory imagery: Theories and applications* (pp. 221–247). New York: Springer.

Hubbard, T. L. (2013b). Do the flash-lag effect and representational momentum involve similar extrapolations? *Frontiers in Psychology, 4*, 290.

Hubbard, T. L. (2013c). Launching, entraining, and representational momentum: Evidence consistent with an impetus heuristic in perception of causality. *Axiomathes, 23*, 633–643.

Hubbard, T. L. (2013d). Phenomenal causality I: Varieties and variables. *Axiomathes, 23*, 1–42.

Hubbard, T. L. (2013e). Phenomenal causality II: Integration and implication. *Axiomathes, 23*, 485–524.

Hubbard, T. L. (2014a). Forms of momentum across space: Representational, operational, and attentional. *Psychonomic Bulletin & Review, 21*, 1371–1403.

Hubbard, T. L. (2014b). The flash-lag effect and related mislocalizations: Findings, properties, and theories. *Psychological Bulletin, 140*, 308–338.

Hubbard, T. L. (2015a). Forms of momentum across time: Behavioral and psychological. *Journal of Mind and Behavior, 36*, 47–82.

Hubbard, T. L. (2015b). The varieties of momentum-like experience. *Psychological Bulletin, 141*, 1081–1119.

Hubbard, T. L. (2017). Toward a general theory of momentum-like effects. *Behavioural Processes, 141*, 50–66.

Hubbard, T. L., & Baird, J. C. (1993). The effects of size, clutter, and complexity on the vanishing point distance in visual imagery. *Psychological Research/Psychologische Forschung, 55*, 223–236.

Hubbard, T. L., & Bharucha, J. J. (1988). Judged displacement in apparent vertical and horizontal motion. *Perception & Psychophysics, 44*, 211–221.

Hubbard, T. L., & Blessum, J. A. (2001). A structural dynamic of form: Displacements in memory for the size of an angle. *Visual Cognition, 8*, 725–749.

Hubbard, T. L., Blessum, J. A., & Ruppel, S. E. (2001). Representational momentum and Michotte's (1946/1963) "launching effect" paradigm. *Journal of Experimental Psychology: Learning, Memory, and Cognition, 27*, 294–301.

Hubbard, T. L., & Courtney, J. R. (2010). Cross-modal influences on representational momentum and representational gravity. *Perception, 39*, 851–862.

Hubbard, T. L., & Favretto, A. (2003). Naïve impetus and Michotte's "tool effect": Evidence from representational momentum. *Psychological Research, 67*, 134–152.

Hubbard, T. L., Hutchison, J. L., & Courtney, J. R. (2010). Boundary extension: Findings and theories. *The Quarterly Journal of Experimental Psychology, 63*, 1467–1494.

Hubbard, T. L., Kumar, A. M., & Carp, C. L. (2009). Effects of spatial cueing on representational momentum. *Journal of Experimental Psychology: Learning, Memory, and Cognition, 35*, 666–677.

Hubbard, T. L., & Lange, M. (2010). Prior probabilities and representational momentum. *Visual Cognition, 18*, 1063–1087.

Hubbard, T. L., Matzenbacher, D. L., & Davis, S. E. (1999). Representational momentum in children: Dynamic information and analogue representation. *Perceptual and Motor Skills, 88*, 910–916.

Hubbard, T. L., & Motes, M. A. (2002). Does representational momentum reflect a distortion of the length or the endpoint of a trajectory? *Cognition, 82*, B89–B99.

Hubbard, T. L., & Motes, M. A. (2005). An effect of context on whether memory for initial position exhibits a Fröhlich effect or an onset repulsion effect. *Quarterly Journal of Experimental Psychology, 58*, 961–979.

Hubbard, T. L., & Ruppel, S. E. (1999). Representational momentum and the landmark attraction effect. *Canadian Journal of Experimental Psychology, 53*, 242–256.

Hubbard, T. L., & Ruppel, S. E. (2000). Spatial memory averaging, the landmark attraction effect, and representational gravity. *Psychological Research, 64*, 41–55.

Hubbard, T. L., & Ruppel, S. E. (2002). A possible role of naive impetus in Michotte's "Launching Effect:" Evidence from representational momentum. *Visual Cognition, 9*, 153–176.

Hubbard, T. L., & Ruppel, S. E. (2011). Effects of spatial cuing on the onset repulsion effect. *Attention, Perception, & Psychophysics, 73*(7), 2236–2248.

Hubbard, T. L., & Ruppel, S. E. (2013). A Fröhlich effect and representational gravity in memory for auditory pitch. *Journal of Experimental Psychology: Human Perception and Performance, 39*, 1153–1164.

Hubbard, T. L., & Ruppel, S. E. (2014). An effect of contrast and luminance on visual representational momentum for location. *Perception, 43*, 754–766.

Hubbard, T. L., & Ruppel, S. E. (2017a). The (dynamic) mind in the cave: Representational space of cave paintings and petroglyphs. *Empirical Studies of the Arts, 35*, 67–92.

Hubbard, T. L., & Ruppel, S. E. (2017b). Perceived causality, force, and resistance in the absence of launching. *Psychonomic Bulletin & Review*, 24, 591, 596.

Hubbard, T. L., Ruppel, S. E., & Courtney, J. R. (2005). The force of appearance: Gamma movement, naive impetus, and representational momentum. *Psicológica (Valencia), 26*, 209–228.

Hubel, D. H., & Wiesel, T. N. (1959). Receptive fields of single neurons in the cat's striate cortex. *Journal of Physiology, 148*, 574–591.

Hubel, D. H., & Wiesel, T. N. (1962). Receptive fields, binocular interaction and functional architecture in the cat's visual cortex. *Journal of Physiology, 160*, 106–154.

Huber, S., Klein, E., Moeller, K., & Willmes, K. (2016). Spatial-numerical and ordinal positional associations coexist in parallel. *Frontiers in Psychology, 7:438*

Huberle, E., & Karnath, H.-O. (2010). Saliency modulates global perception in simultanagnosia. *Experimental Brain Research, 204*(4), 595–603.

Hudson, M., & Jellema, T. (2011). Resolving ambiguous behavioral intentions by means of involuntary prioritization of gaze processing. *Emotion, 11*, 681–686.

Hudson, M., Liu, C. H., & Jellema, T. (2009). Anticipating intentional actions: The effect of eye gaze direction on the judgment of head rotation. *Cognition, 112*, 423–434.

Hudson, M., Nicholson, T., Ellis, R., & Bach, P. (2016). I see what you say: Prior knowledge of other's goals automatically biases the perception of their actions. *Cognition, 146*, 245–250.

Hudson, M., Nicholson, T., Simpson, W. A., Ellis, R., & Bach, P. (2016). One step ahead: The perceived kinematics of others' actions are biased toward expected goals. *Journal of Experimental Psychology: General, 145*, 1–7.

Hume, D., & Selby-Bigge, L. A. (1967). *A treatise of human nature: Reprinted from the original edition in three volumes and edited, with an analytical index, by LA Selby-Bigge*. Oxford: Clarendon Press.

Hund, A. M., & Plumert, J. M. (2002). Delay-induced bias in children's memory for location. *Child Development, 73*(3), 829–840.

Hund, A. M., & Plumert, J. M. (2003). Does information about what things are influence children's memory for where things are? *Developmental Psychology, 39*(6), 939–948.

Hund, A. M., & Plumert, J. M. (2005). The stability and flexibility of spatial categories. *Cognitive Psychology, 50*(1), 1–44.

Hund, A. M., Plumert, J. M., & Benney, C. J. (2002). Experiencing nearby locations together in time: The role of spatiotemporal contiguity in children's memory for location. *Journal of Experimental Child Psychology, 82*(3), 200–225.

Huron, D. (2006). *Sweet anticipation: Music and the psychology of anticipation*. Cambridge, MA: MIT Press.

Huron, D., & Shanahan, D. (2013). Eyebrow movements and vocal pitch height: Evidence consistent with an ethological signal. *Journal of the Acoustical Society of America, 133*, 2947–2952.

Husain, M., Mannan, S., Hodgson, T., Wojciulik, E., Driver, J., & Kennard, C. (2001). Impaired spatial working memory across saccades contributes to abnormal search in parietal neglect. *Brain, 124*, 941–952.

Husain, M., & Nachev, P. (2007). Space and the parietal cortex. *Trends in Cognitive Sciences, 11*(1), 30–36.

Husain, M., & Rorden, C. (2003). Non-spatially lateralized mechanisms in hemispatial neglect. *Nature Reviews Neuroscience, 4*, 26–36.

Hutchins, E. (1995). *Cognition in the wild*. Cambridge, MA: MIT Press.

Hutchison, J. L., Hubbard, T. L., Ferrandino, B., Brigante, R., Wright, J. M., & Rypma, B. (2012). Auditory memory distortion for spoken prose. *Journal of Experimental Psychology: Learning, Memory, and Cognition, 38*(6), 1469–1489.

Huttenlocher, J., Hedges, L. V., & Bradburn, N. M. (1990). Reports of elapsed time: Bounding and rounding processes in estimation. *Journal of Experimental Psychology: Learning, Memory, and Cognition, 16*(2), 196–213.

Huttenlocher, J., Hedges, L. V., Corrigan, B., & Crawford, L. E. (2004). Spatial categories and the estimation of location. *Cognition, 93*(2), 75–97.

Huttenlocher, J., Hedges, L. V., & Duncan, S. (1991). Categories and particulars: Prototype effects in estimating spatial location. *Psychological Review, 98*(3), 352–376.

Huttenlocher, J., Hedges, L. V., Lourenco, S. F., Crawford, L. E., & Corrigan, B. (2007). Estimating stimuli from contrasting categories: Truncation due to boundaries. *Journal of Experimental Psychology: General, 136*(3), 502–519.

Huttenlocher, J., Hedges, L., & Prohaska, V. (1988). Hierarchical organization in ordered domains: Estimating the dates of events. *Psychological Review, 95*(4), 471–484.

Huttenlocher, J., Hedges, L. V., & Prohaska, V. (1992). Memory for day of the week: A 5+2 day cycle. *Journal of Experimental Psychology: General, 121*(3), 313–325.

Huttenlocher, J., Hedges, L. V., & Vevea, J. L. (2000). Why do categories affect stimulus judgment? *Journal of Experimental Psychology: General, 129*(2), 220–241.

Huttenlocher, J., Newcombe, N., & Sandberg, E. H. (1994). The coding of spatial location in young children. *Cognitive Psychology, 27*(2), 115–147.

Iacoboni, M., Woods, R. P., Brass, M., Bekkering, H., Mazziotta, J. C., & Rizzolatti, G. (1999). Cortical mechanisms of human imitation. *Science, 286*, 2526–2528.

Ichikawa, M., & Masakura, Y. (2006). Manual control of the visual stimulus reduces the flash-lag effect. *Vision Research, 46*, 2192–2203.

Ichikawa, M., & Masakura, Y. (2010). Reduction of the flash-lag effect in terms of active observation. *Attention, Perception, & Psychophysics, 72*, 1032–1044.

Ichikawa, M., & Masakura, Y. (2013). Effects of consciousness and consistency in manual control of visual stimulus on reduction of the flash-lag effect for luminance change. *Frontiers in Psychology, 4*, 120.

Ignaschenkova, A., Dicke, P. W., Haarmeier, T., & Thier, P. (2004). Neuron-specific contribution of the superior colliculus to overt and covert shifts of attention. *Nature Neuroscience, 7*, 56–64.

Imbo, I., De Brauwer, J., Fias, W., & Gevers, W. (2012). The development of the SNARC effect: Evidence for early verbal coding. *Journal of Experimental Child Psychology, 111*(4), 671–680.

Immordino-Yang, M. H., Chiao, J. Y., & Fiske, A. P. (2010). Neural re-use in the social and emotional brain. *Behavioral and Brain Sciences, 33*(4), 275–276.

Indovina, I., Maffei. V., Bosco, G., Zago, M., Macaluso, E., & Lacquaniti, F. (2005). Representation of visual gravitational motion in the human vestibular cortex. *Science, 308*(5720), 416–419.

Indow, T. (2004). *The global structure of visual space.* River Edge, NJ: World Scientific.

Indow, T., & Watanabe, T. (1988). Alleys on an extensive frontoparallel plane: A second experiment. *Perception, 17*, 647–666.

Intraub, H. (1997). The representation of visual scenes. *Trends in Cognitive Sciences, 1*, 217–222.

Intraub, H. (2002). Anticipatory spatial representation of natural scenes: Momentum without movement? *Visual Cognition, 9*, 93–119.

Intraub, H. (2004). Anticipatory spatial representation of 3D regions explored by sighted observers and a deaf-and-blind-observer. *Cognition*, 94(1), 19–37.

Intraub, H. (2010). Rethinking scene perception: A multisource model. In B. Ross (Ed.), *The Psychology of learning and motivation: Advances in research and theory* (Vol. 52, pp. 231–264). San Diego, CA: Elsevier Academic Press.

Intraub, H. (2012). Rethinking visual scene perception. *Wiley Interdisciplinary Reviews: Cognitive Science, 3*(1), 117–127.

Intraub, H. (2014). Visual scene representation: A spatial-cognitive perspective. In K. Kveraga & M. Bar (Eds.), *Scene vision: Making sense of what we see* (pp. 5–26). Cambridge, MA: MIT Press.

Intraub, H., Bender, R. S., & Mangels, J. A. (1992). Looking at pictures but remembering scenes. *Journal of Experimental Psychology: Learning, Memory and Cognition, 18*(1), 180–191.

Intraub, H., & Bodamer, J. L. (1993). Boundary extension: Fundamental aspect of pictorial representation or encoding artifact? *Journal of Experimental Psychology: Learning, Memory, and Cognition, 19*, 1387–1397.

Intraub, H., Daniels, K. K., Horowitz, T. S., & Wolfe, J. M. (2008). Looking at scenes while searching for numbers: Dividing attention multiplies space. *Perception & Psychophysics, 70*(7), 1337–1349.

Intraub, H., & Dickinson, C.A. (2008). False memory 1/20 of a second later: What the early onset of boundary extension reveals about perception. *Psychological Science, 19*, 1007–1013.

Intraub, H., Gottesman, C. V., & Bills, A. J. (1998). Effects of perceiving and imaging scenes on memory for pictures. *Journal of Experimental Psychology: Learning, Memory, and Cognition, 24*, 186–201.

Intraub, H., Gottesman, C. V., Willey, E. V., & Zuk, I. J. (1996). Boundary extension for briefly glimpsed photographs: Do common perceptual processes result in unexpected memory distortions? *Journal of Memory and Language, 35*(2), 118–134.

Intraub, H., & Hoffman, J. E. (1992). Reading and visual memory: remembering scenes that were never seen. *American Journal of Psychology, 105*(1), 101–114.

Intraub, H., Hoffman, J. E., Wetherhold, C. J., & Stoehs, S. (2006). More than meets the eye: The effect of planned fixations on scene representation. *Perception & Psychophysics, 68*, 759–769.

Intraub, H., Morelli, F., & Gagnier, K. M. (2015). Visual, haptic and bimodal scene perception: Evidence for a unitary representation. *Cognition, 138*, 132–147.

Intraub, H., & Richardson, M. (1989). Wide-angle memories of close-up scenes. *Journal of Experimental Psychology: Learning, Memory, and Cognition, 15*, 179–187.

Ishiai, S., Sugishita, M., Watabiki, S., Nakayama, T., Kotera, M., & Gono, S. (1994). Improvement of left unilateral spatial neglect in a line extension task. *Neurology, 44*, 294–298.

Ishiai, S., Watabiki, S., Lee, E., Kanouchi, T., & Odajima, N. (1994). Preserved leftward movement in left unilateral spatial neglect due to frontal lesions. *Journal of Neurology, Neurosurgery and Psychiatry, 57*, 1085–1090.

Ishihara, M., Keller, P. E., Rossetti, Y., & Prinz, W. (2008). Horizontal spatial representations of time: Evidence for the STEARC effect. *Cortex*, 44, 454–461.

Ishihara, M., Revol, P., Jacquin-Courtois, S., Maye, R., Rode, G., Boisson, D., et al. (2013). Tonal cues modulate line bisection performance: preliminary evidence for a new rehabilitation prospect? *Frontiers in Psychology 4*, 704.

Ishii, M., Seekkuarachchi, H., Tamura, H., & Tang, Z. (2004). 3D flash lag illusion. *Vision Research, 44*, 1981–1984.

Ito, M. (1993). Movement and thought: Identical control mechanisms by the cerebellum. *Trends in Neurosciences, 16*, 448–450.

Ito, M. (2008). Control of mental activities by internal models in the cerebellum. *Nature Reviews Neuroscience, 9*, 304–313.

Ittelson, W. H. (1993). Mirror reversals: Real and perceived. *Perception, 22*, 855–861.

Ivanoff, J., & Klein, R. M. (2001). The presence of a nonresponding effector increases inhibition of return. *Psychonomic Bulletin and Review, 8*, 307–314.

Ivanoff, J., & Klein, R. M. (2003). Orienting of attention without awareness is affected by measurement-induced attentional control settings. *Journal of Vision, 3*(1), 32–40.

Ivanoff, J., & Klein, R. M. (2004). Stimulus-response expectancies and inhibition of return. *Psychonomic Bulletin & Review, 11*, 542–550.

Ivanoff, J., & Klein, R. M. (2006). A speed-accuracy analysis of inhibition of return in go/no-go and choice-RT tasks. *Journal of Experimental Psychology: Human Perception and Performance, 32*, 908–919.

Ivanoff, J., Klein, R. M., & Lupianez, J. (2002). Inhibition of return interacts with the Simon effect: An omnibus analysis and its implications. *Perception & Psychophysics, 64*, 318–327.

Jacobs, L. F., & Schenk, F. (2003). Unpacking the cognitive map: The parallel map theory of hippocampal function. *Psychological Review, 110*(2), 285–315.

Jacoby, L. L., Kelley, C., Brown, J., & Jasechko, J. (1989). Becoming famous overnight: Limits on the ability to avoid unconscious influences of the past. *Journal of Personality and Social Psychology, 56*(3), 326–338.

Jammer, M. (2000). *Concepts of mass in contemporary physics and philosophy*. Princeton, NJ: Princeton University Press.

Jancke, D., & Erlhagen, W. (2010). Bridging the gap: A model of common neural mechanisms underlying the Fröhlich effect, the flash-lag effect, and the representational momentum effect. In R. Nijhawan & B. Khurana (Eds.), *Space and time in perception and action* (pp.

422–440). New York: Cambridge University Press.

Jancke, D., Erlhagen, W., Dinse, H., Akhavan, A., Giese, M., Steinhage, A., et al. (1999). Parametric population representation of retinal location: Neuronal interaction dynamics in cat primary visual cortex. *Journal of Neuroscience, 19*, 9016–9028.

Jans, B., Peters, J. C., & DeWeerd, P. (2010). Visual spatial attention to multiple locations at once: The jury is still out. *Psychological Review, 117*, 637–684.

Janzen, G. (2006). Memory for object location and route direction in virtual large-scale space. *Quarterly Journal of Experimental Psychology, 59*(3), 493–508.

Jarchow, T., & Mast, F. W. (1999). The effect of water immersion on postural and visual orientation. *Aviation, Space, and Environmental Medicine, 70*, 879–886.

Jarraya, M., Amorim, M. A., & Bardy, B. G. (2005). Optical flow and viewpoint change modulate the perception and memorization of complex motion. *Perception & Psychophysics, 67*, 951–961.

Jarrett, C. B., Phillips, M., Parker, A., & Senior, C. (2002). Implicit motion perception in schizotypy and schizophrenia: A representational momentum study. *Cognitive Neuropsychiatry, 7*, 1–14.

Jasmin, K., & Casasanto, D. (2012). The QWERTY effect: How typing shapes the meanings of words. *Psychonomic Bulletin & Review, 19*(3), 499–504.

Jastrow, J. (1900). *Fact and Fable in Psychology.* Boston, MA: Houghton Mifflin.

Jenkin, M. R., Dyde, R. T., Jenkin, H. L., Zacher, J. E., & Harris L. R. (2011). Perceptual upright: The relative effectiveness of dynamic and static images under different gravity states. *Seeing and Perceiving, 24*, 53–64.

Jenkin, H. L., Jenkin, M. R., Dyde, R. T., & Harris, L. R. (2004). Shape-from-shading depends on visual, gravitational, and body-orientation cues. *Perception, 33*(12), 1453–1461.

Jeschonek, S., Pauen, S., & Babocsai, L. (2013). Cross-modal mapping of visual and acoustic displays in infants: The effect of dynamic and static components. *European Journal of Developmental Psychology, 10*, 337–358.

Jewell, G., & McCourt, M. E. (2000). Pseudoneglect: A review and meta-analysis of performance factors in line bisection task. *Neuropsychologia, 38*, 93–110.

Jian, B. J., Shintani, T., Emanuel, B. A., & Yates, B. J. (2002). Convergence of limb, visceral, and vertical semicircular canal or otolith inputs onto vestibular nucleus neurons. *Experimental Brain Research, 144*(2), 247–257.

Jiang, T., Sun, L., & Zhu, L. (2015). The influence of vertical motor responses on explicit and incidental processing of power words. *Consciousness and Cognition, 34*, 33–42.

Jiang, T., & Zhu, L. (2015). Is power-space a continuum? Distance effect during power judgments. *Consciousness and Cognition, 37*, 8–15.

Joanette, Y., Brouchon, M., Gauthier, L., & Samson, M. (1986). Pointing with left vs right hand in left visual field neglect. *Neuropsychologia, 24*(3), 391–396.

Johnston, H. M., & Jones, M. R. (2006). Higher order pattern structure influences auditory representational momentum. *Journal of Experimental Psychology: Human Perception and Performance, 32*, 2–17.

Johnson, M. K., Hashtroudi, S., & Lindsay, D. S. (1993). Source monitoring. *Psychological Bulletin, 114*, 3–28.

Johnson, M. K., & Raye, C. L. (1981). Reality monitoring. *Psychological Review*, 88, 67–85.

Johnson, M. K., Raye, C. L., Wang, A. Y., & Taylor, T. H. (1979). Fact and fantasy: The roles of accuracy and variability in confusing imaginations with perceptual experiences. *Journal of Experimental Psychology: Human Learning and Memory, 5*, 229–240.

Johnson, M. K., Taylor, T. H., & Raye, C. L. (1977). Fact and fantasy: The effects of internally generated events on the apparent frequency of externally generated events. *Memory & Cognition, 5*(1), 116–122.

Jones, B., & Huang, Y. L. (1982). Space–time dependencies in psychophysical judgment of extent and duration: Algebraic models of the tau and kappa effects. *Psychological Bulletin*, *91*, 128–142.

Jones, L. A. (1988). Motor illusions: What do they reveal about proprioception? *Psychological Bulletin*, 103, 72–86.

Jones, S. (2002). *Antonymy: A corpus-based perspective*. London UK: Routledge.

Jonides, J. (1981). Voluntary vs. automatic control over the mind's eye's movement. In J. B. Long & A. D. Baddeley (Eds.), *Attention and performance IX* (pp. 187–203). Hillsdale, NJ: Lawrence Erlbaum Publishers.

Joo, S. J., Shin, K., Chong, S. C., & Blake, R. (2009). On the nature of the stimulus information necessary for estimating mean size of visual arrays. *Journal of Vision*, *9*, 1–12.

Joordens, S., Spalek, T. M., Razmy, S., & van Duijn, M. (2004). A clockwork orange: Compensation opposing momentum in memory for location. *Memory & Cognition*, *32*, 39–50.

Jordan, J. S. (1998). Recasting Dewey's critique of the reflex-arc concept via a theory of anticipatory consciousness: Implications for theories of perception. *New Ideas in Psychology*, *16*(3), 165–187.

Jordan, J. S. (2000, July). Intentionality in perception/action space. In *Attention and Performance XIX: Common Mechanisms in Perception and Action*. Organized by the Wolfgang Prinz & Bernhard Hommel: Max Planck Institute for Psychological Research, Munich Germany.

Jordan, J. S. (2003). Emergence of self and other in perception and action. *Consciousness and Cognition*, *12*, 633–646.

Jordan, J. S. (2008). Wild-agency: Nested intentionalities in neuroscience and archeology. *Philosophical Transactions of the Royal Society B (Biological Sciences)*, *363*, 1981–1991.

Jordan, J. S. (2013). The wild ways of conscious will: What we do, how we do it, and why it has meaning. *Frontiers in Psychology*, *4*, 574.

Jordan, J. S., Bai, J., Cialdella, V., & Schloesser, D. (2015). Foregrounding the context: Cognitive science as the study of embodied context. In E. Dzhafarov & J. S. Jordan (Eds.), *Contextuality from physics to psychology* (pp. 209–228). Berlin: Springer.

Jordan, J. S. & Day, B. (2015). Wild systems theory as a 21st century coherence framework for cognitive science. In T. Metzinger & J. M. Windt (Eds.), *Open MIND: 21(T)*. Frankfurt am Main: MIND Group.

Jordan, J. S., & Ghin, M. (2006). (Proto-) consciousness as a contextually-emergent property of self-sustaining systems, *Mind & Matter*, *4*(1), 45–68.

Jordan, J. S., & Ghin, M. (2007). The role of control in a science of consciousness: Causality, regulation and self-sustainment, *Journal of Consciousness Studies*, *14*(1–2), 177–197.

Jordan, J. S., Gill, D., & Dupuis, R. (2014, November). The influence of control and intentional dynamics on spatial perception. Presented at the 55th Annual Meeting of the Psychonomic Society, Long Beach, CA.

Jordan, J. S., & Heidenreich, B. (2010). The intentional nature of self-sustaining systems. *Mind & Matter*, *8*, 45–62.

Jordan, J. S., & Hunsinger, M. (2008). Learned patterns of action-effect anticipation contribute to the spatial displacement of continuously moving stimuli. *Journal of Experimental Psychology: Human Perception and Performance*, *34*, 113–124.

Jordan, J. S., Kenning, A., Clinton, J., & Durtschi, J. (2010). Spatial perception during control with another: The "other" as potential perturbation. *Fechner Day 2010. Proceedings of the 26th Annual Meeting of the International Society of Psychophysics*, Padua, Italy: The International Society for Psychophysics.

Jordan, J. S., & Knoblich, G. (2004). Spatial perception and control. *Psychonomic Bulletin & Review*, *11*, 54–59.

Jordan, J. S., Stork, S., Knuf, L., Kerzel, D., & Müsseler, J. (2002). Action planning affects spatial localization. In W. Prinz & B. Hommel (Eds.), *Common mechanisms in perception and action: Attention and performance XIX* (pp. 158–176). New York: Oxford University Press.

Jordan, J. S., & Vinson, D. (2012). After nature: On bodies, consciousness, and causality. *Journal of Consciousness Studies, 19*, 229–250.

Jordan, K., & Schiano, D. J. (1986). Serial processing and the parallel-lines illusion: Length contrast through relative spatial separation of contours. *Perception & Psychophysics, 40*, 384–390.

Jordan, K., & Uhlarik, J. (1985). Assimilation and contrast of perceived length depends on temporal factors. *Perception & Psychophysics, 37*, 447–454.

Jost, J. T., Banaji, M. R., & Nosek, B. A. (2004). A decade of system justification theory: Accumulated evidence of conscious and unconscious bolstering of the status quo. *Political Psychology, 25*(6), 881–919.

Joye, Y., & Verpooten, J. (2013). An exploration of the functions of religious monumental architecture from a Darwinian perspective. *Review of General Psychology, 17*(1), 53–68.

Judd, C. H. (1899). A study of gemoetrical illusions. *Psychological Review, 6*, 241–249.

Judge, T. A., & Cable, D. M. (2004). The effect of physical height on workplace success and income: Preliminary test of a theoretical model. *Journal of Applied Psychology, 89*(3), 428–441.

Julian, J. B., Keinath, A. T., Muzzio, I. A., & Epstein, R. A. (2015). Place recognition and heading retrieval are mediated by dissociable cognitive systems in mice. *Proceedings of the National Academy of Sciences, 112*(20), 6503–6508.

Kafaligonul, H., & Oluk, C. (2015). Audiovisual associations alter the perception of low-level visual motion. *Frontiers in Integrative Neuroscience, 9*, 26.

Kafaligönül, H., Patel, S. S., Öğmen, H., Bedell, H. E., & Purushothaman, G. (2010). Perceptual asynchronies and the dual-channel differential latency hypothesis. In R. Nijhawan & B. Khurana (Eds.), *Space and time in perception and action* (pp. 379–395). New York: Cambridge University Press.

Kahan, T. A., Colligan, S. M., & Wiedman, J. N. (2011). Are visual features of a looming or receding object processed in a capacity-free manner? *Consciousness and Cognition, 20*(4), 1761–1767.

Kahneman, D., Treisman, A., & Gibbs, B. J. (1992). The reviewing of object files: Object-specific integration of information. *Cognitive Psychology, 24*(2), 175–219.

Kaiser, M. K., & Mowafy, L. (1993). Optical specification of time-to-passage: Observers' sensitivity to global tau. *Journal of Experimental Psychology: Human Perception and Performance, 19*(5), 1028–1040.

Kanai, R., Sheth, B. R., & Shimojo, S. (2004). Stopping the motion and sleuthing the flash-lag effect: Spatial uncertainty is the key to perceptual mislocalization. *Vision Research, 44*, 2605–2619.

Kanai, R., & Verstraten, F. A. J. (2006). Visual transients reveal the veridical position of a moving object. *Perception, 35*, 453–460.

Kanizsa, G. (1979). *Organization in vision.* New York: Praeger.

Kanten, A. B., & Teigen, K. H. (2015). A magnitude effect in judgments of subjective closeness. *Personality and Social Psychology Bulletin, 41*, 1712–1722.

Kaptein, R. G., & Van Gisbergen, J. A. (2004). Interpretation of a discontinuity in the sense of verticality at large body tilt. *Journal of Neurophysiology, 91*, 2205–2214.

Kaptein, R. G., & Van Gisbergen, J. A. (2005). Nature of the transition between two modes of external space perception in tilted subjects. *Journal of Neurophysiology, 93*(6), 3356–3369.

Karwoski, T. F., Odbert, H. S., & Osgood, C. E. (1942). Studies in synesthetic thinking: II. The role of form in visual responses to music. *The Journal of General Psychology, 26*, 199–222.

Kastner, S., Pinsk, M. A., De Weerd, P., Desimone, R., & Ungerleider, L. G. (1999). Increased activity in human visual cortex during directed attention in the absence of visual stimulation. *Neuron, 22*, 751–761.

Kauffman, S. (1995). *At home in the universe.* New York: Oxford University Press.

Kawato, M., Furukawa, K., & Suzuki, R. (1987). A hierarchical neural-network model for control and learning of voluntary movement. *Biological Cybernetics*, *57*(3), 169–185.

Kay, P., & Kempton, W. (1984). What is the Sapir-Whorf hypothesis? *American Anthropologist*, *86*, 65–79.

Keating, C. F. (1985). Gender and the physiognomy of dominance and attractiveness. *Social Psychology Quarterly*, *48*, 61–70.

Keil, M. S., & Cristobal, G. (2000). Separating the chaff from the wheat: Possible origins of the oblique effect. *Journal of the Optical Society of America. A, Optics, Image Science, and Vision*, *17*, 697–710.

Kelly, J. W., Loomis, J. M., & Beall, A. C. (2004). Judgments of exocentric direction in large-scale space. *Perception*, *33*(4), 443–454.

Kelly, M. H., & Freyd, J. J. (1987). Explorations of representational momentum. *Cognitive Psychology*, *19*, 369–401.

Kelso, J. A. S. (1995). *Dynamic patterns: The self-organization of brain and behavior.* Cambridge, MA: MIT Press.

Kelso, J. A. S. (2003), Cognitive coordination dynamics. In W. Tschacher & J. Dauwalder (Eds.), *Dynamical systems approaches to embodied cognition* (pp. 45–67). Berlin: Springer Verlag.

Kelso, J. A. S., & Engstrøm, D. (2006). *The complementary nature.* Cambridge, MA: MIT Press.

Kemp, M. (1992). *The science of art: Optical themes in western art from Brunelleschi to Seurat.* New Haven, CT: Yale University Press.

Kerkman, D. D., Friedman, A., Brown, N. R., Stea, D., & Carmichael, A. (2003). The development of geographic categories and biases. *Journal of Experimental Child Psychology*, *84*(4), 265–285.

Kerzel, D. (2000). Eye movements and visible persistence explain the mislocalization of the final position of a moving target. *Vision Research*, *40*, 3703–3715.

Kerzel, D. (2002a). Attention shifts and memory averaging. *Quarterly Journal of Experimental Psychology*, *55A*, 425–443.

Kerzel, D. (2002b). Different localization of motion onset with pointing and relative judgements. *Experimental Brain Research*, *145*(3), 340–350.

Kerzel, D. (2002c). The locus of "memory displacement" is at least partially perceptual: Effects of velocity, expectation, friction, memory averaging, and weight. *Perception & Psychophysics*, *64*, 680–692.

Kerzel, D. (2002d). A matter of design: No representational momentum without predictability. *Visual Cognition*, *9*, 66–80.

Kerzel, D. (2003a). Attention maintains mental extrapolation of target position: Irrelevant distractors eliminate forward displacement after implied motion. *Cognition*, *88*, 109–131.

Kerzel, D. (2003b). Centripetal force draws the eyes, not memory of the target, toward the center. *Journal of Experimental Psychology: Learning, Memory, and Cognition*, *29*, 458–466.

Kerzel, D. (2003c). Mental extrapolation of target position is strongest with weak motion signals and motor responses. *Vision Research*, *43*, 2623–2635.

Kerzel, D. (2004). Attentional load modulates mislocalization of moving stimuli, but does not eliminate the error. *Psychonomic Bulletin & Review*, *11*, 848–853.

Kerzel, D. (2006). Why eye movements and perceptual factors have to be controlled in studies on "representational momentum." *Psychonomic Bulletin & Review*, *13*, 166–173.

Kerzel, D. (2010). The Fröhlich effect: Past and present. In R. Nijhawan & B. Khurana (Eds.), *Space and time in perception and action* (pp. 321–337). New York: Cambridge University Press.

Kerzel, D., & Gegenfurtner, K. R. (2003). Neuronal processing delays are compensated in the sensorimotor branch of the visual system. *Current Biology*, *13*, 1975–1978.

Kerzel, D., & Gegenfurtner, K. R. (2004). Spatial distortions and processing latencies in the onset repulsion and Fröhlich effects. *Vision Research*, *44*(6), 577–590.

Kerzel, D., Jordan, J. S., & Müsseler, J. (2001). The role of perception in the mislocalization of the final position of a moving target.

Journal of Experimental Psychology: Human Perception and Performance, 27, 829–840.

Kerzel, D., & Müsseler, J. (2002). Effects of stimulus material on the Fröhlich illusion. *Vision Research*, *42*, 181–189.

Keslo, J. A. S. (1995). *Dynamic patterns: The self-organization of brain and behavior.* Cambridge, MA: MIT Press/Bradford Books.

Kessler, K., Gordon, L., Cessford, K., & Lages, M. (2010). Characteristics of motor resonance produce the pattern of flash-lag effects for biological motion. *PLoS ONE*, *5*, e8258.

Keus, I. M., & Schwarz, W. (2005). Searching for the functional locus of the SNARC effect: Evidence for a response-related origin. *Memory & Cognition*, *33*(4), 681–695.

Khurana, B., Carter, R. M., Watanabe, K., & Nijhawan, R. (2006). Flashlag chimeras: The role of perceived alignment in the composite face effect. *Vision Research*, *46*, 2757–2772.

Khurana, B., & Nijhawan, R. (1995). Extrapolation or attention shift? *Nature*, *378*, 566.

Khurana, B., Watanabe, K., & Nijhawan, R. (2000). The role of attention in motion extrapolation: Are moving objects "corrected" or flashed objects attentionally delayed? *Perception*, *29*, 675–692.

Kim, R., Peters, M. A., & Shams, L. (2012). 0 + 1 > 1: How adding noninformative sound improves performance on a visual task. *Psychological Science*, *23*, 6–12.

Kim, S., Dede, A. J., Hopkins, R.O., & Squire, L. R. (2015). Memory, scene construction, and the human hippocampus. *Proceedings of the National Academy of Sciences USA 112*, 4767–4772.

Kimura, D. (1999). *Sex and cognition.* Cambridge, MA: MIT Press.

King, A. J. (2004). The Superior Colliculus, *Current Biology* 14(9), R335–R338.

King, S. M., Dykeman, C., Redgrave, P., & Dean, P. (1992). Use of a distracting task to obtain defensive head movements to looming visual-stimuli by human adults in a laboratory setting. *Perception*, *21*(2), 245–259.

King, Z. R., Tenhundfeld, N. T., & Witt, J. K. (2018). What you see and what you are told: Feedback does not diminish action-specific perception. *Psychological Research, 82*. 507–519.

Kingdom, F. A., Prins, N., & Hayes, A. (2003). Mechanism independence for texture-modulation detection is consistent with a filter-rectify-filter mechanisms. *Visual Neuroscience*, *20*, 65–76.

Kinsbourne, M., & Jordan, J. S. (2009). Embodied anticipation: A neurodevelopmental interpretation. *Discourse Processes*, *46*, 103–126.

Kirschfeld, K., & Kammer, T. (1999). The Fröhlich effect: A consequence of the interaction of visual focal attention and metacontrast. *Vision Research*, *39*, 3702–3709.

Kitagawa, N., & Ichihara, S. (2002). Hearing visual motion in depth. *Nature*, *416*, 172–174.

Kitajima, N., & Yamashita, Y. (1999). Dynamic capture of sound motion by light stimuli moving in three-dimensional space. *Perceptual and Motor Skills*, *89*, 1139–1158.

Klatzky, R. L. (1998). Allocentric and egocentric spatial representations: Definitions, distinctions, and interconnections. In C. Freksa, C. Habel, & K. F. Wender (Eds.), *Spatial cognition – An interdisciplinary approach to representation and processing of spatial knowledge* (pp. 1–17). Berlin: Springer-Verlag.

Klein, R., Brennan, M., D'Aloisio, A., D'Entremont, B., & Gilani, A. (1987a). *Covert cross-modality orienting of attention.* Unpublished manuscript.

Klein, R. M. (2009). On the control of attention. *Canadian Journal of Experimental Psychology/Revue canadienne de psychologie expérimentale*, *63*(3), 240–252.

Klein, R. M., Brennan, M., & Gilani, A. (1987b, November). Covert cross-modality orienting of attention in space. Paper presented at the annual meeting of the Psychonomics Society, Seattle, WA.

Klein, R. M., Christie, J., & Morris, E. P. (2005). Vector averaging of inhibition of return. *Psychonomic Bulletin & Review*, *12*, 295–300.

Klein, R. M., & Hilchey, M. D. (2011). Oculomotor inhibition of return. In S. Liversedge, I. D. Gilchrist, & S. Everling (Eds.), *The Oxford handbook of eye movements* (pp. 471–492). Oxford: Oxford University Press.

Klein, R. M., & Juckes, T. (1989, October). Can auditory frequency control the direction of visual attention? Paper presented at the Canadian Acoustic Association, Halifax, Nova Scotia. (Abstract published in Proceedings).

Klein, S. (2002). Libet's research on the timing of conscious intention to act: A commentary. *Consciousness and Cognition, 11*, 273–279.

Kleinman, J. T., Newhart, M., Davis, C., Heidler-Gary, J., Gottesman, R. F., & Hillis, A. E. (2007). Right hemispatial neglect: Frequency and characterization following acute left hemisphere stroke. *Brain and Cognition, 64*, 50–59.

Klippel, A. (2012). Spatial information theory meets spatial thinking: Is topology the Rosetta Stone of spatio-temporal cognition? *Annals of the Association of American Geographers, 102*, 1310–1328.

Klippel, A., Knuf, L., Hommel, B., & Freksa, C. (2004). Perceptually induced distortions in cognitive maps. In *Spatial Cognition IV. Reasoning, Action, Interaction* (pp. 204–213). Berlin: Springer Heidelberg.

Kluss, T., Marsh, W. E., Zetzsche, C., & Schill, K. (2015). Representation of impossible worlds in the cognitive map. *Cognitive Processing, 16* (Suppl 1), 271–276.

Knill, D. C., & Richards, W. (1996). *Perception as Bayesian inference*. New York: Cambridge University Press.

Kobayashi, M., Teramoto, W., Hidaka, S., & Sugita, Y. (2012a). Indiscriminable sounds determine the direction of visual motion. *Scientific Reports, 2*, 365.

Kobayashi, M., Teramoto, W., Hidaka, S., & Sugita, Y. (2012b). Sound frequency and aural selectivity in sound-contingent visual motion aftereffect. *PLoS ONE, 7*, e36803.

Koenderink, J. J., & van Doorn, A. J. (1998). Exocentric pointing. In L. R. Harris & M. Jenkin (Eds.), *Vision and action* (pp. 295–313). Cambridge: Cambridge University Press.

Koenderink, J. J., & van Doorn, A. J. (2008). The structure of visual spaces. *Journal of Mathematical Imaging and Vision, 31*, 171–187.

Koenderink, J. J., van Doorn, A. J., Kappers, A. M. L., Doumen, M. J. A., & Todd, J. T. (2008). Exocentric pointing in depth. *Vision Research, 48*, 716–723.

Koenderink, J. J., van Doorn, A. J., Kappers, A. M. L., & Lappin, J. S. (2002). Large-scale visual frontoparallels under full-cue conditions. *Perception, 31*, 1467–1475.

Koenderink, J. J., van Doorn, A. J., & Lappin, J. S. (2000). Direct measurement of the curvature of visual space. *Perception, 29*, 69–79.

Koffka K. (1935). *Principles of Gestalt psychology*. New York: Harcourt Brace.

Köhler, W. (1938). Physical Gestalten. In W. D. Ellis (Ed.), *A source book of Gestalt psychology* (pp. 17–54). London: Routledge & Kegan Paul. (Original work published 1920).

Köhler, W. (1920). Die physischen Gestalten in Ruhe und im stationären Zustand. Eine natur-philosophische Untersuchung. Braunschweig. Germany: Friedr. Vieweg und Sohn. (Translated extract reprinted as "Physical Gestalten." In W. D. Ellis [Ed.]. [1938]. *A source book of Gestalt psychology* [pp. 17–54]. London: Routledge & Kegan Paul Ltd.)

Köhler, W. (1940). *Dynamics in psychology*. New York: Liveright.

Kolb, H. (2007). Feedback loops in the retina. Retrieved from www.ncbi.nlm.nih.gov/books/NBK11514/

Konečni, V. J. (2001). The golden section in the structure of 20th-century painting. Rivista Psicologia dell'Arta, *Visual Arts Research, 22*, 27–42.

Konečni, V. J. (2005). On the "golden section." *Visual Arts Research, 31*, 76–87.

Konen, C. S., & Kastner, S. (2008). Two hierarchically organized neural systems for object information in human visual cortex. *Nature Neuroscience, 11*(2), 224–231.

Koriat, A., Goldsmith, M., & Pansky A. (2000). Toward a psychology of memory accuracy. *Annual Review of Psychology, 51*, 481–537.

Körding, K. P., Beierholm, U., Ma, W. J., Quartz, S., Tenenbaum, J. B., & Shams, L. (2007). Causal inference in multisensory perception. *PLoS ONE, 2*, e943.

Kornblum, S., Hasbroucq, T., & Osman, A. (1990). Dimensional overlap: Cognitive basis for stimulus–response compatibility – A model and taxonomy. *Psychological Review, 97*, 253–270.

Kornblum, S., & Lee, J.-W. (1995). Stimulus–response compatibility with relevant and irrelevant stimulus dimensions that do and do not overlap with the response. *Journal of Experimental Psychology: Human Perception and Performance, 21*, 855–875.

Kosovicheva, A. A., Maus, G. W., Anstis, S., Cavanagh, P., Tse, P. U., & Whitney, D. (2012). The motion-induced shift in the perceived location of a grating also shifts its aftereffect. *Journal of Vision, 12*(8): 1–14.

Kosslyn, S. M. (1987). Seeing and imagining in the cerebral hemispheres: A computational approach. *Psychological Review, 94*(2), 148–175.

Kosslyn, S. M., Koenig, O., Barrett, A., Cave, C. B., Tang, J., & Gabrieli, J. D. (1989). Evidence for two types of spatial representations: Hemispheric specialization for categorical and coordinate relations. *Journal of Experimental Psychology: Human Perception and Performance, 15*(4), 723–735.

Kosslyn, S. M., Pick, H. L., & Fariello, G. R. (1974). Cognitive maps in children and men. *Child Development, 45*, 707–716.

Kourtzi, Z., & Kanwisher, N. (2000). Activation in human MT/MST by static images with implied motion. *Journal of Cognitive Neuroscience, 12*, 48–55.

Kozhevnikov, M., & Hegarty, M. (2001). Impetus beliefs as default heuristics: Dissociation between explicit and implicit knowledge about motion. *Psychonomic Bulletin & Review, 8*, 439–453.

Koziol, L., Budding, D., & Chidekel, D. (2011). From movement to thought: Executive function, embodied cognition, and the cerebellum. *The Cerebellum, 11*, 505–525.

Koziol, L., & Lutz, J. (2013). From movement to thought: The development of executive function. *Applied Neuropsychology: Child, 2*(2), 104–115.

Kragen, B. P. (2005). *The art of composition*. Allworth Communications.

Kranjec, A. (2013). Thought is a material: Talking with Mel Bochner about space, art, and language. *Journal of Cognitive Neuroscience, 25*(12), 2015–2024.

Kranjec, A., & Chatterjee, A. (2010). Are temporal concepts embodied? A challenge for cognitive neuroscience. *Frontiers in Psychology* (Special Issue: Embodied and grounded cognition) 1:240.

Kranjec, A., Ianni, G., & Chatterjee, A. (2013). Schemas reveal spatial relations to a patient with simultanagnosia. *Cortex, 49*(7), 1983–1988.

Kranjec, A., Lupyan, G., & Chatterjee, A. (2014). Categorical biases in perceiving spatial relations. *PLoS ONE, 9*(5), e98604.

Kreegipuu, K., & Allik, J. (2003). Perceived onset time and position of a moving stimulus. *Vision Research, 43*, 1625–1635.

Kreegipuu, K., & Allik, J. (2004). Confusion of space and time in the flash-lag effect. *Perception, 33*, 293–306.

Kreindel, E., & Intraub, H. (2017). Anticipatory scene representation in preschool children's recall and recognition memory. *Developmental Science*. doi: 10.1111/desc.12444

Krekelberg, B. (2001). The persistence of position. *Vision Research, 41*, 529–539.

Krekelberg, B. (2003). Sound and vision. *Trends in Cognitive Sciences, 7*, 277–279.

Krekelberg, B. (2008). Perception of direction is not compensated for neural latency. *Behavioral and Brain Sciences, 31*(2), 208–209.

Krekelberg, B., & Lappe, M. (1999). Temporal recruitment along the trajectory of moving objects and the perception of position. *Vision Research, 39*, 2669–2679.

Krekelberg, B., & Lappe, M. (2000a). A model of the perceived relative positions of moving objects based upon a slow averaging process. *Vision Research, 40*, 201–215.

Krekelberg, B., & Lappe, M. (2000b). The position of moving objects. *Science, 289*, 1107a.

Krekelberg, B., & Lappe, M. (2001). Neuronal latencies and the position of moving objects. *Trends in Neurosciences, 24*, 335–339.

Krekelberg, B., & Lappe, M. (2002). Response: Untangling spatial from temporal illusions. *Trends in Neurosciences, 25*, 294.

Krüger, H., Collins, T., Pressnitzer, D., Kang, H., Teki, S., & Patrick, C. (2015). Evidence for the common coding of location in auditory and visual space. *Journal of Vision, 15*(12), 368.

Krumhansl, C. L. (1979). The psychological representation of musical pitch in a tonal context. *Cognitive Psychology, 11*, 346–374.

Krumhansl, C. L. (2001). *Cognitive foundations of musical pitch*. Oxford: Oxford University Press.

Kuang, S., & Zhang, T. (2014). Smelling directions: Olfaction modulates ambiguous visual motion perception. *Scientific Reports*, 4, 5796.

Kubovy, M. (1986). *The psychology of perspective and Renaissance art*. Cambridge: Cambridge University Press.

Kuipers, B., & Byun, Y. T. (1991). A robot exploration and mapping strategy based on a semantic hierarchy of spatial representations. *Robotics and Autonomous Systems, 8*(1), 47–63.

Kumar, D., & Srinivasan, N. (2013). Hierarchical control and sense of agency: Differential effects of control on implicit and explicit measures of agency. Proceedings of 35th Annual Meeting of the Cognitive Science Society, Berlin, Germany.

Kunde, W. (2001). Response–effect compatibility in manual choice–reaction tasks. *Journal of Experimental Psychology: Human Perception and Performance, 27*, 387–394.

Künnapas, T. M. (1955). An analysis of the "vertical–horizontal illusion." *Journal of Experimental Psychology, 49*, 134–140.

Künnapas, T. M. (1957). The vertical-horizontal illusion and the visual field. *Journal of Experimental Psychology, 53*, 405–407.

Kushiro, K., Taga, G., & Watanabe, H. (2007). Frame of reference for visual perception in young infants during change of body position. *Experimental Brain Research, 183*, 523–529.

Kwon, O.-C., Tadin, D., & Knill, D. C. (2015). Unifying account of visual motion and position perception. *PNAS, 112*, 8142–8147.

Kwon, T., & Kim, T. (2012). The effect of skill level on perceptual judgment. *European Journal of Scientific Research, 91*, 184–187.

La Scaleia, B., Lacquaniti, F., & Zago, M. (2014). Neural extrapolation of motion for a ball rolling down an inclined plane. *PLoS ONE*, 9, e99837.

La Scaleia, B., Zago, M., & Lacquaniti, F. (2015). Hand interception of occluded motion in humans: a test of model-based vs. on-line control. *Journal of Neurophysiology, 114*(3), 1577–1592.

Laberge, D., Auclair, L., & Siéroff, E. (2000). Preparatory attention: Experiment and theory. *Consciousness and Cognition, 9*, 396–434.

Lackner, J. R., & DiZio, P. (2000). Human orientation and movement control in weightless and artificial gravity environments. *Experimental Brain Research, 130*(1), 2–26.

Lacquaniti, F. (1997). Frames of reference in sensorimotor coordination. In F. Boller & J. Grafman (Eds.), *Handbook of neuropsychology* (vol. 11, pp. 27–64). Amsterdam: Elsevier.

Lacquaniti, F., Bosco, G., Gravano, S., Indovina, I., La Scaleia, B., Maffei, V., & Zago, M. (2014). Multisensory integration and internal models for sensing gravity effects in primates. *Biomedical Research International, 2014*, 615854.

Lacquaniti, F., Bosco, G., Gravano, S., Indovina, I., La Scaleia, B., Maffei, V., & Zago, M. (2015). Gravity in the brain as a reference for space and time perception. *Multisensory Research*. doi:10.1163/22134808-00002471

Lacquaniti, F., & Maioli, C. (1989). The role of preparation in tuning anticipatory and reflex responses during catching, *Journal of Neuroscience*, *9*, 134–148.

Lakens, D., Semin, G. R., & Foroni, F. (2011). Why your highness needs the people. *Social Psychology*, *42*(3), 205–213.

Lakoff, G., & Johnson, M. (1980). *Metaphors we live by*. Chicago: University of Chicago Press.

Landau, B., & Jackendoff, R. (1993). Whence and whither in spatial language and spatial cognition? *Behavioral and Brain Sciences*, *16*(2), 255–265.

Landau, M. J., Meier, B. P., & Keefer, L. A. (2010). A metaphor-enriched social cognition. *Psychological Bulletin*, *136*(6), 1045–1067.

Lappe, M., & Krekelberg, B. (1998). The position of moving objects. *Perception*, *27*, 1437–1449.

Lappin, J. S., Shelton, A. L., & Rieser, J. J. (2006). Environmental context influences visually perceived distance. *Attention, Perception, & Psychophysics*, *68*(4), 571–581.

Larson, S. (2012). *Musical forces: Motion, metaphor, and meaning in music*. Bloomington: Indiana University Press.

Latto, R., Brain, D., & Kelly, B. (2000). An oblique effect in aesthetics: Homage to Mondrian (1872–1944). *Perception*, *29*, 981–987.

Latto, R., & Russell-Duff, K. (2002). An oblique effect in the selection of line orientation by twentieth century painters. *Empirical Studies of the Arts*, *20*, 49–60.

Laurens, J., Meng, H., & Angelaki, D. E. (2013) Neural representation of orientation relative to gravity in the macaque cerebellum. *Neuron*, *80*, 1508–1518.

Laurienti, P. J., Kraft, R. A., Maldjian, J. A., Burdett, J. H., & Wallace, M. T. (2004). Semantic congruence is a critical factor in multisensory behavioral performance. *Experimental Brain Research*, *158*, 405–414.

Lawson, R. (2010). People cannot locate the projection of an object on the surface of a mirror. *Cognition*, *115*, 336–342.

Lawson, R. (2012). Mirrors, mirrors on the wall . . . the ubiquitous multiple reflection error. *Cognition*, *122*, 1–11.

Lawson, R., & Bertamini, M. (2006). Errors in judging information about reflections in mirrors. *Perception*, *35*, 1265–1288.

Lawson, R., Bertamini, M., & Liu, D. (2007). Overestimation of the projected size of objects on the surface of mirrors and windows. *Journal of Experimental Psychology: Human Perception and Performance*, *33*, 1027–1044.

Lawton, C. A. (1994). Gender differences in way-finding strategies: Relationship to spatial ability and spatial anxiety. *Sex Roles*, *30*(11–12), 765–779.

Lawton, C. A. (1996). Strategies for indoor wayfinding: The role of orientation. *Journal of Environmental Psychology*, *16*, 59–68.

Lawton, C. A., & Kallai, J. (2002). Gender differences in wayfinding strategies and anxiety about wayfinding: A cross-cultural comparison. *Sex Roles*, *47*(9–10), 389–401.

Leder, H., Belke, B., Oeberst, A., & Augustin, D. (2004). A model of aesthetic appreciation and aesthetic judgments. *British Journal of Psychology*, *95*, 489–508.

Leder, H., Carbon, C., & Ripsas, A. (2006). Entitling art: Influence of title information on understanding and appreciation of paintings. *Acta Psychologica*, *121*, 176–198.

Lee, D. N. (1976). Theory of visual control of braking based on information about time-to-collision. *Perception*, *5*(4), 437–459.

Lee, D. N. (1998). Guiding movement by coupling taus. *Ecological Psychology*, *10*(3-4), 221-250.

Lee, D. N., & Reddish, P. E. (1981). Plummeting gannets – A paradigm of ecological optics. *Nature*, *293*(5830), 293–294.

Lee, T. C. P., Khuu, S. K., Li, W., & Hayes, A. (2008). Distortion in perceived image size accompanies flash lag in depth. *Journal of Vision*, *8*(11), 20.

Lee, Y., Lee, S., Carello, C., & Turvey, M. T. (2012). An archer's perceived form scales the "hitableness" of archery targets. *Journal of Experimental Psychology: Human Perception and Performance*, *38*(5), 1125–1131.

Lenggenhager, B., Loetscher, T., Kavan, N., Pallich, G., Brodtmann, A., Nicholls, M.

E. R., & Brugger, P. (2012). Paradoxical extension into the contralesional hemispace in spatial neglect. *Cortex, 48*, 1320–1328.

Leslie, A. M. (1994). ToMM, ToBy, and Agency: Core architecture and domain specificity. In L. Hirschfeld & S. Gelman (Eds.), *Mapping the mind: Domain specificity in cognition and culture* (pp. 119–148). New York: Cambridge University Press.

Lessard, D. A., Linkenauger, S. A., & Proffitt, D. R. (2009). Look before you leap: Jumping ability affects distance perception. *Perception, 38*, 1863–1866.

Leung, A. K., Qiu, L., Ong, L., & Tam, K. P. (2011). Embodied cultural cognition: Situating the study of embodied cognition in socio-cultural contexts. *Social and Personality Psychology Compass, 5*(9), 591–608.

Levelt, W. J. M. (1982a). Cognitive styles in the use of spatial direction terms. In R. J. Jarvella & W. Klein (Eds.), *Speech, place, and action* (pp. 251–268). Chichester: Wiley.

Levelt, W. J. M. (1982b). Linearization in describing spatial networks. In S. Peters & E. Saarinen (Eds.), *Processes, beliefs, and questions* (pp. 199–220). Dordrecht: Reidel.

Levelt, W. J. M. (1984). Some perceptual limitations on talking about space. In A. J. Van Doorn, W. A. Van de Grind, & J. J. Koenderink (Eds.), *Limits in perception* (pp. 323–358). Utrecht: VNU Science Press.

Levelt, W. J. M. (1989). *Speaking: From intention to articulation.* Cambridge, MA: MIT Press.

Leventhal, A. G. (1983). Relationship between preferred orientation and receptive field position of neurons in cat striate cortex. *Journal of Comparative Neurology, 220*, 476–483.

Levi, D. M., Klein, S. A., & Yap, Y. L. (1988). Weber's law for position: Unconfounding the role of separation and eccentricity. *Vision Research, 28*, 597–603.

Levin, C. A., & Haber, R. N. (1993). Visual angle as a determinant of perceived interobject distance. *Attention, Perception, & Psychophysics, 54*(2), 250–259.

Levine, M., Marchon, I., & Hanley, G. (1984). The placement and misplacement of you-are-here maps. *Environment and Behavior, 16*(2), 139–157.

Levinson, S. (1996) Frames of reference and Molyneux's question: Cross-linguistic evidence. In P. Bloom, M. A. Peterson, L. Nadel, & M. Garrett (Eds.), *Space and language* (pp. 109–169). Cambridge, MA: MIT Press.

Levinson, S. C. (2003). *Space in language and cognition: Explorations in cognitive diversity.* Cambridge: Cambridge University Press.

Levy, J. (1976). Lateral dominance and aesthetic preference. *Neuropsychologia, 14*, 431–445.

Lewis, C. F., & McBeath, M. K. (2004). Bias to experience approaching motion in a three-dimensional virtual environment. *Perception, 33*(3), 259–276.

Lewis, J. E., & Neider, M. B. (2015). Fixation not required: Characterizing oculomotor attention capture for looming stimuli. *Attention, Perception, & Psychophysics, 77*(7), 2247–2259.

Lewis, J. W., Beauchamp, M. S., & DeYoe, E. A. (2000). A comparison of visual and auditory motion processing in human cerebral cortex. *Cerebral Cortex, 10*, 873–888.

Lewkowicz, D. J., & Minar, N. J. (2014). Infants are not sensitive to synesthetic cross-modality correspondences: A comment on Walker et al. (2010). *Psychological Science, 25*, 832–834.

Leyssen, M. H. R., Linsen, S., Sammartino, J., & Palmer, S. E. (2012). Aesthetic preference for spatial composition in multiobject pictures. *I-Perception, 3*, 25–49.

Leyton, M. (1989). Inferring causal history from shape. *Cognitive Science, 13*, 357–387.

Leyton, M. (1992). *Symmetry, causality, mind.* Cambridge, MA: MIT Press.

Li, B., Peterson, M. R., & Freeman, R. D. (2003). Oblique effect: A neural basis in the visual cortex. *Journal of Neurophysiology, 90*, 204–217.

Li, C., & Chen, T. (2009). Aesthetic visual quality assessment of paintings. *IEEE Journal of Selected Topics in Signal Processing, 3*, 236–252.

Li, Z., Phillips, J., & Durgin, F. H. (2011). The underestimation of egocentric distance: Evidence from frontal matching tasks. *Attention, Perception, & Psychophysics, 73*, 2205–2217.

Liberman, N., & Förster, J. (2009). The effect of psychological distance on perceptual level of construal. *Cognitive Science, 33*(7), 1330–1341.

Liberman, N., Trope, Y., & Stephan, E. (2007). Psychological distance. In A. W. Kruglanski & E. T. Higgins (Eds.), *Social psychology: A handbook of basic principles* (2nd ed, pp. 353–383). New York: Guilford Press.

Libet, B. (1985). Unconscious cerebral initiative and the role of conscious will in voluntary action. *Behavioral and Brain Sciences, 8*, 529–566.

Libet, B., Gleason, C. A., Wright, E. W., & Pearl, D. K. (1983). Time of conscious intention to act in relation to onset of cerebral activity (readiness-potential): The unconscious initiation of a freely voluntary act. *Brain, 106*, 623–642.

Libet, B., Wright, E. W., Jr., & Gleason, C. A. (1982). Readiness-potentials preceding unrestricted "spontaneous" vs. pre-planned voluntary acts. *Electroencephalography and Clinical Neurophysiology, 54*, 322–335.

Lidji, P., Kolinsky R., Lochy A., & Morais J. (2007). Spatial associations for musical stimuli: A piano in the head? *Journal of Experimental Psychology: Human Perception and Performance, 33*, 1189–1207.

Lima, S. L., Blackwell, B. F., DeVault, T. L., & Fernandez-Juricic, E. (2015). Animal reactions to oncoming vehicles: A conceptual review. *Biological Reviews, 90*(1), 60–76.

Lin, J. Y., Franconeri, S., & Enns, J. T. (2008). Objects on a collision path with the observer demand attention. *Psychological Science, 19*(7), 686–692.

Linares, D., & López-Moliner, J. (2007). Absence of flash-lag when judging global shape from local positions. *Vision Research, 47*, 357–362.

Linares, D., López-Moliner, J., & Johnston, A. (2007). Motion signal and the perceived positions of moving objects. *Journal of Vision, 7*(7), 1.

Lindauer, M. S. (1969). The orientation of form in abstract art. *Proceedings of the American Psychological Association, 4*, 475–476.

Lindell, A. K. (2013). The silent social/emotional signals in left and right cheek poses: A literature review. *Laterality: Asymmetries of Body, Brain and Cognition, 18*(5), 612–624.

Lindell, A. K. & Nicholls, M. E. R. (2003). The Cortex Forum – Cortical representation of the fovea: Implications for visual half-field research. *Cortex, 39*, 111–117.

Lindemann, O., Abolafia, J. M., Pratt, J., & Bekkering, H. (2008). Coding strategies in number space: Memory requirements influence spatial–numerical associations. *Quarterly Journal of Experimental Psychology, 61*(4), 515–524.

Lindsay, S. (2008). Source monitoring. In H. L. Roediger, III (Ed.), *Cognitive psychology of memory*, Vol. 2 of Learning and memory: A comprehensive reference, 4 vols. (J. Byrne Editor) (pp. 325–348). Oxford: Elsevier.

Link, T., Moeller, K., Huber, S., Fischer, U., & Nuerk, H.-C. (2013). Walk the number line – An embodied training of numerical concepts. *Trends in Neuroscience and Education, 2*(2), 74–84.

Link, T., Nuerk, H.-C., & Moeller, K. (2014). On the relation between the mental number line and arithmetic competencies. *Quarterly Journal of Experimental Psychology, 67*(8), 1597–1613.

Linkenauger, S. A., Bulthoff, H. H., & Mohler, B. J. (2015). Virtual arm's reach influences perceived distance but only after experience reaching. *Neuropsychologia, 70*, 393–401.

Linkenauger, S. A., Leyrer, M., Buelthoff, H. H., & Mohler, B. J. (2013). Welcome to wonderland: The influence of the size and shape of a virtual hand on the perceived size and shape of virtual objects. *PLoS ONE, 8*(7), e68594.

Linkenauger, S. A., Ramenzoni, V., & Proffitt, D. R. (2010). Illusory shrinkage and growth: Body-based rescaling affects the perception of size. *Psychological Science, 21*(9), 1318–1325.

Linkenauger, S. A., Witt, J. K., & Proffitt, D. R. (2011). Taking a hands-on approach:

Apparent grasping ability scales the perception of object size. *Journal of Experimental Psychology: Human Perception and Performance*, *37*(5), 1432–1441.

Lipinski, J., Simmering, V. R., Johnson, J. S., & Spencer, J. P. (2010). The role of experience in location estimation: Target distributions shift location memory biases. *Cognition*, *115*(1), 147–153.

Lipshits, M., Bengoetxea, A., Cheron, G., & McIntyre, J. (2005). Two reference frames for visual perception in two gravity conditions. *Perception*, *34*, 545–555.

Lipshits, M., & McIntyre, J. (1999). Gravity affects the preferred vertical and horizontal in visual perception of orientation. *NeuroReport*, *10*, 1085–1089.

Lisi, M., & Cavanagh, P. (2015) Dissociation between the perceptual and saccadic localisation of moving objects. *Current Biology*, *25*, 235–240.

Liu, E. H., Mercado, E., & Church, B. A. (2011). Multidimensional processing of dynamic sounds: More than meets the ear. *Attention, Perception, & Psychophysics*, *73*(8), 2624–2638.

Liu, G. B., & Pettigrew, J. D. (2003). Orientation mosaic in barn owl's visual Wulst revealed by optical imaging: Comparison with cat and monkey striate and extra-striate areas. *Brain Research*, *961*, 153–158.

Liu, T., Heeger, D. J., & Carrasco, M. (2006). Neural correlates of the visual vertical meridian asymmetry. *Journal of Vision*, *6*, 1294–1301.

Livio, M. (2003). *The golden ratio: The story of phi, the world's most astonishing number*. New York: Broadway Books.

Lloyd, R. (1989). The estimation of distance and direction from cognitive maps. *American Cartographer*, *16*, 109–122.

Lloyd, R., & Heivly, C. (1987) Systematic distortions in urban cognitive maps. *Annals of the Association of American Geographers*, *77*, 191–207.

Lobmaier, J. S., & Mast, F. W. (2007). The Thatcher illusion: Rotating the viewer instead of the picture. *Perception*, *36*, 537–546.

Locher, P., & Nagy, Y. (1996). Vision spontaneously establishes the percept of pictorial balance. *Empirical Studies of the Arts*, *14*, 17–31.

Locher, P. J., Cornelis, E., Wagemans, J., & Stappers, P. J. (2001). Artists' use of compositional balance for creative visual displays. *Empirical Studies of the Arts*, *19*, 213–227.

Locher, P. J., & Nodine, C. (1987). Symmetry catches the eye. In J. K. O'Regan & A. Levy-Schoen (Eds.), *Eye movements: Physiology to cognition* (pp. 353–361). Amsterdam: Elsevier.

Locher, P., & Nodine, C. (1989). The perceptual value of symmetry. *Computers & Mathematics with Applications*, *17*, 475–484.

Locher, P. J., Smith, J. K., & Smith, L. F. (2001). The influence of presentation format and viewer training in the visual arts on the perception of pictorial and aesthetic qualities in paintings. *Perception*, *30*, 449–465.

Locher, P. J., & Stappers, P. J. (2002). Factors contributing to the implicit dynamic quality of static abstract designs. *Perception*, *32*, 1093–1107.

Locher, P. J., Stappers, P. J., & Overbeeke, K. (1998). The role of balance as an organizing design principle underlying adults' compositional strategies for creating visual displays. *Acta Psychologica*, *99*, 141–161.

Loetscher, T., & Brugger, P. (2009). Random number generation in neglect patients reveals enhanced response stereotypy, but no neglect in number space. *Neuropsychologia*, *47*, 276–279.

Loetscher, T., Nicholls, M. E. R., Towse, J. N., Bradshaw, J. L., & Brugger, P. (2010). Lucky numbers: Spatial neglect affects physical, but not representational, choices in a Lotto task. *Cortex*, *46*, 685–690.

Loetscher, T., Schwarz, U., Schubiger, M., & Brugger, P. (2008). Head turns bias the brain's internal random generator. *Current Biology*, *18*(2), R60–R62.

Loftus, A. M., Nicholls, M. E., Mattingley, J. B., Chapman, H. L., & Bradshaw, J. L. (2009). Pseudoneglect for the bisection of mental number lines. *Quarterly Journal Experimental Psychology*, *62*(5), 925–945.

Loftus, E. F., & Marburger, W. (1983). Since the eruption of Mt. St. Helens, has anyone beaten you up? Improving the accuracy of retrospective reports with landmark events. *Memory & Cognition, 11*, 114–120.

Longo, M. R., & Lourenco, S. F. (2006). On the nature of near space: Effects of tool use and the transition to far space. *Neuropsychologia*, 44, 977–981.

Loomis, J. M., Da Silva, J. A., Fujita, N., & Fukusima, S. S. (1992). Visual space perception and visually directed action. *Journal of Experimental Psychology: Human Perception and Performance, 18*(4), 906–921.

Loomis, J. M., Da Silva, J. A., Philbeck, J. W., & Fukusima, S. S. (1996). Visual perception of location and distance. *Current Directions in Psychological Science, 5*, 72–77.

Loomis, J. M., & Philbeck, J. W. (1999). Is the anisotropy of perceived 3-D shape invariant across scale? *Attention, Perception, & Psychophysics, 61*, 397–402.

Loomis, J. M., & Philbeck, J. W. (2008). Measuring perception with spatial updating and action. In R. L. Klatzky, M. Behrmann, & B. MacWhinney (Eds.), *Embodiment, ego-space, and action* (pp. 1–43). Mahwah, NJ: Erlbaum.

Lopez, C., Bachofner, C., Mercier, M., & Blanke, O. (2009). Gravity and observer's body orientation influence the visual perception of human body postures. *Journal of Vision, 9*(1), 1–14.

Lopez, C., & Blanke, O. (2011). The thalamocortical vestibular system in animals and humans. *Brain Research Reviews, 67*(1–2), 119–146.

López-Moliner, J., & Linares, D. (2006). The flash-lag is reduced when the flash is perceived as a sensory consequence of our action. *Vision Research, 46*, 2122–2129.

Lourenco, S. F., Bonny, J. W., & Schwartz, B. L. (2016). Children and adults use physical size and numerical alliances in third-party judgments of dominance. *Frontiers in Psychology, 6*, 1–10.

Louwerse, M. M., & Jeuniaux, P. (2010). The linguistic and embodied nature of conceptual processing. *Cognition, 114*(1), 96–104.

Low, A., Lang, P. J., Smith, J. C., & Bradley, M. M. (2008). Both predator and prey emotional arousal in threat and reward. *Psychological Science, 19*(9), 865–873.

Lu, A., Zhang, H., He, G., Zheng, D., & Hodges, B. H. (2014). Looking up to others: Social status, Chinese honorifics, and spatial attention. *Canadian Journal of Experimental Psychology, 68*, 77–83.

Luck, S. J., & Vogel, E. K. (1997). The capacity of visual working memory for features and conjunctions. *Nature, 390*, 279–281.

Ludwig, V. U., Adachi, I., & Matzuzawa, T. (2011). Visuoauditory mappings between high luminance and high pitch are shared by chimpanzees (Pan troglodytes) and humans. *Proceedings of the National Academy of Sciences of the USA, 108*, 20661–20665.

Lundholm, H. (1921). The affective tone of lines: Experimental researches. *Psychological Review, 28*, 43–60.

Luneburg, R. K. (1947). *Mathematical analysis of binocular vision*. Princeton, NJ: Princeton University Press.

Luneburg, R. K. (1950). The metric of binocular visual space. *Journal of the Optical Society of America, 40*, 627–642.

Lunven, M., Thiebaut De Schotten, M., Bourlon, C., Duret, C., Migliaccio, R., Rode, G., & Bartolomeo, P. (2015). White matter lesional predictors of chronic visual neglect: A longitudinal study. *Brain, 138*(3), 746–760.

Lupón-Bas, N., Torrents-Gómez, A., Cardona, G., Da Silva, J. A., & Aznar-Casanova, J. A. (2014). New evidence of visual space anisotropy with autostereograms. *Psychology & Neuroscience, 7*(3), 261–267.

Lupyan, G. (2008). The conceptual grouping effect: Categories matter (and named categories matter more). *Cognition, 108*(2), 566–577.

Lupyan, G., Thompson-Schill, S. L., & Swingley, D. (2010). Conceptual penetration of visual processing. *Psychological Science, 21*(5), 682–691.

Luria, A. R. (1959). Disorders of "simultaneous perception" in a case of bilateral occipito-parietal brain injury. *Brain, 82*(3), 437–449.

Lynch, K. (1960). *The image of the city.* Cambridge, MA: MIT Press.

Maass, A., Pagani, D., & Berta, E. (2007). How beautiful is the goal and how violent is the fistfight? Spatial bias in the interpretation of human behavior. *Social Cognition, 25*(6), 833–852.

Maass, A., & Russo, A. (2003). Directional bias in the mental representation of spatial events: nature or culture? *Psychological Science, 14*(4), 296–301.

Maass, A., & Suitner, C. (2011). Spatial constraints of social cognition. *Social Psychology, 42*(3), 159–164.

Maass, A., Suitner, C., Favaretto, X., & Cignacchi, M. (2009). Groups in space: Stereotypes and the spatial agency bias. *Journal of Experimental Social Psychology, 45*(3), 496–504.

Maass, A., Suitner, C., & Nadhmi, F. (2014). What drives the spatial agency bias? An Italian–Malagasy–Arabic comparison study. *Journal of Experimental Psychology: General, 143*(3), 991–996.

Macaluso, E., Frith, C. D., & Driver, J. (2002). Supramodal effects of covert spatial orienting triggered by visual or tactile events. *Journal of Cognitive Neuroscience, 14*, 389–401.

Macevoy, S. P., & Epstein, R. A. (2007). Position selectivity in scene- and object-responsive occipitotemporal regions. *Journal of Neurophysiology, 98*, 2089–2098.

Mach, E. (1959). *The analysis of sensations.* New York: Dover.

Macmillan, N. A., & Creelman, C. D. (2008). *Detection theory: A user's guide* (2nd Ed.). New York: Psychology Press.

MacNeilage, P. R., Banks, M. S., Berger, D. R., & Bulthoff, H. H. (2007). A Bayesian model of the disambiguation of gravitoinertial force by visual cues. *Experimental Brain Research, 179*, 263–290.

Maeda, F., Kanai, R., & Shimojo, S. (2004). Changing pitch induced visual motion illusion. *Current Biology, 14*, R990–R991.

Maglio, S. J., Trope, Y., & Liberman, N. (2013). The common currency of psychological distance. *Current Directions in Psychological Science, 22*, 278–282.

Maguire, E. A., Intraub, H., & Mullally, S. L. (2016). Scenes, spaces, and memory traces: What does the hippocampus do? *Neuroscientist, 22*(5), 432–439.

Mai, L., Le, H., Niu, Y., & Liu, F. (2011). Rule of third detection from photograph. In *Proceedings of the IEEE International Symposium on Multimedia (ISM)* (pp. 91–96). Dana Point, CA.

Maiche, A., Budelli, R., & Gómez-Sena, L. (2007). Spatial facilitation is involved in flash-lag effect. *Vision Research, 47*, 1655–1661.

Maier, J. X., Chandrasekaran, C., & Ghazanfar, A. A. (2008). Integration of bimodal looming signals through neuronal coherence in the temporal lobe. *Current Biology, 18*(13), 963–968.

Maier, J. X., & Ghazanfar, A. A. (2007). Looming biases in monkey auditory cortex. *Journal of Neuroscience, 27*(15), 4093–4100.

Maier, J. X., Neuhoff, J. G., Logothetis, N. K., & Ghazanfar, A. A. (2004). Multisensory integration of looming signals by Rhesus monkeys. *Neuron, 43*(2), 177–181.

Mainwaring, S. D., Tversky, B., Ohgishi, M., & Schiano, D. J. (2003). Descriptions of simple spatial scenes in English and Japanese. *Spatial Cognition and Computation, 3*, 3–42.

Maki, R. H. (1981) Categorization and distance effects with spatial linear orders. *Journal of Experimental Psychology: Human Learning and Memory, 7*, 15–32.

Mamassian, P., & de Montalembert, M. (2010). A simple model of the vertical-horizontal illusion. *Vision Research, 50*(10), 956–962.

Mamassian, P., & Landy, M. S. (2001). Interaction of visual prior constraints. *Vision Research, 41*, 2653–2668.

Mandler, G. (1967). Organization and memory. In K. W. Spence & J. T. Spence (Eds.), *The psychology of learning and motivation* (Vol. 1). New York: Academic Press.

Mandler, J. M. (1992). How to build a baby: II. Conceptual primitives. *Psychological Review, 99*(4), 587–604.

Mannan, S. K., Mort, D. J., Hodgson, T. L., Driver, J., Kennard, C., & Husain, M. (2005). Revisiting previously searched locations in visual neglect: Role of right parietal and frontal lesions in misjudging old locations as new. *Journal of Cognitive Neuroscience*, *17*, 340–354.

Mannion, D. J., McDonald, J. S., & Clifford, C. W. (2010). Orientation anisotropies in human visual cortex. *Journal of Neurophysiology*, *103*, 3465–3471.

Manser, M. P., & Hancock, P. A. (1996). Influence of approach angle on estimates of time-to-contact. *Ecological Psychology*, *8*(1), 71–99.

Mansfield, R. J. W., & Ronner, S. F. (1978). Orientation anisotropy in monkey visual cortex. *Brain Research*, *149*, 229–234.

Mao, L., Zhou, B., Zhou, W., & Han, S. (2007). Neural correlates of covert orienting of visual spatial attention along vertical and horizontal dimensions. *Brain Research*, *1136*, 142–153.

Mapelli, D., Rusconi, E., & Umiltà, C. (2003). The SNARC effect: An instance of the Simon effect? *Cognition*, *88*(3), B1–B10.

Marangolo, P., Piccardi, L., & Rinaldi, M. C. (2003). Dissociation between personal and extrapersonal neglect in a crossed aphasia study. *Neurocase*, *9*(5), 414–420.

Mareschal, I., Morgan, M. J., & Solomon, J. A. (2010). Cortical distance determines whether flankers cause crowding or the tilt illusion. *Journal of Vision*, *10*(8), 1–14.

Marin, D., Pitteri, M., Della Puppa, A., Meneghello, F., Biasutti, E., Priftis, K., & Vallesi, A. (2016). Mental time line distortion in right-brain-damaged patients: Evidence from a dynamic spatiotemporal task. *Neuropsychology*, *30*(3), 338–345.

Mark, V. W., Kooistra, C. A., & Heilman, K. M. (1988). Hemispatial neglect affected by non-neglected stimuli. *Neurology*, *38*(8), 1207–1211.

Markman, A. B., & Gentner, D. (1993). Splitting the difference: A structural alignment view of similarity. *Journal of Memory and Language*, *32*, 517–535.

Markovits, H., & Benenson, J. F. (2010). Males outperform females in translating social relations into spatial positions. *Cognition*, *117*(3), 332–340.

Marks, L. E. (2004). Cross-modal interactions in speeded classification. In G. A. Calvert, C. Spence, & B. E. Stein (Eds.), *Handbook of multisensory processes* (pp. 85–105). Cambridge, MA: MIT Press.

Marr, D. (1982). *Vision*. San Francisco: Freeman.

Marsh, A. A., Yu, H. H., Schechter, J. C., & Blair, R. J. R. (2009). Larger than life: Humans' nonverbal status cues alter perceived size. *PLoS ONE*, *4*(5), e5707.

Marsh, E. B., & Hillis, A. (2008). Dissociation between egocentric and allocentric visuospatial and tactile neglect in acute stroke. *Cortex*, *44*, 1215–1220.

Marshall, J. C., & Halligan, P. W. (1990). Line bisection in a case of visual neglect: Psychophysical studies with implications of theory. *Cognitive Neuropsychology*, *7*, 107–130.

Masson, N., Pesenti, M., & Dormal, V. (2016). Duration and numerical estimation in right brain-damaged patients with and without neglect: Lack of support for a mental time line. *British Journal of Psychology*, *107*(3), 467–483.

Masson, N., Pesenti, M., & Dormal, V. (2017). Impact of optokinetic stimulation on mental arithmetic. *Psychonological Research*, *81*, 840–849.

Mast, F., & Jarchow, T. (1996). Perceived body position and the visual horizontal. *Brain Research Bulletin*, *40*, 393–397.

Mateeff, S., & Hohnsbein, J. (1988). Perceptual latencies are shorter for motion towards the fovea than for motion away. *Vision Research*, *28*, 711–719.

Mateeff, S., Hohnsbein, J., & Noack, T. (1985). Dynamic visual capture: Apparent auditory motion induced by a moving visual target. *Perception*, *14*, 721–727.

Mather, G. (2012). Aesthetic judgment of orientation in modern art. *i-Perception*, *3*, 18–24.

Mathys, C. D., Lomakina, E. L., Daunizeau, J., Iglesias, S., Brodersen, K. H., Friston, K. J., &

Stephan, K. E. (2014). Uncertainty in perception and the Hierarchical Gaussian Filter. *Frontiers in Human Neuroscience, 8.* doi:10.3389/fnhum.2014.00825

Mattingley, J. B., Bradshaw, J. L., & Bradshaw, J. A. (1995). The effects of unilateral visuospatial neglect on perception of Muller-Lyer illusory figures. *Perception, 24,* 415–433.

Mattingley, J. B., David, G., & Driver, J. (1997). Preattentive filling-in of visual surfaces in parietal extinction. *Science, 275,* 671–674.

Maturana, H., & Varela, F. (1980) *Autopoiesis and cognition: The realization of the living.* Boston, MA: Reidel.

Maus, G. W., & Nijhawan, R. (2006). Forward displacements of fading objects in motion: The role of transient signals in perceiving position. *Vision Research, 46,* 4375–4381.

Maus, G. W., & Nijhawan, R. (2009). Going, going, gone: Localizing abrupt offsets of moving objects. *Journal of Experimental Psychology: Human Perception and Performance, 35,* 611–626.

Maus, G. W., Weigelt, S., Nijhawan, R., & Muckli, L. (2010). Does area V3A predict positions of moving objects? *Frontiers of Psychology, 1,* 186.

Mayne R. (1974). A systems concept of the vestibular organs. In H. Kornhuber (Ed.), *Handbook of sensory physiology: Vestibular system: Psychophysics, applied aspects and general interpretations* (Vol. 6, pp. 493–580). Berlin: Springer-Verlag.

Mazur, A., Mazur, J., & Keating, C. (1984). Military rank attainment of a West Point class: Effects of cadets' physical features. *American Journal of Sociology, 90*(1), 125–150.

McBeath, M. K. (1990). The rising fastball: Baseball's impossible pitch. *Perception, 19*(4), 545–552.

McBeath, M. K., Brimhall, S. E., Miller, T. S., & Holloway, S. R. (2010). Naïve curvilinear impetus bias occurs for locomotion. *Journal of Vision, 10*(7), 1021.

McBeath, M. K., & Dye, R. A. (2015). Naïve psychophysics of physics teachers. *Abstracts of the Psychonomic Society, 20,* 18.

McBeath, M. K., & Morikawa, K. (1997). Forward-facing motion biases for rigid and nonrigid biologically likely transformations. *Perceptual and Motor Skills, 85,* 1187–1193.

McBeath, M. K., Morikawa, K., & Kaiser, M. K. (1992). Perceptual bias for forward-facing motion. *Psychological Science, 3,* 362–367.

McBeath, M. K., Nathan, A. M., Bahill, A. T., & Baldwin, D. G. (2008). Paradoxical pop-ups: Why are they hard to catch? *American Journal of Physics, 76*(8), 723–729.

McBeath, M. K., & Neuhoff, J. G. (2002). The Doppler effect is not what you think it is: Dramatic pitch change due to dynamic intensity change. *Psychonomic Bulletin & Review, 9*(2), 306–313.

McBeath, M. K., Schiano, D. J., & Tversky, B. (1997). Three-dimensional bilateral symmetry assumed in judging figural identity and orientation. *Psychological Science, 8*(3), 217–223.

McBeath, M. K., Shaffer, D. M., & Kaiser, M. K. (1995a). How baseball outfielders determine where to run to catch fly balls. *Science, 268* (5210), 569–573.

McBeath, M. K., Shaffer, D. M., & Kaiser, M. K. (1995b). Play ball. *Science 268*(5218), 1681–1685.

McBeath, M. K., Shaffer, D. M., & Kaiser, M. K. (1996). On catching fly balls. *Science, 273* (5272), 256–259.

McBeath, M. K., & Shepard, R. N. (1989). Apparent motion between shapes differing in location and orientation: A window technique for estimating path curvature. *Perception & Psychophysics, 46*(4), 333–337.

McBeath, M. K., & Sugar, T. G. (2005). Natural selection of asymmetric traits operates at multiple levels. *Behavioral and Brain Sciences, 28*(4), 605–606.

McCarthy, L., & Olsen, K. N. (2017). A "looming bias" in spatial hearing? Effects of acoustic intensity and spectrum on categorical sound source localization. *Attention, Perception, & Psychophysics, 79*(1), 352–362.

McCloskey, M. (1983a). Intuitive physics. *Scientific American, 248,* 122–130.

McCloskey, M. (1983b). Naive theories of motion. In D. Gentner & A. Stevens (Eds.), *Mental models* (pp. 299–324). Hillsdale, NJ: Erlbaum.

McCloskey, M., Caramazza, A., & Green, B. (1980). Curvilinear motion in the absence of external forces: Naive beliefs about the motion of objects. *Science, 210*(4474), 1139–1141.

McCloskey, M., Washburn, A., & Felch, L. (1983). Intuitive physics: The straight-down belief and its origin. *Journal of Experimental Psychology: Learning, Memory, and Cognition, 9*(4), 636–649.

McCloud, S. (1994). *Understanding comics: The invisible art*. Northampton, MA: William Morrow Paperbacks.

McCourt, M. E., & Jewell, G. (1999). Visuospatial attention in line bisection: Stimulus modulation of pseudoneglect. *Neuropsychologia, 37*, 843–855.

McDonough, L., Choi, S., & Mandler, J. M. (2003). Understanding spatial relations: Flexible infants, lexical adults. *Cognitive Psychology, 46*, 229–259.

McDunn, B. A., Siddiqui, A. P., & Brown, J. M. (2014). Seeking the boundary of boundary extension. *Psychonomic Bulletin & Review, 21*, 370–375. doi: 10.3758/s13423-013-0494-0

McFie, J., Piercy, M., & Zangwill, O. (1950). Visual-spatial agnosia associated with lesions of the right cerebral hemisphere. *Brain, 73*(2), 167–190.

McGeorge, P., Beschin, N., Colnaghi, A., Rusconi, M. L., & Della Sala, S. (2007). A lateralized bias in mental imagery: evidence for representational pseudoneglect. *Neuroscience Letters, 421*, 259–263.

McGeorge, P., Beschin, N., & Della Sala, S. (2006). Representing target motion: The role of the right hemisphere in the forward displacement bias. *Neuropsychology, 20*, 708–715.

McGuire, A., Gillath, O., & Vitevitch, M. (2016). Effects of mental resource availability on looming task performance. *Attention, Perception, & Psychophysics, 78*(1), 107–113.

McGurk, H., & MacDonald, J. (1976). Hearing lips and seeing voices. *Nature, 264*, 746–748.

McIntyre, J., Zago, M., Berthoz, A., & Lacquaniti, F. (2001). Does the brain model Newton's laws? *Nature Neuroscience, 4*, 693–694.

McKee, S. P., & Welch, L. (1992). The precision of size constancy. *Vision Research, 32*(8), 1447–1460.

McKyton, A., & Zohary, E. (2007). Beyond retinotopic mapping: The spatial representation of objects in the human lateral occipital complex. *Cerebral Cortex, 17*, 1164–1172.

McLeod, R. W., & Ross, H. E. (1983). Optic-flow and cognitive-factors in time-to-collision estimates. *Perception, 12*(4), 417–423.

McManus, I. C. (1980). The aesthetics of simple figures. *British Journal of Psychology, 71*, 505–524.

McManus, I. C. (2005). Symmetry and asymmetry in aesthetics and the arts. *European Review, 13*, 157–180.

McManus, I. C., Cook, R., & Hunt, A. (2010). Beyond the golden section and normative aesthetics: Why do individuals differ so much in their aesthetic preference for rectangles? *Psychology of Aesthetics, Creativity, and the Arts, 4*, 113–126.

McManus, I. C., Edmondson, D., & Rodgers, J. (1985). Balance in pictures. *British Journal of Psychology, 76*, 311–324.

McManus, I. C., & Humphrey, N. K. (1973). Turning the left cheek. *Nature, 243*, 271–272.

McManus, I. C., & Kitson, C. M. (1995). Compositional geometry in pictures. *Empirical Studies of the Arts, 13*, 73–94.

McManus, I. C., Stöver, K., & Kim, D. (2011). Arnheim's Gestalt theory of visual balance: Examining the compositional structure of art photographs and abstract images. *I-Perception, 2*, 615–647.

McManus, I. C., & Thomas, P. (2007). Eye centering in portraits: A theoretical and empirical evaluation. *Perception, 36*, 167–182.

McManus, I. C., & Weatherby, P. (1997). The golden section and the aesthetics of form and composition: A cognitive model. *Empirical Studies of the Arts, 15*, 209–232.

McManus, I. C., & Wu, W. (2013). "The square is . . . bulky, heavy, contented, plain, good-natured stupid. . .": A cross-cultural study of the aesthetics and meanings of rectangles. *Psychology of Aesthetics, Creativity, and the Arts, 7*, 130–139.

McNamara T. P. (1986). Mental representations of spatial relations. *Cognitive Psychology, 18*(1), 87–121.

McNamara, T. P. (1991). Memory's view of space. In G. H. Bower (Ed.), *The psychology of learning and motivation: Advances in research and theory* (Vol. 27, pp. 147–186). San Diego, CA: Academic Press.

McNamara, T. P., & Diwadkar, V. A. (1997). Symmetry and asymmetry of human spatial memory. *Cognitive Psychology, 34*(2), 160–190.

McWhinnie, H. J. (1986). A review of the use of symmetry, the golden section, and dynamic symmetry in contemporary art. *Leonardo, 19*, 241–245.

Medin, D. L., Goldston, R. L., & Gentner, D. (1990). Similarity involving attributes and relations: Judgments of similarity and difference are not inverses. *Psychological Science, 1*, 64–69.

Meier, B. P., & Dionne, S. (2009). Downright sexy: Verticality, implicit power, and perceived physical attractiveness. *Social Cognition, 27*(6), 883–892.

Meier, B. P., Hauser, D. J., Robinson, M. D., Friesen, C. K., & Schjeldahl, K. (2007). What's "up" with God? Vertical space as a representation of the divine. *Journal of Personality and Social Psychology, 93*(5), 699–710.

Meier, B. P., & Robinson, M. D. (2004). Why the sunny side is up. Associations between affect and vertical position. *Psychological Science, 15*, 243–247.

Meilinger, T., Frankenstein, J., & Bülthoff, H. H. (2013). Learning to navigate: Experience versus maps. *Cognition, 129*(1), 24–30.

Meilinger, T., Frankenstein, J., & Bülthoff, H. H. (2014, November). When in doubt follow your nose – a wayfinding strategy. *Frontiers in Psychology, 5*, 1–7.

Meilinger, T., Franz, G., & Bülthoff, H. H. (2012). From isovists via mental representations to behaviour: First steps toward closing the causal chain. *Environment and Planning B-Planning & Design, 39*(1), 48–62.

Meilinger, T., Hölscher, C., Büchner, S. J., & Brösamle, M. (2007). How much information do you need? Schematic maps in wayfinding and self localisation. In T. Barkowsky, M. Knauff, G. Ligozat, & D. R. Montello (Eds.), *Spatial cognition v reasoning, action, interaction* (Vol. 4387, pp. 381–400). Berlin: Springer.

Melamed, T., & Bozionelos, N. (1992). Managerial promotion and height. *Psychological Reports, 71*(2), 587–593.

Melara, R. D., & Marks, L. E. (1990). Interaction among auditory dimensions: Timbre, pitch, and loudness. *Perception & Psychophysics, 48*, 169–178.

Melara, R. D., & O'Brien, T. P. (1987). Interaction between synesthetically corresponding dimensions. *Journal of Experimental Psychology: General, 116*, 323–336.

Meng, J. C., & Sedgwick, H. A. (2001). Distance perception mediated through nested contact relations among surfaces. *Attention, Perception, & Psychophysics, 63*, 1–15.

Menon, T., Sim, J., Fu, J. H. Y., Chiu, C. Y., & Hong, Y. Y. (2010). Blazing the trail versus trailing the group: Culture and perceptions of the leader's position. *Organizational Behavior and Human Decision Processes, 113*(1), 51–61.

Menzel, R., Geiger, K., Chittka, L., Joerges, J., Kunze, J., & Müller, U. (1996). The knowledge base of bee navigation. *Journal of Experimental Biology, 199*(1), 141–146.

Merfeld, D., Zupan, L., & Peterka, R. (1999). Humans use internal models to estimate gravity and linear acceleration. *Nature, 398*, 615–618.

Merleau-Ponty, M. (1945). *The phenomenology of perception*. Paris: Editions Gallimard.

Mettinger, A. (1994). *Aspects of semantic opposition in English*. Oxford: Clarendon Press.

Metzger W. (1930). Optische Untersuchungen am Ganzfeld: II. Zur Phanomenologie des homogenen Ganzfelds [Optical researches on the total field [of vision]. II. On the phenomenology of homogeneous total fields]. *Psychologische Forshung, 13*, 6–29.

Metzger, W. (1932). Versuch einer gemeinsamen Theorie der Phänomene Fröhlichs und Hazelhoffs und Kritik ihrer Verfahren zur Messung der Empfindungszeit [An attempt to unify the theories of Fröhlich and Hazelhoff's phenomena and a review of their method of measuring the time of sensation]. *Psychologische Forschung, 16*(1), 176–200.

Meyer, G. F., & Wuerger, S. M. (2001). Cross-modal integration of auditory and visual motion signals. *Neuroreport, 12*, 2557–2560.

Meyer, G. F., Wuerger, S. M., Röhrbein, F., & Zetzsche, C. (2005). Low-level integration of auditory and visual motion signals requires spatial co-localization. *Experimental Brain Research, 166*, 538–547.

Meyer, L. B. (1956). *Emotion and meaning in music*. Chicago: University of Chicago Press.

Michaels, C. F., & Carello, C. (1981). *Direct perception*. Englewood Cliffs, NJ: Prentice-Hall.

Michotte, A. (1963). *The perception of causality*. T. Miles & E. Miles (Trans.), New York: Basic Books. (Original work published 1946)

Mikellidou, K., & Thompson, P. (2013). The vertical-horizontal illusion: Assessing the contributions of anisotropy, abutting, and crossing to the misperception of simple line stimuli. *Journal of Vision, 13*(8), 1–11.

Miles, L. K., Tan, L., Noble, G. D., Lumsden, J., & Macrae, C. N. (2011). Can a mind have two time lines? Exploring space-time mapping in Mandarin and English speakers. *Psychonomic Bulletin & Review, 18*, 598–604.

Milgram, S., & Jodelet, D. (1976). Psychological maps of Paris. In H. Proshansky, W. Ittelson, & L. Rivlin (Eds.), *Environmental psychology* (2nd Ed., pp. 104–112). New York: Holt, Rinehart and Winston.

Millar, S., & Al-Attar, Z. (2000). Vertical and bisection bias in active touch. *Perception, 29*, 481–500.

Miller, E. F., & Graybiel, A. (1962). Counterrolling of the human eyes produced by head tilt with respect to gravity. *Acta Otolaryngology, 54*, 479–501.

Miller, G. A., & Fellbaum, C. (1991). Semantic networks of English. *Cognition, 41*, 197–229.

Miller, J. O. (1982). Divided attention: Evidence for coactivation with redundant signals. *Cognitive Psychology, 14*, 247–279.

Miller, J. O. (1991). Channel interaction and the redundant targets effect in bimodal divided attention. *Journal of Experimental Psychology: Human Perception and Performance, 17*, 160–169.

Miller, R. J. (2007). Another slant on the oblique effect in drawings and paintings. *Empirical Studies of the Arts, 27*, 25–42.

Millis, K. (2001). Making meaning brings pleasure: The influence of titles on aesthetic experiences. *Emotion, 1*, 320–329.

Milner, A. D., Brechmann, M., & Pagliarini, L. (1992). To halve and to halve not: An analysis of line bisection judgements in normal subjects. *Neuropsychologia, 30*, 515–526.

Milner, A. D., & Goodale, M. A. (1995). *The visual brain in action*. New York: Oxford University Press.

Milner, A. D., Harvey, M., Roberts, R. C., & Forster, S. V. (1993). Line bisection errors in visual neglect: Misguided action or size distortion? *Neuropsychologia, 31*, 39–49.

Mishkin, M., Ungerleider, L. G., & Macko, K. A. (1983). Object vision and spatial vision: Two cortical pathways. *Trends in Neurosciences, 6*, 414–417.

Mittelstaedt, H. (1983). A new solution to the problem of the subjective vertical. *Naturwissenschaften, 70*, 272–281.

Mittelstaedt, H. (1996). Somatic graviception. *Biological Psychology, 42*(1), 53–74.

Moeller, K., Neuburger, S., Kaufmann, L., Landerl, K., & Nuerk, H. C. (2009). Basic number processing deficits in developmental dyscalculia: Evidence from eye tracking. *Cognitive Development, 24*(4), 371–386.

Mohr, C., & Leonards, U. (2007). Rightward bisection errors for letter lines: The role of semantic information. *Neuropsychologia, 45*, 295–304.

Møller, A. P. (1992). Female swallow preference for symmetric male sexual ornaments. *Nature*, *357*, 238–240.

Monacelli, A. M., Cushman, L. A., Kavcic, V., & Duffy, C. J. (2003). Spatial disorientation in Alzheimer's disease. *Neurology*, *61*(11), 1491–1497.

Montello, D. R. (1991). Spatial orientation and the angularity of urban routes: A field study. *Environment and Behavior*, *23*, 47–69.

Montello, D. R. (1998). A new framework for understanding the acquisition of spatial knowledge in large-scale environments. In M. J. Egenhofer & R. G. Golledge (Eds.), *Spatial and temporal reasoning in geographic information systems* (pp. 143–154). New York: Oxford University Press.

Montello, D. R. (2003). Regions in geography: Process and content. In M. Duckham, M. F. Goodchild, & M. F. Worboys (Eds.), *Foundations of geographic information science* (pp. 173–189). London: Taylor & Francis.

Montello, D. R. (2005). Navigation. In P. Shah & A. Miyake (Eds.), *The Cambridge handbook of visuospatial thinking* (pp. 257–294). Cambridge: Cambridge University Press.

Montgomery, J., Kusano, K. D., & Gabler, H. C. (2014). Age and gender differences in time to collision at braking from the 100-car naturalistic driving study. *Traffic Injury Prevention*, *15*, S15–S20.

Moore, B. C. J. (2003). *An introduction to the psychology of hearing* (5th Ed.) London: Academic Press.

Moore, C. M., & Enns, J. T. (2004). Object updating and the flash-lag effect. *Psychological Science*, *15*, 866–871.

Morein-Zamir, S., Soto-Faraco, S., & Kingstone, A. (2003). Auditory capture of vision: Examining temporal ventriloquism. *Cognitive Brain Research*, *17*, 154–163.

Morgan, M. J. (1977). *Molyneux's Question: Vision, touch and the philosophy of perception*. Cambridge: Cambridge University Press.

Morgan, M. J. (1990). Hyperacuity. In D. Regan (Ed.), *Spatial vision* (pp. 87–113). London: Macmillan.

Morgan, M. J. (1992). On the scaling of size judgments by angular cues. *Vision Research*, *32*, 1433–1455.

Morgan, M. J. (1996). Visual illusions. In V. Bruce (Ed.), *Unsolved mysteries of the mind* (pp. 29–58). Hove, East Sussex: Earlbaum UK.

Morgan, M. J. (1999). The Poggendorff illusion: A bias in the estimation of the orientation of virtual lines by second-stage filters. *Vision Research*, *39*(14), 2361–2380.

Morgan, M. (2003). *The space between our ears: How the brain represents the visual space*. London: Weidenfeld & Nicolson.

Morgan, M. J. (2005). The visual computation of 2-D area by human observers. *Vision Research*, *45*, 2564–2570.

Morgan, M. J. (2015). Dissociating perceptual and occulomotor localization of moving objects. *Current Biology*, *25*, R825–R844.

Morgan, M. J., & Baldassi, S. (1997). How the human visual system encodes the orientation of a texture, and why it makes mistakes. *Current Biology*, *7*(12), 999–1002.

Morgan, M. J., & Casco, C. (1990). Spatial filtering and spatial primitives in early vision: An explanation of the Zollner-Judd class of geometrical illusion. *Proceedings of the Royal Society B: Biological Sciences*, *242*, 1–10.

Morgan, M. J., & Dillenburger, B. (2016) Geometric features underlying the perception of colinearity. *Vision Research*, *128*, 83–94.

Morgan, M., Dillenburger, B., Raphael, S., & Solomon, J. A. (2012). Observers can voluntarily shift their psychometric functions without losing sensitivity. *Attention, Perception, & Psychophysics*, *74*(1), 185–193.

Morgan, M. J., & Glennerster, A. (1991) Efficiency of locating centres of dot-clusters by human observers. *Vision Research*, *31*, 2075–2083.

Morgan, M. J., Grant, S., Melmoth, D., & Solomon, J. A. (2015). Tilted frames of reference have similar effects on the perception of the gravitational vertical and the planning of saccadic eye movement.

Experimental Brain Research, 233, 2115–2125

Morgan, M., Hole, G. J., & Glennerster, A. (1990). Biases and sensitivities in geometrical illusions. *Vision Research, 30*, 1793–1810.

Morgan, M. J., Hole, G. J., & Ward, R. M. (1990). Evidence for positional encoding in hyperacuity. *Journal of the Optical Society of America, 7*, 297–304.

Morgan, M. J., & Hotopf, W. H. (1989). Perceived diagonals in grids and lattices. *Vision Research, 29*(8), 1005–1015.

Morgan, M. J., Medford, A., & Newsome, P. (1995). The orthogonal orientation shift and spatial filtering. *Perception, 24*(5), 513–524.

Morgan, M. J., & Moulden, B. (1986). The Munsterberg figure and twisted cords. *Vision Research, 26*, 1793–1800.

Morgan, M. J., Raphael, S., Tibber, M. S., & Dakin, S.C. (2014). A texture-processing model of the "visual sense of number". *Proceedings of the Royal Society B 281*, 20141137.

Morgan, M. J., & Regan, D. (1987). Opponent model for line interval discrimination: Interval and vernier acuity compared. *Vision Research, 27*, 107–118.

Morgan, M. J., & Ward, R. M. (1985). Spatial and spatial-frequency primitives in spatial-interval discrimination. *Journal of the Optical Society of America, 2*, 1205–1210.

Morgan, M. J., Watamanuik, S. N., & McKee, S. P. (2000). The use of an implicit standard for measuring discrimination thresholds. *Vision Research, 40*(17), 109–117.

Morgan, M. J., & Watt, R. J. (1989). The Weber's Law for position is not an artefact of eccentricity. *Vision Research, 29*, 1457–1462.

Morikawa, K. (1999). Symmetry and elongation of objects influence perceived direction of translational motion. *Perception & Psychophysics, 61*(1), 134–143.

Morikawa, K., & McBeath, M. K. (1992). Lateral motion bias associated with reading direction. *Vision Research, 32*(6), 1137–1141.

Morrison, J. B., & Tversky, B. (2005). Bodies and their parts. *Memory & Cognition, 33*, 696–709.

Morrongiello, B. A., Hewitt, K. L., & Gotowiec, A. (1991). Infants discrimination of relative distance in the auditory modality – approaching versus receding sound sources. *Infant Behavior & Development, 14*(2), 187–208.

Morsella, E. (2005). The function of phenomenal states: Supramodular interaction theory. *Psychological Review, 112*, 1000–1021.

Mort, D. J., Malhotra, P., Mannan, S. K., Rorden, C., Pambakian, A., Kennard, C., et al. (2003). The anatomy of visual neglect. *Brain, 126*, 1986–1997.

Moser, E. I., Kropff, E., & Moser, M.-B. (2008). Place cells, grid cells, and the brain's spatial representation system. *Annual Review of Neuroscience, 31*, 69–89.

Mossbridge, J. A., Grabowecky, M., & Suzuki, S. (2011). Changes in auditory frequency guide visual-spatial attention. *Cognition, 121*, 133–139.

Motes, M. A., Hubbard, T. L., Courtney, J. R., & Rypma, B. (2008). A principal components analysis of dynamic spatial memory biases. *Journal of Experimental Psychology: Learning, Memory, and Cognition, 34*, 1076–1083.

Mou, W., Fan, Y., McNamara, T. P., & Owen, C. (2008). Intrinsic frames of reference and egocentric viewpoints in scene recognition. *Cognition, 106*, 750–769.

Mou, W., McNamara, T. P., Rump, B., & Xiao, C. (2006). Roles of egocentric and allocentric spatial representations in locomotion and reorientation. *Journal of Experimental Psychology: Learning, Memory, and Cognition, 32*, 1274–1290.

Mou, W., McNamara, T. P., Valiquiette, C. M., & Rump, B. (2004). Allocentric and egocentric updating of spatial memories. *Journal of Experimental Psychology: Learning, Memory, and Cognition, 30*, 142–157.

Mudd, S. A. (1963). Spatial stereotypes of four dimensions of pure tone. *Journal of Experimental Psychology, 66*, 347–352.

Mudd, S. A. (1965). Experimental evaluation of binary pure-tone auditory displays. *Journal of Applied Psychology, 49*, 112–121.

Mudd, S. A., & McCormick, J. (1960). The use of auditory cues in a visual search task. *Journal of Applied Psychology, 44*, 184–188.

Muelenz, C., Hecht, H., & Gamer, M. (2010). Testing the egocentric mirror-rotation hypothesis. *Seeing and Perceiving, 23*(5–6), 373–383.

Mueller, A. S., & Timney, B. (2014). Effects of radial direction and eccentricity on acceleration perception. *Perception, 43*(8), 805–810.

Mueller, U., & Mazur, A., (2001). Evidence of unconstrained directional selection for male tallness. *Behavioral Ecology and Sociobiology, 50*(4), 302–311.

Mullally, S. L., Intraub, H., & Maguire, E. A. (2012). Attenuated boundary extension produces a paradoxical memory advantage in amnesic patients. *Current Biology, 22*(4), 261–268.

Müller, D., & Schwarz, W. (2007). Is there an internal association of numbers to hands? The task set influences the nature of the SNARC effect. *Memory & Cognition, 35*(5), 1151–1161.

Müller, G. E. (1931). Erklärung der Erscheinungen eines mit konstanter Geschwindigkeit bewegten Lichtstreifens, insbesondere auch des Pihl-Fröhlichschen Phänomens [Explanation of the appearances of a lit bar moving at constant velocity, in particular the phenomenon of Pihl-Fröhlich]. *Zeitschrift für Sinnesphysiologie, 62*, 167–202.

Muller-Lyer, F. C. (1889). Optische Urteilstauschungen. *Arch Fur Physiologie*, 263–270.

Muller-Lyer, F. C. (1896a). Zur Lehre von den optischen Täuschungen: Über Kontrast und Konfluxion. *Zeitschrift fur Psychologie und Physiologie der Sinnesorgane, 9*, 1–16.

Muller-Lyer, F. C. (1896b). Über Kontrast und Konfluxion. (Zweiter Artikel). *Zeitschrift fur Psychologie und Physiologie der Sinnesorgane, 10*, 421–431.

Multhaup, K. S., Munger, M. P. & Smith, K. C. (2016). Boundary extension is sensitive to hand position in young and older adults. Journals of Gerontology Series B: Psychological Sciences and Social Sciences. Advance online publication. *doi:10.1093/geronb/gbw011*

Munger, M. P. (2015). The ministry of silly walks' report: Representational momentum sensitive to awkwardness following action, not single posture. *Visual Cognition, 23*, 796–808.

Munger, M. P., Dellinger, M. C., Lloyd, T. G., Johnson-Reid, K., Tonelli, N. J., Wolf, K., & Scott, J. M. (2006). Representational momentum in scenes: Learning spatial layout. *Memory & Cognition, 34*, 1557–1568.

Munger, M. P., & Minchew, J. H. (2002). Parallels between remembering and predicting an object's location. *Visual Cognition, 9*, 177–194.

Munger, M. P., & Multhaup, K. S. (2016). No imagination effect on boundary extension. *Memory & Cognition, 44*, 73–88.

Munger, M. P., & Owens, T. R. (2004). Representational momentum and the flash-lag effect. *Visual Cognition, 11*, 81–103.

Munger, M. P., Owens, T. R., & Conway, J. E. (2005). Are boundary extension and representational momentum related? *Visual Cognition, 12*, 1041–1056.

Munger, M. P., Solberg, J. L., & Horrocks, K. K. (1999). The relationship between mental rotation and representational momentum. *Journal of Experimental Psychology: Learning, Memory, and Cognition, 25*, 1557–1568.

Munger, M. P., Solberg, J. L., Horrocks, K. K., & Preston, A. S. (1999). Representational momentum for rotations in depth: Effects of shadings and axis. *Journal of Experimental Psychology: Learning, Memory, and Cognition, 25*, 157–171.

Murakami, I. (2001a). The flash-lag effect as a spatiotemporal correlation structure. *Journal of Vision, 1*(2), 6.

Murakami, I. (2001b). A flash-lag effect in random motion. *Vision Research, 41*, 3101–3119.

Murphy, L. (2003). *Semantic relations and the lexicon: Antonyms, synonyms and other semantic paradigms.* Cambridge: Cambridge University Press.

Murray, G. R., & Schmitz, J. D. (2011). Caveman politics: Evolutionary leadership preferences and physical stature. *Social Science Quarterly*, *92*(5), 1–21.

Müsseler, J., & Aschersleben, G. (1998). Localizing the first position of a moving stimulus: The Fröhlich effect and an attention-shifting explanation. *Perception & Psychophysics*, *60*, 683–695.

Müsseler, J., & Kerzel, D. (2004). The trial context determines adjusted localization of stimuli: Reconciling the Fröhlich and Onset Repulsion Effects. *Vision Research*, *44*(19), 2201–2206.

Müsseler, J., & Neumann, O. (1992). Apparent distance reduction with moving stimuli (tandem effect): Evidence for an attentional-shifting model. *Psychological Research*, *54*, 246–266.

Müsseler, J., Stork, S., & Kerzel, D. (2002). Comparing mislocalizations with moving stimuli: The Fröhlich effect, the flash-lag, and representational momentum. *Visual Cognition*, *9*, 120–138.

Müsseler, J., Stork, S., & Kerzel, D. (2008). Localizing the onset of moving stimuli by pointing or relative judgment: Variations in the size of the Fröhlich effect. *Vision Research*, *48*, 611–617.

Müsseler, J., & Tiggelbeck, J. (2013). The perceived onset position of a moving target: Effects of trial contexts are evoked by different attentional allocations. *Attention, Perception, & Psychophysics*, *75*(2), 349–357.

Näätänen, R. (1990). The role of attention in auditory information processing as revealed by event-related potentials and other brain measures of cognitive function. *Behavioral and Brain Sciences*, *13*, 201–288.

Näätänen, R., Gaillard, A. W. K., & Mäntysalo, S. (1978). Early selective attention effect on evoked potential reinterpreted. *Acta Psychologica*, *42*, 313–329.

Nachmias, J. (2008). Judging spatial properties of simple figures. *Vision Research*, *48*, 1290–1296.

Nachson, I., Argaman, E., & Luria, A. (1999). Effects of directional habits and handedness on aesthetic preference for left and right profiles. *Journal of Cross-Cultural Psychology*, *30*, 106–114.

Nagai, M., Kazai, K., & Yagi, A. (2002). Larger forward displacement in the direction of gravity. *Visual Cognition*, *9*, 28–40.

Nagai, M., & Saiki, J. (2005). Illusory motion and representational momentum. *Perception & Psychophysics*, *67*, 855–866.

Nagai, M., Suganuma, M., Nijhawan, R., Freyd, J. J., Miller, G., & Watanabe, K. (2010). Conceptual influence on the flash-lag effect and representational momentum. In R. Nijhawan & B. Khurana (Eds.), *Space and time in perception and action* (pp. 366–378). New York: Cambridge University Press.

Nagai, M., & Yagi, A. (2001). The pointedness effect on representational momentum. *Memory & Cognition*, *29*, 91–99.

Naito, T., Sato, H., & Osaka, N. (2010). Direction anisotropy of human motion perception depends on stimulus speed. *Vision Research*, *50*(18), 1862–1866.

Nakamoto, H., Mori, S., Ikudome, S., Unenaka, S., & Imanaka, K. (2015). Effects of sport expertise on representational momentum during timing control. *Attention, Perception, & Psychophysics*, *77*, 961–971.

Namba, J., & Baldo, M. V. C. (2004). The modulation of the flash-lag effect by voluntary attention. *Perception*, *33*, 621–631.

Nardini, M., Begus, K., & Mareschal, D. (2013). Multisensory uncertainty reduction for hand localization in children and adults. *Journal of Experimental Psychology: Human Perception and Performance*, *39*(3), 773–787.

Nardini, M., Jones, P., Bedford, R., & Braddick, O. (2008). Development of cue integration in human navigation. *Current Biology*, *18*(9), 689–693.

Nasr, S., & Tootell, R. B. H. (2012). A cardinal orientation bias in scene-selective visual cortex. *Journal of Neuroscience*, *32*(43), 14921–14926.

Nathan, M. B., Shaki, S., Salti, M., & Algom, D. (2009). Numbers and space: Associations and dissociations. *Psychonomic Bulletin & Review*, *16*(3), 578–582.

Naughtin, C. K., Mattingley, J. B., & Dux, P. E. (2016). Distributed and overlapping neural substrates for object individuation and identification in visual short-term memory. *Cerebral Cortex, 26*(2), 566–575.

Naughtin, C. K., Tamber-Rosenau, B. J., & Dux, P. E. (2014). The neural basis of temporal individuation and its capacity limits in the human brain. *Journal of Neurophysiology, 111*(3), 499–512.

Nava, E., Grassi, M., & Turati, C. (2016). Audio-visual, visuo-tactile and audio-tactile correspondences in preschoolers. *Multisensory Research, 29*, 93–111.

Nelson, T. M., & MacDonald, G. A. (1971). Lateral organization, perceived depth, and title preference in pictures. *Perceptual and Motor Skills, 33*, 983–986.

Neuhoff, J. G. (1998). Perceptual bias for rising tones. *Nature, 395*(6698), 123–124.

Neuhoff, J. G. (1999). Perception of changes in loudness – reply. *Nature, 398*(6729), 673–674.

Neuhoff, J. G. (2001). An adaptive bias in the perception of looming auditory motion. *Ecological Psychology, 13*, 87–110.

Neuhoff, J. G. (2016). Looming sounds are perceived as faster than receding sounds. *Cognitive Research: Principles and Implications 1*, 15.

Neuhoff, J. G., Hamilton, G. R., Gittleson, A. L., & Mejia, A. (2014). Babies in traffic: Infant vocalizations and listener sex modulate auditory motion perception. *Journal of Experimental Psychology-Human Perception and Performance, 40*(2), 775–783.

Neuhoff, J. G., Long, K. L., & Worthington, R. C. (2012). Strength and physical fitness predict the perception of looming sounds. *Evolution and Human Behavior, 33*(4), 318–322.

Neuhoff, J. G., & McBeath, M. (1996). The Doppler illusion: The influence of dynamic intensity change on perceived pitch. *Journal of Experimental Psychology: Human Perception and Performance, 22*(4), 970–985.

Neuhoff, J. G., McBeath, M., & Wanzie, W. (1999). Dynamic frequency change influences loudness perception: A central, analytic process. *Journal of Experimental Psychology: Human Perception and Performance, 25*(4), 1050–1059.

Neuhoff, J. G., Planisek, R., & Seifritz, E. (2009). Adaptive sex differences in auditory motion perception: Looming sounds are special. *Journal of Experimental Psychology: Human Perception and Performance, 35*(1), 225–234.

Neumann, O., & Müsseler, J. (1990b). Visuelles Fokussieren: Das Wetterwart-Modell und einige seiner Anwendungen [Visual focusing: The weather-station model and its applications] In C. Meinecke & L. Kehrer (Eds.), *Bielefelder Beiträge zur Kognitionspsychologie* (pp. 77–108). Göttingen: Hogrefe.

Neumann, O., & Scharlau, I. (2006). Visual attention and the mechanism of metacontrast. *Psychological Research, 71*, 626–633.

Neuroskeptic. (2012). The nine circles of scientific hell. *Perspectives on Psychological Science, 7*(6), 643–644.

Newcombe, N. S., & Chiang, N. C. R. (2007). Learning geographical information from hypothetical maps. *Memory & Cognition, 35*(5), 895–909.

Newcombe, N., Huttenlocher, J., Sandberg, E., Lie, E., & Johnson, S. (1999). What do misestimations and asymmetries in spatial judgement indicate about spatial representation? *Journal of Experimental Psychology: Learning, Memory, and Cognition, 25*(4), 986–996.

Newcombe, N, & Liben, L. (1982) Barrier effects in the cognitive maps of children and adults. *Journal of Experimental Child Psychology, 34*, 46–58.

Newcombe, N. S., Levine, S. C., & Mix, K. S. (2015), Thinking about quantity: The intertwined development of spatial and numerical cognition. *WIREs Cognitive Science, 6*, 491–505.

Newman, D. G. (2007). An overview of spatial disorientation as a factor in aviation accidents and incidents. Canberra, ACT, Australia: Australian Transport Safety Bureau, Research and Analysis Report B2007/0063.

Newsome, L. R. (1972). Visual angle and apparent size of objects in peripheral vision. *Perception & Psychophysics, 12*, 300–304.

Nicholls, M. E. R., Bradshaw, J. L. & Mattingley, J. B. (1999). Free-viewing perceptual asymmetries for the judgement of brightness, numerosity and size. *Neuropsychologia, 37*(3), 307–14.

Nicholls, M. E. R., Clode, D., Lindell, A. K., & Wood, A. G. (2002). Which cheek to turn? The effect of gender and emotional expressivity on posing behaviour. *Brain and Cognition, 48*, 480–484.

Nicholls, M. E. R., Clode, D., Wood, S. J., & Wood, A. G. (1999). Laterality of expression in portraiture: Putting your best cheek forward. *Proceedings of the Royal Society of London (Section B), 266*, 1517–1522.

Nicholls, M. E. R., Loftus, A., Mayer, K., & Mattingley, J. (2007). Things that go bump in the right: The effect of unimanual activity on rightward collisions. *Neuropsychologia, 45*, 1122–1126.

Nicholls, M. E. R., Mattingley, J. B., Berberovic, N., Smith, A., & Bradshaw, J. L. (2004). An investigation of the relationship between free-viewing perceptual asymmetries for vertical and horizontal stimuli. *Cognitive Brain Research, 19*, 289–301.

Nicholls, M. E. R., Wolfgang, B. J., Clode, D., & Lindell, A. K. (2002). The effect of left and right poses on the expression of facial emotion. *Neuropsychologia, 40*, 1662–1665.

Nicolay, C. W., & Walker, A. L. (2005). Grip strength and endurance: Influences of anthropometric variation, hand dominance, and gender. *International Journal of Industrial Ergonomics, 35*(7), 605–618.

Nicoletti, R., Anzola, G. P., Luppino, G., Rizzolatti, G., & Umiltà, C. (1982). Spatial compatibilità effects on the same side of the body midline. *Journal of Experimental Psychology: Human Perception and Performance, 8*, 664–673.

Nicoletti, R., & Umiltà, C. (1989). Splitting visual space with attention. *Journal of Experimental Psychology: Human Perception and Performance, 15*, 164–169.

Nicoletti, R., & Umiltà, C. (1994). Attention shifts produce spatial stimulus codes. *Psychological Research, 56*(3), 144–150.

Nieder, A., & Dehaene, S. (2009). Representation of number in the brain. *Annual Review of Neuroscience, 32*, 185–208.

Nielsen, K. E., Intriligator, J., & Barton, J. J. S. (1999). Spatial representation in the normal visual field. *Neuropsychologia, 37*, 267–277.

Nieman, D., Nijhawan, R., Khurana, B., & Shimojo, S. (2006). Cyclopean flash-lag illusion. *Vision Research, 46*, 3909–3914.

Nijhawan, R. (1994). Motion extrapolation in catching. *Nature, 370*, 256–257.

Nijhawan, R. (1997). Visual decomposition of colour through motion extrapolation. *Nature, 386*, 66–69.

Nijhawan, R. (2001). The flash-lag phenomenon: Object motion and eye movements. *Perception, 30*, 263–282.

Nijhawan, R. (2002). Neural delays, visual motion and the flash-lag effect. *Trends in Cognitive Sciences, 6*(9), 387–393.

Nijhawan, R. (2008). Visual prediction: Psychophysics and neurophysiology of compensation for time delays. *Behavioral and Brain Sciences, 31*, 179–198.

Nijhawan, R., Khurana, B., Kamitani, Y., Watanabe, K., & Shimojo, S. (1998). Eye-movement based extrapolation leads to decomposition of color. *Investigative Ophthalmology and Visual Science, 39*, 1045.

Nijhawan, R., & Kirschfeld, K. (2003). Analogous mechanisms compensate for neural delays in the sensory and the motor pathways: Evidence from motor flash-lag. *Current Biology, 13*, 749–753.

Nijhawan, R., Watanabe, K., Khurana, B., & Shimojo, S. (2004). Compensation of neural delays in visual–motor behavior: No evidence for shorter afferent delays for visual motion. *Visual Cognition, 11*, 275–298.

Ninio, J. (2014). Geometrical illusions are not always where you think they are: a review of some classical and less classical illusions, and ways to describe them. [Review]. *Frontiers in Human Neuroscience*, 8, 856.

Ninio, J., & O'Regan, J. K. (1999). Characterisation of the misalignment and misangulation components in the Poggendorff and corner–Poggendorff illusions. *Perception, 28*(8), 949–964.

Nishimura, A., & Yokosawa, K. (2009). Effects of laterality and pitch height of an auditory accessory stimulus on horizontal response selection: The Simon effect and the SMARC effect. *Psychonomic Bulletin & Review, 16,* 666–670.

Noël, B., van der Kamp, J., Weigelt, M., & Memmert, D. (2015). Asymmetries in spatial perception are more prevalent under explicit than implicit attention. *Consciousness and Cognition, 34,* 10–15.

Noguchi, Y., & Kakigi, R. (2008). Knowledge-based correction of flashlag illusion. *Journal of Cognitive Neuroscience, 20,* 513–525.

Noordzij, M. L., Neggers, S. F. W., Ramsey, N. F., & Postma, A. (2008). Neural correlates of locative prepositions. *Neuropsychologia, 46,* 1576–1580.

Norman, J. (2002). Two visual systems and two theories of perception: An attempt to reconcile the constructivist and ecological approaches. *Behavioral and Brain Science, 25,* 73–144.

Notebaert, W., Gevers, W., Verguts, T., & Fias, W. (2006). Shared spatial representations for numbers and space: The reversal of the SNARC and the Simon effects. *Journal of Experimental Psychology: Human Perception and Performance, 32*(5), 1197–1207.

Notebaert, W., Soetens, E., & Melis, A. (2001). Sequential analysis of a Simon task-evidence for an attention-shift account. *Psychological Research, 65,* 170–184.

Nuerk, H.-C., Patro, K., Cress, U., Schild, U., Friedrich, C. K., & Göbel, S. M. (2015). How space-number associations may be created in preliterate children: Six distinct mechanisms. *Frontiers in Psychology, 6:215.*

Núñez, R., & Cooperrider, K. (2013). The tangle of space and time in human cognition. *Trends in Cognitive Sciences, 17*(5), 220–229.

Núñez, R., & Fias, W. (2017). Ancestral mental number lines: What is the evidence? *Cognitive Science, 41,* 2262–2266. doi: 10.1111/cogs.12296

Nussbaum, C. O. (2007). *The musical representation: Meaning, ontology, and emotion.* Cambridge, MA: MIT Press.

Nys, G. M. S., Zandvoort, M. J. E. v., Worp, H. B. v. d., Kappelle, L. J., & Haan, E. H. F. d. (2006). Neuropsychological and neuroanatomical correlates of perseverative responses in subacute stroke. *Brain, 129*(8), 2148–2157.

Oaksford, M., & Chater, N. (2001). The probabilistic approach to human reasoning. *Trends in Cognitive Sciences, 5*(8), 349–357.

Occelli, V., Spence, C., & Zampini, M. (2009). Compatibility effects between sound frequencies and tactually stimulated locations on the hand. *Neuroreport, 20,* 793–797.

O'Connell, R. G., Schneider, D., Hester, R., Mattingley, J. B., & Bellgrove, M. A. (2011). Attentional load asymmetrically affects early electrophysiological indices of visual orienting. *Cerebral Cortex, 21*(5), 1056–1065.

Odum, H. T. (1988). Self-organization, transformity, and information. *Science, 242,* 1132–1139.

Ogawa, A., & Macaluso, E. (2013). Audio-visual interactions for motion perception in depth modulate activity in visual area V3A. *Neuroimage, 71,* 158–167.

Öğmen, H., Patel, S. S., Bedell, H. E., & Camuz, K. (2004). Differential latencies and the dynamics of the position computation process for moving targets, assessed with the flash-lag effect. *Vision Research, 44,* 2109–2128.

O'Keefe, J. (1991). An allocentric spatial model for the hippocampal cognitive map. *Hippocampus, 1*(3), 230–235.

O'Keefe, J. (1993). Kant and the sea-horse: An essay in the neurophilosophy of space. In N. Eilan, R. McCarthy, & B. Brewer (Eds.), *Spatial representation: Problems in philosophy and psychology* (pp. 43–64). Malden, MA: Blackwell Publishing.

O'Keefe, J., & Nadel, L. (1978). *The hippocampus as a cognitive map.* Oxford: Clarendon.

Oliveri, M., Magnani, B., Filipelli, A., Avanzi, S., & Frassinetti, F. (2013). Prismatic adaptation effects on spatial representation of time in neglect patients. *Cortex, 49,* 120–130.

Olsen, K. N., & Stevens, C. J. (2010). Perceptual overestimation of rising intensity: Is stimulus continuity necessary? *Perception, 39*(5), 695–704.

Olsen, K. N., Stevens, C. J., & Tardieu, J. (2010). Loudness change in response to dynamic acoustic intensity. *Journal of Experimental Psychology: Human Perception and Performance, 36*(6), 1631–1644.

Olson, C. R. (2003). Brain representation of object-centered space in monkeys and humans. *Annual Review of Neuroscience, 26*, 331–354.

Oman, C. (2007). Spatial orientation and navigation in microgravity. In F. W. Mast & L. Jancke (Eds.), *Spatial processing in navigation, imagery and perception* (pp. 209–247). New York: Springer.

Ooi, T. L., Wu, B., & He, Z. J. (2001). Distance determined by the angular declination below the horizon. *Nature, 414*, 197–200.

Oppel, J. (1855). Ueber geometrisch-opeische tauschungen, *Jber. phys. Ver. Frankfurt*, 34–47.

Orchard-Mills, E., Van der Burg, E., & Alais, D. (2016). Crossmodal correspondence between auditory pitch and visual elevation affects temporal ventriloquism. *Perception, 45*, 409–424.

O'Regan, J. K. (1992) Solving the "real" mysteries of visual perception: The world as an outside memory. *Canadian Journal of Psychology, 46*, 461–488.

Orne, M. T. (1682). On the social psychology of the psychological experiment: With particular reference to demand characteristics and their implications. *American Psychologist, 17*(11), 776–783.

Osiurak, F., Morgado, N., & Palluel-Germain, R. (2012). Tool use and perceived distance: When unreachable becomes spontaneously reachable. *Experimental Brain Research, 218* (2), 331–339.

Ouellet, M., Santiago, J., Funes, M. J., & Lupianez, J. (2010). Thinking about the future moves attention to the right. *Journal of Experimental Psychology: Human Perception and Performance, 36*(1), 17–24.

Ouellet, M., Santiago, J., Israeli, Z., & Gabay, S. (2010). Is the future the right time? *Experimental Psychology, 57*, 308–314.

Ozkan, K., & Braunstein, M. L. (2009) Predominance of ground over ceiling surfaces in binocular rivalry. *Attention, Perception, & Psychophysics, 71*(6), 1305–1312.

Ozkan, K., & Braunstein, M. L. (2010). Background surface and horizon effects in the perception of relative size and distance. *Visual Cognition, 18*, 229–254.

Padakannaya, P., Devi, M. L., Zaveria, B., Chengappa, S. K., & Vaid, J. (2002). Directional scanning effect and strength of reading habit in picture naming and recall. *Brain and Cognition, 48*, 484–490.

Paillard, J. (1991). Motor and representational framing of space. In J. Paillard (Ed.), *Brain and space* (pp. 163–182). New York: Oxford University Press.

Paladino, M. P., & Castelli, L. (2008). On the immediate consequences of intergroup categorization: Activation of approach and avoidance motor behavior toward ingroup and outgroup members. *Personality and Social Psychology Bulletin, 34*(6), 755–768.

Palmer, S. E. (1991). On goodness, Gestalt groups, and Garner: Local symmetry subgroups as a theory of figural goodness. In G. Lockhead & J. R., Pomerantz (Eds.), *The perception of structure: Essays in honor of Wendell R. Garner* (pp. 23–39). Washington, DC: American Psychological Association.

Palmer, S. E. (1999). *Vision science: Photons to phenomenology*. Cambridge, MA: MIT Press.

Palmer, S. E., Gardner, J. S., & Wickens, T. D. (2008). Aesthetic issues in spatial composition: Effects of position and direction on framing single objects. *Spatial Vision, 21*, 421–449.

Palmer, S. E., & Guidi, S. (2011). Mapping the perceptual structure of rectangles through goodness-of-fit ratings. *Perception, 40*, 1428–1446.

Palmer, S. E., & Hemenway, K. (1978). Orientation and symmetry: Effects of multiple, rotational, and near symmetries. *Journal of Experimental Psychology: Human Perception and Performance, 4*, 691–702.

Palmer, S. E., Rosch, E., & Chase, P. (1981). Canonical perspective and the perception of objects. In J. Long & A. Baddeley (Eds.),

Attention and Performance IX (pp. 135–151). Hillsdale, NJ: Erlbaum.

Palmer, S. E., Schloss, K. B., Sammartino, J. (2012). Hidden knowledge in aesthetic judgments: Preference for color and spatial composition. In A. P. Shimamura & S. E. Palmer (Eds.), *Aesthetic science: Connecting minds, brains, and experience* (pp. 189–222). Oxford: Oxford University Press.

Palmer, S. E., Schloss, K. B., Sammartino, J. (2013). Visual aesthetics and human preference. *Annual Review of Psychology, 64*, 77–107.

Paradis, C., Hudson, J. & Magnusson, U. (2013). *The construal of spatial meaning: Windows into conceptual space*. Oxford: Oxford University Press.

Paradis, C., & Willners, C. (2011). Antonymy: From conventionalization to meaning-making. *Review of Cognitive Linguistics, 9*(2), 367–391.

Pardo, J., Fox, P., & Raichle, M. (1991). Localization of a human system for sustained attention by positron emission tomography. *Nature, 349*, 61–64.

Parise, C. V. (2016). Crossmodal correspondences: Standing issues and experimental guidelines. *Multisensory Research, 29*, 7–28.

Parise, C. V., Harrar, V., Ernst, M. O., & Spence, C. (2013). Cross-correlation between auditory and visual signals promotes multisensory integration. *Multisensory Research, 26*, 307–316.

Parise, C. V., Knorre, K., & Ernst, M. O. (2014) Natural auditory scene statistics shapes human spatial hearing. *Proceedings of the National Academy of Science USA, 111*, 6104-6108.

Parise, C., & Spence, C. (2008). Synaesthetic congruency modulates the temporal ventriloquism effect. *Neuroscience Letters, 442*, 257–261.

Parise, C. V., Spence, C., & Ernst, M. O. (2012). When correlation implies causation in multisensory integration. *Current Biology, 22*, 46–49.

Pariyadath, V., & Eagleman, D. M. (2007). The effect of predictability on subjective duration. *PLoS ONE, 2*(11), e1264.

Pariyadath, V., & Eagleman, D. M. (2008). Brief subjective durations contract with repetition. *Journal of Vision, 8*(16), 1–6.

Park, J. H., Faulkner, J., & Schaller, M. (2003). Evolved disease-avoidance processes and contemporary anti-social behavior: Prejudicial attitudes and avoidance of people with physical disabilities. *Journal of Nonverbal Behavior, 27*(2), 65–87.

Park, S., Intraub, H., Yi, D. J., Widders, D., & Chun, M. M. (2007). Beyond the edges of a view: Boundary extension in human scene-selective visual cortex. *Neuron, 54*(2), 335–342.

Parker. A., & Hawken, M. (1985) Capabilities of monkey cortical cells in spatial-resolution tasks. *Journal of the Optical Society of America, A, 2*, 1101–1114.

Parkes, L., Lund, J., Angelluci, A., Solomon, J., & Morgan, M. J. (2001). Compulsory averaging of crowded orientation signals in human vision. *Nature Neuroscience, 4*, 739–744.

Parkhurst, D. J., & Niebur, E. (2003). Scene content selected by active vision. *Spatial Vision, 16*, 125–154.

Parkinson, C., Kohler, P. J., Sievers, B., & Wheatley, T. (2012). Associations between auditory pitch and visual elevation do not depend on language: Evidence from a remote population. *Perception, 41*, 854–861.

Pashler, H., Coburn, N., & Harris, C. R. (2012). Priming of social distance? Failure to replicate effects on social and food judgments. *PLoS ONE, 7*(8), e42510.

Pasqualotto, A., Taya, S., & Proulx, M. J. (2014). Sensory deprivation: Visual experience alters the mental number line. *Behavioural Brain Research, 261*, 110–113.

Patching, G. R., & Quinlan, P. T. (2002). Garner and congruence effects in the speeded classification of bimodal signals. *Journal of Experimental Psychology: Human Perception and Performance, 28*, 755–775.

Patro, K., Fischer, U., Nuerk, H.-C., & Cress, U. (2016a). How to rapidly construct a spatial–numerical representation in preliterate children (at least temporarily). *Developmental Science, 19*(1), 126–144.

Patro, K., Nuerk, H.-C., & Cress, U. (2016b). Mental number line in the preliterate brain: The role of early directional experiences. *Child Development Perspectives, 10*(3), 172–177.

Patro, K., Nuerk, H.-C., Cress, U., & Haman, M. (2014). How number-space relationships are assessed before formal schooling: A taxonomy proposal. *Frontiers in Psychology, 5:419.*

Pavan, A., Cuturi, L. F., Maniglia, M., Casco, C., & Campana, G. (2011). Implied motion from static photographs influences the perceived position of stationary objects. *Vision Research, 51*, 187–194.

Pavlova, M., & Sokolov, A. (2000). Orientation specificity in biological motion perception. *Perception & Psychophysics, 62*(5), 889–899.

Pavlovskaya, M., Bonneh, Y., Soroker N., & Hochstein, S. (2011). Neglect field objects impact statistical property report in patients with unilateral spatial neglect. *Perception*, 40, 122.

Pavlovskaya, M., Bonneh, Y., Soroker, N., & Hochstein, S. (2010). Processing visual scene statistical properties in patients with unilateral spatial neglect. *Journal of Vision, 10*, 280.

Pavio, A. (1986). *Mental representations: A dual coding approach.* Oxford: Oxford University Press.

Peacocke, C. (2009). The perception of music: Sources of significance. *The Modern Schoolman, 86*, 239–260.

Pedley, P. E., & Harper, R. S. (1959). Pitch and the vertical localization of sound. *The American Journal of Psychology, 72*, 447–449.

Peers, P. V., Cusack, R., & Duncan, J. (2006). Modulation of spatial bias in the dual task paradigm: Evidence from patients with unilateral parietal lesions and controls. *Neuropsychologia*, 44, 1325–1335.

Peers, P., Ludwig, C., Rorden, C., Cusack, R., Bonfiglioli, C., et al. (2005). Attentional functions of parietal and frontal cortex. *Cerebral Cortex, 15*(10), 1469–1484.

Peirce, C. S. (1955). *Philosophical writings of Peirce.* New York: Dover.

Pellicano, A., Lugli, L., Baroni, G., & Nicoletti, R. (2009). The Simon effect with conventional signals. A time-course analysis. *Experimental Psychology, 56*, 219–227.

Pérez González, C. (2012). *Local portraiture: Through the lens of the 19th-century Iranian photographers.* Leiden: Leiden University Press.

Perfors, A., Tenenbaum, J. B., Griffiths, T. L., & Xu, F. (2011). A tutorial introduction to Bayesian models of cognitive development. *Cognition, 120*(3), 302–321.

Perri, R., Bartolomeo, P., & Gainotti, G. (2000). Lack of impairments on leftward and rightward line extension tasks in neglect patients. *International Journal of Neuroscience, 103*, 101–113.

Perrott, D. R., Costantino, B., & Cisneros, J. (1993). Auditory and visual localization performance in a sequential discrimination task. *Journal of the Acoustical Society of America, 93*, 2134–2138.

Perry, L. K., Smith, L. B., & Hockema, S. A. (2008). Representational momentum and children's sensori-motor representations of objects. *Developmental Science, 11*, F17–F23.

Peru, A., & Pinna, G. (1997). Right personal neglect following a left hemisphere stroke. A case report. *Cortex, 33*, 585–590.

Peterhans, E., von der Heydt, R., & Baumgartner, G. (1986) Neuronal responses to illusory contour stimuli reveal stages of visual cortical processing. In J. D. Pettigrew, K. J. Sanderson, & W. R. Levick (Eds.), *Visual neuroscience* (pp. 343–351). Cambridge: Cambridge University Press.

Petersen, S. E., & Posner, M. I. (2012). The attention system of the human brain: 20 years after. *Annual Review of Neuroscience, 35*, 73–89.

Petersik, J. T., & Thiel, R. L. (2010). Detecting sudden changes in dynamic rotation displays. *Seeing and Perceiving, 23*(3), 241–261.

Peterson, M. A., Gillam, B., & Sedgwick, H. A. (Eds.). (2006). *In the mind's eye: Julian Hochberg on the perception of pictures, films, and the world.* Oxford: Oxford University Press.

Peyrin, C., Michel, C. M., Schwartz, S., et al. (2010). The neural substrates and timing of top-down process during coarse-to-fine categorization of visual scenes: a combined fMRI and ERP study. *Journal of Cognitive Neuroscience, 22*(12), 2768–2780.

Pfister, R., Schroeder, P. A., & Kunde, W. (2013). SNARC struggles: Instant control over spatial-numerical associations. *Journal of Experimental Psychology: Learning, Memory, and Cognition, 39*(6), 1953–1958.

Philbeck, J. W., & Loomis, J. M. (1997). Comparison of two indicators of visually perceived egocentric distance under full-cue and reduced-cue conditions. *Journal of Experimental Psychology: Human Perception and Performance, 23*, 72–85.

Philbeck, J. W., & Witt, J. K. (2015). Action-specific influences on perception and post-perceptual processes: Present controversies and future directions. *Psychological Bulletin, 141*(6), 1120–1144.

Phillips, J. C., & Ward, R. (2002). S-R correspondence effects of irrelevant visual affordance: Time course and specificity of response activation. *Visual Cognition, 9*, 540–558.

Phillips, R. J., & Noyes, L. (1982). An investigation of visual clutter in the topographic base of a visual map. *Cartographic Journal, 19*, 122–132.

Piéron, H. (1935). Le processus du métacontraste [The process involved in "after-contrast"]. *Journal de Psychologie Normale et Pathalogique, 32*, 1–24.

Pinkerton, E., & Humphrey, N. K. (1974). The heaviness of colors. *Nature, 250*, 164–165.

Pinna, B., & Albertazzi, L. (2011). From grouping to visual meanings: A new theory of perceptual organization. In L. Albertazzi, G. van Tonder, & D. Vishwanath (Eds.), *Information in perception* (pp. 287–344). Cambridge, MA: MIT Press.

Piotrowski, A. S., & Jakobson, L. S. (2011). Representational momentum in older adults. *Brain and Cognition, 77*, 106–112.

Pirenne, M. H. (1970). *Optics, painting and photography.* Cambridge: Cambridge University Press.

Pittenger, J. B., & Shaw, R. E. (1975). Aging faces as viscal-elastic events: Implications for a theory of nonrigid shape perception. *Journal of Experimental Psychology: Human Perception and Performance, 1*(4), 374–382.

Pittenger, J. B., Shaw, R. E., & Mark, L. S. (1979). Perceptual information for the age level of faces as a higher order invariant of growth. *Journal of Experimental Psychology: Human Perception and Performance, 5*(3), 478–493.

Pizlo, Z., Li, Y., Sawada, T., & Steinman, R. M. (2014). *Making a machine that sees like us.* New York: Oxford University Press.

Plumhoff, J., & Schirillo, J. A. (2009). Mondrian, eye movements, and the oblique effect. *Perception*, 38, 719–731.

Poffenberger, A. T., & Barrows, B. E. (1924). The feeling value of lines. *Journal of Applied Psychology, 8*, 187–205.

Ponsot, E., Meunier, S., Kacem, A., Chatron, J., & Susini, P. (2015). Are rising sounds always louder? Influences of spectral structure and intensity-region on loudness sensitivity to intensity-change direction. *Acta Acustica United with Acustica, 101*(6), 1083–1093.

Ponsot, E., Susini, P., & Meunier, S. (2015). A robust asymmetry in loudness between rising- and falling-intensity tones. *Attention Perception & Psychophysics, 77*(3), 907–920.

Popper, A. N., & Fay, R. R. (1997). Evolution of the ear and hearing: Issues and questions. *Brain Behavior and Evolution, 50*(4), 213–221.

Portin, K., Vanni, S., Virsu, V. & Hari, R. (1999). Stronger occipital cortical activation to lower than upper visual field: Neuromagnetic recordings. *Experimental Brain Research, 124*(3), 287–294.

Portugali, Y. (1993). *Implicate relations: Society and space in the Israeli-Palestinian conflict.* Amsterdam: Kluwer.

Posner, M. I. (1978). *Chronometric explorations of mind.* Hillsdale, NJ: Erlbaum.

Posner, M. I. (1980). Orienting of attention. *Quarterly Journal of Experimental Psychology, 32*, 3–25.

Posner, M. I., & Cohen, Y. (1984). Components of visual orienting. In H. Bouma & D. G.

Bouwhuis (Eds.), *Attention and performance X: Control of language processes* (pp. 531–556). Hove, UK: Lawrence Erlbaum Associates Ltd.

Posner, M. I., Rafal, R. D., Choate, L., & Vaughan, J. (1985). Inhibition of return: Neural basis and function. *Cognitive Neuropsychology, 2*, 211–228.

Post, R. B., O'Malley, M. D., Yeh, T. L., & Bethel, J. (2006). On the origin of vertical line bisection errors. *Spatial Vision, 19*(6), 505–527.

Poulton, E. C. (1989). *Bias in quantifying judgments*. Hillsdale, NJ: Erlbaum Associates.

Pratt, C. C. (1930). The spatial character of high and low tones. *Journal of Experimental Psychology, 66*, 347–352.

Pratt, J., Spalek, T. M., & Bradshaw, F. (1999). The time to detect targets at inhibited and noninhibited locations: Preliminary evidence for attentional momentum. *Journal of Experimental Psychology: Human Perception and Performance, 25*, 730–746.

Predebon, J. (2006). Decrement of the Muller-Lyer and Poggendorff illusions: The effects of inspection and practice. *Psychological Research, 70*(5), 384–394.

Pressey, A. W. (2014). Assimilation theory, attention and asymmetry in the Müller-Lyer illusion: Quantitative predictions. *Perceptual and Motor Skills, 119*, 509–529.

Preuss, N., Harris, L. R., & Mast, F. W. (2013). Allocentric visual cues influence mental transformation of bodies. *Journal of Vision, 13*(12), 14.

Previc, F. H. (1990). Functional specialization in the lower and upper visual fields in humans: Its ecological origins and neurophysiological implications. *Behavioral and Brain Sciences, 13*, 519–565.

Previtali P., de Hevia, M. D., & Girelli, L. (2010). Placing order in space: The SNARC effect in serial learning. *Experimental Brain Research, 201*, 599–605.

Priest, S. (1991). *Theories of the mind: A compelling investigation into the ideas of leading philosophers on the nature of the mind and its relation to the body*. New York: Mariner Books.

Priftis, K., Zorzi, M., Meneghello, F., Marenzi, R., & Umiltà, C. (2006). Explicit versus implicit processing of representational space in neglect: Dissociations in accessing the mental number line. *Journal of Cognitive Neuroscience, 18*(4), 680–688.

Prime, D. J., & Jolicoeur, P. (2009). On the relationship between occipital cortex activity and inhibition of return. *Psychophysiology, 46*, 1278–1287.

Prime, D. J., & Ward, L. M. (2006). Cortical expressions of inhibition of return. *Brain Research, 1072*, 161–174.

Prinz, W. (2002). Experimental approaches to imitation. In A. N. Meltzoff, & W. Prinz (Eds.), *The imitative mind: Development, evolution, and brain bases* (pp. 143–163). Cambridge: Cambridge University Press.

Prinzmetal, W. & Gettleman, L. (1993). Vertical-horizontal illusion: One eye is better than two. *Attention, Perception & Psychophysics, 53*(1), 81–88.

Prior, E., Pargetter, R., & Jackson, F. (1982). Three theses about dispositions. *American Philosophical Quarterly, 19*, 251–257.

Proctor, R. W., & Cho, Y. S. (2006). Polarity correspondence: A general principle for performance of speeded binary classification tasks. *Psychological Bulletin, 132*, 416–442.

Proctor, R. W., & Reeve, T. G. (Eds.). (1989). *Stimulus-response compatibility: An integrated perspective*. Amsterdam: Elsevier.

Proctor, R. W., & Vu, K.-P. L. (2002). Eliminating, magnifying, and reversing spatial compatibilità effects with mixed location-relevant and irrelevant trials. In W. Prinz & B. Hommel (Eds.), *Common mechanisms in perception and action: Attention and performance* (Vol. 19, pp. 443–473). Oxford: Oxford University Press

Proctor, R. W., & Vu, K.-P. L.(2006). *Stimulus-response compatibility principles*. Boca Raton, FL: Taylor & Francis.

Proctor, R. W., Vu, K.-P. L., & Marble, J. G. (2003). Eliminating spatial compatibility effects for location-relevant trials by intermixing location-irrelevant trials. *Visual Cognition, 10*, 15–50.

Proctor, R. W., Vu, K.-P. L., & Nicoletti, R. (2003). Does right–left prevalence occur for the Simon effect? *Perception & Psychophysics, 65*, 1318–1329.

Proffitt, D. R. (2006). Embodied perception and the economy of action. *Perspectives on Psychological Science, 1*, 110–122.

Proffitt, D. R. (2009). Affordances matter in geographical slant perception. *Psychonomic Bulletin & Review, 16*(5), 970–972.

Proffitt, D. R., Bhalla, M., Gossweiler, R., & Midgett, J. (1995). Perceiving geographical slant. *Psychonomic Bulletin & Review, 2*(4), 409–428.

Proffitt, D. R., & Linkenauger, S. A. (2013). Perception viewed as a phenotypic expression. In W. Prinz, M. Beisert, & A. Herwig (Eds.), *Action science: Foundations of an emerging discipline* (pp. 171–198). Cambridge, MA: MIT Press.

Proffitt, D. R., Stefanucci, J., Banton, T., & Epstein, W. (2003). The role of effort in perceiving distance. *Psychological Science, 14*(2), 106–112.

Purushothaman, G., Patel, S. S. Bedell, H. E., & Öğmen, H. (1998). Moving ahead through differential visual latency. *Nature, 396*, 424.

Put, K., Baldo, V. C. M., Cravo, A. M., Wagemans, J., & Helsen, W. F. (2013). Experts in offside decision making learn to compensate for their illusory perceptions. *Journal of Sport & Exercise Psychology, 35*, 576–584.

Pylyshyn, Z. W. (1999). Is vision continuous with cognition? The case for cognitive impenetrability of visual perception. *Behavioral and Brain Sciences, 22*, 341–423.

Pylyshyn, Z. W. (2003). *Seeing and visualizing: It's not what you think*. Cambridge, MA: MIT Press.

Pyoun, H. K., Sargent, J., Dopkins, S., & Philbeck, J. (2013). Two-category place representations persist over body rotations. *Memory & Cognition, 41*(8), 1132–1143.

Qu, Z., Song, Y. & Ding, Y. L. (2006). Asymmetry between the upper and lower visual fields: An event-related potential study. *Chinese Science Bulletin, 51*(5), 536–541.

Quadflieg, S., Etzel, J. A, Gazzola, V., Keysers, C., Schubert, T. W., Waiter, G. D., & Macrae, C. N. (2011). Puddles, parties, and professors: Linking word categorization to neural patterns of visuospatial coding. *Journal of Cognitive Neuroscience, 23*(10), 2636–2649.

Quattrone, G. A., (1986). On the perception of a group's variability. In S. Worchei & W. Austin (Eds.), *The psychology of intergroup relations* (2nd ed., pp. 25–48). Chicago: Nelson-Hall.

Quek, G. L., & Finkbeiner, M. (2016). The upper-hemifield advantage for masked face processing: Not just an attentional bias. *Attention, Perception, & Psychophysics, 78*(1), 52–68.

Quinn, P. C., Cummins, M., Kase, J., Martin, E., & Weisman, S. (1996). Development of categorical representations for above and below spatial relations in 3- to 7-month-old infants. *Developmental Psychology, 32*, 942–950.

Quinn, P. C., & Intraub, H. (2007). Perceiving "outside the box" occurs early in development: Evidence for boundary extension in three- to seven-month-old infants. *Child Development, 78*(1), 324–334.

Rafal, R. D., Egly, R., & Rhodes, D. (1994). Effects of inhibition of return on voluntary and visually guided saccades. *Canadian Journal of Experimental Psychology, 48*, 284–300.

Raiguel, S. E., Lagae, L., Gulyàs, B., & Orban, G. A. (1989). Response latencies of visual cells in macaque areas V1, V2 and V5. *Brain Research, 493*, 155–159.

Raine, A., Reynolds, C., Venables, P. H., Mednick, S. A., & Farrington, D. P. (1998). Fearlessness, stimulation-seeking, and large body size at age 3 years as early predispositions to childhood aggression at age 11 years. *Archives of General Psychiatry, 55*(8), 745–751.

Ramachandran, V. S. (1988). Shape from shading. *Nature, 331*, 163–166.

Ramachandran, V. S. (2011). *The tell-tale brain*. New York: W. W. Norton.

Ramachandran, V. S., & Hirsten, W. (1999). The science of art: A neurological theory of aesthetic experience. *Journal of Consciousness Studies, 6*, 15–51.

Ramachandran, V. S., & Seckel, E. (2012). Neurology of visual aesthetics: Indian nymphs, modern art, and sexy beaks. In A. P. Shimamura & S.E. Palmer (Eds.), *Aesthetic science: Connecting minds, brains, and experience*. Oxford, UK: Oxford University Press.

Ramani, G. B., & Siegler, R. S. (2008). Promoting broad and stable improvements in low-income children's numerical knowledge through playing number board games. *Child Development, 79*(2), 375–394.

Rao, R. P. N., Eagleman, D. M., & Sejnowski, T. J. (2001). Optimal smoothing in visual motion perception. *Neural Computation, 13*, 1243–1253.

Rao, H., Han, S., Jiang, Y., Xue, Y., Gu, H., Cui, Y., & Gao, D. (2004). Engagement of the prefrontal cortex in representational momentum: An fMRI study. *NeuroImage, 23*, 98–103.

Rapcsak, S. Z., Cimino, C. R., & Heilman, K. M. (1988). Altitudinal neglect. *Neurology, 38*, 277–281.

Raphael, S., & Morgan, M.J. (2016) The computation of relative numerosity, size and density. *Vision Research, 124*, 15–23.

Rastelli, F., Tallon-Baudry, C., Migliaccio, R., Toba, M. N., Ducorps, A., Pradat-Diehl, P., et al. (2013). Neural dynamics of neglected targets in patients with right hemisphere damage. *Cortex, 49*(7), 1989–1996.

Raudsepp, J. (2001a). *Horizontal-vertical illusion: Functional and causal issues*. Umeä: Umeä University.

Raudsepp, J. (2001b). Peripersonal horizontal-vertical illusions: towards comparable kinesthetic and visual task. *Journal of Human Movement Studies, 41*, 263–286.

Raudsepp, J., & Djupsjöbacka, M. (2005). Handgrip maximum force and the visual horizontal vertical illusion. *Perception, 34*, 421–428.

Re, D. E., Dzhelyova, M., Holzleitner, I. J., Tigue, C. C., Feinberg, D. R., & Perrett, D. I. (2012). Apparent height and body mass index influence perceived leadership ability in three-dimensional faces. *Perception, 41*(12), 1477–1485.

Reber, R. (2012). Processing fluency, aesthetic pleasure, and culturally shared taste. In A. P. Shimamura & S. E. Palmer (Eds.), *Aesthetic science: Connecting minds, brains, and experience* (pp. 223–249). Oxford: Oxford University Press.

Reber, R., Schwartz, N., & Winkielman, P. (2004). Processing fluency and aesthetic pleasure: Is beauty in the perceiver's processing experience? *Personality and Social Psychology Review, 8*, 364–382.

Redden, R. S., Hilchey, M. D., & Klein, R. M. (2015a). *Using speed and accuracy and the Simon effect to explore two forms of inhibition of return*. Manuscript under review.

Redden, R. S., Hilchey, M. D., & Klein, R. M. (2015b). *Oculomotor inhibition of return: Evidence against object-centered representation*. Manuscript under review.

Redden, R. S., Hilchey, M. D., & Klein, R. M. (2016). Peripheral stimuli generate different forms of inhibition of return when participants make prosaccades versus antisaccades to them. *Attention, Perception, & Psychophysics, 78*(8), 2283–2291.

Reed, C. L., Stone, V., Bozova, S., & Tanaka, J. (2003). The body inversion effect. *Psychological Science, 14*, 302–308.

Reed, C. L., & Vinson, N. G. (1996). Conceptual effects on representational momentum. *Journal of Experimental Psychology: Human Perception and Performance, 22*, 839–850.

Regan, D., & Hamstra, S. J. (1993). Dissociation of discrimination thresholds for time to contact and for rate of angular expansion. *Vision Research, 33*(4), 447–462.

Regier, T., & Kay, P. (2009). Language, thought, and color: Whorf was half right. *Trends in Cognitive Sciences, 13*(10), 439–446.

Ren, P., Nicholls, M. E., Ma, Y. Y., & Chen, L. (2011). Size matters: Non-numerical magnitude affects the spatial coding of response. *PLoS ONE, 6*(8), e23553.

Rengachary, J., He, B., Shulman, G., & Corbetta, M. (2011). A behavioral analysis of spatial neglect and its recovery after stroke. *Frontiers in Human Neuroscience*, 5(29), doi: 10.3389/fnhum.2011.00029.

Repp, B. H., & Knoblich, G. (2009). Performed or observed keyboard actions affect pianists' judgements of relative pitch. *The Quarterly Journal of Experimental Psychology, 62*, 2156–2170.

Restle, F., & Levinson, M. (1971) Method of constant stimuli: Invalidity to the third power. *Percepion & Psychophysics, 9*, 312–314.

Rhee, I., Shin, M., Hong, S., Lee, K., & Chong, S. (2008). On the levy-walk: Nature of human mobility. In *IEEE INFOCOM 2008 – IEEE Conference on Computer Communications* (pp. 924–932). Phoenix, AZ: IEEE.

Richards, W. (1975). Visual space perception. In E. C. Carterette & M. P. Friedman (Eds.), *Handbook of Perception, vol. 5: Seeing* (pp. 351–386). New York: Academic Press.

Richardson, D. C., Spivey, M. J., Barsalou, L. W., & McRae, K. (2003). Spatial representations activated during real-time comprehension of verbs. *Cognitive Science, 27*(5), 767–780.

Richter, H. O., Wennberg, P., & Raudsepp, J. (2007). The effects of inverting prisms on the horizontal–vertical illusion: A systematic effect of downward gaze. *Experimental Brain Research, 183*, 9–15.

Riddoch, M. J., & Humphreys, G. W. (1983). The effect of cueing on unilateral neglect. *Neuropsychologia, 21*, 589–599.

Riello, M., & Rusconi, E. (2011). Unimanual SNARC effect: Hand matters. *Frontiers in Psychology, 2:372.*

Rieser, J. J., Pick, H. L., Ashmead, D. H., & Garing, A. E. (1995). Calibration of human locomotion and models of perceptual-motor organization. *Journal of Experimental Psychology: Human Perception and Performance, 21*(3), 480–497.

Riestra, A. R., Womack, K. B., Crucian, G. P. & Heilman, K. M. (2002). Is the middle between both halves? Midpoint location and segment size estimation in neglect. *Neurology, 59*, 1580–1584.

Riordan, B., & Jones, M. N. (2011). Redundancy in perceptual and linguistic experience: Comparing feature-based and distributional models of semantic representation. *Topics in Cognitive Sciences, 3*(2), 303–345.

Riskind, J. H., Kleiman, E. M., Seifritz, E., & Neuhoff, J. G. (2014). Influence of anxiety, depression and looming cognitive style on auditory looming perception. *Journal of Anxiety Disorders, 28*(1), 45–50.

Rizk, J. K., Chappell, M., & Hine, T. J. (2009). Effect of motion smoothness on the flash-lag illusion. *Vision Research, 49*, 2201–2208.

Rizzolatti, G., Fadiga, L., Gallese, V., & Fogassi, L. (1996). Premotor cortex and the recognition of motor actions. *Cognitive Brain Research, 3*(2), 131–141.

Rizzolatti, G., Riggio, L., Dascola, I., & Umiltà, C. (1987). Reorienting attention across the horizontal and vertical meridians: Evidence in favor of a premotor theory of attention. *Neuropsychologia, 25*, 31–46.

Rizzolatti, G., Riggio, L., & Sheliga, B. M. (1994). Space and selective attention. In C. Umiltà & M. Moscovitch (Eds.), *Attention and performance XV* (pp. 231–265). Cambridge, MA: MIT Press.

Roach, N. W., Heron, J., & McGraw, P. V. (2006). Resolving multisensory conflict: A strategy for balancing the costs and benefits of audio-visual integration. *Proceedings of the Royal Society of London, Series B: Biological Sciences, 273*, 2159–2168.

Roberston, L. C., Mirjam, E., & Knight, R. (2003). Grouping influences in unilateral visual neglect. *Journal of Clinical and Experimental Neuropsychology, 25*(3), 297–307.

Robertson, I. H. (2001). Do we need the "lateral" in unilateral neglect? Spatially nonselective attention deficits in unilateral neglect and their implications for rehabilitation. *NeuroImage, 14*, 85–90.

Robinson, J. O. (1972). *The psychology of visual illusion*. London: Hutchinson.

Rock, I. (1983). *The logic of perception*. Cambridge, MA: MIT Press.

Roese, N. J. (1997). Counterfactual thinking. *Psychological Bulletin, 121*, 133–148.

Roffler, S. K., & Butler, R. A. (1968). Factors that influence the localization of sound in the vertical plane. *Journal of the Acoustical Society of America, 43*, 1255–1259.

Rogers, B. J., & Bradshaw, M. F. (1993). Vertical disparities, differential perspective and binocular stereopsis. *Nature, 361*, 253–255.

Rogers, S. (1996). The horizon-ratio relation as information for relative size in pictures. *Perception & Psychophysics, 58*, 142–152.

Rogers, Y., & Lindley, S. (2004). Collaborating around vertical and horizontal large interactive displays: Which way is best? *Interacting with Computers, 16*(6), 1133–1152.

Roidl, E., Siebert, F. W., Oehl, M., & Hoeger, R. (2013). Introducing a multivariate model for predicting driving performance: The role of driving anger and personal characteristics. *Journal of Safety Research, 47*, 47–56.

Rorden, C., Berger, M. F., & Karnath, H. O. (2006). Disturbed line bisection is associated with posterior brain lesions. *Brain Research, 1080*, 17–25.

Rorden, C., & Brett, M. (2000). Stereotaxic display of brain lesions. *Behavioural Neurology, 12*, 191–200.

Rorden, C., Fruhmann Berger, M., & Karnath, H. (2006). Disturbed line bisection is associated with posterior brain lesions. *Brain Research, 1080*(1), 17–25.

Rorden, C., Karnath, H. O., & Bonilha, L. (2007). Improving lesion-symptom mapping. *Journal of Cognitive Neuroscience, 19*, 1081–1088.

Rosch, E. (1975). Cognitive reference points. *Cognitive Psychology, 7*, 532–547.

Rosch, E., (1978). Principles of categorization. In E. Rosch & B. B. Lloyd (Eds.), *Cognition and categorization* (pp. 27–48). Hillsdale, NJ: Erlbaum.

Rosenberg, A., & Angelaki, D. E. (2014). Gravity influences the visual representation of object tilt in parietal cortex. *Journal of Neuroscience, 34*, 14170–14180.

Rosenbloom, P. S., & Newell, A. (1987). An integrated computational model of stimulus–response compatibility and practice. In G. H. Bower (Ed.), *The psychology of learning and motivation* (Vol. 21, pp. 1–52). San Diego, CA: Academic Press.

Rosenblum, L. D., Carello, C., & Pastore, R. E. (1987). Relative effectiveness of 3 stimulus variables for locating a moving sound source. *Perception, 16*(2), 175–186.

Rosenblum, L. D., Gordon, M. S., & Wuestefeld, A. P. (2000). Effects of performance feedback and feedback withdrawal on auditory looming perception. *Ecological Psychology, 12*(4), 273–291.

Rosenblum, L. D., Wuestefeld, A. P., & Saldana, H. M. (1993). Auditory looming perception – influences on anticipatory judgments. *Perception, 22*(12), 1467–1482.

Rosenholtz, R., Li, Y., & Nakano, L. (2007). Measuring visual clutter. *Journal of Vision, 7*(2), 1–22.

Rosenthal, R. (1979). The "file drawer problem" and tolerance for null results. *Psychological Bulletin, 86*, 638–641.

Ross, H., & Plug, K. (2002). *The mystery of the moon illusion*. Oxford: Oxford University Press.

Rossetti, Y., Jacquin-Courtois, S., Rode, G., Ota, H., Michel, C., & Boisson, D. (2004). Does action make the link between number and space representation? Visuo-manual adaptation improves number bisection in unilateral neglect. *Psychological Science, 15*(6), 426–430.

Rotman, G., Brenner, E., & Smeets, J. B. J. (2002). Spatial but not temporal cueing influences the mislocalization of a target flashed during smooth pursuit. *Perception, 31*, 1195–1203.

Rotman, G., Brenner, E., & Smeets, J. B. J. (2004). Mislocalization of targets flashed during smooth pursuit depends on the change in gaze direction after the flash. *Journal of Vision, 4*(7), 4.

Rotman, G., Brenner, E., & Smeets, J. B. J. (2005). Flashes are localized as if they were moving with the eyes. *Vision Research, 45*, 355–364.

Roulston, B. W., Self, M. W., & Zeki, S. (2006). Perceptual compression of space through position integration. *Proceedings of the Royal Society: Series B. Biological Sciences, 273*, 2507–2512.

Rousseaux, M., Honoré, J., Vuilleumier, P., & Saj, A. (2013). Neuroanatomy of space, body, and posture perception in patients with right hemisphere stroke. *Neurology*, *81*(15), 1291–1297.

Rowland, E., & Durant, S. (2014). High temporal frequency adaptation compresses time in the flash-lag illusion. *Vision Research*, *105*, 130–136.

Rubichi, S., Nicoletti, R., Iani, C., & Umiltà, C. (1997). The Simon effect occurs relative to the direction of an attention shift. *Journal of Experimental Psychology: Human Perception and Performance*, *23*, 1353–1364.

Rubin, E. (1930). Kritisches und Experimentelles zur "Empfindungszeit" Fröhlichs [Review and experiments regarding Fröhlich's time of sensation]. *Psychologische Forschung*, *13*(1), 101–112.

Ruch, T. C., & Fulton, J. F. (1960). *Medical physiology and biophysics*. Philadelphia: W. B. Saunders Company.

Rueckert, L., & Grafman, J. (1996). Sustained attention deficits in patients with right frontal lesions. *Neuropsychologia*, *34*(10), 953–963.

Rugani, R., Vallortigara, G., Priftis, K., & Regolin, L. (2015). Number-space mapping in the newborn chick resembles humans' mental number line. *Science*, *347*(6221), 534–536.

Runeson, S. (1974). Constant velocity: Not perceived as such. *Psychological Research*, *37*, 3–23.

Ruppel, S. E., Fleming, C. N., & Hubbard, T. L. (2009). Representational momentum is not (totally) impervious to error feedback. *Canadian Journal of Experimental Psychology*, *63*, 49–58.

Rusconi, E., Giordano, B. L., Casey, A., Umiltà, C., & Butterworth, B. (2006). Vertical and horizontal dimensions in the spatial representation of pitch height. In M. Baroni, A. R. Addessi, R. Caterina, & M. Costa (Eds.), *Proceedings of the 9th International Conference on Music Perception and Cognition (ICMPC2006)*. Bologna, Italy.

Rusconi, E., Kwan, B., Giordano, B. L., Umiltà, C., & Butterworth, B. (2005). The mental space of pitch height. *NYAS*, *1060*, 195–197.

Rusconi, E., Kwan, B., Giordano, B. L., Umiltà, C., & Butterworth, B. (2006). Spatial representation of pitch height: The SMARC effect. *Cognition*, 99, 113–129.

Russell, P. A. (2000). The aesthetics of rectangle proportion: Effects of judgment scale and context. *American Journal of Psychology*, 113, 27–42.

Sackeim, H. A., & Gur, R. C. (1978). Lateral asymmetry in intensity of emotional expression. *Neuropsychologia*, *16*(4), 473–481.

Sadaghiani, S., Maier, J. X., & Noppeney, U. (2009). Natural, metaphoric, and linguistic auditory direction signals have distinct influences on visual motion processing. *Journal of Neuroscience*, *29*, 6490–6499.

Sadalla, E. K., Burroughs, W. J., & Staplin, L. J. (1980). Reference points in spatial cognition. *Journal of Experimental Psychology: Human Learning and Memory*, *6*(5), 516–528.

Sadalla, E. K., & Magel, S. G. (1980). The perception of traversed distance. *Environment and Behavior*, *12*, 65–79.

Sadalla, E. K., & Montello, D. R. (1989). Remembering changes in direction. *Environment and Behavior*, *21*, 346–363.

Sadalla, E. K., & Staplin, L.J. (1980a) An information storage model for distance cognition. *Environment and Behavior*, *12*, 183–193.

Sadalla, E. K., & Staplin, L. J. (1980b). lloyd. *Environment and Behavior*, *12*, 167–182.

Sagi, D., & Julesz, B. (1985). "What" and "where" in vision. *Science*, *228*, 1217–1219.

Sailor, K. M., & Antoine, M. (2005). Is memory for stimulus magnitude Bayesian? *Memory & Cognition*, *33*(5), 840–851.

Saj, A., Fuhrman, O., Vuilleumier, P., & Boroditsky, L. (2014). Patients with left spatial neglect also neglect the "left side" of time. *Psychological Science*, *25*, 207–214.

Sakata, H., Tsutsui, K., & Taira, M. (2005). Toward an understanding of the neural processing for 3D shape perception. *Neuropsychologia*, *43*, 151–161.

Salgado-Montejo, A., Marmolejo-Ramos, F., Alvarado, J. A., Arboleda, J. C., Suarez, D. R.,

& Spence, C. (2016). Drawing sounds: Representing tones and chords spatially. *Experimental Brain Research, 234,* 3509–3522.

Sammartino, J., & Palmer, S. E. (2012a). Aesthetic issues in spatial composition: Effects of vertical position and perspective on framing single objects. *Journal of Experimental Psychology: Human Perception and Performance, 38,* 865–879.

Sammartino, J., & Palmer, S. E. (2012b). Aesthetic issues in spatial composition: Representational fit and the role of sematic context. *Perception, 41,* 1434–1457.

Sampaio, C., & Wang, R. F. (2009). Category-based errors and the accessibility of unbiased spatial memories: A retrieval model. *Journal of Experimental Psychology: Learning, Memory, and Cognition, 35*(5), 1331–1337.

Sampaio, C., & Wang, R. F. (2010). Overcoming default categorical bias in spatial memory. *Memory & Cognition, 38*(8), 1041–1048.

Sanabria, D., Spence, C., & Soto-Faraco, S. (2007). Perceptual and decisional contributions to audiovisual interactions in the perception of apparent motion: A signal detection theory. *Cognition, 102,* 299–310.

Sanborn, A. N., Mansinghka, V. K., & Griffiths, T. L. (2013). Reconciling intuitive physics and Newtonian mechanics for colliding objects. *Psychological Review, 120,* 411–437.

Sandberg, E. H., Huttenlocher, J., & Newcombe, N. (1996). The development of hierarchical representation of two-dimensional apace. *Child Development, 67*(3), 721–739.

Santiago, J, & Lakens, D. (2015). Can conceptual congruency effects between number, time, and space be accounted for by polarity correspondence? *Acta Psychologica, 156,* 179–191.

Santiago, J., Lupiáñez, J., Pérez, E., & Funes, M. J. (2007). Time (also) flies from left to right. *Psychonomic Bulletin & Reviews, 14,* 512–516.

Santiago, J., Román, A., Ouellet, M., Rodríguez, N., & Pérez-Azor, P. (2010). In hindsight, life flows from left to right. *Psychological Research, 74*(1), 59–70.

Santiago, J., Román, A., & Ouellet, M. (2011). Flexible foundations of abstract thought: A review and a theory. In A. Maas & T. Schubert (Eds.), *Spatial dimensions of social thought* (pp. 39–108). Berlin: Mouton de Gruyter.

Sareen, P., Ehinger, K. A., & Wolfe, J. M. (2014). Through the looking-glass: Objects in the mirror are less real. *Psychonomic Bulletin & Review, 22*(4), 980–986.

Sarich, D., Chappell, M., & Burgess, C. (2007). Dividing attention in the flash-lag illusion. *Vision Research, 47,* 544–547.

Sasaki, Y., Rajimehr, R., Kim, B. W., Ekstrom, L. B., Vanduffel, W., & Tootell, R. B. (2006). The radial bias: A different slant on visual orientation sensitivity in human and nonhuman primates. *Neuron, 51,* 661–670.

Sato, K., & Yamawaki, Y. (2014). Role of a looming-sensitive neuron in triggering the defense behavior of the praying mantis Tenodera aridifolia. *Journal of Neurophysiology, 112*(3), 671–682.

Saunders, J., & Knill, D. C. (2001). Perception of 3D surface orientation from skew symmetry. *Vision Research, 41*(24), 3163–3185.

Saunders, J. A., & Backus, B. T. (2006). Perception of surface slant from oriented textures. *Journal of Vision, 6*(9), 882–897.

Saunders, J. A., & Backus, B. T. (2007). Both parallelism and orthogonality are used to perceive 3D slant of rectangles from 2D images. *Journal of Vision, 7*(6), 1–11.

Sauvan, X. M., & Peterhans, E. (1999). Orientation constancy in neurons of monkey visual cortex. *Visual Cognition, 6,* 43–54.

Savardi, U., & Bianchi I. (2009). The spatial path to contrariety. In U. Savardi (Ed.), *The perception and cognition of contraries* (pp. 63–92). Milan, IT: McGraw-Hill.

Savardi, U., Bianchi, I., & Bertamini, M. (2010). Naive prediction of orientation and motion in mirrors: From what we see to what we expect reflections to do. *Acta Psychologica, 134*(1), 1–15.

Savelsbergh, G. J., Whiting, H. T., & Bootsma, R. J. (1991). Grasping tau. *Journal of*

Experimental Psychology: Human Perception and Performance, 17(2), 315–322.

Saygin, A. P., Driver, J., & de Sa, V. R. (2008). In the footsteps of biological motion and multisensory perception. *Psychological Science, 19*, 469–485.

Scarisbrick, D. J., Tweedy, J. R., & Kuslansky, G. (1987). Hand preference and performance effects in line bisection. *Neuropsychologia, 25*, 695–699.

Schacter, D. L., & Addis, D. R. (2007). Constructive memory: The ghosts of past and future. *Nature, 445*(7123), 27.

Schaffer, J. (2003). Is there a fundamental level? *Noûs, 37*(3), 498–517.

Schall, J. D., Perry, V. H., & Leventhal, A. G. (1986). Retinal ganglion cell dendritic fields in old-world monkeys are oriented radially. *Brain Research, 368*, 18–23.

Scharine, A. A., & McBeath, M. K. (2002) Right handers and Americans favor turning to the right. *Human Factors, 44*(1), 248–256.

Scheef, L., Boecker, H., Daamen, M., Fehse, U., Landsberg, M. W., Granath, D. O., et al. (2009). Multimodal motion processing in area V5/MT: Evidence from an artificial class of audio-visual events. *Brain Research, 1252*, 94–104.

Scheier, C. R., Nijhawan, R., & Shimojo, S. (1999). Sound alters visual temporal resolution. *Investigative Ophthalmology & Visual Science, 40*, S792.

Schenkenberg, T., Bradford, D. C., & Ajax, E. T. (1980). Line bisection and unilateral visual neglect in patients with neurologic impairment. *Neurology, 30*(5), 509–517.

Schiano, D. J., McBeath, M. K., & Chambers, K. W. (2008). Regularity of symmetry verticality guides perceptual judgments of objects. *American Journal of Psychology, 121* (2), 209–227.

Schiano, D., & Tversky, B. (1992). Structure and strategy in viewing simple graphs. *Memory & Cognition, 20*, 12–20.

Schiff, W. (1965). Perception of impending collision – a study of visually directed avoidant behavior. *Psychological Monographs, 79*(11), 1–26.

Schiff, W., Caviness, J. A., & Gibson, J. J. (1962). Persistent fear responses in rhesus monkeys to the optical stimulus of "looming". *Science, 136*, 982–983.

Schiff, W., & Oldak, R. (1990). Accuracy of judging time to arrival – effects of modality, trajectory, and gender. *Journal of Experimental Psychology: Human Perception and Performance, 16*(2), 303–316. doi:10.1037/0096-1523.16.2.303

Schiller, F., Eloka, O., & Franz, V. H. (2016). Using key distance to clarify a theory on the SNARC. *Perception, 45*, 196–221.

Schinazi, V. R., Nardi, D., Newcombe, N. S., Shipley, T. F., & Epstein, R. A. (2013). Hippocampal size predicts rapid learning of a cognitive map in humans. *Hippocampus, 23*, 515–528.

Schirillo, J. A. (2000). Hemispheric asymmetries and gender influence Rembrandt's portrait orientations. *Neuropsychologia, 38*, 1593–1606.

Schlag, J., Cai, R. H., Dorfman, A., Mohempour, A., & Schlag-Rey, M. (2000). Extrapolating movement without retinal motion. *Nature, 403*, 38–39.

Schmahmann, J. D. (2001). The cerebrocerebellar system: Anatomic substrates of the cerebellar contribution to cognition and emotion. *International Review of Psychiatry, 13*(4), 247–260.

Schmiedchen, K., Freigang, C., Nitsche, I., & Rubsamen, R. (2012). Crossmodal interactions and multisensory integration in the perception of audio-visual motion – a free-field study. *Brain Research, 1466*, 99–111.

Schmiedchen, K., Freigang, C., Rubsamen, R., & Richter, N. (2013). A comparison of visual and auditory representational momentum in spatial tasks. *Attention, Perception, & Psychophysics, 75*, 1507–1519.

Schmitt, D. P., & Pilcher, J. J. (2004). Evaluating evidence of psychological adaptation: How do we know one when we see one? *Psychological Science, 15*(10), 643–649.

Schnall, S., Harber, K. D., Stefanaucci, J. K., & Proffitt, D. R. (2008) Social support and the perception of geographical slant. *Journal of*

Experimental Social Psychology, 44, 1246–1255.

Schnall, S., Zadra, J. R., & Proffitt, D. R. (2010). Direct evidence for the economy of action: Glucose and the perception of geographical slant. *Perception, 39*, 464–482.

Schober, M. F. (1995). Speakers, addressees, and frames of reference: Whose effort is minimized in conversations about locations? *Discourse Processes, 20*, 219–247.

Schouten, B., Troje, N. F., Vroomen, J., & Verfaillie, K. (2011). The effect of looming and receding sounds on the perceived in-depth orientation of depth-ambiguous biological motion figures. *PLoS ONE, 6*, e14725.

Schubert, L., Schubert, T. W., & Topolinski, S. (2013). The effect of spatial elevation on respect depends on merit and medium. *Social Psychology, 44*(2), 147–159.

Schubert, T., & Maass, A. (Eds.). (2011). *Spatial schemas in social thought*. Berlin: Mouton de Gruyter.

Schubert, T. W. (2005). Your highness: Vertical positions as perceptual symbols of power. *Journal of Personality and Social Psychology, 89*, 1–21.

Schubert, T. W., & Otten, S. (2002). Overlap of self, ingroup, and outgroup: Pictorial measures of self-categorization. *Self and Identity, 1*(4), 353–376.

Schubert, T. W., Waldzus, S., & Giessner, S. R. (2009). Control over the association of power and size. *Social Cognition, 27*(1), 1–19.

Schubert, T. W., Waldzus, S., & Seibt, B. (2008). The embodiment of power and communalism in space and bodily contact. In G. R. Semin & E. R. Smith (Eds.), *Embodied grounding: Social, cognitive, affective, and neuroscientific approaches* (pp. 160–183). New York: Cambridge University Press.

Schubert, T. W., Waldzus, S., & Seibt, B. (2010). More than a metaphor: How the understanding of power is grounded in experience. In T. W. Schubert & A. Maass (Eds.), *Spatial dimensions of social thought* (pp. 177–213). Berlin: Mouton de Gruyter.

Schuler, J. R., Bockisch, C. J., Straumann, D., & Tarnutzer, A. A. (2010). Precision and accuracy of the subjective haptic vertical in the roll plane. *BMC Neuroscience*, 11(1), 83.

Schwartz, B., Tesser, A., & Powell, E. (1982). Dominance cues in nonverbal behavior. *Social Psychology Quarterly, 45*(2), 114–120.

Scocchia, L., Actis-Grosso, R., de'Sperati, C., Stucchi, N., & Baud-Bovy, G. (2009). Observer's control of the moving stimulus increases the flash-lag effect. *Vision Research, 49*, 2363–2370.

Scott, G., & Hellige, J. (1998). Hemispheric asymmetry for word naming: Effects of frequency and regularity of pronunciation. *Laterality, 3*, 343–371.

Scott, T. R., Lavender, A. D., McWhirt, R. A., & Powell, D. A. (1966). Directional asymmetry of motion aftereffect. *Journal of Experimental Psychology, 71*(6), 806–815.

Scruton, R. (1983). *The aesthetic understanding*. South Bend, IN: St Augustine's Press.

Seamon, J. G., Schlegel, S. E., Hiester, P. M., Landau, S. M., & Blumenthal, B. F. (2002). Misremembering pictured objects: People of all ages demonstrate the boundary extension illusion. *American Journal of Psychology, 115*(2), 151–167.

Sedgwick, H. A. (1986). Space perception. In K. R. Boff, L. Kaufman, & J. P. Thomas (Eds.), *Handbook of human perception and performance: Sensory processes* (pp. 21.1–21.57). New York: Wiley.

Sedgwick, H. A. (2001). Visual space perception. In E. B. Goldstein (Ed.), *Blackwell handbook of perception* (pp. 128–167). Oxford: Blackwell.

Seifritz, E., Neuhoff, J. G., Bilecen, D., Scheffler, K., Mustovic, H., Schachinger, H., et al. (2002). Neural processing of auditory looming in the human brain. *Current Biology, 12*(24). doi:10.1016/s0960-9822(02)01356-8

Seitz, A. R., Kim, R., van Wassenhove, V., & Shams, L. (2007). Simultaneous and independent acquisition of multisensory and unisensory associations. *Perception, 36*, 1445–1453.

Sekuler, R., Sekuler, A. B., & Lau, R. (1997). Sound alters visual motion perception. *Nature, 385*, 308.

Senior, C., Barnes, J., Giampietro, V., Simmons, A., Bullmore, E. T., Brammer, M., & David, A. S. (2000). The functional neuroanatomy of implicit-motion perception or representational momentum. *Current Biology, 10*, 16–22.

Senior, C., Ward, J., & David, A. S. (2002). Representational momentum and the brain: An investigation of the functional necessity of V5/MT. *Visual Cognition, 9*, 81–92.

Senot, P., Zago, M., Lacquaniti, F., & McIntyre, J. (2005). Anticipating the effects of gravity when intercepting moving objects: Differentiating up and down based on nonvisual cues. *Journal of Neurophysiology, 94*, 4471–4480.

Senot, P., Zago, M., Le Séac'h, A., Zaoui, M., Berthoz, A., Lacquaniti, F., & McIntyre, J. (2012). When up is down in 0g: How gravity sensing affects the timing of interceptive actions. *Journal of Neuroscience, 32*(6), 1969–1973.

Serences, J. T., Yantis, S., Culberson, A., & Awh, E. (2004). Preparatory activity in visual cortex indexes distractor suppression during covert spatial orienting. *Journal of Neurophysiology, 92*, 3538–3545.

Sereno, A. B., & Maunsell, J. H. (1998). Shape selectivity in primate lateral intraparietal cortex. *Nature, 395*(6701), 500–503.

Sereno, M., Pitzalis, S., & Martinez, A. (2001). Mapping of contralateral space in retinotopic coordinates by a parietal cortical area in humans. *Science, 294*, 1350–1354.

Sereno, M. I., Dale, A. M., Reppas, J. B., et al. (1995). Borders of multiple visual areas in human revealed by functional magnetic resonance imaging. *Science, 268*, 889–893.

Sergent, J. (1982). Theoretical and methodological consequences of variations in exposure duration in visual laterality studies. *Perception & Psychophysics, 31*(5), 451–461.

Sergent, J. (1995). Hemispheric contribution to face processing: Patterns of convergence and divergence. In R. Davidson & K. Hughdahl (Eds.), *Brain asymmetry* (pp. 157–181). Cambridge, MA: MIT Press.

Serrano-Pedraza, I., Brash, C., & Read, J. C. (2013). Testing the horizontal-vertical stereo anisotropy with the critical-band masking paradigm. *Journal of Vision, 13*(11), 1–15.

Serrano-Pedraza, I., & Read, J. C. A. (2010). Multiple channels for horizontal, but only one for vertical corrugations? A new look at the stereo anisotropy. *Journal of Vision, 10*(12), 1–11.

Seuss, Dr. (1953). *The Sneetches and other stories*. New York: Redbook.

Shadmehr, R., & Krakauer, J. W. (2008). A computational neuroanatomy for motor control. *Experimental Brain Research, 185*, 359–381.

Shaffer, D. M., Krauchunas, S. M., Eddy, M., & McBeath, M. K. (2004). How dogs navigate to catch frisbees. *Psychological Science, 15*(7), 437–441.

Shaffer, D. M., & McBeath, M. K. (2002). Baseball outfielders maintain a linear optical trajectory when tracking uncatchable fly balls. *Journal of Experimental Psychology: Human Perception and Performance, 28*(2), 335–348.

Shaffer, D. M., & McBeath, M. K. (2005). Naïve beliefs in baseball: Systematic distortion in perceived time of apex for fly balls. *Journal of Experimental Psychology: Learning, Memory, and Cognition, 31*(6), 1492–1501.

Shaffer, D. M., McBeath, M. K., & Roy, W. L., & Krauchunas, S. M. (2003). A linear optical trajectory informs fielders where to run to the side to catch fly balls. *Journal of Experimental Psychology: Human Perception and Performance, 29*(6), 1244–1250.

Shaffer, D. M., McManama, E., Swank, C., & Durgin, F. H. (2013). Sugar and space? Not the case: Effects of low blood glucose on slant estimation are mediated by beliefs. *i-Perception, 4*, 147–155.

Shaki, S., & Fischer, M. H. (2008). Reading space into numbers – a cross-linguistic comparison of the SNARC effect. *Cognition, 108*(2), 590–599.

Shaki, S., & Fischer, M. H. (2012). Multiple spatial mappings in numerical cognition. *Journal of Experimental Psychology: Human Perception and Performance, 38*(3), 804–809.

Shaki, S., & Fischer, M. H. (2014). Random walks on the mental number line. *Experimental Brain Research*, *232*(1), 43–49.

Shaki, S., Fischer, M. H., & Göbel, S. M. (2012). Direction counts: A comparative study of spatially directional counting biases in cultures with different reading directions. *Journal of Experimental Child Psychology*, *112*(2), 275–281.

Shaki, S., Fischer, M. H., & Petrusic, W. M. (2009). Reading habits for both words and numbers contribute to the SNARC effect. *Psychonomic Bulletin & Review*, *16*, 328–331.

Shaki, S., Petrusic, W. M., & Leth-Steensen, C. (2012). SNARC effects with numerical and non-numerical symbolic comparative judgments: Instructional and cultural dependencies. *Journal of Experimental Psychology: Human Perception and Performance*, *38*, 515–530.

Shallice, T., Mussoni, A., D'Agostino, S., & Skrap, M. (2010). Right posterior cortical functions in a tumour patient series. *Cortex*, *46*(9), 1178–1188.

Shams, L., Kamitani, Y., & Shimojo, S. (2000). Illusions: What you see is what you hear. *Nature*, *408*, 788.

Shams, L., Kamitani, Y., & Shimojo, S. (2002). Visual illusion induced by sound. *Cognitive Brain Research*, *14*, 147–152.

Shams, L., & Kim, R. (2010). Crossmodal influences on visual perception. *Physics of Life Reviews*, *7*, 269–284.

Shams, L., Ma, W. J., & Beierholm, U. (2005). Sound-induced flash illusion as an optimal percept. *Neuroreport*, *16*, 1923–1927.

Shannon, C. E. (1948). A mathematical theory of communication. *Bell Systems Technical Journal*, *27*, 279–423.

Shanon, B. (1983). Answers to where-questions. *Discourse Processes*, *6*, 319–352.

Shapiro, K., Hillstrom, A. P., & Husain, M. (2002). Control of visuotemporal attention by inferior parietal and superior temporal cortex. *Current Biology*, *12*, 1320–1325.

Shaw, B. K., McGowan, R. S., & Turvey, M. T. (1991). An acoustic variable specifying time-to-contact. *Ecological Psychology*, *3*(3), 253–261.

Shaw, E. A. G. (1982). External ear response and sound localization. In R. Gatehouse (Ed.), *Localization of sound: Theory and applications* (pp. 30–41). Groton, CT: Amphora.

Shaw, R., Turvey, M. T., & Mace, W. (1982). Ecological psychology: The consequence of a commitment to realism. *Cognition and the Symbolic Processes*, *2*, 159–226.

Shayan, S., Ozturk, O., & Sicoli, M. A. (2011). The thickness of pitch: Crossmodal metaphors in Farsi, Turkish, and Zapotec. *The Senses and Society*, *6*, 96–105.

Shelton, A. L., & McNamara, T. P. (1997). Multiple views of spatial memory. *Psychonomic Bulletin & Review*, *4*, 102–106.

Shelton, P. A., Bowers, D. & Heilman, K. M. (1990). Peripersonal and vertical neglect. *Brain*, *113*, 191–205.

Shen, M., Zhou, J., Gao, T., Liang, J., & Shui, R. (2007). The perceived position of a moving object is not the result of position integration. *Vision Research*, *47*, 3088–3095.

Shepard, R. N. (1982). Geometrical approximations to the structure of musical pitch. *Psychological Review*, *89*, 305–333.

Shepard, R. N. (1984). Ecological constraints on internal representation: Resonant Kinematics of perceiving, imaging, thinking, and dreaming. *Psychological Review*, *91*(4), 417–447.

Shepard, R. N. (1994). Perceptual-cognitive universals as reflections of the world. *Psychonomic Bulletin & Review*, *1*(1), 2–28.

Shepard, R. N. (2001). Perceptual-cognitive universals as a reflection of the world. *Behavioral and Brain Sciences*, *24*, 581–601.

Sheth, B. R., Nijhawan, R., & Shimojo, S. (2000). Changing objects lead briefly flashed ones. *Nature Neuroscience*, *3*, 489–495.

Shi, Z., & de'Sperati, C. (2008). Motion-induced positional biases in the flash-lag configuration. *Cognitive Neuropsychology*, *25*, 1027–1038.

Shimojo, S., & Shams, L. (2001). Sensory modalities are not separate modalities: Plasticity and interactions. *Current Opinion in Neurobiology*, *11*, 505–509.

Shipley, T. (1964). Auditory flutter-driving of visual flicker. *Science, 145*, 1328–1330.

Shipley, T. F. (2003). The effect of object and event orientation on perception of biological motion. *Psychological Science, 14*, 377–380.

Shirai, N., Birtles, D., Wattam-Bell, J., Yamaguchi, M. K., Kanazawa, S., Atkinson, J., & Braddick, O. (2009). Asymmetrical cortical processing of radial expansion/contraction in infants and adults. *Developmental Science, 12*(6), 946–955.

Shirai, N., Kanazawa, S., & Yamaguchi, M. K. (2004). Asymmetry for the perception of expansion/contraction in infancy. *Infant Behavior & Development, 27*(3), 315–322.

Shirai, N., Kanazawa, S., & Yamaguchi, M. K. (2006). Anisotropic motion coherence sensitivities to expansion/contraction motion in early infancy. *Infant Behavior & Development, 29*(2), 204–209.

Sholl, M. J. (1987). Cognitive maps as orienting schemata. *Journal of Experimental Psychology: Learning, Memory, and Cognition, 13*, 615–628.

Shi, Z., & de'Sperati, C. (2008). Motion-induced positional biases in the flash-lag configuration. *Cognitive Neuropsychology, 25*, 1027–1038.

Shi, Z., & Nijhawan, R. (2008). Behavioral significance of motion direction causes anisotropic flash-lag, flash-drag, flash-repulsion, and movement-mislocalization effects. *Journal of Vision, 8*(7), 24.

Shioiri, S., Yamamoto, K., Oshida, H., Matsubara, K., & Yaguchi, H. (2010). Measuring attention using flash-lag effect. *Journal of Vision, 10*(10), 10.

Shulman, G., Popen, D., Astafiev, S., McAvoy, M., Snyder, A., & Corbetta, M. (2010). Right hemisphere dominance during spatial selective attention and target detection occurs outside the dorsal frontoparietal network. *Journal of Neuroscience, 30*, 3640–3651.

Shum, M. S., Bailenson, J., Hwang, S., Piland, L., & Uttal, D. (1998). Road climbing: Principles of route choice. In *Proceedings of the 20th Annual Conference of the Cognitive Science Society*. Mahwah, NJ: Erlbaum.

Siegler, R. S., & Opfer, J. E. (2003). The development of numerical estimation evidence for multiple representations of numerical quantity. *Psychological Science, 14*(3), 237–250.

Silver, M., Ress, D., & Heeger, D. (2005). Topographic maps of visual spatial attention in human parietal cortex. *Journal of Neurophysiology, 94*(2), 1358–1371.

Silverman, I., Choi, J., & Peters, M. (2007). The Hunter-Gatherer theory of sex differences in spatial abilities: Data from 40 countries. *Archives of Sexual Behavior, 36*(2), 261–268.

Silvia, P. J., & Barona, C. M. (2009). Do people prefer curved objects? Angularity, expertise, and aesthetic preference. *Empirical Studies of the Arts, 27*, 25–42.

Simmering, V. R., Schutte, A. R., & Spencer, J. P. (2008). Generalizing the dynamic field theory of spatial cognition across real and developmental time scales. *Brain Research, 1202*, 68–86.

Simon, J. R. (1990). The effects of an irrelevant directional cue on human information processing. In R. W. Proctor & T. G. Reeve (Eds.), *Stimulus-response compatibility: An integrated perspective* (pp. 31–86). Amsterdam: North-Holland.

Simon, J. R., Acosta, E., Mewaldt, S. P., & Speidel, C. R. (1976). The effect of an irrelevant directional cue on choice-reaction time: Duration of the phenomenon and its relation to stages of processing. *Perception & Psychophysics, 19*, 16–22.

Simon, J. R., & Craft, J. L. (1971). Communicating directional information with an auditory display. *Journal of Applied Psychology, 55*, 241–243.

Simon, J. R., Mewaldt, S. P., Acosta, E, & Hu, J. M. (1976). Processing auditory information: interaction of two spatial stereotypes. *Journal of Applied Psychology, 61*, 354–358.

Simon, J. R., & Small, Jr., A. M. (1969). Processing auditory information: Interference from an irrelevant cue. *Journal of Applied Psychology, 53*, 433–435.

Sinai, M. J., Ooi T. L., & He, Z. J. (1998). Terrain influences the accurate judgement of distance. *Nature, 395*(6701), 497–500.

Sinico, M., Parovel, G., Casco, C., & Anstis, S. (2009). Perceived shrinkage of motion paths. *Journal of Experimental Psychology: Human Perception and Performance, 35*(4), 948–957.

Skarratt, P. A., Cole, G. G., & Gellatly, A. R. H. (2009). Prioritization of looming and receding objects: Equal slopes, different intercepts. *Attention, Perception, & Psychophysics, 71*(4), 964–970.

Skarratt, P. A., Gellatly, A. R. H., Cole, G. G., Pilling, M., & Hulleman, J. (2014). Looming Motion Primes the Visuomotor System. *Journal of Experimental Psychology: Human Perception and Performance, 40*(2), 566–579.

Skinner, B. F. (1976). *About behaviorism*. New York: Vintage Books.

Slobin, D. (1996). From "thought and language" to "thinking for speaking." In J. Gumperz & S. Levinsohn (Eds.), *Rethinking linguistic relativity* (pp. 70–96). New York: Cambridge University Press.

Sloutsky, V. M. (2003). The role of similarity in the development of categorization. *Trends in Cognitive Sciences, 7*(6), 246–251.

Sluzenski, J., Newcombe, N., & Ottinger, W. (2004). Changes in reality monitoring and episodic memory in early childhood. *Developmental Science, 7*(2), 225–245.

Smith, A. T., Singh, K. D., & Greenlee, M. W. (2000). Attentional suppression of activity in the human visual cortex. *NeuroReport, 11*, 271–277.

Smith, D. T., & Schenk, T. (2012). The premotor theory of attention: Time to move on? *Neuropsychologia, 50*, 1104–1114.

Smith, L. F. (2014). The science and aesthetics of astronomical images. *Psychology of Aesthetics, Creativity, and the Arts, 8*, 506–513.

Smith, P. K., & Trope, Y. (2006). You focus on the forest when you're in charge of the trees: Power priming and abstract information processing. *Journal of Personality and Social Psychology, 90*(4), 578–596.

Snow, J. C., & Mattingley, J. B. (2006). Goal-driven selective attention in patients with right hemisphere lesions: How intact is the ipsilesional field? *Brain, 129*, 168–181.

So, W. C., Shum, P. L. C., & Wong, M. K. Y. (2015). Gesture is more effective than spatial language in encoding spatial information. *Quarterly Journal of Experimental Psychology, 68*(12), 2384–2401.

Soechting, J. F., & Flanders, M. (1992). Moving in three-dimensional space: Frames of reference, vectors, and coordinate systems. *Annual Review of Neuroscience, 15*(1), 167–191.

Solomon, J. A. (2009). The history of dipper functions. *Attention, Perception, & Psychophysics, 71*, 435–443.

Solomon, J. A. (Ed.). (2011). *Fechner's legacy in psychology: 150 years of elementary psychophysics*. Leiden: Brill.

Solomon, J. A., & Morgan, M. J. (2006) Stochastic re-calibration: Contextual effects on perceived tilt. *Proceedings of the Royal Society, B, 273*, 2681–2686.

Solomon, J. A., Morgan, M. J., & Chubb, C. (2011) Efficiences for the statistics of size discrimination. *Journal of Vision*, 11, 13.

Soto, D., & Humphreys, G. W. (2009). Semantically induced distortions of visual awareness in a patient with Balint's syndrome. *Cognition, 110*(2), 237–241.

Soto-Faraco, S., Kingstone, A., & Spence, C. (2003). Multisensory contributions to the perception of motion. *Neuropsychologia, 41*, 1847–1862.

Soto-Faraco, S., Lyons, J., Gazzaniga, M., Spence, C., & Kingstone, A. (2002). The ventriloquist in motion: Illusory capture of dynamic information across sensory modalities. *Cognitive Brain Research, 14*, 139–146.

Soto-Faraco, S., Spence, C., & Kingstone, A. (2004). Cross-modal dynamic capture: Congruency effects in the perception of motion across sensory modalities. *Journal of Experimental Psychology: Human Perception and Performance, 30*, 330–345.

Soto-Faraco, S., Spence, C., & Kingstone, A. (2005). Assessing automaticity in the audiovisual integration of motion. *Acta Psychologica, 118*, 71–92.

Spanò, G., Intraub, H., & Edgin, J. O. (2017). Testing the "Boundaries" of boundary extension: Anticipatory scene representation across development and disorder. *Hippocampus*, *27*, 726–739.

Spelke, E., Lee, S. A., & Izard, V. (2010). Beyond core knowledge: Natural geometry. *Cognitive Science*, *34*(5), 863–884.

Spence, C. (2010). Crossmodal spatial attention. *Annals of the New York Academy of Sciences (The Year in Cognitive Neuroscience)*, *1191*, 182–200.

Spence, C. (2011). Crossmodal correspondences: A tutorial review. *Attention, Perception, & Psychophysics*, *73*, 971–995.

Spence, C. (2013). Just how important is spatial coincidence to multisensory integration? Evaluating the spatial rule. *Annals of the New York Academy of Sciences*, *1296*, 31–49.

Spence, C., & Deroy, O. (2013). How automatic are crossmodal correspondences? *Consciousness and Cognition*, *22*, 245–260.

Spence, C., Deroy, O., & Bremner, A. (2013). Questioning the utility of the concept of amodality: Towards a revised framework for understanding crossmodal relations. *Multisensory Research*, *26*(Suppl.), 57.

Spence, J. T., Helmreich, R. L., & Stapp, J. (1974). The personal attributes questionnaire: A measure of sex-role stereotypes and masculinity and femininity. *Journal Supplement Abstract Service: Catalog of Selected Documents in Psychology*, *4*, 43–44.

Spencer, J. P., & Hund, A. M. (2002). Prototypes and particulars: Geometric and experience-dependent spatial categories. *Journal of Experimental Psychology: General*, *131*(1), 16–37.

Spetch, M. L., Friedman, A., Bialowas, J., & Verbeek, E. (2010). Contributions of category and fine-grained information to location memory: When categories don't weigh in. *Memory & Cognition*, *38*(2), 154–162.

Stafford, L. (2010). Geographic distance and communication during courtship. *Communication Research*, *37*, 275–297.

Stefanucci, J. K., & Proffitt, D. R. (2009). The roles of altitude and fear in the perception of height. *Journal of Experimental Psychology: Human Perception and Performance*, *35*(2), 424–438.

Stefanucci, J. K., Proffitt, D. R., Banton, T., & Epstein, W. (2005). Distances appear different on hills. *Perception & Psychophysics*, *67*(6), 1052–1060.

Steinman, B. A., Steinman, S. B., & Lehmkuhle, S. (1995). Visual attention mechanisms show a center-surround organization. *Vision Research*, *35*, 1859–1869.

Stekelenburg, J. J., & Vroomen, J. (2005). An event-related potential investigation of the time-course of temporal ventriloquism. *NeuroReport*, *16*, 641–644.

Stern, E., & Leiser, D. (1988). Levels of spatial knowledge and urban travel modeling. *Geographical Analysis*, *20*(2), 140–155.

Stevens, S. S. (1935). The relation of pitch to intensity. *The Journal of the Acoustical Society of America*, *6*, 150–154.

Stevens, A., & Coupe, P. (1978). Distortions in judged spatial relations. *Cognitive Psychology*, *10*(4), 422–437.

Stidwill, D., & Fletcher, R. (2011). *Normal binocular vision: Theory, investigation and practical aspects*. Oxford: Wiley-Blackwell.

Stieger, S., & Swami, V. (2015). Time to let go? No automatic aesthetic preference for the golden ratio in art pictures. *Psychology of Aesthetics, Creativity, and the Arts*, *9*, 91–100.

Stigler, R. (1910). Chronophotische Studien über den Umgebungskontrast [Chronophotic studies of the vicinity contrast]. *Pflügers Archiv für die gesamte Physiologie*, *134*, 365–435.

Stigmar, G. (1970). Observations on vernier and stereo acuity with special reference to their relationship. *Acta Ophthalmologica (Copenhagen)*, *48*(5), 979–998.

Stoffregen, T. A., & Bardy, B. G. (2001). On specification and the senses. *Behavioral and Brain Sciences*, *24*, 195–261.

Stone, J. P., & McBeath, M. K. (2010). Gender differences in distance estimates when exposed to multiple routes. *Environment and Behavior*, *42*(4), 469–478.

Stork, S., & Müsseler, J. (2004). Perceived localizations and eye movements with action-generated and computer-generated vanishing points of moving stimuli. *Visual Cognition*, *11*, 299–314.

Streeck, J., & Jordan, J. S. (2009). Communication as a dynamical self-sustaining system: The importance of time-scales and nested contexts, *Communication Theory, 19*, 445–464.

Strybel, T. Z., & Vatakis, A. (2004). A comparison of auditory and visual apparent motion presented individually and with crossmodal moving distractors. *Perception, 33*, 1033–1048.

Stulp, G., Buunk, A. P., Verhulst, S., & Pollet, T. V. (2012a). High and mighty: Height increases authority in professional refereeing. *Evolutionary Psychology, 10*(3), 588–601.

Stulp, G., Buunk, A. P., Verhulst, S., & Pollet, T. V. (2012b). Tall claims? Sense and nonsense about the importance of height of US presidents. *The Leadership Quarterly, 24*(1), 1–13.

Stulp, G., Buunk, A. P., Verhulst, S., & Pollet, T. V. (2013). Tall claims? Sense and nonsense about the importance of height of US presidents. *The Leadership Quarterly, 24*(1), 159–171.

Stulp, G., Buunk, A. P., Verhulst, S., & Pollet, T. V. (2015). Human height is positively related to interpersonal dominance in dyadic interactions. *PLoS ONE, 10*(2), 1–18.

Stumpf, C. (1883). *Tonpsychologie*. Leipzig: Verlag von S. Hirzel.

Stumpf, C. (1911/2012). *The origins of music* (David Trippett, Ed. and Trans.). Oxford: Oxford University Press.

Sturmer, B., Leuthold, H., Soetens, E., Schröter, H., & Sommer, W. (2002). Control over location-based response activation in the Simon task: Behavioral and electrophysiological evidence. *Journal of Experimental Psychology: Human Perception and Performance, 28*, 1345–1363.

Suavansri, K., Falchook, A. D., Williamson, J. B., & Heilman, K. M. (2012). Right up there: Hemispatial and hand asymmetries of altitudinal pseudoneglect. *Brain and Cognition, 79*, 216–220.

Sugar, D. G., & McBeath, M. K. (2001). Robotic modeling of mobile ball-catching as a tool for understanding biological interceptive behavior. *Behavioral and Brain Sciences, 24*(6), 1078–1080.

Sugar, T. G., McBeath, M. K., Suluh, A., & Mundhra, K. (2006). Mobile robot interception using human navigational principles: Comparison of active versus passive tracking algorithms. *Autonomous Robots, 21*(1), 43–54.

Sugar, T. G., McBeath, M. K., & Wang, Z. (2006). A unified fielder theory for interception of moving objects either above or below the horizon. *Psychonomic Bulletin & Review, 13*(5), 908–917.

Sugovic, M., Turk, P., & Witt, J. K. (2016). Perceived distance and obesity: It's what you weigh, not what you think. *Acta Psychologica, 165*, 1–8.

Suitner, C., & Giacomantonio, M. (2012). Seeing the forest from left to right: How Construal Level affects the Spatial Agency Bias. *Social Psychological and Personality Science, 3*, 180–185.

Suitner, C., & Maass, A. (2007). Positing bias in portraits and self-portraits: Do female artists make different choices? *Empirical Studies of the Arts, 25*, 71–95.

Suitner, C., & Maass, A. (2016). Spatial agency bias: Representing people in space. *Advances in Experimental Social Psychology, 53*, 245–301.

Suitner, C., Maass, A., Bettinsoli, M. L., Carraro, L., & Kumar, S. (2017). Left-handers' struggle in a rightward wor(l)d: The relation between horizontal spatial bias and effort in directed movements. *Laterality: Asymmetries of Body, Brain and Cognition, 22*(1) 60–89.

Suitner, C., Maass, A., & Ronconi, L. (2017). From spatial to social asymmetry: Spontaneous and conditioned associations of gender and space. *Psychology of Women Quarterly, 41*(1), 46–64.

Suitner, C., & McManus, I. C. (2011). Aesthetic asymmetries, spatial agency, and art history: A social psychological perspective. In T. Schubert & A. Maass (Eds.), *Spatial dimensions of social thought* (pp. 277–302). Berlin & New York: Mouton de Gruyter.

Sumi, S. (1984). Upside-down presentation of the Johansson moving light-spot pattern. *Perception, 13*, 283–286.

Sussman, A. L. (2001). Reality monitoring of performed and imagined interactive events: Developmental and contextual effects. *Journal of Experimental Child Psychology, 79*(2), 115–138.

Suzuki, S., & Cavanagh, P. (1997). Focused attention distorts visual space: An attentional repulsion effect. *Journal of Experimental Psychology: Human Perception and Performance, 23*(2), 443–463.

Swartz, P., & Hewitt, D. (1970). Lateral organization in pictures and aesthetic preference. *Perceptual and Motor Skills, 30*, 991–1007.

Tabata, T., & Okuda, S. (2000). Mirror reversal simply explained without recourse to psychological processes. *Psychonomic Bulletin & Review, 7*(1), 170–173.

Tagliabue, M., Zorzi, M., & Umiltà, C. (2002). Cross-modal remapping influences the Simon effect. *Memory & Cognition, 30*, 18–23.

Tagliabue, M., Zorzi, M., Umiltà, C., & Bassignani, F. (2000). The role of LTM links and STM links in the Simon effect. *Journal of Experimental Psychology: Human Perception and Performance, 26*, 648–670.

Tajadura-Jimenez, A., Vakali, M., Fairhurst, M. F., Mandringin, A., Bianchi-Berthouze, N., & Deroy, O. (2017). Contingent sounds change the mental representation of one's finger length. *Scientific Reports, 7*, 5748.

Tajadura-Jimenez, A., Valjamae, A., Asutay, E., & Vastfjall, D. (2010). Embodied auditory perception: The emotional impact of approaching and receding sound sources. *Emotion, 10*(2). doi:10.1037/a0018422

Tajfel, H., Billig, M. G., Bundy, R. P., & Flament, C. (1971). Social categorization and intergroup behaviour. *European Journal of Social Psychology, 1*(2), 149–178.

Tajfel, H. (1974). Social identity and intergroup behaviour. *Social Science Information/sur les sciences sociales, 13*(2), 65–93.

Takano, Y. (1998). Why does a mirror image look left–right reversed? A hypothesis of multiple processes. *Psychonomic Bulletin & Review, 5*(1), 37–55.

Takeuchi, T. (1997). Visual search of expansion and contraction. *Vision Research, 37*(15), 2083–2090.

Talgar, C. P., & Carrasco, M. (2002). Vertical meridian asymmetry in spatial resolution: visual and attentional factors. *Psychonomic Bulletin & Review, 9*(4), 714–722.

Talmy, L. (1983). How language structures space. In H. Pick & L. Acredolo (Eds.), *Spatial orientation: Theory, research and application* (pp. 225–282). New York: Plenum Press.

Talmy, L. (2000). *Towards a cognitive semantics: Concept structuring systems*. Cambridge, MA: MIT Press.

Tan, S., & Dixon, P. (2011). Repetition and the SNARC effect with one-and two-digit numbers. *Canadian Journal of Experimental Psychology/Revue canadienne de psychologie expérimentale, 65*(2), 84–97.

Tanner, W. P., & Swets, J. A. (1954). A decision-making theory of visual detection. *Psychological Review, 61*(6), 401–409.

Tarnutzer, A. A., Bockisch, C., Straumann, D., & Olasagasti, I. (2009). Gravity dependence of subjective visual vertical variability. *Journal of Neurophysiology, 102*(3), 1657–1671.

Taya, S., & Miura, K. (2010). Cast shadow can modulate the judged final position of a moving target. *Attention, Perception, & Psychophysics, 72*, 1930–1937.

Taylor, H. A., & Tversky, B. (1996). Perspective in spatial descriptions. *Journal of Memory and Language, 35*(3), 371–391.

Taylor, J. E. T., Witt, J. K., & Sugovic, M. (2011). When walls are no longer barriers: Perception of wall height in parkour. *Perception, 40*(6), 757–760.

Taylor, N. M., & Jakobson, L. S. (2010). Representational momentum in children born preterm and at term. *Brain and Cognition, 72*, 464–471.

Taylor, T. L., & Ivanoff, J. (2003). The interplay of stop signal inhibition and inhibition of return. *Quarterly Journal of Experimental Psychology, 56A*, 1349–1371.

Taylor, T. L., & Klein, R. M. (2000). Visual and motor effects in inhibition of return. *Journal of Experimental Psychology: Human Perception and Performance, 26*, 1639–1656.

Taylor-Covill, G. A. H., & Eves, F. F. (2013). Slant perception for stairs and screens: Effects of sex and fatigue in a laboratory environment. *Perception, 42*(4), 459–469.

Taylor-Covill, G. A. H., & Eves, F. F. (2014). When what we need influences what we see: Choice of energetic replenishment is linked with perceived steepness. *Journal of*

Experimental Psychology: Human Perception and Performance, 40(3), 915–919.

Taylor-Covill, G. A. H., & Eves, F. F. (2016). Carrying a biological "backpack": Quasi-experimental effects of weight status and body fat change on perceived steepness. *Journal of Experimental Psychology: Human Perception and Performance, 42*(3), 331–338.

Teghtsoonian, M., & Teghtsoonian, R. (1969). Scaling apparent distance in natural indoor settings. *Psychonomic Science, 16*(6), 281–283.

Teghtsoonian, R., Teghtsoonian, M., & Canevet, G. (2005). Sweep-induced acceleration in loudness change and the "bias for rising intensities." *Perception & Psychophysics, 67*(4), 699–712.

Teneggi, C., Canzoneri, E., di Pellegrino, G., & Serino, A. (2013). Social modulation of peripersonal space boundaries. *Current Biology, 23*(5), 406–411.

Tenhundfeld, N. L., & Witt, J. K. (2017). Distances on hills look farther than distances on flat ground: Evidence from converging measures. *Attention, Perception, & Psychophysics, 79*(4), 1165–1181.

Teramoto, W., Hidaka, S., Gyoba, J., & Suzuki, Y. (2010). Auditory temporal cues can modulate visual representational momentum. *Attention, Perception, & Psychophysics, 72*, 2215–2226.

Teramoto, W., Hidaka, S., & Sugita, Y. (2010). Sounds move a static visual object. *PLoS ONE, 5*, e12255.

Teramoto, W., Hidaka, S., Sugita, Y., Sakamoto, S., Gyoba, J., Iwaya, Y., & Suzuki, Y. (2012). Sounds can alter the perceived direction of a moving visual object. *Journal of Vision, 12*, 11.

Teramoto, W., Kobayashi, M., Hidaka, S., & Sugita, Y. (2013). Vision contingent auditory pitch aftereffects. *Experimental Brain Research, 229*, 97–102.

Teramoto, W., Manaka, Y., Hidaka, S., Sugita, Y., Miyauchi, R., Sakamoto, S., Iwaya, Y. & Suzuki, Y. (2010). Visual motion perception induced by sounds in vertical plane. *Neuroscience Letters, 479*, 221–225.

Terhardt, E. (1974). On the perception of periodic sound fluctuations (roughness). *Acustica, 30*, 201–213.

Thiebaut de Schotten, M., Dell'Acqua, F., Forkel, S., Simmons, A., Vergani, F., Murphy, D., et al. (2011). A lateralized brain network for visuospatial attention. *Nature Neuroscience, 14*(10), 1245–1246.

Thiebaut de Schotten, M., Tomaiuolo, F., Aiello, M., Merola, S., Silvetti, M., Lecce, F., et al. (2012). Damage to white matter pathways in sub-acute and chronic spatial neglect: A group study and two single-case studies with complete virtual "in-vivo" tractography dissection. *Cerebral Cortex, 24*(3), 691–706.

Thiebaut de Schotten, M., Urbanski, M., Duffau, E., Volle, E., Levy, R., Dubois, B., et al. (2005). Direct evidence for a parietal-frontal pathway subserving spatial awareness in humans. *Science, 309*, 2226–2228.

Thomas, B. (1971). *Geometry in pictorial composition*. Newcastle, UK: Oriela Press.

Thomas, C., Kveraga, K., Huberle, E., Karnath, H.-O., & Bar, M. (2012). Enabling global processing in simultanagnosia by psychophysical biasing of visual pathways. *Brain, 135*(5), 1578–1585.

Thomas, N. A., Burkitt, J. A., & Saucier, D. M. (2006). Photographer preference or image bias? An investigation of posing bias in mammalian and non-mammalian species. *Laterality, 11*, 350–354.

Thomas, N. A., & Elias, L. J. (2011). Upper and lower visual field differences in perceptual asymmetries. *Brain Research, 1387*, 108–115.

Thompson, P. (1980). Margaret Thatcher: A new illusion. *Perception, 9*, 483–484.

Thomsen, L., & Carey, S. (2013). Core cognition of social relations. In M. R. Banaji & S. Gelman (Eds.), *Navigating the social world: What infants, children, and other species can teach us* (pp. 17–22). New York: Oxford University Press.

Thomsen, L., Frankenhuis, W. E., Ingold-Smith, M., & Carey, S. (2011). Big and mighty: Preverbal infants mentally represent social dominance. *Science, 331*(6016), 477–480.

Thorndyke, P. W. (1981). Distance estimation from cognitive maps. *Cognitive Psychology, 13*, 526–550.

Thorndyke, P. W., & Hayes-Roth, B. (1982) Differences in spatial knowledge acquired from maps and navigation. *Cognitive Psychology, 14*, 560–589.

Thornton, I. M. (2002). The onset repulsion effect. *Spatial Vision, 15*(2), 219–243.

Thornton, I. M. (2014). Representational momentum and the human face: An empirical note. *Xjena Online – Journal of the Malta Chamber of Scientists.* doi: 10.7423/XJENZA.2014.2.09

Thornton, I. M., & Hayes, A. E. (2004). Anticipating action in complex scenes. *Visual Cognition, 11*, 341–370.

Tian, Y., Huang, Y., Zhou, K., Humphreys, G. W., Riddoch, M. J., & Wang, K. (2011). When Connectedness Increases Hemispatial Neglect. *PLoS ONE, 6*(9), e24760.

Tibber, M. S., Anderson, E. J., Melmoth, D. R., Rees, G., & Morgan, M. J. (2009). Common cortical loci are activated during visuospatial interpolation and orientation discrimination judgements. *PLoS ONE, 4*(2), e4585.

Tibber, M. S., Melmoth, D. R., & Morgan, M. J. (2008). Biases and sensitivities in the Poggendorff effect when driven by subjective contours. *Invesigative Ophthalmology & Visual Science, 49*(1), 474–478.

Tillman, B., & Bharucha, J. J. (1999). Perceiving and learning harmonic structure: Some news from MUSACT. *International Journal of Computing Anticipatory Systems, 4*, 289–300.

Tinbergen, N., Impekoven, M., & Franck, D. (1967). An experiment on spacing-out as a defence against predation. *Behaviour, 28*, 307–321.

Tlauka, M. (2002). The processing of numbers in choice-reaction tasks. *Australian Journal of Psychology, 54*(2), 94–98.

Toba, M.-N., Cavanagh, P., & Bartolomeo, P. (2011). Attention biases the perceived midpoint of horizontal lines. *Neuropsychologia, 49*(2), 238–246.

Todd, J. T. (1981). Visual information about moving-objects. *Journal of Experimental Psychology-Human Perception and Performance, 7*(4), 795–810.

Tolman, E. C. (1948). Cognitive maps in rats and men. *Psychological Review, 55*, 189–208.

Tomassini, A., Morgan, M. J., & Solomon, J. S. (2010) Orientation uncertainly reduces perceived obliquity. *Vision Research, 50*, 541–547.

Torralbo, A., Santiago, J., & Lupiáñez, J. (2006). Flexible conceptual projection of time onto spatial frames of reference. *Cognitive Science, 30*, 745–757.

Toth, C., & Kirk, A. (1996). A normal bias towards a pictorially defined top in line bisection. *Canadian Journal of Neurological Science, 23*, 110–113.

Tozawa, J. (2010). Role of texture gradient in the perception of relative size. *Perception, 39*(5), 641–660.

Tozawa, J., & Oyama, T. (2006). Effects of motion parallax and perspective cues on perceived size and distance. *Perception, 35*(8), 1007–1023.

Tresilian, J. R. (1993). Four questions of time to contact – a critical examination of research on interceptive timing. *Perception, 22*(6), 653–680.

Tresilian, J. R. (1994). Approximate information-sources and perceptual variables in interceptive timing. *Journal of Experimental Psychology: Human Perception and Performance, 20*(1), 154–173.

Tresilian, J. R. (1995). Perceptual and cognitive-processes in time-to-contact estimation – analysis of prediction-motion and relative judgment tasks. *Perception & Psychophysics, 57*(2), 231–245.

Tresilian, J. R. (1999). Visually timed action: Time-out for "tau"? *Trends in Cognitive Sciences, 3*(8), 301–310.

Troje, N. F. (2003). Reference frames for orientation anisotropies in face recognition and biological-motion perception. *Perception, 32*, 201–210.

Troje, N. F., & Westhoff, C. (2006). The inversion effect in biological motion perception: Evidence for a "life detector?" *Current Biology, 16*, 821–824.

Trope, Y., & Liberman, N. (2003). Temporal construal. *Psychological Review, 110*, 403–421.

Trope, Y., & Liberman, N. (2010). Construal-level theory of psychological distance. *Psychological Review, 117*(2), 440–463.

Tsal, Y., & Bareket, T. (1999). Effects of attention on localization of stimuli in the visual field. *Psychonomic Bulletin & Review*, *6*(2), 292–296.

Tsal, Y., & Bareket, T. (2005). Localization judgments under various levels of attention. *Psychonomic Bulletin & Review, 12*, 559–566.

Tsal, Y., Meiran, N., & Lamy, D. (1995). Towards a resolution theory of visual attention. *Visual Cognition*, *2*, 313–330.

Tseng, R. (2003). *The sceptical idealist: Michael Oakeshott as a critic of the Enlightenment (Vol. 1)*. Charlottesville, VA: Imprint Academic.

Tucker, M., & Ellis, R. (1998). On the relations between seen objects and components of potential actions. *Journal of Experimental Psychology: Human Perception and Performance, 24*, 830–846.

Turvey, M. T., & Carello, C. (1986). The ecological approach to perceiving-acting: A pictorial essay. *Acta Psychologica, 63*(1–3), 133–155.

Turvey, M. T., & Shaw, R. E. (1999). Ecological foundations of cognition. I. Symmetry and specificity of animal-environment systems. *Journal of Consciousness Studies, 6*(11–12), 95–110.

Tversky, A. (1977). Features of similarity. *Psychological Review, 84*, 327–352.

Tversky, A., & Gati, I. (1978). Studies of similarity. In E. Rosch & B. B. Lloyd (Eds.), *Cognition and categorization* (pp. 79–98). Hillsdale, NJ: Erlbaum.

Tversky, B. (1981). Distortions in memory for maps. *Cognitive Psychology, 13*(3), 407–433.

Tversky, B. (1993). Cognitive maps, cognitive collages, and spatial mental models. In A. U. Frank & I. Campari (Eds.), *Spatial information theory: A theoretical basis for GIS* (pp. 14–24). Berlin: Springer-Verlag.

Tversky, B. (1996). Spatial perspective in descriptions. In P. Bloom, M. A. Peterson, L. Nadel, & M. Garrett (Eds.), *Language and space* (pp. 463–491). Cambridge, MA: MIT Press.

Tversky, B. (1998). Three dimensions of spatial cognition. In M. A. Conway, S. E. Gathercole, & C. Cornoldi (Eds.), *Theories of memory II* (pp. 259–275). Hove, East Sussex: Psychological Press.

Tversky, B. (2000). Remembering space. In E. Tulving & F. I. M. Craik (Eds.), *Handbook of Memory* (pp. 363–378). New York: Oxford University Press.

Tversky, B. (2001). Spatial schemas in depictions. In M. Gattis (Ed.), *Spatial schemas and abstract thought* (pp. 79–111). Cambridge, MA: MIT Press.

Tversky, B. (2003). Navigating by mind and by body. In C. Freksa, W. Brauer, C. Habel, & K. F. Wender (Eds.), *Spatial cognition III: Routes and navigation, human memory and learning, spatial representation and spatial reasoning* (pp. 1–10). Berlin: Springer Verlag.

Tversky, B. (2005a). Functional significance of visuospatial representations. In P. Shah & A. Miyake (Eds.), *Handbook of higher-level visuospatial thinking* (pp. 1–34). Cambridge: Cambridge University Press.

Tversky, B. (2005b). Visuospatial reasoning. In K. Holyoak & R. Morrison (Eds.), *The Cambridge handbook of thinking and reasoning* (pp. 209–241). Cambridge: Cambridge University Press.

Tversky, B. (2009). Spatial cogntion: Embodied and situated. In P. Robbins & M. Aydede (Eds.), *The Cambridge handbook of situated cognition* (pp. 201–216). Cambridge: Cambridge University Press.

Tversky, B. (2011a). Spatial thought, social thought. In T. Schubert & A. Maass (Eds.), *Spatial schemas in social thought* (pp. 75–38). Berlin: Mouton de Gruyter.

Tversky, B. (2011b). Visualizations of thought. *Topics in Cognitive Science, 3*, 499–535.

Tversky, B. (2016). Lines: Orderly and messy. In Y. Portugali & E. Stolk (Eds.), *Complexity, cognition, urban planning, and design* (pp. 237–250). Dordrecht: Springer.

Tversky, B., Kugelmass, S., & Winter, A. (1991). Cross-cultural and developmental trends in graphic productions. *Cognitive Psychology, 23*, 515–557.

Tversky, B., & Hard, B. M. (2009). Embodied and disembodied cognition: Spatial perspective-taking. *Cognition, 110*(1), 124–129.

Tversky, B., & Hemenway, K. (1983). Categories of scenes. *Cognitive Psychology, 15*, 121–149.

Tversky, B., Lee, P. U., & Mainwaring, S. (1999). Why speakers mix perspectives. *Journal of Spatial Cognition and Computation, 1*, 399–412.

Tversky, B., & Schiano, D. J. (1989). Perceptual and conceptual factors in distortions in memory for graphs and maps. *Journal of Experimental Psychology: Human Perception and Performance, 118*(4), 387–398.

Tversky, B., & Schiano, D. J. (1997). Distortions in visual memory: Reply to Engebretson and Huttenlocher (1996). *Journal of Experimental Psychology: General, 126*(3), 312–314.

Tversky, B., Zacks, J. M., & Hard, B. M. (2008). The structure of experience. In T. Shipley & J. M. Zacks (Eds.), *Understanding events* (pp. 436–464). Oxford: Oxford University.

Tyler, C. W. (1973). Periodic vernier acuity. *Journal of Physiology, 228*(3), 637–647.

Tyler, C. W. (1998). Painters centre one eye in portraits. *Nature, 392*, 877–878.

Tyler, C. W. (2011). Paradoxical perception of surfaces in the Shepard tabletop illusion. *I-Perception, 2*(2), 137–141.

Tyler, C. W., & Nakayama, K. (1984). Size interactions in the perception of orientation. In L. Spillman & B. R. Wooton (Eds.), *Sensory experience, adaptation and perception* (pp. 529–546). Hillsdale, NJ: Lawrence Erlbaum.

Tyll, S., Bonath, B., Schoenfeld, M. A., Heinze, H. J., Ohl, F. W., & Noesselt, T. (2013). Neural basis of multisensory looming signals. *Neuroimage, 65*, 13–22.

Udo De Haes, H. A. (1970). Stability of apparent vertical and ocular counter-torsion as a function of lateral tilt. *Perception & Psychophysics, 8*, 137–142.

Ulrich, R., & Maieborn, C. (2010). Left–right coding of past and future in language: The mental timeline during sentence processing. *Cognition, 117*(2), 126–138.

Umiltà, C. (1991). Problems of the salient-features coding hypothesis: Comment on Weeks and Proctor. *Journal of Experimental Psychology: General, 120*, 83–86.

Umiltà, C., & Nicoletti, R. (1985). Attention and coding effects in S-R compatibility due to irrelevant spatial cues. In M. I. Posner & O. S. M. Marin (Eds.), *Attention and performance XI* (pp. 457–471). Hillsdale, NJ: Erlbaum.

Umiltà, C., & Nicoletti, R. (1990). Spatial stimulus–response compatibility. In R. W. Proctor & T. G. Reeve (Eds.), *Stimulus–response compatibility: An integrated perspective* (pp. 89–143). Amsterdam: North-Holland.

Ungerleider, L. G., & Mishkin, M. (1982). Two cortical visual systems. In D. J. Ingle, M. A. Goodale, & R. J. W. Mansfield (Eds.), *Analysis of visual behavior* (pp. 549–586). Cambridge, MA: MIT Press.

Uono, S., Sato, W., & Toichi, M. (2010). Brief report: Representational momentum for dynamic facial expressions in pervasive developmental disorder. *Journal of Autism and Developmental Disorders, 40*, 371–377.

Uono, S., Sato, W., & Toichi, M. (2014). Reduced representational momentum for subtle dynamic facial expressions in individuals with autism spectrum disorders. *Research in Autism Spectrum Disorders, 8*, 1090–1099.

Urbanski, M., & Bartolomeo, P. (2008). Line bisection in left neglect: The importance of starting right. *Cortex, 44*, 782–793.

Uttal, D. H., Friedman, A., Hand, L. L., & Warren, C. (2010). Learning fine-grained and category information in navigable real-world space. *Memory & Cognition, 38*(8), 1026–1040.

Vagnoni, E., Lourenco, S. F., & Longo, M. R. (2012). Threat modulates perception of looming visual stimuli. *Current Biology, 22* (19), R826–R827.

Vagnoni, E., Lourenco, S. F., & Longo, M. R. (2015). Threat modulates neural responses to looming visual stimuli. *European Journal of Neuroscience, 42*(5), 2190–2202.

Vaid, J., Singh, M., Sakhuja, T., & Gupta, G. C. (2002). Stroke direction asymmetry in figure drawing: Influence of handedness and reading/writing habits. *Brain and Cognition, 48*, 597–602.

Vallacher, R. R., Wegner, D. M., & Somoza, M. P. (1989). That's easy for you to say! Action identification and speech fluency. *Journal of Personality and Social Psychology, 56*(2), 199–208.

Vallar, G. (2001). Extrapersonal visual unilateral spatial neglect and its neuroanatomy. *Neuroimage, 14*(1), 52–58.

Vallar, G., Daini, R., & Antonucci, G. (2000). Processing of illusion of length in spatial hemineglect: A study of line bisection. *Neuropsychologia, 38*(7), 1087–1097.

Vallar, G., & Perani, D. (1986). The anatomy of unilateral neglect after right-hemisphere stroke lesions. A clinical/CT-scan correlation study in man. *Neuropsychologia, 24*(5), 609–622.

Vallesi, A., Binns, M. A., & Shallice, T. (2008). An effect of spatial–temporal association of time. *Cognition, 107*, 501–527.

Vallesi, A., Weisblatt, Y., Semenza, C., & Shaki, S. (2014). Cultural modulations of space-time compatibility effects. *Psychonomic Bulletin & Review, 21*, 666–669.

van Beers, R. J., Wolpert, D. M., & Haggard, P. (2001). Sensorimotor integration compensates for visual localization errors during smooth pursuit eye movements. *Journal of Neurophysiology, 85*, 1914–1922.

van de Grind, W. (2002). Physical, neural, and mental timing. *Consciousness and Cognition, 11*, 241–264.

van der Hoort, B., & Ehrsson, H. H. (2014). Body ownership affects visual perception of object size by rescaling the visual representation of external space. *Attention, Perception, & Psychophysics, 76*(5), 1414–1428.

van der Hoort, B., Guterstam, A., & Ehrsson, H. H. (2011). Being Barbie: The size of one's own body determines the perceived size of the world. *PLoS ONE, 6*(5), e20195.

Van der Lubbe, R. H. J., & Abrahamse, E. L. (2011). The premotor theory of attention and the Simon effect. *Acta Psychologica, 136*, 259–264.

Van der Lubbe, R. H. J., Abrahamse, E. L., & De Kleine, E. (2012). The premotor theory of attention as an account of the Simon effect. *Acta Psychologica, 140*, 25–34.

Van der Lubbe, R. H. J., & Verleger, R. (2002). Aging and the Simon task. *Psychophysiology, 39*, 100–110.

van Dijck, J. P., & Fias, W. (2011). A working memory account for spatial–numerical associations. *Cognition, 119*(1), 114–119.

van Dijck, J. P., Gevers, W., & Fias, W. (2009). Numbers are associated with different types of spatial information depending on the task. *Cognition, 113*(2), 248–253.

van Dijck, J. P., Gevers, W., Lafosse, C., Doricchi, F., & Fias, W. (2011). Non-spatial neglect for the mental number line. *Neuropsychologia, 49*, 2570–2583.

van Ee, R., & Erkelens, C. J. (2000). Is there an interaction between perceived direction and perceived aspect ratio in stereoscopic vision? *Perception & Psychophysics, 62*(5), 910–926.

van Elk, M., van Schie, H. T., & Bekkering, H. (2010). From left to right: Processing acronyms referring to names of political parties activates spatial associations. *Quarterly Journal of Experimental Psychology, 63*, 2202–2219.

van Gelder, T. J. (1998) The dynamical hypothesis in cognitive science, *Behavioral and Brain Sciences, 21*, 1–14.

van Hasselt, P. C. (1972). *The centrifugal control of retinal function*. Benschop: Nijimegen.

Van Opstal, A. J., & Van Gisbergen, J. A. (1987). Skewness of saccadic velocity profiles: A unifying parameter for normal and slow saccades. *Vision Research, 27*(5), 731–745.

Van Orden, G. C., & Holden, J. G. (2002) Intentional contents and self-control. *Ecological Psychology, 14*, 87–109.

van Quaquebeke, N., & Giessner, S. R. (2010). How embodied cognitions affect judgments: Height-related attribution bias in football foul calls. *Journal of Sport & Exercise Psychology, 32*(1), 3–22.

van Vugt, P., Fransen, I., Creten, W., & Paquier, P. (2000). Line bisection performances of 650 normal children. *Neuropsychologia, 38*(6), 886–895.

Vandervert, L. (1995) Chaos theory and the evolution of consciousness and mind: A thermodynamic-holographic resolution to the mind-body problem. *New Ideas in Psychology, 13*(2), 107–127.

Vartanian, O., Martindale, C., Podsaidlo, J., Overbay, S., & Borkum, J. (2005). The link between composition and balance in masterworks vs. paintings of lower artistic quality. *British Journal of Psychology, 96*, 493–503.

Vartanian, O., & Nadal, M. (2007). A biological approach to a model of aesthetic experience. In L. Dorfman, C. Martindale, & V. Petrov (Eds.), *Aesthetics and innovation* (pp. 429–444). Newcastle, UK: Cambridge Scholars Publishing.

Vaziri-Pashkam, M., & Cavanagh, P. (2011). Effect of speed overestimation on flash-lag effect at low luminance. *i-Perception, 2*, 1063–1075.

Vera-Constán, F. Rodríguez-Cuadrado, S., Romero-Rivas, C., Puigcerver, Fernández-Prieto, I., & Navarra, J. (submitted). *Seeing music: The perception of melodic "ups and downs" modulates the spatial processing of visual stimuli.*

Verfaillie, K., & Daems, A. (2002). Representing and anticipating human actions in vision. *Visual Cognition, 9*, 217–232.

Verfaillie, K., De Troy, A., & Van Rensbergen, J. (1994). Transsaccadic integration of biological motion. *Journal of Experimental Psychology: Learning, Memory, and Cognition, 20*, 649–670.

Verfaillie, K., & d'Ydewalle, G. (1991). Representational momentum and event course anticipation in the perception of implied periodical motions. *Journal of Experimental Psychology: Learning, Memory, and Cognition, 17*, 302–313.

Verschuure, J., & Van Meeteren, A. A. (1975). The effect of intensity on pitch. *Acta Acustica, 32*, 33–44.

Viarouge, A., Hubbard, E. M., & Dehaene, S. (2014). The organization of spatial reference frames involved in the SNARC effect. *The Quarterly Journal of Experimental Psychology, 67*(8), 1484–1499.

Vickers, J. N. (1992). Gaze control in putting. *Perception, 21*(1), 117–132.

Vingerhoets, R. A. A., De Vrijer, M., van Gisbergen, J. A. M., & Medendorp, W. P. (2009). Fusion of visual and vestibular tilt cues in the perception of visual vertical. *Journal of Neurophysiology, 101*, 1321–1333.

Vinson, D. W., Jordan, J. S., & Hund, A. M. (2017). Perceptually walking in another's shoes: Goals and memories constrain spatial perception. *Psychological Research, 81*(1), 66–74.

Vinson, N. G., & Reed, C. L. (2002). Sources of object-specific effects in representational momentum. *Visual Cognition, 9*, 41–65.

Vishton, P. M., Reardon, K. M., & Stevens, J. A. (2010). Timing of anticipatory muscle tensing control: Responses before and after expected impact. *Experimental Brain Research, 202*(3), 661–667.

Viswanathan, G. M. (2011). *The physics of foraging: An introduction to random searches and biological encounters.* Cambridge: Cambridge University Press.

Viswanathan, G. M., Afanasyev, V., Buldyrev, S. V., Havlin, S., da Luz, M. G. E., Raposo, E. P., & Stanley, H. E. (2001). Lévy flights search patterns of biological organisms. *Physica A: Statistical Mechanics and its Applications 295* (1–2), 85–88.

Viswanathan, G. M., Buldyrev, S. V., Havlin, S., da Luz, M. G., Raposo, E. P., & Stanley, H. E. (1999). Optimizing the success of random searches. *Nature, 401*(6756), 911–914.

von Hecker, U., Klauer, K. C., & Sankaran, S. (2013). Embodiment of social status: Verticality effects in multilevel rank-orders. *Social Cognition, 31*, 374–389.

von Hecker, U., Klauer, K. C., Wolf, L., & Fazilat-Pour, M. (2016). Spatial processes in linear ordering. *Journal of Experimental Psychology: Learning, Memory, and Cognition, 42*, 1003–1033.

von Muhlenen, A., & Lleras, A. (2007). No-onset looming motion guides spatial attention. *Journal of Experimental Psychology: Human*

Perception and Performance, 33(6), 1297–1310.

Voyer, D., Voyer, S., & Bryden, M. P. (1995). Magnitude of sex-differences in spatial abilities – a metaanalysis and consideration of critical variables. *Psychological Bulletin, 117*(2), 250–270.

Vreven, D., & Verghese, P. (2005). Predictability and the dynamics of position processing in the flash-lag effect. *Perception, 34*, 31–44.

Vroomen, J., & de Gelder, B. (2004). Temporal ventriloquism: Sound modulates the flash-lag effect. *Journal of Experimental Psychology: Human Perception and Performance, 30*, 513–518.

Vu, K.-P., Minakata, K., & Ngo, M. K. (2014). Influence of auditory and audiovisual stimuli on the right-left prevalence effect. *Psychological Research, 78*, 400–410.

Vu, K.-P., Proctor, R. W., & Pick, D. F. (2000). Vertical versus horizontal spatial compatibility: Right-left prevalence with bimanual responses. *Psychological Research, 64*, 25–40.

Vuilleumier, P., Ortigue, S., & Brugger, P. (2004). The number space and neglect. *Cortex, 40*(2), 399–410.

Vuilleumier, P., & Rafal, R. D. (2000). A systematic study of visual extinction: Between- and within-field deficits of attention in hemispatial neglect. *Brain, 123*, 1263–1279.

Vuilleumier, P., Valenza, N., Mayer, E., Reverdin, A., & Landis, T. (1998). Near and far visual space in unilateral neglect. *Annals of Neurology, 43*, 406–410.

Wagemans, J. (1995). Detection of visual symmetries. *Spatial Vision, 9*, 9–32.

Wagemans, J., Elder, J. H., Kubovy, M., Palmer, S. E., Peterson, M. A., Singh, M., & von der Heydt, R. (2012). A century of Gestalt psychology in visual perception: I. Perceptual grouping and figure–ground organization. *Psychological Bulletin, 138*, 1172–1271.

Wagner, M. (1985). The metric of visual space. *Perception & Psychophysics, 38*, 483–495.

Wagner, M. (2006). *The geometries of visual space*. London: Erlbaum.

Wagner, M., & Gambino, A. J. (2015). Variations in the anisotropy and affine structure of visual space: A geometry of visibles with a third dimension. *Topoi – An International Review of Philosophy*. Advance online publication. doi: http://link.springer.com/article/10.1007/s11245-015-9303-x.

Walker, B. N., & Ehrenstein, A. (2000). Pitch and pitch change interact in auditory displays. *Journal of Experimental Psychology: Applied, 6*, 15–30.

Walker, P. (2012). Cross-sensory correspondences and cross talk between dimensions of connotative meaning: Visual angularity is hard, high-pitched, and bright. *Attention, Perception, & Psychophysics, 74*, 1792–1809.

Walker, P. (2015). Depicting visual motion in still images: Forward leaning and a left to right bias for lateral movement. *Perception, 44*, 111–128.

Walker, P., Bremner, J. G., Mason, U., Spring, J., Mattock, K., Slater, A., & Johnson, S. P. (2010). Preverbal infants' sensitivity to synesthetic cross-modality correspondences. *Psychological Science, 21*, 21–25.

Walker, P., Bremner, J. G., Mason, U., Spring, J., Mattock, K., Slater, A., & Johnson, S. P. (2014). Preverbal infants are sensitive to cross-sensory correspondences: Much ado about the null results of Lewkowicz and Minar (2014). *Psychological Science, 25*, 835–836.

Walker, P., & Smith, S. (1986). The basis of Stroop interference involving the multimodal correlates of auditory pitch. *Perception, 15*, 491–496.

Walker, R. (1987). The effects of culture, environment, age, and musical training on choices of visual metaphors for sound. *Perception & Psychophysics, 42*, 491–502.

Walker, R., Findlay, J. M., Young, A. W., & Lincoln, N. (1996). Saccadic eye movements in object-based neglect. *Cognitive Neuropsychology, 13*, 569–615.

Walker, R., & Young, A. W. (1996). Object-based neglect: An investigation of the contributions of eye movements and perceptual completion. *Cortex, 32*(2), 279–285.

Walker-Andrews, A. S., & Lennon, E. M. (1985). Auditory-visual perception of changing distance by human infants. *Child Development, 56*(3), 544–548.

Walsh, P. D. (1996). Area-restricted search and the scale dependence of path quality discrimination. *Journal of Theoretical Biology, 183*(4), 351–361.

Walsh, V. (2003). A theory of magnitude: Common cortical metrics of time, space and quantity. *Trends in Cognitive Sciences, 7*(11), 483–488.

Wang, W., McBeath, M. K., & Sugar, T. (2015a). Optical angular constancy is maintained as a navigational control strategy when pursuing robots moving along complex pathways. *Journal of Vision, 15*(3), 16.

Wang, W., McBeath, M. K., & Sugar, T. (2015b). Navigational strategy used to intercept fly balls under real-world conditions with moving visual background fields. *Attention, Perception, & Psychophysics, 77*(2), 613–625.

Wang, Z., & Klein, R. M. (2010) Searching for inhibition of return in visual search: A review. *Vision Research, 50*, 220–228.

Wansard, M., Bartolomeo, P., Bastin, C., Segovia, F., Gillet, S., Duret, C., and Meulemans, T. (2015). Support for distinct subcomponents of spatial working memory: a double dissociation between spatial-simultaneous and spatial-sequential performance in unilateral neglect. *Cognitive Neuropsychology, 32*, 14–28.

Wansard, M., Meulemans, T., Gillet, S., Segovia, F., Bastin, C., Toba, M. N., et al. (2014). Visual neglect: Is there a relationship between impaired spatial working memory and re-cancellation? *Experimental Brain Research, 232*(10), 3333–3343.

Ward, L. M., Porac, C., Coren, S. & Girgus, J. S. (1977). The case for misapplied constancy scaling: Depth associations elicited by illusion configurations. *American Journal of Psychology, 90*, 609–620.

Ward, R., Goodrich, S., & Driver, J. (1994). Grouping reduces visual extinction: Neuropsychological evidence for weight-linkage in visual selection. *Visual Cognition, 1*(1), 101–129.

Warren, W. H., Rothman, D. B., Schnapp, B. H., & Ericson, J. D. (2017). Wormholes in virtual space: From cognitive maps to cognitive graphs. *Cognition, 166*, 152–163.

Wascher, E., Schatz, U., Kuder, T., & Verleger, R. (2001). Validity and boundary conditions of automatic response activation in the Simon task. *Journal of Experimental Psychology: Human Perception and Performance, 27*, 731–751.

Wasner, M., Moeller, K., Fischer, M. H., & Nuerk, H.-C. (2014). Aspects of situated cognition in embodied numerosity: The case of finger counting. *Cognitive Processing, 15*(3), 317–328.

Watanabe, K. (2004). Visual grouping by motion precedes the relative localization between moving and flashed stimuli. *Journal of Experimental Psychology: Human Perception and Performance, 30*, 504–512.

Watanabe, K., Nijhawan, R., Khurana, B., & Shimojo, S. (2001). Perceptual organization of moving stimuli modulates the flash-lag effect. *Journal of Experimental Psychology: Human Perception and Performance, 27*, 879–894.

Watanabe, K., & Yokoi, K. (2006). Object-based anisotropies in the flash-lag effect. *Psychological Science, 17*, 728–735.

Watanabe, K., & Yokoi, K. (2007). Object-based anisotropic mislocalization by retinotopic motion signals. *Vision Research, 47*, 1662–1667.

Watson, C., Cardillo, E., Bromberger, B., & Chatterjee, A. (2014). The specificity of action knowledge in sensory and motor systems. *Frontiers in Psychology, 5*, 494.

Watt, R. J. (1984). Further evidence concerning the analysis of curvature in human foveal vision. *Vision Research, 24*, 251–253.

Watt, R. J. (1988). *Visual processing: Computational, psychological and cognitive research*. London: Lawrence Erlbaum Associates.

Watt, R. J. (1991). *Understanding vision*. London: Academic Press.

Watt, R. J., & Andrews, D. P. (1982). Contour curvature analysis: Hyperacuities in the discrimination of detailed shape. *Vision Research, 22*, 449–460.

Watt, R. J., & Morgan, M. J. (1984). Spatial filters and the localization of luminance changes in human vision. *Vision Research, 24*, 1387–1397.

Watts, R. G., & Bahill, A. T. (2000). *Keep your eye on the ball: Curve balls, knuckleballs, and fallacies of baseball.* Patomac Books.

Wearden, J. H., Norton, R., Martin, S., & Montford-Bebb, O. (2007). Internal clock processes and the filled-duration illusion. *Journal of Experimental Psychology: Human Perception and Performance, 33*, 716–729.

Wedell, D. H., Fitting, S., & Allen, G. L. (2007). Shape effects on memory for location. *Psychonomic Bulletin & Review, 14*(4), 681–686.

Weeks, D. J., & Proctor, R. W. (1990). Salient-features coding in the translation between orthogonal stimulus–response dimensions. *Journal of Experimental Psychology: General, 119*, 355–366.

Weger, U. W., & Pratt, J. (2008). Time flies like an arrow: Space-time compatibility effects suggest the use of a mental timeline. *Psychonomic Bulletin & Review, 15*, 426–430.

Weintraub, D. J., Krantz, D. H., & Olson, T. P. (1980). The Poggendorff illusion: Consider all the angles. *Journal of Experimental Psychology: Human Perception and Performance, 6*(4), 718–725.

Weintraub, D. J., & Schneck, M. K. (1986). Fragments of Delboef and Ebbinghaus illusions. *Perception & Psychophysics, 40*, 147–158.

Weis, T., Estner, B., van Leeuwen, C., & Lachmann, T. (2016). SNARC (spatial–numerical association of response codes) meets SPARC (spatial–pitch association of response codes): Automaticity and interdependency in compatibility effects. *Quarterly Journal of Experimental Psychology, 69*(7), 1366–1383.

Weisberg, S. M., & Newcombe, N. S. (2016). How do (some) people make a cognitive map? Routes, places and working memory. *Journal of Experimental Psychology: Learning, Memory, and Cognition, 42*, 768–785.

Weisberg, S. M., Schinazi, V. R., Newcombe, N. S., Shipley, T. F., & Epstein, R. A. (2014). Variations in cognitive maps: Understanding individual differences in navigation. *Journal of Experimental Psychology: Learning, Memory, and Cognition, 40*, 669–682.

Welch, R. B., DutionHurt, L. D., & Warren, D. H. (1986). Contributions to audition and vision to temporal rate perception. *Perception & Psychophysics, 39*, 294–300.

Welch, R. B., & Warren, D. H. (1980). Immediate perceptual response to intersensory discrepancy. *Psychological Bulletin, 88*, 638–667.

Welch, R. B., & Warren, D. H. (1986). Intersensory interactions. In K. R. Boff, L. Kaufman, & J. P. Thomas (Eds.), *Handbook of perception and human performance* (Vol. 2, pp. 25.1–25.36). New York: Wiley.

Wen, W., & Kawabata, H. (2013). The best route is not always the easiest one: Spatial references in heuristics of route choice. *Psychology, 4*(9), 704–710.

Wenderoth, P. (1994). The salience of vertical symmetry. *Perception, 23*, 221–236.

Wenderoth, P., White, D., & Beh, H. (1978). The effects of peripheral and central fixation on a Poggendorff-like vernier alignment task. *Perception & Psychophysics, 24*(4), 377–386.

Werner, S., & Schmidt, K. (2000). Environmental reference systems for large-scale spaces. *Spatial Cognition and Computation, 1*, 447–473.

Wesp, R., Cichello, P., Gracia, E. B., & Davis, K. (2004). Observing and engaging in purposeful actions with objects influences estimates of their size. *Perception & Psychophysics, 66*(8), 1261–1267.

Westheimer, G. (1981). Visual hyperacuity. *Progress in Sensory Physiology, 1*, 2–29.

Westheimer, G. (2003). The distribution of preferred orientations in the peripheral visual field. *Vision Research, 43*, 53–57.

Westheimer, G. (2005). Anisotropies in peripheral Vernier acuity. *Spatial Vision, 18*, 159–167.

Westheimer, G., & McKee, S. P. (1977). Spatial configurations for hyperacuity. *Vision Research, 17*, 941–947.

Westphal-Fitch, G., Oh, J., & Fitch, W. T. (2013). Studying aesthetics with the method of production: Effects of context and local symmetry. *Journal of Aesthetics, Creativity, and the Arts*, 7, 13–26.

Weyl, H. (1952). *Symmetry*. Princeton, NJ: Princeton University Press.

Whipp, B. J., & Ward, S. A. (1992). Will women soon outrun men. *Nature, 355*(6355), 25–25.

White, E. J., Shockley, K., & Riley, M. A. (2013). Multimodally specified energy expenditure and action-based distance judgments. *Psychonomic Bulletin & Review, 20*(6), 1371–1377.

White, H., Minor, S. W., Merrell, J., & Smith, T. (1993). Representational momentum effects in the cerebral hemispheres. *Brain and Cognition, 22*, 161–170.

White, P. A. (2011). Visual impressions of forces between objects: Entraining, enforced disintegration, and shattering. *Visual Cognition, 19*, 635–674.

White, P. A. (2012). The experience of force: The role of haptic experience of forces in visual perception of object motion and interactions, mental simulation, and motion-related judgments. *Psychological Bulletin, 138*, 589–615.

Whitney, D. (2002). The influence of visual motion on perceived position. *Trends in Cognitive Sciences, 6*, 211–216.

Whitney, D. (2008). Visuomotor extrapolation. *Behavioral and Brain Sciences, 31*(2), 220–221.

Whitney, D., & Cavanagh, P. (2000a). The position of moving objects. *Science, 289*, 1107a.

Whitney, D., & Cavanagh, P. (2000b). Motion distorts visual space: Shifting the perceived position of remote stationary objects. *Nature Neuroscience, 3*, 954–959.

Whitney, D., & Cavanagh, P. (2002). Surrounding motion affects the perceived locations of moving stimuli. *Visual Cognition, 9*, 139–152.

Whitney, D., Cavanagh, P., & Murakami, I. (2000). Temporal facilitation for moving stimuli is independent of changes in direction. *Vision Research, 40*, 3829–3839.

Whitney, D., & Murakami, I. (1998). Latency difference, not spatial extrapolation. *Nature Neuroscience, 1*, 656–657.

Whitney, D., Murakami, I., & Cavanagh, P. (2000). Illusory spatial offset of a flash relative to a moving stimulus is caused by differential latencies for moving and flashed stimuli. *Vision Research, 40*, 137–149.

Wickelgren, W. A. (1968). Sparing of short-term memory in an amnesic patient: Implications for strength theory of memory. *Neuropsychologia, 6*, 235–244.

Wiegand, K., & Wascher, E. (2005). Dynamic aspects of S-R correspondence: Evidence for two mechanisms involved in the Simon effect. *Journal of Experimental Psychology: Human Perception and Performance, 31*, 453–464.

Wickelgren, W. A. (1977). Speed-accuracy tradeoff and information processing dynamics. *Acta Psychologica, 41*(1), 67–85.

Wiegand, K., & Wascher, E. (2007). The Simon effect for vertical S-R relations: Changing the mechanism by randomly varying the S-R mapping rule? *Psychological Research, 71*, 219–233.

Wiener, J. M., Büchner, S., & Hölscher, C. (2009). Taxonomy of human wayfinding: A knowledge-based approach. *Spatial Cognition and Computation, 9*, 152–165.

Wiener, J. M., de Condappa, O., Harris, M. A., & Wolbers, T. (2013). Maladaptive bias for extrahippocampal navigation strategies in aging humans. *Journal of Neuroscience, 33*, 6012–6017.

Wiener, J. M., Hölscher, C., Büchner, S., & Konieczny, L. (2012). Gaze behaviour during space perception and spatial decision making. *Psychological Research, 76*(6), 713–729.

Wiener, J. M., & Mallot, H. A. (2003). "Fine-to-coarse" route planning and navigation in regionalized environments. *Spatial Cognition & Computation: An Interdisciplinary Journal, 3*(4), 331–358.

Wiener, J. M., Schnee, A., & Mallot, H. A. (2004). Use and interaction of navigation strategies in regionalized environments. *Journal of Environmental Psychology*, *24*(4), 475–493.

Wilkins, A. J., Shallice, T., & McCarthy, R. (1987). Frontal lesions and sustained attention. *Neuropsychologia*, *25*, 359–365.

Williams, D. W., & Sekuler, R. (1984). Coherent global motion percepts from stochastic local motions. *Vision Research*, *24*, 55–62.

Williams, L. E., & Bargh, J. A. (2008). Keeping one's distance: The influence of spatial distance cues on affect and evaluation. *Psychological Science*, *19*(3), 302–308.

Williams, P. A., & Enns, J. T. (1996). Pictorial depth and framing have independent effects on the horizontal-vertical illusion. *Perception*, *25*(8), 921–926.

Williamson, V. J., Cocchini, G., & Stewart, L. (2011). The relationship between pitch and space in congenital amusia. *Brain and Cognition*, *76*, 70–76.

Wilson A., & Bingham, G. P. (2001). Dynamics, not kinematics, is an adequate basis for perception. *Behavioral and Brain Sciences*, *24*, 709–710.

Wilson, H. R. (1986) Responses of spatial mechanisms can explain hyperacuity. *Vision Research*, *26*, 453–469.

Wilson, M., Lancaster, J., & Emmorey, K. (2010). Representational momentum for the human body: Awkwardness matters, experience does not. *Cognition*, *116*, 242–250.

Wilson, P. R. (1968). Perceptual distortion of height as a function of ascribed academic status. *Journal of Social Psychology*, *74*, 97–102.

Wilton, R. N. (1979) Knowledge of spatial relations: The specification of information used in making inferences. *Quarterly Journal of Experimental Psychology*, *31*, 133–146.

Winawer, J., Huk, A. C., & Boroditsky, L. (2008). A motion aftereffect from still photographs depicting motion. *Psychological Science*, *19*, 276–283.

Winawer, J., Witthoft, N., Frank, M. C., Wu, L., Wade, A. R., & Boroditsky, L. (2007). Russian blues reveal effects of language on color discrimination. *Proceedings of the National Academy of Sciences*, *104*(19), 7780–7785.

Winner, E., Dion, J., Rosenblatt, E., & Gardner, H. (1987). Do lateral or vertical reversal affect balance in paintings? *Visual Arts Research*, *13*, 1–9.

Winter, B., Matlock, T., Shaki, S., & Fischer, M. H. (2015). Mental number space in three dimensions. *Neuroscience & Biobehavioral Reviews*, *57*, 209–219.

Witkin, H. A. (1949). Perception of body position and of the position of the visual field. *Psychological Monographs: General and Applied*, *63*(7), i–46.

Witt, J. K. (2011a). Action's effect on perception. *Current Directions in Psychological Science*, *20*(3), 201–206.

Witt, J. K. (2011b). Tool use influences perceived shape and perceived parallelism, which serve as indirect measures of perceived distance. *Journal of Experimental Psychology: Human Perception and Performance*, *37*(4), 1148–1156.

Witt, J. K. (2015). Awareness is not a necessary characteristic of a perceptual effect: Commentary on Firestone (2013). *Perspectives on Psychological Science*, *10*(6), 865–872.

Witt, J. K. (2017). Action potential influences spatial perception: Evidence for genuine top-down effects on perception. *Psychonomic Bulletin & Review*, *24*(4), 999–1021.

Witt, J. K., & Dorsch, T. E. (2009). Kicking to bigger uprights: Field goal kicking performance influences perceived size. *Perception*, *38*(9), 1328–1340.

Witt, J. K., Linkenauger, S. A., Bakdash, J. Z., & Proffitt, D. R. (2008). Putting to a bigger hole: Golf performance relates to perceived size. *Psychonomic Bulletin & Review*, *15*(3), 581–585.

Witt, J. K., Linkenauger, S. A., & Proffitt, D. R. (2012). Get me out of this slump! Visual illusions improve sports performance. *Psychological Science*, *23*(4), 397–399.

Witt, J. K., Linkenauger, S. A., & Wickens, C. D. (2016). Action-specific effects in perception and their potential applications. *Journal of Applied Research in Memory and Cognition, 5*(1), 69–76.

Witt, J. K., & Proffitt, D. R. (2005). See the ball, hit the ball: Apparent ball size is correlated with batting average. *Psychological Science, 16*(12), 937–938.

Witt, J. K., & Proffitt, D. R. (2008). Action-specific influences on distance perception: A role for motor simulation. *Journal of Experimental Psychology: Human Perception and Performance, 34*(6), 1479–1492.

Witt, J. K., Proffitt, D. R., & Epstein, W. (2004). Perceiving distance: A role of effort and intent. *Perception, 33*(5), 577–590.

Witt, J. K., Proffitt, D. R., & Epstein, W. (2005). Tool use affects perceived distance, but only when you intend to use it. *Journal of Experimental Psychology: Human Perception and Performance, 31*(5), 880–888.

Witt, J. K., Proffitt, D. R., & Epstein, W. (2010). When and how are spatial perceptions scaled? *Journal of Experimental Psychology: Human Perception and Performance, 36*(5), 1153–1160.

Witt, J. K., & Riley, M. A. (2014). Discovering your inner Gibson: Reconciling action-specific and ecological approaches to perception-action. *Psychonomic Bulletin & Review, 21*(6), 1353–1370.

Witt, J. K., Schuck, D. M., & Taylor, J. E. T. (2011). Action-specific effects underwater. *Perception, 40*(5), 530–537.

Witt, J. K., & Sugovic, M. (2010). Performance and ease influence perceived speed. *Perception, 39*(10), 1341–1353.

Witt, J. K., & Sugovic, M. (2012). Does ease to block a ball affect perceived ball speed? Examination of alternative hypotheses. *Journal of Experimental Psychology: Human Perception and Performance, 38*(5), 1202–1214.

Witt, J. K., & Sugovic, M. (2013a). Catching ease influences perceived speed: Evidence for action-specific effects from action-based measures. *Psychonomic Bulletin & Review, 20*, 1364–1370.

Witt, J. K., & Sugovic, M. (2013b). Response bias cannot explain action-specific effects: Evidence from compliant and non-compliant participants. *Perception, 42*, 138–152.

Witt, J. K., Sugovic, M., & Dodd, M. D. (2016). Action-specific perception of speed is independent of attention. *Attention, Perception, & Psychophysics, 78*(3), 880–890.

Witt, J. K., Sugovic, M., Tenhundfeld, N. T., & King, Z. R. (2016). An action-specific effect on perception that avoids all pitfalls. *Behavioral and Brain Sciences, 39*, e261.

Witt, J. K., Taylor, J. E. T., Sugovic, M., & Wixted, J. T. (2015). Signal detection measures cannot distinguish perceptual biases from response biases. *Perception, 44*(3), 289–300.

Witt, J. K., Taylor, J. E. T., Sugovic, M., & Wixted, J. T. (2016). Further clarifying signal detection theoretic interpretations of the Müller-Lyer and sound-induced flash illusions. *Journal of Vision, 16(11)*, 19 (1–7).

Witt, J. K., Tenhundfeld, N. L., & Tymoski, M. J. (2018). Is there a chastity belt on perception? *Psychological Science, 29*(1), 139–146.

Wohlgemuth, A. (1911). On the after-effect of seen movement. *British Journal of Psychology, Monograph Supplement, 1*, 1–117.

Wojciulik, E., Husain, M., Clarke, K., & Driver, J. (2001). Spatial working memory deficit in unilateral neglect. *Neuropsychologia, 39*, 390–396.

Wojciulik, E., Rorden, C., Clarke, K., Husain, M., & Driver, J. (2004). Group study of an "undercover" test for visuospatial neglect: Invisible cancellation can reveal more neglect than standard cancellation. *Journal of Neurology, Neurosurgery and Psychiatry, 75*, 1356–1358.

Wojtach, W. T., Sung, K., Truong, S., & Purves, D. (2008). An empirical explanation of the flash-lag effect. *Proceedings of the National Academy of Sciences of the United States of America, 105*, 16338–16343.

Wolbers, T., & Wiener, J. M. (2014). Challenges for identifying the neural mechanisms that support spatial navigation: The impact of spatial scale. *Frontiers in Human Neuroscience, 8*, 571.

Wolfe, U., Maloney, L. T., & Tam, M. (2005). Distortions of perceived length in the frontoparallel plane: Tests of perspective theories. *Perception & Psychophysics*, *67*(6), 967–979.

Wölfflin, H. (1994). Prolegomena zu einer Psychologie der Architektur [Prolegomena to a psychology of architecture]. In H. F. Mallgrave & E. Ikonomou (Eds.), *Empathy, form, and space: Problems in German aesthetics 1873–1893* (pp. 149–190). Santa Monica, CA: Getty Center for the History of Art and the Humanities (original work published 1886).

Wolfe, U., Maloney, L. T., & Tam, M. (2005). Distortions of perceived length in the frontoparallel plane: Tests of perspective theories. *Perception & Psychophysics*, *67*(6), 967–979.

Wolpert, D. M., Doya, K., & Kawato, M. (2003). A unifying computational framework for motor control and social interaction. *Philosophical Transactions of the Royal Society B Biological Sciences*, *358*, 593–602.

Wood, G., Nuerk, H.-C., & Willmes, K. (2006a). Crossed hands and the SNARC effect: A failure to replicate Dehaene, Bossini and Giraux (1993). *Cortex*, *42*(8), 1069–1079.

Wood, G., Nuerk, H.-C., & Willmes, K. (2006b). Variability of the SNARC effect: Systematic interindividual differences or just random error? *Cortex*, *42(8)*, 1119–1123.

Wood, G., Vine, S. J., & Wilson, M. R. (2013). The impact of visual illusions on perception, action planning, and motor performance. *Attention, Perception, & Psychophysics*, *75*, 830–834.

Wood, G., Willmes, K., Nuerk, H.-C., & Fischer, M. H. (2008). On the cognitive link between space and number: A meta-analysis of the SNARC effect. *Psychology Science Quarterly*, *50*(4), 489–525.

Woodworth, R. S. (1938). *Experimental psychology*. New York: Holt.

Wu, B., Ooi, T. L., & He, Z. J. (2004). Perceiving distances accurately by a directional process of integrating ground information. *Nature*, *428*, 73–77.

Wu, D. H., Waller, S., & Chatterjee, A. (2007). The functional neuroanatomy of thematic role and locative relational knowledge. *The Journal of Cognitive Neuroscience*, *19*, 1542–1555.

Wuerger, S. M., Hofbauer, M., & Meyer, G. F. (2003). The integration of auditory and visual motion signals at threshold. *Perception & Psychophysics*, *65*, 1188–1196.

Wühr, P., Fasold, F., & Memmert, D. (2015). Soccer offside judgments in laypersons with different types of static displays. *PLoS ONE*, *10*(8), e0133687.

Xiao, Y. J., & Van Bavel, J. J. (2012). See your friends close and your enemies closer: social identity and identity threat shape the representation of physical distance. *Personality and Social Psychology Bulletin*, *38*(7), 959–972.

Xu, Y. (2007). The role of the superior intraparietal sulcus in supporting visual short-term memory for multifeature objects. *The Journal of Neuroscience*, *24*(43), 11676–11686.

Xu, Y. (2009). Distinctive neural mechanisms supporting visual object individuation and identification. *Journal of Cognitive Neuroscience*, *21*(3), 511–518.

Yang, T. L., Dixon, M. W., & Proffitt, D. R. (1999). Seeing big things: Overestimation of heights is greater for real objects than for objects in pictures. *Perception*, *28*(4), 445–467.

Yang, Z., & Purves, D. (2003). A statistical explanation of visual space. *Nature Neuroscience*, *6*, 632–640.

Yap, A. J., Mason, M. F., & Ames, D. R. (2013). The powerful size others down: The link between power and estimates of others' size. *Journal of Experimental Social Psychology*, *49*, 591–594.

Yau, J. M., Olenczak, J. B., Dammann, J. F., & Bensmaia, S. J. (2009). Temporal frequency channels are linked across audition and touch. *Current Biology*, *19*, 561–566.

Yeshurun, Y., & Carrasco, M. (1999). Spatial attention improves performance in spatial resolution tasks. *Vision Research*, *39*, 293–306.

Yilmaz, M., & Meister, M. (2013). Rapid innate defensive responses of mice to looming visual stimuli. *Current Biology*, *23*(20), 2011–2015.

Yin, R. K, (1969) Looking at upside-down faces. *Journal of Experimental Psychology, 81*, 141–145.

Yoshikawa, S., & Sato, W. (2006). Enhanced perceptual, emotional, and motor processing in response to dynamic facial expressions of emotion. *Japanese Psychological Research, 48*, 213–222.

Yoshikawa, S., & Sato, W. (2008). Dynamic facial expressions of emotion induce representational momentum. *Cognitive, Affective & Behavioral Neuroscience, 8*, 25–31.

Yost, W. A. (2007). *Fundamentals of hearing* (5th ed.). London: Academic Press.

Zago, M., Bosco, G., Maffei, V., Iosa, M., Ivanenko, Y. P., & Lacquaniti, F. (2004). Internal models of target motion: Expected dynamics overrides measured kinematics in timing manual interceptions, *Journal of Neurophysiology, 91*, 1620–1634.

Zago, M., Bosco, G., Maffei, V., Iosa,M., Ivanenko, Y. P., & Lacquaniti, F. (2005). Fast adaptation of the internal model of gravity for manual interceptions: Evidence for event-dependent learning, *Journal of Neurophysiology, 93*, 1055–1068.

Zago, M., La Scaleia, B., Miller, W. L., & Lacquaniti, F. (2011). Coherence of structural visual cues and pictorial gravity paves the way for interceptive actions. *Journal of Vision, 11*(10), 13.

Zago, M., & Lacquaniti, F. (2005). Internal model of gravity for hand interception: Parametric adaptation to zero-gravity visual targets on Earth. *Journal of Neurophysiology, 94*, 1346–1357.

Zago, M., McIntyre, J., Senot, P., & Lacquaniti, F. (2008). Internal models and prediction of visual gravitational motion, *Vision Research, 48*, 1532–1538.

Zago, M., McIntyre, J., Senot, P., & Lacquaniti, F. (2009). Visuo-motor coordination and internal models for object interception, *Experimental Brain Research, 192*, 571–604.

Zajonc, R. B. (1968). Attitudinal effects of mere exposure. *Journal of Personality and Social Psychology, 9*, 1–27.

Zanolie, K., van Dantzig, S., Boot, I., Wijnen, J., Schubert, T. W., Giessner, S. R., & Pecher, D. (2012). Mighty metaphors: Behavioral and ERP evidence that power shifts attention on a vertical dimension. *Brain and Cognition, 78*, 374–389.

Zanutti, L. (1976). A new explanation for the Poggendorff illusion. *Perception & Psychophysics, 20*, 29–32.

Zbikowski. L. (1998). *Metaphor and music theory: Reflections from cognitive science.* Music Theory Online 4:1. Santa Barbara, CA: Society for Music Theory.

Zebian, S. (2005). Linkages between number concepts, spatial thinking, and directionality of writing: The SNARC effect and the reverse SNARC effect in English and Arabic monoliterates, biliterates, and illiterate Arabic speakers. *Journal of Cognition and Culture, 5*, 165–190.

Zeki, S. (1999). Art and the brain. *Journal of Consciousness Studies, 6*, 76–96.

Zetzsche, C., Wolter, J., Galbraith, C., & Schill, K. (2009). Representation of space: Image-like or sensorimotor? *Spatial Vision, 22*(5), 409–424.

Zhang, J., & Kornblum, S. (1997). Distributional analysis and De Jong, Liang, and Lauber's (1994) dual-process model of the Simon effect. *Journal of Experimental Psychology: Human Perception and Performance, 23*, 1543–1551.

Zhao, M., & Warren, W. H. (2015a). Environmental stability modulates the role of path integration in human navigation. *Cognition, 142*, 96–109.

Zhao, M., & Warren, W. H. (2015b). How you get there from here: Interaction of visual landmarks and path integration in human navigation. *Psychological Science, 26*(6), 915–924.

Zhou, H., Ge, Y., Wang, L., Zhang, P., & He, S. (2015). Feedback signal contributes to the flash grab effect: Evidence from fMRI and ERP study. *Journal of Vision, 15*(12), 1248–1284.

Zhou, L., Yan, J. J., Liu, Q., Li, H., & Xie, C. X. (2007). Visual and auditory information specifying an impending collision of an approaching object. *Human-Computer*

Interaction, Pt 2, Proceedings, 4551, 720–729.

Zorzi, M., Priftis, K., & Umiltà, C. (2002). Brain damage: Neglect disrupts the mental number line. *Nature, 417*(6885), 138–139.

Zorzi, M., & Umiltà, C. (1995). A computational model of the Simon effect. *Psychological Research, 58*, 193–205.

Zuckerkandl, V. (1956). *Sound and symbol* (Willard R. Trask, Trans.). New York: Pantheon Books.

Zurek, D. B., Perkins, M. Q., & Gilbert, C. (2014). Dynamic visual cues induce jaw opening and closing by tiger beetles during pursuit of prey. *Biology Letters, 10*(11). doi:10.1098/rsbl.2014.0760

Index